Introduction to
VLSI Circuits and Systems

John P. Uyemura
Professor of Electrical and Computer Engineering
Georgia Institute of Technology

WILEY-
INDIA

INTRODUCTION TO VLSI CIRCUITS AND SYSTEMS

Authorized reprint by Wiley India (P.) Ltd., 4435/7, Ansari Road, Daryaganj, New Delhi 110 002.

Bicentennial logo designer: Richard J. Pacifico

Reprint : 2012

Printed at : Satyam Enterprises, Delhi

ISBN : 978-81-265-0915-7

To the Gang

Bear, Bunny, Cow, Bear 2, Polar Bear, the Baby Bears, Polar Bear Gone to Paris, Malinda, Elise, and all of the others (including Valerie and Christine, of course)

植 村

Preface

VLSI has become a major driving force in modern technology. It provides the basis for computing and telecommunications, and the field continues to grow at an amazing pace.

This book was written as a text that covers the foundations of digital CMOS VLSI systems design. It was written for use in a first course at the senior level in the fields of electrical engineering, computer engineering, and computer science. It can also be used in a first-year graduate course that is supplemented with journal articles. The core topics can be covered in a 10-week course, but there is enough material for a 15-week offering, or a 2-quarter sequence with readings from current literature.

No previous background VLSI or CMOS is required. It is assumed that the reader has taken the core curriculum in electrical or computer engineering, with courses in digital logic design and electrical circuit analysis being the most important. Electronics starts at a basic level, so that computer science majors should be able to follow the discussions. The book is *introductory* in that it does not assume any prior exposure to the field. It is accessible for self-study by both students and professionals alike. Even though discussions start from basics, they evolve to high-level treatments of advanced topics.

Coverage

After a brief introduction to some jargon and the concept of design hierarchies in **Chapter 1**, the treatment in divided into three main parts.

Part 1, entitled *Silicon Logic*, introduces the ideas of logic design in a CMOS technology. It takes the student from the ideas of logic design to designing CMOS networks. **Chapter 2** introduces MOSFETs as simple logic-controlled switches, and then concentrates on the design of CMOS static logic gates at the Boolean level. A first look at the physical design is provided in **Chapter 3**, which views an integrated circuit as a set of patterned material layers that are used to control the flow of signals. This provides a transition from a switch-level description down to the physical level. Silicon chip fabrication is covered in **Chapter 4**, and examines both general and specific aspects of the manufacturing process. The discussion culminates in **Chapter 5**, where physical design and layout are examined in detail. Design hierarchies and cell libraries discussions link the material to the usage of CAD tools. By the end of Part 1, students can implement basic logic designs in CMOS and understand how they are translated to a silicon environment.

Electronic aspects of CMOS are presented in Part 2, *The Logic-Electronics Interface*. The approach emphasizes simple explanations and switch models for system design level. MOSFETs are characterized in **Chapter 6** using square-law equations, which are then used to develop linear RC switching models. **Chapter 7** covers the electrical properties of CMOS logic circuits. High-speed CMOS cascades are introduced in **Chapter 8**. The treatment covers the classical methods, but includes a section on Logical Effort. Some advanced CMOS logic families are examined in **Chapter 9**; the material is specialized, but appropriate for some groups.

The remaining chapters are contained in Part 3, *The Design of VLSI Systems*. This material was chosen to link logic networks, physical layout, and electronics together into the single discipline of VLSI systems design. **Chapter 10** is an introduction to Verilog® HDL, which was chosen as a basis for the high-level descriptions used in some parts of the book. Students already familiar with VHDL tend to adapt to Verilog very quickly. **Chapter 11** analyzes a few digital library components from a viewpoint that links high-level HDL descriptions to the logic, circuit, and physical implementations. The purpose of this approach is to illustrate how high-level abstractions are written, and then examine the many options that are available in CMOS VLSI to actually build the components. The same idea is used to present adder and multiplier networks in **Chapter 12**. These two chapters emphasize the interaction among the various levels in the design hierarchy. **Chapter 13** covers memories circuits and architectures. Chip-level physical design considerations are presented in **Chapter 14**. Interconnect modeling, crosstalk, floorplaning, and routing are discussed. Power distribution schemes, I/O circuits, and low-power design are also introduced. **Chapter 15** deals with system-level design. Clock drivers and distribution trees are presented as are system-level datapath techniques such as pipelining and bit-slice design. The book ends with a brief introduction to reliability and testing in **Chapter 16**.

A concerted effort has been made to present VLSI as an interdisciplinary field that relies on many specialists working together at every level. Emphasis is placed on illustrating the interaction among the different viewpoints, and how design choices made at one level affect all of the other groups.

Philosophy

The term "VLSI" means different things to different people. It consists of many distinct specialties that interact in a unique manner. A true "VLSI designer" may work in one particular area, but understands how units interact to form a system. I have taught both the undergraduate and graduate-level VLSI courses at Georgia Tech for several years. The view presented here is one that I have found to be effective in introducing students with varying backgrounds to the field.

Every topic can be found discussed elsewhere with more detail. The differences between this text and specialized treatments are in the context, the level, and the applications. This book provides a quasi-uniform basis for learning the main aspects of the VLSI. The topics were selected because of their importance to seeing an overall view of the field. The analyses are detailed enough to provide useful results without having to be a specialist.

Students who choose to work in the field will concentrate studies in one or two areas. For example, the author teaches courses in digital electronics at both the senior and graduate level that go much deeper than the present text. Reference books are listed at the end of every chapter for deeper studies of the topics. No attempt was made to include a listing of journal articles, as that is more appropriate for a specialized text.

Use as a Text

The material in each chapter corresponds to 3 to 6 one-hour lectures. The actual time required to cover a particular section or chapter varies with the background of the students. Problems have been provided at the end of most chapters, and a Solutions Manual is available to instructors adopting the text. The URL for the book web site is

www.wiley.com/college/Uyemura

The material in some chapters emphasizes practical design and layout issues, and does not easily yield "homework-type" problems. These topics are best addressed in design projects.

The order of the topics has been chosen to introduce the concepts from the bottom up (Parts 1 and 2), and then jump to the system level in Part 3. This tends to work well with electrical engineering and hardware-oriented computer engineering students. System-oriented design courses will find it more natural to start with the Verilog discussion in Chapter 10 to emphasize high-level system design and abstraction. Physical design and electronics are used as a basis for implementation. This makes the material more accessible to computer science and system-oriented computer engineering majors.

Several course outlines have been developed around the book. Three one-semester formats for a senior level course are shown. Each provides a suggested order of presentation on a chapter-by-chapter basis. Deviations will be necessary depending upon the students' background and the instructor's interests. Part 1 topics should be presented in order. Chapters 3–5 can be covered very quickly. Topics in Part 2 require more time due to the electronics content. Chapters 6 and 7 are important to the rest of the book; Chapters 8 and 9 emphasize advanced electronic circuit design, making them optional. The material in Part 3 is mixed. Chapter 10 on Verilog can be covered in 3 or 4 lectures. The remaining chapters

General Introduction	Systems Orientation	Circuits Emphasis
Part 1 Chapters 2-5	**Part 3** Chapters 10, 15	**Part 1** Chapters 2-5
Part 2 Chapters 6-8	**Part 1** Chapters 2-5	**Part 2** Chapters 6-9
Part 3 Chapters 10-11, 14-15	**Part 2** Chapters 6, 7	**Part 3** Chapters 12-15
	Part 3 Chapters 14, 16	

emphasize specific topics, and do not have to be followed in order. Chapters 14 and 15 should be included in a systems design course.

CAD Tools

CAD tools are discussed in a generic manner without any reference to a particular toolset. That being said, it should be noted that the accompanying CD contains limited-feature versions of the AIM-SPICE and MicroCap6 SPICE circuit analysis programs, and the Silos III Verilog simulator. These can be used to provide hands-on experience in circuit simulation and Verilog coding for PC users. The current contents of the CD are listed in the README.TXT file, which also has loading instructions.

Design Projects

A design project is an excellent vehicle for learning VLSI in a classroom environment. At Georgia Tech, the initial lab assignments deal with layout, DRC, extraction, and circuit simulation. These are stored in a library for eventual use in designing a large chip. Specifications change every year, spurred on by the rapid advances in technology.

The design project itself is always done by groups. Each group consists of 5 to 20 students. Design projects are assigned by the instructor. Examples of past projects are basic microprocessors, MIMD array processors, telecommunication interfaces, pipelined datapaths, and DSP networks.

The project assignment consists of only a system level specification. The details are left entirely to the student engineering group. They must produce a high-level HDL model that is used to translate down to the behavioral, logic, circuit, and layout levels. A full CAD suite is made available, with the exception of a synthesis tool. This forces the group to use the cell library that they designed in the early lab exercises. The design project is meant to expose the students to as many aspects of VLSI as possible.

Acknowledgments

Mark Berrafato and Charity Robey of John Wiley & Sons provided much inspiration for me during the early stages of this project. Bill Zobrist, my editor, never seemed to lose faith that the book would be eventually completed. He maintained his support even after the birth of his first child Ian, somehow balancing time between his family and his job! Jenny Welter and Susannah Barr kept track of every detail, large and small, that kept the project running. In addition to designing the cover for the book, Maddy Lesure helped me solve graphics problems that enhanced the appearance. Finally, my hat is off to my production editor, Christine Cervoni, for the spectacular job she did coordinating the project and checking every detail!

Several reviewers provided helpful comments that influenced the final form of the book. Particular thanks are due to Professors Krishnendu Chakrabarty (Duke University), Mona Zaghloul (George Washington University), Ralph Teeing-Cummings (The Johns Hopkins University), and Giovanni De Micheli (Stanford University). I would like to thank the several hundred Georgia Tech ECE students who have taken my chip design courses over the past few years. The countless hours they spent on the design projects allowed me to see how well the lecture material translated into practical application. Their feedback on the course and manuscript has been a great help in reworking the presentation for the next offering. Michael Robinson in particular did an exceptionally thorough reading of several chapters. Tony Alvarez (Cypress Semiconductor) and Brian Butka (IDT) went the extra mile to provide several of the die photos in the book.

I would like to thank Dr. Roger Webb, Chair of the School of Electrical & Computer Engineering at Georgia Tech, for his continuing support of my writing projects. Professors Bill Sayle and Joe Hughes have always managed to accommodate my teaching requests that allowed me to get involved with the VLSI offerings. Conversations with Professors John Buck and Glenn Smith always boost my morale.

Finally, I would like to once again thank my wife Melba, and my daughters Christine and Valerie for their endless patience and support throughout this (and every) project. Although I cannot repay them for the hours I spent writing this book, perhaps a short trip to France will help. *Au revoir!*

John P. Uyemura
Atlanta, GA
April, 2001

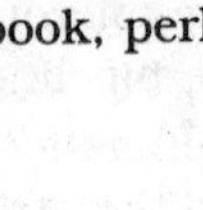

Table of Contents

Part 3 - The Design of VLSI Systems

An Overview of VLSI

1

VLSI is an acronym that stands for **very-large-scale integration**. This somewhat nebulous term is used to collectively refer to the many fields of electrical and computer engineering that deal with the analysis and design of very dense electronic integrated circuits. Although a strict definition is difficult to come by, one commonly used metric is to say that a VLSI contains more than a million (10^6) or so switching devices or logic gates. Early in the first decade of the 21st century, the actual number of **transistors** (the switching devices) has exceeded 100 million (10^8) for the more complex designs on a piece of silicon (a **chip**), which is typically about 1 centimeter on a side.

This book has been written to provide an understanding of the basics of **digital VLSI chip design**. Emphasis is placed on presenting the details of translating a system specification to a small piece of silicon. The treatment is very technical with many details. Some statements and analyses will appear immediately obvious, while others may not make sense until later chapters. This occurs because the field of VLSI engineering encompasses several distinct "areas of specialization" that mesh together in a unique manner. The most difficult aspect of learning VLSI is seeing the common theme that links the areas together. Once this is accomplished, you are on your way to understanding one of the most fascinating fields of modern times.

1.1 Complexity and Design

Engineering a VLSI chip is an extremely complex task. When attempting to describe the field to a non-technical group, the idea of the "VLSI design funnel" shown in Figure 1.1 helps break the ice. This views the process as one where we provide the basic necessities such as money, an idea, and

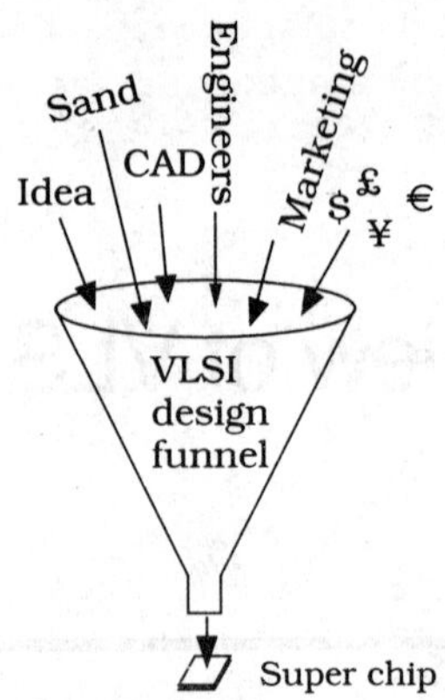

Figure 1.1 The VLSI design funnel

marketing information and dump them all into a "magic technology funnel." Adding a pile of sand as a raw material produces the super chip at the bottom that will sell millions of units and hopefully revolutionize the world. And maybe make someone rich. Of course, engineers and scientists are needed somewhere in the process, but they just put the things together. Unfortunately, the process is slightly more complicated than portrayed in this example.

Any system that is composed of millions of elements is inherently difficult to understand. One human mind cannot process information of the complexity that is required for the design and implementation. Creating a **design team** provides a realistic approach to approaching a VLSI project, as it allows each person to study small sections of the system. In a modern design, hundreds of engineers, scientists, and technicians may be working different parts of the design. However, since the team is working on a single project, it is important that each team member have some understanding of where their work falls within the overall scheme. This is accomplished by means of the **design hierarchy**, where the chip is viewed at many different "levels" from the abstract to the physical implementation. Every level is important, and each has subdivisions that can evolve into a lifetime career.

In our treatment of VLSI, we will continually stress the fact that the field is inherently multidisciplinary in nature. Specialists in an uncountable number of areas are needed to produce a working functional design. Computer architects must interact with code writers and logic designers, and they must be able to comprehend some of the problems of circuit design and silicon processing. Electronics experts must move beyond circuits to see how their units will affect the system. And everyone depends upon the computer-aided design tools and the support groups that perform the 10,000 or so other tasks not described here. If this description makes the field sound complicated, that's because it is. VLSI is not a simple discipline to understand. But it is possible to learn the basics in a reasonable amount of time. Persons who end up working in the area usually gravitate there because one or more aspects catch their interest and fall within their background.

Now that we have an appreciation of what is involved, let us move to a better description of the design process. An overview with the major steps in the sequence is shown in Figure 1.2. The starting point of a VLSI design is the system specification. At this point, the product is defined in both general and specific terms that provide design targets such as functions, speed, size, etc., for the entire project. This is the "Top" level of the design hierarchy. The system specifications are used to create an abstract, high-level model. Digital design is usually based on some type of **hardware description language** (HDL) that allows abstract modeling of the operation. VHDL and Verilog™ are the most common HDLs in practice, but several others (including C and C++) are used. The abstract model contains information on the behavior of each block and the interaction among the blocks in the system. The model is subjected to extensive verification steps where the design is checked and rechecked to ensure that it is correct.

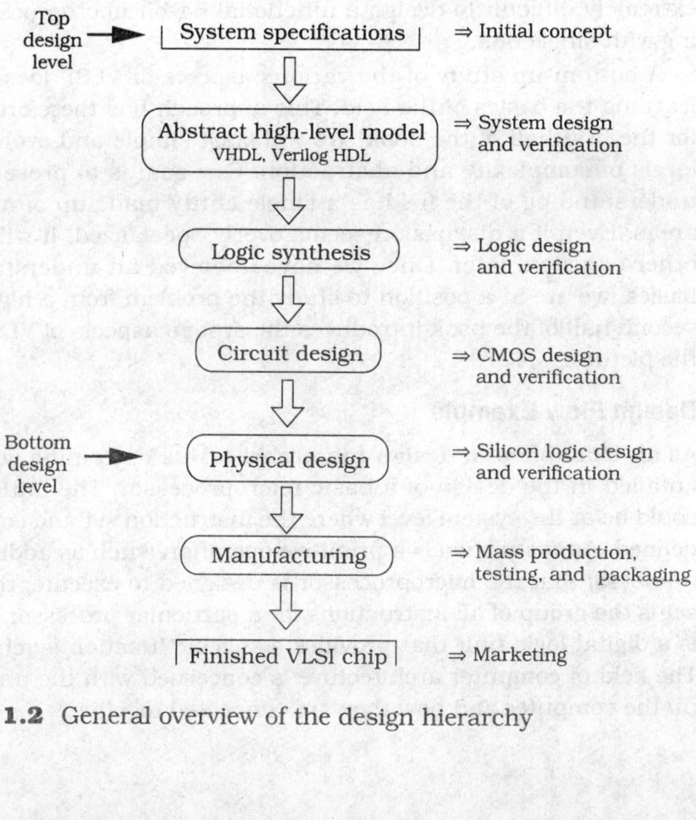

Figure 1.2 General overview of the design hierarchy

The next step in the process is called **synthesis**. The abstract logic model is used to provide the **logical design** of the network by specifying the primitive gates and units needed to build each unit. This then forms the basis for transferring the design to the **electronic circuit level** where transistors are used as switches and Boolean variables are treated as varying voltage signals. To create a transistor, we move down another level to that of **physical design**. At this level, the network is built on a tiny area on a slice of silicon using a complex mapping scheme that translated transistors and wires into extremely fine-line patterns of metals and other materials. The physical design level constitutes the Bottom of the design hierarchy. After the design process is completed, the project moves on to the manufacturing line. The final result is a finished electronic VLSI chip.

When we start at the system level specification, the design process is called the **top-down** approach. The initial work is quite abstract and theoretical and there is no direct connection to silicon until many steps have been completed. The reverse approach starts at the silicon or circuit level and builds primitive units such as logic gates, adders, and registers as the first steps. These are combined to obtain larger and more complex logic blocks, which are then used as building blocks in even larger designs. This **bottom-up** approach is acceptable for small projects, but the complexity of modern VLSI designs makes it impractical; it is extremely difficult to design a functional 64-bit microprocessor by starting with single bits.

A bottom-up study of the various aspects of VLSI does work well for learning the basics of the field. This approach has therefore been chosen for the first half of the book. We will start simple and evolve into higher levels of complexity and abstraction. Our goal is to present a coherent understanding of the field as a single entity made up of many different areas. Even if a discussion seems overly specialized, it will be linked to other concepts later. Once we have achieved an understanding of the basics, we are in a position to study the problem from a higher level. The second half of the book introduces the system aspects of VLSI to complete the picture.

1.1.1 Design Flow Example

As an example of a design hierarchy, let us determine what would be entailed in the design of a basic microprocessor. The initial conception could be at the system level where the instruction set and components are defined. An *instruction* is a primitive operation (such as adding two binary numbers) that the microprocessor is designed to execute; the instruction set is the group of all instructions for a particular processor. A component is a digital logic unit that provides a specific function (such as addition). The field of computer architecture is concerned with the units that make up the computer and how they are connected together.

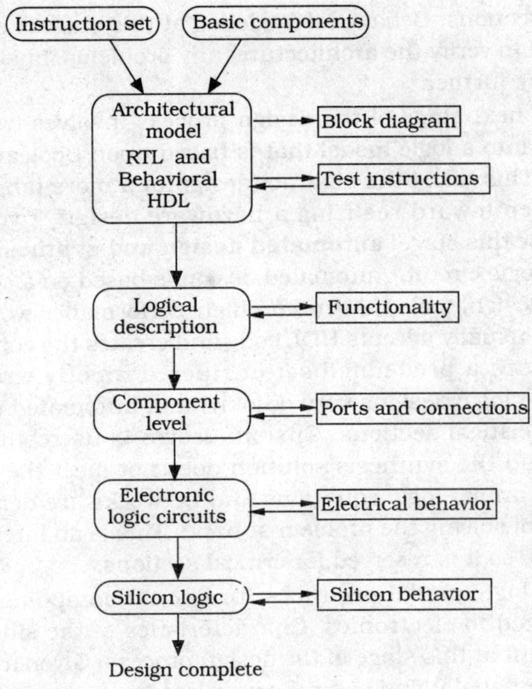

Figure 1.3 A simple design flow for a microprocessor

A basic design flow for the problem is shown in Figure 1.3. The instruction set and component group can be used to construct a high-level model of the architecture. At this level, the behavior of the system is described in an abstract manner that ignores the low-level details needed to actually build the network. For example, we may define an addition event by writing

$$\text{Register\_X} \leftarrow A + B$$

which is translated as saying that the sum of A and B is transferred to a storage device named Register_X. High-level abstractions of this type can be used to define the processor architecture, and are commonly known as the register-transfer level (RTL) description. RTL models describe the operation of the system without reference to specific components. When written with an HDL, it can be used to test instructions and verify the architectural behavior Abstract design allows us to construct a block diagram for the system.

RTL code can be translated to an equivalent description that contains more detail about the operation and behavior components. The operation of each block can be summarized at the HDL **behavioral level**, where the emphasis is on the large-scale behavior of the blocks as they interact with other sections. Behavioral modeling at this stage is extremely critical as it is used to verify the architecture; any problems must be solved before progressing further.

The next stage of the design process involves translating the system blocks into a logic model that is based upon Boolean equations and logic gates. This takes the abstract design to a more tangible level, and is the first step toward realizing a hardware design. Two approaches can be used for this stage: **automated design and synthesis**, or **custom design** of the logic circuit. Automated design is based on a set of CAD (computer-aided design) tools that run on high-performance workstations. A synthesis tool usually accepts HDL code and creates the corresponding logic network with a predefined set of rules. Properly written HDL code can produce logic designs very quickly, and automated synthesis is used for all noncritical sections. Custom design is used when special problems arise and the synthesis solution does not meet the necessary specifications. Various logic equations and networks are derived and tested as a means of solving the problem at hand. This is an intense, time-consuming process, so it is reserved for critical sections.

The logic model produces functional components, which are then translated to electronics. Characteristics of the silicon circuits become important at this stage of the design process. Given a large-scale function, one can usually find several equivalent logic expressions; all produce the same output, but will use different equations and gates. Recall, for example, how a Karnaugh map is used to simplify logic equations. Silicon VLSI is complicated by the fact that each type of logic gate or circuit has distinct characteristics, and we must often search for circuits that are faster or smaller than what can be obtained using an obvious solution. A sophisticated synthesis tool can take HDL code and provide suggested designs for both the logic gates and the silicon circuits. However, the toolsets have not yet reached the level where they are powerful enough to produce the "best" design, whatever that may mean.

After the logic network and circuits have been designed, the next step is to use the information to produce an integrated circuit at the physical design level. This is accomplished in a series of steps where transistors are defined as 3-dimensional structures on a chip of silicon, and are then placed and wired using another set of graphical CAD tools. Once this is accomplished, the designs are tested and verified, and then used to create a database that allows the manufacturing line to actually build the electronic chip. Fabricating a VLSI chip is itself a complex specialized field. Once started, it may take several weeks to produce the final circuits.

The specifics of the procedures are much more complicated than what is portrayed in the simple flowchart. It does, however, illustrate the essence of a top-down design flow. VLSI design is concerned with filling in the details needed to produce a manufactured chip that functions as designed with high reliability and a long lifetime. And can be sold at a profit, of course.

1.1.2 VLSI Chip Types

At the engineering level, digital VLSI chips are classified by the approach used to implement and build the circuit. A **full-custom** design is one where every circuit is custom designed for the project. This is an extremely tedious and time-consuming process that makes it impractical for designing an entire system.

Application-specific integrated circuits (ASICs) allow digital designers to create ICs for a particular application. ASICs are very popular for prototyping or low-volume production runs. They are designed using an extensive suite of CAD tools that portray the system design in terms of standard digital logic constructs: state diagrams, function tables, and logic diagrams. Usually, an ASIC designer does not need any knowledge of the underlying electronics or the actual structure of the silicon chip. Design automation CAD tools are responsible for taking the logic design and building most of the chip. One drawback of ASIC design is that all characteristics such as speed are set by the architectural design; the designer does not have access to the electronics, so delay times cannot be changed. Modern ASICs have evolved to a high level of sophistication, and are generally capable of providing solutions to a large class of problems.

A **semi-custom** design is in between that of a full-custom and an ASIC-type circuit. The majority of the chip is designed using a group of primitive predefined cells as building blocks. Each cell provides a basic function, such as a logic operation or a storage circuit, and the master design resides in a database collection called a library. A cell entry contains all of the information needed to create the circuit on silicon. If it is not possible to meet the system specifications using the cell library, then the semi-custom approach permits the designer to engineer a solution by creating alternate silicon circuits that have the desired characteristics. These are used only in small sections where the problems occur. For example, floating point circuits in microprocessors can be extremely complex, so that some sections may require custom design to meet the clocking budget. Variations of semi-custom design are used for most high-performance chips.

1.2 Basic Concepts

The objective of this book is to present the field of VLSI in its entirety. Overall, VLSI design is a system design discipline. Many aspects can be taught without any reference to the underlying silicon circuits. System solutions can be generated using the CAD tools, and the necessary data turned over to the manufacturing group for production. While this approach produces functional solutions, it makes many of the details invisible to the designer. Simplifying the design process is important. However, many of the most powerful techniques and ideas of VLSI reside at lower levels and are therefore lost. Circuits work, but they are not as fast or as small as they could have been.

VLSI should be thought of as a single discipline that deals with the conception, design, and manufacture of complex integrated circuits. Many system-level concepts are based on the characteristics of electronic circuits that are made at the silicon level. When Carver Mead of Caltech pioneered the field in the 1970's, one of the most important foundations for VLSI arose from his observation that digital electronic integrated circuits could be viewed as a set of geometrical patterns on the surface of a silicon chip. Groups of patterns represented different logic functions and were repeated many times in the system. Complexity could thus be dealt with using the concept of repeated patterns that were fitted together in a structured manner. Signal flow and data movement could be followed by tracing the paths of the metallic "lines" that carried electricity. It was possible to write Boolean expressions that could be directly translated to geometrical patterns on silicon in a well-defined manner. The microphotograph of a CMOS chip in Figure 1.4 shows many of these features in a finished device. Note in particular the repeated patterns and ordered placement of rectangular lines, polygons, and groups of geometric patterns.[1] Mead's observation (and a huge amount of work) has structured VLSI into the important field it is today. The importance of gaining an overall unified view of VLSI becomes clear.

VLSI design encompasses many practical aspects of digital system design. One is the fact that even the most powerful **system on a chip** (SOC) must be interfaced to other components to create an operational unit. This is achieved by placing the silicon circuitry in the center of a rectangular piece of material, and then providing some type of scheme that allows external wires to contact it. Figure 1.5 shows the use of **bonding pads**, which are square metal sections where wires can be bonded and connected to the package in which the chip is mounted. A more advanced technique developed by IBM is called the C4 technology; it allows metal "bumps" to be located across the surface area. Contact to the package is established by "flipping" the chip so that the bumps are on the bottom and can be aligned to a wiring grid. Regardless of the approach, there is a limit on the actual size of the chip.

In an ideal world, we could make the chip as large as desired. This would allow increasingly complex systems to be designed without any bounds. Unfortunately, it is not possible to manufacture a functional

[1] This is a section of a binary adder network designed at Georgia Tech. Each group of patterns adds two bits and produces sum and carry outputs.

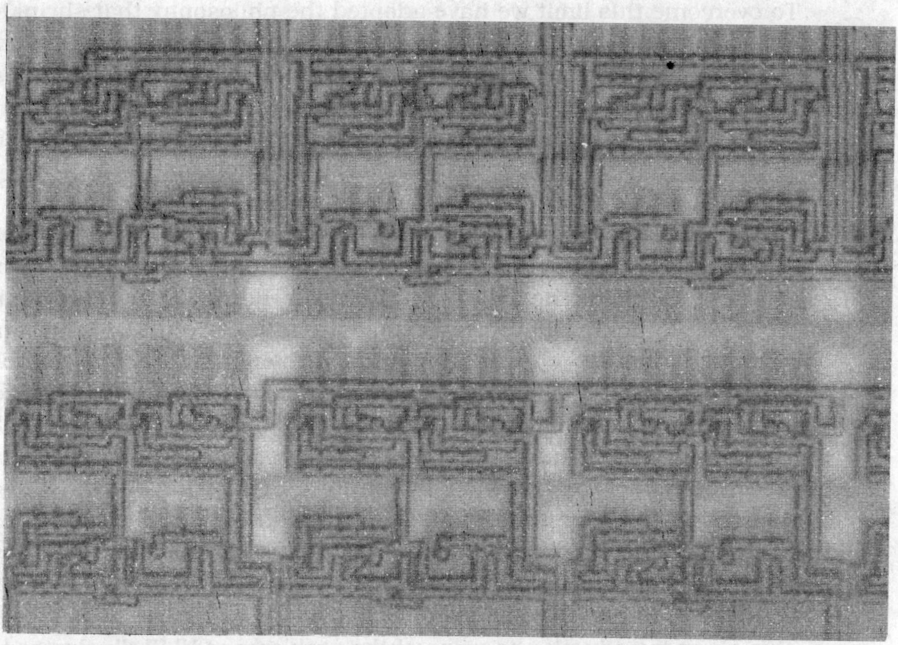

Figure 1.4 Micrograph of a section of a digital CMOS integrated circuit

design because of defects in the silicon crystal structure that cannot be avoided. The larger the area of the circuit, the higher the probability that a defect will occur. Even a single bad transistor or connection renders the circuit nonfunctional, so we attempt to keep the overall size of the chip small. We note in passing that other problems in manufacturing also limit the size of the chip.

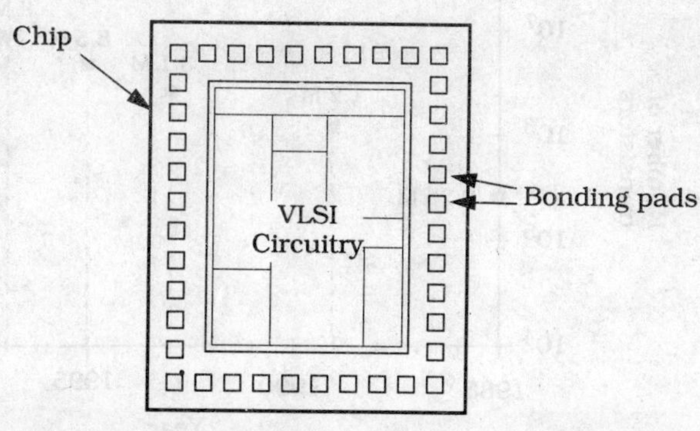

Figure 1.5 Bonding pad frame for interfacing

To overcome this limit we have adopted the philosophy that shrinking the size of a transistor will allow more devices to be placed in a given chip area. Technologically, this is a very difficult problem. Precision design and manufacturing techniques give devices where the smallest dimension is around 1.3×10^{-7} meters. At this level, we change our measurement metric to the micrometer (μm), or **micron** for short, such that 1 μm = 10^{-6} m = 10^{-4} cm, and refer to the technology as a 0.13-μm process.

One of the classical predictions in VLSI transistor densities is known as **Moore's Law**. Gordon Moore, one of the cofounders of the Intel Corporation, visualized in the 1970's that chip building technology would improve very quickly. He projected that the number of transistors on a chip would double about every 18 months. Although there have been variations due to technological problems or economic slowdowns, Moore's Law has proved amazingly close to actual trends. Figure 1.6 shows a plot of device count as a function of year for a group of randomly selected microprocessor chips from major vendors. There have always been debates as to how long the transistor count can continue to increase at this rate due to technological limitations in reducing the size. Regardless of the actual slope, however, it seems clear that VLSI design will remain a powerful force for many years to come.

This short introduction to some of the problems of VLSI illustrates the vast nature of the field itself. The role of a VLSI design group is to create a large, complex system on a tiny piece of silicon. The group faces constraints at every level, from the abstract modeling and timing down to building a chip with millions of transistors. Project status presentations, engineering summaries, and critical deadlines are always present.

Welcome to the exciting world of VLSI!

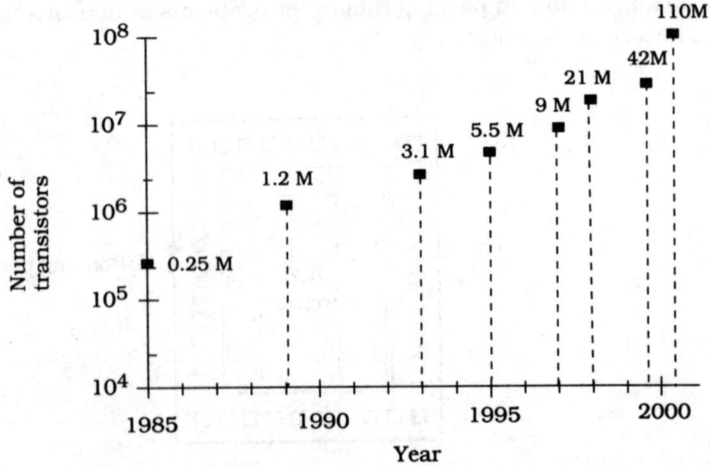

Figure 1.6 Device count by year

1.3 Plan of the Book

The book has been divided into three main sections. For self-study, it is best to follow them sequentially, although it is not necessary to read every section in a first reading.

Part 1 is entitled *Silicon Logic*, and includes Chapters 2 through 5. The material examines the techniques for designing logic networks in silicon. It concentrates on introducing transistor logic circuits and how they translate to patterns on a silicon chip. Details of the CMOS processing sequence are presented, and applied to realistic chip design. After completing Part 1, the reader will be able to design a myriad of CMOS logic gates at both the circuit and silicon levels.

The electronics of VLSI are covered in Part 2, which is entitled *The Logic-Electronics Interface*. Part 2 includes Chapters 6 through 9, with the more advanced concepts in Chapters 8 and 9. Transistor switching characteristics are presented and then used to analyze digital electronic logic gates. The treatment is quite detailed, but it concentrates on the important basics that affect system performance and switching speeds. Completing this section of the book will provide a solid understanding of the relationship between logic design and electrical characteristics.

System-level problems are addressed in Part 3, *The Design of VLSI Systems*, which includes Chapters 10 through 16. The basics of Verilog® HDL are presented as the vehicle for system-level modeling. Many VLSI logic components such a multiplexors, adders, and memories are studied in Chapters 11 through 13. Large-scale chip design issues are addressed in Chapters 14 and 15. The book concludes with an introduction to digital testing in the final chapter.

An effort has been made to present the material in a readable, coherent manner that concentrates on explaining the details. This is particularly true in Part 1, where the reader is exposed to subject matter that is not typically found in other courses.

So, without further delay, let us begin our trek into the world of very-large scale integration.

1.4 General References

[1] Dan Clein, **CMOS IC Layout**, Newnes Publishing Co., Boston, 2000.

[2] Randy H. Katz, **Contemporary Logic Design**, Benjamin-Cummings Publishing Co., Redwood City, CA, 1994.

[3] Ken Martin, **Digital Integrated Circuit Design**, Oxford University Press, New York, 2000.

[4] Jan Rabaey, **Digital Integrated Circuits**, Prentice-Hall, Upper Saddle River, NJ, 1996.

[5] Michael John Sebastian Smith, **Application-Specific Integrated Circuits**, Addison-Wesley Longman Inc., Reading, MA, 1997.

[6] John P. Uyemura, **A First Course in Digital Systems Design**, Brooks-Cole Publishers, Pacific Grove, CA, 2000.

[7] John P. Uyemura, **CMOS Logic Circuit Design**, Kluwer Academic Press, Norwell, MA, 1999.

[8] John P. Uyemura, **Physical Design of CMOS Integrated Circuits Using L-Edit®**, PWS /Brooks-Cole Publishers, Pacific Grove, CA, 1995.

[9] M. Michael Vai, **VLSI Design**, CRC Press. Boca Raton, FL, 2001.

[10] Neil H.E. Weste and Kamran Eshraghian, **Principles of CMOS VLSI Design**, 2nd ed., Addison-Wesley Publishing Co., Reading, MA, 1993.

[11] Wayne Wolf, **Modern VLSI Design**, 2nd ed., Prentice-Hall PTR, Upper Saddle River, NJ, 1998.

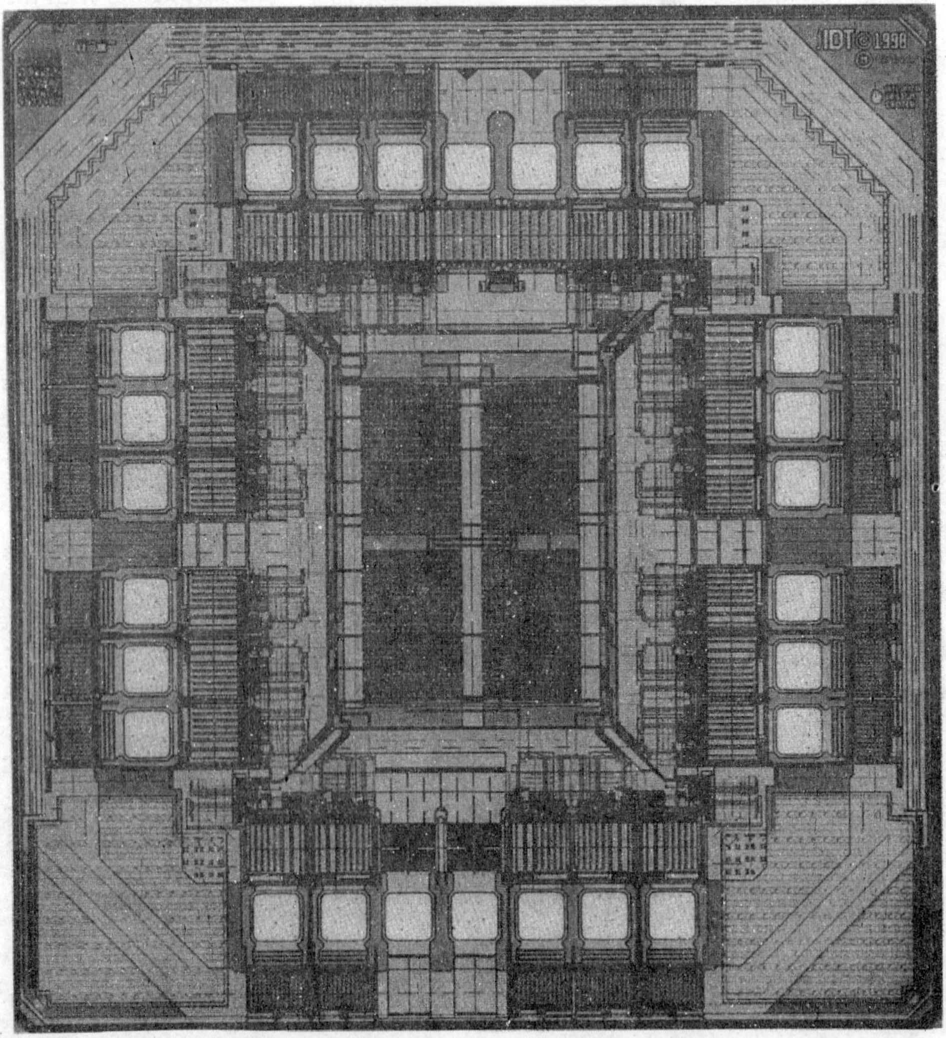

Part 1

Silicon Logic

Logic Design with MOSFETs 2

CMOS integrated circuits use bi-directional devices called MOSFETs as logic switches. This chapter examines the logical characteristics of MOSFETs and develops techniques for building digital networks.

2.1 Ideal Switches and Boolean Operations

All digital designs are based on primitive logic operations. The first task in our study of VLSI will be to create electronic logic gates that can be used as building blocks in complex switching networks.

Logic gates are created by using sets of **controlled switches**. The characteristics of an **assert-high** controlled switch are illustrated by the drawings in Figure 2.1. In this idealized situation, the state of the switch (**open** or **closed**) is determined by the value of the control variable A. In Figure 2.1(a), the control bit has the value of $A = 0$ which is defined to give an open switch. This means that there is no relationship between the two variables x and y as represented by the gap between the left and right sides. The opposite case is a closed switch where we visualize the top portion of the switch being "pushed down" as shown in Figure 2.1(b). This condition occurs when $A = 1$ and connects the two sides of the switches

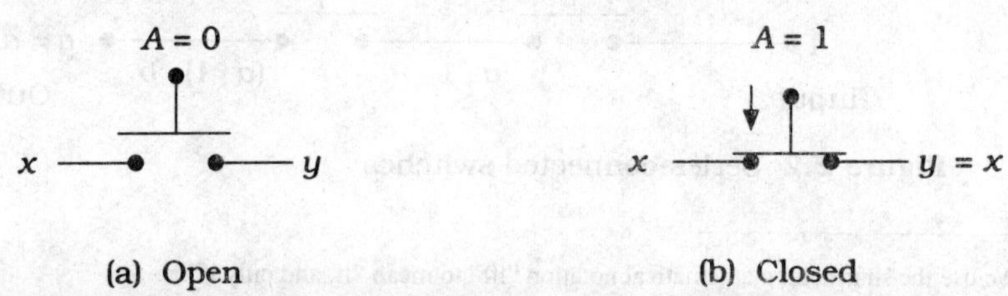

Figure 2.1 Behavior of an assert-high switch

so that

$$y = x \tag{2.1}$$

is valid. If we interpret the left side variable x as the input and the right side as the output, then we can say that the condition $A = 1$ allows the input variable to flow through the switch and establish the value of the output. This is called an "assert-high" switch because a high control bit of $A = 1$ is required to close the circuit.

A different approach to characterizing the behavior of the switch is to write the logic equation[1]

$$y = x \cdot A \quad \text{iff} \quad A = 1 \tag{2.2}$$

By itself, the relationship between x and y is undefined if $A = 0$. Although this appears to be a serious deficiency, in practice we will avoid the problem by using additional switches to define the value of y for this case.

Let us now proceed to create a logic network by combining the concept of an ideal switch with a voltage source. Suppose that we take two switches that are controlled by the independent variables a and b and connect them as shown by the diagram in Figure 2.2. The two switches are said to be **in series** with each other. As we trace the signal path through the first switch, equation (2.2) shows that the output (directly after the switch) is given by $a \cdot 1$ as indicated on the drawing. This acts as the input to the second switch, so that applying equation (2.2) again yields

$$g = (a \cdot 1) \cdot b = a \cdot b \tag{2.3}$$

for the output. This is easy to interpret using a qualitative analysis: both switches must be closed with $a = 1$ AND $b = 1$ to allow the input 1 to reach the output and result in $g = 1$. The circuit appears to provide the AND2 operation.[2] However, note that equation (2.2) is valid only if the control bit has a value of 1; it if is 0, then there is no direct relationship between the left and right sides of the switch. There are three other possi-

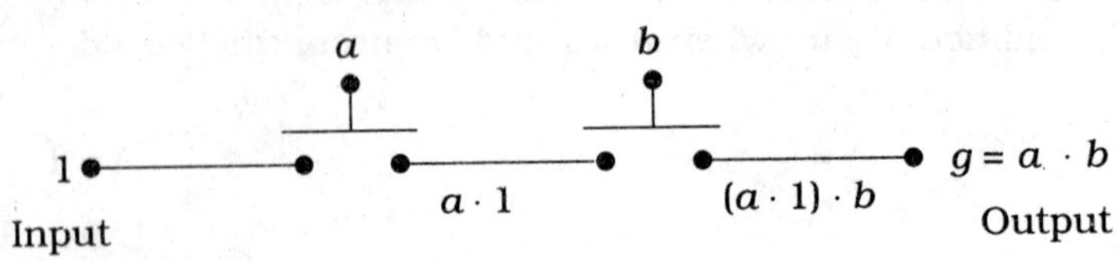

Figure 2.2 Series-connected switches

---

1 We use the shorthand mathematical notation "iff" to mean "if, and only if".

2 We denote a 2-input AND operation as an AND2. This type of notation will be used for all gates. For exam ple, an OR2 operation implies a 2-input OR.

bilities for the two inputs:

$$(a, b) = (1, 0), (0, 1), (0, 0) \tag{2.4}$$

Any of these input combinations should result in a logical output of $g = 0$ but the logic equations say that g is undefined.

Before proceeding any further, let us clarify our approach to portraying logic networks. In general, the switch drawings will be called **schematic diagrams** since they show the "scheme" used in the wiring. We extend this terminology to include diagrams that contain electronic devices. Since we want to keep the drawings relatively compact and neat looking, wiring lines will often cross one another in the drawing. When this occurs, we will adopt the convention shown in Figure 2.3. In Figure 2.3(a), Wire 1 and Wire 2 are assumed to be totally separate. The signal a on Wire 1 has no relationship to the signal b on Wire 2. If we wish to create a connection, we will use a "dot" as in Figure 2.3(b). In this case, the two wires are connected so that placing a signal a on one of the lines results in the same value on all points of both lines.

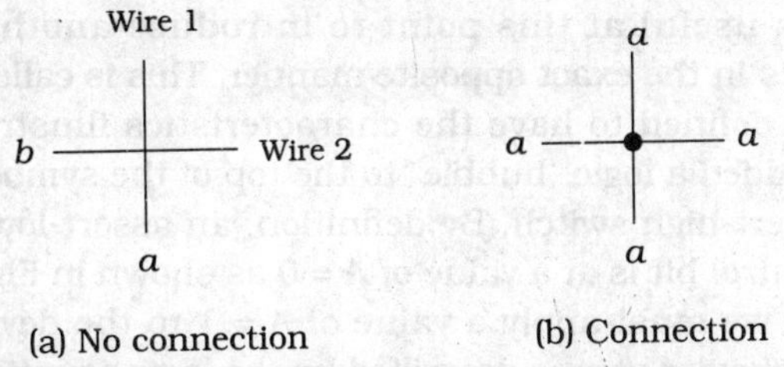

Figure 2.3 Connection convention used in schematic diagrams

Let us examine another circuit that has the same problem. Figure 2.4 shows two switches that are controlled by the independent variables a and b, but the two are wired **in parallel** with one another; this means that the left (input) sides are connected together and the right (output) sides are connected together. The output f can be constructed by recog-

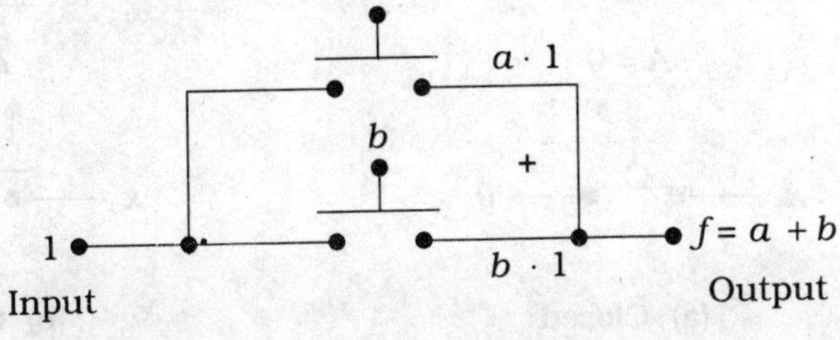

Figure 2.4 Parallel-connected switches

nizing that according to equation (2.2), the top switch produces an output of $a \cdot 1$ iff $a = 1$, while the lower switch produces an output of $b \cdot 1$ iff $b = 1$. Both expressions are shown at the appropriate points in the diagram. We conclude that if either $a = 1$ OR $b = 1$ (or both), then the output is described by the single expression

$$g = a + b \tag{2.5}$$

which appears to be the OR2 operation at this point in the analysis. Parallel-connected switches can thus be used to OR variables together; this is indicated on the diagram by including the "+" between the switches. Note, however, that if $a = 0$ and $b = 0$ at the same time, then the output g of the switching network is undefined. It thus fails to provide the entire OR2 function.

The preceding examples illustrate that switches have characteristics that can be used as a basis for implementing logic operations. However, since the logic equation (2.2) is valid only if the switch is closed, we were not able to obtain complete AND and OR gates as neither network could produce a logic 0 output.

It is useful at this point to introduce another type of switch that behaves in the exact opposite manner. This is called an **assert-low** switch and is defined to have the characteristics illustrated in Figure 2.5. We have added a logic "bubble" to the top of the symbol to distinguish it from an assert-high switch. By definition, an assert-low switch is closed when the control bit is at a value of $A = 0$ as shown in Figure 2.5(a). To open the switch we must apply a value of $A = 1$ to the device as in Figure 2.5(b). This behavior can be described by the logic equation

$$y = x \cdot \overline{A} \qquad \text{iff } A = 0 \tag{2.6}$$

In this case, the value of y is not defined if $A = 1$. Comparing the two types of switches we see that they behave in a complementary manner.

As an example of how this type of switch can be used, consider the series-connected pair in Figure 2.6. Tracing the signal path from the input through the first switch gives an output of $\overline{a} \cdot 1$ which is valid iff $a = 0$. This acts as the input to the second switch so that the output of the

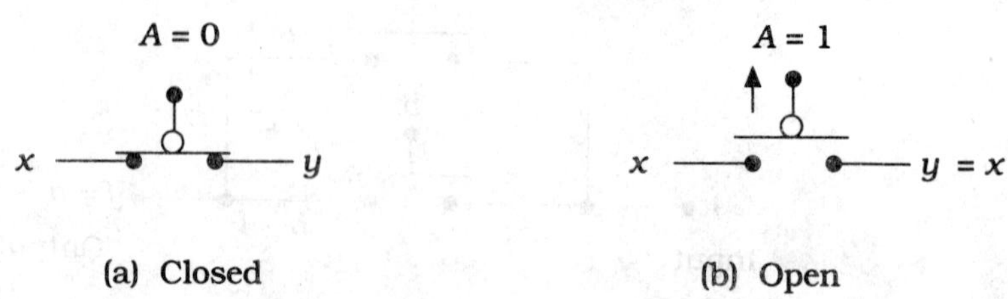

Figure 2.5 An assert-low switch

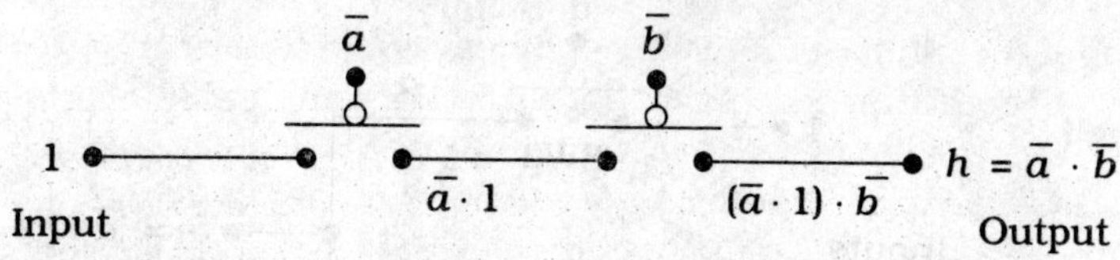

Figure 2.6 Series-connected complementary switches

series chain is given by

$$h = (\bar{a} \cdot 1) \cdot \bar{b}$$
$$= \overline{a + b} \tag{2.7}$$

where we have used the DeMorgan relation to write the second line. This looks like the NOR2 operation. However, since the second switch must be closed with $b = 0$, this result is correct only if both $a = 0$ and $b = 0$. If either a or b is a 1, then g is undefined. We thus have the same type of problem experienced in our earlier examples.

Let us now progress to the idea of using both types of switches in a single network. We will provide both logic 1 and logic 0 inputs in an effort to produce an output that is defined for all possible input combinations. In Figure 2.7, the assert-high switch SW1 is used to connect a logic 0 input to the output y, while the assert-low switch SW2 connects a logic 1 input to y. The variable a controls both switches. Since the two are in parallel, we may write the OR relation between the upper and lower branches to give the output in the form

$$y = \bar{a} \cdot 1 + a \cdot 0 \tag{2.8}$$

The operation of the circuit can be understood by specifying a value for a. If $a = 0$, then SW1 is open and SW2 is closed which gives

$$y = \bar{0} \cdot 1 + 0 \cdot 0 = 1 \tag{2.9}$$

If $a = 1$, then SW1 is closed and SW2 is open. Substituting into the expression we have

$$y = \bar{1} \cdot 1 + 1 \cdot 0 = 0 \tag{2.10}$$

This circuit thus eliminates the problem of an undefined voltage. Moreover, since logically $a \cdot 0 = 0$, the expression reduces to

$$y = \bar{a} \tag{2.11}$$

In other words, this circuit implements the NOT operation.

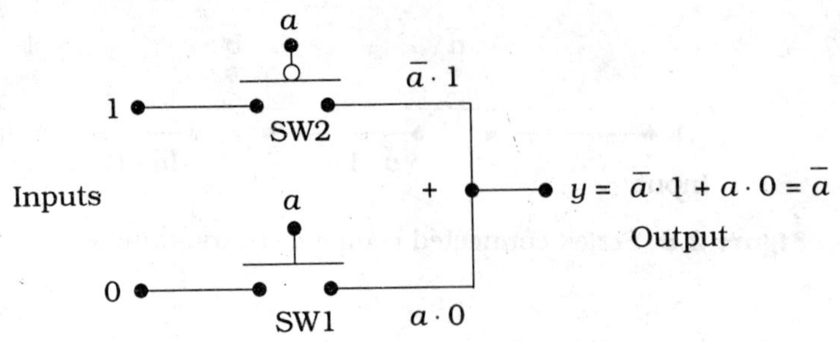

Figure 2.7 A switch-based NOT gate

$$y = \text{NOT}(a) = \bar{a} \tag{2.12}$$

This demonstrates that using two switches with opposite characteristics allows us to build a network with well-defined results.

The NOT circuit in Figure 2.7 is based on the behavior of a 2:1 multiplexor as shown in Figure 2.8. The MUX uses the input a to select between input 0 (that has a "1" applied to it) when $a = 0$, or input 1 (that has a "0" applied to it) when $a = 1$. The output is given by the expression

$$y = \bar{a} \cdot 1 + a \cdot 0 \tag{2.13}$$

which reduces to $y = \bar{a}$. A close examination of the switching circuit in Figure 2.7 verifies that there is a one-to-one correspondence with the 2:1 multiplexor.

2.2 MOSFETs as Switches

MOSFETs are electronic devices that are used to direct and control logic signals in high-density digital IC design.[3] The acronym "MOSFET" stands

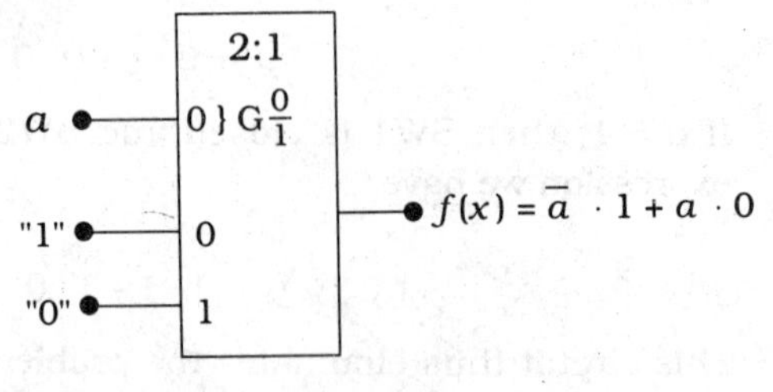

Figure 2.8 A MUX-based NOT gate

"MOSFET" is pronounced as *moss-fet*.

for **metal-oxide-semiconductor field-effect transistor**, but we will not worry about the details just yet. In many ways, MOSFETs behave like the idealized switches introduced in the previous section. There are important differences that must be taken into account before they are used. These arise from the fact that MOSFETs must obey circuit equations and their ultimate performance is limited by the laws of physics. In this section we will concentrate on creating switching models for the devices. The more complicated aspects of current flow will be discussed in later chapters.

Complementary MOS (**CMOS**) uses two types of MOSFETs to create logic networks. One type is called an n-channel MOSFET (or nFET for short), and uses negatively charged electrons for electrical current flow. The circuit symbol for an nFET is shown in Figure 2.9(a). The **gate** terminal acts as the control electrode for the device. Applying a voltage on the gate electrode determines the current flow between the **drain** and **source** terminals. The other type of transistor is called a p-channel MOSFET or pFET. It uses positive charges for current flow and has the circuit symbol drawn in Figure 2.9(b). The only graphical difference between the nFET and pFET symbols is the inversion bubble at the gate. As with the nFET, the voltage applied to the gate determines the current flow between the source and drain terminals. Do not confuse the **gate terminal** of a MOSFET with a **logic gate**, as the two "gates" are not related. The context of the discussion always helps clarify the usage.

MOSFETs are intrinsically electronic devices. To use them as logic-controlled switches, we must first define how to translate between Boolean values and electrical parameters. This is accomplished by using voltages that exist on the chip when we apply an external power supply. In the most general case, two power supply voltages V_{DD} and V_{SS} are defined as shown in Figure 2.10. The reference terminal is taken to be the **ground** connection (which is at 0 volts) between the two sources so that the chip receives both a positive power voltage V_{DD} and a negative power supply voltage V_{SS}. Early generations of silicon MOS logic circuits used both positive and negative supply voltages. However, modern designs require only a single positive voltage V_{DD} and the ground connection; common values are V_{DD} = 5 V and 3.3 V or lower. The remaining source is set to a value of

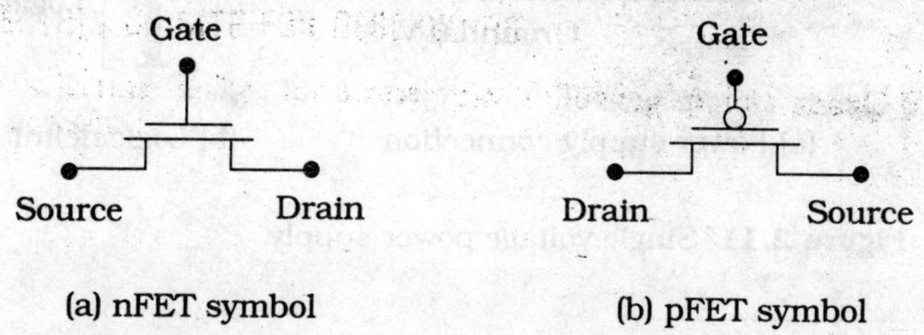

Figure 2.9 Symbols used for nFETs and pFETs

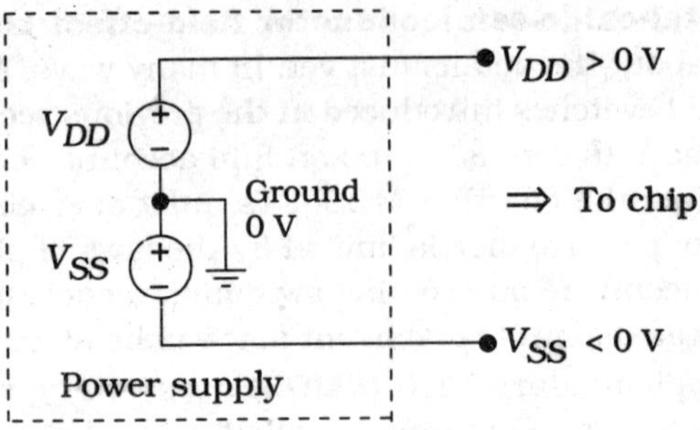

Figure 2.10 Dual power supply voltages

V_{SS} = 0 V, which results in the power supply network portrayed in Figure 2.11(a).[4] We will assume that all of our circuits use only a single positive voltage source V_{DD}. In practice it is still common to use V_{SS} to denote the lowest voltage in the circuit such that V_{SS} has an implied value of 0 V.

We can now define the relationship between logic variables and voltages. Recall that Boolean variables are discrete; a binary variable *x* can have the value of *x* = 0 or *x* = 1 only. At the circuit level we represent the variable *x* using a voltage V_x such that

$$0 \le V_x \le V_{DD} \tag{2.14}$$

gives the normal range of values with a power supply providing 0 V and V_{DD} directly to the circuit. The **positive logic convention** then defines the **ideal** logic 0 and logic 1 voltages as

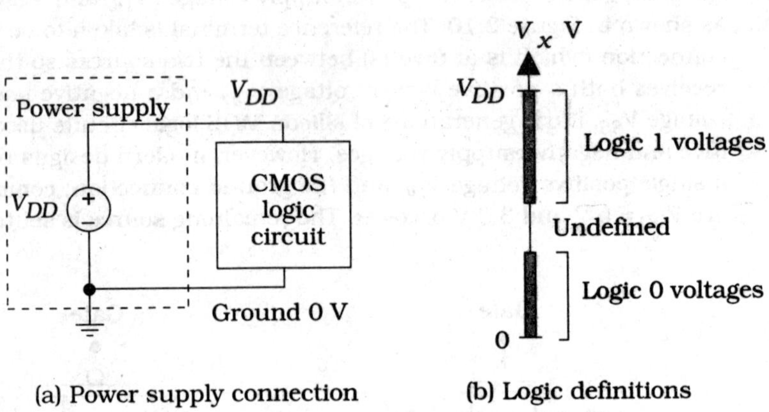

Figure 2.11 Single voltage power supply

4 The unit of volt is denoted by V in the text.

$$x = 0 \text{ means that } V_x = 0 \text{ V}$$
$$x = 1 \text{ means that } V_x = V_{DD} \quad (2.15)$$

Realistic circuits are more lenient and allow us to use a *range* of voltages for both logic 0s and logic 1s as portrayed in Figure 2.11(b). In general,

- Low voltages correspond to logic 0 values
- High voltages correspond to logic 1 values

The transition region between the highest logic 0 voltage and the lowest logic 1 voltage is undefined in that it does not represent either a 0 or a 1. The actual extent of both voltage ranges is determined by the characteristics of the logic circuits and will be dealt with later.

With the logic-voltage conversion defined, let us now examine the switching characteristics of MOSFETs. Ideally, an nFET behaves like an assert-high switch. This is shown in Figure 2.12 where A is the logic variable applied to the gate. If $A = 0$ corresponding to a low voltage, then the nFET acts like an open switch and there is no relationship between the left and right sides; this is illustrated in Figure 2.12(a). Increasing the gate voltage to a high value is the same as changing to $A = 1$. This results in a closed switch as shown in Figure 2.12(b). As with the assert-high switch, this can be described by the logic equation

$$y = x \cdot A \quad (2.16)$$

which is valid iff $A = 1$.

The pFET is exactly opposite in that it behaves like an assert-low switch. In Figure 2.13(a), the signal applied to the gate has a logical value of $A = 1$ corresponding to a high voltage. This gives an open circuit and there is no direct relationship between x and y. If the gate voltage is reduced to give $A = 0$, then the pFET acts as a closed switch. This allows us to write the ideal relationship

$$y = x \cdot \overline{A} \quad (2.17)$$

which is valid so long as $A = 0$ is true; this condition is shown in Figure 2.13(b).

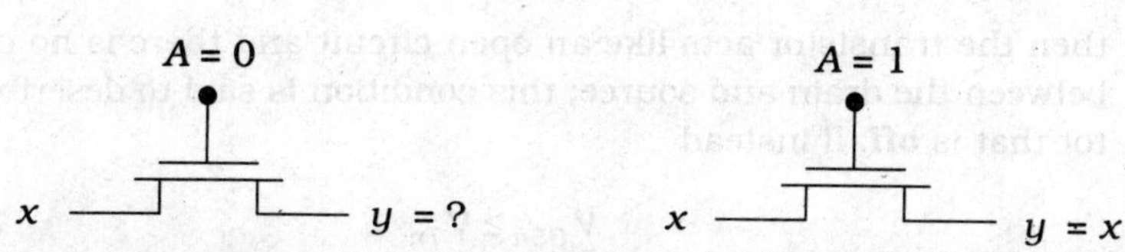

(a) Open (b) Closed

Figure 2.12 nFET switching characteristics

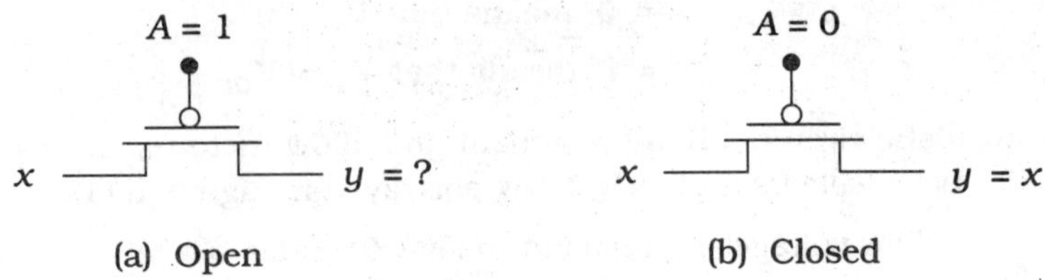

Figure 2.13 pFET switching characteristics

MOSFETs allow us to design logic circuits using the techniques of assert-high and assert-low switching networks. However, FETs are physical devices that do not behave exactly like the ideal switch models above. This is not a severe problem so long as we understand the differences and learn the limitations.

FET Threshold Voltages

The switching equations assume that the binary variable A applied to the gate of a FET is either a 0 or a 1. The corresponding voltage V_A is a physical quantity and does not behave in such a discrete manner. Moreover, we want to define a range of voltages for both cases of $A = 0$ and $A = 1$ to aid in the design of working circuits. Every MOSFET has a characterizing parameter called the **threshold voltage** V_T that helps us define the important gate voltage ranges. The specific value of V_T is established during the manufacturing process, and is thus taken to be a given value to the VLSI designer. One complicating factor is that nFETs and pFETs have different threshold voltages.

An nFET is characterized by a threshold voltage V_{Tn} that is a positive number with values around V_{Tn} = 0.5 V to 0.7 V being typical. The meaning of V_{Tn} can be understood by referring to the parameters shown in Figure 2.14(a). First note that the drain terminal has been identified as the one closest to the power supply V_{DD}, while the source terminal has been connected to ground (0 V). The gate-source voltage V_{GSn} shown in the drawing is the important parameter that determines whether the nFET acts as an open or closed switch. In particular, if

$$V_{GSn} \leq V_{Tn} \tag{2.18}$$

then the transistor acts like an open circuit and there is no current flow between the drain and source; this condition is said to describe a transistor that is **off**. If instead

$$V_{GSn} \geq V_{Tn} \tag{2.19}$$

then the nFET drain and source are connected and the equivalent switch is closed. A transistor that conducts current is said to be **on**. This behav-

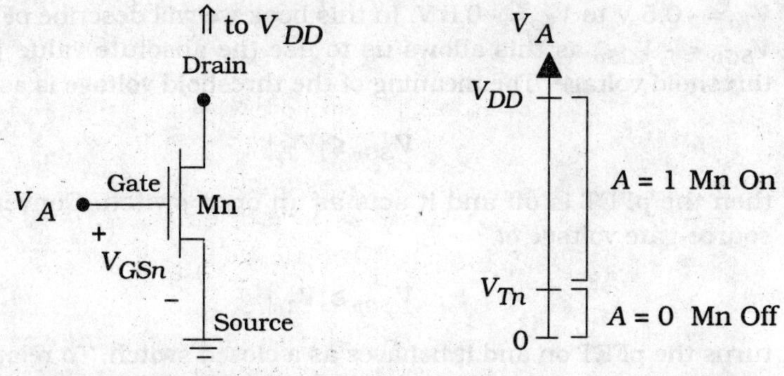

(a) Gate-source voltage (b) Logic translation

Figure 2.14 Threshold voltage of an nFET

ior allows us to create the voltage diagram shown in Figure 2.14(b) to define the voltage V_A that is associated with the binary variable A. In particular, we note that

$$V_A = V_{GSn} \tag{2.20}$$

This shows that $A = 0$ corresponds to values of $V_A \leq V_{Tn}$, while $A = 1$ implies that $V_A \geq V_{Tn}$. These relations establish the voltage ranges needed to control the nFET.

A pFET behaves in a **complementary** manner. Consider the transistor shown in Figure 2.15(a). For the pFET, the source terminal has been connected to the power supply V_{DD} while the drain is the side closest to ground; this is opposite to that used for the nFET. In this device, the source-gate voltage V_{SGp} is the important applied voltage. By convention, the pFET threshold voltage V_{Tp} is referenced to the gate-source voltage

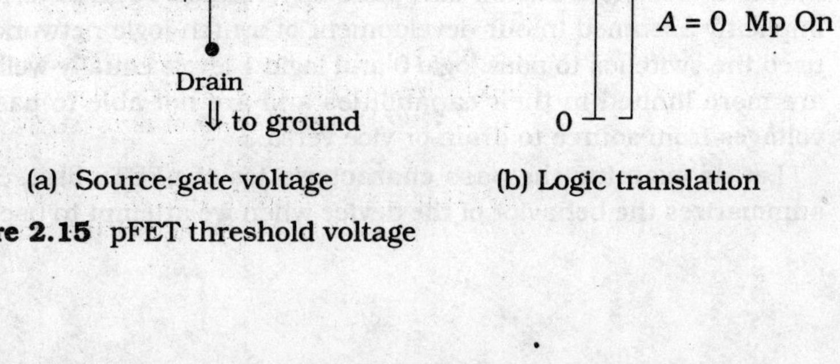

(a) Source-gate voltage (b) Logic translation

Figure 2.15 pFET threshold voltage

V_{GSp} and is a negative number with typical values in the range of about V_{Tp} = - 0.5 V to V_{Tp} = - 0.8 V. In this book we will describe pFETs by using V_{SGp} = - V_{GSp} as this allows us to use the absolute value $|V_{Tp}|$ for the threshold voltage. The meaning of the threshold voltage is as follows. If

$$V_{SGp} \le |V_{Tp}| \tag{2.21}$$

then the pFET is off and it acts as an open switch. Conversely, a large source-gate voltage of

$$V_{SGp} \ge |V_{Tp}| \tag{2.22}$$

turns the pFET on and it behaves as a closed switch. To relate this behavior to the applied voltage V_A we first sum voltages to write

$$V_A + V_{SGp} = V_{DD} \tag{2.23}$$

Thus,

$$V_A = V_{DD} - V_{SGp} \tag{2.24}$$

shows that a low value of V_A implies a large V_{SGp} and the pFET is on. Similarly, if V_A is large then V_{SGp} is small and the pFET is off. This gives us the logic 0 and logic 1 ranges summarized in Figure 2.15(b). Note that the transition between a logic 0 and a logic 1 is at

$$V_{DD} - |V_{Tp}| \tag{2.25}$$

since this corresponds to the source-gate voltage where the device turns on.

It is important to note that the logic 0 and logic 1 voltage ranges of V_A are different for the two types of FETs. One way around this problem is to note that there are regions of overlap for both A = 0 and A = 1 values that can be used if a uniform definition is needed. The ideal values of

$$\begin{aligned} V_A &= 0 \text{ V} \\ V_A &= V_{DD} \end{aligned} \tag{2.26}$$

are, however, valid for both devices.

Pass Characteristics

An ideal electrical switch can pass any voltage applied to it. This was implicitly assumed in our development of switch logic networks where we used the switches to pass logic 0 and logic 1 levels equally well. MOSFETs are more limited in their capabilities and are not able to pass arbitrary voltages from source to drain or vice versa.

Let us examine the pass characteristics of nFETs first. Figure 2.16 summarizes the behavior of the device when we attempt to use it to pass a

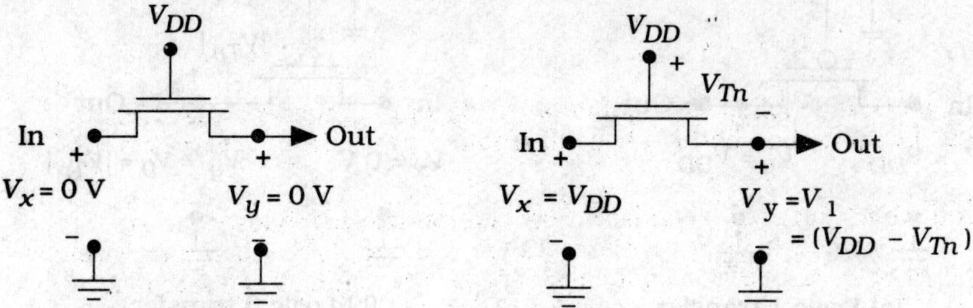

(a) Logic 0 transfer

(b) Logic 1 transfer

Figure 2.16 nFET pass characteristics

voltage from left to right. Applying V_{DD} to the gate insures that the nFET is on and the device acts like a closed switch. In Figure 2.16(a) a logic 0 voltage of $V_x = 0$ V is applied to the left side. This results in an output voltage of $V_y = 0$ V as desired. Increasing the input voltage results in that value being transmitted to the output side. However, a problem occurs if we apply an ideal logic 1 input voltage of $V_x = V_{DD}$ as shown in Figure 2.16(b). In this case, the output voltage V_y is reduced to a value

$$V_1 = V_{DD} - V_{Tn} \tag{2.27}$$

which is less than the input voltage V_{DD}. This is referred to as a **threshold voltage loss**. It arises from the fact that the minimum value of the gate-source voltage need to maintain an on state is

$$V_{GSn} = V_{Tn} \tag{2.28}$$

Using the Kirchhoff voltage law, this subtracts from the voltage V_{DD} that is applied to the *gate* as shown in the drawing.[5] Since the transmitted voltage V_y is less than the ideal logic 1 value of V_{DD}, we say that the nFET can only pass a **weak** logic 1. In the same terminology, the nFET is said to pass a **strong** logic 0 since it is capable of producing an output voltage of $V_y = 0$ V without any problems. In general, the nFET can pass a voltage in the range $[0, V_1]$, but nothing above V_1.

A pFET has opposite pass characteristics. To examine the pFET properties we apply a logic 0 to the gate by grounding it. Figure 2.17 shows the circuits for both input values. Figure 2.17(a) portrays the case where $V_x = V_{DD}$ corresponding to a logic 1 input. The output voltage is

$$V_y = V_{DD} \tag{2.29}$$

[5] Kirchhoff's voltage law (KVL) says that the algebraic sum of voltages around a closed loop is 0.

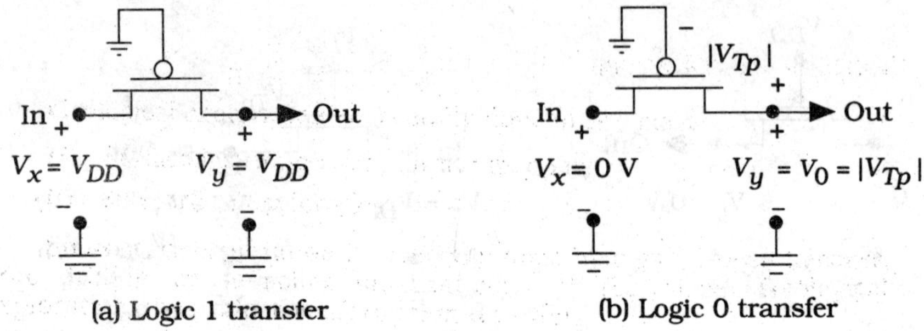

Figure 2.17 pFET pass characteristics

which is an ideal logic 1 level. The pFET is therefore capable of passing a strong logic 1 voltage. The problem arises when we attempt to pass an ideal logic 0 voltage of $V_y = 0$ V and is presented in Figure 2.17(b). In this case, the transmitted voltage can only drop to a minimum value of

$$V_y = |V_{Tp}| \tag{2.30}$$

This is also due to a threshold effect. In order to keep the pFET on requires a minimum source-gate voltage of

$$V_{SGp} = |V_{Tp}| \tag{2.31}$$

as shown in the drawing. Since the gate is at 0 V, this represents a rise to a voltage of $|V_{Tp}|$, which is turn affects the output. Obviously, the pFET transmits a weak logic 0 voltage. In summary, a pFET can pass a voltage in the range $[V_{DD}, V_0]$, but nothing below V_0.

Let us restate the results of the above discussion:

- nFETs pass strong logic 0 voltages, but weak logic 1 values
- pFETs pass strong logic 1 voltages, but weak logic 0 levels

Complementary MOS (CMOS) circuits are designed to account for the transmission levels. In particular, we can write down the following rules as a basis for our design:

1. Use pFETs to pass logic 1 voltages of V_{DD}
2. Use nFETs to pass logic 0 voltages of $V_{SS} = 0$ V

These allow us to build circuits that can pass the ideal logic voltages 0 V and V_{DD} to the output terminal. We will find, however, that ideal levels are not always needed in practice.

2.3 Basic Logic Gates in CMOS

The concept of a general CMOS digital logic gate can be understood by referring to the drawing in Figure 2.18. In this example, *a*, *b*, and *c* are the

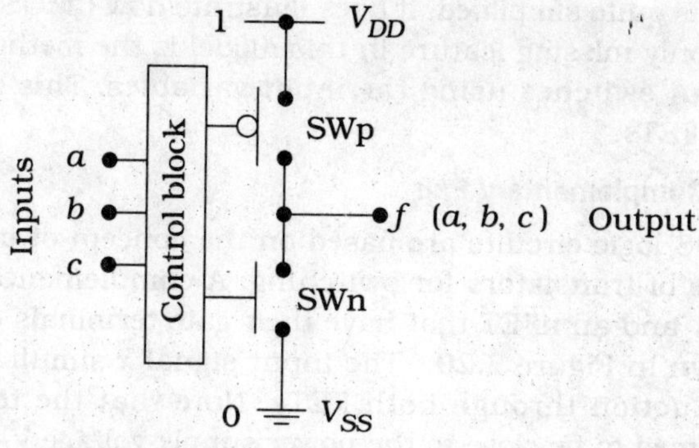

Figure 2.18 General CMOS logic gate

input bits that combine to give the output function bit $f(a, b, c)$. Since this is by definition a digital circuit, all of the quantities are restricted to values of 0 or 1. Digital logic circuits are nonlinear networks that use transistors as electronic switches to divert *one* of the supply voltages V_{DD} or 0 V to the output. This corresponds to a logical result of $f = 1$ or $f = 0$. Internally, we may view the output network of the gate as consisting of two switches SWp (an assert-low device) and SWn (an assert-high device) as shown. These are wired in to insure that one switch is closed while the other switch is open.

The operation of the general logic gate is shown in Figure 2.19 for both output possibilities. In Figure 2.19(a), the upper switch is closed while the lower switch is open. This connects the output to the power supply and yields a value of $f = 1$. The opposite situation is shown in Figure 2.19(b): the upper switch is open and the lower switch is closed. Because the output is now connected to $V_{SS} = 0$ V, the logical result is $f = 0$. Although this

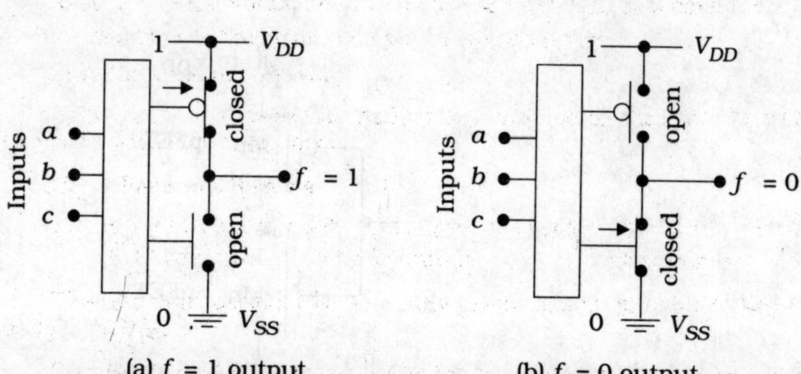

Figure 2.19 Operation of a CMOS logic gate

view is quite simplified, it does illustrate how CMOS logic circuits operate. The only missing feature in this model is the method used to control the output switches using the input variables. This is accomplished with MOSFETs.

The Complementary Pair

CMOS logic circuits are based on the concept of using **complementary pairs** of transistors for switching. A complementary pair consists of a pFET and an nFET that have their gate terminals connected together as shown in Figure 2.20. The input signal x simultaneously controls the conduction through both FETs. Note that the top of the pFET Mp is assumed to be close to the power supply voltage V_{DD}, while the nFET Mn is close to the ground (V_{SS}). The behavior of the complementary pair is easily understood by observing the state of each FET for the two possible input values as in Figure 2.21. An input of $x = 0$ turns Mp on while the nFET Mn is off and acts like an open switch; this is shown in Figure 2.21(a). The opposite case shown in Figure 2.21(b) is where $x = 1$. Now the pFET MP is off while Mn is on. The name "complementary" is derived from this operation: when one FET is on, the other is off. The important aspect of this behavior is that the nFET and pFET are electrical opposites, which translates directly into a coherent switching scheme.

Now that we have seen the overall structure of CMOS logic gates and the idea of a complementary pair, we have all of the concepts needed to create and analyze basic logic gate circuits.

2.3.1 The NOT Gate

The NOT or INVERT function is often considered the simplest Boolean operation. It has an input x and produces an output $f(x)$ of

$$f(x) = \text{NOT}(x) = \bar{x} \tag{2.32}$$

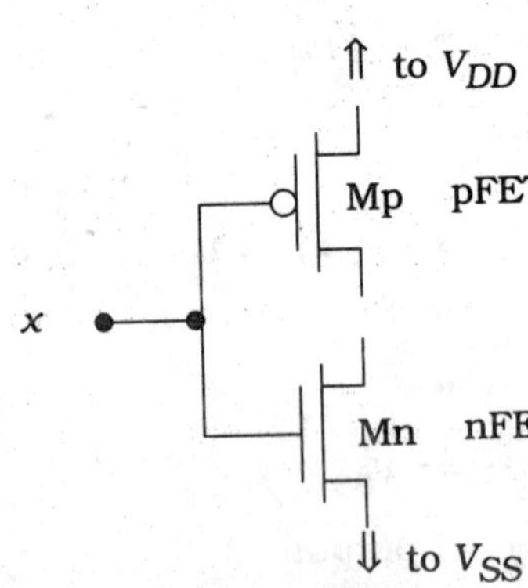

Figure 2.20 A CMOS complementary pair

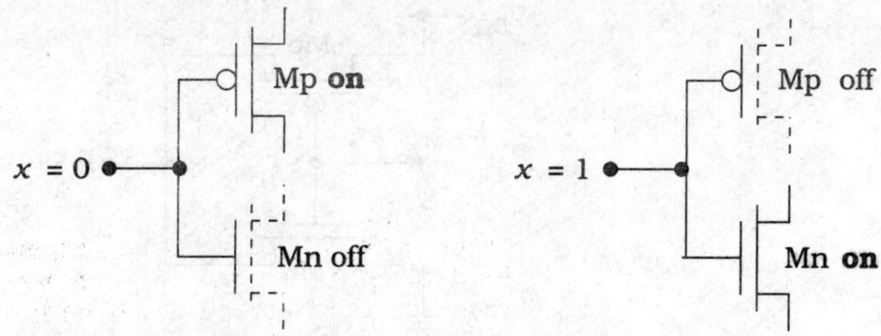

(a) $x = 0$ input (b) $x = 1$ input

Figure 2.21 Operation of the complementary pair

such that

$$\begin{aligned} &\text{If } x = 0 \quad \text{then } \bar{x} = 1 \\ &\text{If } x = 1 \quad \text{then } \bar{x} = 0 \end{aligned} \tag{2.33}$$

defines the notation. The logic symbol and truth table are provided in Figure 2.22 for future reference.

A CMOS NOT gate is shown in Figure 2.23. This has been constructed using the same idea as for the switch-based circuit discussed earlier in the context of Figure 2.7. The circuit uses a complementary pair of MOSFETs such that the input variable x controls both transistors.

The operation follows directly from the properties of the complementary pair. If the input x has a value of 0, then pFET MP is on and the nFET Mn is off. As shown in Figure 2.24(a), this connects the output node to the power supply voltage V_{DD}, giving an output of $\bar{x} = 1$. Conversely, if $x = 1$ then Mp is off and Mn is on. The output is then connected to the ground node and gives $\bar{x} = 0$ as verified by the circuit in Figure 2.24(b). It is clear that this simple circuit does indeed provide the NOT operation. This can be verified analytically by applying the FET logic rules to write the output f as

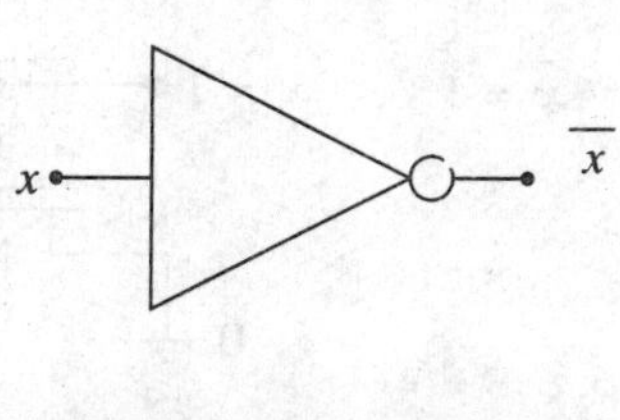

x	$\bar{x}$
0	1
1	0

(a) Logic symbol (b) Truth table

Figure 2.22 NOT gate

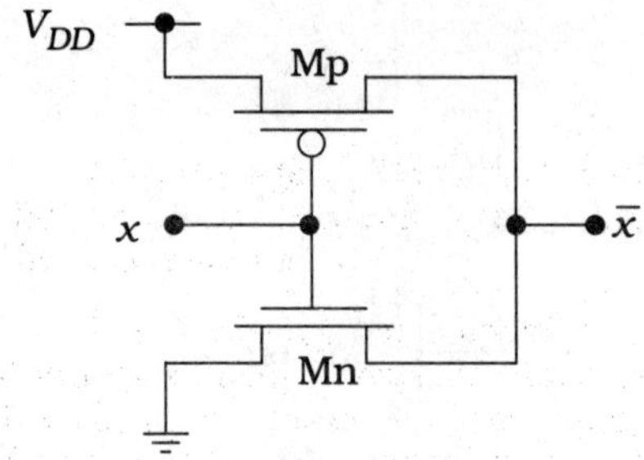

Figure 2.23 CMOS NOT gate

$$f = \bar{x} \cdot 1 + x \cdot 0 \tag{2.34}$$

where the first term describes Mp and the second term is due to Mn. Simplifying gives

$$f = \bar{x} \tag{2.35}$$

as expected.

One of the most important characteristics of the CMOS NOT gate is the manner in which the complementary FET pair insures that, for a given input logic state of $x = 0$ or 1, the output is connected to either V_{DD} or ground and gives a well-defined value. This circuit specifically avoids the possibilities where (i) both FETs are off at the same time, or (ii) both FETs are on at the same time; either situation would give an ill-defined output.

2.3.2 The CMOS NOR Gate

Now that we have seen the basic NOT gate, let us extend the concepts to create a 2-input NOR gate using the same principles. These are

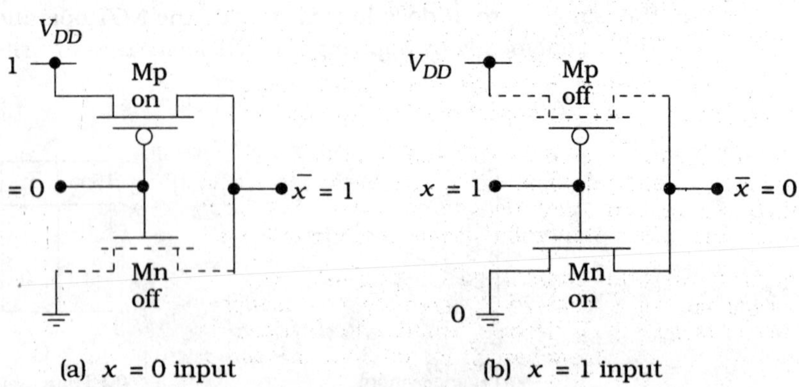

Figure 2.24 Operation of the CMOS NOT gate

- Use a complementary nFET/pFET pair for each input
- Connect the output node to the power supply V_{DD} through pFETs
- Connect the output node to ground through nFETs, and
- Insure that the output is always a well-defined high or low voltage

This set of guidelines helps us design logic circuits that have input and output characteristics which are compatible with the NOT gate.

The logic symbol and truth table for the NOR2 gate are provided in Figure 2.25.[6] With input variables x and y, the NOR2 produces the output

$$g(x, y) = \overline{x + y} \tag{2.36}$$

such that a 1 at either input gives $g = 0$. Only the input combination $(x, y) = (1,1)$ yields an output of $g = 1$.

One way to synthesize the NOR2 operation at the logic design level is to use a 4:1 MUX as shown in Figure 2.26(a). Path selection is obtained using the input pair (x, y) such that every combination gives either a 1 or a 0 to the output. The Boolean expression for the output of the MUX is

$$g(x, y) = \bar{x} \cdot \bar{y} \cdot 1 + \bar{x} \cdot y \cdot 0 + x \cdot \bar{y} \cdot 0 + x \cdot y \cdot 0 \tag{2.37}$$

which reduces to the desired form

$$g(x, y) = \overline{x + y} \tag{2.38}$$

using the DeMorgan theorem. A voltage-equivalent circuit is obtained by replacing the binary quantities with voltages, and results in the circuit shown in Figure 2.26(b). In the notation of the drawing, V_x and V_y respectively represent the Boolean variables x and y. This information provides the basis for constructing the CMOS NOR2 circuit.

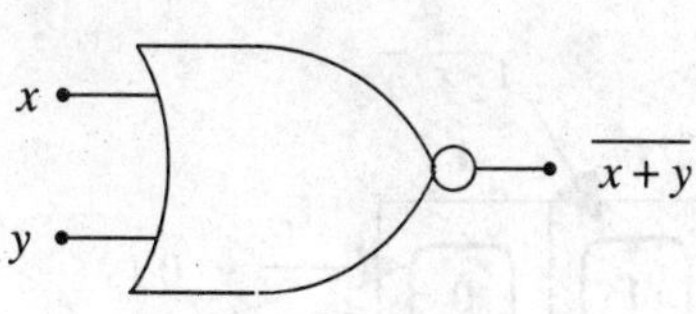

(a) Logic symbol

x	y	$\overline{x+y}$
0	0	1
0	1	0
1	0	0
1	1	0

(b) Truth table

Figure 2.25 NOR logic gate

6 The terminology "NOR2" means a 2-input NOR gate.

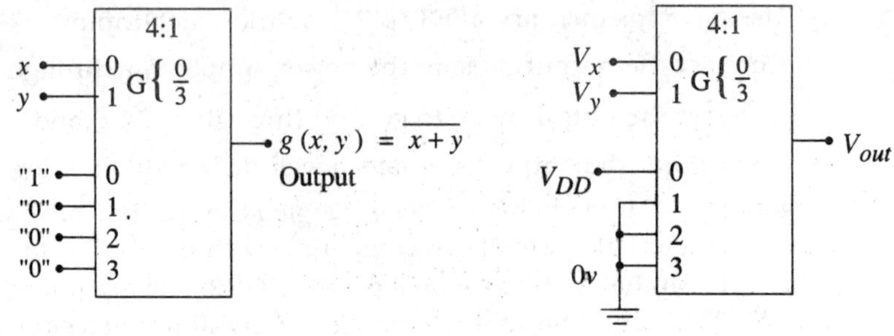

(a) Logic diagram (b) Voltage network

Figure 2.26 NOR2 operation using a 4:1 multiplexor

One approach to building the logic gate is to use the Karnaugh map drawn in Figure 2.27. CMOS generally produces **inverting** logic because our gates are constructed using the NOT circuit as a basis. This creates the situation where we are generally interested in the occurrence of both 1's and 0's when dealing with K-maps. In particular, note that we have created two 0-groupings in the drawing. The map allows us to write the logic expression in the form

$$g(x, y) = \bar{x} \cdot \bar{y} \cdot 1 + x \cdot 0 + y \cdot 0 \tag{2.39}$$

and work backward to construct the circuit. Each term represents a FET path to the output. The first term connects the output to 1 (the power supply V_{DD}) and is controlled by complements of the input variables in a series-connected AND arrangement. The second and third terms represent two independent nFET paths between the output and 0 (ground). Combining these statements results in the CMOS NOR2 circuit shown in

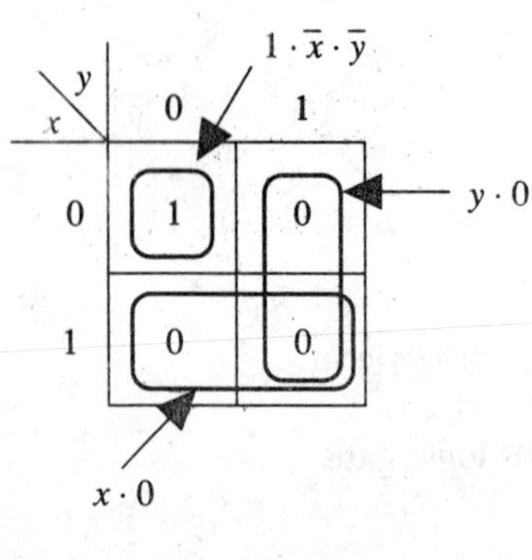

Figure 2.27 NOR2 gate Karnaugh map

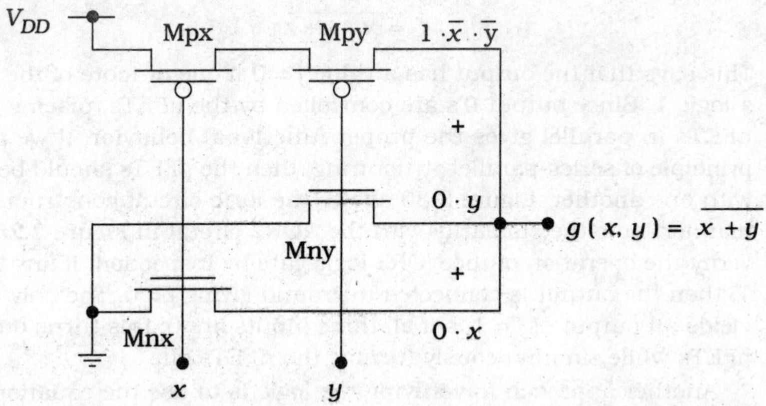

Figure 2.28 CMOS NOR2 gate

Figure 2.28; the one-to-one correspondence between each line in the equation and the circuit is obvious.

To verify that this circuit does have the proper electrical behavior, we may construct the table shown in Figure 2.29. This shows the state (on or off) of every FET for each of the four input possibilities. Tracing the output connections for each possibility easily shows that the switching circuit is consistent with the truth table.

x	y	Mpx	Mpy	Mnx	Mny	g
0	0	on	on	off	off	1
0	1	on	off	off	on	0
1	0	off	on	on	off	0
1	1	off	off	on	on	0

Figure 2.29 Operational summary of the NOR2 gate

The electrical structure of the NOR2 gate also illustrates another important point in the manner in which the FETs are wired together. Note that the two pFETs Mpx and Mpy are connected in series such that both must be on to establish a conducting path from V_{DD} to the output. The nFETs Mnx and Mny, on the other hand, are wired in parallel so that a connection between the output and ground is created if either nFET is on. This is called a **series-parallel** transistor arrangement; the principle allows us to design more complex gates.

As an example, let us construct a 3-input NOR (NOR3) gate using the NOR2 topology as a guideline. Let us label the inputs as x, y, and z. Each input is connected to the gate of a complementary nFET/pFET pair. The logical output expression for the gate is given by

$$f = \overline{x + y + z} \tag{2.40}$$

This says that the output has a value $f = 0$ if one or more of the inputs is a logic 1. Since output 0's are controlled by the nFETs, placing the three nFETs in parallel gives the proper functional behavior. If we apply the principle of series-parallel structuring, then the pFETs should be in series with one another. Figure 2.30 shows the logic circuit constructed in this manner; note the similarity with the NOR2 circuit in Figure 2.28. We can verify the operation of the NOR3 logic gate by inspection: if any input is a 1, then the output is connected to ground giving $f = 0$. The only case that yields an output of $f = 1$ is if all three inputs are 0; this turns on all three pFETs while simultaneously turning the nFETs off.

Another approach to verifying the logic is to use the equations of FET switches and derive the MUX equations. The top branch in Figure 2.30 is through a series group of three pFETs as described by the term

$$1 \cdot \bar{x} \cdot \bar{y} \cdot \bar{z} \tag{2.41}$$

where we recognize that the power supply voltage V_{DD} is equivalent to a logic 1. Each of the three nFET branches consists of a single FET passing the ground to the output. Since a ground is a logic 0, we can OR the four branches together to give a complete output expression of

$$f = 1 \cdot \bar{x} \cdot \bar{y} \cdot \bar{z} + 0 \cdot x + 0 \cdot y + 0 \cdot z \tag{2.42}$$

The nFET terms insure that the output voltage of the circuit is 0 V whenever one or more of the inputs is 1. Logically, however, they evaluate to 0 leaving the final form

$$f = 1 \cdot \bar{x} \cdot \bar{y} \cdot \bar{z} = \overline{x + y + z} \tag{2.43}$$

where we have use a DeMorgan relation in the reduction. This shows that

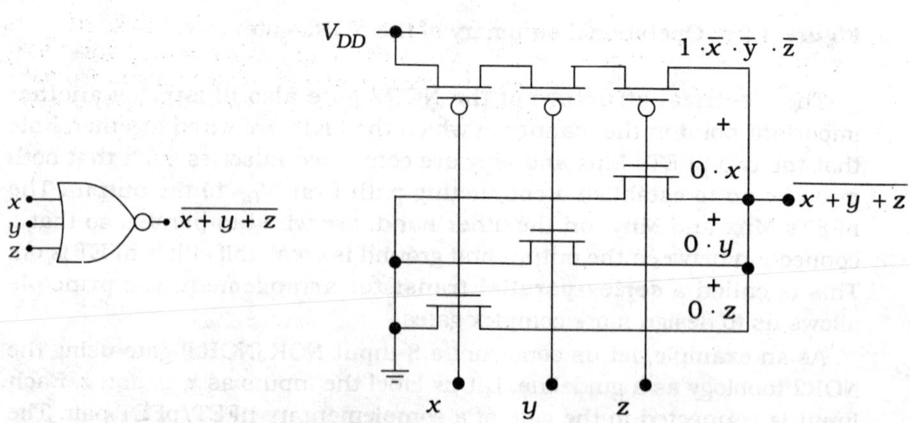

Figure 2.30 A NOR3 gate in CMOS

the circuit does in fact provide the NOR3 operation.

In principle, one may use the same arguments to construct multiple input NOR gates in CMOS such as a NOR4 or a NOR6. This technique is easy to apply and yields functional logic circuits. For VLSI applications, however, the choice of logic circuits is based on more than just the ability to provide a logic operation. Hardware characteristics such as switching speed and area consumption on the silicon chip must be taken into account. In this chapter, we will concentrate solely on forming logic functions through the circuit topology. More detailed considerations will be discussed later in the text.

2.3.3 The CMOS NAND Gate

Let us next construct the CMOS circuit for the NAND2 gate with the logical symbol and behavior summarized in Figure 2.31. This gate is characterized by an output that is 0 unless both of the inputs are 1. The truth table can be used to build the 4:1 MUX implementation drawn in Figure 2.32(a) such that the output is described by

$$h(x, y) = \bar{x} \cdot \bar{y} \cdot 1 + \bar{x} \cdot y \cdot 1 + x \cdot \bar{y} \cdot 1 + x \cdot y \cdot 0 \tag{2.44}$$

The voltage-equivalent network in Figure 2.32(b) is now somewhat obvious.

As with the NOR2 gate, it is useful to examine the Karnaugh map for the NAND2 function. Figure 2.33 shows the map along with two groupings that simplify the cases where $h = 1$. Using these reductions, our expression can be rewritten to read

$$h(x, y) = \bar{x} \cdot 1 + \bar{y} \cdot 1 + x \cdot y \cdot 0 \tag{2.45}$$

Translating each term to FET groups yields the CMOS circuit shown in Figure 2.34. This gives the NAND2 function as can be verified by the oper-

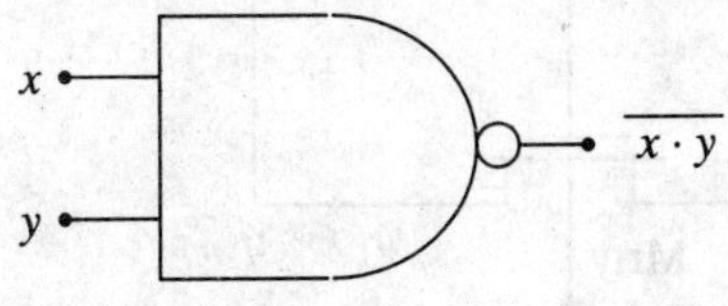

(a) Logic symbol

x	y	$\overline{x \cdot y}$
0	0	1
0	1	1
1	0	1
1	1	0

(b) Truth table

Figure 2.31 NAND2 logic gate

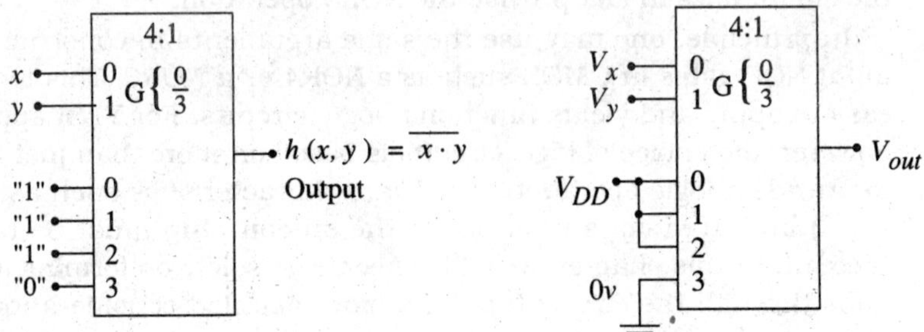

(a) Logic diagram (b) Voltage network

Figure 2.32 NAND2 operation using a 4:1 multiplexor

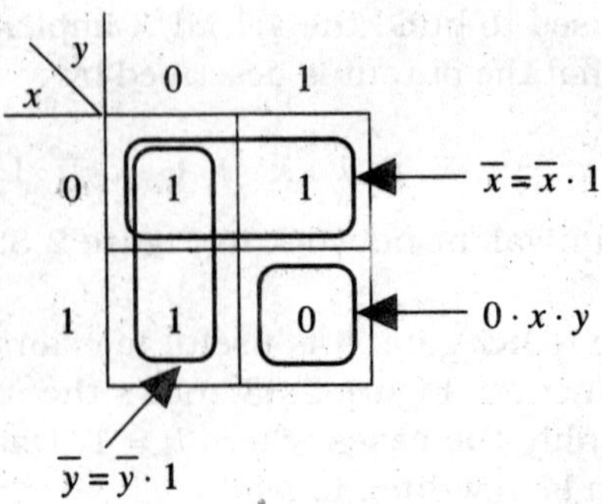

Figure 2.33 NAND2 K-map

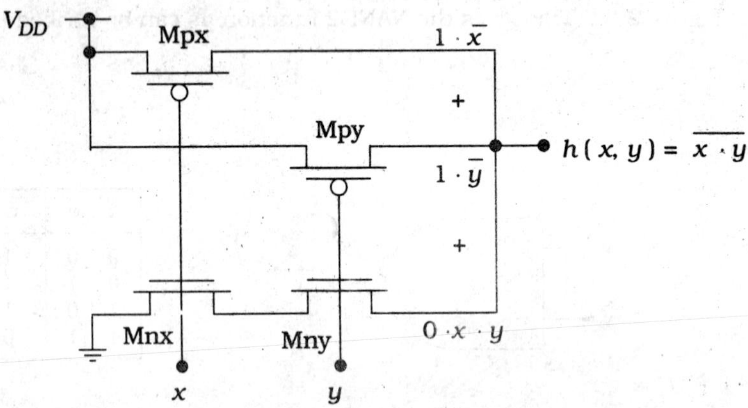

Figure 2.34 CMOS NAND2 logic circuit

x	y	Mpx	Mpy	Mnx	Mny	h
0	0	on	on	off	off	1
0	1	on	off	off	on	1
1	0	off	on	on	off	1
1	1	off	off	on	on	0

Figure 2.35 Operational summary of the NAND2 circuit

ation summarized in the table of Figure 2.35. An important characteristic of the NAND2 gate is that it uses two parallel-connected pFETs, while the nFETs are in series. This is exactly opposite to the structure of the NOR2 gate.

A NAND3 gate can be created using the same topology. It requires three sets of complementary pairs, each driven by a separate input. The nFETs are placed in series, while the pFETs are wired in parallel. This gives the gate shown in Figure 2.36. To verify the operation of the circuit, note that all three inputs must be 1's to provide a conduction path between the output and ground. If any one (or more) of the inputs is a 0, then the corresponding nFET is off while the pFET it drives acts like a closed switch; this gives a logic 1 voltage of V_{DD} at the output.

Switch logic analysis can also be applied by treating the circuit as a multiplexor. The series-connected nFET chain at the bottom of the circuit is described by the logic term

$$0 \cdot x \cdot y \cdot z \tag{2.46}$$

Each pFET branch consists of a single transistor that acts like a closed switch when a 0 is applied. Performing the OR operation among the four

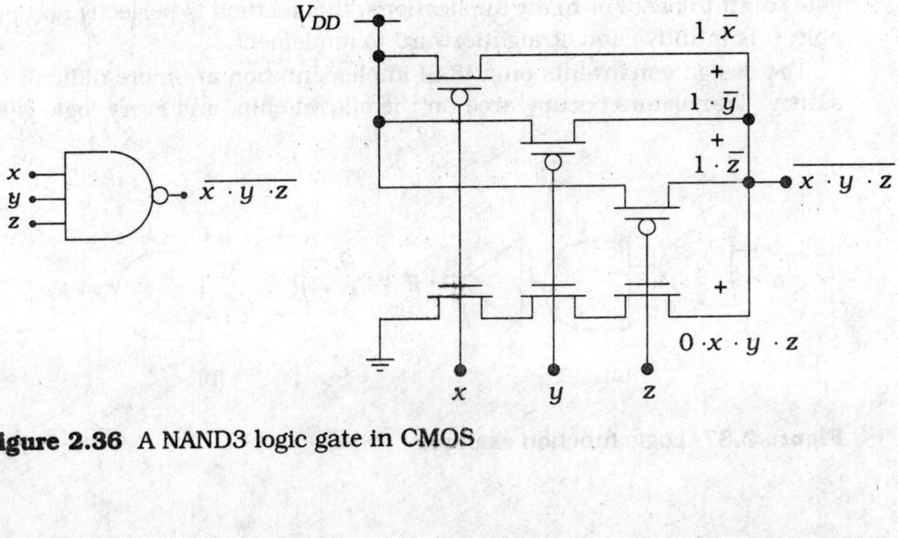

Figure 2.36 A NAND3 logic gate in CMOS

branches gives

$$0 \cdot x \cdot y \cdot z + 1 \cdot \bar{x} + 1 \cdot \bar{y} + 1 \cdot \bar{z} \tag{2.47}$$

Eliminating the 0 terms and using DeMorgan reduction gives the output function as

$$\overline{x \cdot y \cdot z} \tag{2.48}$$

which is the NAND3 function. This technique can be extended to design the CMOS circuitry for NAND gates with more inputs.

2.4 Complex Logic Gates in CMOS

One of the most powerful aspects of building logic circuits in CMOS is the ability to create a single circuit that provides several primitive operations (NOT, AND, and OR) in an integrated manner. These will be called **complex** or **combinational logic gates** in our discussion. Complex logic gates are very useful in VLSI system-level design.

To illustrate the main idea of a complex logic gate, consider the Boolean expression

$$F(a, b, c) = \overline{a \cdot (b + c)} \tag{2.49}$$

The simplest way to construct the logic network for this function is to use one OR-gate, one AND-gate, and one NOT-gate as shown in Figure 2.37(a). Alternately, we might simplify the network to that in Figure 2.37(b) if a NAND2 gate is available. If we build the electronic equivalent of either implementation, then the traditional approach would be to use a one-to-one map: each gate requires one electronic logic circuit. For the first case (a), this would require three separate gates, while (b) reduces the gate count to two. For many applications, this method is perfectly acceptable; it is intuitive and straightforward to implement.

The design constraints on a VLSI implementation are more difficult to satisfy. Transistors occupy area on the silicon chip, and every logic gate

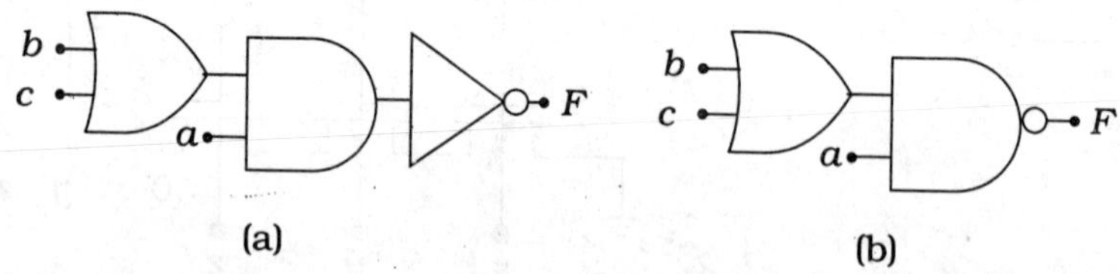

Figure 2.37 Logic function example

uses transistors. Since the gate count on a VLSI chip can easily exceed several hundred thousand, we often look for techniques that reduce the number of gates and/or FETs while still performing the required logic. In the present discussion, we will achieve this objective by building a single logic gate that implements the entire function.

Let us investigate the characteristics of the function F in more detail by applying DeMorgan expansions to the function to write

$$\begin{aligned} F &= \overline{a \cdot (b+c)} \\ &= \bar{a} + \overline{(b+c)} \\ &= [\bar{a} + (\bar{b} \cdot \bar{c})] \cdot 1 \end{aligned} \tag{2.50}$$

The last step is simply ANDing the result with a logical 1. Expanding gives

$$F = \bar{a} \cdot 1 + (\bar{b} \cdot \bar{c}) \cdot 1 \tag{2.51}$$

which is in a form that can be used to build the pFET switching circuit shown in Figure 2.38. The correspondence can be verified by checking each term. The first term implies a pFET connected between the power supply (V_{DD}) and the output that is controlled by the input a. The second term is identical in form to that encountered for the NOR2 gate. It represents two series-connected pFETs (with control variables b and c) that connect the power supply to the output.

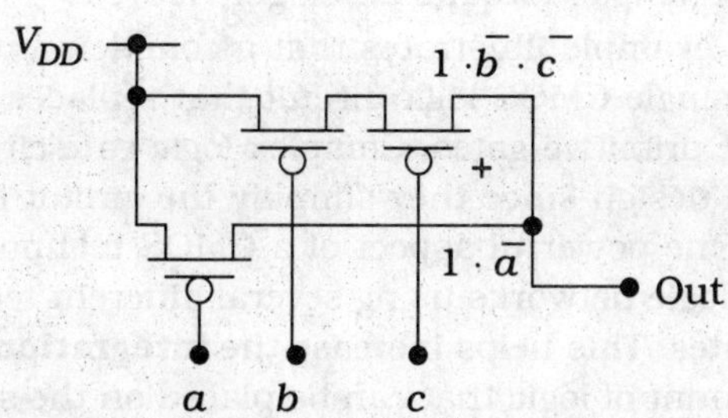

Figure 2.38 pFET circuit for F from equation (2.51)

The pFET circuit alone is not sufficient to create a functional electronic network. We must add an nFET array that gives $F = 0$ when necessary. The original form of the function in equation (2.49) shows that $F = 0$ occurs when

$$a = 1 \text{ AND } (b + c) = 1$$

This is equivalent to writing the output expression

$$0 \cdot [a \cdot (b+c)] \tag{2.52}$$

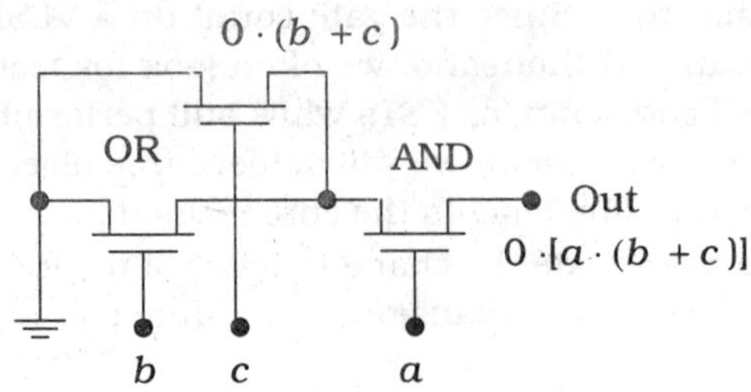

Figure 2.39 nFET logic circuit for *F*

which can in turn be used to describe the nFET array shown in Figure 2.39. Two parallel-connected nFETs that are controlled by *b* and *c* give the OR operation. This group is in series with the *a*-input nFET to produce the AND. The logic can be verified by the Karnaugh map groupings shown in Figure 2.40. Simplification to using a single *a*-input nFET occurs because of the common term encompassed by the two groups.

The completed CMOS logic gate is built by combining the nFET and pFET circuits, and results in the circuit of Figure 2.41. We have rotated the orientation of the FETs by 90 degrees to arrive at the finished schematic. This is the most common way to draw CMOS logic circuits since it makes series and parallel-connected FETs more obvious. The equivalence of the circuit can be verified by tracing out each branch and comparing it with the simpler circuits developed above.

This example illustrates that a complex function can be implemented with a single CMOS logic circuit that replaces a cascade made up of two or more primitive gates. Complex logic gate circuits can be more efficient in VLSI design since they simplify the circuit requirements and the logic flows. One powerful aspect of a CMOS technology is that it allows us to design logic networks using several different techniques, such as complex logic gates. This helps increase the **integration density**, which measures the amount of logic that can be placed on the silicon chip.

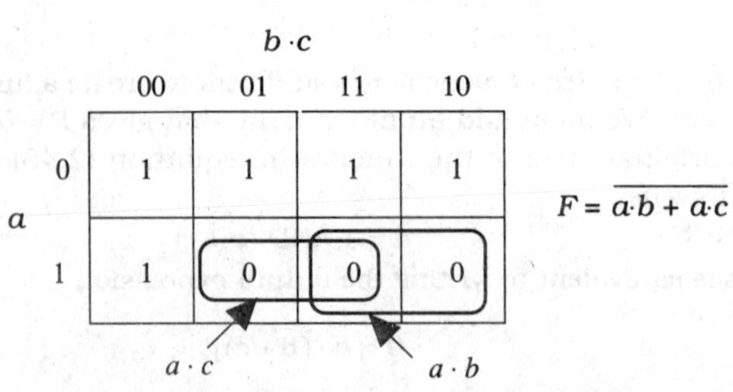

Figure 2.40 Karnaugh map grouping for the nFET circuit

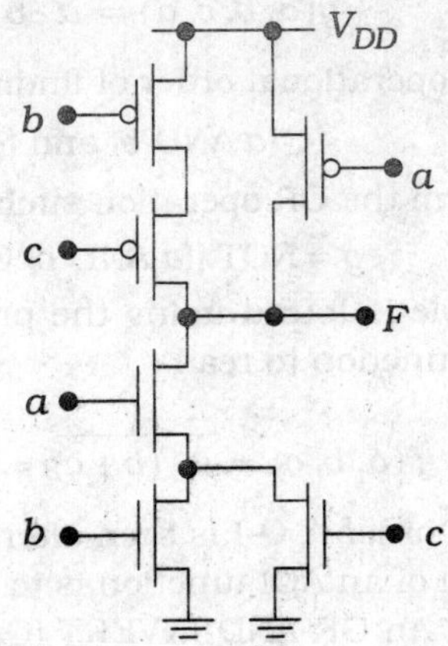

Figure 2.41 Finished complex CMOS logic gate circuit

2.4.1 Structured Logic Design

A structured approach to designing complex logic gates can be developed by focusing on the circuit characteristics. CMOS logic gates are intrinsically **inverting**; this means that the output always produces a NOT operation acting on the input variables. The simple inverter in Figure 2.42 illustrates the origin of this property. If the input a is a logic 1, then the nFET is ON and the pFET is OFF. The nFET passes the logic 0 (ground) to the output, giving $\overline{a}$ there. This characteristic was also observed in the NAND and NOR circuits.

The inverting nature of CMOS logic circuits allows us to construct logic circuits for AOI and OAI logic expressions using a structured approach. An AOI logic function is one that implements the operations in the order AND then OR then NOT (invert). For example,

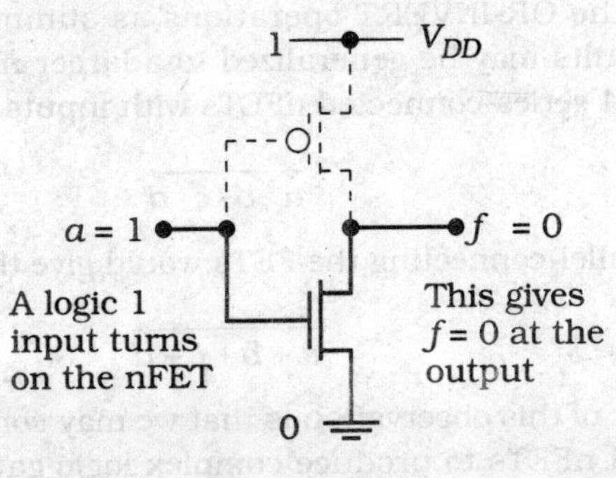

Figure 2.42 Origin of the inverting characteristic of CMOS gates

$$g(a, b, c, d) = \overline{a \cdot b + c \cdot d} \tag{2.53}$$

has an implied operational order of finding

(*a* AND *b*) and (*c* AND *d*)

and then perform the OR operation such that the final result is

g = NOT [(*a* AND *b*) OR (*c* AND *d*)]

Another example is found using the preceding CMOS gate example by expanding the function to read

$$f(a, b, c) = \overline{a \cdot (b + c)} = \overline{a \cdot b + a \cdot c} \tag{2.54}$$

The operational order A-O-I is seen after distributing the terms. An alternate description of an AOI function is to say that it is an inverted sum-of-products (SOP). An OR-AND-INVERT (OAI) function reverses the order of the AND and OR operations. An example of an OAI form is

$$h(x, y, z, w) = \overline{(x + y) \cdot (z + w)} \tag{2.55}$$

since this implies that we first calculate

(*x* OR *y*) along with (*w* OR *z*)

and then

h = NOT [(*x* OR *y*) AND (*w* OR *z*)]

to evaluate the value of *h*. An OAI form is equivalent to an inverted product-of-sums (POS) expression.

CMOS switching characteristics provide a natural means for implementing inverting logic forms such as AOI and OAI. The technique is based on using nFET and pFET arrays in a consistent manner. Complex logic gates of this type allow the designer to compress three or more primitive operations into a single logic gate. Consider first the logic formation properties of nFETs. From the NAND analysis, we learned that nFETs in series provide AND-INVERT logic; this is shown in Figure 2.43(a). Similarly, the NOR gate analysis showed us that parallel connected nFETs produce the OR-INVERT operations as summarized in Figure 2.43(b). These results may be generalized to a larger number of transistors. For example, 4 series-connected nFETs with inputs *a*, *b*, *c*, *d* would produce

$$\overline{a \cdot b \cdot c \cdot d} \tag{2.56}$$

while parallel-connecting the FETs would give the OR-INVERT operation

$$\overline{a + b + c + d} \tag{2.57}$$

The power of this observation is that we may combine series- and parallel-connected nFETs to produce complex logic gates. An example of this is shown in Figure 2.44. This array consists of parallel-connected groups,

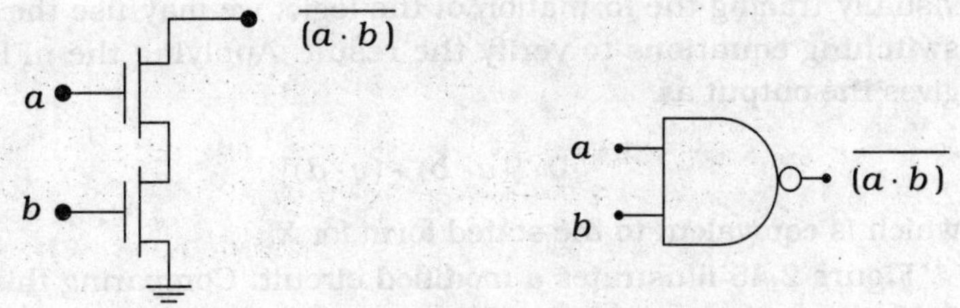

(a) Series-connected nFETs

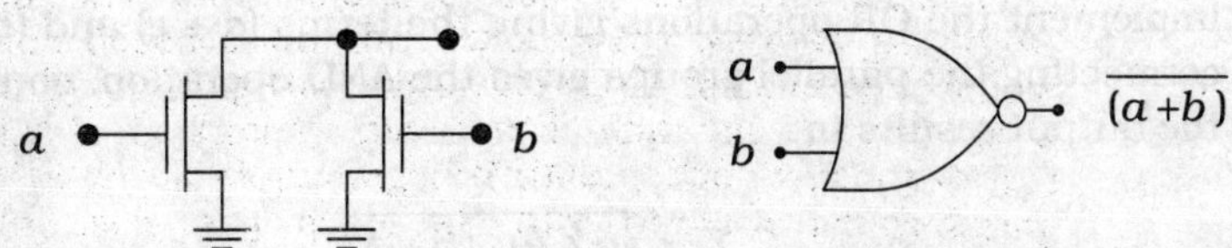

(b) Parallel-connected nFETs

Figure 2.43 nFET logic formation

with each group made of 2 series-connected nFETs. The transistors on the left side form the AND operation $(a \cdot b)$ while the right group of nFETs yields $(c \cdot d)$; the parallel connection of the two groups gives the OR operation, while the final output from the gate yields the NOT. We thus see that the function is described by

$$X = \overline{(a \cdot b) + (c \cdot d)} \tag{2.58}$$

which is an AOI expression that is represented by the logic circuit shown next to the circuit. It is important to note that the NOT operation is viewed at the exit point of the logic (i.e., only for the function X). The AND operation is provided by series-connected nFETs, while the OR is accomplished by using a parallel-connected group. Although this approach is based on

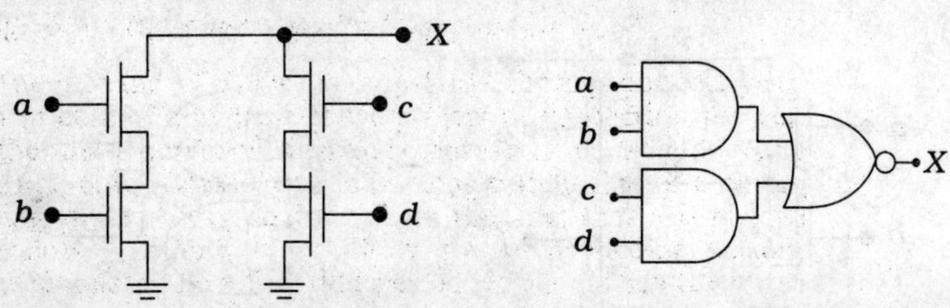

Figure 2.44 nFET AOI circuit

visually tracing the formation of the logic, we may use the formalism of switching equations to verify the result. Applying the nFET equations gives the output as

$$0 \cdot [(a \cdot b) + (c \cdot d)] \tag{2.59}$$

which is equivalent to the stated form for X.

Figure 2.45 illustrates a modified circuit. Comparing this with Figure 2.43 shows that a connection has been added such that now the upper two transistors (with inputs a and e) are in parallel with one another. Similarly, the nFETs with inputs b and f are in parallel. Both parallel groups implement the OR operations giving the terms $(a + e)$ and $(b + f)$. Series-connecting the parallel groups gives the AND operation, so that inverting the output results in

$$Y = \overline{(a+e) \cdot (b+f)} \tag{2.60}$$

which has OAI form. To verify this result, use the switch-level equations to write

$$0 \cdot [(a+e) \cdot (b+f)] \tag{2.61}$$

This is equivalent to the expression in equation (2.60) for Y.

Now recall that a CMOS logic gate uses nFETs to pass a 0 to the output, and pFETs to pass a logic 1. Since pFETs complement nFETs, we can construct the logic formation characteristics summarized in Figure 2.46. The parallel-connected pFETs shown in Figure 2.46(a) are described by the logic equation

$$1 \cdot (\bar{x} + \bar{y}) = 1 \cdot \overline{(x \cdot y)} \tag{2.62}$$

which is the AND-NOT operation sequence. To obtain the OR-NOT operation, we must use series-connected pFETs as in Figure 2.46(b). In this case, the logic is formed from switching equations as

$$1 \cdot \bar{x} \cdot \bar{y} = 1 \cdot \overline{(x+y)} \tag{2.63}$$

which verifies the statement.

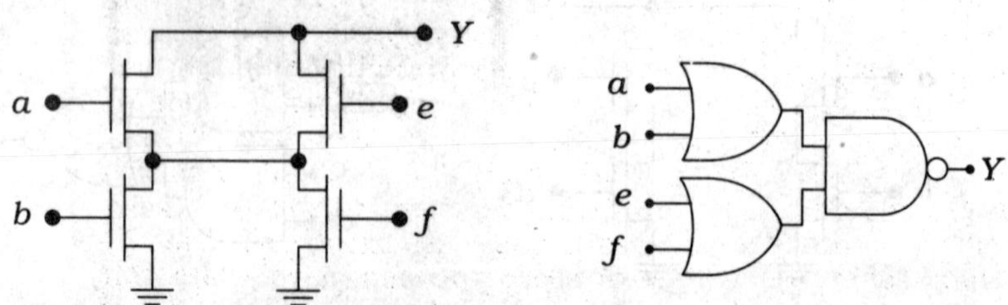

Figure 2.45 nFET OAI network

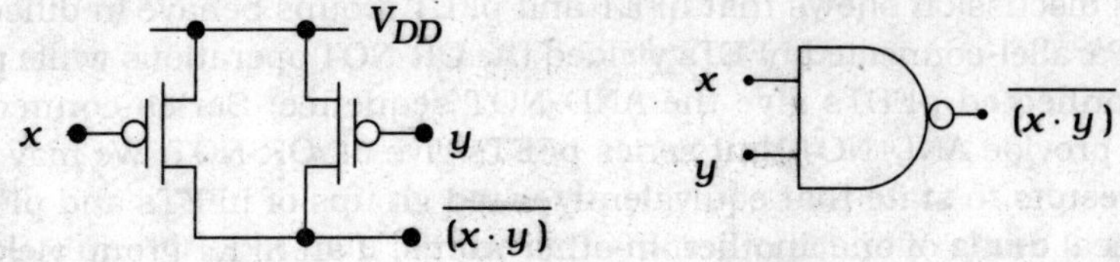

(a) Parallel-connected pFETs

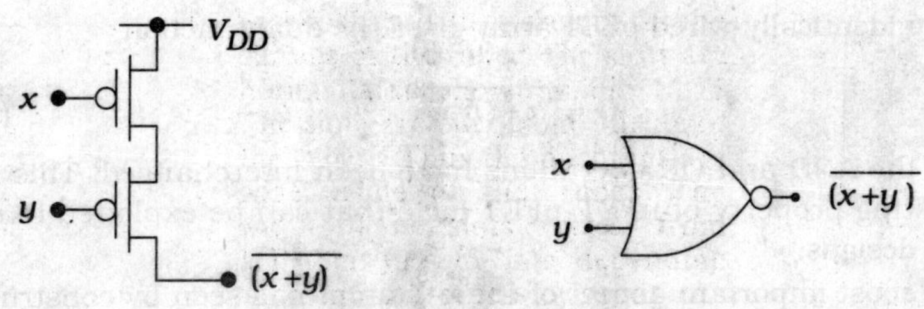

(b) Series-connected pFETs

Figure 2.46 pFET logic formation

Let us examine the pFET array needed for the AOI function

$$X = \overline{(a \cdot b) + (c \cdot d)} \tag{2.64}$$

discussed earlier for the nFET circuit shown in Figure 2.44. Using the pFET rules results in the network illustrated in Figure 2.47(a). Similarly, the OAI function

$$Y = \overline{(a + e) \cdot (b + f)} \tag{2.65}$$

yields the pFET array in Figure 2.47(b).

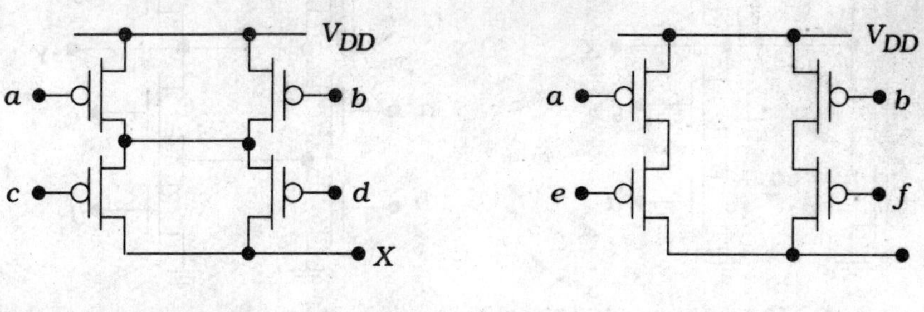

(a) pFET AOI circuit (b) pFET OAI circuit

Figure 2.47 pFET arrays for AOI and OAI gates

This discussion shows that nFET and pFET groups behave in different ways. Parallel-connected nFETs yielded the OR-NOT operations while parallel-connected pFETs give the AND-NOT sequence. Series-connected nFETs provide AND-NOT, but series pFETs give us OR-NOT. We may use these results to state that equivalently wired groups of nFETs and pFETs are logical **duals** of one another. In other words, if an nFET group yields a function of the form

$$g = \overline{a \cdot (b + c)} \tag{2.66}$$

then an identically-wired pFET array gives the dual function

$$G = \overline{a + (b \cdot c)} \tag{2.67}$$

where the AND and OR operations have been interchanged. This is an interesting property of nFET-pFET logic that can be exploited in some CMOS designs.

The most important aspect of these examples is seen by constructing the complete CMOS circuit for each; both are shown in Figure 2.48. Consider first the AOI circuit in Figure 2.48(a). The nFETs with inputs a and b are in series, while the corresponding pFETs are wired in parallel. This scheme is also applied to the FETs with input variables c and d. Finally, the nFET group with inputs (a, b) is in parallel with the input group (c, d), so the corresponding pFET groups are in series. This is another example of series-parallel structuring of the nFET-pFET arrays. The OAI circuit in Figure 2.48(b) exhibits the same features. In this case, the nFETs with

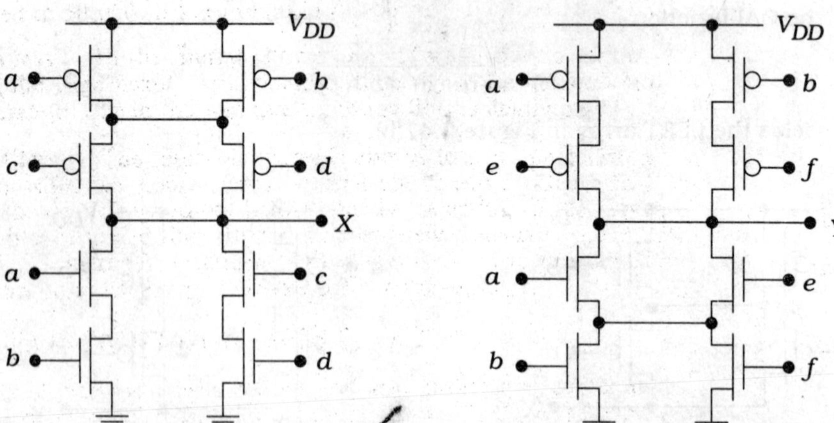

(a) AOI circuit (b) OAI circuit

Figure 2.48 Complete CMOS AOI and OAI circuits

inputs *a* and *e* are in parallel, as are the nFETs with inputs *b* and *f*. The pFET group with inputs *a* and *e* are wired in series; the same comment holds for the pFETs driven by *b* and *f*. Finally, since the nFET (*a*, *e*) group is in series with the (*b*, *f*) group, the corresponding pFETs groups are in parallel. This may be used to construct any AOI or OAI circuit in CMOS.

Example 2.1

Consider the complex function

$$X = \overline{a + b \cdot (c + d)} \tag{2.68}$$

The nFET circuit can be constructed by using the following arrangements:

Group 1: nFETs with inputs *c* and *d* are in parallel;

Group 2: an nFET with input *b* is in series with Group 1;

Group 3: an nFET with input *a* is in parallel with the Group 1-Group 2 circuit.

The circuit in Figure 2.49 shows each group explicitly. The pFETs are arranged using series-parallel structuring. Each group of pFETs can be associated with the nFET group that has the same inputs such that

Group 1: pFETs with inputs *c* and *d* are in series;

Group 2: a pFET with input *b* is in parallel with Group 1 pFETs;

Group 3: a pFET with input *a* is in series with the Group 1-Group 2 pFETs.

The equivalent logic diagram for the circuit is shown in Figure 2.50.

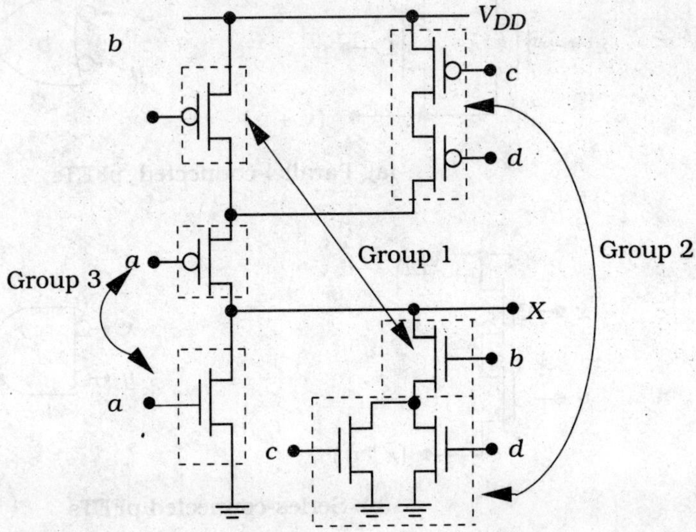

Figure 2.49 AOI circuit for Example 2.1

Tracing the data flow from the inputs to the output shows that the gate has OAOI structuring. This is just an AOI circuit with an additional OR input.

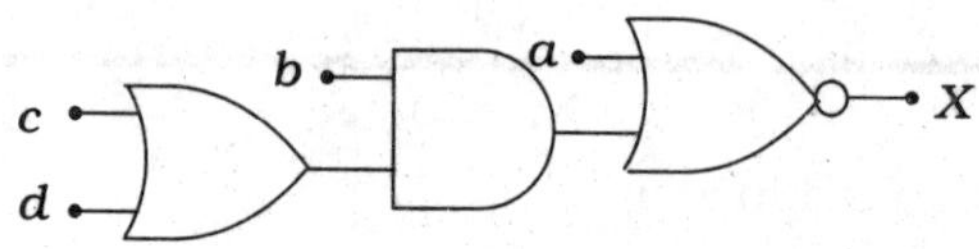

Figure 2.50 Equivalent logic diagram for Example 2.1

Bubble Pushing

Series-parallel wiring of complex CMOS logic circuits can be designed using an approach that is based on logic diagrams. The procedure is obtained by applying the DeMorgan rules to the pFET relations illustrated in Figure 2.46. Recall that pFETs can be modeled as assert-low switches. Let us therefore model pFET groups as logic gates that have assert-low inputs. This leads to the modified logic associations shown in Figure 2.51. In Figure 2.51(a), we apply the DeMorgan rule to write

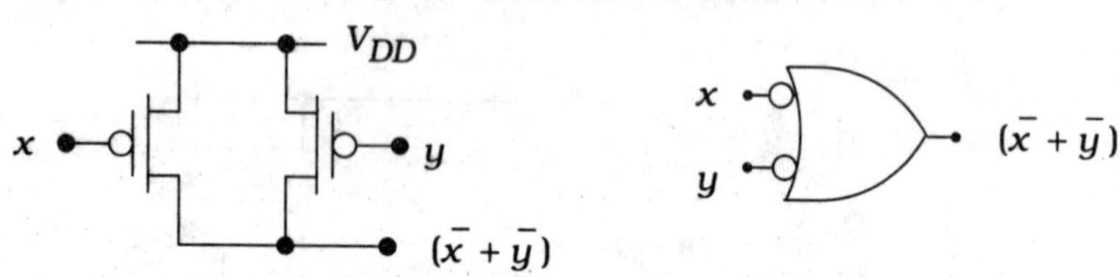

(a) Parallel-connected pFETs

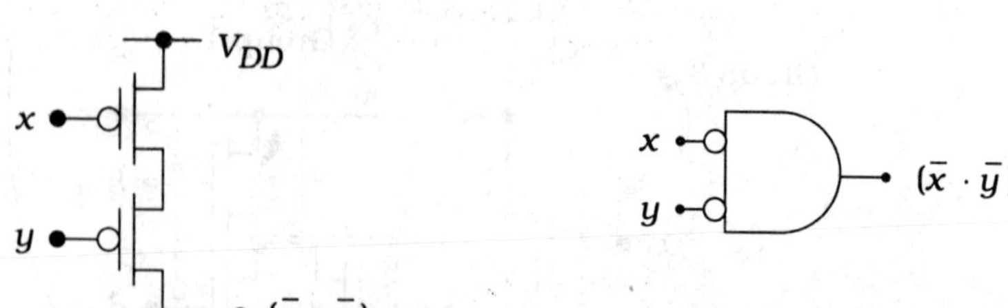

(b) Series-connected pFETs

Figure 2.51 Assert-low models for pFETs

$$1 \cdot (\overline{x \cdot y}) = 1 \cdot (\bar{x} + \bar{y}) \tag{2.69}$$

so that parallel-connected pFETs may be viewed as an OR operation with assert-low (bubbled) inputs. In the same manner, the series-connected pFETs in Figure 2.51(b) provide the AND operation with assert-low inputs as verified by the identity

$$1 \cdot (\overline{x + y}) = 1 \cdot (\bar{x} \cdot \bar{y}) \tag{2.70}$$

Both operations can be represented graphically by the operations shown in Figure 2.52 where we visualize pushing the bubble backward through the gate to the inputs to create the dual operation with assert-low input ports.

The procedure for designing the transistor circuitry for a CMOS logic gate can be summarized by the following steps.

- Construct the logic diagram using basic AOI or OAI structuring. Deeper nesting, such a OAOI and AOAI, is allowed.
- Use the gate-nFET relations summarized in Figure 2.43 to construct the nFET logic circuit between the output and ground.
- To obtain the topology of the pFET array, start with the original logic diagram and push the bubble back toward the inputs using the DeMorgan rules. Continue the backward pushing until every input is bubbled. The pFET circuitry between the output and VDD is then obtained using the rules in Figure 2.51.

Note that both the nFETs and the pFETs are wired such that parallel-connected transistors give the OR operation, while series-connected FETs provide the AND operation. The only difference between the two is that nFETs are assert-high devices while pFETs are assert-low (bubbled-input) switches.

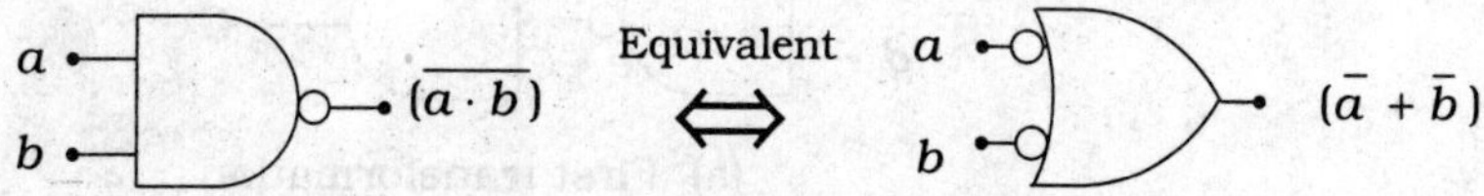

(a) NAND - OR

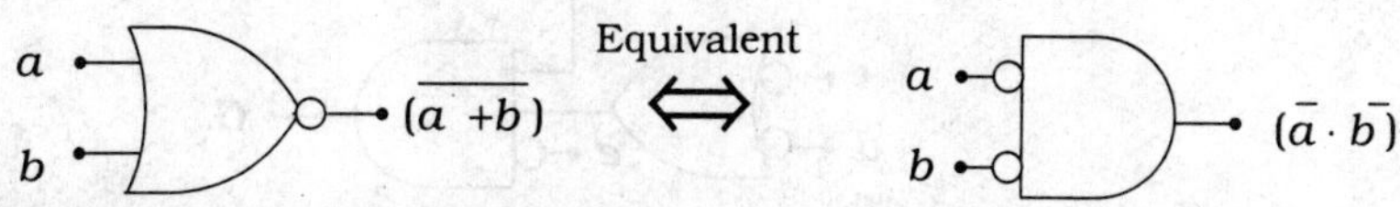

(b) NOR - AND

Figure 2.52 Bubble pushing using DeMorgan rules

Example 2.2

Consider the logic diagram shown in Figure 2.53. This provides us with a map for building the nFET logic array. We see that the nFETs with inputs *a* and *b* are in series (due to the AND gate), as are the nFETs with inputs *c* and *d*. These series-connected groups are in parallel with an nFET that has the input *e* since they are OR'ed at the output. The NOT operation (in the NOR gate) is automatic in the nFET array.

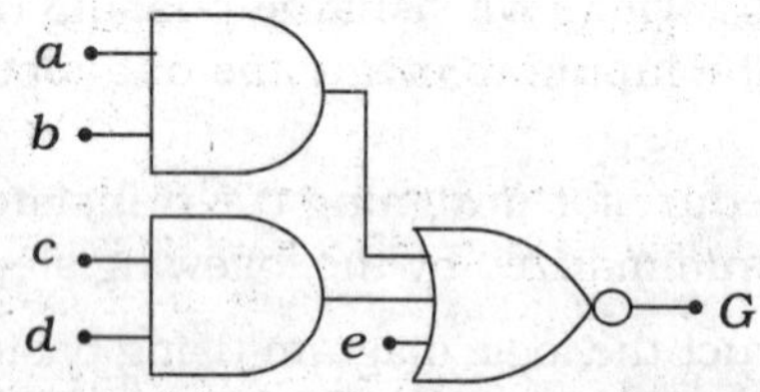

Figure 2.53 AOI logic diagram for bubble-pushing example

To obtain the wiring of the pFETs, we push the bubble back as shown in Figure 2.54. The first step is to transform the output NOR gate into an AND gate with assert-low inputs; this results in the intermediate diagram drawn in Figure 2.54(a). Pushing the bubbles back through the AND gates gives assert-low OR gates as in Figure 2.54(b). This shows that the

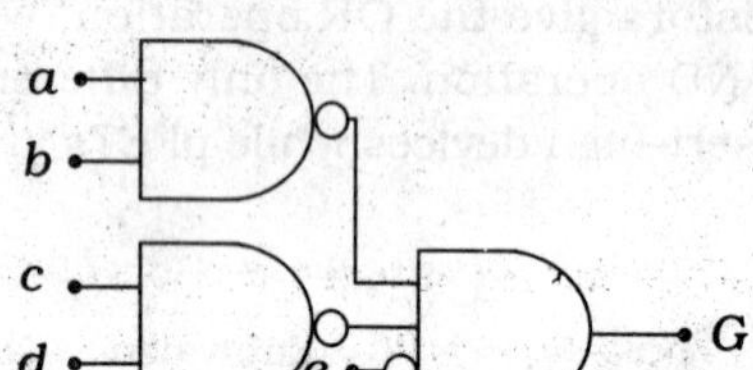

(a) First transformation

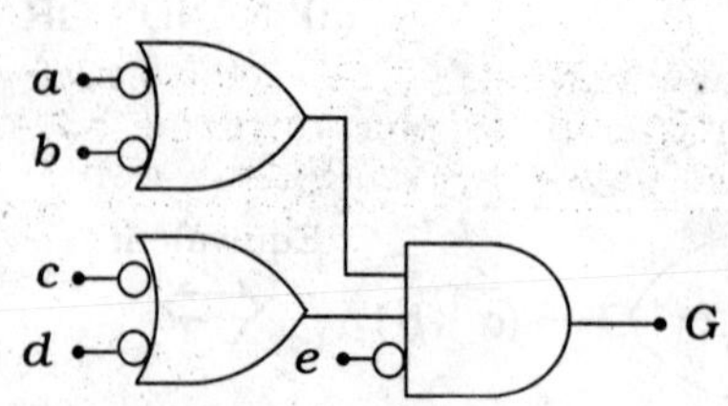

(b) Final form

Figure 2.54 Bubble pushing to obtain the topology of the pFET array

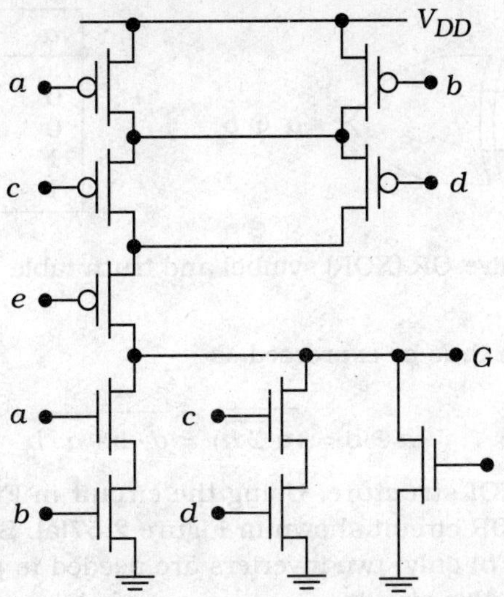

Figure 2.55 Final circuit for the bubble-pushing example

pFET array consists of

- Two pFETs with inputs *a* and *b* wired in parallel
- Two pFETs with inputs *c* and *d* wired in parallel
- One pFET with an input *e* that is in series with the two groups above.

The final circuit is drawn in Figure 2.55. It is worth the effort to trace through the construction procedure. And, it is important to remember that the CMOS logic gate implements the entire function *G* portrayed in the logic diagram. It is not possible to break the circuit down into more primitive logic.

2.4.2 XOR and XNOR Gates

An important example of using an AOI circuit is constructing Exclusive-OR (XOR) and Exclusive-NOR circuits. These often-used gates are constructed from logic primitives. Figure 2.56 gives the circuit symbol and truth table for the XOR. Reading the logic 1 outputs gives the standard SOP equation

$$a \oplus b = \bar{a} \cdot b + a \cdot \bar{b} \tag{2.71}$$

from the second and third lines. This is not in AOI form. However, if we read the 0 output lines, then the XNOR expression is

$$\overline{a \oplus b} = a \cdot b + \bar{a} \cdot \bar{b} \tag{2.72}$$

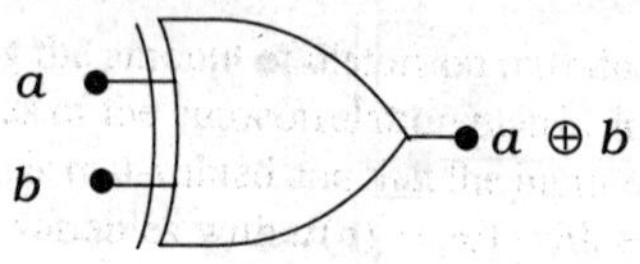

a	b	$a \oplus b$
0	0	0
0	1	1
1	0	1
1	1	0

Figure 2.56 Exclusive-OR (XOR) symbol and truth table

The XOR can thus be expressed as

$$a \oplus b = \overline{(\overline{a \oplus b})} = \overline{a \cdot b + \bar{a} \cdot \bar{b}} \tag{2.73}$$

which has AOI structure. Using the circuit in Figure 2.48(a) gives the basic AOI XOR circuit shown in Figure 2.57(a). Since the XOR gate has inputs of (a, b) only, two inverters are needed to provide the 4-input set $(a, b, \bar{a}, \bar{b})$ in this circuit.

To obtain an XNOR circuit, we just complement the XOR SOP equation to write

$$\overline{a \oplus b} = \bar{a} \cdot b + a \cdot \bar{b} \tag{2.74}$$

Interchanging a and $\bar{a}$ in the XOR circuit thus gives the XNOR gate in Figure 2.57(b). Switching the b and $\bar{b}$ variables would have given the same result.

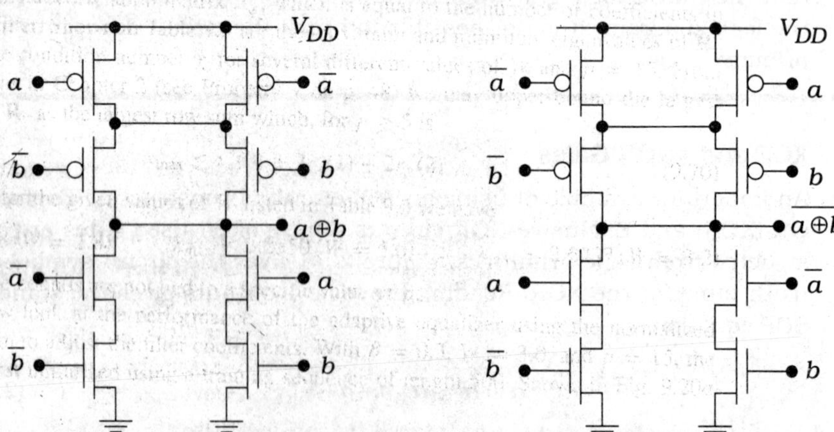

(a) Exclusive-OR (b) Exclusive-NOR

Figure 2.57 AOI XOR and XNOR gates

2.4.3 Generalized AOI and OAI Logic Gates

Standard logic design is often simplified using generalized multiple-input AOI and OAI logic gates. This is particularly true in ASIC-type circuits that rely on predesigned logic circuits. A straightforward nomenclature for distinguishing among various input configurations is developed in Figure 2.58. The network in Figure 2.58(a) has an AOI pattern with 2 inputs to each AND gate; it is therefore called an AOI22 gate. Similarly, the logic pattern in Figure 2.58(b) is called an AOI 321, where a "1" label implies an input that bypasses the AND gates and is connected directly to an OR gate. The third example, shown in Figure 2.58(c), is termed an OAI221 gate using the same convention. The CMOS circuits are easily designed using series-parallel wiring or bubble pushing.

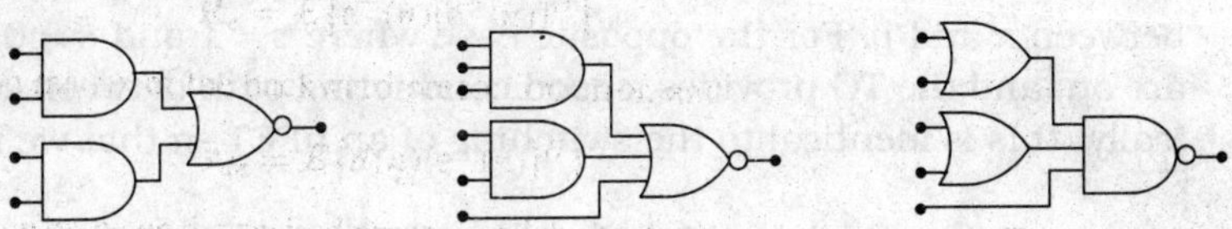

(a) AOI22 gate (b) AOI321 gate (c) OAI221 gate

Figure 2.58 General naming convention

Generalized complex logic gates provide a uniform basis for creating different logic operations using a generic gate. As a simple example, consider the AOI22 gate shown in Figure 2.59(a). This provides an output of

$$\text{AOI22}(a, b, c, d) = \overline{a \cdot b + c \cdot d} \tag{2.75}$$

To create an XOR circuit, we can define the inputs as shown in Figure 2.59(b), which allows us to write

$$a \oplus b = \text{AOI22}(a, b, \bar{a}, \bar{b}) \tag{2.76}$$

Using the same reasoning, the XNOR function can be obtained using

$$\overline{a \oplus b} = \text{AOI22}(a, \bar{b}, \bar{a}, b) \tag{2.77}$$

This illustrates how generic logic gates can be used in random logic design.

2.5 Transmission Gate Circuits

A CMOS **transmission gate** is created by connecting an nFET and pFET in parallel as shown in Figure 2.60(a). The nFET Mn is controlled by the signal s, while the pFET Mp is controlled by the complement $\bar{s}$. When wired in this manner, the pair acts as a good electrical switch between the

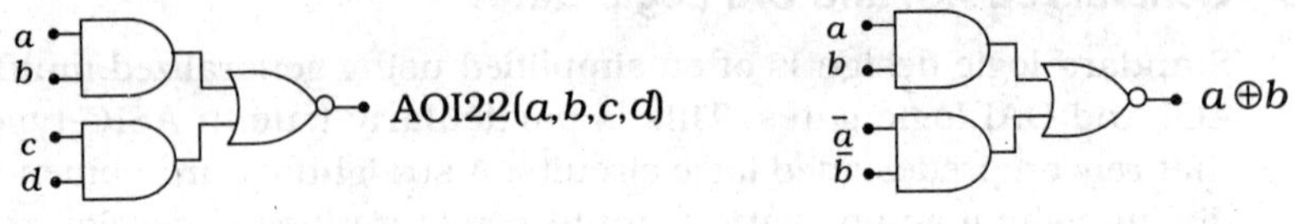

(a) General AOI22 gate (b) XOR

Figure 2.59 Application of an AOI22 gate

input and the output variables x and y, respectively.

The operation of the switch can be understood by analyzing the two cases for s. If $s = 0$, the nFET is OFF; since $\bar{s} = 1$, the pFET is also OFF, so that the TG acts as an open switch. In this case, there is no relationship between x and y. For the opposite case where $s = 1$ and $\bar{s} = 0$, both FETs are on, and the TG provides a good conducting path between x and y. Logically, this is identical to the switching of an nFET so that we may write

$$y = x \cdot s \quad \text{iff} \quad s = 1 \tag{2.78}$$

This assumes that x is the input and y is the output. However, the TG is classified as a **bi-directional** switch. The TG symbol in Figure 2.60(b) is based on this observation. It is created using two back-to-back arrows indicating that the data can flow in either direction. Control is achieved by s and $\bar{s}$; the bubble indicates the connection to the pFET gate.

Transmission gates are useful because they can transmit the entire voltage range [0, V_{DD}] from left to right (or vice versa). This is due to the parallel connection of the transistors. Zero voltage levels are transmitted by the nFET, while the pFET is responsible for transmitting the power supply voltage V_{DD}. The main drawback of using TGs in modern VLSI is that they require two FETs and an implied inverter that takes s and produces $\bar{s}$.

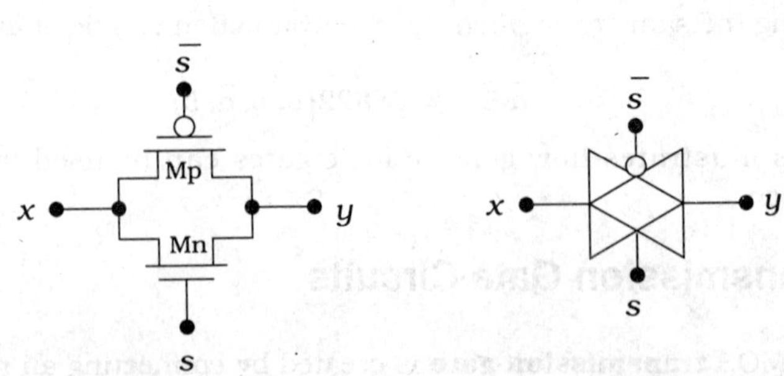

(a) Transistor circuit (b) Symbol

Figure 2.60 Transmission gate (TG)

2.5.1 Logic Design

Transmission gate logic design has been used extensively in CMOS design for many years. The simplicity of the switching and the ability to transmit the entire range of voltages made it attractive for many applications. TG circuits are found in many ASIC structures, making them worth studying in more detail.

Multiplexors

The ideal-switch characteristics of TGs make them useful for creating some rather unique circuits. An example is the 2-to-1 MUX shown in Figure 2.61. The operation of the circuit is summarized in the table. When the selector signal has a value $s = 0$, TG0 is closed and TG1 is open, so that P_0 is transmitted to the output. If $s = 1$, the situation is reversed with TG0 open and TG1 closed; in this case, $F = P_1$. Combining these results gives

$$F = P_0 \cdot \bar{s} + P_1 \cdot s \tag{2.79}$$

which is the required equation. Note that the use of a pair of TGs eliminates the possibility of having a floating (disconnected) output since one TG is always closed while the other will be open. The 2-to-1 architecture can be extended to a 4:1 network by using the 2-bit selector word $(s_1\ s_0)$ that has values of (0 0), (0 1), (1 0), and (1 1). Each input line (P_0, P_1, P_2, P_3) will have two TGs in its path such that the output is

$$F = P_0 \cdot \overline{s_1} \cdot \overline{s_0} + P_1 \cdot \overline{s_1} \cdot s_0 + P_2 \cdot s_1 \cdot \overline{s_0} + P_3 \cdot s_1 \cdot s_0 \tag{2.80}$$

For example, the P_0 path will have TGs that are closed with $(s_1\ s_0) = (0\ 0)$. The construction of the network is left as an exercise for the reader.

The 2:1 MUX can be modified to produce other useful functions. One is illustrated in Figure 2.62(a). The input to the top TG is a; this is inverted so that $\bar{a}$ enters the lower TG. Variable b and its complement are used to control the TGs. When $b = 0$, the upper TG is closed and a is passed to the

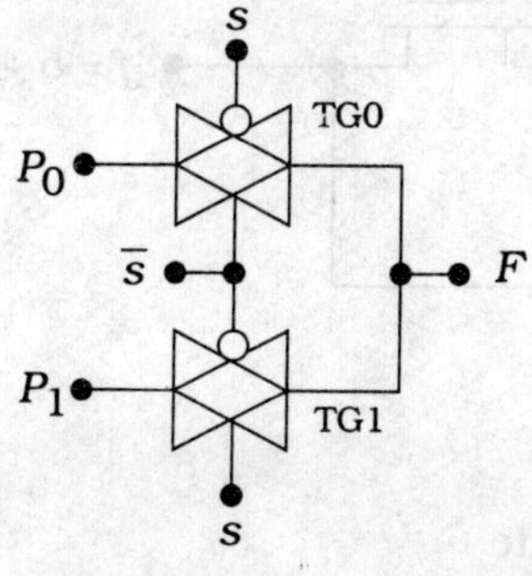

s	TG0	TG1	F
0	Closed	Open	P_0
1	Open	Closed	P_1

Figure 2.61 A TG-based 2-to-1 multiplexor

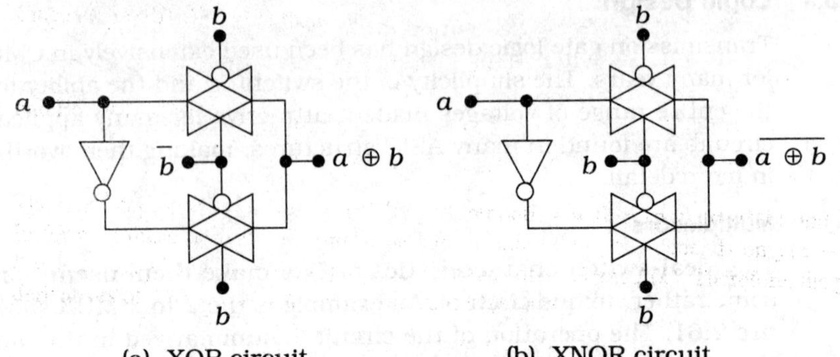

Figure 2.62 TG-based exclusive-OR and exclusive-NOR circuits

output, while $b = 1$ closes the lower TG and steers $\overline{a}$ to the output. This gives

$$a \cdot \overline{b} + \overline{a} \cdot b = a \oplus b \tag{2.81}$$

i.e., the circuit provides the XOR (exclusive-OR) function. The expression can be verified using the 2:1 MUX result. An XNOR function

$$\overline{a \oplus b} = a \cdot b + \overline{a} \cdot \overline{b} \tag{2.82}$$

is obtained if we interchange b and $\overline{b}$. The circuit for this simple modification is shown in Figure 2.62(b).

OR Gate

Transmission gate characteristics can be used to create the simple OR circuit shown in Figure 2.63; this is useful since complementary CMOS gates can only provide the NOR operation. The operation of the circuit can

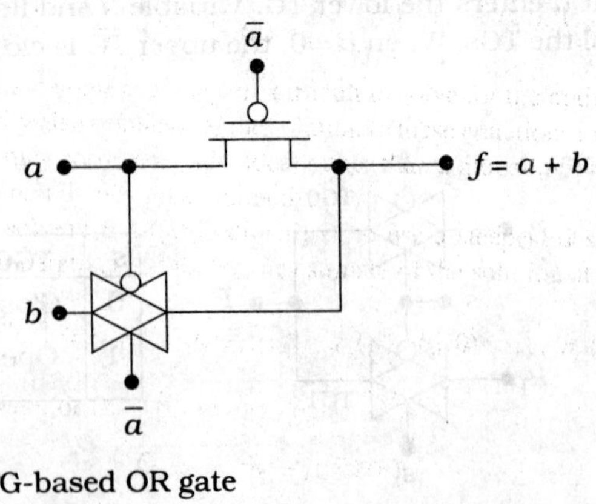

Figure 2.63 A TG-based OR gate

be understood by examining the effect that a has on the switches. If $a = 0$, then the pFET is OFF (since $\overline{a} = 1$ drives it into cutoff) while the TG acts as a closed switch. This gives an output of $f = b$. If $a = 1$, then the pFET is ON and the value of $f = a = 1$ is transmitted to the output. Thus, the output is $f = 1$ if either input is a 1, which establishes the OR operation. We can alternately use logic equations for the TG and the pFET to write the output as

$$\begin{aligned} f &= a \cdot (\overline{\overline{a}}) + \overline{a} \cdot b \\ &= a + \overline{a} \cdot b \\ &= a + b \end{aligned} \tag{2.83}$$

where the last step follows by absorption. This verifies the simpler bit-by-bit analysis.

Alternate XOR/XNOR Circuits

Mixing TGs and FETs as in the OR gate circuit gives rise to many variations for the design of basic logic gates. Many of these designs are for exclusive-OR and equivalence (XNOR) functions due to their importance in adders and error detection/correction algorithms.

An example of this type of circuit is the XNOR network in Figure 2.64. This uses the input pair $(b, \overline{b})$ to control the transmission gate. To understand the operation, remember that the output of an XNOR gate is 1 if and only if the inputs are equal. Suppose that $b = 1$; the TG acts as a closed switch and a is transmitted to the output to give $g = a$. For this case, the output is a 1 iff $a = 1$. The circuit operates differently if $b = 0$. Now, the TG is off and a is directed toward the gates of the Mp/Mn pair. Since $b = 0$ is applied to the source of the nFET Mn and $\overline{b} = 1$ is connected to the source (upper side) of the pFET, the $(b, \overline{b}) = (0, V_{DD})$ pair provides power to the FETs, resulting in an inverter! For this case, the output is $g = \overline{a}$, so that g is 1 iff $a = 0$. This establishes the circuit as an XNOR gate as stated. Interchanging b and $\overline{b}$ gives an XOR gate.

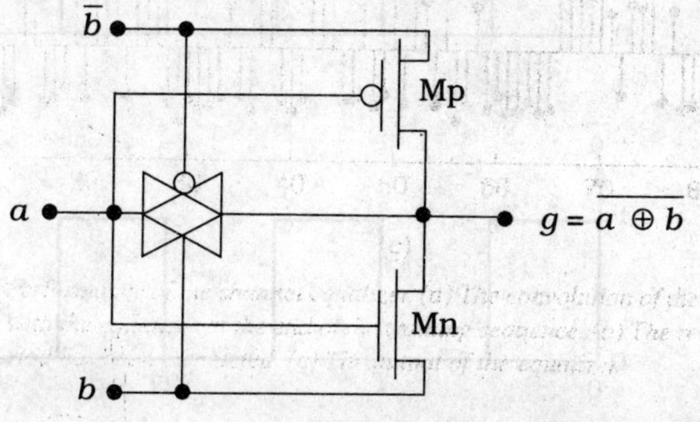

Figure 2.64 An XNOR gate that uses both TGs and FETs

2.6 Clocking and Dataflow Control

Synchronous digital design relies on the ability to control the flow of data using a clocking signal ϕ. The switching characteristics of TGs can be used to provide a simple approach to system clocking. Since complementary signals are required to switch a TG, both ϕ and $\overline{\phi}$ are used in this type of design; waveforms are shown in Figure 2.65. The period T is the time in seconds needed for one complete cycle. The frequency f is defined by

$$f = \frac{1}{T} \tag{2.84}$$

and has units of Hertz [Hz] = [1/sec], where 1 hertz means one cycle is completed in 1 second. We will assume that the clock is at a logic 1 value of one-half of the period, and at a logic 0 value for the remaining half of the period.

Let us examine the effect of applying the complementary clocks to a transmission gate. Figure 2.66(a) shows that when a value of $\phi = 1$ is applied to the nFET and $\overline{\phi} = 0$ to the pFET, the TG is On and acts like a closed switch. Reversing these values as in Figure 2.66(b) gives an open switch. Under static conditions, the value of y would not be known when the switch is opened. However, the electrical characteristics of CMOS allow us to temporarily hold the value of $y = x$ for a very short time t_{hold}; typically, t_{hold} is less than 1 second. If we use a high-frequency clock then the periodic open-closed change occurs at every half clock cycle. The node can hold the previous value so long as $(T/2) < t_{hold}$. This provides an accurate time base for controlling data flow in a complex network.

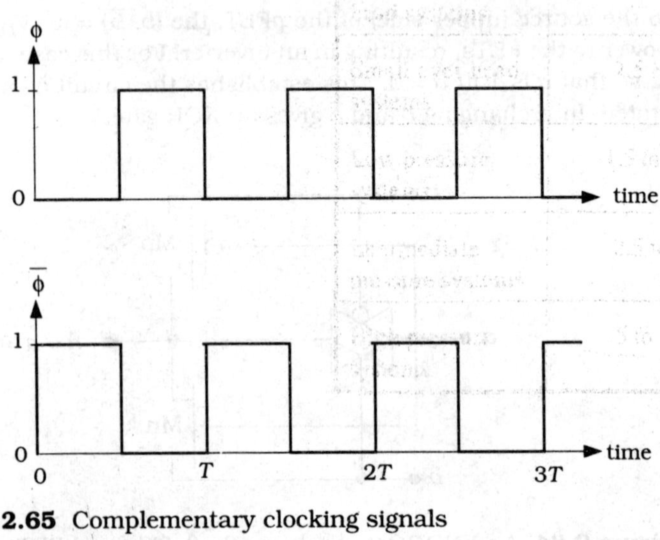

Figure 2.65 Complementary clocking signals

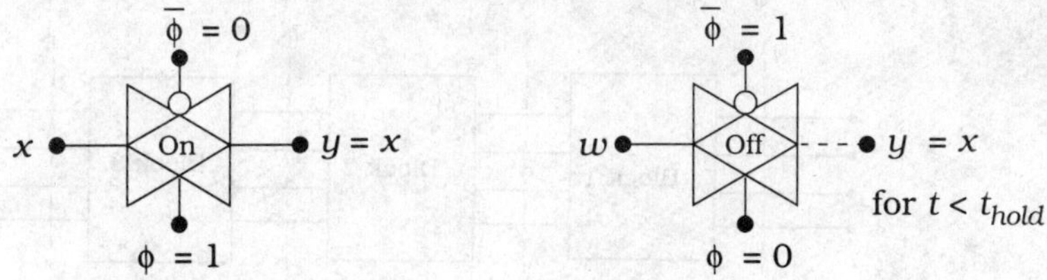

(a) Closed switch (b) Open switch

Figure 2.66 Behavior of a clocked TG

To use clocked TGs for data flow control, we place oppositely phased TGs at the inputs and outputs of logic blocks. A gate-level example is shown in Figure 2.67. The inputs on the left side are admitted when the clock is high with $\phi = 1$; the first group of logic gates evaluates the input bits and produce outputs *f* and *g* during this time. Since the output TGs are off, the outputs are held until the clock changes to $\phi = 0$. When this happens, *f* and *g* are allowed to enter the next group of logic gates, which results in *F*, *G*, and *H*. These are held at the outputs until they are transferred out when the clock returns to the value $\phi = 1$. This shows how data flow through the system is synchronized by the clocked TG.

Data flow can be visualized using system level block timing diagrams as in Figure 2.68. Each clock plane is shown graphically by a dashed line with either ϕ or $\overline{\phi}$ next to it. These represent a clock-controlled TG at every input. When the variable is true (equal to 1), then data is allowed to pass the plane from one side to the other. Otherwise the data is held on the left side until a clock transition takes place. With the labeling shown, this

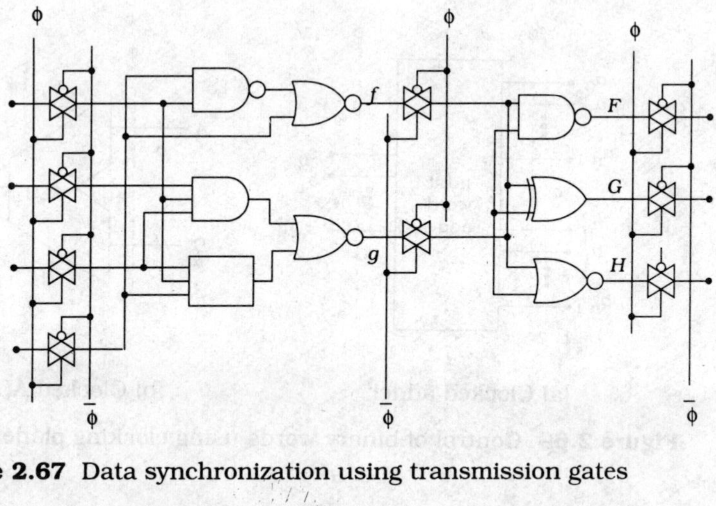

Figure 2.67 Data synchronization using transmission gates

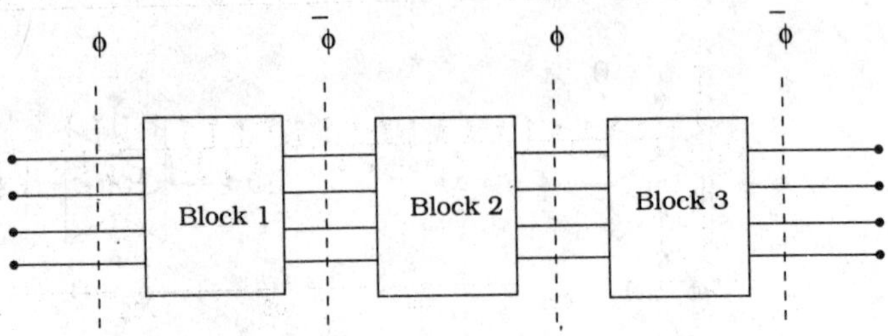

Figure 2.68 Block-level system timing diagram

says that a clock of $\phi = 1$ allows inputs into logic block 1. The outputs are transferred to logic block 2 when the clock changes to $\phi = 0$ and so on. In this scheme, data moves through a logic block every half cycle. Since the logic blocks are arbitrary, it can be used as the basis for building very complex logic chains. It also allows us to synchronize the operations performed on each bit of an n-bit binary word.

A synchronized word adder is illustrated in Figure 2.69(a). The input words $a_{n-1}...a_0$ and $b_{n-1}...b_0$ are controlled by the ϕ-clock plane, while the sum $s_{n-1}...s_0$ is transferred to the output when $\phi = 0$. Every bit in a word is transmitted from one point to another at the same time, which allows us to track the data flow through the system. This is extended to a larger scale with the ALU (arithmetic and logic unit) example in Figure 2.69(b). Inputs A and B are "gated" into the ALU by the ϕ -plane control; the result word Out is transferred to the next stage when $\overline{\phi} = 1$, i.e., $\phi = 0$. This illustrates the power of using clocked data transfer in VLSI design.

Clocked transmission gates synchronize the flow of signals, but the lines themselves cannot store the values for times longer than t_{hold}, which is very small. A storage element such as a latch is needed to obtain long-

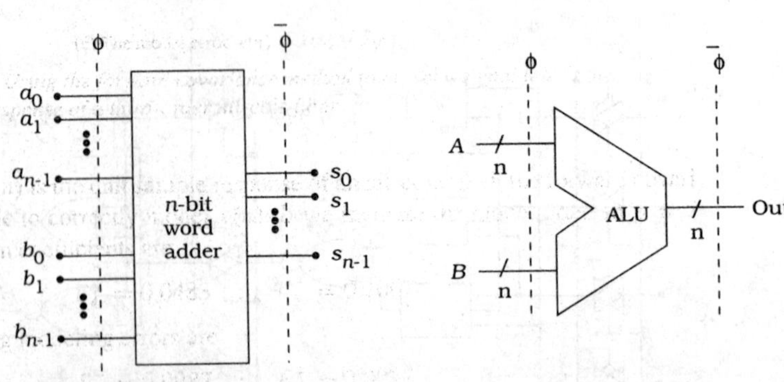

(a) Clocked adder (b) Clocked ALU

Figure 2.69 Control of binary words using clocking planes

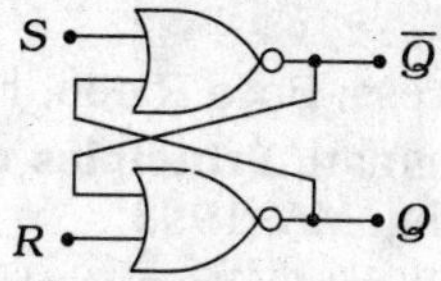

(a) Logic diagram

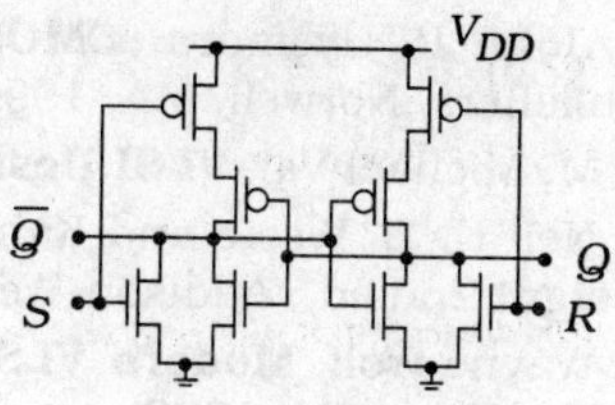

(b) CMOS circuit

Figure 2.70 SR latch

term storage of a data bit. Figure 2.70(a) shows the logic diagram for a simple NOR-based SR-latch. The CMOS circuit in Figure 2.70(b) is obtained by wiring two NOR2 gates together.

Clock control can be added to the circuit by inserting AND gates at the inputs to arrive at the modified logic diagram in Figure 2.71(a). This only allows changes in the inputs when $\phi = 1$. A compact CMOS circuit can be obtained by observing that two identical CMOS AOI circuits can be used to create the circuit in Figure 2.71(b). Designing a CMOS circuit using a logic diagram as a starting point thus becomes a straightforward process. This makes CMOS easy to adapt to. The challenge arises in making the circuits as fast and as compact as possible.

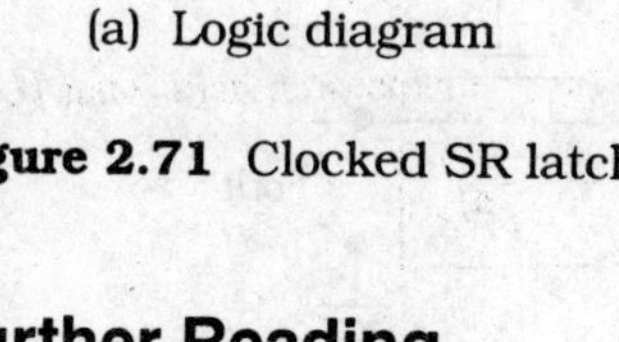

(a) Logic diagram

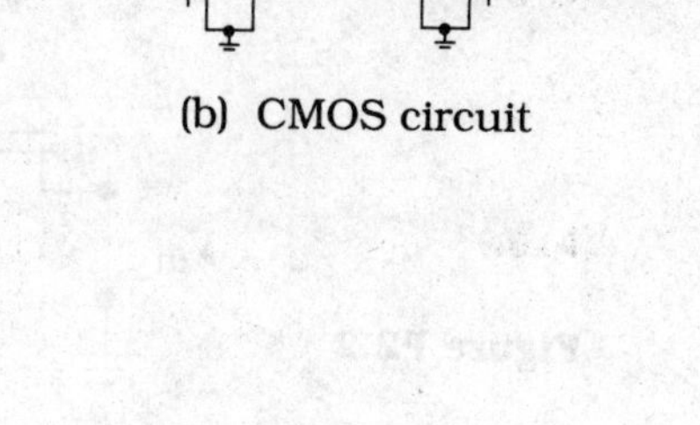

(b) CMOS circuit

Figure 2.71 Clocked SR latch

2.7 Further Reading

[1] Ken Martin, **Digital Integrated Circuit Design**, Oxford University Press, New York, 2000.

[2] Michael John Sebastian Smith, **Application Specific Integrated Circuits**, Addison-Wesley, Reading, MA, 1997.

[3] John P. Uyemura, **A First Course in Digital Systems Design**, Brooks-Cole Publishers, Monterey, CA, 2000.

[4] John P. Uyemura, **CMOS Logic Circuit Design**, Kluwer Academic Publishers, Norwell, MA, 1999.

[5] M. Michael Vai, **VLSI Design**, CRC Press, Boca Raton, FL, 2001.

[6] Neil H. E. Weste and Kamran Eshraghian, **Principles of CMOS VLSI Design**, 2nd ed., Addison-Wesley, Reading, MA, 1993.

[7] Wayne Wolf, **Modern VLSI Design**, 2nd ed., Prentice-Hall PTR, Upper Saddle River, NJ, 1998.

2.8 Problems

[2.1] Suppose that V_{DD} = 5 V and V_{Tn} = 0.7 V. Find the output voltage V_{out} of the nFET in Figure P2.1 for the following input voltage values: (a) V_{in} = 2 V; (b) V_{in} = 4.5 V; (c) V_{in} = 3.5 V; (d) V_{in} = 0.7 V.

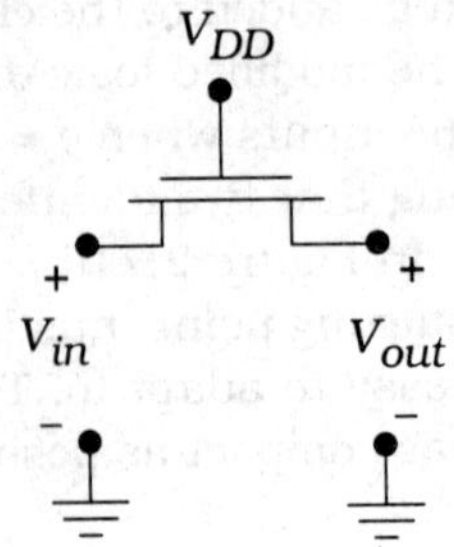

Figure P2.1

[2.2] Consider the two-FET chain in Figure P2.2. The power supply is set to a value of V_{DD} = 3.3 V and the nFET threshold voltage is V_{Tn} = 0.55 V. Find the output voltage V_{DD} at the right side of the chain for the following values of V_{in}: (a) V_{in} = 2.9 V; (b) V_{in} = 3.0 V; (c) V_{in} = 1.4 V; (d) V_{in} = 3.1V.

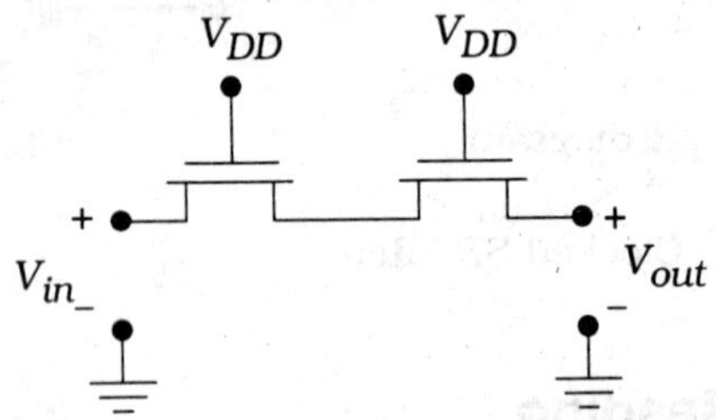

Figure P2.2

[2.3] The output of an nFET is used to drive the gate of another nFET as shown in Figure P2.3. Assume that V_{DD} = 3.3 V and V_{Tn} = 0.60 V. Find the output voltage V_{out} when the input voltages are at the following values: (a) V_a = 3.3 V and V_b = 3.3 V; (b) V_a = 0.5 V and V_b = 3.0 V; (c) V_a = 2.0 V and V_b = 2.5 V; (d) V_a = 3.3 V and V_b = 1.8 V.

[2.4] Design a NAND3 gate using an 8:1 MUX.

[2.5] Design a NOR3 gate using an 8:1 MUX as a basis.

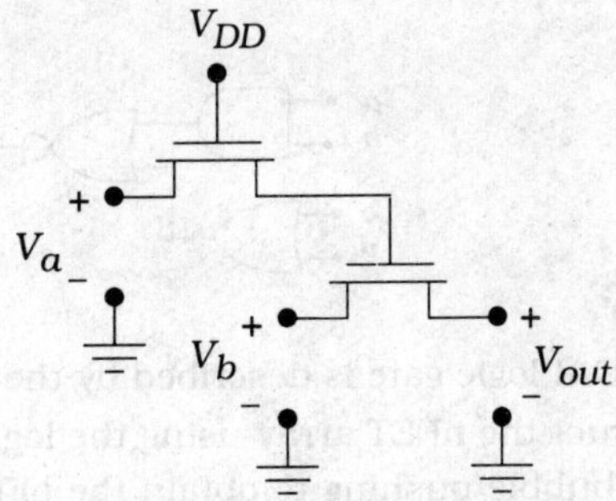

Figure P2.3

[2.6] Consider the 2-input XOR function $a \oplus b$.

(a) Design an XOR gate using a 4:1 MUX.

(b) Modify the circuit in (a) to produce a 2-input XNOR.

(c) A full adder accepts inputs a, b, and c and calculates the sum bit

$$s = a \oplus b \oplus c \tag{2.85}$$

Use your MUX-based gates to design a circuit with this output.

[2.7] Design a CMOS logic gate for the function

$$f = \overline{a \cdot b + a \cdot c + b \cdot d} \tag{2.86}$$

using the smallest number of transistors.

[2.8] Design a CMOS circuit for the OAI expression

$$h = \overline{(a+b) \cdot (a+c) \cdot (b+d)} \tag{2.87}$$

Use the smallest number of transistors in your design.

[2.9] Construct the CMOS logic gate for the function

$$g = \overline{x \cdot (y+z) + y} \tag{2.88}$$

Start with the minimum-transistor nFET network, and then apply bubble pushing to find the pFET wiring.

[2.10] Design a CMOS logic gate circuit that implements

$$F = \overline{a + b \cdot c + a \cdot b \cdot c} \tag{2.89}$$

using series-parallel logic. The objective is to minimize the transistor count.

[2.11] Consider the logic described by the diagram in Figure P2.4. A single, complex logic CMOS gate is to be designed for F.

(a) Construct the nFET array using the logic diagram.

(b) Apply bubble pushing to obtain the pFET logic. Then construct the pFET array using the rules.

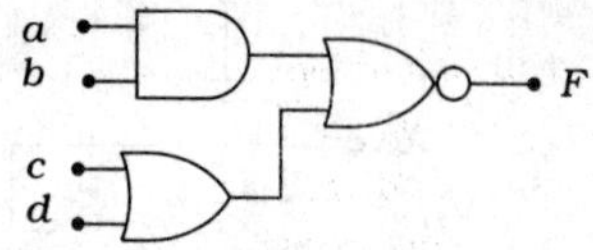

Figure P2.4

[2.12] An AOAI logic gate is described by the schematic in Figure P2.5.

(a) Construct the nFET array using the logic diagram.

(b) Apply bubble pushing to obtain the pFET logic. Use the diagram to construct the pFET array using the pFET rules.

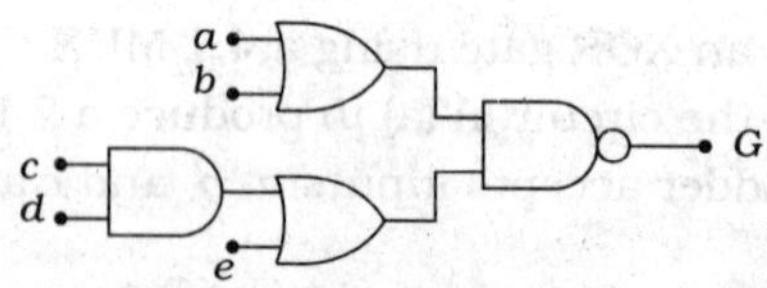

Figure P2.5

[2.13] A pFET logic array is shown in Figure P2.6. Construct the logic diagram using the pFET logic equations. Then construct the nFET circuit.

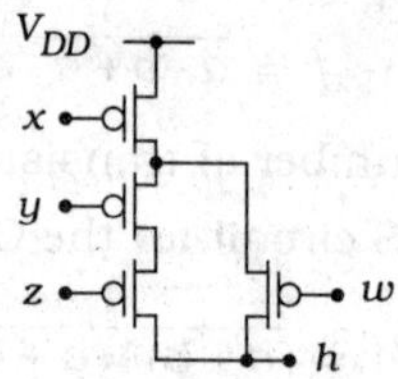

Figure P2.6

[2.14] Design the 4:1 multiplexor circuit that implements the function in equation (2.80) by using TG switches.

[2.15] Use an AOI22 gate to design a 2:1 MUX. Inverters are permitted in your design.

[2.16] Design a 4:1 MUX using three 2:1 TG multiplexors.

[2.17] A CPU clock ϕ has a frequency 2.1 GHz. What is the period T?

[2.18] Suppose that the hold time for a TG is given as t_{hold} = 120 milliseconds (ms). What is the smallest clock frequency that can be used to clock the data flow using a scheme such as that shown in Figure 2.67?

Physical Structure of CMOS Integrated Circuits

3

CMOS integrated circuits are electronic switching networks that are created on small area of a silicon wafer using a complex set of physical and chemical processes. A primary task of the VLSI designer is to translate circuit schematics into silicon form. This process is called **physical design** and is one aspect that separates the field of VLSI from general digital engineering. In this chapter we will examine the structure of a CMOS integrated circuit as seen at the microscopic silicon level in the design hierarchy.

3.1 Integrated Circuit Layers

A silicon integrated circuit can be viewed as a collection of patterned material layers, with each layer having specific conduction properties. The layers may be **metals** that conduct current very well, or they may be **insulators** that block the flow of current. Another material used to create layers is the element **silicon**. It is classified as a **semiconductor**, which means that it is a "partial" conductor. We sometimes refer to both metals and silicon as "conductors," but it is important to distinguish between the two.

An integrated circuit is made by stacking different layers of materials in a specific order to form three-dimensional structures that collectively act as an electronic switching network. Each layer has a predefined pattern that is specified in the system design process. The idea can be understood by referring to Figure 3.1, which portrays two separated layers. The bottom layer is a "sheet" of insulator on a base material called the "substrate." Above this is a patterned material layer of metal that is labeled "Layer M1." The pattern consists of two parallel **lines** of material which are to be placed on top of the insulator in the positions indicated by the

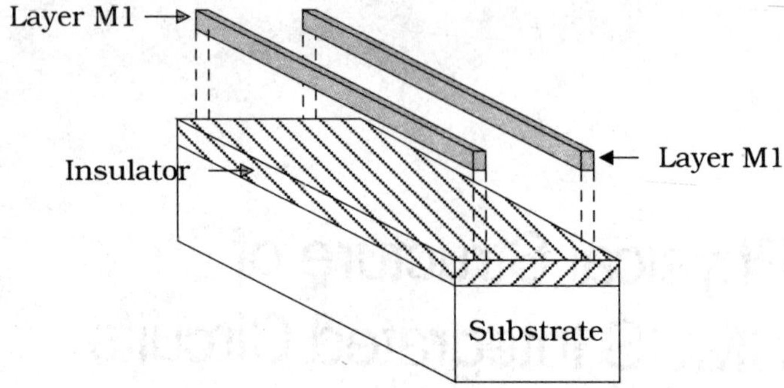

Figure 3.1 Two separate material layers

dashed lines. Figure 3.2 shows the structure after the stacking operation is finished. The end view is illustrated in Figure 3.2(a) and shows the two lines of Layer M1 on top of the insulating layer. Figure 3.2(b) provides the top view which shows that the two lines are parallel. The insulator is shown explicitly in this drawing, but we often take its existence as being implied and omit it from top views. Physically this is acceptable because the insulator itself is usually a layer of **silicon dioxide** (SiO_2), which is generically known as **quartz glass** and is visually transparent. Although quite simple, this example provides one main feature of layering in a silicon integrated circuit (IC): patterned conducting layers on top of a glass insulator. Complex VLSI chips employ several conducting layers of aluminum or copper with this type of structuring.

The concepts introduced in the preceding drawings can be extended by adding more layers. Suppose that we want to place another metal pattern on top of the structure shown in Figure 3.2. First we coat the surface with another layer of insulating glass to keep it from coming in contact with Layer M1, and then subject it to a **chemical-mechanical planarization** (CMP) sequence. In CMP, the surface is etched and "sanded" to provide a

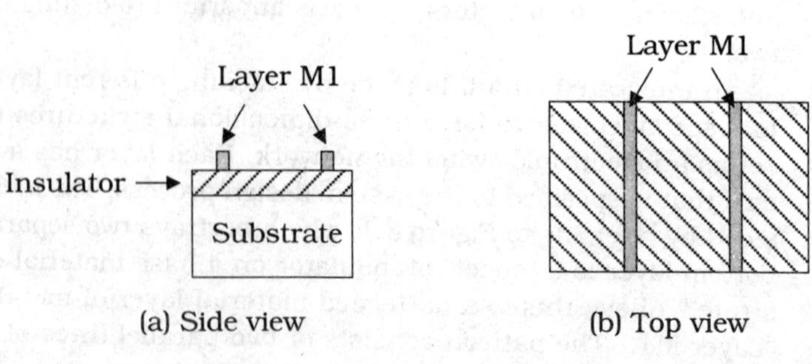

Figure 3.2 Layers after the stacking process is completed

flat surface for the next layer. Next, we coat the surface with the second metal (Layer M2) to arrive at the structure drawn in Figure 3.3. The side view in Figure 3.3(a) shows the added insulator that covers Layer M1 and the second metal Layer M2 on top. This illustrates the stacking order of the various layers but does not show that the two metal layers have different patterns. The top view in Figure 3.3(b) provides the distinguishing features of the patterning. In particular, we see that the Layer M2 pattern is a single metal line that is perpendicular to the parallel lines of Layer M1. The line has been drawn so that it covers Layer M1 patterns when the two cross. Also note that we have not shown any of the insulating layers explicitly in the top view, but it is important to remember that the two do not touch.

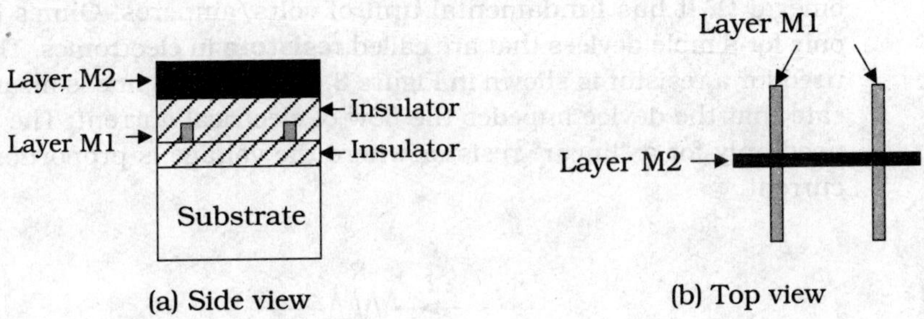

Figure 3.3 Addition of another insulator and a second metal layer

Combining the top and end views of an integrated circuit allows us to visualize the three-dimensional structure. Some important points that arise from this example are

- The side view illustrates the order of the stacking
- Insulating layers separate the two metal layers so that they are electrically distinct
- The patterning of each layer is shown by a top view perspective

The stacking order is established in the manufacturing process and cannot be altered by the VLSI designer. However, creating the pattern for each layer is a critical part of the chip design sequence as it defines the locations and sizes of all MOSFETs and specifies how the transistors are connected together.

3.1.1 Interconnect Resistance and Capacitance

Logic gates communicate with each other by signal flow paths from one point to another. At the integrated circuit level, this is accomplished by using patterned metal lines as wires to conduct electrical currents. These lines are generically referred to as **interconnects**. While this seems like a straightforward translation, the level of electric current flow is governed by the physical characteristics of the material and the dimensions of the line. This implies that the signal transfer speed is directly affected by the physical implementation of the wiring, making it a very important aspect of chip design.

Applying a voltage V (in units of volts V) to a patterned metal line creates a flow of current I (in amperes A) through it. For a simple conductor such as a metal, the relationship between the voltage and the current is given by Ohm's law

$$V = IR \tag{3.1}$$

where R is a constant of proportionality called the **resistance**. The unit of resistance is the **Ohm** and is denoted by the Greek uppercase letter omega: Ω. It has fundamental unit of volts/amperes. Ohm's law holds only for simple devices that are called **resistors** in electronics. The symbol used for a resistor is shown in Figure 3.4. The jagged line is meant to indicate that the device impedes the flow of electrical current. The symbol is used only for a "linear" resistor where the voltage is proportional to the current.

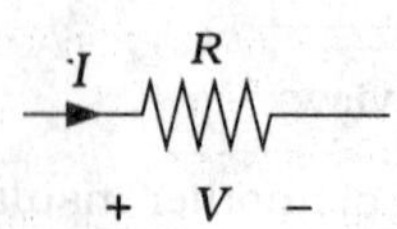

Figure 3.4 Symbol for a linear resistor

Now recall that a Boolean variable x is represented by a voltage V_x in a CMOS circuit. When applied to a patterned metal line, this voltage causes a current I_x to flow; the actual value of I_x is determined by the **line resistance** R_{line} and a few other electrical parameters. R_{line} is measured in units of ohms and is classified as a **parasitic** (unwanted) electrical element that cannot be avoided. Resistance impedes the flow of electrical signals, so that the value of R_{line} should be kept as small as possible.

The value of R_{line} for a given line can be calculated using the geometry shown in Figure 3.5. The length of the line is denoted by l and is measured in units of centimeters [cm]. The cross-sectional area A (with units of cm^2) is the product of the width w and the thickness t of the layer:

$$A = wt \tag{3.2}$$

The **conductivity** σ shown in the drawing is a characteristic of the material used in the layer. It is measured in units of $[\Omega\text{-cm}]^{-1}$ and represents how easily current flows: a large value of σ means that the layer conducts very well. Metals have large conductivities, while insulators have very small values of σ; we assume that the numerical value of σ is known. With

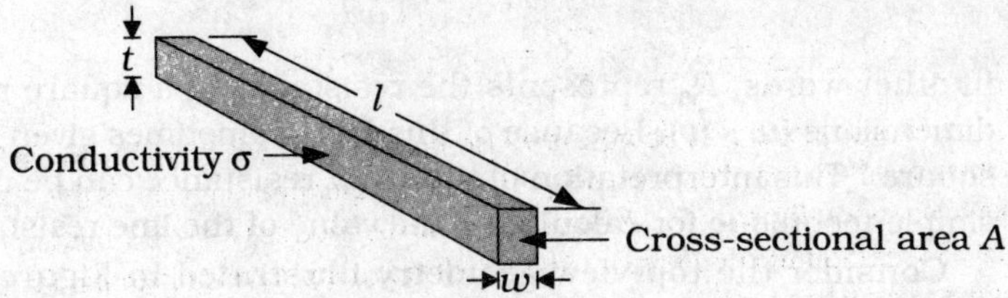

Figure 3.5 Geometry of a conducting line

these parameters, the line resistance in units of ohms is calculated from the expression

$$R_{line} = \frac{l}{\sigma A} \tag{3.3}$$

This shows the important relations that R_{line} is proportional to the length l of the line, and inversely proportional to the cross-sectional area A.

The **resistivity** ρ is the inverse of the conductivity such that

$$\rho = \frac{1}{\sigma} \tag{3.4}$$

ρ has units of [Ω-cm]. A high resistivity implies a low conductivity. The formula for the line resistance becomes

$$R_{line} = \rho\frac{l}{A} \tag{3.5}$$

by simple substitution.

A VLSI designer cannot control the values of t or σ, as these are established by the manufacturing process. Because of this, it is useful to rewrite the equation in the form

$$R_{line} = \left(\frac{1}{\sigma t}\right)\left(\frac{l}{w}\right) \tag{3.6}$$

so that the process-related terms are grouped together. This can be used to define the **sheet resistance** R_s of the line as

$$R_s = \frac{1}{\sigma t} = \frac{\rho}{t} \tag{3.7}$$

It is easily verified that R_s has units of ohms [Ω]. The sheet resistance of a layer is very useful because of two reasons. First, we find that it can be directly measured in the laboratory without knowing the actual values of σ or t. The second reason is due to the observation that a line with a length of $l = w$ has a line resistance of

$$R_{line} = R_s\left(\frac{w}{w}\right) = R_s \tag{3.8}$$

In other words, R_s represents the resistance of a square region with top dimensions ($w \times w$). Because of this R_s is sometimes given units of "Ω per square." This interpretation of the sheet resistance can be used to derive a simple technique for calculating the value of the line resistance.

Consider the top-view geometry illustrated in Figure 3.6(a), which identifies one square. By definition, the square has an end-to-end resistance of R_s. The square can be used to construct the equivalent line illustrated in Figure 3.6(b) by stringing many squares in a linear fashion. To calculate the total end-to-end resistance R_{line}, we note that each square has a resistance of R_s and that a string of resistors in series is equivalent to a single resistor with a value equal to the sum of the individual resistances. If there are a total of n squares, then we may write that

$$R_{line} = R_s n \tag{3.9}$$

where

$$n = \frac{l}{w} \tag{3.10}$$

gives the total number of squares from one end to the other. Note that n is not restricted to integer values; fractional contributions are permitted as shown on the right-hand side of the line.

This analysis demonstrates that, for a given layer, the line resistance depends upon the ratio (l/w) of the patterned line. The importance of this result is based on the qualitative observation that the speed of a transmitted signal along a patterned line is affected by the value of R_{line}. A small

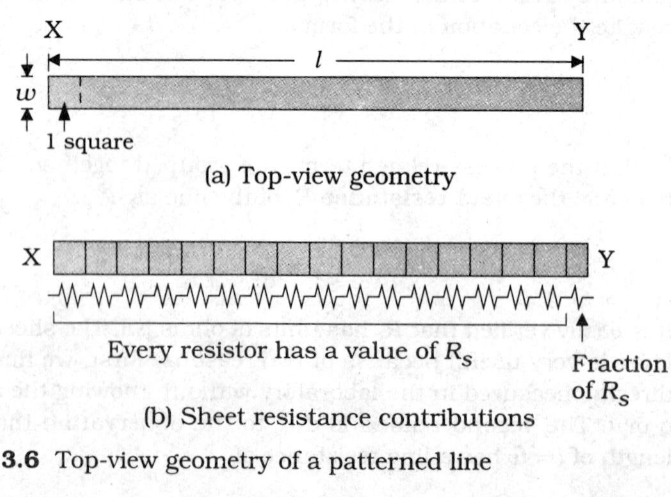

Figure 3.6 Top-view geometry of a patterned line

value of R_{line} allows for a high level of current flow, and is desirable for high-speed designs. We will quantify these statements later.

Interconnect lines also exhibit the property of **capacitance**, which is the ability to store electric charge and energy. In electronics, the element that stores charge is called a **capacitor**, and has the circuit symbol shown in Figure 3.7. It is characterized by the capacitance value C such that the charge Q on the positive side of the device is given by

$$Q = CV \tag{3.11}$$

where V is the voltage; this is balanced by a negative charge $-Q$ on the other plate. The unit of capacitance is the **farad** [F] where 1 F is defined as 1 coulomb/volt. Since electric current is defined by the time derivative $I = (dQ/dt)$, differentiating gives the I-V equation

$$I = C\frac{dV}{dt} \tag{3.12}$$

for the device.

Capacitance exists between any two conducting bodies that are electrically separated. For the interconnect line, the conductor is isolated from the semiconductor substrate by an insulating layer of silicon dioxide glass. The capacitance depends on the geometry of the line. Consider the structure shown in Figure 3.8 where T_{ox} is the thickness of the oxide between the interconnect line and the substrate in units of cm. Using basic physics, the line capacitance is given by the *parallel-plate* formula

$$C_{line} = \frac{\varepsilon_{ox} wl}{T_{ox}} \tag{3.13}$$

and is measured in units of farads. In this equation, wl is the area of the interconnect in cm^2 as seen from the top. The parameter ε_{ox} is the permittivity of the insulating oxide with units of F/cm; ε_{ox} is determined by the composition of the oxide.

Capacitance will be examined in more detail later in this chapter. For our present purposes, it is sufficient to note that the interconnect line exhibits both parasitic resistance R_{line} [Ω] and capacitance C_{line} [F]. Forming the product of these two quantities gives

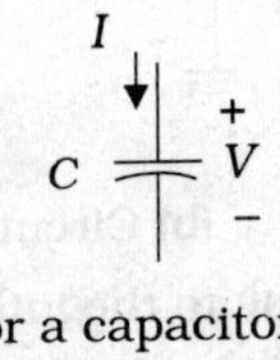

Figure 3.7 Circuit symbol for a capacitor

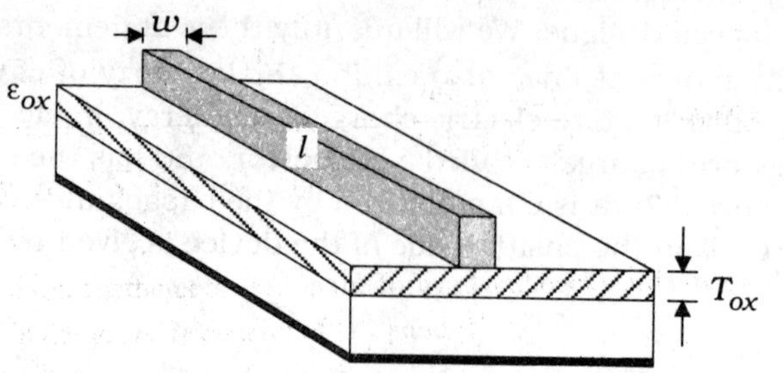

Figure 3.8 Geometry for calculating the line capacitance

$$\tau = R_{line}C_{line} \tag{3.14}$$

where τ has units of seconds [s] and is called a **time constant**. In high-speed digital circuits, signals on an interconnect line are delayed by τ, which places a limiting factor on the speed of the network. This is illustrated in Figure 3.9. In the physical layout of Figure 3.9(a), the output signal $v_s(t)$ from a NOT gate is connected to an interconnect line leading to the next gates in a logic chain. The voltage at the end of the interconnect is labeled as $v(t)$. The parasitic elements R_{line} and C_{line} are used to model the interconnect circuit as shown in Figure 3.9(b). With this simple circuit, $v(t)$ is the voltage across the capacitor. If the output voltage $v_s(t)$ of NOT gate makes a voltage transition from a 0 to a 1 level as shown in the waveform plot, then $v(t)$ also rises in the same manner. However, the capacitor voltage is delayed by a time constant τ and the shape of the

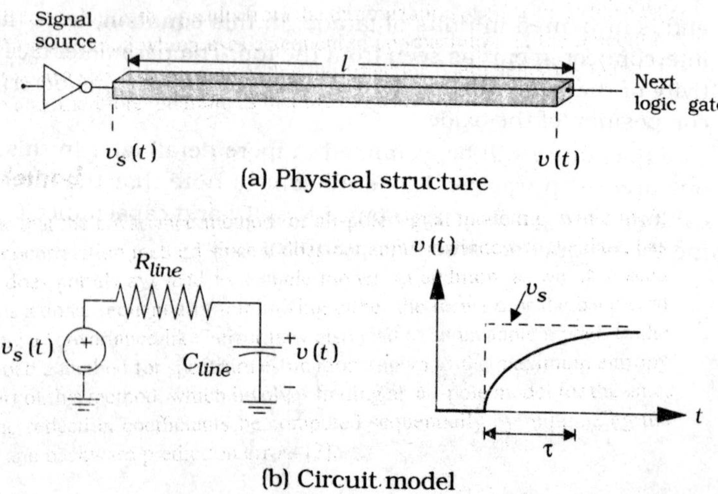

Figure 3.9 Time delay due to the interconnect time constant

waveform is not as sharp as the source.

Many aspects of VLSI processing are directed toward minimizing both R_{line} and C_{line}; circuit designers are then faced with creating the fastest possible switching network within the limits of the interconnect delay. This simple discussion gives us a first look at the signal transmission characteristics of interconnect lines. The problems associated with interconnect delays are critically important in high-density VLSI chip designs.

3.2 MOSFETs

Our discussion in the previous chapter emphasized the technique of designing logic gates from MOSFET switches. To build the circuit on silicon, we need to first understand what a MOSFET looks like at the physical level. We may then proceed to study how logic gates can be designed.

An integrated MOSFET is a small area set of two basic patterned layers that together act like a controlled switch. To determine what the layering scheme should look like, recall the circuit symbol for an nFET shown in Figure 3.10(a). This schematic symbol was designed to resemble the physical structure of the FET itself. Each terminal provides an electrical "entry point" to a patterned feature on one of the layers that makes up the transistor at the chip level. The terminals have been labeled as the gate, the source, and the drain, and each provides access to the device. From the analysis in the previous chapter, we know that the gate electrode acts as the control terminal in that the voltage applied to it determines whether the switch is open or closed. In electrical terms, the voltage applied to the gate determines the electrical current flow between the source and drain terminals.

Our task at this point is to use the concept of integrated circuit layers to create a silicon FET. In Figure 3.10(b) we have drawn a simple representation of the nFET using conducting layers. The vertical line represents the gate layer and divides another layer into source and drain regions that correspond to the schematic symbol. This simplified view is

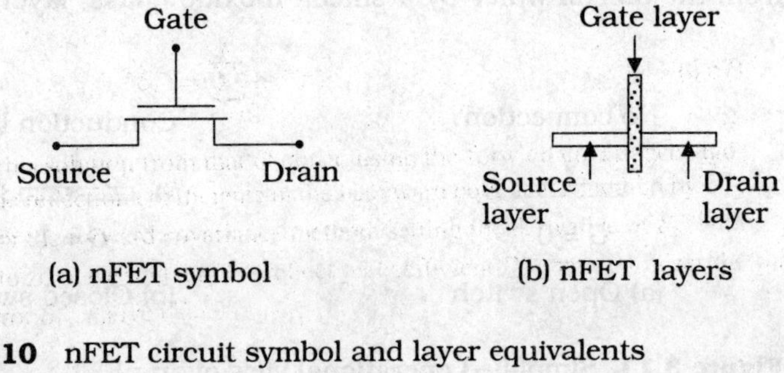

Figure 3.10 nFET circuit symbol and layer equivalents

sufficient to understand the physical structure and operation of an integrated transistor.

We may use this drawing to determine the operational characteristics needed in the physical structure. Let us assume that a signal G is applied to the gate and study the behavior of the nFET. If $G = 0$ the source and drain are not connected electrically. This is shown in Figure 3.11(a) where we have removed the gate layer to more clearly illustrate the behavior of the device. In this case an open circuit exists so that the two sides are electrically separate; this means that there is no relationship between A and B. If we instead apply a gate signal of $G = 1$, then the nFET acts as a closed switch and the source and drain sides are electrically connected. This is illustrated in Figure 3.11(b). A conduction layer that bridges the gap has been formed, yielding the logic expression

$$B = A \tag{3.15}$$

Assuming that the drain and source are formed on the same layer, then this behavior can be used to deduce that

- The gate signal G is responsible for the absence or presence of the conducting region between the drain and source regions

This is, in fact, how a MOSFET works. The voltage V_G applied to the gate is used to electrically create a conduction path that allows current to flow between the drain and source sections of the transistor.

Now that we have seen how IC layers can be used to create a MOSFET, let us examine the physical structure of the transistor in more detail. Figure 3.12 shows the layers involved in creating a generic FET. The drain and source regions are patterned into a silicon wafer; the wafer is equivalent to the substrate introduced previously in Figure 3.1. Although the drain and source regions are on the same layer, they are physically separated from one another by a distance L; L has units of centimeters [cm] and is called the **channel length** of the FET. The width W of the drain and source regions is called the **channel width** and also has units of centimeters. The **aspect ratio** of the FET is defined as (W/L) and is the most important parameter to the VLSI designer. The gate layer is separated from the silicon wafer by a silicon dioxide (glass) layer that acts as an

Figure 3.11 Simplified operational view of an nFET

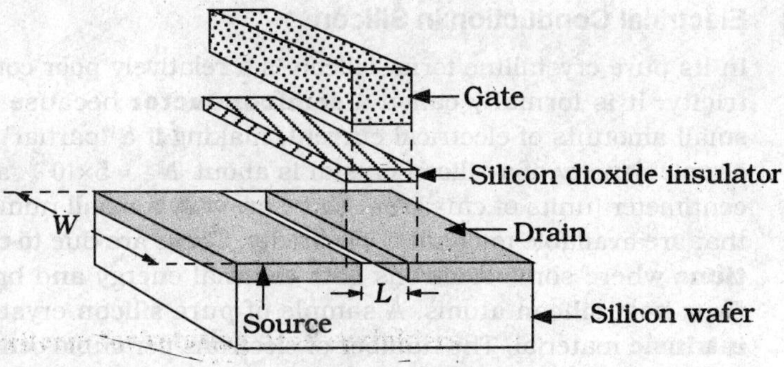

Figure 3.12 Layers used to create a MOSFET

insulator. The dashed vertical lines in the drawing show the alignment of the layers after the stacking process is completed.

Stacking the layers results in the 3-dimensional structure characterized by the drawing in Figure 3.13. The view of Figure 3.13(a) shows a cross-section of the layering scheme. The silicon dioxide layer has been renamed as the **gate oxide** as it resides directly underneath the gate region. The channel length L is shown explicitly in the drawing. The top view in Figure 3.13(b) is identical in form to the simple FET drawing we created in Figure 3.10(b). It shows the drain and source layer being separated by the gate pattern. The only major difference is that the simple drawing was concerned with layers and conduction paths and did not specify sizes.

The basic structure of nFETs and pFETs is the same as that portrayed in Figure 3.13. The difference between the two devices is in the nature of the layers used for the drain and source regions. Both use patterned layers in silicon, but the nFET layer is made to have an excess number of negatively charged electrons, while the pFET drain-source layer has an excess number of positive charges in it. Let us take a short excursion into the world of silicon physics to see how this is accomplished.

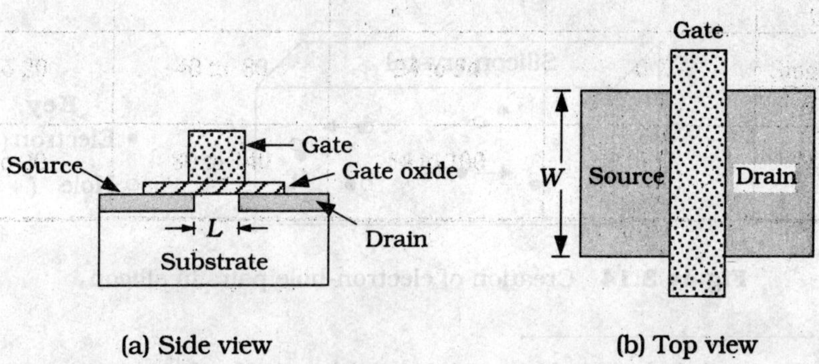

Figure 3.13 Views of a MOSFET

3.2.1 Electrical Conduction in Silicon

In its pure crystalline form, silicon is a relatively poor conductor of electricity. It is formally called a **semiconductor** because it can conduct small amounts of electrical current, making it a "partial" conductor. The atomic density of a silicon crystal is about $N_{Si} \approx 5\times10^{22}$ atoms per cubic centimeter (units of cm^{-3}), but there are only a small number of electrons that are available to conduct electricity. These are due to **thermal excitations** where some electrons gain thermal energy and break away from their host silicon atoms. A sample of pure silicon crystal is said to be **intrinsic** material. The number of electrons per cubic centimeter that are free to carry current is denoted by the symbol n_i and is called the **intrinsic carrier density**; the term "carrier" is short for "charge carrier," meaning that the particle has charge. The value of n_i is a function of the temperature T. At **room temperature** (T = 27° C = 300 K), the intrinsic density is given by

$$n_i \approx 1.45\times10^{10}\ \text{cm}^{-3} \tag{3.16}$$

so that only a small fraction of the electrons in crystal are available for conduction. The value of n_i increases with increasing temperature since more thermal energy is added to the structure. However, the number of free electrons remains small compared to that in a metal.

If we analyze the bonding structure of pure crystal silicon we find that most of the electrons are confined to orbits around the atomic nuclei of the atoms. When an electron gains sufficient thermal energy to break away from its host atom, it may move around in the crystal as a **free** (or mobile) electron. When an electron leaves its atomic site, it leaves behind an empty covalent bond that is called a **hole**; this is illustrated in Figure 3.14. The hole represents the absence of an electron, and may be treated as a "particle" with properties that are opposite to those of electrons. In particular, since the electron has a negative charge of $-q$ associated with it, the hole carries a positive charge of value $+q$ that allows it to participate in the current flow process.[1] Although the particles are independent of

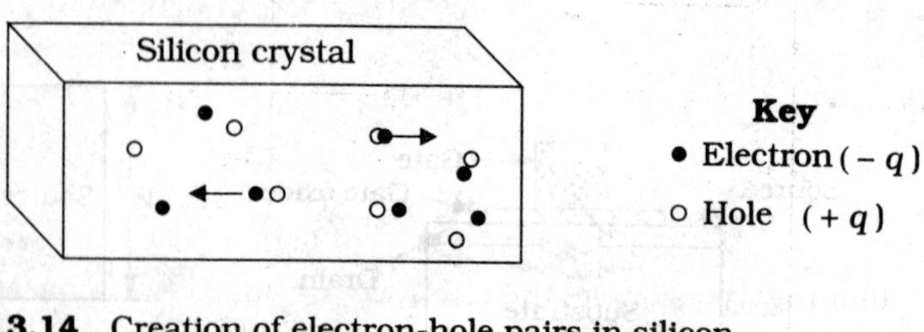

Figure 3.14 Creation of electron-hole pairs in silicon

[1] The numerical value of the fundamental charge unit is $q = 1.602 \times 10^{-19}$ coulombs.

one another, they constitute an **electron-hole pair** when they are created.

The ability of a material to conduct electrical current depends upon the number of freely moving charged particles that are available. Let us introduce two variables that provide this information. We define n to be the number of free electrons per cubic centimeter and p to be the number of free holes per cubic centimeters; both n and p have units of cm^{-3}. In a sample of pure silicon, the only way that a hole is created is by freeing an electron from its host atom. We thus see that

$$n = p = n_i \tag{3.17}$$

holds for our sample. The product of the two values gives

$$np = n_i^2 \tag{3.18}$$

which is a statement of the **mass-action law** that governs the relative numbers of electrons and holes if no currents are flowing. This is valid for any semiconductor in thermal equilibrium, which is equivalent to having zero current flow.

Pure silicon does not conduct current very well, but this may be changed by purposely adding small amounts of impurity atoms, called **dopants**, to create a **doped** sample. The idea is to enhance either the number of electrons, or the number of holes, to aid in the current flow process. The population of free electrons can be increased by adding arsenic (As) or phosphorus (P) atoms to the crystal. The resulting sample is called **n-type** material because it has an excess of negatively charged electrons. When used as dopants, both arsenic and phosphorus "donate" free electrons to the crystal, and are said to act as **donor atoms** or simply **donors**. The number of donors added to one cubic centimeter is given the symbol N_d, with a typical range of values for N_d of around 10^{16} to 10^{19} cm^{-3}. Each donor atoms adds a free electron to the crystal so that we may compute the electron density from

$$n_n \approx N_d \text{ cm}^{-3} \tag{3.19}$$

where the notation n_n means the electron density in an n-type sample. The number of holes in the n-type sample, which we will denote by p_n, is given by the mass-action law as

$$p_n \approx \frac{n_i^2}{N_d} \text{ cm}^{-3} \tag{3.20}$$

In an n-type sample, the electrons are called the **majority carriers** while the holes are called the **minority carriers** due to their relative numbers.

Example 3.1

Suppose that the donor doping density is $N_d = 2 \times 10^{17}$ cm^{-3}. The electron density is

$$n_n \approx N_d = 2\times10^{17} \text{ cm}^{-3} \tag{3.21}$$

while the hole concentration is

$$p_n \approx \frac{n_i^2}{N_d} = \frac{(1.45\times10^{10})^2}{2\times10^{17}} \tag{3.22}$$

which gives

$$p_n \approx 1\times10^3 \text{ cm}^{-3} \tag{3.23}$$

Obviously, $n_n \gg p_p$ holds for the sample.

The opposite-polarity material is called **p-type** and is created by adding boron (B) atoms to the crystal. A p-type material has more positively charged holes than negatively charged electrons. Boron is used because every impurity atom induces a free hole into the bonding scheme. Since a hole can "accept" an electron, boron is called an **acceptor** dopant, and the number of acceptors added per cubic centimeter is denoted by the symbol N_a. The acceptor density has approximately the same range as that state for donors (about 10^{14} to 10^{19} cm^{-3}), but the effect is exactly opposite: adding boron enhances the concentration of holes p_p in the p-type semiconductor. To calculate the carrier densities we use

$$p_p \approx N_a \qquad n_p \approx \frac{n_i^2}{N_a} \tag{3.24}$$

and refer to the holes as the majority carriers, while the electrons are the minority carriers since $p_p > n_p$. Both p_p and n_p have units of cm^{-3}.

The conductivity σ of a semiconductor region with carrier densities n and p is given by

$$\sigma = q(\mu_n n + \mu_p p) \tag{3.25}$$

where μ_n and μ_p are called the electron and hole **mobilities**, respectively, with units of cm^2/V-sec. Qualitatively, the mobilities are parameters that indicate "how mobile" a particle is. A small value of μ indicates that it is difficult for the particle to move, while a large value of μ implies relatively free motion. For intrinsic silicon, the room temperature mobilities are

$$\mu_n = 1360 \qquad \mu_p = 480 \tag{3.26}$$

which gives a conductivity of $\sigma \approx 4.27 \times 10^{-6}$ [Ω-cm]$^{-1}$ or $\rho \approx 2.34 \times 10^{5}$ [Ω-cm]. For comparison purposes, we note that quartz glass, which is an excellent insulator, has a resistivity ρ of about 10^{12} [Ω-cm].

If we specialize to an n-type sample where $n_n >> p_n$, then we may usually approximate the conductivity as

$$\sigma \approx q\mu_n n_n \tag{3.27}$$

Similarly, the conductivity of a p-type region is often estimated by

$$\sigma \approx q\mu_p p_p \tag{3.28}$$

For the present discussion, however, the most important point to remember is that an n-type region is dominated by negatively charged electrons, while a p-type region has mostly positively charged holes.

Example 3.2

Consider a sample of silicon that is doped p-type with boron added at a density of 10^{15} cm^{-3}. The majority charge carriers are holes with a density of

$$p_p = 10^{15} \text{ cm}^{-3} \tag{3.29}$$

while the minority carrier electron density is

$$n_p \approx \frac{(1.45\times10^{10})^2}{10^{15}} = 2.2\times10^{5} \text{ cm}^{-3} \tag{3.30}$$

For this sample, the mobilities are given by $\mu_n \approx 1350$ cm^2/V-sec and $\mu_p \approx 450$ cm^2/V-sec. The conductivity is

$$\begin{aligned}\sigma &= (1.6\times10^{-19})[(1350)(2.2\times10^{5}) + (450)(10^{15})] \\ &= 0.072 \quad [\Omega\text{-cm}]^{-1}\end{aligned} \tag{3.31}$$

which is equivalent to a resistivity of

$$\rho = \frac{1}{0.08} = 13.9[\Omega\text{-cm}] \tag{3.32}$$

A quick check on the values shows that $\mu_p p_p >> \mu_n n_p$ for this example. In general, the resistivity of silicon samples is on the order of 1 to 10 Ω-cm.

This example shows that the doping level is the most important factor in determining the conductivity in n-type or p-type silicon. Increasing the doping density creates more charged particles that aid in the conduction process. However, a large number of impurity atoms creates more barriers that the particles must pass, making them less mobile. This is called **impurity scattering**, and is described by writing the mobility μ as a function of the total doping density N. In general, $\mu(N)$ decreases with increasing N. An empirical equation for this effect is (see Reference [3])

$$\mu = \mu_1 + \frac{\mu_2 - \mu_1}{1 + \left(\frac{N}{N_{ref}}\right)^{\alpha}} \tag{3.33}$$

where μ_1, μ_2, N_{ref}, and α are constants. For electrons, the room temperature silicon values are approximately $\mu_1 = 92$ cm^2/V-sec, $\mu_2 = 1380$ cm^2/V-sec, $N_{ref} = 1.3 \times 10^{17}$ cm^{-3}, and $\alpha = 0.91$. The corresponding hole values are $\mu_1 = 47.7$ cm^2/V-sec, $\mu_2 = 495$ cm^2/V-sec, $N_{ref} = 6.3 \times 10^{16}$ cm^{-3}, and $\alpha = 0.76$. The decrease in mobility with increasing doping is called a **second-order effect** in device physics. Although it is tempting to ignore impurity scattering in simple calculations, doing so can introduce significant errors. A final comment is that for a given doping level N,

$$\mu_n > \mu_p \tag{3.34}$$

This means the electrons can move more easily than holes. Physically, this can be visualized by assuming that electrons are true particles in the classical sense, while holes are seen as the "absence of particles."

The above analysis assumes that only donors N_d *or* acceptors N_a will be present in the sample. In CMOS processing, however, most doped regions have both donors *and* acceptors. The polarity is established by the dominant species. To create an n-type region, we need $N_d > N_a$ so that the donors outnumber the acceptors. The carriers are computed by

$$n_n \approx N_d - N_a \qquad p_n \approx \frac{n_i^2}{(N_d - N_a)} \tag{3.35}$$

with the electrons in the majority. For a p-type region, we need $N_a > N_d$ such that the carrier densities are given by

$$p_p \approx N_a - N_d \qquad n_p \approx \frac{n_i^2}{(N_a - N_d)} \tag{3.36}$$

for the majority carrier holes p_p and minority carrier electrons n_p. To calculate the mobility, we use the total doping density $N = N_a + N_d$ in equation (3.33). The conductivity is still calculated from

$$\sigma = q(\mu_n n + \mu_p p) \tag{3.37}$$

since only the values are altered. One special case is where $N_d = N_a$. Since every electron released by a donor is matched by a hole in an acceptor, the material looks like an intrinsic same with $n = p = n_i$. This is called

total compensation. Note that the mobility will be smaller than the intrinsic value since the number of dopants is not zero.

When an n-type region touches a p-type region, a very special interface is formed. This **pn junction** allows electrical conduction in only one direction, from the p-side to the n-side. If we attempt to force current from the n-side to the p-side, the junction blocks it and acts like an open switch. The properties of a pn junction are summarized in Figure 3.15. In electronics, this feature is used to make a device called a **diode**. The characteristic of allowing current to flow in only one direction is called **rectification**.

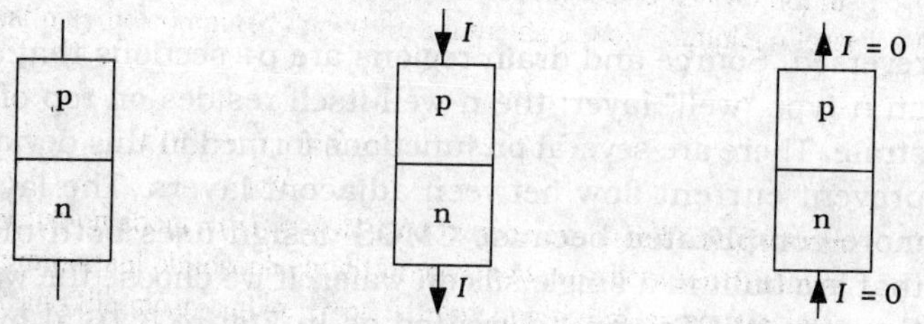

Figure 3.15 Formation and characteristics of a pn junction

3.2.2 nFETs and pFETs

With the distinction between n-type and p-type regions established, we can now define the structures of nFET and pFETs. This is a very simple task: the polarity of a FET (n or p) is determined by the polarity of the drain and source regions. The device is designed so that the conducting layer shown in Figure 3.11(b) has the same polarity as the drain and source regions when the device is conducting. An nFET uses n-type drain and source regions, while a pFET has p-type drain and source regions. These are shown in Figure 3.16(a) and (b), respectively. Metal contacts have been added to illustrate how we can connect the drain and source regions to other parts of the circuit.

Let us examine the nFET first. The drain and source regions are labeled as "n+" to indicate that they are **heavily doped**. This means that the donor doping density N_d is relatively large, with a typical value around $N_d = 10^{19}$ cm^{-3}. The substrate layer (at the bottom) is now specified to be p-type with an implied boron doping density of N_a; a reasonable value for the acceptor doping would be $N_a = 10^{15}$ cm^{-3}. Note that pn junctions are formed between the n+ regions and the p-type substrate. These are used to block the current flow between the substrate and the top n+ layers of the device as discussed in the context of Figure 3.15.

The pFET has the same structure as the nFET but the polarities are

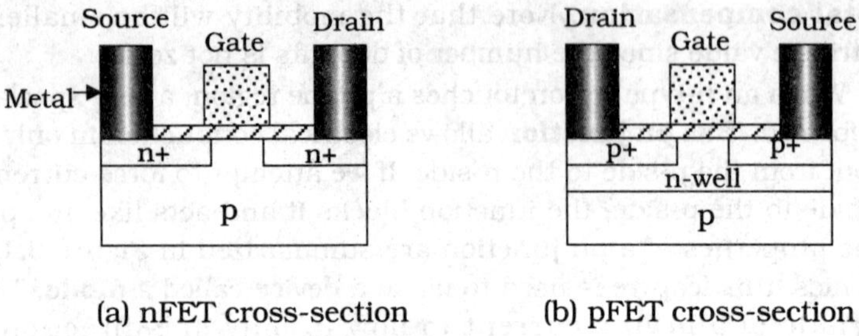

Figure 3.16 nFET and pFET layers

reversed. Source and drain regions are p+ sections that are embedded in an n-type "well" layer; the n-well itself resides on top of the p-type substrate. There are several pn junctions formed in this device; all are used to prevent current flow between adjacent layers. The layering scheme is more complicated because CMOS design uses both nFETs and pFETs that are built in a single silicon wafer. If we choose the wafer to be p-type, then the nFETs can be created as in Figure 3.16(a) by just adding n+ regions. However, if we add pFET p+ regions directly to the p-substrate, then we lose the needed structure of the layering: p+ into an n-region. Since no pn junction is formed, we cannot control the current flow. To correct for this problem, the n-well layer is used to build the pFET as shown. This assures us that the transistors have opposite electrical characteristics.

3.2.3 Current Flow in a FET

MOSFETs are used as voltage controlled switches in CMOS logic circuits. Applying a signal to the gate electrode results in either an open or closed switch as we saw previously in Figure 3.11 for an nFET. The creation of the conducting layer underneath the gate is due to the property of the capacitance that is built into the gate region of the MOSFET itself. A simple parallel-plate capacitor from basic physics is shown in Figure 3.17.

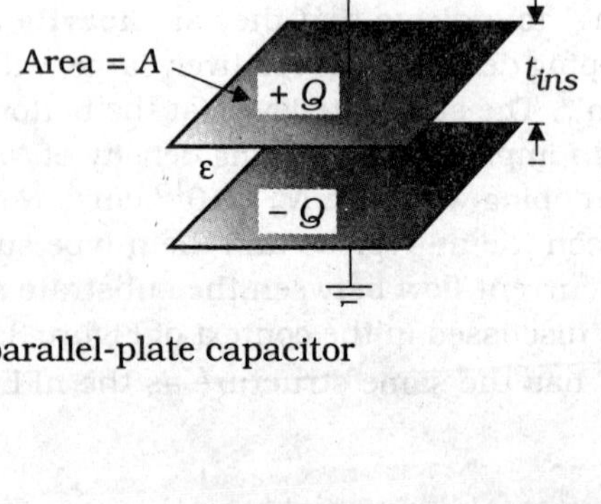

Figure 3.17 A parallel-plate capacitor

This device consists of two identical metal plates that are separated by an insulator with a thickness t_{ins} cm. The plates have an area A in units of cm^2. A capacitor stores electric charge Q on the plates as indicated in the drawing. With a voltage difference of V applied across the plates, the charge is given by

$$Q = CV \tag{3.38}$$

where C is the capacitance. For a parallel-plate structure, the basic formula for the capacitance is well known as

$$C = \frac{\varepsilon A}{t_{ins}} \tag{3.39}$$

where ε is the permittivity of the insulator in units of F/cm. The value of ε depends upon the material used to separate the plate.[2] The most important observation to be made is that applying a positive voltage V to the upper plate **induces** a negative charge $-Q$ on the lower plate.

Let us examine an nFET in more detail by referring to the drawing in Figure 3.18. The central region of the device is designed to be a capacitor. The gate oxide layer is the insulating glass between the gate (which acts as the upper plate) and the p-type substrate (which acts as the lower plate). In the early days of MOS, the gate was made out of aluminum (Al), which is a metal. The layering thus gave rise to the acronym "MOS" for metal-oxide-semiconductor. In modern processing, the gate material is **polycrystal silicon**, which is usually called **polysilicon** or just **poly**.[3] Although the gate material is no longer a metal, the acronym has never been changed with any success and still remains in use today.[4]

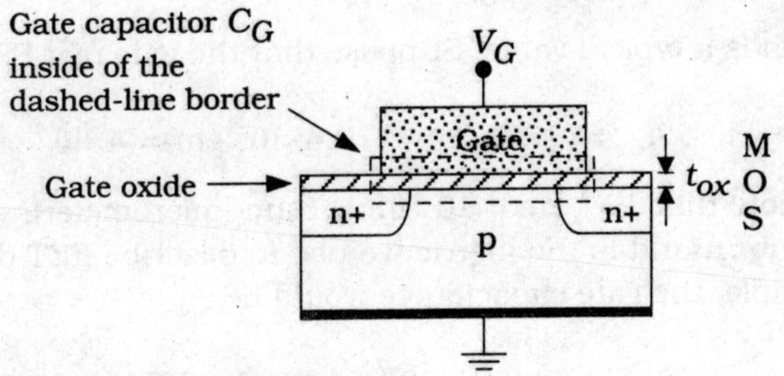

Figure 3.18 The gate capacitance in an n-channel MOSFET

2 Physically, the permittivity is a measure of the electric energy storage capacity of the material.

3 Polycrystal consists of small regions of silicon crystals, called crystallites; the material is discussed in more detail in Chapter 4.

4 The most common substitute is the IGFET, which stands for insulated-gate FET.

To describe the MOS structure we introduce the **oxide capacitance**

$$C_{ox} = \frac{\varepsilon_{ox}}{t_{ox}} \tag{3.40}$$

which has units of F/cm^2. Comparing this to the parallel-plate formula in equation (3.39) shows that the area has been left out. This is done on purpose so that C_{ox} can be applied to any device in the circuit. If the gate has an area A_G cm^2, then the total **gate capacitance** of the FET is

$$C_G = C_{ox}A_G \tag{3.41}$$

which has units of farads. In this formula, ε_{ox} is the permittivity of the glass insulating layer. In modern technology, silicon dioxide (SiO_2) is used for almost all silicon MOSFETs and the permittivity is given by

$$\varepsilon_{ox} = 3.9\varepsilon_0 \tag{3.42}$$

where $\varepsilon_o \approx 8.854 \times 10^{-14}$ F/cm is the permittivity of free space. The oxide thickness t_{ox} is a critical parameter in CMOS. For reasons that will be seen later, a thin oxide (small t_{ox}) is desirable. Modern processing lines have $t_{ox} \leq 10$ nm = 100 Å with advanced fabrication facilities providing the capabilities to create an oxide with less than half this thickness.[5]

Example 3.3

Consider a gate oxide that has a thickness of t_{ox} = 50 Å = 50 × 10^{-8} cm. The oxide capacitance per unit area is

$$C_{ox} = \frac{(3.9)(8.854\times10^{-14})}{50\times10^{-8}} = 6.91\times10^{-7} \quad \text{F/cm}^2 \tag{3.43}$$

which is a typical value. Suppose that the gate of a FET has an area

$$A_G = (1\times10^{-4}\text{cm})\times(0.4\times10^{-4}\text{cm}) = 4\times10^{-9}\text{cm}^2 \tag{3.44}$$

We note that 10^{-4} cm = 10^{-6} m = 1 μm (micrometer), which is often called 1 **micron** and is the metric we use to describe FET dimensions. For this example, the gate capacitance would be

$$C_G = (6.91\times10^{-7})(4\times10^{-9}) = 2.76\times10^{-15}\ \text{F} \tag{3.45}$$

Defining 1 **femtofarad** (fF) as 1 fF = 10^{-15} F, the gate capacitance is

$$C_G = 2.76\ \text{fF} \tag{3.46}$$

---

5 One Angstrom (Å) is 10^{-10} cm = 10^{-8} m.

which is typical for a modern device. The electronics expert will notice that this is much smaller than the typical capacitance values encountered in the everyday world.

The above discussion illustrates that the gate of a MOSFET is really one side of an MOS capacitor. Applying a voltage on the gate causes a charge layer of the opposite polarity to form on the opposite side of the capacitor, i.e., in the silicon region directly beneath the gate oxide. If we apply a positive voltage to the gate, then a negative electron layer is created in the silicon. Conversely, using a voltage that is negative with respect to the rest of the device creates a positively charged layer of holes in the silicon. The formation of thin layers of charge in the silicon is possible because it is a semiconductor material where the number of charge carriers depends upon the local electrical conditions. Armed with this observation, the mechanism of current flow becomes easy to visualize.

Consider first an nFET as shown in Figure 3.19. The drain and source regions are n-type, but they are physically separated by a section of the p-type substrate. In Figure 3.19(a) the gate voltage has the value of 0 V so that no charge is induced underneath the gate oxide. The top view of the silicon provided on the right side shows that the drain and source are separated, so that no current can flow between them. This is due to the current-blocking properties of the pn junctions, and is analogous to having an open switch. Applying a positive voltage to the gate as in Figure 3.19(b), the capacitive MOS structure induces a layer of negatively

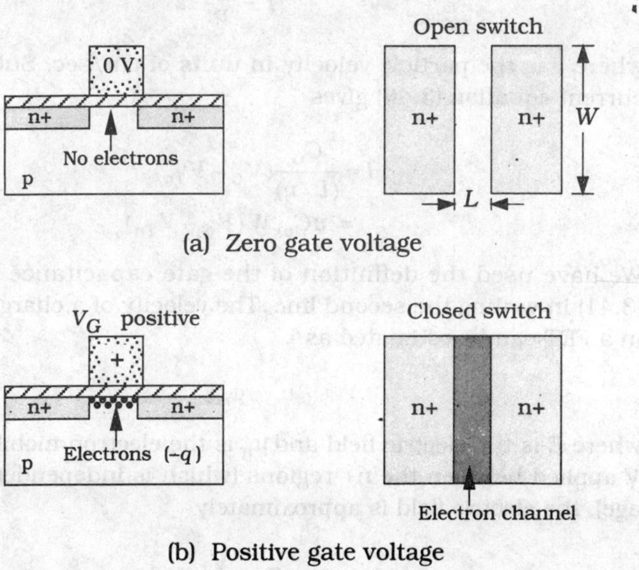

Figure 3.19 Controlling current flow in an nFET

charged electrons underneath the gate oxide. The electron layer establishes an electrical connection between the drain and source region as shown. The electrons form a "channel" for current to flow through the nFET, and the device acts like a closed switch. The formation of the layer requires that the gate voltage be larger than the **threshold voltage** V_{Tn} that was introduced in Chapter 2. A typical value for the nFET threshold voltage is $V_{Tn} = 0.70$ V. This parameter is established by the fabrication sequence, and is always assumed to be a known value at the VLSI design level.

The channel charge in units of coulombs is given by

$$Q_c = -C_G(V_G - V_{Tn}) \tag{3.47}$$

where V_G is the gate voltage and C_G is the gate capacitance. The voltage difference $(V_G - V_{Tn})$ is used because no charge forms until V_G reaches V_{Tn}. The negative sign in the equation indicates that the channel consists of negatively charged electrons. The current I flowing through the channel can be written as

$$I = \frac{|Q_c|}{\tau_t} \text{ C/sec} \tag{3.48}$$

where we have introduced τ_t as the channel **transit time** in units of seconds. Physically, τ_t is the average time needed for an electron to move from one n+ region to the other, and can be calculated from

$$\tau_t = \frac{L}{v} \tag{3.49}$$

where v is the particle velocity in units of cm/sec. Substituting into the current equation (3.48) gives

$$\begin{aligned} I &\approx \frac{C_G}{(L/v)}(V_G - V_{Tn}) \\ &= vC_{ox}W(V_G - V_{Tn}) \end{aligned} \tag{3.50}$$

We have used the definition of the gate capacitance C_G from equation (3.41) in writing the second line. The velocity of a charged particle moving in a FET can be estimated as

$$v \approx \mu_n E \tag{3.51}$$

where E is the electric field and μ_n is the electron mobility. With a voltage V applied between the n+ regions (which is independent of the gate voltage), the electric field is approximately

$$E = \frac{V}{L} \tag{3.52}$$

with units of volts/cm. Substituting these relations into equation (3.50) gives

$$I \approx \mu_n C_{ox}\left(\frac{W}{L}\right)(V_G - V_{Tn})V \tag{3.53}$$

as our first approximation for the current. The linear resistance R_n of the device can be calculated by taking the ratio

$$R_n = \frac{V}{I} = \frac{1}{\beta_n(V_G - V_{Tn})} \tag{3.54}$$

where we have defined the parameter

$$\beta_n = \mu_n C_{ox}\left(\frac{W}{L}\right) \tag{3.55}$$

called the **device transconductance**[6] that has units of A/V^2. With this model we can view the nFET as a device that acts either as an open switch with no channel and $R \to \infty$, or a closed switch with a resistance of R_n between the drain and source sides.

One fine point that needs to be mentioned is that the mobility μ_n used in the MOSFET analysis is the value at the "surface" of the silicon, and is therefore called the **surface mobility**. This is different from that calculated using equation (3.33) which is valid for the **bulk mobilities**, i.e., the value inside the material. A simple estimate is that the surface mobility is about (1/2) the bulk mobility value. In practice, measured values from the laboratory are used to design circuits.

A deeper analysis will show that MOSFETs are intrinsically **non-linear devices** in that the current I through a FET is a non-linear function of the voltage V across it. This relationship will be examined in more detail in Chapter 6. For simple modeling, however, we often treat the transistor as a linear resistor with a value

$$R_n = R_{c,n}\left(\frac{L}{W}\right) \tag{3.56}$$

where

$$R_{c,n} = \frac{1}{\mu_n C_{ox}(V_G - V_{Tn})} \tag{3.57}$$

is the equivalent sheet resistance for the electron current flow channel.

A pFET behaves in a similar manner, except that all of the polarities

6 In general, a transconductance parameter has units of amps over volt with squares, cubes, etc. A transresistance has basic units of volts/amps.

are reversed. The operation is summarized in Figure 3.20. If the gate voltage is made positive [see Figure 3.20(a)] then only negative charge from the n-type layer exists underneath the gate oxide. Since the drain and source regions are both p-type, they are electrically separated from each other by the n-type region. The transistor thus acts like an open switch. If, on the other hand, we apply a negative voltage on the gate, then a layer of positively charged holes can form underneath the gate oxide as illustrated in Figure 3.20(b). To create the hole layer, the difference in voltage between the highest voltage p+ region and the gate must be greater than the magnitude of the pFET voltage $|V_{Tp}|$. This gives an electrical conduction channel between the p-type source and drain regions so current flows through the transistor. The pFET is then similar to a closed switch. Like the nFET, the pFET also exhibits a resistance that is estimated by

$$R_p = \frac{1}{\beta_p(V_G - |V_{Tp}|)} \tag{3.58}$$

By convention, V_{Tp} is a negative number, so we use $|V_{Tp}|$ to make the formula have the same form as the nFET expression. In this equation,

$$\beta_p = \mu_p C_{ox}\left(\frac{W}{L}\right) \tag{3.59}$$

is the device transconductance for the pFET, with μ_p the hole mobility.

Although our first look at conduction through nFETs and pFETs has

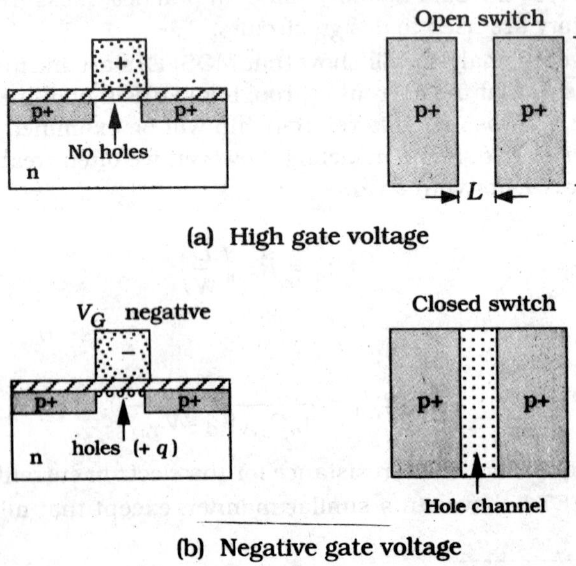

Figure 3.20 Switching behavior of a pFET

been highly simplified, it does relay on a useful visualization for the VLSI designer. Oftentimes a simple model is more useful than a complex one.

3.2.4 Driving the Gate Capacitance

Let us dig a little deeper into the behavior of the MOS capacitor system. It is fundamental to the operation of the FET, but the presence of any capacitance in a CMOS integrated circuit causes signal delays. Figure 3.21 shows the circuit symbol for a capacitor with a capacitance C. The drawing defines a positive current i as flowing into the side of the capacitor that has a positive voltage on it. Note that positive charge $+Q$ is stored on the top plate, while the bottom plate holds a negative charge with a value of $-Q$. The current i flowing into the capacitor as a function of time t is the time rate-of-change of the charge

$$i(t) = \frac{dQ}{dt} \tag{3.60}$$

Since $Q = CV$, we may substitute for the charge and obtain the I - V relation for a capacitor

$$i = C\frac{dV}{dt} \tag{3.61}$$

This tells us several things about how voltage signals behave in a CMOS circuit. First, it is not possible to change the voltage $V(t)$ across a capacitor in an instantaneous manner, i.e., with $dt \rightarrow 0$; this is because (dV/dt) would have to be infinite in value, but it is physically impossible to have an infinite current i. If we apply this conclusion to the gate capacitance C_G of a FET as shown in Figure 3.22, we conclude that the gate voltage V_G cannot be changed without experiencing a delay. This corresponds to the time required to transfer the charge

$$Q = C_G V_G \tag{3.62}$$

on to, or off of, the gate electrode. When combined with the mechanism of switching a FET to open and closed switching states, this implies that the

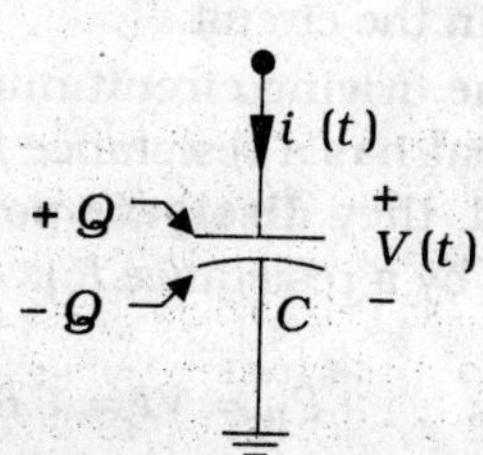

Figure 3.21 Voltage and current in a capacitor

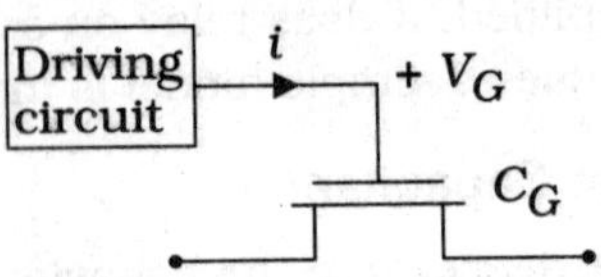

Figure 3.22 Driving the gate of a FET

transistor itself introduces signal delays. The value of the capacitance C_G, determines the amount of charge necessary to change the voltage, so a large capacitance implies a long delay.

The second important observation is that capacitors store electric energy. Charging and discharging C_G corresponds to changing the energy stored in the device, so that switching a transistor on or off requires that we transfer energy from one point to another in a circuit. The power P in units of **watts** [W] is related to the energy E by

$$P = \frac{dE}{dt} \tag{3.63}$$

where E has units of **joules** [J]. By definition, 1 watt of power means that 1 joule of energy has been transferred in 1 second. For an electrical device with voltage V that has a current i flowing into it, the power is given by the product $P = Vi$. Using equation (3.61) for a capacitor gives

$$P = V\left(C\frac{dV}{dt}\right) = \frac{d}{dt}\left(C\frac{V^2}{2}\right) \tag{3.64}$$

so that the electric energy E_e stored in a capacitor with a voltage V is

$$E_e = \frac{1}{2}CV^2 \tag{3.65}$$

When applied to a CMOS switching network, this means that changing the gate voltage of a FET from 0 V to V_{DD} requires energy of

$$E_e = \frac{1}{2}C_G V_{DD}^2 \tag{3.66}$$

for *every* transistor in the circuit.

Now note that the driving circuit must transmit current through an interconnect wire that has a resistance R_{line}. Resistors do not store electric energy. Instead, they **dissipate** power by changing it to heat. The power P_R dissipated by a resistance R is calculated from

$$P_R = Vi = i^2R \tag{3.67}$$

where we have used Ohm's law. This illustrates that flowing currents induce localized heating effects. This applies to every electrical device in the circuit, not just interconnect lines.

These simple observations bring out some critical aspects of VLSI circuit design that will be examined throughout the book. Two immediate considerations that arise are

- Switching delays are due to the physical characteristics of the devices and interconnects.
- Every switching event requires energy transfer in the circuit. This implies that power dissipation will occur within the circuit.

The first consideration implies that the designer must understand the nature of switching delays in order to design a fast digital network. The characteristics of both the FETs and the interconnects influence the overall system speed, so VLSI design deals with the network as a whole. The second statement is more practical. Excessive localized heating may be so severe that it melts the silicon crystal and destroys the chip. This, of course, must be avoided by proper design and the use of heat-removal techniques. If the chip is for a portable unit that uses batteries for a power supply, then the design must reduce the power requirements to extend battery life.

3.3 CMOS Layers

Now that we have seen how patterned layers of materials are used to create nFETs and pFETs, let us move to a higher level and examine the entire structure of a CMOS integrated circuit.

CMOS provides the economic basis for a huge segment of the world computing industry. Many companies compete in the marketplace, with each attempting to provide a more advanced technological base than the other. Because of the rapid evolution of high-density circuit manufacturing techniques in the first years of the 21st century, countless variations in CMOS have been introduced. We will choose a rather simple process to study here and purposely avoid advanced (and, hence, complicated) techniques. In particular, we will concentrate on an **n-well process** as being typical.

First of all, let us define what a "CMOS fabrication process" is. In simplest terms, this refers to the sequence of steps that we use to take a bare "wafer" of silicon to the finished form of an electronic integrated circuit. The details of the fabrication process will be discussed in Chapter 5. For the moment, we will only be concerned with the final structure.

The n-well process starts with a p-type substrate (wafer) that is used as a base layer for building all transistors. nFETs can be fabricated directly in the p-type substrate, while n-well regions are added to accommodate pFETs. The cross-sectional view in Figure 3.23 illustrates the nFET and pFET structures after they have been fabricated on the sub-

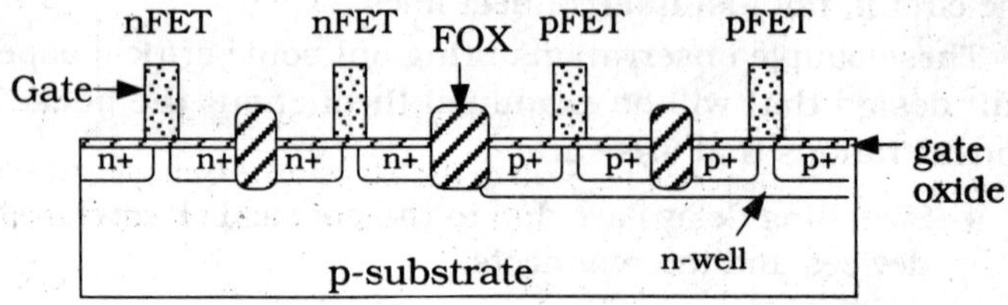

Figure 3.23 MOSFET layers in an n-well process

strate. From this drawing, we can identify the following layer types:

- p-substrate
- n-well
- n+ (nFET drain/source)
- p+ (pFET drain/source)
- gate oxide
- gate (polysilicon)

Note that the term "layer" implies a region with distinct electrical characteristics, even though it may be physically at the same geometrical level as another layer (such as the n+ and p+ layers). The drawing also shows regions labeled as "FOX" which defines **field oxide** sections. **Field regions** are simply recessed insulating glass (silicon dioxide) sections that are inserted in between adjacent FETs to provide **electrical isolation**. The glass acts to insure that there is no current flowing between the two transistors, keeping them electrically separate. Another point that is worth mentioning again is that junctions between n-type and p-type regions have the ability to block current flow. It is therefore assumed that n-regions and p-regions are electrically isolated.[7]

The top view patterning for this example is shown in Figure 3.24. In this drawing, the only layers that are shown explicitly are n-well, n+ (nFET drain/source), p+ (pFET drain/source), and gate (polysilicon). The p-substrate is implied, as are the oxide layers. Note that FOX surrounds every transistor so there is an implied field region everywhere except at transistor sites. FOX regions are rarely shown explicitly, so it is important to remember that every device in the wafer is automatically isolated from every other device.

Once the base transistor layers have been defined, we add conductive metal layers separated by glass insulators to allow for wiring. Modern processes tend to allow for five or more metal interconnect layers to ease the problem of massive wiring in complex circuits. The example in Figure

7 The ability to block current flow requires that the voltage on the n-side be higher than the voltage on the p-side.

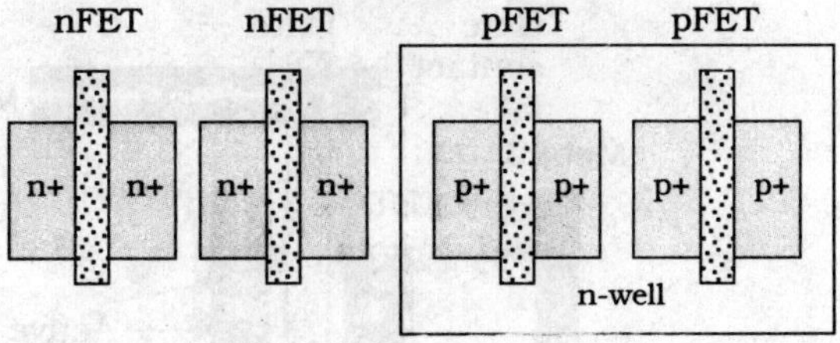

Figure 3.24 Top view FET patterning

3.25 shows two metal layers to illustrate the main points. After the FETs are formed, an oxide layer (Ox1) is deposited over the wafer surface and planarized. A "hole" (called a contact cut) is then etched in the oxide to allow electrical access to drain/source regions. This is shown as the Active contact in the drawing; it is filled with a conducting metal such as tungsten. Metal1 is then deposited on top, followed by another insulating oxide layer (Ox2). We note that Metal1 can also be connected to the gate layer by using an etch hole in Ox1. The second metal layer (Metal2) is then deposited on top of Ox2. Electrical contact between Metal1 and Metal2 is accomplished using a Via, which is a hole etched in Ox2 and filled with a conducting metallic "plug" as shown.

Now that we have seen how metal interconnect layers can be added to the CMOS process, it is important to make the following observations:

- Metal layers are electrically isolated from each other and the transistors by glass
- Electrical contact between adjacent conducting layers requires that we create **contact cuts** and **vias** in the oxide between them

These imply that we can "cross" conducting layers without creating an electrical path between them. Examples of these rules are provided by the layout (top view) drawing in Figure 3.26. Metal2 lines can cross over every other layer; a via is needed to contact Metal1. Metal1 can be connected to

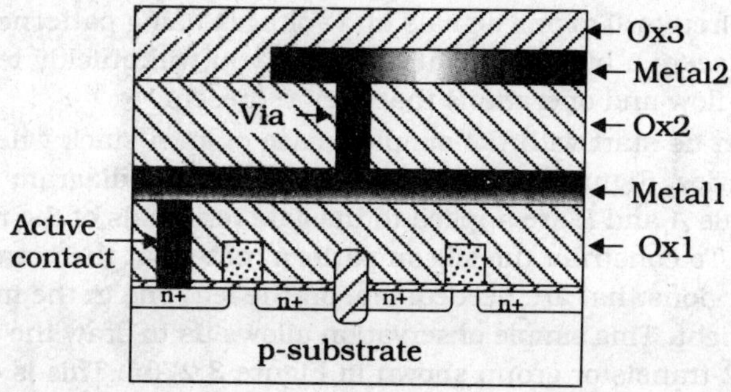

Figure 3.25 Metal interconnect layers

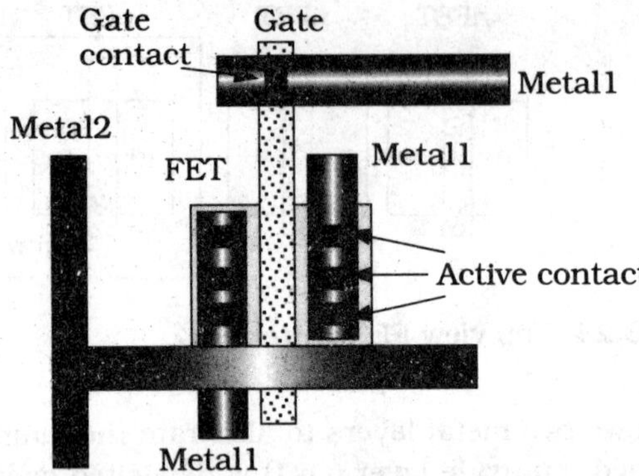

Figure 3.26 Interconnect layout example

the gate using the Gate contact, while an Active contact is used to provide an electrical connection between Metal1 and the drain/source regions of the FET.

CMOS circuits are designed by creating nFETs and pFETs in silicon, and then wiring them together using interconnect lines formed on the conducting layers. One interesting point that may be obvious is that digital CMOS logic circuits consist only of transistors and wires. No other devices are needed, regardless of the complexity of the system. Once we learn how to design basic FETs and add the interconnect wiring, then we will understand the basics of CMOS VLSI!

3.4 Designing FET Arrays

CMOS logic gates are switching networks that are controlled by the input variables. These switching arrays use FETs that are wired together in series and parallel groups in a manner that allows us to create the desired functions. In Chapter 2 we learned how to build logic gates using FETs at the schematic level. These must eventually be translated to silicon patterns for the final design. In VLSI, the patterns themselves become the circuits. Tracing signals and voltages using patterned polygon shapes may seem a bit strange at first, but you will quickly learn to "read" the logic flow and operations that they represent.

Let us start with the simplest case of an n-stack where two nFETs are in series. Figure 3.27(a) shows the schematic diagram for this case. The signals *A* and *B* are applied to the gate terminals of the respective transistors. To construct the silicon pattern, note that there are really only three n+ regions that are needed: one on the left, one in the middle, and one on the right. This simple observation allows us to draw the silicon pattern for the 2-transistor group shown in Figure 3.27(b). This is our first lesson in high-density integration:

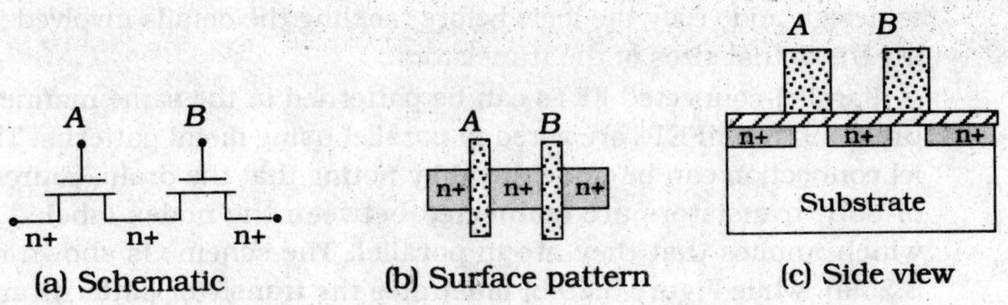

Figure 3.27 Silicon patterning for two series-connected nFETs

- Devices can share patterned regions, which may reduce the layout area or complexity.

In the present case, this says that it is not necessary to first build individual devices and then wire them together. A more efficient design results if we combine n+ regions. The side view shown in Figure 3.27(c) shows that this technique does indeed create a signal path that consists of two transistors. Conduction from left to right occurs only if both transistors are conducting.

This technique can be applied to any group of series-connected FETs. A 3-FET chain is shown in Figure 3.28. Instead of labeling every region in the surface layout drawing, it is usually more convenient to provide a key that associates each fill pattern with a specific material. Color coding the layers is even easier, and is the preferred technique used in computer-based design aids. Metal lines have been added on the left and right sides, along with active contacts (that connect the metal to the n+ regions). These define electrical connections to the nodes x and y shown in the drawing and are required to connect the transistor group to other parts of the circuit. This pattern also shows the channel width W for the three transistors and thus provides more information than the simpler surface patterning scheme used in Figure 3.27. In the initial design stages, the width is not always shown explicitly. At that point in the design, we are usually more interested in the signal flow path and the circuit topology than the details of the transistors. In other words, we want

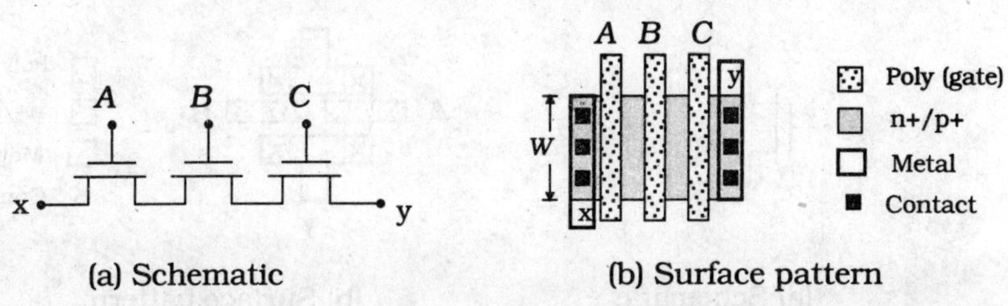

Figure 3.28 Three series-connected nFETs

to design and verify the logic before tackling the details involved in choosing the actual sizes of the transistors.

Parallel-connected FETs can be patterned in the same manner. In Figure 3.29, two nFETs are wired in parallel using metal patterns. The parallel connection can be understood by noting that the drain/source regions of both transistors are connected between the nodes labeled x and y, which implies that they are in parallel. The scheme is shown in Figure 3.29(a), while Figure 3.29(b) illustrates the transistor patterns and wiring scheme. This approach to surface patterning maintains the orientation of the transistor patterns that were used for series-connections groups. This may be desirable in that a uniform layout philosophy may lead to a higher packing density on the silicon surface.

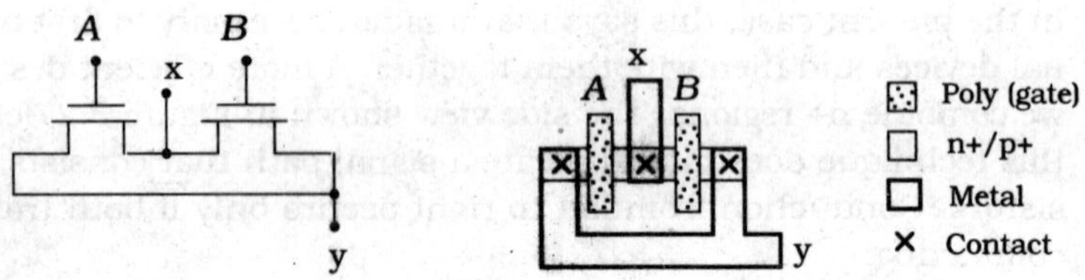

(a) Schematic (b) Surface pattern

Figure 3.29 Parallel-connected FET patterning

An alternate layout strategy for parallel FETs is shown in Figure 3.30. This uses vertical drain-source orientations for the transistors. In this approach, two FETs are created with separate n+ regions. The parallel connection is accomplished using metal interconnects to give the nodes x and y shown in the drawing. While these two techniques maintain the same orientation for both FETs (horizontal or vertical), this is not mandatory. Only the wiring and the resulting electrical connections are important. Separated transistors usually require more area than those that share drain/source regions, so this type of scheme is restricted to special situations.

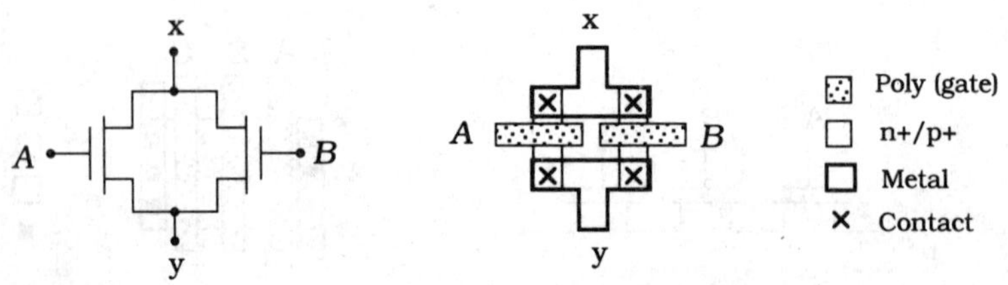

(a) Schematic (b) Surface pattern

Figure 3.30 Alternate layout strategy for parallel FETs

3.4.1 Basic Gate Designs

Now that we have seen the basic ideas involved in CMOS layout, let us examine the surface patterns used for CMOS logic gates in silicon. In the simplified view used in this section, patterned lines on conducting layers are viewed as paths that "steer" electrical current and establish voltages. The widths of the lines are not important at this level; only the topology of the network is needed to trace the logic. This approach is very useful in the initial stages of creating a CMOS layout design as it allows one to play with the location and orientation of the devices to see how well they pack together.

Consider first a NOT gate. Figure 3.31(a) shows how the circuit is wired using transistors Mn and Mp as a complementary pair. The silicon implementation is shown in Figure 3.31(b). The layout has been structured so that there is a visual one-to-one correspondence with the circuit. Some of the important aspects are that

- Both the power supply (VDD) and ground (Gnd) are routed using the Metal layer
- n+ and p+ regions are denoted using the same fill pattern. The difference is that pFETs are embedded within an n-well boundary
- Contacts are needed from Metal to n+ or p+ since they are at different levels in the structure

The ability to trace the logic operation on the layout is a useful skill to develop. In this case, the input x controls the poly gate. When $x = 0$, Mp acts like a closed switch while Mn is open, giving an output of VDD, i.e., $\overline{x} = 1$. Conversely, an input of $x = 1$ forces Mn into conduction while Mp is open. This connects Gnd to the output, and is equivalent to $\overline{x} = 0$.

An alternate layout is shown in Figure 3.32. In this case, the NOT gate has been drawn like a 2:1 multiplexor. While the operation is entirely equivalent, a one-to-one translation results in FETs that are at right

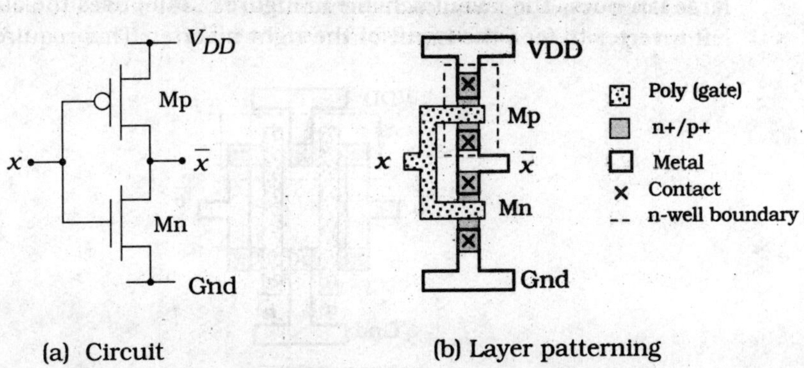

Figure 3.31 Translating a NOT gate circuit to silicon

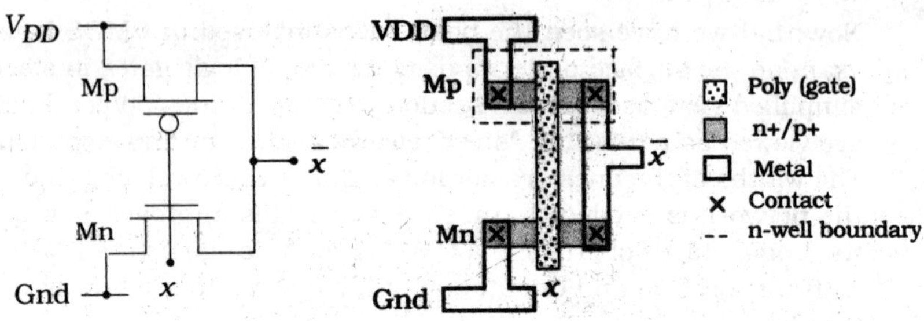

(a) Circuit (b) Layer patterning

Figure 3.32 Alternate layout for a NOT gate

angles to those in Figure 3.31. This illustrates the fact that different geometrical layouts can be used to implement CMOS circuits. Variations in the layout strategy are not important until the actual sizes of the patterns are taken into account. This aspect of physical design is discussed later in the book.

One goal of physical design is to minimize the area of the overall chip. This can be accomplished at many levels using various techniques. One example is shown in Figure 3.33 where two NOT circuits share the VDD and Gnd connections. The left inverter has an input of a and produces $\bar{a}$, while the right circuit inverts b to $\bar{b}$. It is easy to visualize the area savings over the brute-force approach that uses two separate circuits. Of course, the design must have the need for two inverters close together in the logic chain. The same layout may be used as a basis for creating the non-inverting buffer in Figure 3.34. This uses two series-connected inverters as shown in Figure 3.34(a) to provide the logic. Although an input of a produces the same Boolean logical value for a, the buffer provides electrical reshaping of the signal and provides additional "drive strength" for large fan-outs. The layout scheme in Figure 3.34(b) uses the output of the left inverter to feed the input of the right inverter. This requires a metal-

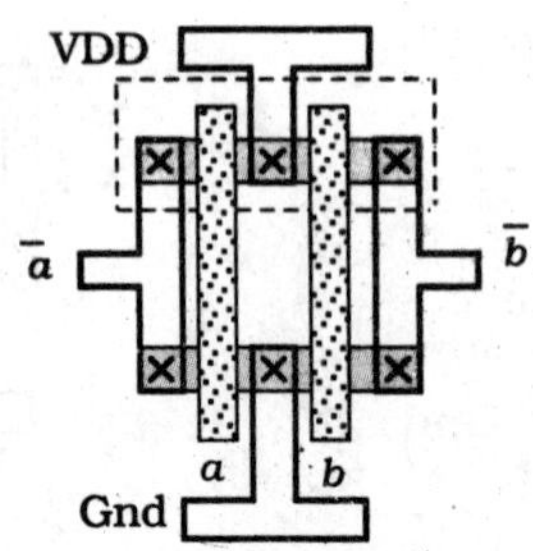

Figure 3.33 Two NOT gates that share power supply and ground

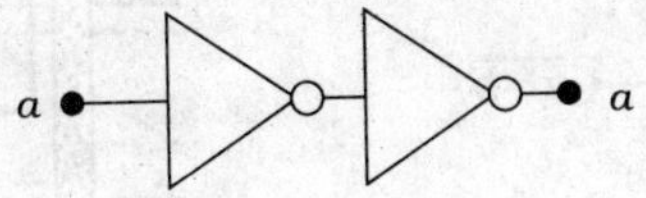

(a) Logic diagram

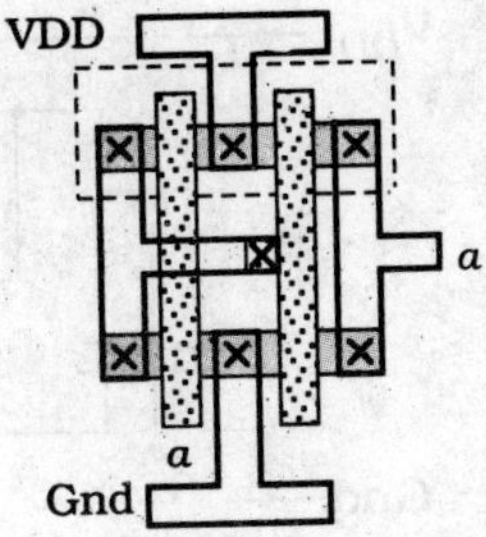

(b) Layout

Figure 3.34 Non-inverting buffer

to-poly contact between the two stages as shown.[8] In addition, the scheme makes use of the fact that metal can cross over the input poly gate without creating an electrical connection.

The transmission gate problem illustrates some of the interconnect routing problems that arise in layout. The logic diagram in Figure 3.35(a) shows a TG with an input x and an output y. Since a transmission gate has only two FETs, it is very simple to design at the physical level. The complicating factor is the inverter that takes the switching signal S and must produce $\overline{S}$ to drive the pFET side of the TG. The NOT gate must be connected to the power supply and ground, but the nFET and pFET of the TG may be located as needed. One solution is shown in Figure 3.35(b). This uses inverter FETs with a tall n+ region so that metal TG input line carrying x may be crossed over it. The complementary switching signal $\overline{S}$ is taken directly from the inverter and fed to the TG pFET.

Once we have laid the foundation for simple layout, it may be used for

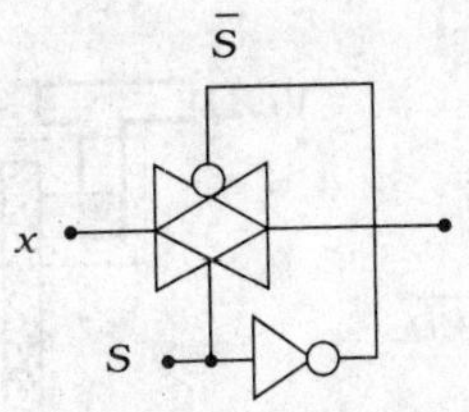

(a) Logic diagram

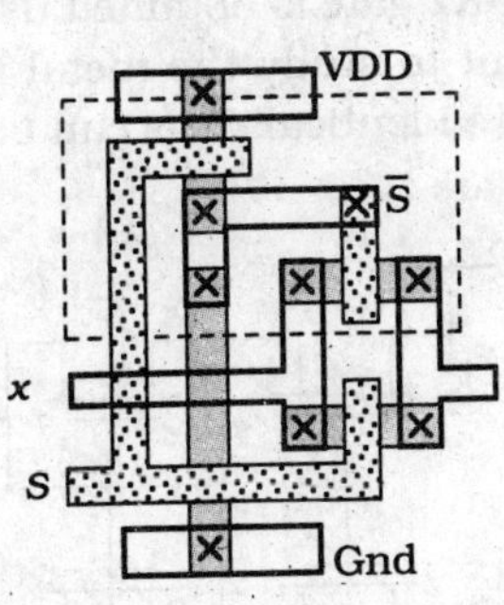

(b) Layout

Figure 3.35 Layout of a transmission gate with a driver

8 This is called a **poly contact** and defines an oxide cut.

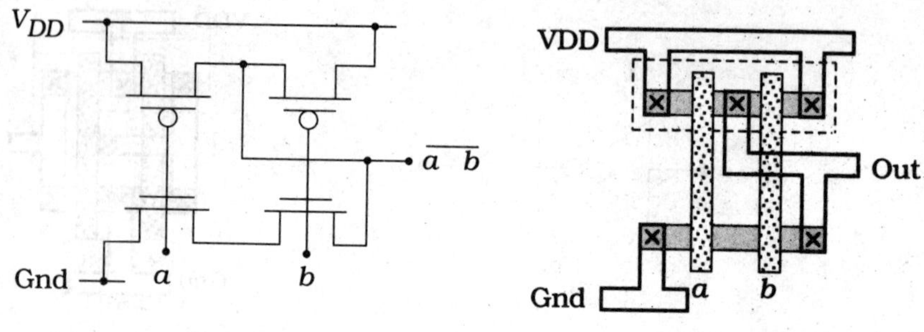

(a) Circuit (b) Layer design

Figure 3.36 NAND2 layout

more complex gates. Figure 3.36(a) shows a NAND2 circuit that has been drawn in a manner that leads to the patterning in Figure 3.36(b). The two nFETs are in series and can be laid out using the method shown in Figure 3.27. Since the gates (with inputs a and b) run in a vertical direction, the parallel-connected pFETs can be added using the technique introduced earlier in Figure 3.29 which achieves the parallel connection by Metal wiring. This allows us to maintain simple gate poly lines shown. The same approach may be used to construct the NOR2 gate. As shown in Figure 3.37(a), the FET arrangement is opposite with the nFETs in parallel and the pFETs in series. The resulting layout in Figure 3.37(b) follows the same philosophy as for the NAND2 gate wiring.

The similarity between the NAND2 and NOR2 layouts can be seen by decomposing the structures into transistors and wiring. The basic FET arrangement for both gates is shown in Figure 3.38(a). To obtain a NAND2 gate, we use the metal wiring pattern provided in Figure 3.38(b); the NOR2 gate is obtained using the wiring in Figure 3.38(c). If you take a moment to study the metal patterns for the two gates, you will see that they are identical! This can be verified by drawing an imaginary horizontal

(a) Circuit (b) Layer design

Figure 3.37 NOR2 gate design

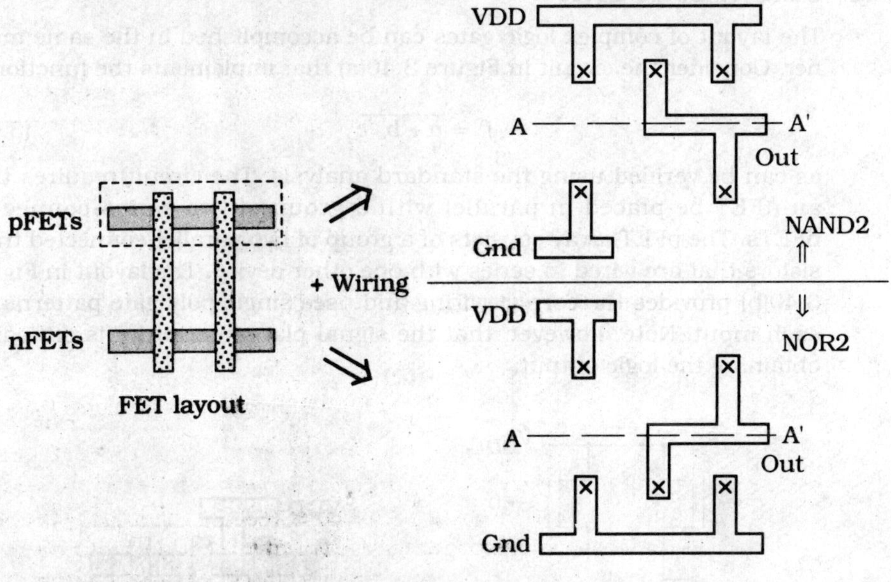

Figure 3.38 NAND2-NOR2 layout comparison

line through the center of one, and then rotating the pattern around it (i.e., flip it vertically). This illustrates how the AND-OR property of duality translates into a layout symmetry.

These layout techniques can be extended to gates with 3 or more inputs. A NOR3 gate is shown in Figure 3.39(a). This uses 3 series-connected pFETs and 3 parallel-connected nFETs. If we "flip" the metal pattern, then we obtain the NAND3 circuit in Figure 3.39(b). On paper, 4-input gates can also be designed in the same manner. However, the electrical switching time of 4-input NAND and NOR gates is relatively slow, which often precludes their usage.

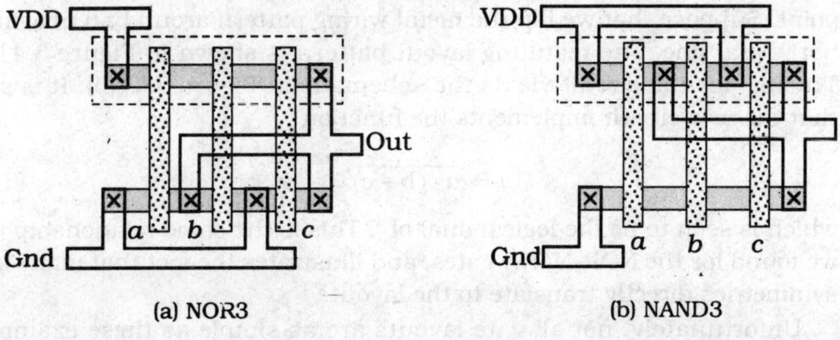

Figure 3.39 Layout for 3-input gates

3.4.2 Complex Logic Gates

The layout of complex logic gates can be accomplished in the same manner. Consider the circuit in Figure 3.40(a) that implements the function

$$f = \overline{a + b \cdot c} \tag{3.68}$$

as can be verified using the standard analysis. The circuit requires that an nFET be placed in parallel with a group of two series-connected nFETs. The pFET array consists of a group of two parallel-connected transistors that are wired in series with one other device. The layout in Figure 3.40(b) provides the correct wiring and uses single poly gate patterns for each input. Note, however, that the signal placement order is critical to obtaining the logic output.

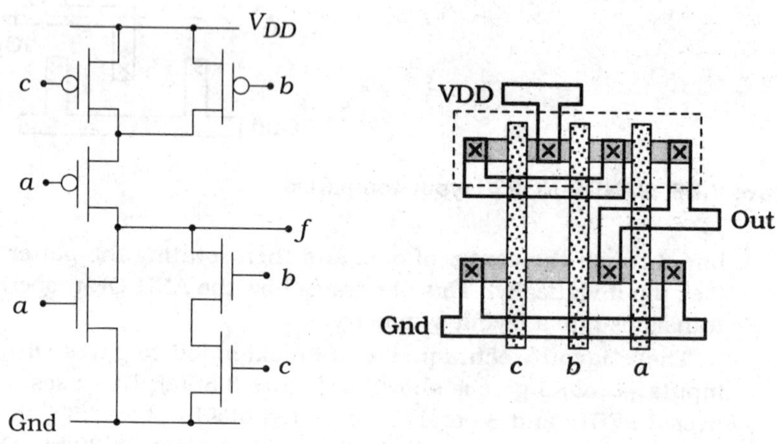

(a) Circuit (b) Patterning

Figure 3.40 Extension of layout technique to a complex logic gate

An interesting variation of the layout demonstrates another important point. Suppose that we flip the metal wiring pattern around an imaginary horizontal line. The resulting layout pattern is shown in Figure 3.41(a). Tracing out the circuit yields the schematic in Figure 3.41(b). It is seen that the new circuit implements the function

$$g = \overline{a \cdot (b + c)} \tag{3.69}$$

which is seen to be the logical dual of *f*. This is the same relationship that we found for the NOR-NAND gates, and illustrates the fact that many logic symmetries directly translate to the layout.

Unfortunately, not all gate layouts are as simple as these examples. Many require much thought and may involve trial-and-error sketches to

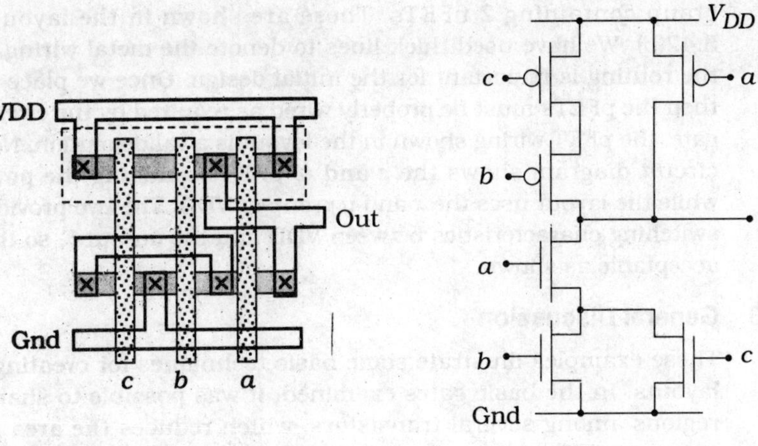

(a) Pattern (b) Circuit

Figure 3.41 Creation of the dual network

accomplish the finished design. Consider the general AOI expression

$$F = \overline{x \cdot y + z \cdot w} \tag{3.70}$$

that can be implemented using the circuit in Figure 3.42(a). If we want to maintain the layout strategy where we use a vertical-running poly line for each input, then we start with 4 gate lines with VDD and Gnd lines. To minimize the area, we would like to share n+ and p+ regions. The nFET patterning is easy since it consists of two groups in parallel, with each

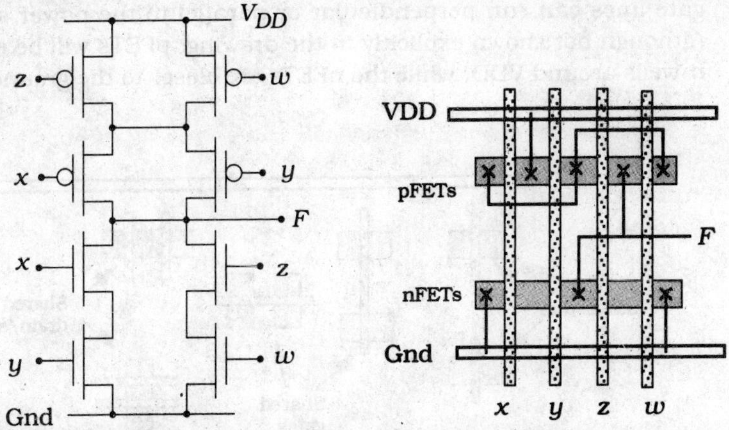

(a) Circuit (b) Layout wiring

Figure 3.42 A general 4-input AOI gate

group containing 2 nFETs. These are shown in the layout of Figure 3.42(b). We have used thick lines to denote the metal wiring, since only the routing is important for the initial design. Once we place the nFETs, then the pFETs must be properly wired as required by the circuit. For this gate, the pFET wiring shown in the layout is a valid solution. Note that the circuit diagram shows the *z* and *w* pFETs touching the power supply while the layout uses the *x* and *y* group at VDD. The two provide the same switching characteristics between VDD and the output *F*, so the layout is acceptable as shown.

3.4.3 General Discussion

These examples illustrate some basic techniques for creating gate-level layouts. In the basic gates examined, it was possible to share n+ or p+ regions among several transistors, which reduces the area and wiring complexity. This is not always possible, especially in complicated arrangements. Various approaches to handling FET placement and wiring have been developed over the years, and are worth discussing here.

Consider the general problem of placing transistors into a CMOS circuit. Experience has shown that regular patterns and arrays will yield the best packing density, and randomly placed polygons should be avoided when possible. In general, every logic gate requires a power supply (VDD) and ground (VSS) connection, which will run as horizontal metal lines in our examples without loss of generality. This leads to the basic framework illustrated in Figure 3.43. All FETs are placed in between the two power rails. In the drawing, transistors are shown as individual devices, groups with shared gate poly lines, and groups with shared drain/source regions. The latter case is the most area-efficient placement, but it may not always be possible to link transistors. The drawing also shows that gate lines can run perpendicular or parallel to the power supply rails. Although not shown explicitly in the drawing, pFETs will be embedded in n-wells around VDD, while the nFETs are closer to the ground rail.

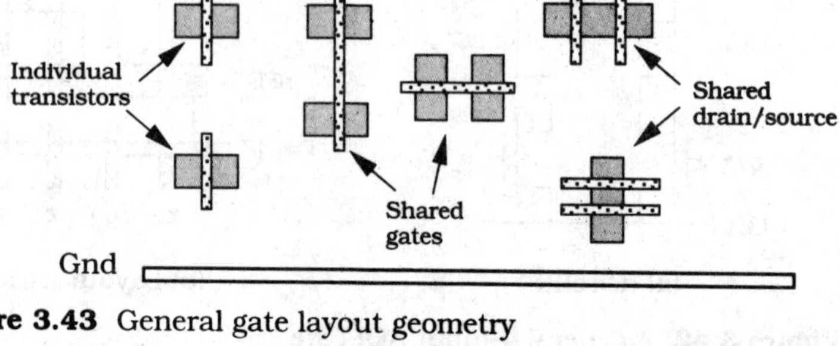

Figure 3.43 General gate layout geometry

One approach to layout is based on the concept of simple **stick diagrams** where each layer is represented by a distinct color, and the routing consists of colored lines that obey the rules of chip formation. A simple example of a stick diagram is shown in Figure 3.44. To save printing costs and keep the price of the book as low as possible, the drawing is monochrome and layers have been represented by different line features such as varying the linewidth or using a dashed line. The key is used to translate the lines into corresponding layers. The most commonly used colors for each layer are listed in the drawing. They are

- Polysilicon (gates): Red
- Doped n+/p+ (active): Green
- N-Well: Yellow (varies)
- Metal1: Blue
- Metal2: Grey (varies)
- Contacts: Black X's

Armed with a set of colored pencils, the layout designer can easily create and verify trial layouts for eventual transferal to silicon. Some of the simple rules associated with colored stick diagrams are as follows.

- A red line crossing a green line creates a transistor
- Red over green inside a yellow border region is a pFET; otherwise it is an nFET
- Red may cross blue or grey
- Blue may cross red, green, or gray
- Grey may cross red, green, or blue
- Transistor contacts must be placed from blue to green
- Vias must be specified to contact blue to grey
- A (poly) contact must be used to connect blue to red

This simple set of rules provides the basics of stick diagram layout. The

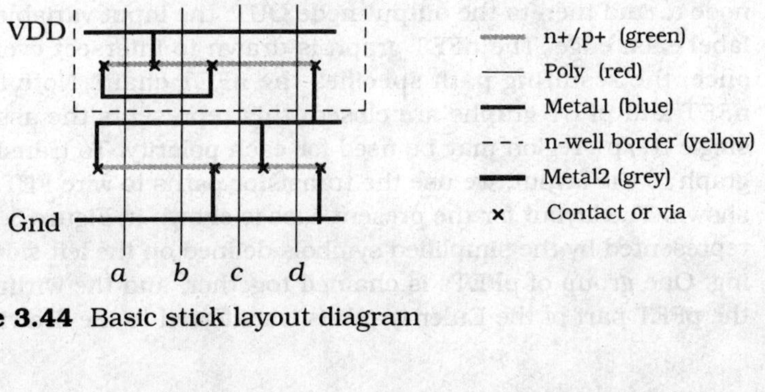

Figure 3.44 Basic stick layout diagram

accompanying CD provides a more detailed discussion of stick diagrams using a color on-screen presentation. Stick diagrams are often used to perform quick layouts or to study large complex routing problems.

A more structured technique is to apply graph theory to the problem of transistor placement and logic gate layout. Figure 3.45 defines the basic components of a graph element that represents a FET. In this approach, the drain and source nodes *x* and *y* of the transistor translate to connection nodes called **vertices**. The transistor itself is represented by an **edge** that corresponds to the signal flow path. Any CMOS circuit can be translated into an equivalent graph consisting of edges and vertices.

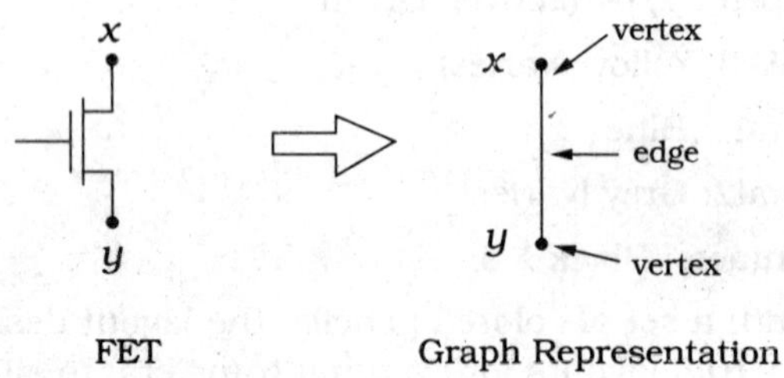

Figure 3.45 Representation of a FET in graph theory

Euler graphs aid in the placement and wiring of circuits where the transistors have shared drain/source regions. To construct an Euler graph, start with the CMOS circuit diagram and select a starting vertex (node). If it is possible to trace the entire graph without passing over an edge more than once, then it is possible to use common n+/p+ regions for nFETs/pFETs. The resulting graph can then be used directly to create the layout strategy.

An example of this process is shown in Figure 3.46. The circuit in Figure 3.46(a) can be traced as shown. The path starts at the vertex shown and follows the arrow to the end vertex, and passes over every edge only once; this defines an **Eulerian path**. Since the path exists, we can use it to construct the Euler graph in Figure 3.46(b). The graph consists of intersecting pFET and nFET graphs. The pFET graph links VDD to the node α, and then to the output node OUT; the input variables are used to label each edge. The nFET graph is drawn to intersect every pFET edge once; the resulting path specifies the nFET chain. Note that both the nFET and pFET graphs are closed; this represents the assertion that a single n+/p+ region may be used for each polarity. To translate the Euler graph to the layout, we use the transistor paths to wire FETs in the order shown. The layout for the present case is shown in Figure 3.47. FETs are represented by the simplified symbols defined on the left side of the drawing. One group of pFETs is chained together, and the wiring specified in the pFET part of the Euler graph is transferred to the drawing. Similarly,

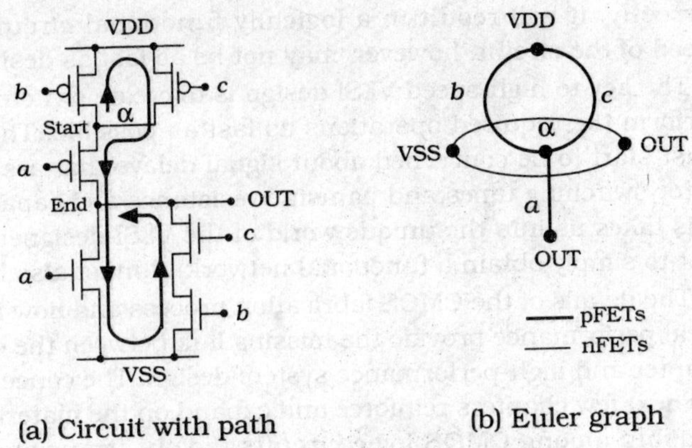

(a) Circuit with path (b) Euler graph

Figure 3.46 Construction of an Euler graph

we use a common n+ region for the nFETs and follow the wiring in the nFET portion of the Euler graph. This may then be translated into the final layout.

If an Euler path cannot be found, then it means that it is not possible to use FET chains to build the circuit. Two or more groups of transistors will be needed, and the layout is much more complex. Design automation tools have been developed to help in some aspects of gate layout, but an experienced layout designer is still considered necessary in critical applications. Many layout specialists have backgrounds in graphics or art, and are able to produce amazingly compact designs that are not obvious to the rest of the group.

3.4.4 Summary

In this chapter we have seen the basics of translating FET logic circuits to silicon. The layout considerations presented here are sufficient to create complex logic networks using a set of standard MOSFETs as building blocks. In many designs it is possible to simply place reasonably sized transistors as specified by the layout, and wire them together. If done

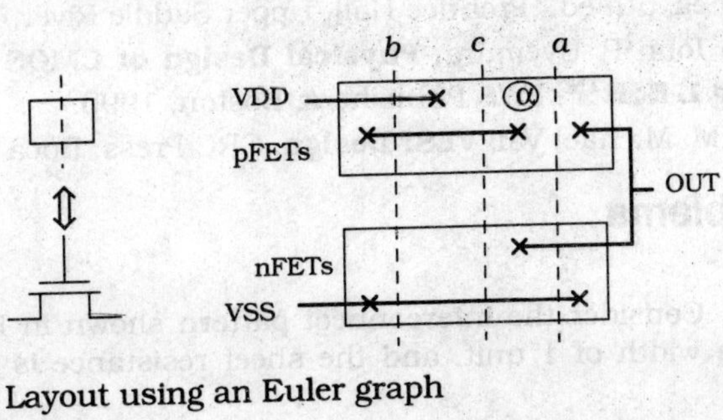

Figure 3.47 Layout using an Euler graph

correctly, it will result in a logically functional circuit. The switching speed of the circuit, however, may not be as fast as desired.

The key to high-speed VLSI design is to create switching networks that perform the required operations as fast as possible. This means that we must start to be concerned about signal delays that are induced by transistor switching times and parasitic resistance and capacitance elements. This takes us into the unique world of the VLSI designer. We are not content to simply obtain a functional network; it must also be fast!

The details of the CMOS fabrication process and how it affects the electrical performance provide the missing link between the discussion in this chapter and high-performance system design. The concepts introduced in the next few chapters reinforce and expand on the material here. The relationship among CMOS logic circuits, layout, transistors, and systems design is quite natural.

3.5 References for Further Reading

[1] H. B. Bakoglu, **Circuits, Interconnections, and Packaging for VLSI**, Addison-Wesley, Reading, MA, 1990.

[2] Dan Clein, **CMOS IC Layout**, Newnes, Woburn, MA, 2000.

[3] Richard S. Muller and Theodore I. Kamins, **Device Electronics for Integrated Circuits**, 2nd. ed., John Wiley & Sons, New York, 1986.

[4] Robert F. Pierret, **Semiconductor Device Fundamentals**, Addison-Wesley, Reading, MA, 1996.

[5] Bryan Preas and Michael Lorenzetti (eds.), **Physical Design Automation of VLSI Systems**, Benjamin/Cummings Publishing Company, Menlo Park, CA, 1988.

[6] M. Sarrafzadeh and C. K. Wong, **An Introduction to VLSI Physical Design**, McGraw-Hill, New York, 1996.

[7] Naveed Sherwani, **Algorithms for VLSI Physical Design Automation**, Kluwer Academic Publishers, Norwell, MA, 1993.

[8] Jasprit Singh, **Semiconductor Devices**, John Wiley & Sons, New York, 2001.

[9] Ben G. Streetman and Sanhay Banerjee, **Solid State Electronic Devices**, 5th ed., Prentice Hall, Upper Saddle River, NJ, 1998.

[10] John P. Uyemura, **Physical Design of CMOS Integrated Circuits Using L-Edit™**, PWS Publishers, Boston, 1995.

[11] M. Michael Vai, **VLSI Design**, CRC Press, Boca Raton, FL, 2001.

3.6 Problems

[3.1] Consider the interconnect pattern shown in Figure P3.1. The line has a width of 1 unit, and the sheet resistance is $R_s = 25\ \Omega$. Find the resistance from A to B if each corner square contributes a factor of 0.625 of a "straight-path" square.

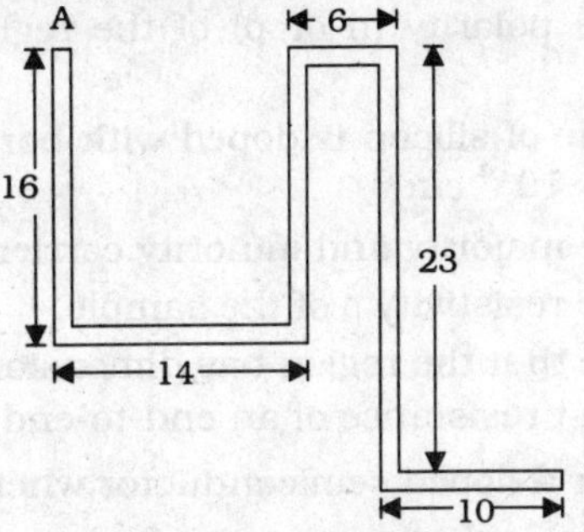

Figure P3.1

[3.2] An interconnect line can be made in either of two layers. If a gate polysilicon layer is selected, the sheet resistance is 25 Ω; for this case, the interconnect will have a width of 0.5 μm and a length of 27.5 μm. A metal layer can also be used. It has a sheet resistance of 0.08 Ω. The metal line has a width of 0.8 μm but requires a different routing length of 32.4 μm.

Calculate the line resistance R_{line} for each case and determine the lower resistance alternate. What is the percentage increase in resistance if the larger resistance line is used instead?

[3.3] An interconnect line is made from a material that has a resistivity of $\rho = 4\ \mu\Omega$-cm. The interconnect is 1200 Å thick, where 1 Angstrom (Å) is 10^{-8} cm. The line has a width of 0.6 μm.

(a) Calculate the sheet resistance R_s of the line.

(b) Find the line resistance for a line that is 125 μm long.

[3.4] Consider equation (3.14) for the interconnect time constant τ. Prove that τ has units of seconds by expressing ohms and farads in fundamental MKS units and reducing.

[3.5] An interconnect line runs over an insulating oxide layer that is 10,000 Å thick. The line has a width of 0.5 μm and is 40 μm long. The sheet resistance is known to be 25 Ω.

(a) Find the line resistance R_{line}.

(b) Find the line capacitance C_{line}. Use $\varepsilon_{ox} = 3.453 \times 10^{-13}$ F/cm, and express your answer in femtofarads (fF) where 1 fF = 10^{-15} F.

(c) Find the time constant τ for the line in units of picoseconds (ps) where 1 ps = 10^{-12} sec.

[3.6] A sample of silicon is doped with arsenic with $N_d = 4 \times 10^{17}$ cm^{-3}.

(a) Find the majority carrier density.

(b) Find the minority carrier density.

(c) Calculate the electron and hole mobilities and then find the conductivity of the sample.

[3.7] A region of silicon is doped with both phosphorus and boron. The P-doping is $N_d = 2 \times 10^{16}$ cm^{-3} while the B-doping level is $N_a = 6 \times 10^{18}$ cm^{-3}. Determine the polarity (n or p) of the region, and find the carrier densities.

[3.8] A sample of silicon is doped with boron atoms at an acceptor density of $N_a = 4 \times 10^{14}$ cm^{-3}.

(a) Find the majority and minority carrier densities.

(b) Find the resistivity ρ of the sample.

(c) Suppose that the region has dimensions of 2 μm × 0.5 μm × 100 μm. Find the largest resistance of an end-to-end block of the region.

[3.9] Consider a doped semiconductor where

$$\sigma = q(\mu_n n + \mu_p p) \tag{3.71}$$

and $np = n_i^2$. Suppose we wish to minimize the conductivity.

(a) Use the mass-action law to write in terms of p only.

(b) Compute the derivative $(d\sigma/dp)$ and set it equal to 0 to find the hole concentration that minimizes σ.

(c) Noting that $\mu_n > \mu_p$, what polarity (n-type or p-type) is required for the highest resistivity? Then use your equations to find the doping type and density that give the highest resistivity.

[3.10] An n-channel MOSFET has a mobility value of $\mu_n = 560$ cm^2/V-sec and uses a gate oxide with a thickness of $t_{ox} = 90$ Å. The gate voltage is given as $V_G = 2.5$ V, and the threshold voltage is 0.65 V.

(a) Calculate the value of C_{ox} in units of F/cm^2.

(b) Find the process transconductance k'_n.

(c) Find the device transconductance β_n if the FET has a channel length of 0.25 μm and a channel width of 2 μm.

[3.11] Use equation (3.57) for R_n to find the units of the electron mobility μ_n. Then suppose that $\mu_n = 500$ cm^2/V-sec and $(V_G - V_{Tn}) = (3.3 - 0.7)$ V is known.

(a) Find the nFET resistance if $W = 10$ μm, $L = 0.5$ μm, and $t_{ox} = 10$ nm.

(b) Find R_n if the channel width is increased to a value of $W = 22$ μm while the channel length remains the same.

[3.12] A pFET is described by $\mu_p = 220$ cm^2/V-sec and $(V_G - |V_{Tp}|) = (3.3 - 0.8)$ V, $W = 14$ μm, $L = 0.5$ μm, and $t_{ox} = 11.5$ nm. Find the pFET resistance R_p of the device.

[3.13] Consider a process that has an oxide thickness of $t_{ox} = 9.5$ nm. The particle mobilities are given as $\mu_n = 540$ and $\mu_p = 220$ cm^2/V-sec. An nFET and a pFET are made, both with $W = 12$ μm, $L = 0.35$ μm. Both have gate voltages of $V_G = 3.3$ V, while the threshold voltages are $V_{Tn} = 0.65$ V and $V_{Tp} = -0.74$ V.

(a) Find the values of R_n and R_p for the two transistors.

(b) Suppose that we want to keep the nFET the same size, but increase the width of the pFET to the point where $R_p = 0.8\ R_n$. Find the required width of the pFET.

[3.14] Design a CMOS logic gate that provides the function

$$Out = \overline{x \cdot (y \cdot z + z \cdot w)} \tag{3.72}$$

Then perform the basic layout of circuit.

[3.15] Design the circuit and layout for a CMOS gate that implements the function

$$F = \overline{a \cdot b \cdot c + a \cdot d} \tag{3.73}$$

using the fewest number of transistors and a compact layout style.

[3.16] Consider the OAI logic function

$$g = \overline{(a+b) \cdot (c+d) \cdot e} \tag{3.74}$$

Design the CMOS logic gate and then construct a basic layout for the circuit.

[3.17] Expand the function g given in equation (3.74) [Problem 3.16 above] into AOI form. Then design the CMOS logic circuit and layout.

[3.18] Examine the stick diagram in Figure 3.44. Is this a functional logic gate? If so, determine the logic operation it provides.

[3.19] Consider the logic function

$$g = \overline{a \cdot b \cdot c + d} \tag{3.75}$$

(a) Design the CMOS logic gate that provides this function.
(b) Is it possible to find an Euler graph for the circuit? If so, construct the graph and use it to perform a stick-level layout. If not, find a layout strategy for the gate.

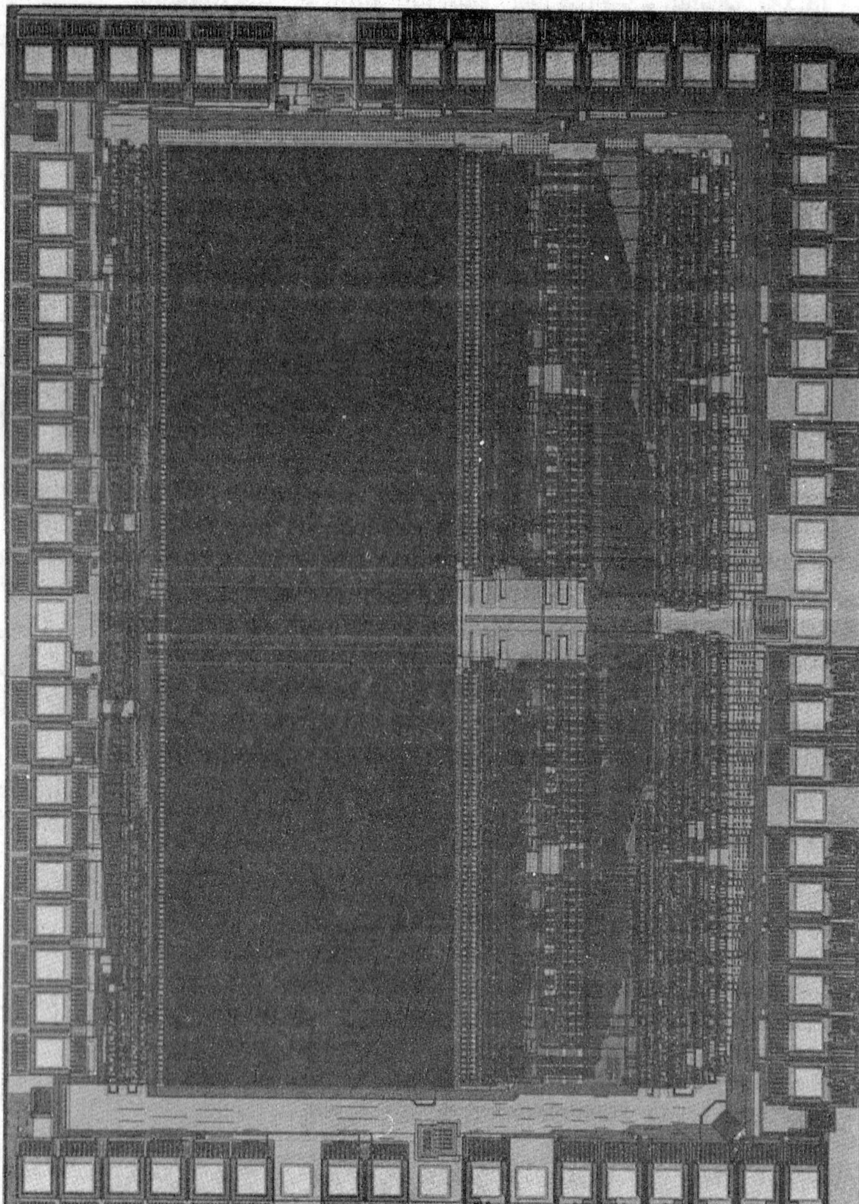

Fabrication of CMOS Integrated Circuits

4

An integrated circuit consists of several patterned layers of materials that are used to form transistors and provide electrical interconnections for the circuit. In a modern process, the minimum feature size is less than about 0.12 μm, which allows for a tremendous packing density. Individual chips with more than 100 million FETs are becoming commonplace. The techniques needed to fabricate silicon chips of this sophistication have been developed over several decades at an enormous cost. In fact, silicon has been characterized as being the most studied element on earth!

Now that we have an understanding of the physical structure of CMOS integrated circuits we may progress to studying how the circuits are fabricated in the manufacturing process. Our treatment will focus on those aspects of silicon chip fabrication that are important to VLSI design.

4.1 Overview of Silicon Processing

Silicon integrated circuits are created on larger circular sheets of silicon called wafers. They are typically 100–300 mm in diameter, and about 0.4–0.7 mm thick. A large silicon circuit is about 1 cm on a side so that many individual circuits can be made on a single wafer. The location of a circuit is called a die site, with the number of sites per wafer depending upon the size of each site and relative to the overall surface area of the wafer. Figure 4.1 portrays a wafer with individual sites. The flat is used as a reference plane to form an imaginary grid that is used to place the individual sites. Some wafers will have additional flats that are coded to provide information about the crystal orientation at a glance.

Starting with a bare polished surface, the wafer is subjected to thousands of individual steps in the manufacturing processes. The most

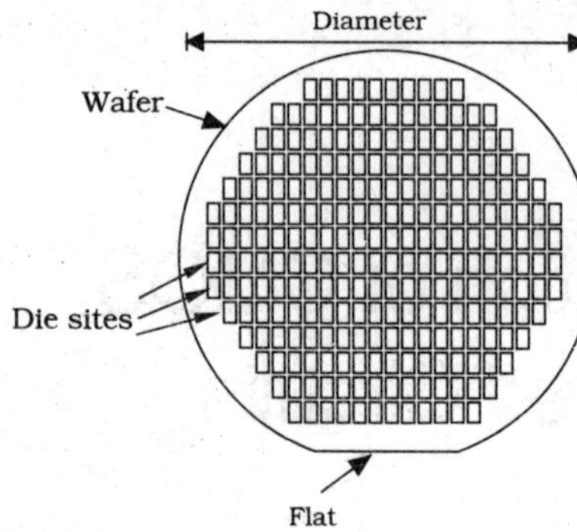

Figure 4.1 Silicon wafer showing die sites

important steps in the sequence are for creating and patterning the layers of materials needed in the CMOS structures. Most of the remaining steps are cleaning and rinsing of the wafer. The manufacturing capacity of a chip factory is usually measured by the number of **wafer starts** per week, i.e., how many fresh wafers are introduced into the fabrication sequence. Wafers are processed in groups, and it takes several weeks for a lot to make it through the entire processing line.

Unfortunately, not every site on the wafer turns out to be a functional circuit. This is due to many factors that may arise in the manufacturing line that are inherent due to the complexity of the processing sequence. To describe this problem, we introduce the concept of the fabrication **yield** Y such that

$$Y = \frac{N_G}{N_T} \times 100\% \tag{4.1}$$

where N_G is the number of good (functional) sites and N_T is the total number of sites. A yield of $Y = 85\%$ means that 85% of the chips operate as they should and can be sold to customers. High yield values are, of course, desirable to help insure the economic stability of the company. However, yield enhancement is a complex problem that requires countless hours of thinking and experimenting with the process line.

Yield analysis is based on predicting the yield of a particular process, and requires a thorough understanding of all aspects of the silicon processing sequence. One working in this area is faced with increasing the value of Y for a given design. One variable that is critically important to increasing the yield is the area A_{die} of the die. The number of total die sites N_T on a wafer of diameter d is estimated from

$$N_T = \pi \frac{(d - d_e)^2}{4A_{die}} \tag{4.2}$$

where d_e is the wasted edge distance that arises from placing rectangular sites onto a round wafer. Empirical analysis shows that large area die are plagued by smaller yields. A simple expression that describes this is

$$Y = e^{-\sqrt{DA}} \tag{4.3}$$

where A is the area of the die. The parameter D is the **defect density** in units of cm^{-2} and is the average number of defects per cm^2 on the wafer. D represents the "limit of perfection" that can be expected of the silicon wafer; this is due to the fact that every crystal wafer has random imperfections that cannot be eliminated. In modern technology, $D = 1\ cm^{-2}$ is a reasonable value for the defect density limit.

Some physical defects tend to occur in clusters on the wafer. An estimate for the yield when these dominate is given by

$$Y = \left(1 - \frac{A_{die}D}{c}\right)^c \tag{4.4}$$

where c is an empirical parameter that accounts for the clusters. This idea has been used to a write a binomial equation of the form

$$Y = \frac{1}{\left(1 + \frac{A_{die}D}{c}\right)^2} \tag{4.5}$$

Alternately, when several die fail in a large area A_{fail} of the wafer, then the yield has been approximated using the expression

$$Y \approx (1 - g)e^{-A_{die}D} \tag{4.6}$$

where

$$g = \frac{A_{fail}}{A_{wafer}} \tag{4.7}$$

is the fractional area where the defects exist.

Yield analysis is a very specialized aspect of VLSI manufacturing. Persons working in the area tend to have strong backgrounds in physics, general and physical chemistry, mathematics, statistics, or engineering (chemical, materials, or electrical), and groups work closely with the manufacturing line and the wafer analysis groups to maximize yield values. It is not possible to solve a problem until it is discovered and defined. The "design of experiments" that can pinpoint problems and lead to solutions becomes very critical.

Economics 101

It is worth examining some important economic factors that deal with the design, manufacture, and marketing of VLSI circuits. Let C_{chip} be the cost of manufacturing a chip, and C_{sell} the selling price. The profit per chip is then given by

$$\text{Profit} = C_{sell} - C_{chip} \tag{4.8}$$

To survive, a product line must result in a value where

$$\text{Profit} > 0 \tag{4.9}$$

While this may seem blatantly obvious, the VLSI designer must recognize that neither C_{chip} nor C_{sell} is easy to compute. The cost of manufacture of the chip includes the materials and salaries of all personnel (design, manufacturing, testing, etc.) plus overhead (electricity, water, taxes, etc.). Increasing the yield reduces the overall costs per unit, so the importance of yield analysis becomes obvious. These factors and many more contribute to C_{chip}.

In modern VLSI, the cost of a state-of-the art chip manufacturing plant is somewhere between $1–3 billion (USD). This includes the land, building, equipment, and start-up costs, but not materials or everyday operations. The cost of the facility must be amortized over the product lines for the lifetime of the plant.

The selling price C_{sell} of every product must include all direct and indirect costs, plus a fraction of the plant debt. The laws of supply and demand also enter the picture: C_{sell} must be at a level that the customers are willing to pay. If a product is in great demand, then C_{sell} may be well above the costs and the design produces a large income. In this case, chips (and products in general) may be sold at a price that is determined entirely by demand; a common phrase that describes this is *whatever the market will bear.* On the other hand, even a great engineering design may fail to gain a following of users, and it will be eventually withdrawn; in this case we encounter the unwanted result that Profit < 0.

Another complication is that C_{sell} tends to decrease in time. Even the "hottest" new microprocessor eventually becomes a cheap bargain basement item. This is not a major problem so long as we have repaid the investment made in engineering costs. Complex VLSI chips are very difficult to design, and the original design can be very expensive. Another helpful factor is that as time progresses,

$$C_{chip} \rightarrow C_{materials} \tag{4.10}$$

where $C_{materials}$ is the cost of materials. Silicon has the advantage of being very cheap, especially when compared to alternates such as gallium arsenide (GaAs). Keeping a product line active for many years greatly enhances the profitability.

This short introduction is designed to help the aspiring VLSI designer understand the overall structure of the industry. Producing a silicon chip with 100 million transistors is much more complicated than starting a "dot-com" web site. It requires financial backing, strong technological support, innovative engineering, and a reliable sales force. The fabrication process is considered to be a major expense in the Profit equation, so we have chosen to study it in some detail in this chapter. Design and engineering costs are almost as high, and are discussed in the rest of the book.

4.1.1 Outline of the Chapter

We have introduced the viewpoint that a silicon chip is a set of patterned material layers. When the layers are properly stacked, the resulting three-dimensional structures are controlled switches (transistors) that are wired together to implement logic operations.

In this chapter, we first examine the most important material layers that are used in silicon processing. This includes oxides, doped silicon regions, and metals. A few chemical reactions are presented along with a brief description of how the layer is actually grown or deposited. We then move on and study how a layer is physically patterned to have the proper shape and size needed for wire and transistors. This allows us to progress to the steps used to fabricate a basic CMOS circuit.

The main objective of the treatment is to provide an understanding of the basics and how they relate to the physical design of a VLSI circuit. Many persons with strong science backgrounds become fascinated with silicon processing, and establish an entire career in the field.

4.2 Material Growth and Deposition

An integrated circuit is created by stacking layers of various materials in a prespecified sequence. Both the electrical properties of the material and the geometrical patterns of the layer are important in establishing the characteristics of devices and networks.

Most layers are created first, and then patterned using the lithographic sequence described in the next section. Doped silicon layers are the exception to this rule, as they are created with the desired shapes by using the lithographic process to define where the dopants can enter the silicon. In this section we will examine some basic processing steps used in silicon VLSI processing.

4.2.1 Silicon Dioxide

Silicon dioxide (SiO_2) is a critically important material in IC processing because

- It is an excellent electrical insulator
- It adheres well to most materials
- It can be "grown" on a silicon wafer or deposited on top of the wafer

SiO_2 is generically known as quartz glass, or simply "glass," and is used for the gate oxide in a MOSFET, in addition to numerous other applications.

There are two types of SiO_2 layers found in VLSI circuits, with the distinction being how they are created. A **thermal oxide** is formed by the reaction

$$Si + O_2 \rightarrow SiO_2 \tag{4.11}$$

using heat as a catalyst. The unique aspect of a thermal oxide is that the silicon (Si) required for the reaction is obtained from the silicon wafer itself. This is illustrated in Figure 4.2(a) where oxygen molecules O_2 are passed over the surface of the wafer where the reaction takes place. This literally "grows" the glass layer with the results shown in Figure 4.2(b). The final thickness of the oxide is denoted as x_{ox} in the drawing, and depends on the temperature, crystal orientation, and growth time. Since silicon atoms from the surface of the wafer are used by the reaction, a layer of silicon with a thickness

$$x_{Si} \approx 0.46\ x_{ox} \tag{4.12}$$

is consumed. An equivalent (and useful) viewpoint is that the surface of the silicon is "recessed" from its original location.

Although pure oxygen yields high-quality oxide layers, it is relatively slow. A faster growth rate is obtained using water (H_2O) in the form of steam via the reaction

$$Si + 2H_2O \rightarrow SiO_2 + 2H_2 \tag{4.13}$$

which is called "wet oxidation." In practice, mixtures of O_2 and steam are used, along with nitrogen as a carrier gas and other chemicals such as chlorine (Cl).

Thermal oxide is a form of a **native oxide**, i.e., one that is created when the surface is exposed to an oxygenated atmosphere. If you take a bare silicon wafer and place it in air, a thin native oxide layer will form. Increasing the temperature enhances the growth rate. Silicon oxidation temperatures are typically in the range of about 850–1100 °C.

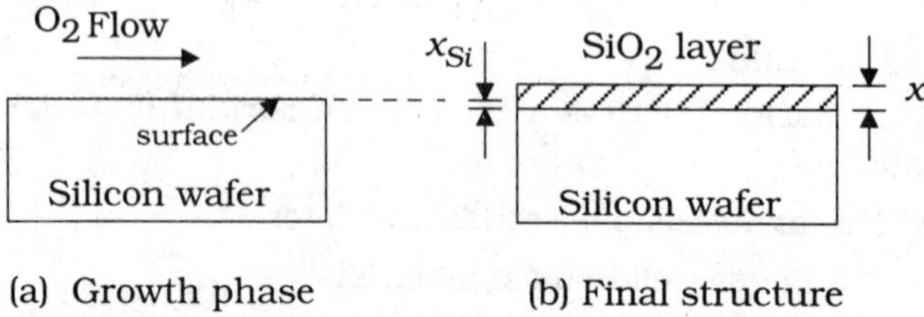

Figure 4.2 Thermal oxide growth

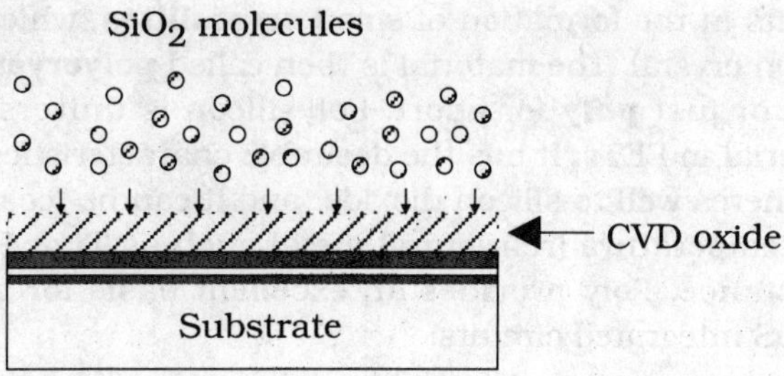

Figure 4.3 CVD oxide process

Most oxide layers in VLSI circuits are well above the wafer surface and no silicon is available for thermal oxide growth. In this case, we create SiO_2 molecules using gaseous reactions, and then **deposit** them onto the surface to provide an oxide coating. The process is shown schematically in Figure 4.3. A chemical reaction using silane (SiH_4) such as

$$SiH_4 \text{ (gas)} + 2O_2 \text{ (gas)} \rightarrow SiO_2\text{(solid)} + 2H_2O \text{ (gas)} \qquad (4.14)$$

can be used to produce the SiO_2 molecules above the wafer. This technique is called **chemical vapor deposition** (CVD) and the resulting layers are often called **CVD oxides**. The thickness of the oxide layer is controlled using the growth rate and deposition time. It is possible to perform the deposition at low temperatures, giving rise to the name **LTO** (low-temperature oxides). Also, it is sometimes advantageous to dope the glass. For example, phosphorus doping yields "P-doped glass" which helps certain types of planarization steps.

4.2.2 Silicon Nitride

Another useful material is silicon nitride Si_3N_4, which is often just called "nitride" when the context is clear. The reaction

$$3SiH_4\text{(gas)} + 4NH_3\text{(gas)} \rightarrow Si_3N_4 \text{ (solid)} + 12H_2\text{(gas)} \qquad (4.15)$$

illustrates one technique. Nitrides are unique in that they act as strong barriers to most atoms. This makes them ideal for use as an **overglass** layer, which is a final protective coating on a chip, since it keeps contaminants from reaching the sensitive silicon circuits. Silicon nitride is used in a fabrication sequence that electrically isolates adjacent FETs (as will be discussed later). And, they have a relatively high dielectric constant $\varepsilon_N \approx 7.8\ \varepsilon_o$, which makes them candidates for insulating ON (oxide-nitride) "sandwich" insulators in various capacitor structures such as those used in DRAM (dynamic random-access memory) cells.

4.2.3 Polycrystal Silicon

If we deposit silicon atoms on top of an amorphous SiO_2 layer, the silicon attempts to crystallize but can't find a crystal structure for reference. This results in the formation of small **crystallites**, which are small regions of silicon crystal. The material is then called **polycrystal silicon** or **polysilicon**, or just **poly** for short. Polysilicon is universally used as the gate material in FETs. It has the desirable characteristics that it can be doped, it adheres well to silicon dioxide, and it can be "coated" with a high-melting-temperature (refractory) metal such as Ti or Pt to reduce the sheet resistance. Poly provides an excellent basis for building MOSFETs in CMOS integrated circuits.

A basic reaction using silane is

$$SiH_4 \rightarrow Si + 2H_2 \tag{4.16}$$

which is performed at a temperature around 500–600° C. Poly deposition techniques have evolved during recent years in the fabrication of **stacked capacitors** used in advanced dynamic random-access memory (DRAM) cells. These are examined in Section 13.3 of Chapter 13.

4.2.4 Metals

Aluminum (Al) is the most common metal used for interconnect wiring in integrated circuits. It can be evaporated by heating in a vacuum chamber with the resulting flux used to coat the wafer. Al has good adhesion characteristics and is easy to pattern. Its popularity is understandable.

Aluminum has a bulk resistivity of about ρ = 2.65 μΩ-cm. An aluminum interconnect line that is 0.1 μm thick has a sheet resistance of about

$$R_s = \frac{\rho}{t} = \frac{2.56\times10^{-6}}{10^{-5}} = 0.265\Omega \tag{4.17}$$

However, aluminum exhibits a problem called **electromigration**. High current flow densities tend to literally move atoms from one end of an interconnect line, creating pits called **voids**. The atoms pile up at the other end in microscopic structures called **hillocks**. These are illustrated schematically in Figure 4.4. Hillocks and voids can lead to failure and

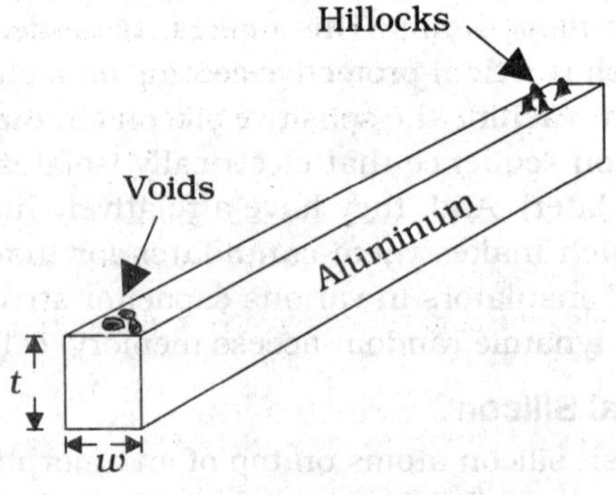

Figure 4.4 Visualization of electromigration effects in aluminum

much research has been devoted toward studying this problem. A common solution is to mix copper with the aluminum during the metal deposition step. This reduces electromigration effects, but increases the resistivity to values around ρ = 3.5 μΩ-cm. The sheet resistance is increased proportionately.

At the physical design level, we avoid excessive formation of hillocks and voids by controlling the current density J A/cm^2 flowing in the interconnect. For an interconnect line with thickness t and width w, the current density is given by

$$J = \frac{I}{A} \tag{4.18}$$

where I is the current in amperes, and $A = wt$ is the cross-sectional area in units of square centimeters. Layout designers cannot alter the thickness t of the layer since it is established in the processing line. Electromigration is thus controlled by specifying the minimum linewidth w needed to keep J below a maximum value J_{max}. This is our first example of a layout **design rule** that specifies a minimum dimension of a feature for a particular situation. We will investigate design rules more thoroughly in the later sections of this chapter.

MOS had its beginnings in metal-gate technology where the "M" truly stood for metal, and aluminum was the choice for the gate layer. The drawback of using Al for a transistor gate is that its low melting temperature prohibits the use of high-temperature processing steps once it is deposited on the wafer. As processing technology continued to improve with increasingly complex processing sequences, this became a limiting factor. Transistors using polysilicon gates were developed and are now standard in CMOS. A significant problem with silicon gates is that even heavily doped poly has a high sheet resistance with values around R_s = 25–50 ohms. To overcome this, the poly is coated with a thin layer of a **refractory** (high-temperature) metal such as titanium (Ti), tungsten (W), or platinum (Pt). This combination is called a **silicide** and the poly-metal mixture is usually treated as a single layer in the design. This will be shown explicitly in the CMOS processing sequence described later. Tungsten is also commonly used for plugs in vias to connect metal layers.

Copper (Cu) has recently been introduced as a replacement to aluminum. Since its resistivity is about one-half the value of Al, it gives smaller sheet resistances. At the device level, the difference is not important. However, the reduction in sheet resistance is significant when copper is used for long, system-level interconnect lines. The improvement in technology does not come easily. Standard patterning techniques cannot be used on copper layers; specialized techniques had to be developed. The use of copper will be discussed in Section 4.4.1.

4.2.5 Doped Silicon Layers

The silicon wafer is the starting point for the CMOS fabrication process. It is defined to be n-type or p-type during the crystal growth and acts as the basis substrate for the entire circuit structure. By our definition, a doped silicon layer is a patterned n- or p-type section of the wafer surface. Even though silicon layers don't always "stack" in the usual sense, we will maintain this terminology to be consistent.

The key to creating doped layers in the substrate is to introduce donor or acceptor atoms into the wafer that can be eventually incorporated into the silicon crystal. In modern CMOS, this is accomplished by a technique called **ion implantation** where the atoms are first ionized in a chamber, then accelerated to high energies in a particle accelerator. The beam is passed through a mass separation unit that selects the desired charge species using a magnetic field. The overall system is shown in Figure 4.5.

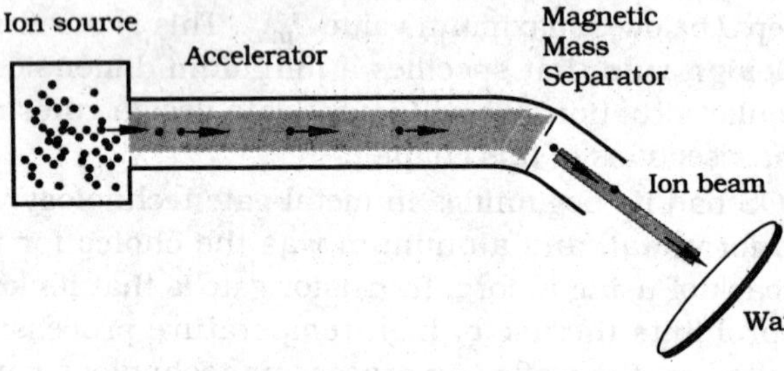

Figure 4.5 Basic sections of an ion implanter

The fast moving ions are literally smashed into the substrate at typical energies around 100–200 keV. The ions come to rest after several collisions with electrons and nuclei in the silicon wafer. This is illustrated schematically in Figure 4.6. The slowing mechanism damages the crystal

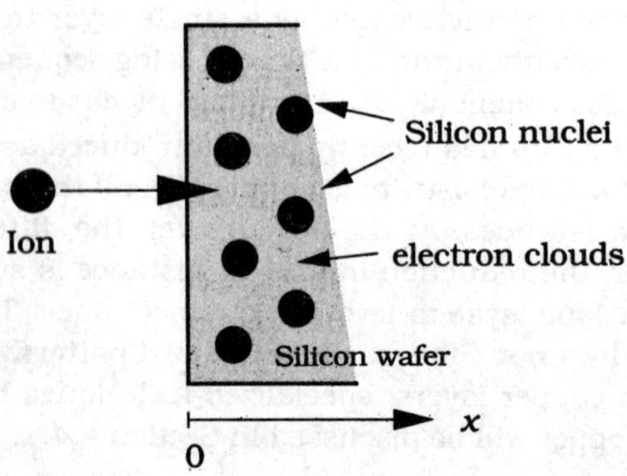

Figure 4.6 The ion stopping process

and leaves the dopants in random locations. To heal the crystal and set the dopants into proper locations within the crystal lattice, the wafer is heated in an **anneal** step. The dopants redistribute a little during the annealing step because of a process known as **particle diffusion**; diffusion is simply the collective heat-induced motion of particles that are concentrated in a small region that makes the particle spread out.

The ion distribution into the silicon can be approximated to first order using the Gaussian form

$$N_{ion}(x) = N_p e^{-\frac{1}{2}\left(\frac{x-R_p}{\Delta R_p}\right)^2} \tag{4.19}$$

with units of cm^{-3}; the surface of the wafer is defined by $x = 0$. This function is shown in Figure 4.7. The quantity R_p is called the **projected range**, and is the average depth of an implanted ion. The value of R_p depends on the incident energy, the species, and the crystal orientation, and can range from about 0.1 μm to as deep as 1 μm. The peak density N_p occurs at $x = R_p$. The standard deviation is denoted as the **straggle** ΔR_p; this represents the variation in the stopping depth of the individual ions due to the statistical nature of the energy loss process. More accurate models of the implant profile employ Pearson Type IV distributions and numerical simulations.

The number of implanted ions is usually described by the implant dose D_I defined by

$$D_I = \int_{\text{All } x} N_{ion}(x)\,dx \tag{4.20}$$

which has units of ions per cm^2 (or just cm^{-2}). This can be very accurately measured using charge counters. The dose is often used when analyzing the macroscopic electrical characteristics of MOS capacitors.

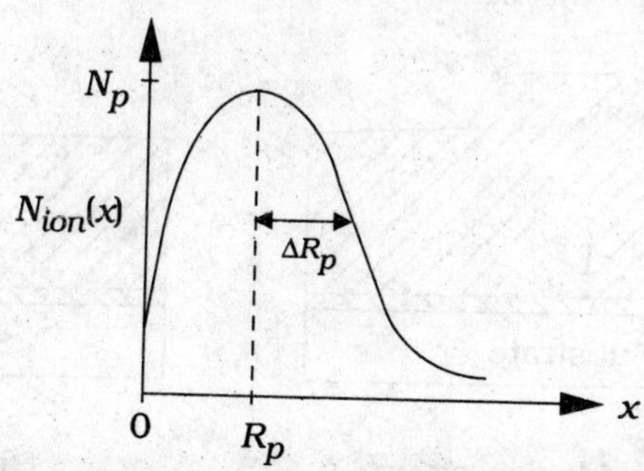

Figure 4.7 Gaussian implant profile

4.2.6 Chemical-Mechanical Polishing

Suppose that we deposit and pattern a polysilicon feature and then deposit silicon dioxide over it. As shown in Figure 4.8(a), the top surface of the deposited oxide would have a "hill" due to the underlying polysilicon line. If we deposit a metal interconnect layer on top, it will follow the surface contour, but may have to be wider and thicker to account for the variations. If we continue to add metal layers, the surface will get increasingly rough and may lead to breaks in fine line features and other problems. Non-planarized surfaces are not really a problem when only one or two metal interconnect lines are used. However, in modern CMOS processes where five or more interconnect layers are commonplace, techniques to planarize the surface have become mandatory.

Chemical-mechanical polishing (CMP) uses a combination of chemical etching and mechanical "sanding" to produce planar surfaces on silicon wafers. When applied to the oxide it results in a flat surface as portrayed in Figure 4.8(b). CMP steps are included at selective points in the CMOS fabrication sequence where it is important to have a flat working surface. This includes metal deposition steps, and the application of the photoresist used in the lithographic sequence discussed in the next section.

4.3 Lithography

We have defined an integrated circuit as a 3-dimensional set of patterned layers. One of the most critical problems in modern CMOS fabrication is the technique used to create a pattern on each layer with submicron features to a material layer. This is achieved using the process of **photolithography** where we optically project the shadow of the pattern onto the surface of the chip, and then employ photographic-type techniques to transfer the pattern to the surface. The same process is used to make

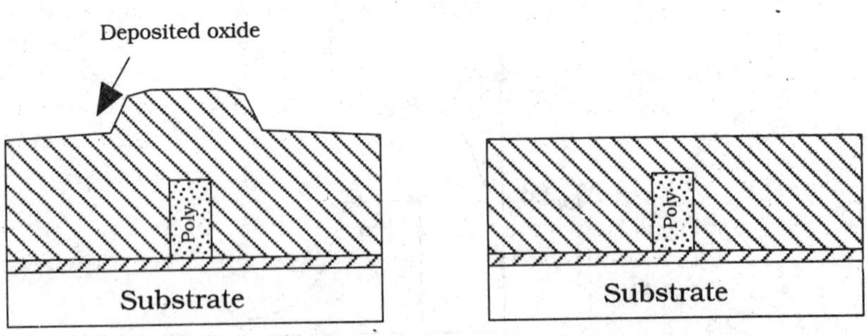

(a) After oxide deposition

(b) After CMP

Figure 4.8 Surface planarization

printed circuit boards, but chip fabrication allows resolutions smaller than 0.12 μm. Lithography has evolved into a complex discipline that has the responsibility of continuing to shrink feature sizes. The overview presented here is sufficient to understand the main points and its relationship to VLSI system design.

The photolithographic process starts with the desired pattern definition for a layer. This is in the form of a computer database file that is created during the chip layout phase of the design. The data is used to create a piece of high-quality glass that has the pattern defined using a metal such as chromium. This is called a **reticle** (or **mask**) and is typically about 5–10× the size of the actual chip. The reticle thus consists of two types of regions: transparent (no metal) and opaque (where there is metal). The components of a reticle are illustrated schematically in Figure 4.9. When light is used to illuminate the reticle, it projects the shadow of the reticle onto the surface of the chip.

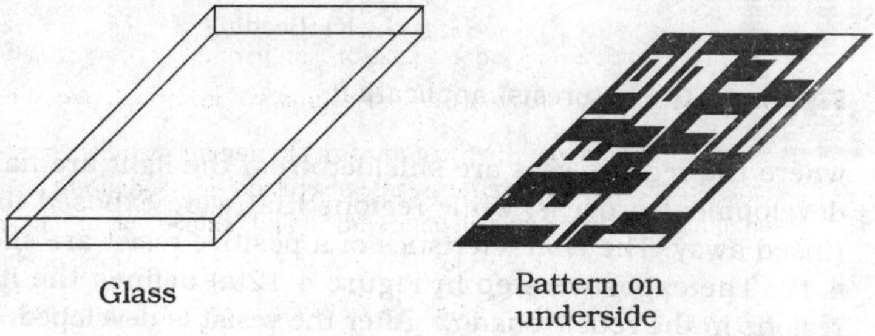

Figure 4.9 A reticle is a glass plate with a chromium pattern

To transfer the reticle pattern to the surface of a silicon region, we first coat the wafer with a light-sensitive liquid plastic material called **photoresist** (or simply **resist**). The process is shown schematically in Figure 4.10. Figure 4.10(a) depicts liquid photoresist being sprayed onto a spinning wafer that is held in place by a vacuum chuck. Spinning the wafer allows centrifugal force to coat the entire surface, which results in a reasonably uniform coating as in Figure 4.10(b). The exception is around the edges of the wafer, where surface tension causes a **beading effect** as illustrated in Figure 4.10(c). This restricts the useful region of the wafer to the interior portions away from the edges.

Resist acts much like photographic film in that it is sensitive to light. VLSI resists react to the ultraviolet (UV) region of the spectrum where the photon energies are highest and the wavelengths are the shortest. This is the **exposure step** of the lithographic process; Figure 4.11 shows the main idea. After exposure is completed, the photoresist layer is developed using a chemical rinse. Most VLSI processes use **positive photoresists**

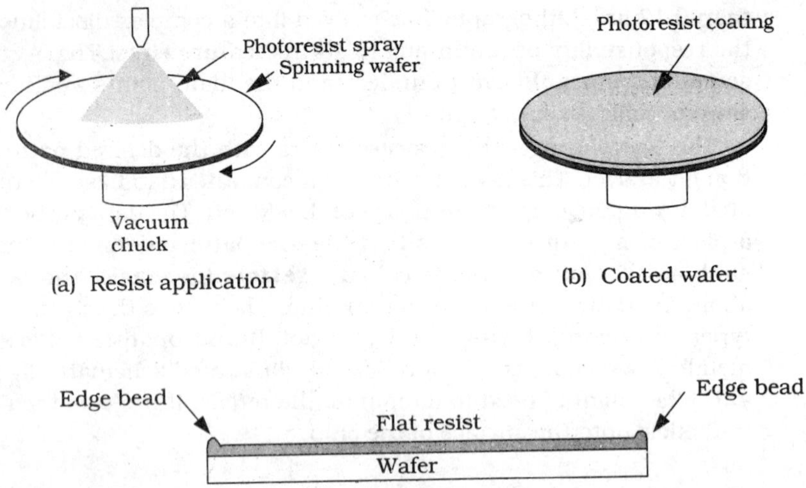

(a) Resist application (b) Coated wafer

(c) Beading

Figure 4.10 Photoresist application

where the regions that are shielded from the light are hardened in the development process, while regions that were exposed to the light are rinsed away. The characteristics of a positive resist are shown in Figure 4.12. The exposure step in Figure 4.12(a) defines the light and dark regions in the reticle shadow. After the resist is developed, hardened layers remain in the regions that were shielded from the light; this is illustrated in Figure 4.12(b). Negative photoresist has opposite characteristics: illuminated regions harden while shielded regions are soluble and are rinsed away.

The hardened resist layer is used to protect underlying regions from

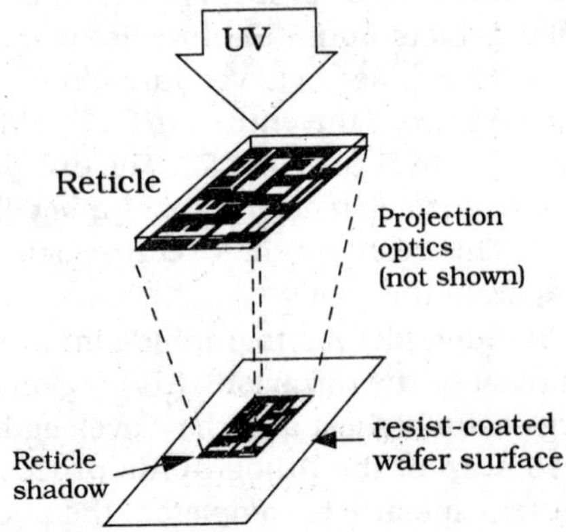

Figure 4.11 Exposure step

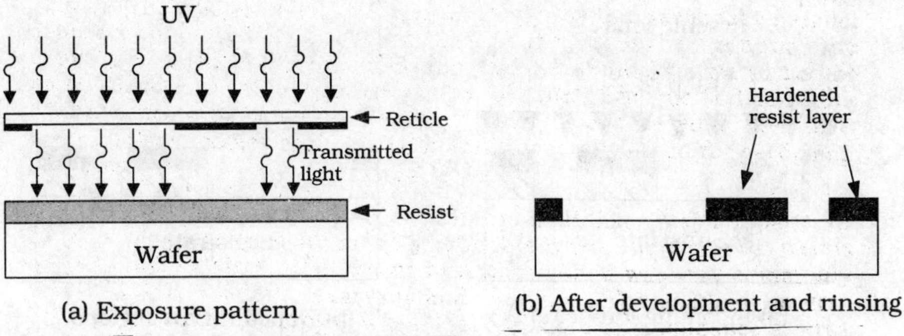

Figure 4.12 Characteristics of positive photoresist

the **etching** process. This is where the surface of the wafer is subjected to a gaseous plasma that is formed from an inert gas such as argon (Ar) and has reactant chemicals in it; overall, this is called a **reactive-ion etch** (RIE). The chemicals and plasma are chosen to attack and remove the material layer not shielded by the hardened photoresist. The resist itself can withstand the etchant mixture for the duration of the process. An example is shown in Figure 4.13. In Figure 4.13(a), a resist pattern is created on top of an oxide layer. The etching step removes oxide in the unprotected regions, so that the oxide has the same pattern as the resist; this is illustrated in Figure 4.13(b). This technique can be used to pattern any material layer above the wafer surface, including polysilicon, CVD oxides, and metals.[1] It allows us to transfer patterns from a computer layout design to the physical silicon level, thus creating the physical implementation of a logic network.

Doped silicon regions are also patterned using the lithographic process but the sequence is different. In this case, we grow an oxide layer on the wafer and then use lithography to etch down to the silicon surface; this is identical to the cross-section that was shown as Figure 4.13(b). The

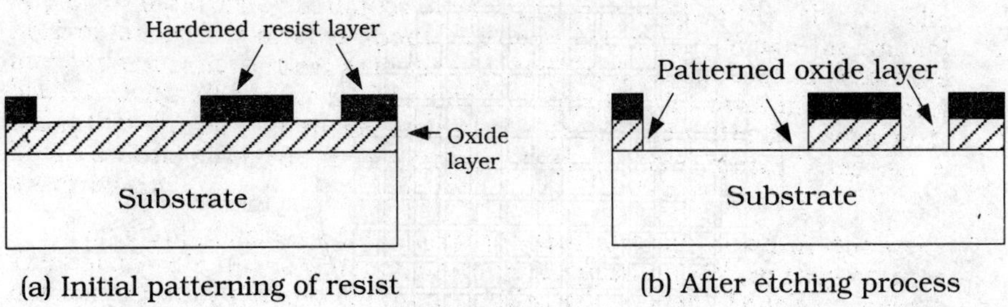

Figure 4.13 Etching of an oxide layer

[1] Copper is an exception as it is patterned using a different technique.

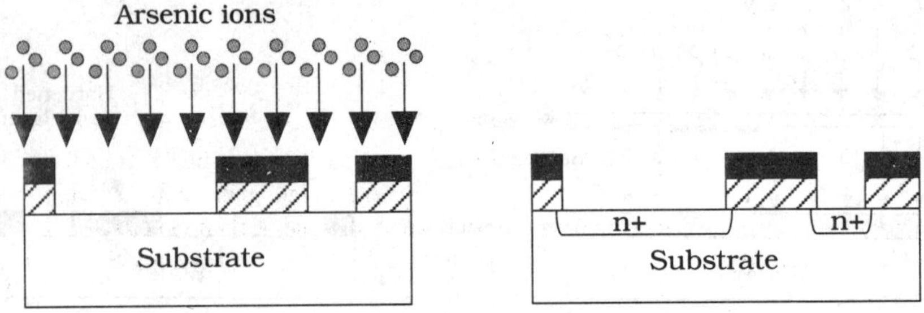

(a) Incoming ion beam (b) Doped n-type regions

Figure 4.14 Creation of doped silicon patterns

resist-oxide layers are then used to shield the silicon from an ion implantation step. Figure 4.14(a) shows that an incoming beam of arsenic ions covers the entire surface, but the dopants can enter the silicon only where the oxide has been etched away. The resulting n+ regions are thus defined by the oxide openings. Note that the widths of the n+ patterns are slightly larger than the oxide openings. This is due to an effect called **lateral doping** that arises from dopant diffusion during the annealing step. Lateral effects can limit the resolution of a narrow-line printing system.

Although we have shown only a single pattern in our examples, the manufacturing processes use larger wafers that accommodate many individual chip sites. Each site is individually exposed using a **step-and-repeat process**; a **wafer stepper** is an apparatus that holds the wafer and allows accurate movement to align the optics to each site, one at a time. After a site is exposed, the mechanism "steps" the wafer to the next site. This sequence produces a wafer with a large number of identical sites as illustrated in Figure 4.15. The **test site** locations contain various

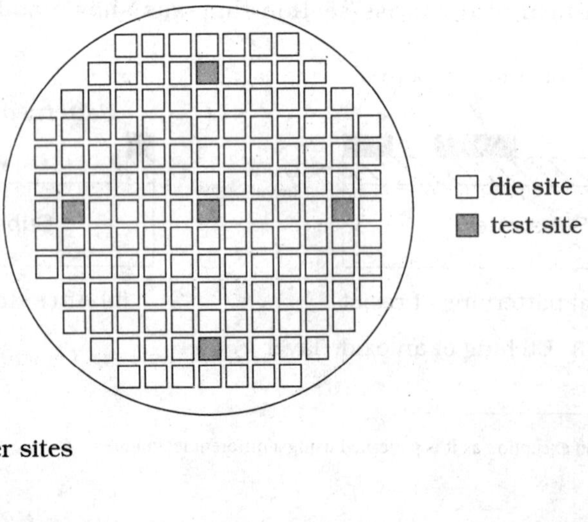

Figure 4.15 Wafer sites

test structures and circuits, such as MOS capacitors, doped regions of silicon, MOSFETs, and simple circuits. These are included to allow the wafer to be electrically tested during various phases of the manufacturing sequence. **Wafer probes** are sets of very small metallic probes that can contact regions on the wafer to allow these tests. The readings provide information on how well the manufacturing flow is progressing and also give critical data on electrical parameters needed for circuit design. It is important to include several test sites that are representative of all regions of the wafer, as nonuniform temperatures, gas flow densities, and other parameters vary across the wafer and affect the electrical characteristics.

The lithographic sequence is repeated for every masking step needed to build the integrated circuit. It is important to note that the first masking step defines the basic outline of the chip patterns; subsequent masking steps must pattern layers that have correct spacing relative to the features already created on the substrate. Correct alignment of a mask with the patterns on other masks is critical to the yield. Mask misalignment can cause the entire chip to be nonfunctional. Accurate alignment is achieved using **registration targets**, which are geometrical patterns that are created on a base layer solely to help align later masking steps. As the layers build, more sets of registration marks are required.

4.3.1 Clean Rooms

The lithographic process is very sensitive to dust particles. If a speck of dust lands on the photoresist, it will interfere with the exposure and development and may lead to a defect. Similarly, if a dust particle lands on the reticle in the focal plane of the optics, it will be imaged down to the wafer site. Events such as these decrease the yield, and are especially critical in submicron geometries.

Many procedures have been developed to deal with these problems. Lithography is performed in a **clean room** environment that uses HEPA (high-efficiency particulate air) filters to remove dust particles. HEPA filters must be able to be 99.97% effective in removing particles with diameters of 0.5 μm or larger. A **Class X** clean room means that there are less than X particles per cubic foot with diameters greater than 0.5 microns; modern facilities have a Class 1 or better rating in critical work areas. To insure this level of cleanliness, workers must take air showers and wear special suits that cover all parts of the body before entering the area; these are generically referred to as "bunny suits" because of their appearance. Alternately, the entire flow may be automated and all movement performed by robots.

Lithographic areas are lighted by yellow light since it does not affect the UV-sensitive photoresist. To keep dust particles on the reticle from ruining the image, a thin layer of transparent plastic is placed above the reticle to catch dust and keep it off of the reticle surface. This is called a

pellicle, and is placed far enough above the reticle to keep the dust out of the image plane of the projection optics.

Many other features of the processing environment are included to insure that functional chips can be produced. Many scientists, engineers, and technicians are required to design, maintain, and update the processing areas. Touring an advanced chip fabrication facility is usually an overwhelming show of VLSI technology.

4.4 The CMOS Process Flow

Modern CMOS processing is, by all definitions, a "technological marvel." Starting literally with sand, the manufacturing line produces tiny rectangular slices that provide the computing power for the world. Semiconductor manufacturing companies have developed highly advanced processing techniques, and the details of their process flows are highly proprietary. Since a new manufacturing plant costs in excess of a billion dollars, it is no wonder that companies must remain secretive.

In this section we will study the main steps in a "standard" silicon CMOS process. The level of presentation has been chosen to insure that the main points are discussed without going into excessive details. Understanding CMOS processing is important to every VLSI designer, some more so than others. It depends on the task that the engineer is currently involved with. Device and circuit engineers view processing parameters as the fundamental limit to how fast their transistors and circuits can switch. The system architect understands that logic blocks need to be created in silicon, and that the processing dictates area allocations, interconnect levels, delays, clock speeds, and dozens of other system-level considerations. Everyone involved in the design of a VLSI chip is affected.

The initial steps are illustrated in Figure 4.16. It should be noted that the features, especially in the vertical directions, are not drawn to scale as this would obscure some of the important details. The starting point in Figure 4.16(a) is a p+ wafer with a thin p-type **epitaxial layer** of silicon grown on top. The epitaxial layer is created by dropping silicon atoms onto a heated wafer to form a high-quality crystal layer for transistors. The wafer itself acts as the substrate for building the chip, and is not shown explicitly in any of the remaining drawings.

The next step shown in Figure 4.16(b) is the formation of n-well regions using a masking step. This defines the locations of pFETs. In general, every transistor (nFET or pFET) is built in an **active area** of the wafer surface. Active areas are defined by a masking step that patterns a layer of silicon nitride that rests on a thin layer of thermal oxide that is used to relieve the mechanical stress of the crystal surface. Figure 4.16(c) shows the details after the patterning. Active areas are introduced as part of the **electrical isolation** scheme that prevents electrical conduction between

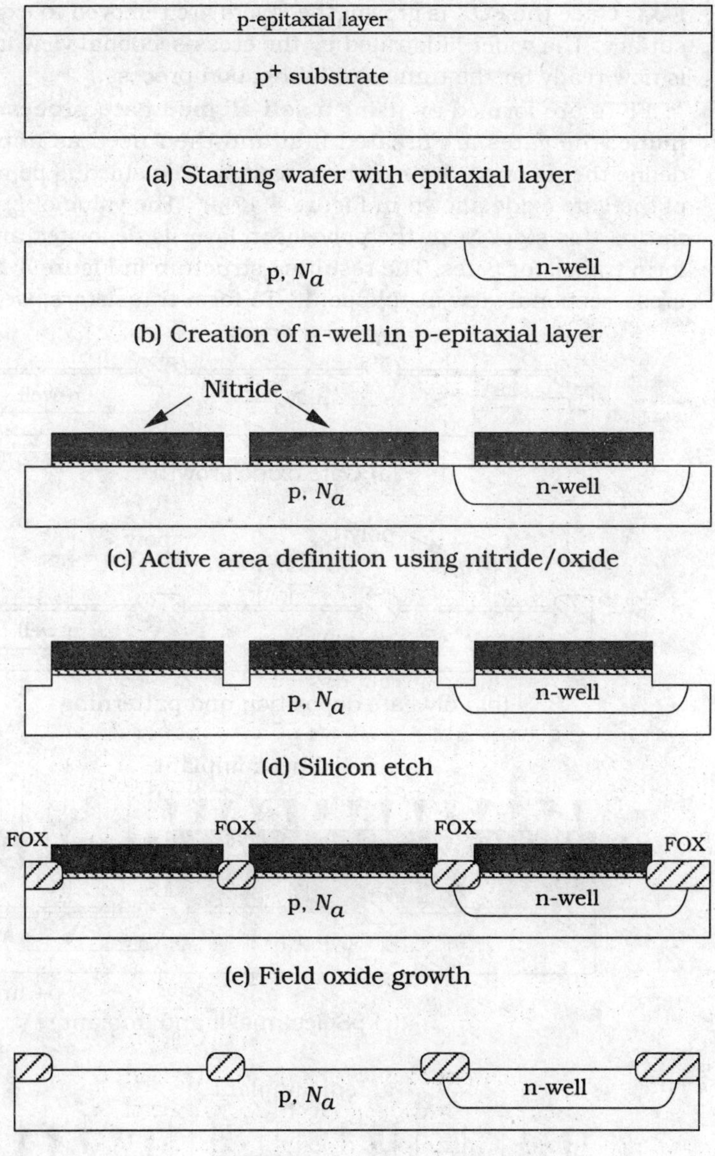

(a) Starting wafer with epitaxial layer

(b) Creation of n-well in p-epitaxial layer

(c) Active area definition using nitride/oxide

(d) Silicon etch

(e) Field oxide growth

(f) Surface preparation

Figure 4.16 Initial sequences in the CMOS fabrication sequence

neighboring devices using recessed regions of glass (oxide) as an insulator. To achieve isolation, the nitride pattern is used to define silicon etched regions shown in Figure 4.16(d). Oxide is then grown or deposited in the etched regions as in Figure 4.16(e). Glass insulation between active areas defines the **field** regions, and the oxide there is called **field oxide** or

FOX. Once the FOX is grown, the layers are removed to expose the silicon surface. The wafer illustrated by the cross-sectional view in Figure 4.16(f) is now ready for the transistor fabrication process.

FETs are formed by using a **self-aligned gate process**. In this technique, the gates are created first and then used as implant masks to define the n+ or p+ drain/source regions. The starting point is the growth of the gate oxide shown in Figure 4.17(a). The value of t_{ox} is established during this step. Next, the polysilicon layer is deposited and patterned to form transistor gates. The resulting structure in Figure 4.17(b) shows the cross-sectional view at this point. To form transistors, we need to create

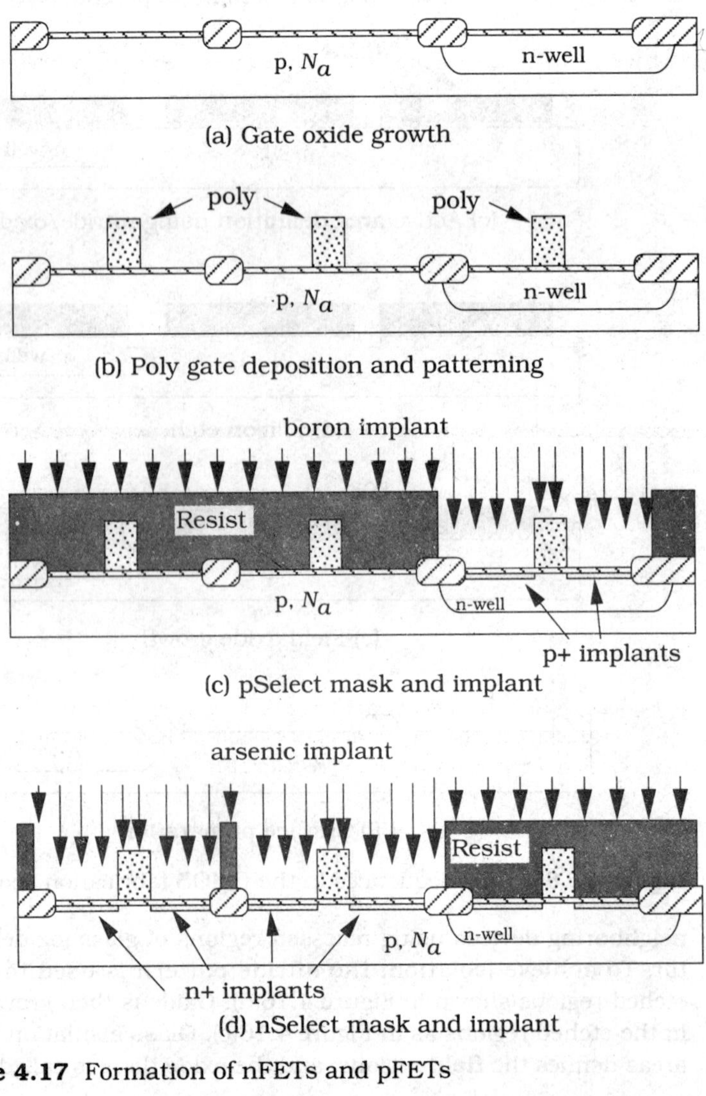

Figure 4.17 Formation of nFETs and pFETs

doped drain and source regions in the silicon. A pFET is created using a **pSelect** masking pattern with a boron ion implant. As illustrated in Figure 4.17(c), the pSelect mask creates a hardened photoresist layer that blocks the implant over nFET locations, but allows the ion beam to hit the pFET region. Ions are absorbed by the gate poly layer, but easily make it through the thin oxide layer to reach the silicon. The term "self-aligned gate" arises from this step. nFETs are formed in a similar manner. An **nSelect** mask is used to block an n-type ion implant from reaching pFET sites. Ions are permitted to bombard nFET locations, creating the n+ regions shown in Figure 4.17(d). At this point, all transistors have been built. Silicided gates can be created by layering a refractory metal on the poly. This lowers the sheet resistance of the poly lines. The remaining steps in the process flow are used to create interconnect layers.

The basic sequence for adding interconnect layers is illustrated in Figure 4.18 for the first layer of metal. CVD oxide is used to coat the surface as in Figure 4.18(a). Electrical contact with the n+ and p+ regions is established by etching holes in the oxide using an **Active Contact** mask. After the cuts are made, they are filled with a metal plug material such as tungsten (W). The resulting structure is shown in Figure 4.18(b). The first layer of metal is deposited and patterned with a Metal1 mask. This mask defines the first level of metal interconnect used to wire the circuit together. The drawing in Figure 4.18(c) illustrates the final view after the first metal has been patterned. Additional layers of metal are added in the same manner. Current processing lines have 5 or more metal interconnect layers (separated by oxide) to aid in complex wiring.

After all of the metal layers have been added, the entire chip is covered with the overglass layer that protects the surface from external contaminants. Silicon nitride is the most common overglass material since it is a dense dielectric that prevents diffusion of unwanted atoms and has good adhesion to metals. It is an insulator, so a via must be etched to gain electrical access to the chip; this requires another masking step. The simplest way to interface the silicon circuitry with the outside world is to use a **pad frame** arrangement where large metal **bonding pads** surround the central chip core area. Wires are attached between the pads and the output pins on the package. Figure 4.19 illustrates the basic idea. The top view in Figure 4.19(a) shows the metal pad (solid line) and an overglass cut (dashed line). The bonding pad itself may be quite large, with 100 μm × 100 μm being used in some processes. The side view in Figure 4.19(b) shows the details of the bond itself. A robotic apparatus is used to place the bond on the pad accurately and string a wire from the chip to the specified pin on the package frame.

4.4.1 Variations

Modern CMOS processing lines use a large number of enhancements to the basic flow described above. These are usually included to provide

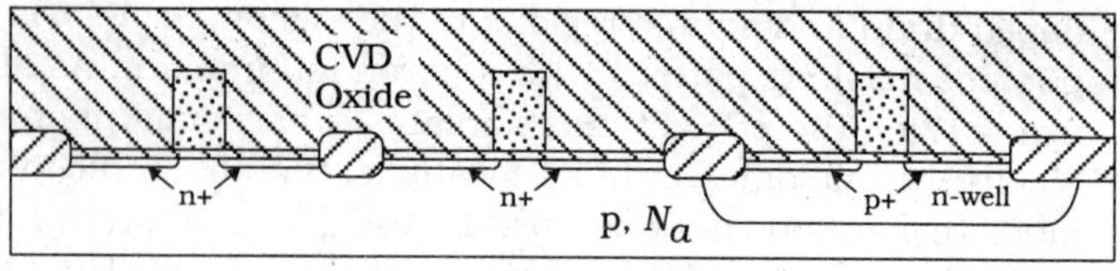

(a) After anneal and CVD oxide

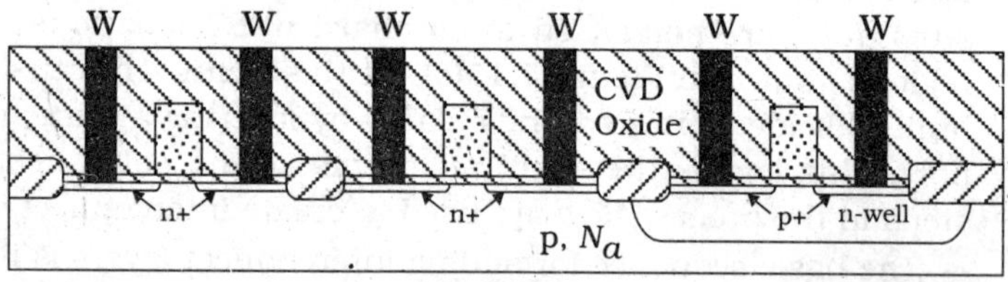

(b) After CVD oxide active contact, W plugs

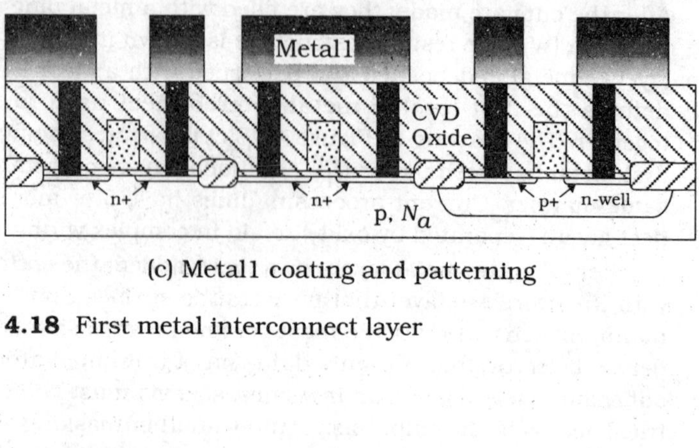

(c) Metal1 coating and patterning

Figure 4.18 First metal interconnect layer

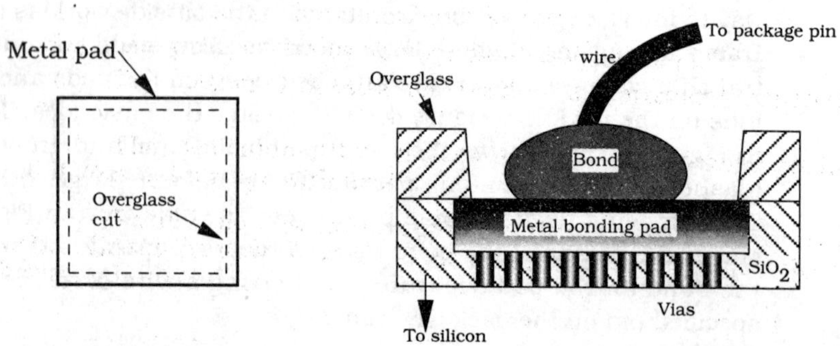

(a) Top view

(b) Side view

Figure 4.19 Bonding pad structure

better electrical characteristics, combat small device or high-density problems, or to enhance the yield. We will examine two additional steps that are now standard: lightly doped drain (LDD) FETs and silicides. In addition, we will take a brief look at how copper interconnect patterns are created.

A lightly doped drain MOSFET is designed to reduce the electric fields in the channel region by providing n- (lightly doped) drain and source regions instead of the usual n+ regions. Theoretically, this reduces the maximum electric field intensity, which in turn increases the reliability of the transistors.[2] LDD structures can be created without an additional mask, so that their presence is usually transparent to the layout designer.

A sequence for creating an LDD FET is portrayed in Figure 4.20. The starting point is shown in Figure 4.20(a). To create an n-channel MOSFET, we start with a low-dose donor doping to create the n- (lightly doped) drain and source regions shown. The next step [illustrated in Figure 4.20(b)] is to deposit an oxide layer over the surface. Note that an oxide layer coats the side (vertical) wall of the poly gate feature. After this, the wafer is subjected to an oxide etching step. When viewed from the top, the sidewall oxides are thicker than the oxide covering the flat portions of the surface. This results in the **sidewall spacers** shown in the drawing of Figure 4.20(c). The spacers are used to block the heavy n+ donor implant in Figure 4.20(d), which keeps the drain and source regions closest to the channel at lightly doped levels. The lateral (horizontal) width of the spacers determines the extent of the n-regions.

Figure 4.21(a) provides an expanded view of a finished LDD nFET with the details of the doped regions shown. This can be used as the basis for studying silicides, which is the second variation from the basic CMOS flow that we will examine. Even heavily doped polysilicon exhibits a sheet resistance of about 25 Ω or more, limiting its use as an interconnect material. To overcome this problem, a refractory metal such as titanium or platinum can be coated over silicon or polysilicon as in Figure 4.21(b). The resulting silicide reduces the sheet resistance of the poly layer without affecting the electrical characteristics of the MOS gate structure; a typical order of magnitude for a silicided poly is $R_s \approx 10$ mΩ. The drain-source n+ silicides reduce the contact resistance when a tungsten plug is used as an active contact. Owing to this fact, silicides have become very common in high-frequency processes. We note in passing that neither Pt nor W by themselves can be used to replace the polysilicon gate (and form a true MOS structure) as they do not adhere to the silicon dioxide insulating layer, but simply "slide off."

The last variation that we will examine is the use of copper (Cu) as an

[2] This is accomplished by using LDD FET to reduce the **hot-electron effects** found in short-channel devices.

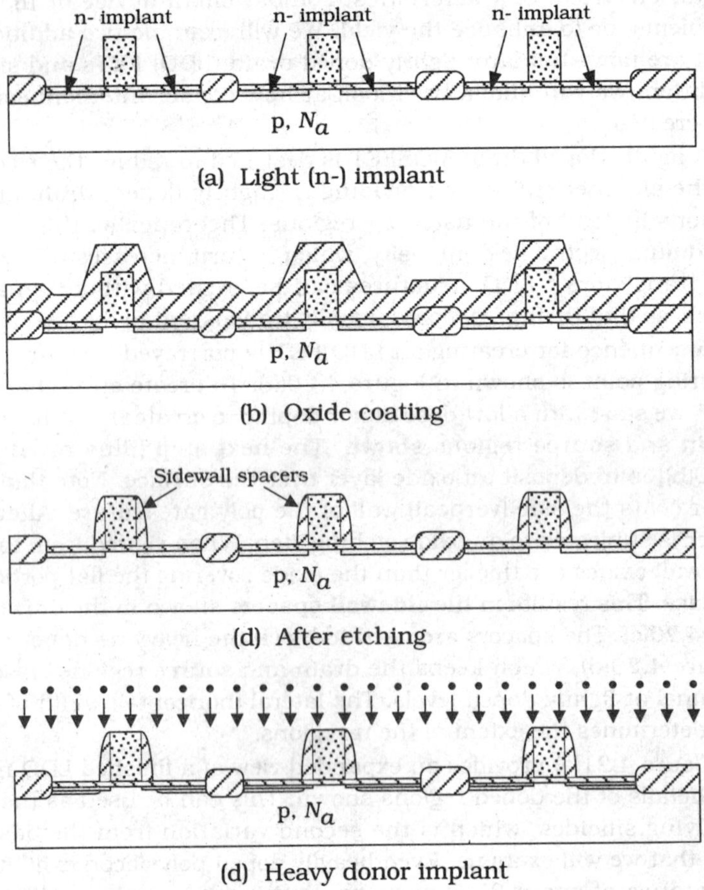

Figure 4.20 Sequence for creating a lightly doped drain nFET

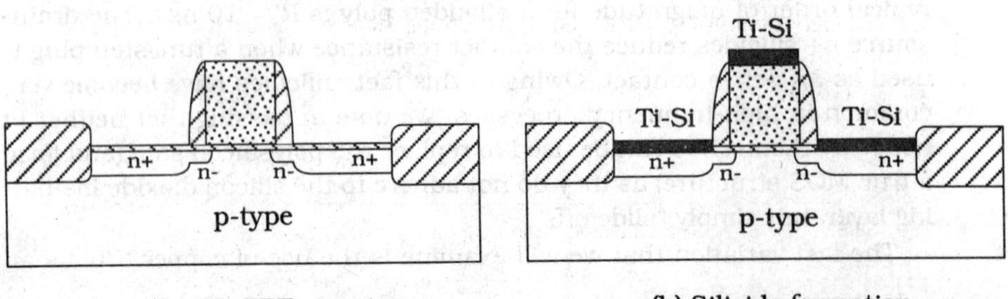

Figure 4.21 LDD nFET with silicided gate and contacts

interconnect material instead of aluminum. It is a well-known fact that the bulk resistivity of copper is $\rho = 1.67\ \mu\Omega\text{-cm}$, which is about one-half that of Al. When used as an interconnect line material, the sheet resistance would be about one-half that of an aluminum line with the same thickness. Copper, however, has proved difficult to introduce into the processing line. It cannot be patterned using the standard sequence of deposition followed by a lithographic step because it is very difficult to etch using standard RIE techniques. Copper diffuses very rapidly through silicon and can alter the electrical characteristics, so it cannot be directly deposited on top of any silicon regions. It also diffuses through silicon dioxide, making the problem even more difficult to deal with. Much research has been directed toward the development of techniques to replace aluminum with low-resistivity interconnect metals. At the present time, copper is being introduced into the majority of new high-speed CMOS lines, making it of interest to the VLSI designer. One of the first VLSI chips to use copper technology was an advanced generation Power PC microprocessor design.

Let us first examine how copper patterns are produced. As mentioned above, dry-etching techniques do not etch copper. Even trace amounts of copper from Al-Cu mixtures are difficult to remove from a chip surface. To get around this problem, we use the **Damascene** process based on the method used in ancient times to inlay gold or silver into an iron sword. The name is taken from the city of Damascus whose artisans were well known for their work. In this technique, the copper pattern is first etched into a silicon dioxide layer; copper is then deposited (using, for example, electroplating) on the surface. The sequence is shown in Figures 4.22(a) and (b). To avoid the etching problem, we subject the

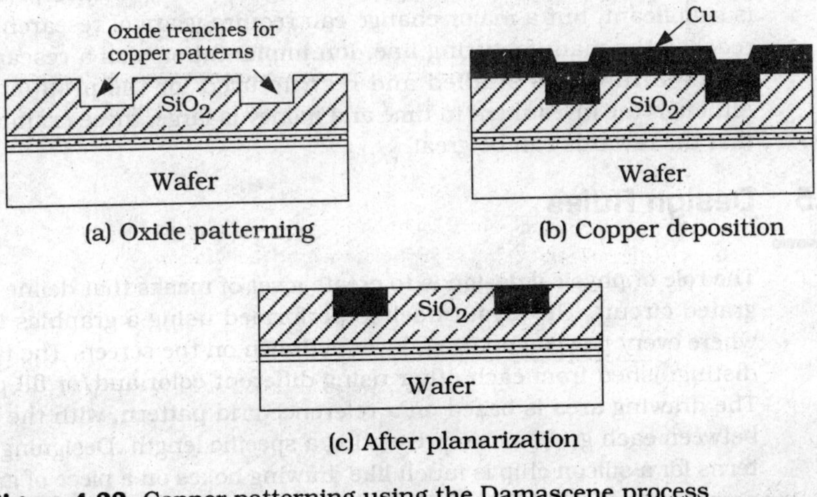

Figure 4.22 Copper patterning using the Damascene process

wafer to a chemical-mechanical polishing (CMP) step that planarizes the surface and removes copper not in a oxide trench. This results in the structure shown in Figure 4.22(c).

Dual-Damascene processes that allow copper vias to be created have also been developed. The basic sequence is the same, except that two oxide etch steps are used to give the general structure portrayed in Figure 4.23. Copper vias have a lower resistance than tungsten, and also avoid the contact resistance introduced by a standard Al-W interface.

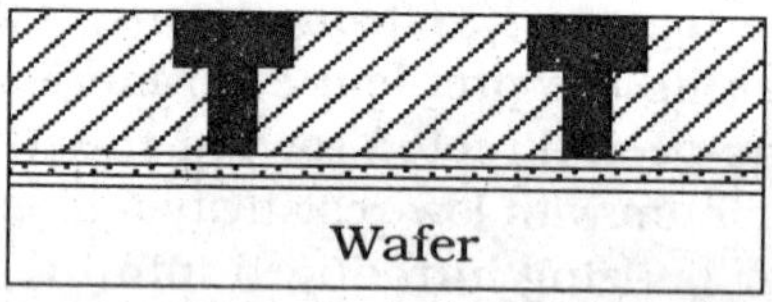

Figure 4.23 Dual-Damascene structure with copper vias

The other major problem with using copper as an interconnect material is the need to prevent it from diffusing into silicon regions. This is achieved by using thin barrier layers to contain the material. Copper has relatively poor adhesion properties, so providing a barrier layer around a copper feature can help. Various barrier materials have been tried with results published in the literature. Included in this group are W, Ti, TiN, Ta, TaN, and TaN_x (where x is the mole fraction of nitrogen). The choice of materials affects the resistivity and sheet resistance of the interconnects, so many trade-offs exist. Moreover, reliability issues and long-term effects still need to be studied in more detail.

The fabrication problems introduced by using copper illustrate the complexity of a modern silicon processing line. Even a small perturbation is significant, but a major change can require years of research before it reaches the manufacturing line. Ion implantation was a research technique until it was studied and restructured for the production line. Although the investment in time and money is large, these examples show that the rewards can be great.

4.5 Design Rules

The role of physical design is to create a set of masks that define the integrated circuit. The layout itself is performed using a graphics CAD tool where every polygon on every layer is drawn on the screen. The layers are distinguished from each other using different color and/or fill patterns. The drawing area is based on a reference grid pattern, with the distance between each grid point representing a specific length. Designing the patterns for a silicon chip is much like drawing boxes on a piece of grid paper using a set of colored pencils. However, just because we can draw something does not mean that it can be fabricated. Every piece of fabrication equipment used in the IC manufacturing process has limited accuracy. A lithographic stepper unit that is designed to image linewidths of 0.25 μm will not operate at 0.18 μm. The same is true for an etching system. Physical limitations at the silicon level also restrict what can be fabricated in the microworld of silicon circuitry.

Topological **design rules** (DRs) are a set of geometrical specifications that dictate the design of the layout masks. A design rule set provides numerical values for minimum dimensions, line spacings, and other geometrical quantities that are derived from the limits of a specific processing line. The design rules must be followed to insure functional structures on the fabricated chip. An example of a design rule specification is shown in Figure 4.24 for the case of two closely spaced polysilicon lines. This drawing is used to show the two parameters

w_p = minimum width of a polysilicon line

$s_{p\text{-}p}$ = minimum poly-to-poly spacing

These are given numerical values in the DR listing; violating these values may lead to a failure. Every layer in the process will have similar quantities assign to it. In our notation,

w = minimum width specification

s = minimum spacing value

d = generic minimum distance

with subscript used to denote the relevant layers. For example,

w_{m1} = minimum width of a metal1 line

$s_{m1\text{-}m1}$ = minimum spacing between metal1 lines

This convention makes it easy to understand each rule as it is introduced and used in layout drawings. In practice, layers are numbered and design rules are usually assigned identifiers that are associated with the layer numbers.

All design rule specifications such as w and s have units of length, with the micron (μm) being the most common metric. For example, a process might specify polysilicon features with minimum width and spacing

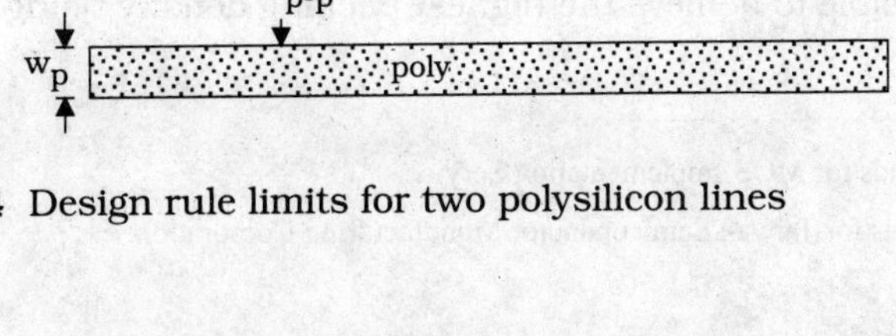

Figure 4.24 Design rule limits for two polysilicon lines

values of

$$w_p = 0.25\ \mu m,\ s_{p\text{-}p} = 0.425\ \mu m$$

The layout grid in the CAD system is usually calibrated to accommodate the necessary resolution. These values are obtained by a careful analysis of relevant parts of the manufacturing line and vary with the process. A set of topological design rules may require a document listing of 100 pages or more, and takes some time to learn. This level of detail is necessary to produce chips with the highest possible packing density.

Design rules change with technological advances, so we have made the decision not to include a specific set in this text. In the United States, the government-sponsored MOSIS group provides foundry access to universities and small companies.[3] The reader is directed to the MOSIS web site at **www.mosis.org** where up-to-date sets can be viewed and downloaded for immediate usage. In this spirit, our discussion will be kept general in nature.

The popularity of modern VLSI applications has introduced the concept of the **silicon foundry**. A foundry provides access to a chip manufacturing process on a pay-by-use basis. On a more global scale, foundries such as TSMC are used by large corporations down to well-funded individuals to create their designs.[4] A foundry allows designers to submit designs using a state-of-the-art process. Since the customer base is wide and varied, most foundry operations allow the submission of designs using a simpler set of design rules that can be easily scaled to different processes. These are called **lambda design rules**.

Lambda design rules are based on a reference metric λ that has units of μm. All widths, spacings, and distances are written in the form

$$\text{Value} = m\lambda \tag{4.21}$$

where m is scaling multipler. For example, we might stipulate that $w = 2\lambda$ and $s = 3\lambda$ for the minimum width and spacing on a layer. The numerical values of w and s are not known until λ itself is specified. If $\lambda = 0.15\ \mu m$, then these would specify that

$$w = 2(0.15) = 0.30\ \mu m$$
$$s = 3(0.15) = 0.45\ \mu m$$

for the design. If the layout is based on a λ-grid, then submitting the design to a different process just means that the numerical value of λ must be changed. The relative dimensions remain the same. The main drawback of using a **scalable** design rule set of this type is that it is not possible to achieve the highest packing density using integer values of m.

3 MOSIS stands for MOS Implementation Service.

4 TSMC stands for Taiwan Semiconductor Manufacturing Corporation.

Design rules can be classified into four main types: minimum width, minimum spacing, surround, and extension. We have already seen minimum width and minimum spacing examples. A surround rule is enforced when a feature must be placed inside of an existing feature on the chip surface. An extension rule is similar in that it requires that portion of the pattern be extended beyond the edge of an existing border.

Let us consider the placement of an active contact as an example of a surround rule. As shown in Figure 4.25(a), the oxide contact cut must be aligned so that it is over the existing (active) n+ region. The corresponding design rule is shown in Figure 4.25(b). The surround spacing $s_{a\text{-}ac}$ between the active area (n+) and active contact edge must be maintained to guard against a misaligned contact cut pattern during the lithographic exposure step.

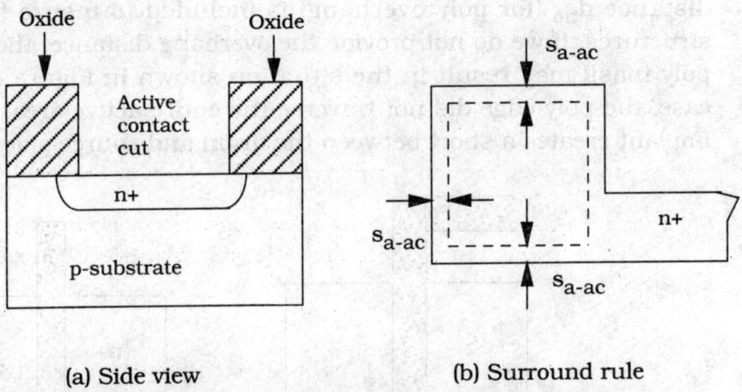

Figure 4.25 Example of a surround design rule

Misalignment problems must be included in the design rule set because it is not possible to project the reticle image to the chip surface with an arbitrary degree of accuracy. The registration marks are in the form of geometrical target patterns on some layers during the processing. The targets are used to align several subsequent patterning steps. When an opaque material layer is deposited, a new set of marks must be introduced. Surround rules are included to compensate for the alignment tolerance of the stepper.

Figure 4.26 illustrates the potential problem with the active contact. Suppose that the contact cut is not aligned to fall within the n+ active region as seen in Figure 4.26(a). After the contact is made and the metal plug added, the cross-sectional view in Figure 4.26(b) shows the existence of a metal-substrate short. This will render the chip nonfunctional.

Extension-type design rules also tend to be based on misalignment problems. Consider the formation of a self-aligned nFET as an example.

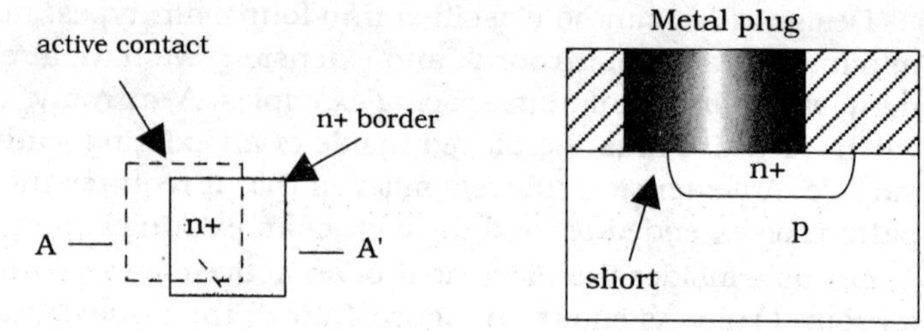

(a) Top view (b) Side view along A-A'

Figure 4.26 Misalignment-induced defect

The polysilicon gate is used as a dopant mask for the n-type ion implant that defines the drain and source regions. In Figure 4.27(a), the extension distance d_{po} (for poly overhang) is included to insure functional FET structures. If we do not provide the overhang distance, then a misaligned poly mask may result in the situation shown in Figure 4.27(b). In this case, the poly edge did not traverse the entire active area, so that the ion implant creates a short between the drain and source sides.

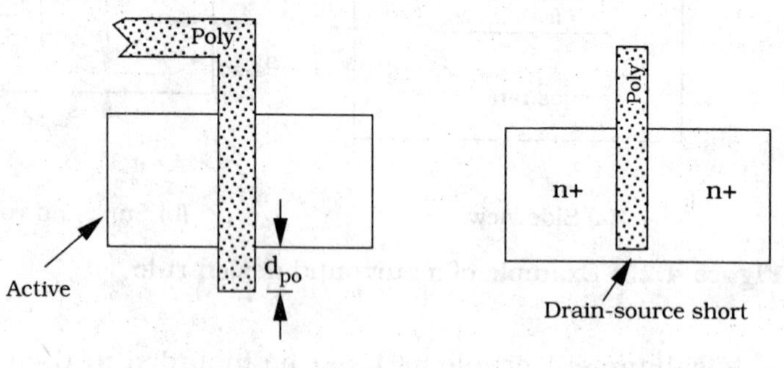

(a) Gate overhang DR (b) Misalignment failure

Figure 4.27 Example of an extend (gate overhang) design rule

4.5.1 Physical Limitations

Some geometrical design rules originate from physical considerations. These enter into the formulation of the design rule set, and may or may not be obvious.

An important aspect is the linewidth limitation of an imaging system. The reticle shadow projected to the surface of the photoresist does not have sharp edges due to optical diffraction. As a simple rule of thumb, a lightwave with an optical wavelength of λ cannot accurately image a feature size much less than that value. UV-sensitive positive photoresists are used because the short wavelengths of the ultraviolet light allow for better resolution of fine linewidths, and positive resists have better development properties than negative resists. In addition, the structure of a reticle is much more complicated than we have alluded to; advanced optical techniques such as phase-shifting structures are used to enhance the resolution.

The etching process introduces another type of problem. When we remove material around a resist edge, both vertical (perpendicular to the wafer surface) and lateral (parallel to the surface) etching occurs. We can characterize the respective etch rates of the two by r_{vert} [μm/min] and r_{lat} [μm/min] and define the **degree of anisotropy** A by

$$A = 1 - \frac{r_{lat}}{r_{vert}} \tag{4.22}$$

The presence of lateral etching in r_{vert} limits the resolution that can be achieved. Figure 4.28(a) shows an oxide layer that is to be patterned by the resist layer on top of it. A pure anisotropic etch profile is shown in Figure 4.28(b). This is characterized by $r_{lat} = 0$ which gives vertical walls and $A = 1$. The result of a pure isotropic etch with $r_{lat} = r_{vert}$ is shown in Figure 4.28(c). Undercutting of the resist due to the lateral etching decreases the resolution that can be used in the design. Another factor that enters the problem is the absorption profile of light by the resist layer itself; this results in the resist edges having finite slopes instead of well-defined vertical shapes.

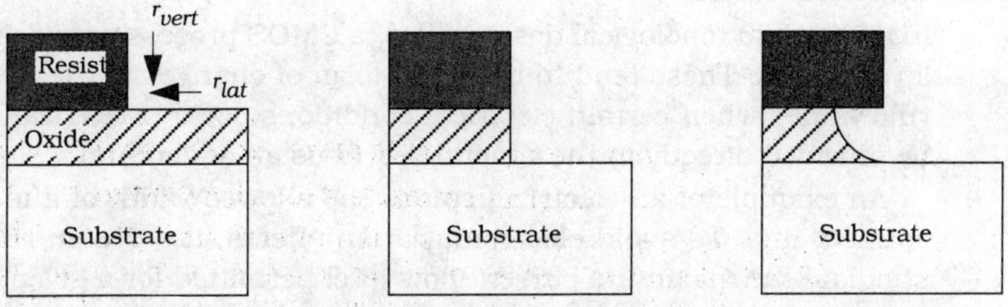

(a) Resist pattern (b) Pure anisotropic etch (c) Isotropic etch

Figure 4.28 Etching profiles

Semiconductor effects in silicon also influence the formulation of design rules. Any time a pn junction is formed it gives rise to what is known as a **depletion region** at the interface. By definition, the depletion region is "depleted" of free electrons and holes because of an electric field that originates from the dopants and forces the charges out. If the depletion regions of adjacent pn junctions touch, then the current blocking characteristics are altered and current can flow between the two. This limits the spacing rule $s_{n\text{-}n}$ shown in Figure 4.29. The drawing also

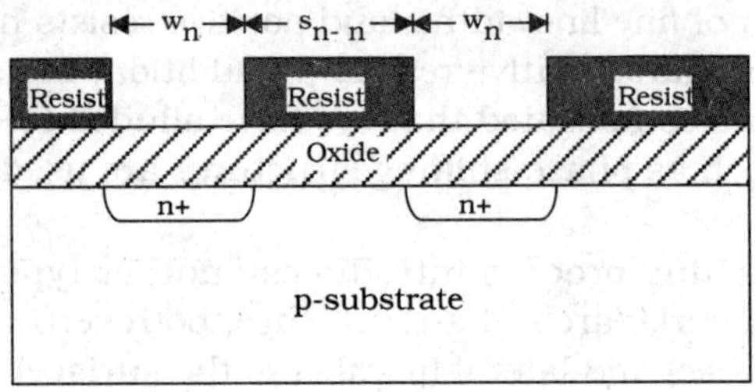

Figure 4.29 Limits on n+ spacings

shows that the minimum linewidth parameter w_n must account for lateral doping and isotropic etching effects.

Another physical problem is the electrical capacitive coupling that occurs between closely spaced conducting lines. This leads to a problem called crosstalk in which a portion of the electrical energy from one line is coupled to another, causing an unwanted perturbation called "noise." This can lead to errors, and is a major problem in high-density designs. Memory chips are particularly sensitive to induced noise of this type. Crosstalk considerations can lead to design rule spacing values that are much larger than the minimum value that can be achieved by the lithography. The problem of crosstalk itself is treated in more detail in Chapter 14.

4.5.2 Electrical Rules

In addition to topological design rules, a CMOS process provides electrical layout rules. These tend to be in the form of changes to the basic design rule values when certain electrical conditions occur. Electrical rules may be provided directly in the general set, or as an addendum.

An example of an electrical rule is the allowed width of a metal interconnect line. To avoid electromigration effects, the design rule set will stipulate the maximum current flow level permitted for a given linewidth. Larger currents require wider lines.

4.6 Further Reading

[1] Stephen A. Campbell, **The Science and Engineering of Microelectronic Fabrication**, Oxford University Press, New York, 1996.

[2] C.Y. Chang and S.M. Sze, **ULSI Technology**, McGraw-Hill, New York, 1996.

[3] James D. Plummer, Michael Deal, and Peter B. Griffin, **Silicon VLSI Technology**, Prentice Hall, Upper Saddle River, NJ, 2000.

Elements of Physical Design

5

In the previous chapter we examined the basic fabrication sequence for manufacturing CMOS integrated circuits. In this chapter we will study the details of translating logic circuits into silicon, which is called **physical design**. Details such as the minimum size specifications allowed for a patterned region become critical. However, the most important lessons in the physical design of VLSI chips revolve around the use of CAD tools and database structures that describe the silicon masks. These give the needed information for creating the chip, and provide the basis for the hierarchical design of large complex logic networks.

5.1 Basic Concepts

Physical design is the actual process of creating circuits on silicon. During this phase of the VLSI design process, schematic diagrams are carefully translated into sets of geometric patterns that are used to define the on-chip physical structures. Every layer in the CMOS fabrication sequence is defined by a distinct pattern. A patterned layer consists of a group of geometrical objects that are generically referred to as **polygons**. This naturally includes rectangles and squares, but allows us to include arbitrarily complex *n*-vertex shapes with specific dimensions. Examples of the types of polygons that occur in a CMOS design are shown in Figure 5.1 where several are superposed to form the overall layout. When stacked into three-dimensional structures, the layers are electrically equivalent to the circuit diagram.

Our study to this point shows that the **topology** of the transistor network establishes the logic function. In other words, the details of how the FETs are wired together (series, parallel, etc.) are sufficient to determine the binary operations of the circuit. Another aspect of logic is that of

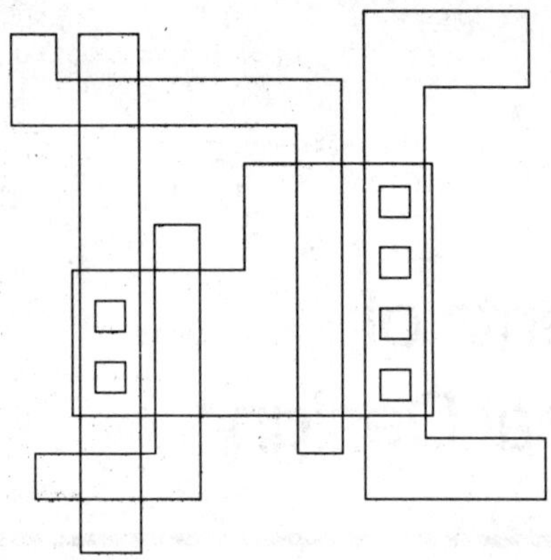

Figure 5.1 Examples of polygons in physical design

switching speed. This is more complicated to analyze, but is crucially important to modern chip design. Although the details will be discussed later, we can summarize the main points here. For a given set of processing parameters, we will find that the electrical characteristics of a logic gate depend on the aspect ratios of the transistors. This is due to both the current flow levels and the parasitic resistance and capacitance of the devices. Physical design must address both of these areas. The patterns must be created to correctly implement the signal flow network, as was discussed in Chapter 3. The complicating factor is that the dimensions of every feature affect the electrical performance of the circuit. In a VLSI chip, the switching speed of some gates will be critical, especially those in long complex logic paths. We will concentrate on studying the basics of circuit layout in this chapter, and delve into the intricacies of high-speed circuits in Part 2 of the book.

The process of physical design is performed using a computer tool called a **layout editor**. This is a graphics program that allows the designer to specify the shape, dimensions, and placement of every polygon on every layer of the chip. Complexity issues are attacked by first designing simple gates and storing their descriptive files in a **library** subdirectory or folder; the gates constitute **cells** in the library. Library cells are used as building blocks by creating copies of the basic cells to construct a larger more complex circuit. This process is called **instancing** of the cells, while a copy of a cell is called an **instance**.

The layout must be an accurate representation of the logic network, but much of the designer's work is directed toward the goal of obtaining a fast circuit in the minimum amount of area. Small changes in the shapes or areas of a polygon will affect the resulting electrical characteristics of the circuit. However, the changes may or may not be significant for the logic chain. An experienced layout designer gains a certain level of intuition that often helps find trouble spots. Circuit simulations also help insure that the layout is accurate and provides a network that meets specifications.

5.1.1 CAD Toolsets

Physical design is based on the use of CAD tools that simplify the procedure and aid in the verification process. The most powerful toolsets are collections of programs that are combined together into an integrated suite environment. There are several packages available, each with its own strengths.

Let us examine what constitutes a basic chip design toolset by listing some of its features. For the physical design process the primary tool is the layout editor described above. This is a graphical interface to a database that allows the user to draw transistors and wiring patterns made up of polygons. Each layer has a distinct color or fill pattern on the screen. The overlap of the boxes or polygons on each layer becomes our view of the transistor. The layout editor creates a database for each layer that describes the patterning on a universal grid. This is eventually used to create the masks needed to pattern the layers in the fabrication sequence.

After a layout is completed, we must run several secondary programs that use database information to determine if our layout is valid. The electrical behavior of the design is simulated by first using an **extraction** routine that translates the polygon patterns and layers into an equivalent electrical network. The output of an extraction routine is a netlist file that can be used in a circuit simulation program; SPICE formats are the most common. Extraction programs provide important geometrical parameters such as the drawn channel width and length for each FET. They also specify how the transistors are wired together. Process-dependent electrical parameters are added to the extract output file to form a complete basis for simulation. Circuit simulation codes such as SPICE are usually included within the toolset (or within a related subdirectory) for easy access. This allows the designer to immediately perform simulations as needed.

A related program that is usually included in the design environment is called **layout versus schematic**, or **LVS** for short. As implied by its name, this program checks the layout against the schematic diagram. This is important to verify that the layout corresponds to the intended circuit. LVS can be performed using either logic diagrams or electronic circuit schematics.

The **design rule checker (DRC)** is a program that uses the layout database and checks every occurrence of the design rule list on the layout. This means, for example, that the width and spacing of every metal line in the layout are checked to insure that they do not violate the minimum specified values. Passing a DRC insures that the design can be fabricated within the limitations of the manufacturing process.

Other tools are provided to help in large designs. **Place and route** routines help the layout designer by automatically finding viable wiring routes between two specified points. This is useful when trying to connect two complex units together. Electrical continuity can be seen using an **electrical rule checker (ERC)** which highlights connecting paths.

This short description of a chip design environment provides us with a starting point for a more detailed discussion of the layout and design of VLSI silicon networks. Our approach will be to stress the fundamental ideas and procedures without going into the details of using any specific set of CAD tools. Once the techniques are understood and mastered, they may be applied to any environment.

5.2 Layout of Basic Structures

Let us start with the sequences that are used to define chip regions. We will base our discussion on the p-substrate (n-well) technology described in Chapter 4.[1] The masking sequence was established as

0. Start with p-type substrate
1. nWell
2. Active
3. Poly
4. pSelect
5. nSelect
6. Active contact
7. Poly contact
8. Metal1
9. Via
10. Metal2
11. Overglass

It is important to remember that oxides are grown or deposited between conducting layers above the substrate. The details needed for chip layout vary with the sequence. However, only minor modifications are needed to extend the ideas here to arbitrary processing lines.

In this section we will study how to design basic structures on the chip such as n+ and p+ regions and MOSFETs using the basic masking sequence. Relevant design rules are introduced for each structure. It is worth remembering that the features on every level have design rule specifications for the minimum width w of a line, and a minimum edge-to-edge spacing s between adjacent polygons. These are illustrated in Figure 5.2.

[1] These techniques are quite general, and may be easily extended to other technologies with minor changes.

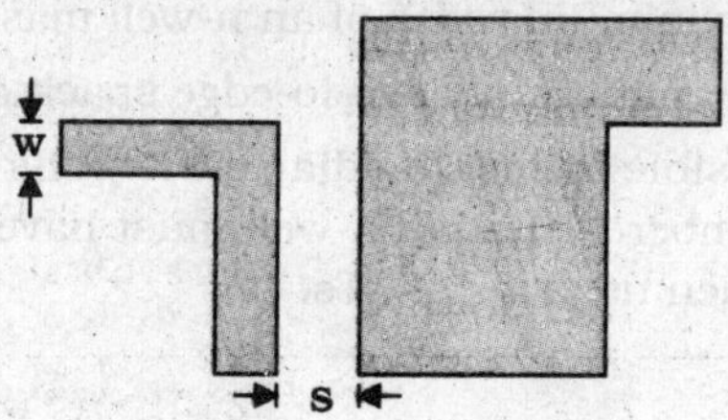

Figure 5.2 Minimum line width and spacing

The actual values for w and s depend on the layer. Design rules apply only to the features on the *mask* (a generic term for the reticle) for that layer. The actual fabricated structure on the chip will have different dimensions For that reason, we will sometimes refer to the layout sizes as the **drawn** values, while the resulting sizes on the finished chip have **effective** or **final** values. This is particularly important in designing FETs.

Our discussion will consider only **Manhattan** geometries where all turns are multiples of 90°. Right-angle layout is the most straightforward to learn, but does not always give the best packing density. Many layout editors allow you to select the angles in an arbitrary manner, but you must be sure that the structures are supported by the fabrication process.

As mentioned in the previous chapter, our discussion will be generic in nature. Detailed and up-do-date sets of design rules for various processes can be obtained from the MOSIS web site at **www.mosis.org**.

5.2.1 n-Wells

An n-well is required at every location where a pFET is to be made. We define these using the nWell mask on which closed polygons represent the placement of the wells. Figure 5.3(a) shows the cross-sectional view of two adjacent n-well regions. The polygons in Figure 5.3(b) constitute the mask set for this part of the chip. The drawing illustrates the two design rules

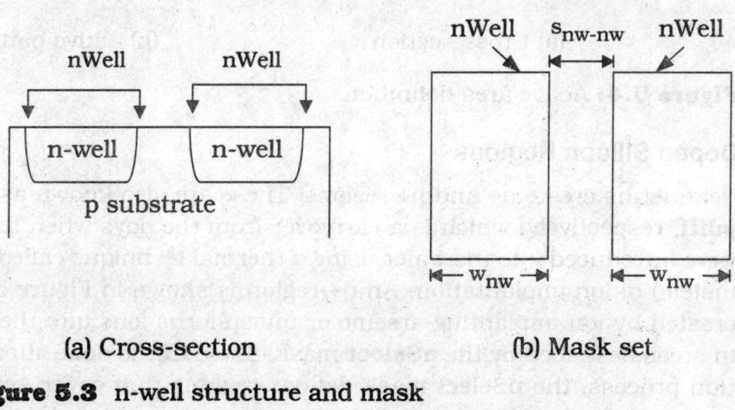

Figure 5.3 n-well structure and mask

w_{nw} = minimum width of an n-well mask feature

$s_{nw\text{-}nw}$ = minimum edge-to-edge spacing of adjacent n-wells

It is often possible to merge adjacent n-wells together into one. Also, it must be remembered that an n-well must have a connection to the power supply V_{DD} when used for pFETs.

5.2.2 Active Areas

Silicon devices are built on active areas of the substrate. Figure 5.4(a) illustrates the cross-sectional view of an active section. After the isolation (field) oxide is grown, an active area is flat and provides access to the top of the silicon wafer. The field oxide (FOX) exists everywhere else on the wafer. Active regions are defined by closed polygons on the **Active** mask. The set of polygons required to define the patterns in Figure 5.4(a) is shown in Figure 5.4(b). The relevant design rule spacings are denoted by

w_a = minimum width of an Active feature

$s_{a\text{-}a}$ = minimum edge-to-edge spacing of Active mask polygons

These are minimum values that must be observed in maximum density designs. The field oxide regions can be derived from the Active mask by the expression

$$\text{FOX} = \text{NOT (Active)} \tag{5.1}$$

This is a symbolic expression that is based on the observation

$$\text{FOX} + \text{Active} = \text{Surface} \tag{5.2}$$

In other words, if a region is not Active, then it is FOX by default.

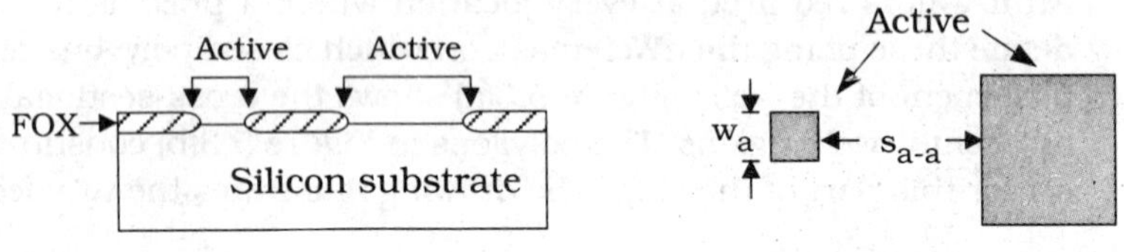

(a) Cross-section (b) Active patterns

Figure 5.4 Active area definition

5.2.3 Doped Silicon Regions

Next, let us create n+ and p+ regions. These are also known as **ndiff** and **pdiff**, respectively, which is a carryover from the days when the dopants were introduced into the wafer using a thermal technique called **diffusion** instead of ion implantation. An n+ region is shown in Figure 5.5(a). It is created by ion implanting arsenic or phosphorus ions into the substrate in areas described by the **nSelect** mask. Since this is done after the isolation process, the nSelect mask defines regions that cover Active areas.

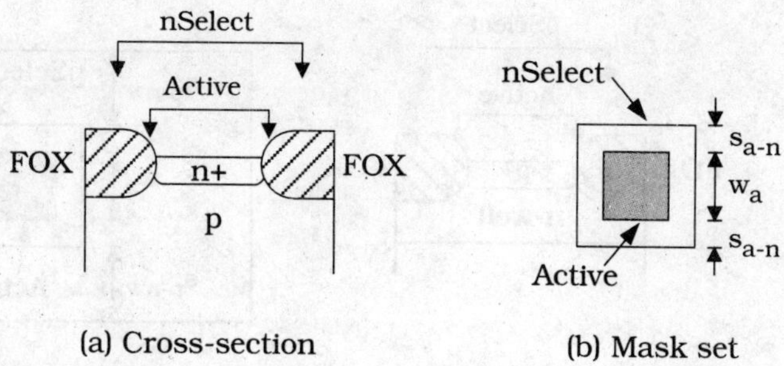

Figure 5.5 Design of n+ regions

The mask set shown in Figure 5.5(b) shows that both the nSelect and Active areas are needed to create the n+ region. Let us use the notation (Mask_name) to imply the set of all polygons on that layer. If only the nSelect and Active masks are included, we may express an n+ region by[2]

$$n+ = (\text{ nSelect }) \cap (\text{ Active }) \tag{5.3}$$

This says that n+ regions are created whenever the Active and nSelect masks intersect, as denoted by the intersection operator $\cap$. Two design rules are illustrated in the drawing. These are

w_a = minimum width of an Active area

s_{a-n} = minimum Active-to-nSelect spacing

where it is implied that spacing distances are measured from edge to edge; this convention will be followed throughout the book. Design rules are usually invariant with respect to direction, so that the same values also apply to the horizontal dimensions.

A p+ region is obtained by ion implanting boron into an active area opening on the wafer. The cross-sectional view in Figure 5.6(a) shows that p+ regions are created in n-well areas in this technology. The active region is made p-type by using the implant defined by the **pSelect** mask. The required mask set is shown in Figure 5.6(b) where the **nWell** mask has been included for completeness. The expression for a p+ region is

$$p+ = (\text{ pSelect }) \cap (\text{ Active }) \cap (\text{ nWell }) \tag{5.4}$$

when only these three masks are considered. This says that p+ regions are created whenever regions on the pSelect and Active masks overlap within an nWell region. The important design rule spacings are shown as

w_a = minimum Active area width

s_{a-p} = minimum Active-to-pSelect spacing

---

2 In the Magic layout editor, **ndiff** and **pdiff** are drawn by a single command so that separate Active and nSelect patterns are not necessary. Note, however, that there are no such things as ndiff and pdiff masks in the process.

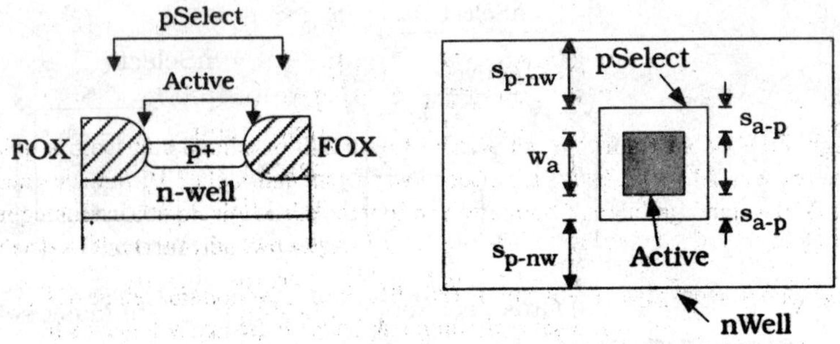

(a) Cross-section (b) Mask set

Figure 5.6 Design of a p+ region

$s_{p\text{-}nw}$ = minimum pSelect-to-nWell spacing

Again, these are specified by the process design rule set.

5.2.4 MOSFETs

Self-aligned MOSFET structures exist every time a poly gate line completely crosses an n+ or p+ region. Physically, the poly line is deposited before the ion implant, and acts to block dopants from entering the silicon. FETs thus require the use of polygons on the **Poly** mask layer. The basic design rules for Poly features are

w_p = minimum poly width

$s_{p\text{-}p}$ = minimum poly-to-poly spacing

The minimum poly linewidth w_p is the same as the drawn channel length for a FET.

Let us construct an nFET first. In Figure 5.7(a) the cross-sectional view shows the n+ and poly layers; the gate oxide between the gate and substrate is not shown explicitly. The top view in Figure 5.7(b) shows the drawn values of the channel length L and the channel width W of the transistor. To construct the mask set, we just add a polygon to the Poly mask that separates the n+ into two regions. This results in the masks shown in Figure 5.8. The implied design rule is

$L = w_p$ = minimum width of a Poly line

The other design rule shown is

d_{po} = minimum extension of Poly beyond Active

which is required to insure the formation of the self-aligned FET if a small registration error occurs in the lithography. This is known as the **gate overhang** distance. Using the figure allows us to write the definition of the central part of the nFET as

$$\text{nFET} = (\text{nSelect}) \cap (\text{Active}) \cap (\text{Poly}) \qquad (5.5)$$

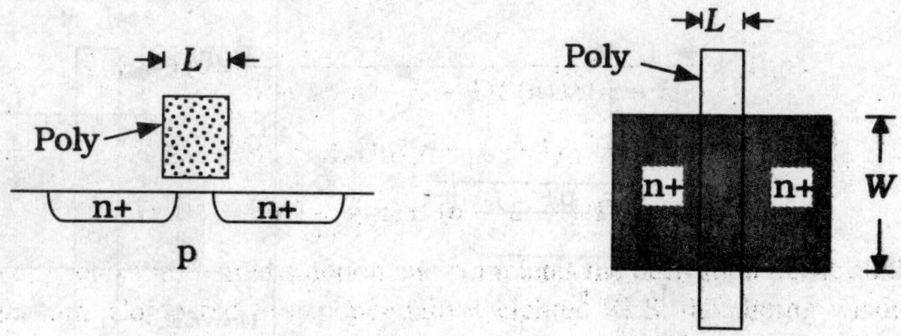

(a) Cross-section (b) Layout view

Figure 5.7 nFET structure

since this is where the channel is formed. The n+ regions are defined by

$$n+ = (\text{ nSelect }) \cap (\text{ Active }) \cap (\text{ NOT [Poly] }) \tag{5.6}$$

This is more precise than the limited definition given previously in equation (5.3) which ignored the existence of the Poly mask.

A pFET is created in the same manner. Figure 5.9(a) shows the cross-sectional view of the device, while the top view in Figure 5.9(b) provides the important channel dimensions L and W as drawn on the masks. It is worth mentioning that the n-well region is surrounded by implied p-substrate; this is shown explicitly in the top-view drawing. The pFET mask set shown in Figure 5.10 has the same basic features as the nFET group. The differences are only in the polarity of the implant (pSelect instead of nSelect) and the presence of nWell surrounding the transistor. The drawn channel length L corresponds to the minimum Poly linewidth, while d_{po} is the gate overhang design rule. The other design rules implied by the drawing have already been discussed. A simple expression for the central pFET region is

$$pFET = (\text{ pSelect }) \cap (\text{ Active }) \cap (\text{ Poly}) \cap (\text{ nWell}) \tag{5.7}$$

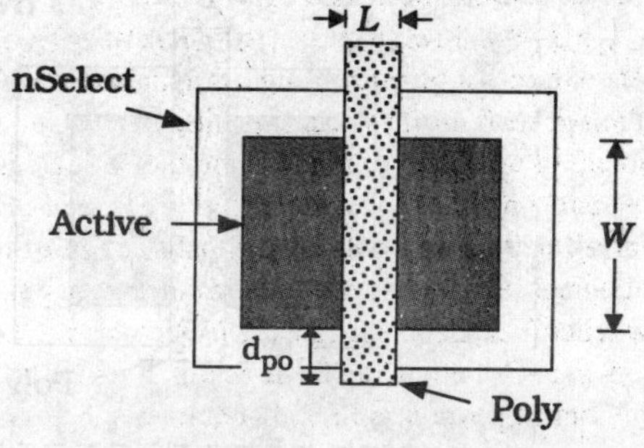

Figure 5.8 Masks for the nFET

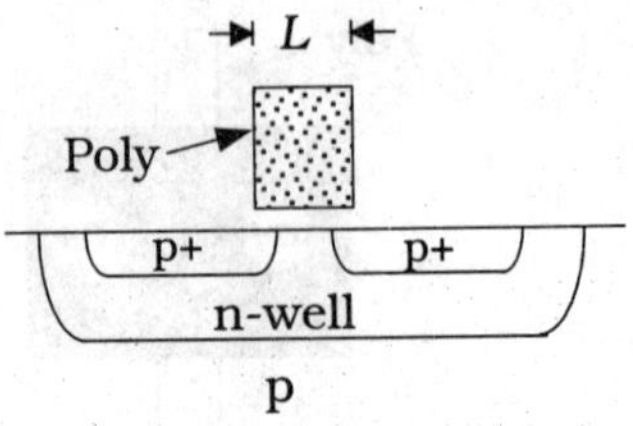

(a) Cross-section

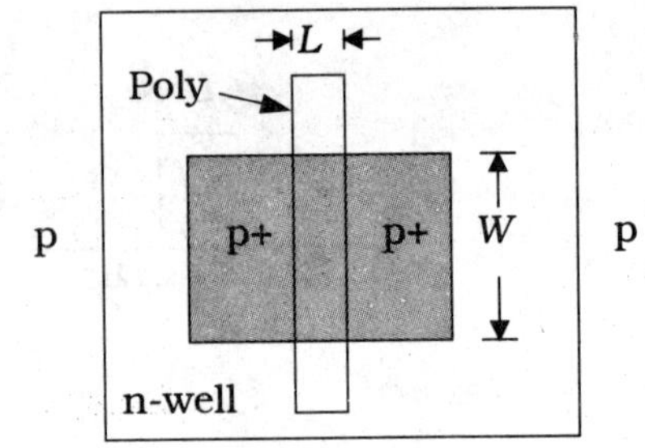

(b) Layout view

Figure 5.9 pFET structure

which expresses the device as the overlap of the four masks. A p+ region is then described by

$$p+ = (\text{ pSelect }) \cap (\text{ Active }) \cap (\text{ nWell }) \cap (\text{ NOT [Poly] }) \tag{5.8}$$

i.e., sections in the device where no poly has been created. This is more precise than the simpler expression in equation (5.4) which ignored the Poly mask.

Drawn and Effective Values in MOSFETs

The critical dimensions of a MOSFET are the channel length L and the channel width W. As we have seen, L is established by the width of the poly gate line. Tracing the fabrication sequence shows that the channel width W is set by the appropriate edge measurement of the Active transistor area, since that region defines where the drain/source ion implant penetrates into the silicon. As mentioned in the previous chapter, design rules dictate the mask layout and represent the drawn dimensions. The final values measured on the chip will be slightly different. The exact relationship is particularly important in the electrical analysis of transistors.

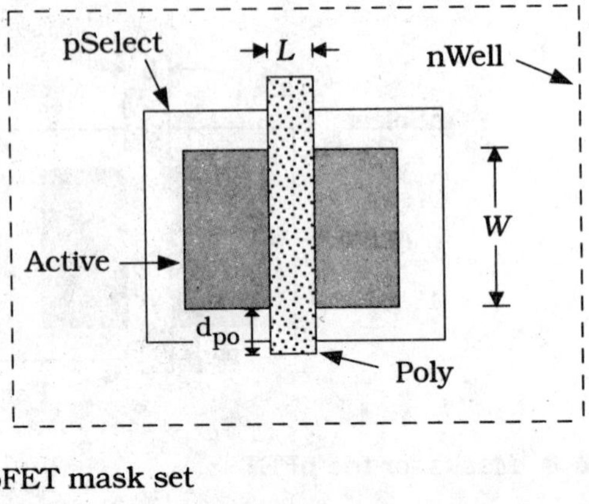

Figure 5.10 pFET mask set

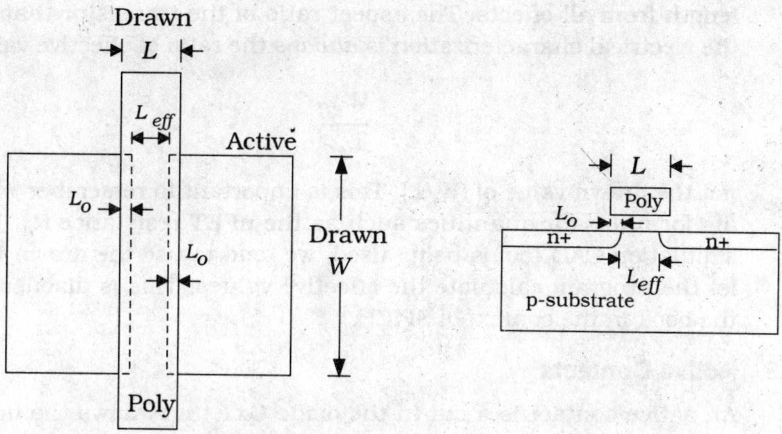

(a) Drawn layout (b) Finished view

Figure 5.11 Drawn and effective dimensions of a MOSFET

Let us examine the FET layout geometry shown in Figure 5.11(a); this general layout applies to both nFETs and pFETs. Consider first the channel length of the device. The drawn value L is the width of the polysilicon line. However, the distance between the n+ regions in the final structure is smaller than L due to lateral doping during the implant annealing step. When the wafer is heated, the dopants on opposite sides move toward one another. The overlap effect is symmetrical and results in the overlap distance L_o on both sides. In the electrical analysis of the transistor, the important distance is the final value between the two n+ regions. When a distinction needs to be made, the final value is referred to as the **electrical** or **effective channel length**. Denoting the effective value by L_{eff} we see that

$$L_{eff} = L - 2L_o \tag{5.9}$$

gives the numerical value. A more general form is

$$L_{eff} = L - \Delta L \tag{5.10}$$

where ΔL is the total reduction in channel length due to overlap and other effects.

The channel width is also smaller than the drawn value due to a reduction of active area by the field oxide growth. This is called active area **encroachment**, and leads to an effective channel width of the form

$$W_{eff} = W - \Delta W \tag{5.11}$$

where W is the drawn value and ΔW is the total reduction in channel length from all effects. The aspect ratio of the transistor that is used in the electrical characterization is *always* the ratio of effective values

$$\frac{W_{eff}}{L_{eff}} \tag{5.12}$$

not the drawn value of (W/L). This is important to remember when applying formulas for quantities such as the nFET resistance R_n. If a circuit simulation CAD tool is being used, we tend to use the drawn values and let the program calculate the effective values. This is discussed later in the book in the context of SPICE.

5.2.5 Active Contacts

An active contact is a cut in the oxide Ox1 that allows the first layer of metal to contact an active n+ or p+ region. This is shown in the cross-sectional view of Figure 5.12(a). These are defined by the **Active Contact** Mask with the general overlay shown in Figure 5.12(b). Since the contact is placed to fall inside of an n+ or p+ region, it is subject to the surround design rule

$s_{a\text{-}ac}$ = minimum spacing between Active and Active Contact

The dimensions of the contact are given by

$d_{ac,\,v}$ = vertical size of the contact

$d_{ac,\,h}$ = horizontal size of the contact

which are exact specifications. A square contact is obtained if

$$d_{ac,\,v} = d_{ac,\,v} = d_{ac} \tag{5.13}$$

but it is not uncommon to have aspect ratios other than 1:1.

5.2.6 Metal1

Metal1 is applied to the wafer after the Ox1 oxide. It is used as interconnect for signals and also for power supply distribution. Figure 5.13(a) shows the cross-sectional view of a first-layer metal line with an active

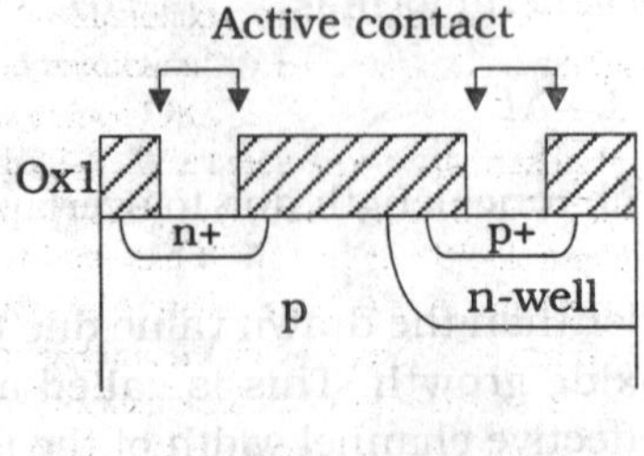

(a) Cross-section

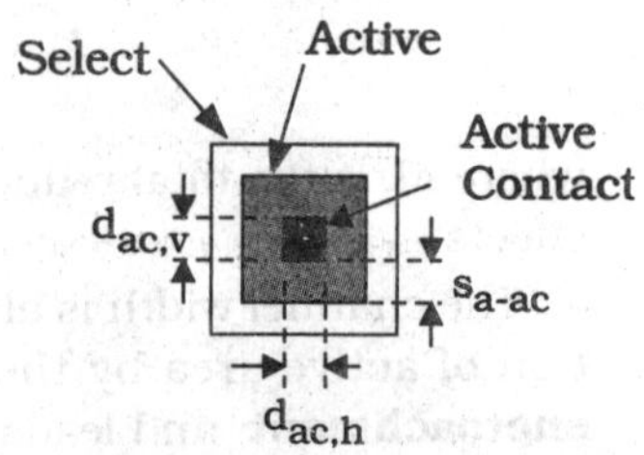

(b) General mask set

Figure 5.12 Active contact formation

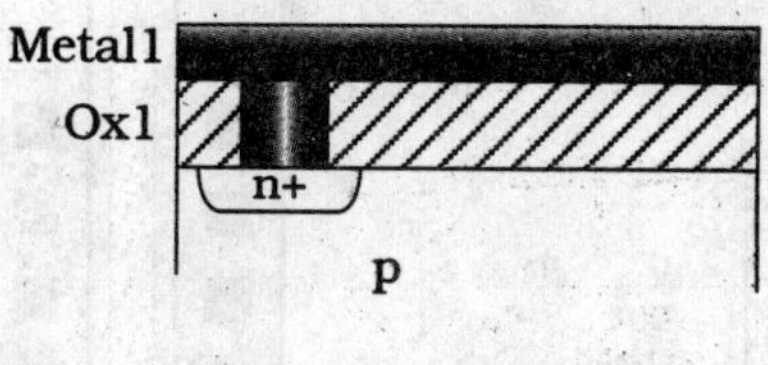

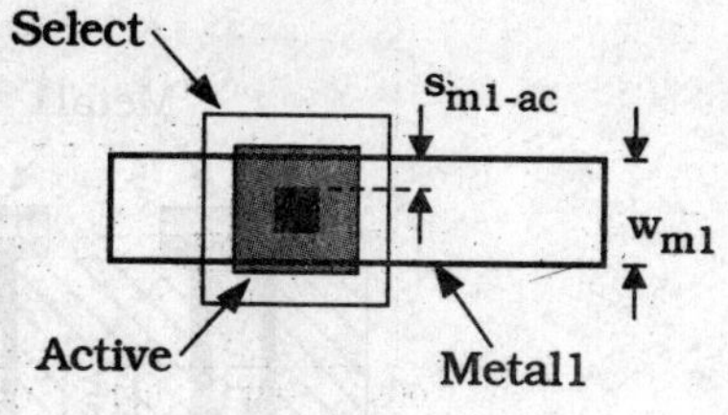

Figure 5.13 Metal1 line with Active Contact

contact to an n+ region. The contact cut through the oxide has been filled with a plug as described in the previous chapter. The mask set for this arrangement is drawn in Figure 5.13(b) with the **Metal1** mask feature overlapping the Active Contact to attain the electrical connection. The two design rules indicated in the drawing are

w_{m1} = minimum width of a Metal1 line

and

$s_{m1\text{-}ac}$ = minimum spacing from Metal1 to Active Contact

In addition, the metal has a minimum spacing rule value of $s_{m1\text{-}m1}$ which is not shown.

Every contact is characterized by a resistance

R_c = contact resistance Ω

due to the metal connections. To limit the overall resistance, it is common to use as many contacts as the design rules permit. An example is shown in Figure 5.14. Since the contacts are all in parallel, the effective resistance of the Metal1-Active connection with N contacts is reduced to

$$R_{c,\,eff} = \frac{1}{N} R_c \tag{5.14}$$

In the example, $N = 16$ so that the effective resistance of the connection is (1/16) the value of a single contact. These also spread the current flow.

Metal1 allows access to the active regions of MOSFETs using the Active

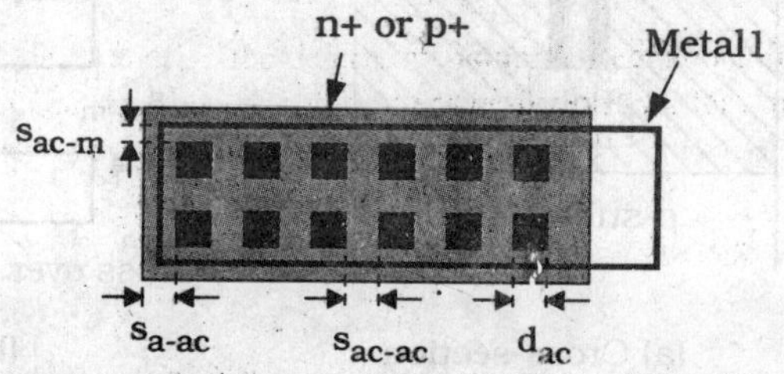

Figure 5.14 Multiple contacts to reduce contact resistance

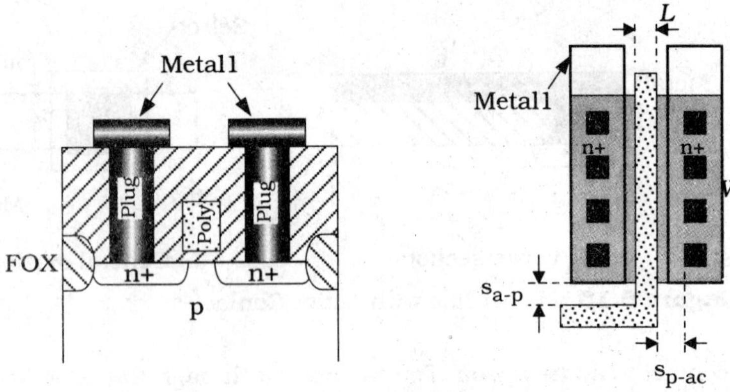

(a) Cross-section (b) Layout

Figure 5.15 Drain and source FET terminals using Metal1

Contact oxide cut. Drain and source terminals are usually at the Metal1 level as shown for an nFET in Figure 5.15(a). The corresponding layout is provided in Figure 5.15(b) where we have included the design rules

$s_{p\text{-}ac}$ = minimum spacing from Poly to Active Contact

$s_{a\text{-}p}$ = minimum spacing from Active to Poly

The first parameter is a surround-type specification to insure that the Active Contact does not destroy any of the polysilicon gate. The second spacing distance $s_{a\text{-}p}$ is due to the self-aligned FET sequence; it insures that the FET has the proper dimensions even if the Poly mask is not perfectly registered with the existing Active pattern on the wafer.

A Poly Contact mask is used to allow electrical connections between Metal1 and the polysilicon gate. Figure 5.16(a) is a cross-sectional view of the contact between the two layers. The Poly Contact mask defines the oxide cut as indicated by the "empty" square shown in the upper part of

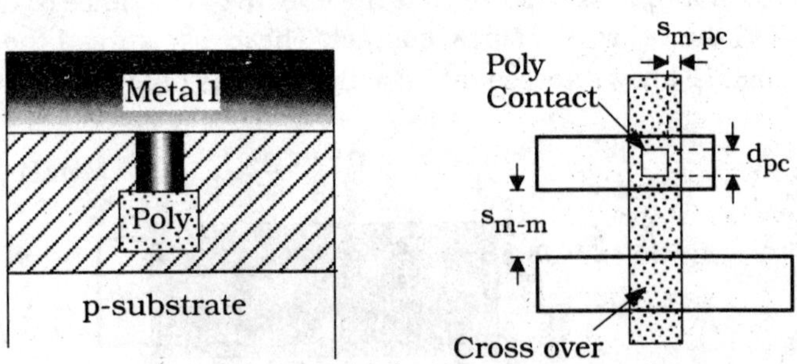

(a) Cross-section (b) Layout

Figure 5.16 Poly Contact

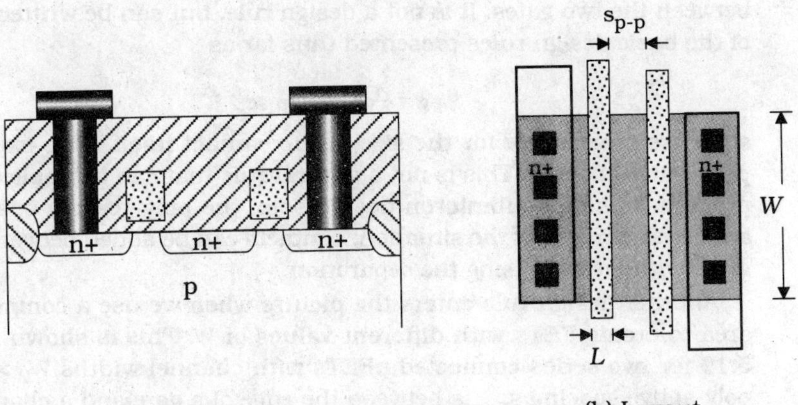

(a) Cross-section (b) Layout

Figure 5.17 Series-connected FETs

the layout in Figure 5.16(b). In the lower portion of the layout, the Metal1 and Poly layers are not connected. This "crossover" characteristic is useful for routing the wiring.

As a final example, let us construct a pair of series-connected FETs. Figure 5.17(a) shows the cross-sectional view for two nFETs. Series wiring is achieved by sharing the central n+ region; since n+ is a reasonable conductor, no additional wiring is needed between the two devices. The layout in Figure 5.17(b) uses this observation: the series transistors are created by parallel Poly lines. The important design rule spacing is

$s_{p\text{-}p}$ = minimum Poly-to-Poly spacing

To obtain a pair of parallel-connected FETs, we add the contacts shown in Figure 5.18. The spacing $s_{g\text{-}g}$ shown in the drawing is the distance

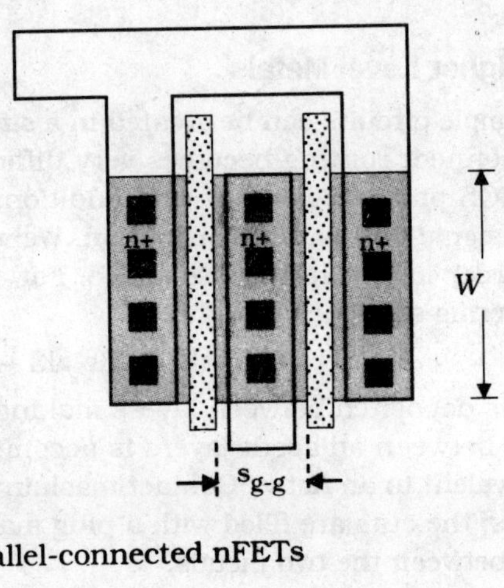

Figure 5.18 Parallel-connected nFETs

between the two gates. It is not a design rule, but can be written in terms of the basic design rules presented thus far as

$$s_{g\text{-}g} = d_{ac} + 2\ s_{p\text{-}ac} \tag{5.15}$$

since we must allow for the size of the contact itself, plus two units of poly-active spacing. This is not a general rule that can be applied to every process. In some submicron design sets, the poly-to-poly spacing $s_{p\text{-}p}$ applies regardless of the situation; contacts can be added between the two gates without increasing the separation.

Another design rule enters the picture when we use a common active area to create FETs with different values of *W*. This is shown in Figure 5.19 for two series-connected nFETs with channel widths $W_2 > W_1$. The poly-active spacing $s_{p\text{-}a}$ is between the edge of a gate and a change in the active border. It must be enforced twice in this design since both FETs see the change.

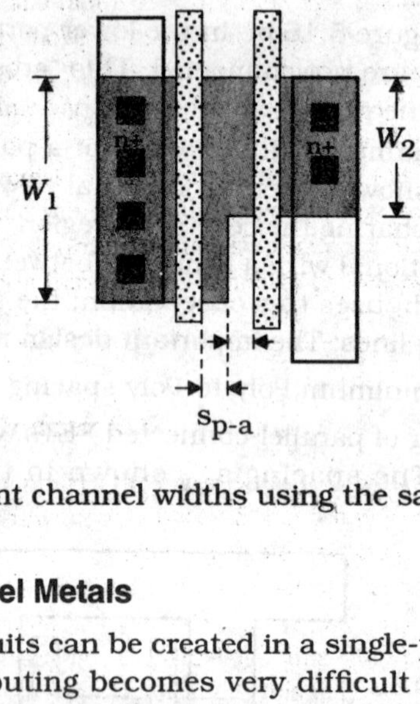

Figure 5.19 Different channel widths using the same active region

5.2.7 Vias and Higher Level Metals

Although simple circuits can be created in a single-poly, single-metal process, interconnect routing becomes very difficult in complex networks. Modern CMOS processes add several additional layers of metal that can be used for signal and power distribution. We will label the layers according to the order in which they are added. For example, in a 4-metal process the layering sequence would be

$$\text{Metal1} \rightarrow \text{Metal2} \rightarrow \text{Metal3} \rightarrow \text{Metal4}$$

CVD oxide is deposited between layers making each electrically distinct. Connection between adjacent layers is accomplished using a Via mask. This is equivalent to an Active Contact mask in that it defines the location of oxide cuts; the cuts are filled with a plug material that gives an electrical contact between the two metals.

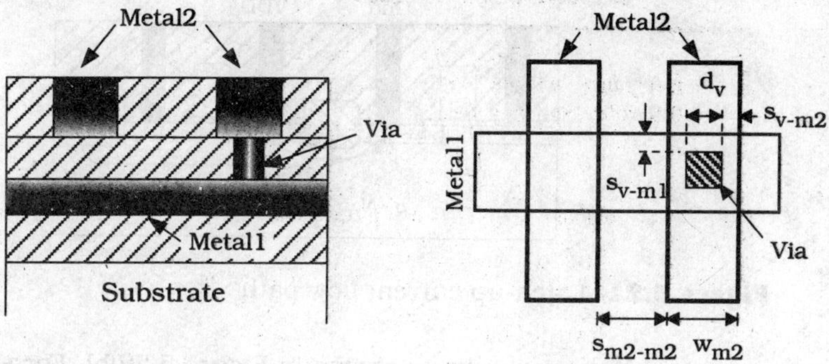

(a) Cross-section (b) Layout

Figure 5.20 Metal1-Metal2 connection using a Via mask

Figure 5.20(a) illustrates the use of a via to connect Metal1 to Metal2. Without a Via (as on the left side of the drawing), the two metal layers are electrically separate. The Via on the right side of the cross-sectional view provides the connection between the two layers. The mask layout is shown in Figure 5.20(b). The new design rule quantities shown are

d_v = dimension of a Via (may be different for vertical direction)

w_{m2} = minimum width of Metal2 feature

$s_{m2\text{-}m2}$ = minimum spacing between adjacent Metal2 features

$s_{v\text{-}m1}$ = minimum spacing between Via and Metal1 edges

$s_{v\text{-}m2}$ = minimum spacing between Via and Metal2 edges

Vias between other metal layers are similar. We note that the values of w_{mj} and $s_{mj\text{-}mj}$ for the j-th metal layer vary for $j > 1$ as the topology and roughness of the wafer surface often dictate that wider lines be used.

5.2.8 Latch-up Prevention

Latch-up is a condition that can occur in a circuit fabricated in a bulk CMOS technology. When a chip is in a state of latch-up it draws a large current from the power supply but does not function in response to input stimuli. A chip may be operating normally and then enter a state of latch-up; in this case, removing and the reconnecting the power supply may restore operations. In the worst-case situation, the chip may enter latch-up when power is applied and never be functional. If the current flow is too large, heat dissipation will destroy the die.

Figure 5.21 shows the current flow path when the chip is in latch-up. Under proper conditions, the path has a very low resistance and can allow large currents to flow. The key to understanding latch-up is noting that the bulk technology gives a **4-layer pnpn** structure between the power supply VDD and ground. This structure, shown in Figure 5.22(a), has the

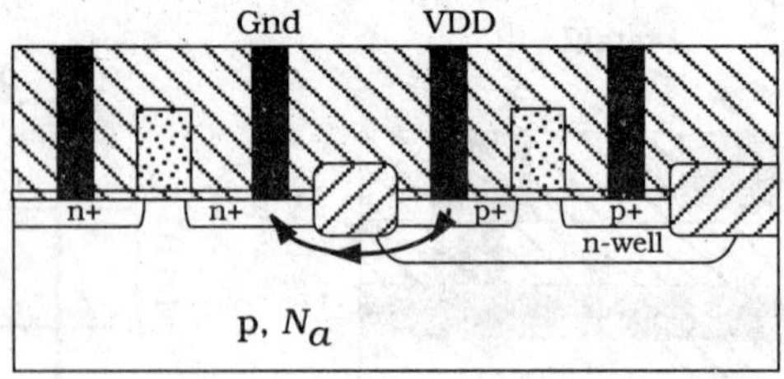

Figure 5.21 Latch-up current flow path

current-voltage dependence shown in Figure 5.22(b). For small voltages V_{DD}, the current I_{DD} is small because of the blocking characteristics of the pn junctions. However, if V_{DD} reaches the **breakover voltage** V_{BO}, the blocking is overwhelmed by internal electric fields. This admits large currents as shown in the drawing, indicating that the chip has entered a latch-up state.

Latch-up prevention starts at the physical design level with various rules used to avoid the formation of the current flow path. One idea is quite simple. Since the current must flow through the n-well and the p-substrate, we can place VDD and ground connections at many different points to steer the current out of the "bad" path. This gives us the general rules

- Include an n-Well contact every time a pFET is connected to the power supply V_{DD}, and
- Include a p-substrate contact every time an nFET is connected to a ground rail.

Since the electrical connections must be made anyway, it is a simple matter to remember to include them. These are illustrated in Figure 5.23. and are very effective for avoiding latch-up. Other techniques have been

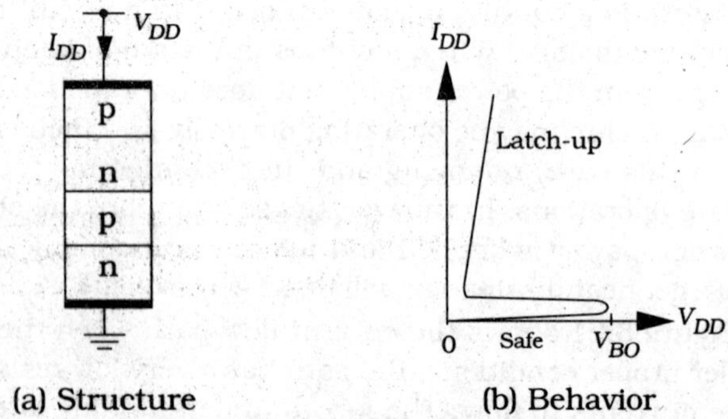

Figure 5.22 Characteristics of a 4-layer pnpn device

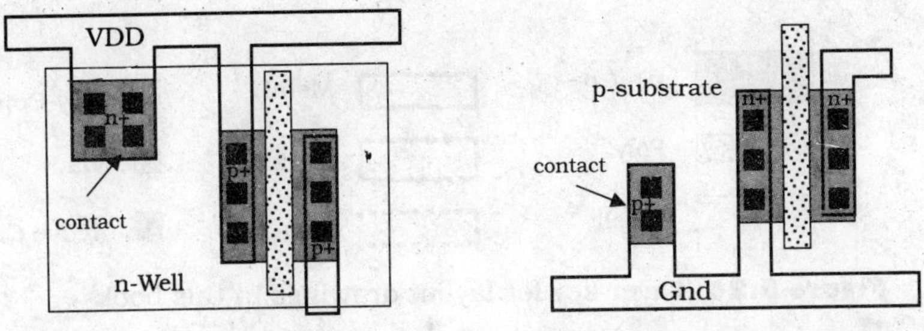

(a) n-Well voltage contact (b) Substrate ground contact

Figure 5.23 n-Well and substrate contacts for latch-up prevention

developed, and one should always check the design rule guidelines on how latch-up is to be avoided.

Non-bulk CMOS technologies that do not build the transistors directly on a silicon substrate avoid latch-up problems by not having the pnpn layering. This is true of silicon-on-insulator (SOI) designs. Alternately, using two separate wells for FETs, an n-well for pFETs and a p-well for nFETs, helps resist the formation of the current flow path. These **twin-tub** technologies are popular in advanced processing lines.

Since latch-up is induced by a high voltage, one must exercise special caution when designing circuits that have high levels of induced electrical "noise" such as a data receiver circuit. Information on avoiding these types of problem is also included in the design rule set. A new designer doesn't always worry about latch-up until a chip fails because of it; from that point on, the problem receives the respect it deserves!

5.2.9 Layout Editors

Several important aspects of layout have been presented in this section. The more critical items are summarized below for future reference.

- n+ is formed whenever Active is surrounded by nSelect; this is also called ndiff.
- p+ is formed whenever Active is surrounded by pSelect; this is also called pdiff.
- an nFET is formed whenever Poly cuts an n+ region into two separate segments.
- a pFET is formed whenever Poly cuts a p+ region into two separate segments.
- No electrical current path exists between conducting layers (n+, p+, Poly, Metal, etc.) unless a contact cut (Active Contact, Poly Contact, or Via) is provided.

These simple observations provide the basis for most of the layout problems we will encounter.

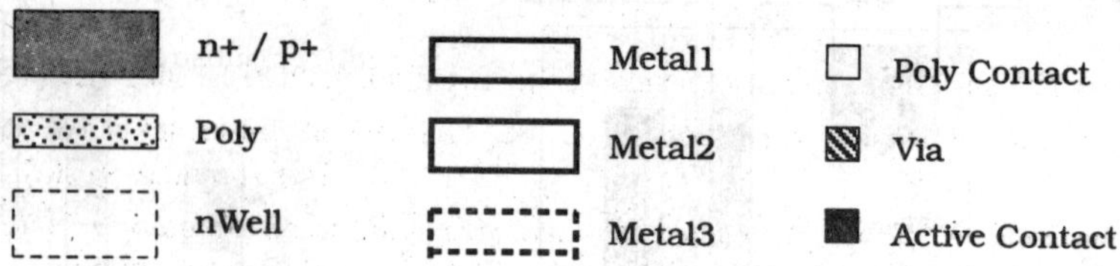

Figure 5.24 Layer key for layout drawings in this book

A layout editor visually distinguishes among the layers by defining different colors and/or fill patterns for each. We have opted to use simple gray-scale and linewidth variations here to save the cost involved in printing the book in color.[3] Figure 5.24 shows the outlines that we will use to identify layers in the book. Note that n+ and p+ regions have the same shading, so that the polarity of a region is implied by where it is located: it will be a p+ layer in an nWell, an n+ section otherwise.

Every layout editor operates in a slightly different manner, but all have the same basic features. In general,

- One enters a polygon by first choosing the desired layer of material and then using the drawing tools to shape the object as needed.
- Layout editors provide a background grid. The distance between each grid point is a specified distance.
- The layers may be drawn in any order, so long as each polygon is properly identified by layer color/name/pattern. The database automatically keeps track of the polygons drawn on each layer.
- The layout pattern is used to create the mask set for the process, and constitute the drawn dimensions.
- Design rules must be obeyed and the spacing must be checked before the drawing is complete.
- Polygons on a given layer may be drawn to touch or overlap. Only the outline is important. This is illustrated in Figure 5.25. The entire layout in Figure 5.25(a) is drawn using rectangles, but results in the finished masks shown in Figure 5.25(b). This simplifies the overall layout process.

Always save your designs in a timely fashion! When the chip is completed, it is usually put into a standard format for transmission to the processing line. Keeping in the spirit of the pioneers of chip design, the process is called **tape-out** because the files were transferred to the fabrication group on magnetic tape. The most common format used is probably the GDS standard which was a standard of one of the early minicomputer-based

3 Which would quadruple the cost of the book!

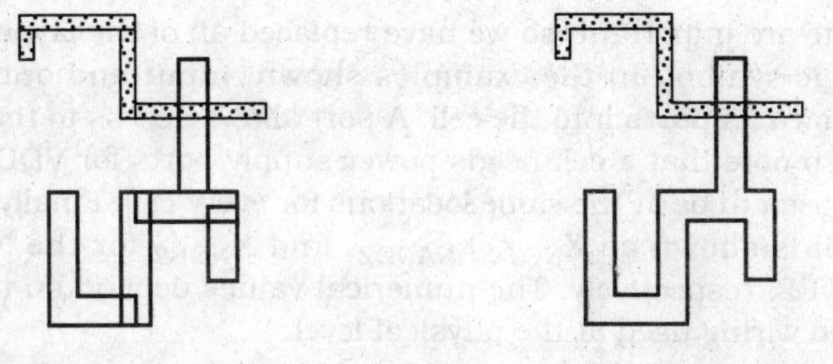

(a) Multiple polygons (b) Mask pattern

Figure 5.25 Drawing complex polygons using rectangles

CAD systems. Academic users often produce files in CIF (Caltech Intermediate Form) format which was developed in the 1970's.

5.3 Cell Concepts

Digital VLSI chips are based on the idea of hierarchical design. Individual transistors are used to build gates, which are then used to create logic cascades and functional blocks, which in turn are used as the basis for even larger units. The basic building blocks in physical design are called **cells**. A cell may be as simple as a FET, or as complex as an arithmetic logic unit (ALU). Regardless of the internal complexity, every cell acts in the same manner: it may be used as a component to create a larger logic network.

The main idea of **cell-based** design is straightforward to visualize. Suppose that we start with a set of CMOS logic gates (NOT, NAND2, NOR2) and design the physical circuit layout for each. At the basic level, we concentrate on placing polygons for each layer with the required sizes. We then "step back" and view the gates as portrayed in Figure 5.26; each block is an independent cell. At this level in the design hierarchy, we do not care about the internal details. Only the external characteristics of a

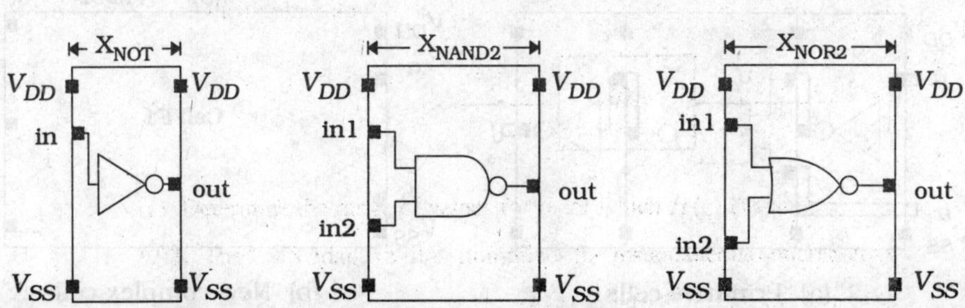

Figure 5.26 Logic gates as basic cells

gate are important, so we have replaced all of the layout by an equivalent logic symbol. In the examples shown, input and output terminals are shown as **ports** into the cell. A port allows access to the interior circuitry. Also note that a cell needs power supply ports for VDD and VSS that are chosen to be at the same locations for every cell. Finally, the width of each cell is shown as X_{NOT}, X_{NAND2}, and X_{NOR2} for the NOT, NAND2, and NOR2, respectively. The numerical values depend on the transistor sizes and wiring used at the physical level.

Once a set of cells are defined, they may be used to create more complex networks. Suppose we want a cell that provides the function

$$f = \bar{a} \cdot b \tag{5.16}$$

This can be created using the simple cascade of two NOT gates and one NAND2 gate in Figure 5.27(a). Metal1 lines have been used to wire the ports of the cells as needed. For example, the output of the first NOT gate is wired to In1 of the NAND2 gate. Once the cascade has been created, we can define a new cell F1 as on Figure 5.27(b). This cell has a total width of

$$2X_{NOT} + X_{NAND2} \tag{5.17}$$

which is just the sum of the widths of the three cells used to construct it. Once defined, the new cell F1 can be used as a building block without decomposing it into the primitive cells that were used to create it. It becomes as basic as the NOT, NAND2, and NOR2 circuits. Using this hierarchical design approach allows us to design and construct extremely complex logic networks. It is, in fact, one of the most important techniques to learn in VLSI.

Let us now turn our attention to the problem of creating a basic collection of cells at the physical level. The first item that we should investigate is the placement of the power supply lines VDD and VSS. The problem is shown in Figure 5.28. Both are shown on the Metal1 layer. The spacing between the two lines is shown as

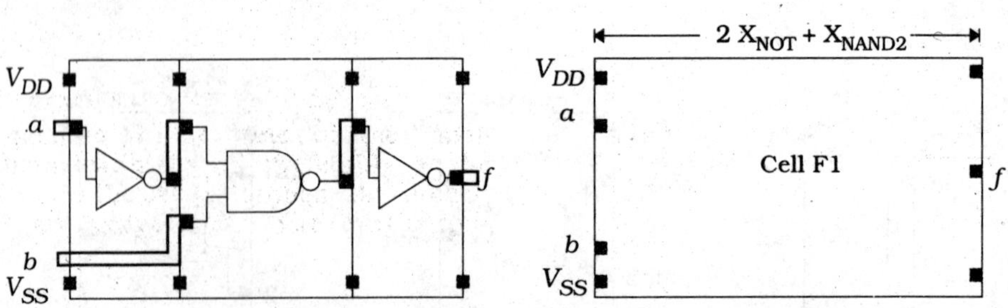

(a) Primitive cells (b) New complex cell

Figure 5.27 Creation of a new cell using basic units

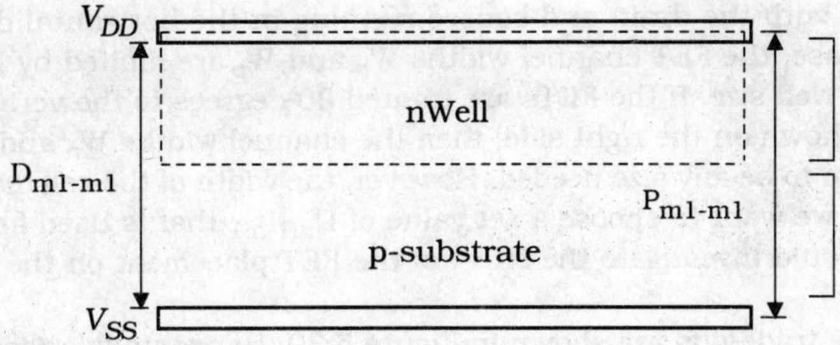

Figure 5.28 VDD and VSS power supply lines

$D_{m1\text{-}m1}$ = Edge-to-edge distance between VDD and VSS and the **pitch**

$P_{m1\text{-}m1}$ = Distance between the middle of the VDD and VSS lines

The two are related by

$$P_{m1\text{-}m1} = D_{m1\text{-}m1} + w_{DD} \tag{5.18}$$

where w_{DD} is the width of the power supply lines.[4] Fabrication specialists often use the pitch specification, while the actual distance D between the edges is more useful for circuit layout. The nWell region that is used for pFETs is placed about the VDD line as shown. The region around VSS is kept as p-substrate since nFETs are connected to it.

Once we have established the VDD and VSS lines, we can proceed to place FETs between them. Figure 5.29 shows two different approaches to transistor orientation. The FETs on the left side of the drawing are ori-

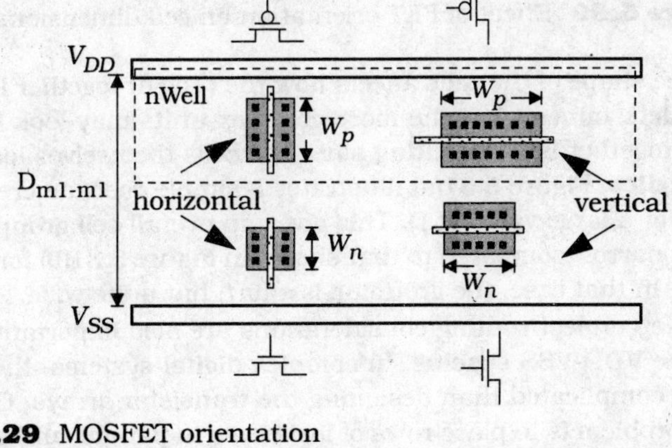

Figure 5.29 MOSFET orientation

4 Note that w_{DD} may be larger than the minimum design rule width w_{m1} allowed for a Metal1 line.

ented with the drain and source running in the horizontal direction. In this case, the FET channel widths W_n and W_p are limited by $D_{m1\text{-}m1}$ and the n-well size. If the FETs are rotated 90 degrees to the vertical orientation shown on the right side, then the channel widths W_n and W_p may be chosen to be any size needed. However, the width of the cell may get large. Since we want to choose a set value of $D_{m1\text{-}m1}$ that is used for every cell, we should investigate the effect of the FET placement on the cell dimensions.

The trade-offs are shown in Figure 5.30. Horizontally oriented transistors are used in Figure 5.30(a). In this case, we would want to make D_1 large enough to accommodate the most complex logic gate needed. Using vertical FETs, the value of D_2 shown in Figure 5.30(b) can be made smaller than D_1. The difference is in the horizontal widths of the cell. In general, we would expect X_2 to be greater than X_1 for a given circuit.

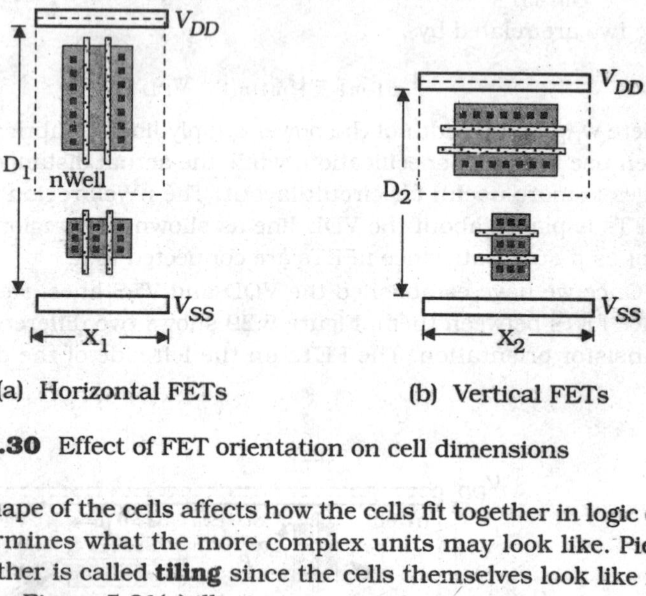

(a) Horizontal FETs (b) Vertical FETs

Figure 5.30 Effect of FET orientation on cell dimensions

The shape of the cells affects how the cells fit together in logic cascades and determines what the more complex units may look like. Piecing the cells together is called **tiling** since the cells themselves look like non-uniform **tiles**. Figure 5.31(a) illustrates a simple cascade created out of four tiles for a large value of D. This gives an overall cell grouping that is relatively narrow compared to that shown in Figure 5.31(b) for a smaller value of D. In that case, the grouping is short, but quite wide.

Interconnect routing considerations are also important considerations for the VDD-VSS spacing. In complex digital systems, the wiring is often more complicated than designing the transistor arrays. One approach to this problem is to place rows of logic cells in parallel and allocate space in between the rows for wiring. The general idea is portrayed in Figure 5.32. Metal1 lines running parallel to the logic rows can be used to route signals as required. Since Metal2 lines can cross over Metal1, vertical lines

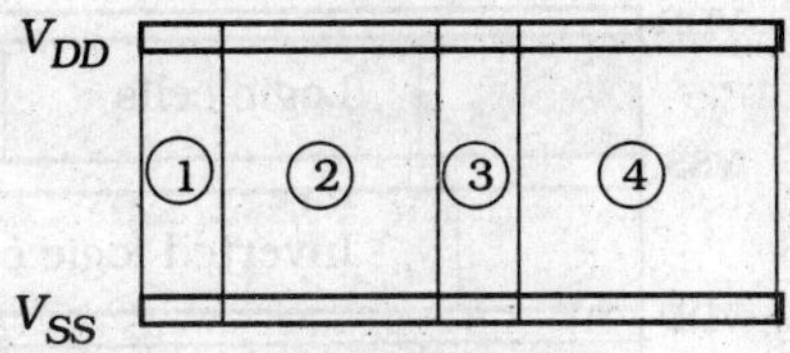

(a) Larger D

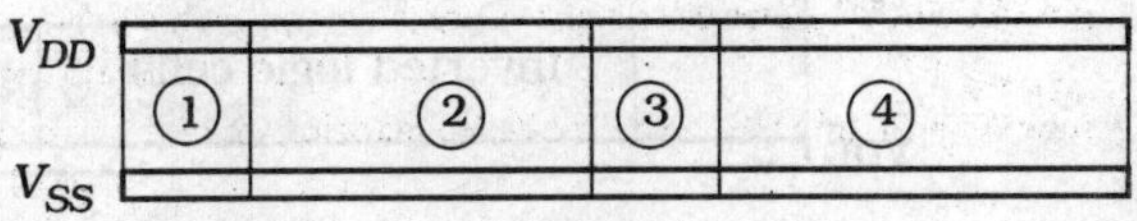

(b) Smaller D

Figure 5.31 Effect of tile shapes on larger cells

can be used to connect logic cells to the Metal1 interconnect as shown. This technique is often found in ASIC designs because it allows a significant amount of freedom for different designs. The main drawback is that the logic density is relatively low compared to close-packed layouts.

An alternate high-density technique is to alternate VDD and VSS power lines and share them with cells above and below. This results in the **Weinberger image** shown in Figure 5.33. The "Inverted logic cells" are defined to be flipped in relation to the rows of "Logic cells" above or below. This is because they have VSS at the top and VDD at the bottom. The

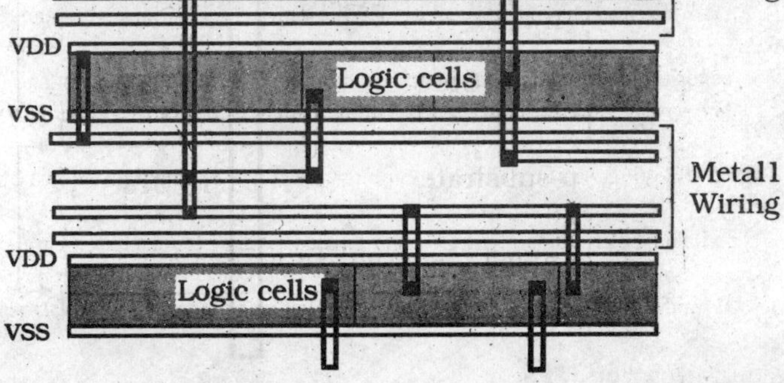

Figure 5.32 Wiring channels

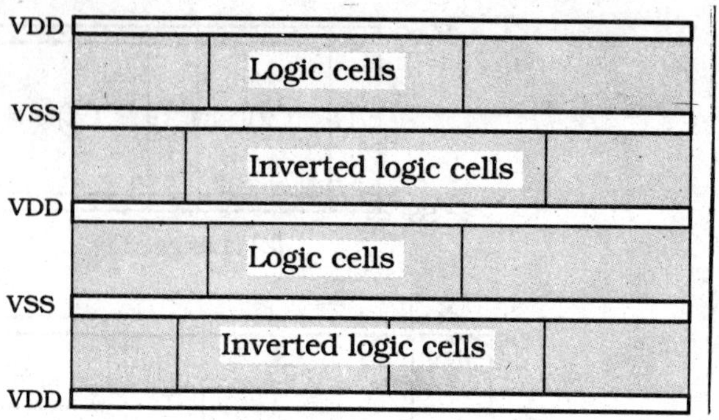

Figure 5.33 Weinberger image array

details of FET placement in a Weinberger image are provided by the close-up in Figure 5.34. The nWell regions surround the VDD rails and allow pFETs to be created above or below the power lines. The nFETs are placed on both sides of the VSS line. Since no space is automatically reserved for wiring, this scheme allows for high-density placement of the cells. The main drawback is that the connections between rows must be accomplished by using Metal2 or higher, since Metal1 is already designated for the power supplies. It may be possible to use horizontal Metal1 interconnect lines within a row if there is sufficient room.

Port Placement

The input and output ports of a cell must be placed at convenient points to facilitate the interconnect wiring. At the basic level, we view logic circuit inputs as being to the gate terminals of MOSFETs, while the outputs

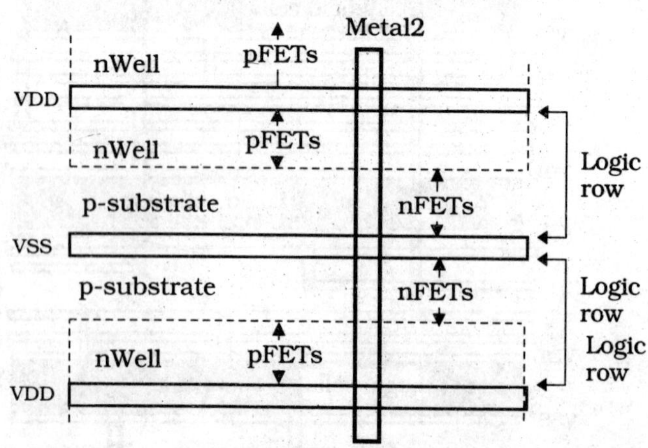

Figure 5.34 FET placement in a Weinberger array

are metal interconnect lines. Since FET gates are at the polysilicon level, we must provide a poly contact to connect the output of a cell to the input of another cell.

Figure 5.35 shows the case where the ports are placed around the periphery of a cell. With this simple view, the input poly lines are on the left side and include a Metal1 pad and poly contact. The output on the right side is at the Metal1 level, which allows cell interconnects to be completed on the same level. Vertical poly inputs are also shown. These are useful if the layout uses wiring channels between cell rows as in Figure 5.32.

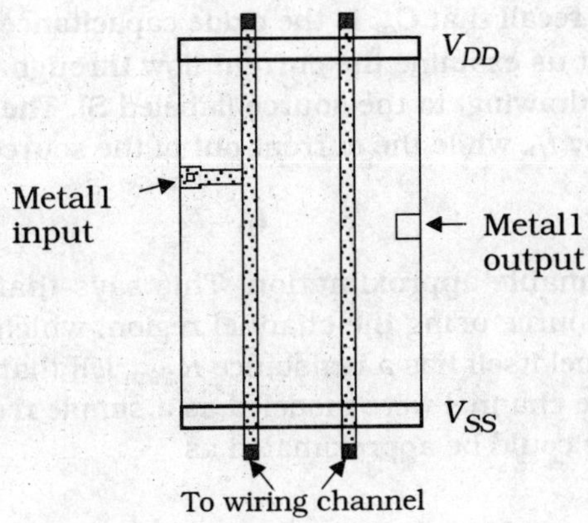

Figure 5.35 Port placement in a cell

There are no *a priori* constraints on the placement of cell ports, and interior ports are also used in practice. The most important factor is to insure that the cells can be wired together as needed in a complex design. Wiring problems have a tendency of appearing at critical times. Careful cell planning and a reliable CAD tool set helps to solve them more efficiently.

Now that we have learned the basics of logic cells, let us study the details of designing a set of CMOS gates at the silicon level. Once we have a reasonable set of gates, we can progress into the next hierarchical design level where we build up more complex units.

5.4 FET Sizing and the Unit Transistor

Field-effect transistors are specified by the aspect ratio (W/L) where W is the channel width and L is the channel length. In modern VLSI, both are on the order of microns [μm], with specific numerical values established in the layout of the masks. These dimensions combine with the processing parameters to give the electrical characteristics of the transistor.

Consider the basic FET drawn in Figure 5.36. The drawn values of the channel length and width are shown explicitly. We may estimate some of the layout-dependent electrical properties of the transistor by using a few simple formulas. First, the area A_G of the gate is defined to be the portion of the poly that is over the channel region. The drawing shows that the area A_G of the gate is given by $A_G = LW$. The gate capacitance C_G seen looking into the gate terminal (labeled as G in the drawing) is then given by

$$C_G = C_{ox}WL \tag{5.19}$$

where we recall that C_{ox} is the oxide capacitance per unit area.

Now let us examine the current flow through the device from the drain (D in the drawing) to the source (labeled S). The current into the drain is denoted by I_D, while the current out of the source is I_S such that

$$I_D \approx I_S \tag{5.20}$$

is a reasonable approximation. This says that the current flows from drain to source using the channel region, which is underneath the gate. The channel itself has a resistance R_{chan} [Ω] that impedes the flow of current. If the channel were modeled as a simple rectangular block, then the resistance could be approximated as

$$R_{chan} = R_{s,c}\left(\frac{L}{W}\right) \tag{5.21}$$

where $R_{s,c}$ is the sheet resistance of the channel region. Unfortunately, FETs are not that simple and computing the drain-to-source resistance is more complicated. The equation does, however, agree with the more rigorous analysis in that it predicts that R_{chan} is inversely proportional to the channel width W:

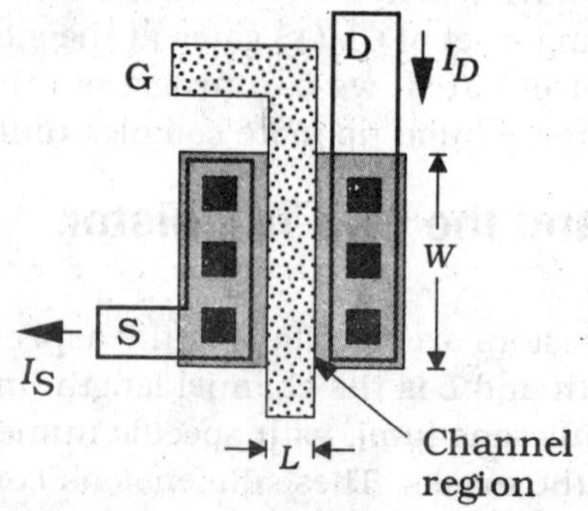

Figure 5.36 Basic geometry of a FET

$$R_{chan} \propto \frac{1}{W} \tag{5.22}$$

This says that increasing W decreases the resistance, which allows more current to flow. The channel dimensions thus establish the resistance and capacitance of a FET.

One other feature is worth mentioning. The primary difference between an nFET and a pFET is the polarity of charge that gives the current. An nFET uses negatively charged electrons, while a pFET relies on positively charged holes. Recall, however, that electrons can move more easily than holes. This is expressed by the relation

$$\mu_n > \mu_p \tag{5.23}$$

that was introduced earlier in Section 3.2 of Chapter 3. In this equation, μ_n and μ_p are the electron and hole mobilities, respectively. A high value of mobility implies that the particle is "more mobile" than a low-mobility particle. Suppose we design an nFET and a pFET with the same aspect ratio (W/L). Since electrons have a higher mobility, the nFET resistance R_n would be smaller than the pFET resistance R_p. Let us define the **mobility ratio** r by

$$r = \frac{\mu_n}{\mu_p} \tag{5.24}$$

In modern CMOS processing, the mobility ratio $r > 1$ is usually between 2 and 3, with the actual value set by the doping densities and other physical considerations. The resistance is inversely proportional to the conductivity, which is proportional to the mobility. We can thus conclude that R_n and R_p for equal size FETs are related by

$$\frac{R_p}{R_n} = r \tag{5.25}$$

This is often stated in the literature by saying that pFETs don't conduct as well as nFETs. Alternately, since electrons travel faster than holes, we conclude that nFETs are faster than pFETs. Both statements assume that the transistors being compared are the same size.

The resistance of a FET can be adjusted by changing the channel width W. Suppose that we have an nFET with an aspect ratio of $(W/L)_n$ that gives a resistance R_n. To design a pFET with the same resistance value $R_p = R_n$, we use an aspect ratio of $(W/L)_p > (W/L)_n$ that compensates for the differences in mobilities. This is accomplished by selecting

$$\left(\frac{W}{L}\right)_p = r\left(\frac{W}{L}\right)_n \tag{5.26}$$

With this design, the resistances are equal. Note, however, that the gate areas are different with $A_{Gp} > A_{Gn}$ due to the increased channel width of the pFET. Assuming that the channel lengths are the same, this gives different gate capacitances such that

$$C_{Gp} = rC_{Gn} \tag{5.27}$$

since the areas are proportional to W.

Example 5.1

Consider an nFET with an aspect ratio of $(W/L)_n = 4$ that is constructed in a process where $r = 2.4$. To create a pFET with the same resistance we must select

$$\left(\frac{W}{L}\right)_p = 2.4(4) = 9.6 \tag{5.28}$$

In practice, we might use the nearest integer value $(W/L)_p = 10$. The gate capacitance of the pFET would be larger than that of the nFET by the same ratio

$$C_{Gp} = 2.4C_{Gn} \tag{5.29}$$

It is also worth mentioning the obvious fact that the pFET will consume more surface area than the nFET.

The electrical characteristics of transistors determine the switching speed of a VLSI circuit. At the physical level, this translates to selecting the aspect ratios $(W/L)_n$ and $(W/L)_p$ for every FET in the circuit. Once the sizes have been determined, the physical design problem revolves around designing the silicon circuit using the specified aspect ratios. Let us concentrate on the physical design problem for now. Many of the remaining sections of this book are concerned with how to choose the transistor sizes for high-speed logic networks.

A useful starting point for circuit layout is to define a **unit transistor.** This is a FET with a specified aspect ratio (W/L) that can be replicated as needed in the layout. Since it only needs to be drawn once, layouts can be completed much more quickly than if the designer had to construct every transistor. Moreover, since the electrical characteristics of device will be known, the switching performance analysis will be straightforward.

One choice for a unit transistor is the **minimum-size MOSFET**. As implied by its name, a minimum-size FET is the smallest transistor that can be created using the design rule set. An example is shown in Figure 5.37. The drawn channel length L is the minimum allowed poly width w_p, while the drawn channel length W is the minimum width w_a allowed for a feature on the Active mask. The aspect ratio for the device is thus

$$\left(\frac{W}{L}\right)_{min} = \frac{w_a}{w_p} \tag{5.30}$$

as can be verified by inspection. The gate capacitance is set as

$$C_G = C_{ox} w_a w_p \tag{5.31}$$

since the gate area is just $A_G = w_a\, w_p$. The minimum-size device is the smallest transistor, so that in theory it allows the highest packing density. However, it does have the largest resistance of any FET, so it may not be the best choice for every circuits.

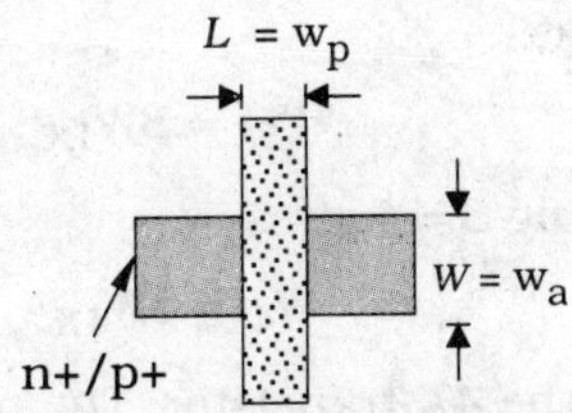

Figure 5.37 Geometry of a minimum-size FET

The basic minimum-size FET shown in Figure 5.37 does not have any contacts. If we add Active Contacts to allow Metal1 connections, then our dimensions may change. Consider the modified layout in Figure 5.38(a). The channel length is still given by $L = w_p$. However, since we have used Active Contact cuts in the oxide, the design rules

d_c = dimension of the contact

$s_{a\text{-}ac}$ = spacing between Active and Active Contact

must be applied. As shown in the drawing, the minimum width is now

$$W = d_c + 2s_{a\text{-}ac} \tag{5.32}$$

In some processes, this value may be the same as $W = w_a$. If not, then the Active region can be enlarged to accommodate the contact as in Figure 5.38(b). This allows us to have $W = w_a < d_c + 2s_{a\text{-}ac}$. Although minimum-size FETs are slow due to their high resistance, they can be useful in situations where slow switching is not a critical concern.

Once a unit FET has been selected, it is useful to allow it to be **scaled** in size. In Figure 5.39, the 1X transistor is used as the reference basis. Larger transistors are obtained by multiplying the width; 2X and 4X versions are shown in the drawing. Altering the size of the transistor changes its resistance and capacitance. Let us denote the resistance and gate capacitance of the 1X device by R_{1X} and C_{1X}, respectively. If the width of

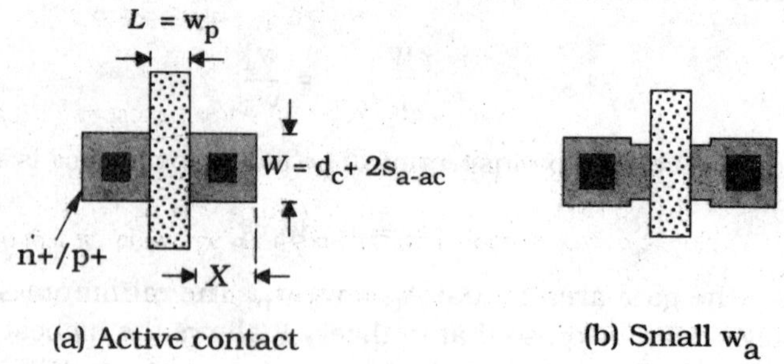

Figure 5.38 Minimum-size FETs with Active Contact features

the 1X device is W_{1X}, then we can create larger FETs using a **scaling factor** $S \geq 1$ such that

$$W_{SX} = SW_{1X} \tag{5.33}$$

For example, setting $S = 4$ gives

$$W_{4X} = 4W_{1X} \tag{5.34}$$

which describes the 4X transistor. The resistance and capacitance of a scaled FET are changed because they depend upon the size of the device. Applying the scaling transformations gives the general values

$$R_{SX} = \frac{R_{1X}}{S} \qquad C_{SX} = SC_{1X} \tag{5.35}$$

For example, the 2X FET has

$$R_{2X} = \frac{R_{1X}}{2} \qquad C_{2X} = 2C_{1X} \tag{5.36}$$

which is easy to remember. Since pFETs have different conduction char-

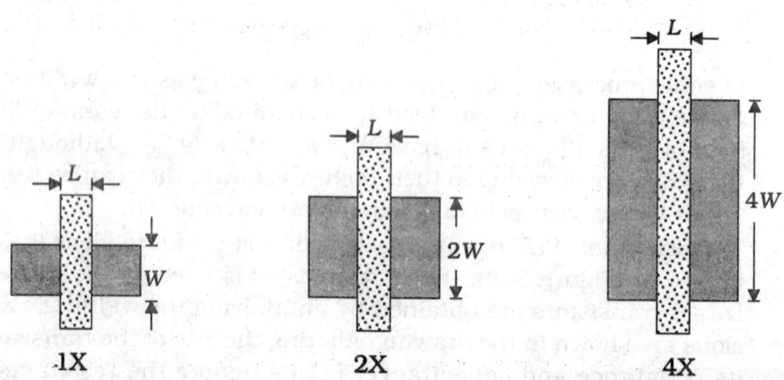

Figure 5.39 Scaling of the unit transistor

acteristics than nFETs, it is common to introduce unit transistors for each type. The scaling relations remain the same regardless of the polarity.

Unit devices are not restricted to individual transistors. It may be useful to define series and parallel groups of FETs as 1X units, and then scale using the same technique. Figure 5.40 shows an example of a series-connected 2-FET chain at 1X and 2X sizes. Since each transistor has been scaled in the same manner, the resistance and capacitance relations are still valid. It is important to note, however, that the total resistance of the series-connected transistors is the sum of the individual resistances. If the resistance of one 1X FET is R_{1X}, then the series group has a resistance of $2R_{1X}$. Since each FET in the 2X circuit has a resistance given by $(R_{1X}/2)$, the resistance of the scaled 2X series pair is

$$2\,(R_{1X}/2) = R_{1X} \tag{5.37}$$

by just adding. Series-connected FETs are usually made larger than individual FETs to reduce the overall end-to-end resistance.

Large transistors often require a bit more thought. There are occasions when aspect ratios reach 100 or greater. A single device with a large channel width W will have a long rectangular shape and may not easily fit into the overall layout. Or, the resistance of the gate material may slow down the signal.

The most common solution is to use a group of parallel-connected transistors. Figure 5.41 shows a group of transistors that is based on a channel width of W. The four gate lines are all connected together, and the wiring is routed to give an effective channel length of $4W$ between sides A and B. One advantage of this approach is that the overall layout geometry can be adjusted to square or nearly square shapes.

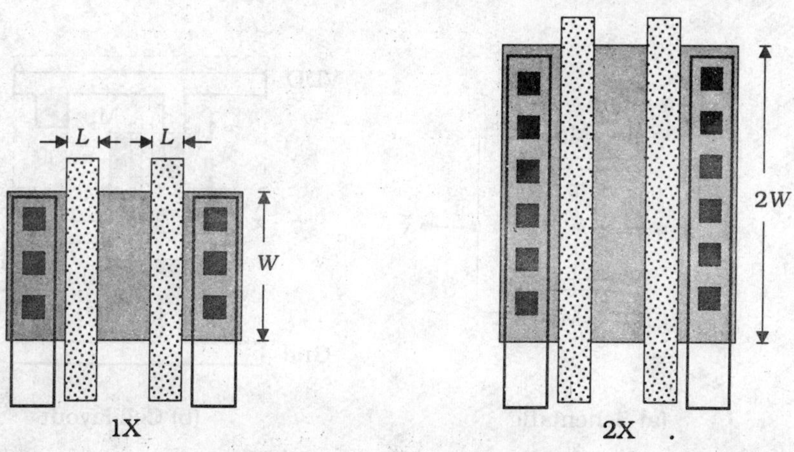

Figure 5.40 Scaling of series-connected FET chain

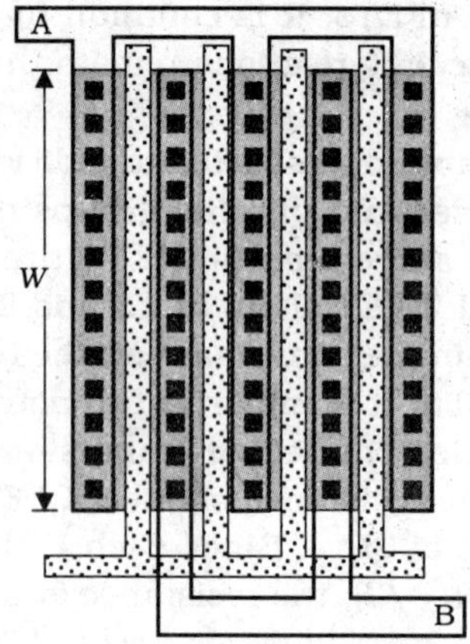

Figure 5.41 A large FET created from parallel-connected transistors

5.5 Physical Design of Logic Gates

Let us now apply the basics of the physical design process to the problem of constructing a set of layouts for basic CMOS logic gates. Each gate is classified as an individual cell. We will concentrate on a unit cell design, since larger cells can be obtained by scaling.

5.5.1 The NOT Cell

The simplest CMOS logic circuit is the inverter that provides the NOT operation. Consider the schematic shown in Figure 5.42(a); the horizontal orientation can be directly translated to the layout in Figure 5.42(b). The layout shows the channel widths W_p and W_n of the FETs. It also shows the important connection of VDD to the nWell (which is the pFET bulk)

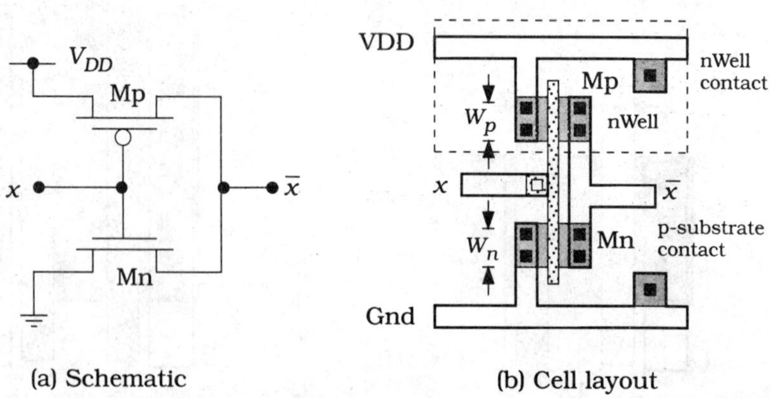

Figure 5.42 NOT gate with horizontal FETs

and Gnd to the p-substrate (which is the nFET bulk). These connections will not always be shown explicitly in our drawings, but must be included in every cell to produce a functional circuit.

Although this simple example illustrates the basic aspects of the layout, the small spacing of the VDD-Gnd Metal1 lines makes it difficult to scale. If we rotate the FETs by 90 degrees, then it is easier to increase the channel widths of the FETs. This is illustrated by the example in Figure 5.43. In Figure 5.43(a), the unit NOT design has aspect ratios of (W/L) for both Mn and Mp. The 2X cell in Figure 5.43(b) uses the same VDD-Gnd pitch, but provides transistors with aspect ratios of $2(W/L)$ by stretching the FETs in the horizontal direction.

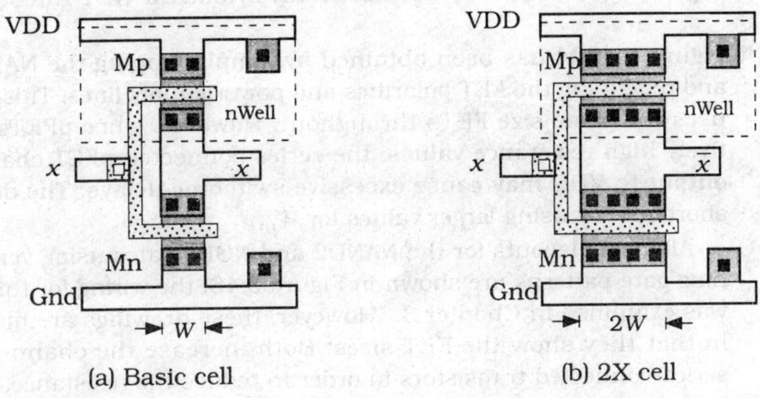

Figure 5.43 NOT layout using vertical FETs

Another example is shown in Figure 5.44. In this design, the pFET is larger than the nFET by a mobility ratio $r = 2.5$. This equalizes the resistance between the output $\bar{x}$ and the two power supply rails VDD and Gnd. Since this has equal nFET and pFET resistances, it is called a **symmetric inverter** (even though it is not geometrically symmetric). The significance of this design will be discussed in Chapter 7 in the context of the electrical design of CMOS logic gates.

5.5.2 NAND and NOR Cells

We can apply the same techniques to design the layout for a NAND gate. Vertical FETs are used in the NAND2 layout shown in Figure 5.45(a). In this design, all transistors have the same aspect ratio. They can be resized as needed, as can the cell itself. If more inputs are used, e.g., as in a NAND3 gate, then the sizing of the nFETs becomes more critical. In this case, the value of W_n should be increased to reduce the series resistance from the output to ground.

A NOR gate may be created in the same manner. The NOR2 layout in

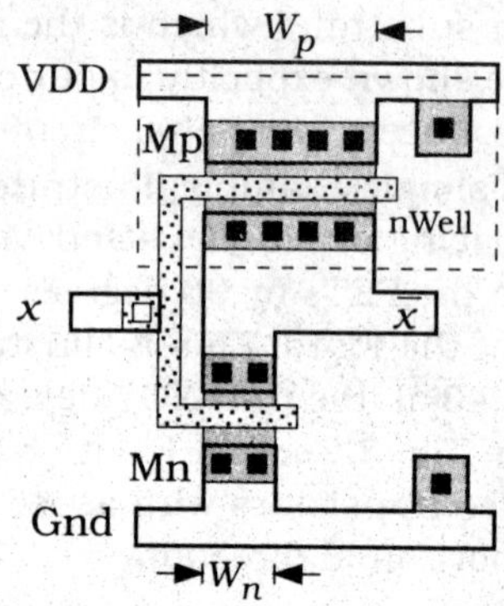

Figure 5.44 Layout for an electrically symmetric NOT gate

Figure 5.45(b) has been obtained by simply flipping the NAND2 layout and redefining the FET polarities and power supply lines. This design also uses the same size FETs throughout. However, since pFETs have relatively high resistance values, the series-connected pFET chain from the output to VDD may cause excessive switching delays. The delay can be shortened by using larger values for W_p.

Alternate layouts for the NAND2 and NOR2 gates using vertically running gate patterns are shown in Figure 5.46; the wiring for this approach was examined in Chapter 3. However, these drawings are more detailed in that they show the FET sizes. Both increase the channel width for series-connected transistors in order to reduce the resistance. The nFETs in the NAND2 gate in Figure 5.46(a) are made wider than the parallel-connected nFETs used in the NOR2 gate of Figure 5.46(b). Similarly, the series pFETs in the NOR2 are wider than the parallel pFETs of the NAND2.

.The actual numerical values of W_n and W_p determine the electrical

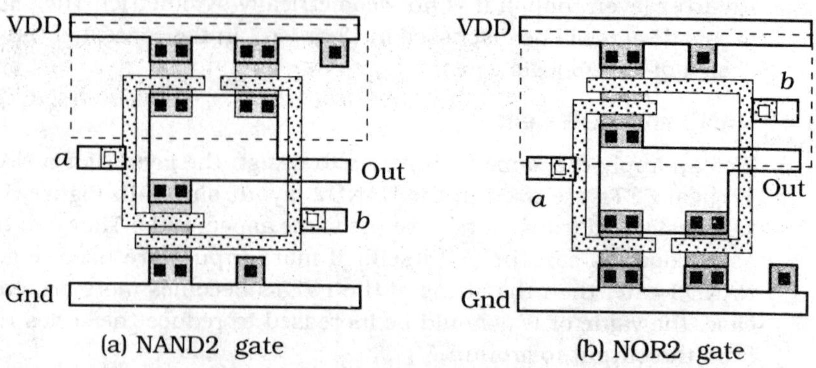

Figure 5.45 NAND2 and NOR2 layouts using vertical FETs

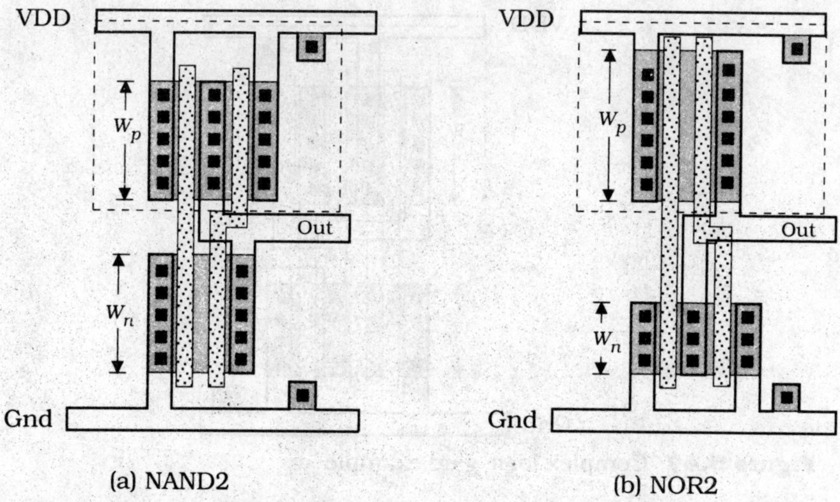

Figure 5.46 Alternate NAND2 and NOR2 cells

characteristics of the gate. In many designs, conveniently sized FETs are used in the layout. The circuits are then simulated to determine their electrical response, and the sizes adjusted if needed. In critical data paths, the values are more important and initial design work concentrates on finding acceptable values. The electrical aspects of logic gate design are discussed in later chapters of the book.

5.5.3 Complex Logic Gates

Complex logic gate layout progresses in the same manner. The routing techniques presented in Chapter 3 give the device placement. In the physical design stage, every transistor size is specified and the FET structures are placed between the VDD and Gnd rails. Series-connected FETs are generally made wider than individual transistors unless they share the same Active area or other considerations are important.

An example of a complex logic gate is shown in Figure 5.47. Since both the nFET and pFET arrays share drain/source regions, single values of W_n and W_p are used for simplicity in layout. Note that the pFETs can be made wider within the given VDD-Gnd spacing to compensate for their higher resistance values.

5.5.4 Generalized Comments on Layout

These examples illustrate the basics for the physical design of logic gates using the following sequence:

- Design the MOSFET logic circuit;

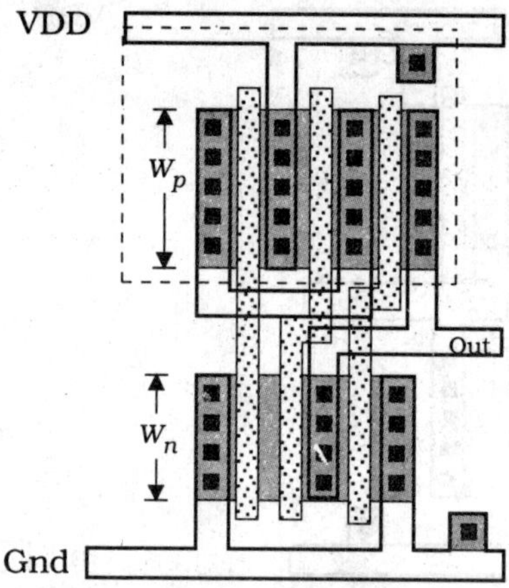

Figure 5.47 Complex logic gate example

- Use the transistor circuit to create a routing diagram where only wiring paths and levels are important;
- Use the routing diagram as the basis for the final physical design of the gate that includes proper sizes for all features and adheres to the design rules.

The final aspect of gate design can be time consuming for the neophyte. Some complex circuits are tricky even for an experienced designer. After a layout is completed, it must be extracted and simulated using the process parameters. Final acceptance of a cell depends on its satisfying all electrical and size specifications.

VLSI designers attempt to place as much circuitry on a given area as possible. At the engineering level, this is accomplished by using the ideas of regular, repeated patterns and mastering the CAD tools. The lessons learned from designing simple cells provide the basis for increasingly complex networks. Design automation tools are becoming quite powerful and more intelligent, and are helping pave the way for designs of incredible complexity.

5.6 Design Hierarchies

VLSI systems are created using the concept of design hierarchies where simple building blocks are used to design more complex units. This nesting continues until the entire chip is complete. The code for a layout editor is structured to provide this type of environment for the chip designer. The key to creating the hierarchy lies in the concept of cells. We define a cell to be a collection of objects that is treated as a single entity. The characteristics of the objects themselves provide the hierarchical viewpoint.

The simplest cells consist of only polygons. The logic gates such as the NOT and NAND2 examples in the previous section fall into this category. A cell with this property is said to be **flat**; this means that every object is independent and not related to any other object. In a flat cell, we can alter any polygon without affecting anything else. To initiate the design process, we create a large number of flat cells and store them in a library. The most primitive library entries are chosen to be transistors and logic gates that can be used as building blocks in more complex designs. Figure 5.48 illustrates the idea. In this simple example, three gate-level designs nor2, nand2, and not are created at the polygon level. Each is then stored as a separate cell in the library. Each cell is independent of the others.

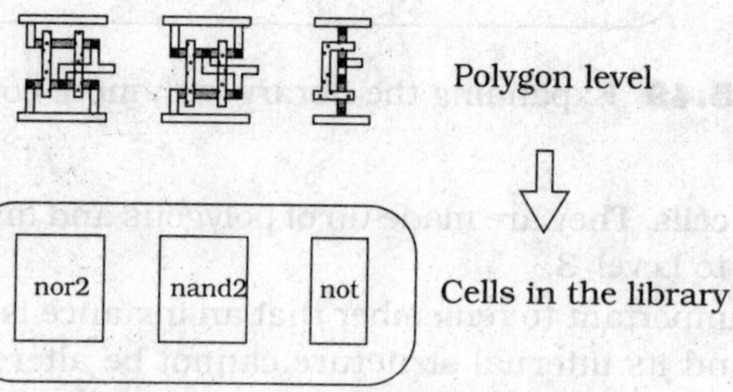

Figure 5.48 Primitive polygon-level library entries

Once the initial library is established, we can use the cell entries in our design by instancing them into our layout. An instance is a copy of the cell in the library. An instanced object cannot be altered in the new layout, as it is always an exact replica of the library entry. The only way to change the characteristics of an instance is to change the library entry. The most important concept to grasp is that the new layout will be a more complex object that can itself be stored as a cell in the library. In Figure 5.49, two new cells named cell_1 and cell_2 are designed using instances of the Primitive Library, plus polygons of their own. We may save the new cells and create a larger library group (Library 1) for use in more complex designs. This process may be repeated as needed. Useful functions are designed into new cells that become a part of the library, and are used to build other cells. The final cell collection chosen for the library should contain the great majority of the cells needed for the design projects.

The concept of cell hierarchy is based on the building of the cell library. Figure 5.50 provides visualization of the scheme. At the most primitive level, the cells consist only of polygons representing the material layers. This is designated as Level_1 in the hierarchy. Level_2 cells consist of polygons and instances of Level_1 cells. The next group is designated as Level_3 cells. These consist of polygons and may contain instances from Level_1 and Level_2 entries. The last group shown in the drawing are the

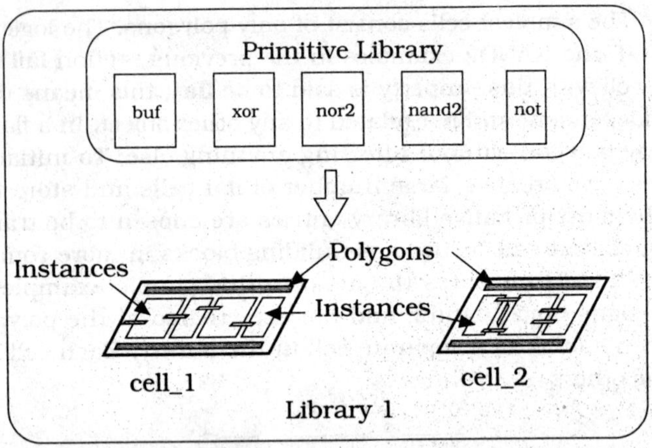

Figure 5.49 Expanding the library with more complex cells

Level_4 cells. They are made up of polygons and instances of any cells from Level_1 to Level_3.

It is important to remember that an instance is only a copy of a simpler entity and its internal structure cannot be altered at a higher level. For example, if a Level_2 cell is instanced into a Level_4 cell, the Level_4 design treats it as being invariant. To alter the Level_2 cell, one must return to the original Level_2 design. Any changes in the cell will propagate to all higher levels where the cell was instanced. In practice, the library is used by a large number of designers, but most users do not

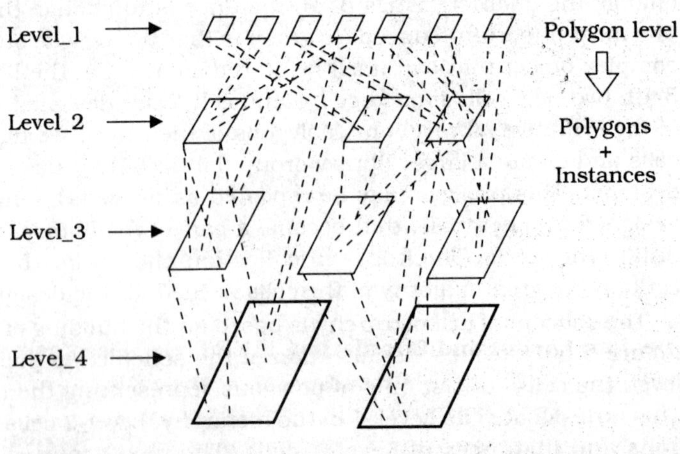

Figure 5.50 Cell hierarchy

have access privileges to the central group of cells. This prevents someone from changing a characteristic that may be critical in another's design.

Although the contents of an instance can be changed, it is possible to decompose it into polygons by the **flatten** command. After a cell is flattened, all references to the original cell are lost and individual features of the circuit can be modified. Figure 5.51 illustrates the effect of the flatten operation. A flattened cell cannot be restored to its original instanced form.

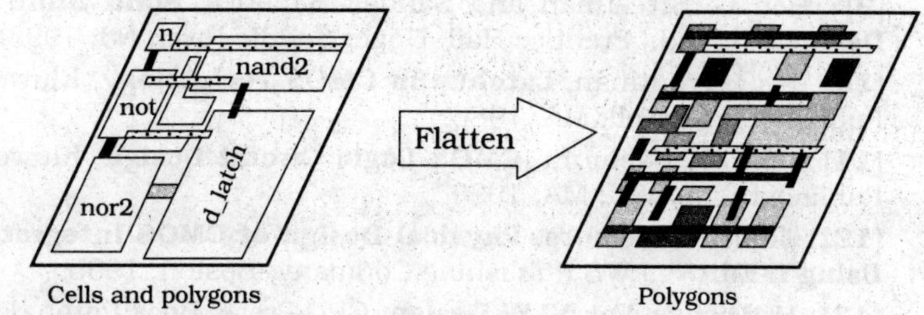

Figure 5.51 Effect of the flatten operation

The concept of design hierarchies is indispensable in VLSI engineering. It allows us to build up complex networks by starting at primitive levels and adding cells as deemed useful. In this manner, various libraries can be built and maintained for use in many different projects. Complex systems are broken down into manageable sections, and the concept of building chips with millions of transistors becomes a reality.

As a closing comment, note that the layout is process dependent. This means that a new library must be built every time a new fabrication plant goes on-line. Unless the new process is radically different, the old cells may be used as a starting point for the new group. Sometimes it is possible to simply scale the dimensions, which is the basis for the concept of cell **reuse**. This helps reduce the time needed for the new chip. Many current designs are created with reuse in mind.

5.7 References for Further Reading

[1] R. Jacob Baker, Harry W. Li, and David E. Boyce **CMOS Circuit Design, Layout and Simulation**, IEEE Press, Piscataway, NJ, 1998.

[2] H. B. Bakoglu, **Circuits, Interconnections, and Packaging for VLSI**, Addison-Wesley, Reading, MA, 1990.

[3] Kerry Bernstein, et al., **High-Speed CMOS Design Styles**, Kluwer Academic Publishers, Norwell, MA, 1998.

[4] Dan Clein, **CMOS IC Layout**, Newnes Publishing Co., Woburn, MA, 2000.

[5] Robert F. Pierret, **Semiconductor Device Fundamentals**, Addison-Wesley, Reading, MA, 1996.

[6] Bryan Preas and Michael Lorenzetti (eds.), **Physical Design Automation of VLSI Systems**, Benjamin/Cummings Publishing Co., Menlo Park, CA, 1988.

[7] M. Sarrafzadeh and C.K. Wong, **An Introduction to VLSI Physical Design**, McGraw-Hill, New York, 1996.

[8] Jasprit Singh, **Semiconductor Devices**, John Wiley & Sons, New York, 2001.

[9] Ben G. Streetman and Sanhay Banerjee, **Solid State Electronic Devices**, 5th ed., Prentice Hall, Upper Saddle River, NJ, 1998.

[10] R. R. Troutman, **Latchup in CMOS Technology**, Kluwer Academic Publishers, Norwell, MA, 1986.

[11] John P. Uyemura, **CMOS Logic Circuit Design**, Kluwer Academic Publishers, Norwell, MA, 1999.

[12] John P. Uyemura, **Physical Design of CMOS Integrated Circuits Using L-Edit™**, PWS Publishing Company, Boston, 1995.

[13] M. Michael Vai, **VLSI Design**, CRC Press, Boca Raton, FL, 2001.

Part 2

The Logic-Electronics Interface

Electrical Characteristics of MOSFETs 6

This chapter centers on MOSFET characteristics and initiates the "electronics" side of VLSI where electrical currents and voltages are the most important quantities. The emphasis, however, is not on studying electronics for its own sake, but to emphasize the link between physical design and logic networks.

6.1 MOS Physics

MOSFETs conduct electrical current by using an applied voltage to move charge from the source side to the drain side of the device. Since the drain and source are physically separate, the flow of charge underneath the gate can occur only if a conduction path, or **channel**, has been created.

Consider the nFET schematic symbol shown in Figure 6.1. The drain current I_{Dn} is controlled by the voltages applied to the device. The primary voltages are identified in the drawing as the gate-source voltage V_{GSn} and the drain-source voltage V_{DSn}. It is important to determine the current versus voltage (I-V) relation

$$I_{Dn} = I_{Dn}(V_{GSn}, V_{DSn}) \tag{6.1}$$

to obtain models for the device operation. Once this is accomplished, we

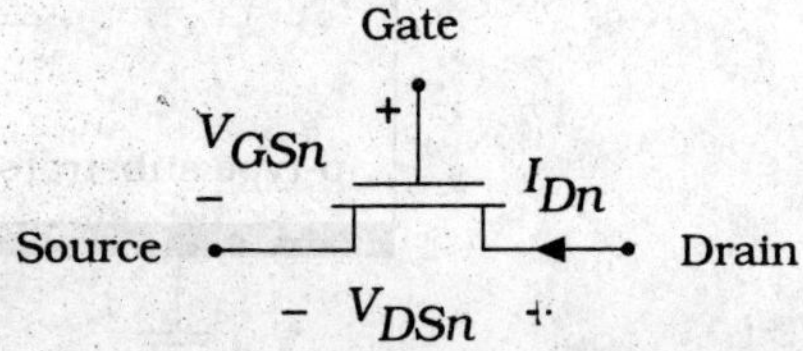

Figure 6.1 nFET current and voltages

will have both a physical understanding and a mathematical model for analyzing and designing CMOS switching networks.

The starting point for our study is the simple MOS structure shown in Figure 6.2; remember that the acronym "MOS" is used to describe any conductor-oxide-semiconductor layering even if the top layer is not a metal. In the present situation, the gate layer is the top conducting layer. This drawing represents the central region of an nFET and provides the physics of how the conduction layer is formed between the drain and source regions. The voltage applied to the gate is denoted by V_G and is assumed to be a positive value with the polarity shown. The oxide layer is taken to be silicon dioxide (SiO_2), which acts as an insulator between the gate and substrate. This gives the oxide capacitance per unit area (in units of F/cm^2) as

$$C_{ox} = \frac{\varepsilon_{ox}}{t_{ox}} \tag{6.2}$$

where t_{ox} is the thickness of the oxide in cm. We recall from Chapter 3 that the permittivity for silicon dioxide is $\varepsilon_{ox} = 3.9\ \varepsilon_o$ such that $\varepsilon_o = 8.854 \times 10^{-14}$ F/cm is the permittivity of free space. Oxide layers in modern CMOS processing are very thin with $t_{ox} < 10$ nm $=10^{-6}$ cm being typical.

The value of C_{ox} determines the amount of electrical coupling that exists between the gate electrode and the p-type silicon region. The effect is most pronounced at the **silicon surface**, i.e., the top of the silicon region. The coupling is described by an electric field E (with units of V/cm) that is created in the insulating oxide layer when a voltage is applied to the gate. The electric field induces charge in the semiconductor and allows us to control the current flow through the FET by varying the gate voltage V_G. This is the origin of the terminology *field-effect*.

To describe the field-effect, we introduce the concept of a surface charge density Q_S that has units of coulombs/square centimeter [C/cm^2].

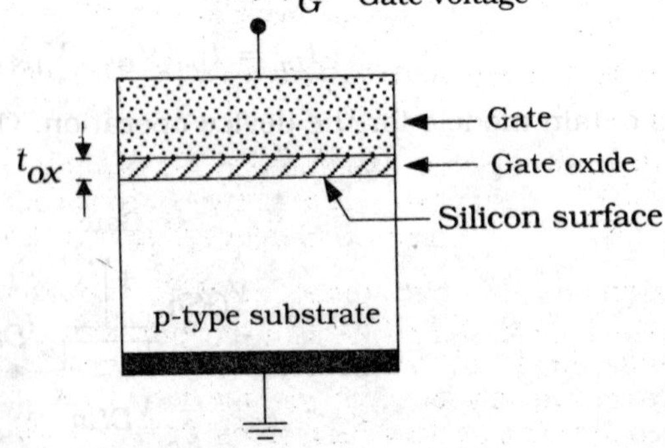

Figure 6.2 Structure of the MOS system

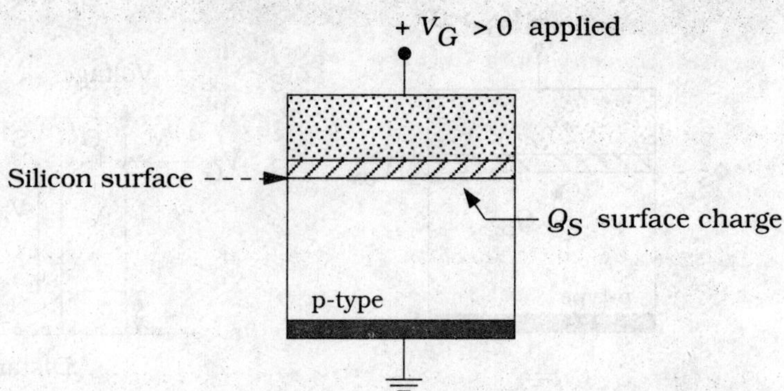

Figure 6.3 Surface charge density Q_S

This is related to the gate voltage by means of

$$Q_S = -C_{ox} V_G \tag{6.3}$$

The concept of the surface charge can be understood using the drawing of Figure 6.3. Physically, Q_S represents the charge density that is seen looking down "into" the semiconductor from the oxide layer. The minus sign is included because a positive V_G induces a negative surface charge density. Although this is a simple-looking equation, MOS physics is complicated by the fact that Q_S represents all of the charge at the semiconductor surface, and the characteristics of the charge depend upon the value of the applied gate voltage.

At the circuit level, the threshold voltage is obtained by applying Kirchhoff's Voltage Law[1] (KVL) to the MOS system shown in Figure 6.4. Assuming that the gate voltage V_G has the polarity shown, KVL gives the expression

$$V_G = V_{ox} + \phi_S \tag{6.4}$$

where V_{ox} is the voltage drop across the oxide layer and ϕ_S is the **surface potential** that represents the voltage at the top of the silicon. The voltages in the MOS system can be plotted as shown, with V_G at the gate and ϕ_S at the silicon surface. The oxide voltage V_{ox} is the difference $(V_G - \phi_S)$ and is the result of a decreasing electric potential inside the oxide as illustrated in the plot. Also note that the voltage in the semiconductor decreases from a value of ϕ_S to $\phi = 0$ in a more gradual manner.

The electric fields in the MOS system are illustrated in Figure 6.5 where we have expanded the vertical dimensions of the oxide to allow us

1 KVL says that the sum of the voltage rises must equal the sum of the voltage drops when a circuit is traced around a closed loop.

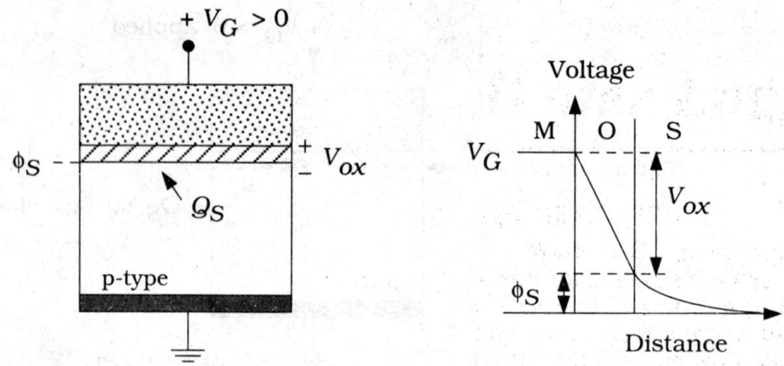

Figure 6.4 Voltages in the MOS system

to see more details. This shows the **oxide electric field** E_{ox} in the insulator pointing away from the higher potential gate electrode. The **surface electric field** E_S also points in the same direction (toward the ground connection), and is the field that controls the surface charge density Q_S at the surface of the semiconductor. This is due to the fact that an electric field exerts a force on a charged particle according to the Lorentz law

$$F = Q_{particle}E \tag{6.5}$$

where $Q_{particle}$ is the charge on the particle with the appropriate sign. Positively charged holes have a charge of $+q$ and the force equation

$$F_h = +qE \tag{6.6}$$

indicates that holes experience a force in the *same direction* as the electric field.[2] Conversely, electrons have a negative charge $-q$ so they experience

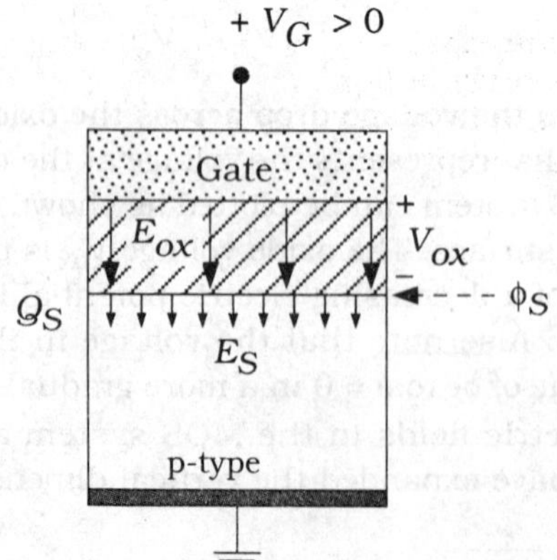

Figure 6.5 MOS electric fields

a force of

$$F_e = -qE \tag{6.7}$$

In this case, the minus sign says that electrons are forced in a direction *opposite* to that of the electric field. With the surface field E_S pointing downward as shown in Figure 6.5, positive charges are forced *away* from the surface while negative charges are attracted *toward* the surface. This explains why the surface charge density consists of negative charge and Q_S itself is a negative number.

The nature of the surface charge depends upon the magnitude of the applied gate voltage. Suppose that V_G starts at 0 V and is then increased to a small positive value, say $V_G = 0.1$ V. The surface field attracts electrons toward the surface while pushing holes downward. This results in a negative charge on the semiconductor surface that is called the **bulk charge** density $Q_B < 0$ with units of C/cm^2. Bulk charge is due to the presence of boron atoms in the p-type substrate. Since a boron acts as an acceptor, it can capture and hold a negatively charged electron. When this happens, it becomes an **ionized dopant** with a net negative charge. Bulk charge is immobile since these ions cannot move. An analysis of the physics gives that

$$Q_B = -\sqrt{2q\varepsilon_{Si}N_a\phi_S} \tag{6.8}$$

where ε_{Si} is the silicon permittivity $\varepsilon_{Si} \approx 11.8\ \varepsilon_o$. For this case the oxide voltage is related to the bulk charge by

$$Q_B = -C_{ox}V_{ox} \tag{6.9}$$

Bulk charge is shown in Figure 6.6, where it is represented by circles with enclosed minus signs. The section from the silicon surface to the bottom of the bulk charge layer is called the **depletion region** because it is "depleted" of free electrons and holes: the holes have been forced away while the electrons have been "absorbed" by the boron dopant atoms. The depth x_d of the depletion layer increases with the applied voltage. This situation defines the "depletion mode of operation" in an MOS system. A depleted MOS structure cannot support the flow of electrical current since bulk charge is trapped by the silicon crystal lattice and cannot move.

If we increase the gate voltage to a special value called the **threshold voltage** V_{Tn}, then we observe a change in the charge properties. As implied by its name, the threshold voltage is the border between two different phenomena. For $V_G < V_{Tn}$, the charge is immobile bulk charge and

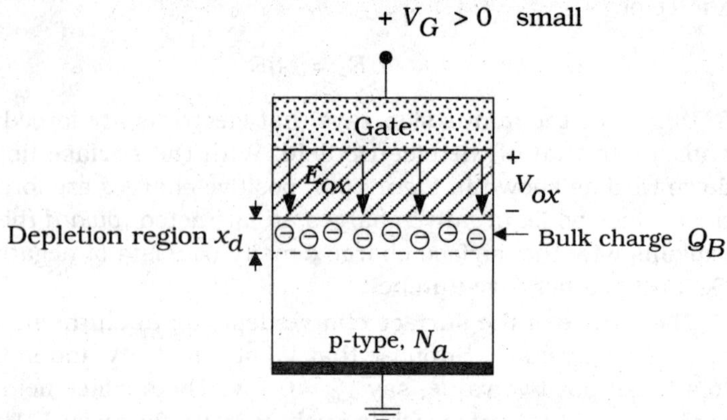

Figure 6.6 Bulk (depletion) charge in the MOS system

$Q_S = Q_B$. However, for $V_G > V_{Tn}$, the charge is made up of two distinct components such that

$$Q_S = Q_B + Q_e < 0 \tag{6.10}$$

where Q_B is the bulk charge but now we observe an electron charge layer that is described by the quantity Q_e C/cm^2. The two components of the surface charge are shown in Figure 6.7. The important point is that electrons are mobile and can move in a lateral direction (parallel to the surface). The electron layer can thus be used as a channel region to construct a MOSFET. The threshold voltage $V_G = V_{Tn}$ represents the value of the gate voltage where Q_e just starts to form. This means that $Q_e = 0$ for $V_G = V_{Tn}$, but Q_e increases for $V_G > V_{Tn}$ according to the capacitor relation

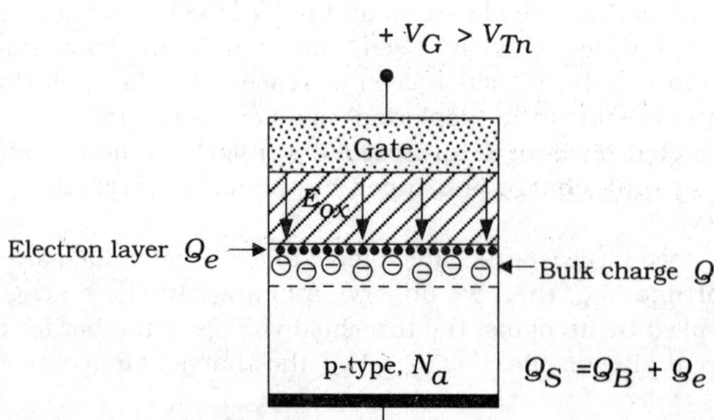

Figure 6.7 Formation of the electron charge layer

$$Q_e = -C_{ox}(V_G - V_{Tn}) \tag{6.11}$$

We must subtract the threshold voltage from V_G to obtain the effective voltage across the insulator after the electron layer has formed. Note that this implies that the bulk charge Q_B does not increase for gate voltages that satisfy $V_G > V_{Tn}$. The negative sign is included to indicate that the electron charge is negative.

The numerical value of the threshold voltage is set in the fabrication process. Typically, it ranges from about V_{Tn} = 0.5 V to V_{Tn} = 0.8 V depending upon the intended type of application for the circuits. In VLSI system design, we assume the V_{Tn} has a value that is specified in the electrical parameters list.

6.1.1 Derivation of the Threshold Voltage[3]

It is not difficult to obtain an approximate expression that illustrates the origin of the numerical value. Recall that KVL gave us the voltage equation

$$V_G = V_{ox} + \phi_S \tag{6.12}$$

A deeper study of the MOS system shows that the electron layer just starts to form when the surface potential reaches a value of

$$\phi_S = 2|\phi_F| \tag{6.13}$$

where $|\phi_F|$ is called the **bulk Fermi potential** which is set by the acceptor doping density N_a of boron in the p-type semiconductor. The analysis gives

$$|\phi_F| = \left(\frac{kT}{q}\right)\ln\left(\frac{N_a}{n_i}\right) \tag{6.14}$$

where k is Boltzmann's constant and T is the temperature in Kelvin. The parameter group (kT/q) is also known as the **thermal voltage** V_{th}, and has a numerical value of $(kT/q) \approx 0.026$ V at room temperature (T = 27° C = 300 K).

With this established, we may write the KVL equation $V_G = V_{Tn}$ as

$$V_{Tn} = V_{ox}\big|_{\phi_S = 2|\phi_F|} + 2|\phi_F| \tag{6.15}$$

Recalling equations (6.8) and (6.9) for Q_B then gives

$$V_{Tn} = \frac{1}{C_{ox}}\sqrt{2q\varepsilon_{Si}N_a(2|\phi_F|)} + 2|\phi_F| \tag{6.16}$$

---

[3] This subsection may be skipped without loss of continuity in the discussion.

This is the threshold voltage for an **ideal MOS** structure in which the oxide is free of all stray charge and the gate and semiconductor materials are identical. A general expression that accounts for a more realistic situation is

$$V_{Tn} = \frac{1}{C_{ox}}\sqrt{2q\varepsilon_{Si}N_a(2|\phi_F|)} + 2|\phi_F| + V_{FB} \tag{6.17}$$

where V_{FB} is called the **flatband voltage** and accounts for both charge in the oxide and different gate and substrate materials.[4] In most modern CMOS processes, V_{FB} is a negative number that gives $V_{Tn} < 0$. Owing to the fact that most CMOS circuits operate with a positive power supply, it is desirable to have a positive threshold voltage with $V_{Tn} > 0$. This is accomplished by introducing another processing step where additional boron ions are implanted into the surface of the region. This alters the threshold voltage equation to read

$$V_{Tn} = \frac{1}{C_{ox}}\sqrt{2q\varepsilon_{Si}N_a(2|\phi_F|)} + 2|\phi_F| + V_{FB} + \frac{qD_I}{C_{ox}} \tag{6.18}$$

where D_I is the implant dose that gives the number of ions implanted per square centimeter; D_I has units of cm^{-2}. The threshold voltage may thus be set by adjusting the implant dose. In some processes, it is also possible to alter the threshold voltage by changing the doping of the gate, which modifies the flatband voltage V_{FB}.

6.2 nFET Current-Voltage Equations

Let us now direct our interest toward the *I-V* characteristics for an n-channel MOSFET. These are determined by the physical structure of the device itself. The nFET consists of an MOS capacitor with n^+ regions added on both sides. The cross-sectional view in Figure 6.8(a) shows how the source and drain n^+ regions are placed with respect to the MOS (gate-oxide-substrate) capacitor. The distance between the edges of the n^+ regions is denoted by L, which is known as the (electrical) **channel length** of the device. L has units of length, and is the smallest feature size in the FET. The labels for the drain and the source are purely arbitrary at this point as the distinction between the two cannot be determined until the voltages are applied. For future reference we will note that in an nFET, the drain is the n^+ side with the higher voltage. A top view of the nFET is provided in Figure 6.8(b). This defines the (electrical) **channel width** W which also has units of length. The dimensionless quantity (W/L) is the

---

4 The name *flatband voltage* arises from an energy band diagram analysis of the system which is beyond the scope of the present treatment.

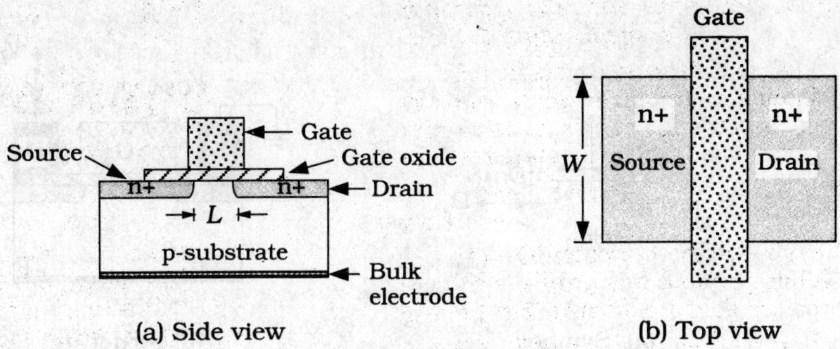

Figure 6.8 Details of the nFET structure

aspect ratio that is used to specify the relative size of a transistor with respect to others in the circuit.

It is important to note that the values used for W and L in this chapter are the electrical or "effective" values, not the drawn values introduced in the previous chapter. This notational convention is used in device physics treatments, and is worthwhile to maintain when the discussion is at the electronics level. To avoid confusion, we will denote the drawn values (used for layout) as L' and W' such that

$$
\begin{aligned}
L &= L' - \Delta L \\
W &= W' - \Delta W
\end{aligned}
\tag{6.19}
$$

give the relationship between the electrical and drawn values; ΔL and ΔW are reduction factors from the processing. All of the equations in this chapter use electrical values W and L, and we maintain this association for the remainder of the book. The actual usage of the two in SPICE will be discussed later to provide final clarification.

The current flow characteristics are found by applying voltages to the physical structure and then analyzing the physics. As shown in Figure 6.9, there is a one-to-one correlation between the voltages represented in the symbol [see Figure 6.9(a)] and those applied to the integrated structure [see Figure 6.9(b)]. The source has been grounded for simplicity. This does not affect the generality of the results, since only relative voltages V_{GSn} and V_{DSn} are used. The program at this point is to determine the dependence of the current I_{Dn} on the voltages.

The key to understanding current flow in an n-channel MOSFET is to note that the MOS structure allows one to control the creation of the electron charge layer Q_e under the gate oxide by using the gate-source voltage V_{GSn}. This is illustrated in Figure 6.10. If $V_{GSn} < V_{Tn}$ then $Q_e = 0$ as illustrated in Figure 6.10(a). Since no electron layer exists, the two n+ regions are physically separated from each other, and no direct current flow path exists between them. From the outside world, an open circuit exists

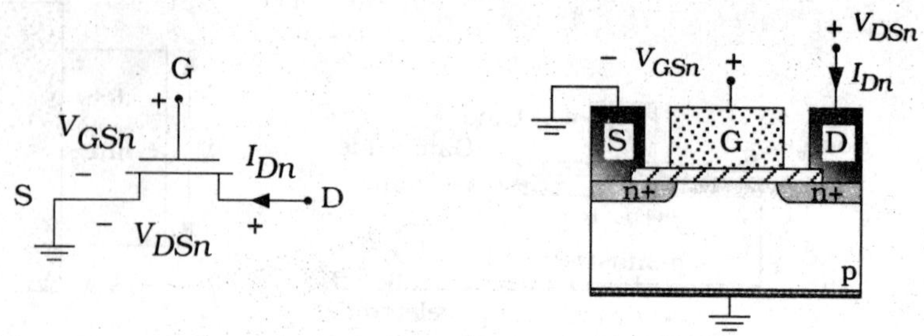

(a) Symbol (b) Structure

Figure 6.9 Currents and voltages for an nFET

between the drain and source terminals, which then says that the current I_{Dn} must be 0. This state of operation is called **cutoff**, and is defined by having $V_{GSn} < V_{Tn}$ with $I_{Dn} = 0$. A cutoff transistor is equivalent to having an open switch between the drain and source terminals.

If, on the other hand, the gate-source voltage is increased to a value $V_{GSn} > V_{Tn}$, the situation changes dramatically. An electron charge layer Q_e is created underneath the gate oxide as shown in Figure 6.10(b). The layer provides an electrical **channel** between the electrons in the drain and source n+ regions and allows current to flow between the two. The presence of a channel defines the **active** mode of operation of the transistor. The numerical value of the current I_{Dn} depends upon both V_{GSn} and V_{DSn}.

Figure 6.11 shows the operational modes of the FET from the viewpoint of layers. These drawings illustrate the device operation as it appears at the silicon surface, i.e., if the gate layer is made transparent. Cutoff with $V_{GSn} < V_{Tn}$ is shown in Figure 6.11(a); since $Q_e = 0$, no channel exists between the drain and source regions and the device acts like an open switch. Figure 6.11(b) illustrates the opposite case where the gate-source voltage satisfies $V_{GSn} > V_{Tn}$ and results in the formation of an

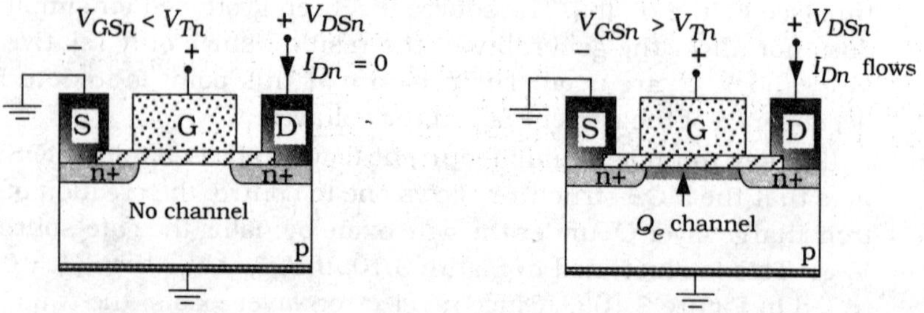

(a) Cutoff (b) Active bias

Figure 6.10 Controlling the channel in an nFET

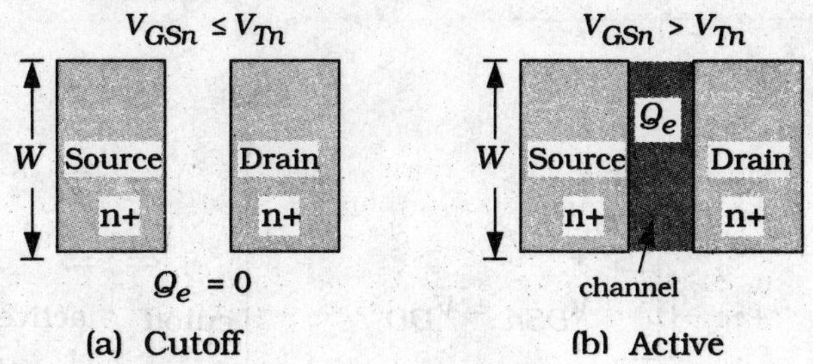

Figure 6.11 Channel formation in an nFET

electron charge layer Q_e. This defines the active mode of operation. The charge layer acts as a conducting channel between the two n+ regions, allowing charge transport between the two.

The behavior described in the preceding paragraphs almost justifies the modeling of an nFET as an assert-high switch that is OPEN with a small gate voltage ($V_{GSn} < V_{Tn}$) and CLOSED with a large gate voltage ($V_{GSn} > V_{Tn}$). In VLSI, the switch model is sufficient for designing logic gates as was demonstrated in Chapter 2. However, the electrical characteristics of FETs deviate substantially from those of an ideal switch. While this consideration does not affect the rules for logic formation with FETs, it does establish fundamental upper limits on the transient response of a CMOS network. Since switching speed is critical in modern chip design, it is worth the effort to dig deeper into the operation of MOSFETs to provide a complete picture of the VLSI design environment. The sizing of transistors provides the link between the physical design and the electronic operation of a logic gate.

To characterize an nFET at the device level, we will adopt the simple procedure where the current is plotted as a function of a voltage. Since there are two voltages (V_{GSn} and V_{DSn}), we will hold one constant while varying the other, and perform two separate experiments to obtain the overall behavior. The first is shown in Figure 6.12, where we have set the drain-source voltage to be the power supply value ($V_{DSn} = V_{DD}$) while we increase V_{GSn} in a positive direction from 0 V. This results in the plot of I_{Dn} vs. V_{GSn} shown. For voltages $V_{GSn} < V_{Tn}$, the transistor is in cutoff and $I_{Dn} = 0$. Increasing the gate-source voltage to values $V_{GSn} > V_{Tn}$ biases the nFET into the active region of operation by forming the electron charge layer Q_e. The drain-source voltage V_{DSn} provides the difference in potential needed to move the charge, which results in the current I_{Dn} flowing through the device. Mathematically, the current can be approximated by the equation

$$I_{Dn} = \frac{\beta_n}{2}(V_{GSn} - V_{Tn})^2 \tag{6.20}$$

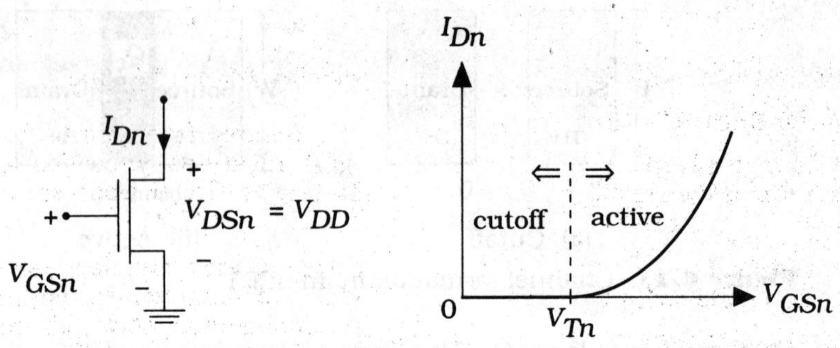

Figure 6.12 *I-V* characteristics as a function of V_{GSn}

which shows a quadratic dependence on the voltage. This defines the **square-law model** of a FET. Although it is only an approximation, it is very useful for calculating the behavior of complex CMOS networks. The factor β_n multiplying the voltage factor is the device transconductance parameter with units of A/V^2. Every nFET has a distinct value of β_n that is determined by its aspect ratio through

$$\beta_n = k'_n\left(\frac{W}{L}\right) \tag{6.21}$$

In this equation, k'_n is the process transconductance parameter that is calculated from

$$k'_n = \mu_n C_{ox} \tag{6.22}$$

It cannot be changed by the VLSI designer. In this equation, μ_n is the electron mobility at the silicon surface. In a silicon MOSFET at room temperature, μ_n is typically around 500 to 580 cm^2/V-sec and is a characteristic of the material.

Note that the process transconductance is proportional to the oxide capacitance per unit area

$$C_{ox} = \frac{\varepsilon_{ox}}{t_{ox}} \tag{6.23}$$

Substituting gives

$$k'_n = \frac{\mu_n \varepsilon_{ox}}{t_{ox}} \tag{6.24}$$

so that a thin oxide (small t_{ox}) gives a large value for k'_n. This increases the sensitivity of the device with respect to the gate voltage, and helps the device switch faster. From the physical viewpoint it can be seen that decreasing t_{ox} increases C_{ox}, which in turn enhances the field-effect.

Example 6.1
Consider an nFET that has a gate oxide thickness of t_{ox} = 12 nm and an electron mobility of μ_n = 540 cm^2/V-sec. The oxide capacitance per cm^2 is

$$C_{ox} = \frac{(3.9)(8.854\times10^{-14})}{1.2\times10^{-6}} = 2.88\times10^{-7} \quad \text{F/cm}^2 \tag{6.25}$$

where we have used t_{ox} = 12 nm = 1.2 × 10^{-6} cm since the permittivity is in units of F/cm. The process transconductance is computed from

$$\begin{aligned} k'_n &= \mu_n C_{ox} \\ &= (540)(2.88\times10^{-7}) \\ &= 1.55\times10^{-4} \quad \text{A/V}^2 \end{aligned} \tag{6.26}$$

or,

$$k'_n = 155 \ \mu\text{A/V}^2 \tag{6.27}$$

If the oxide is reduced to a thickness of t_{ox} = 8 nm, then the process transconductance increases to a value of

$$k'_n = 233 \ \mu\text{A/V}^2 \tag{6.28}$$

indicating a more sensitive device.

Let us now change the voltages to the situation shown in Figure 6.13. In this case, we apply a constant gate-source voltage $V_{GSn} > V_{Tn}$ to the nFET and vary the drain-source voltage V_{DSn}. This gives the plot of I_{Dn} vs. V_{DSn} shown. For small values of V_{DSn}, the current can be estimated by the equation

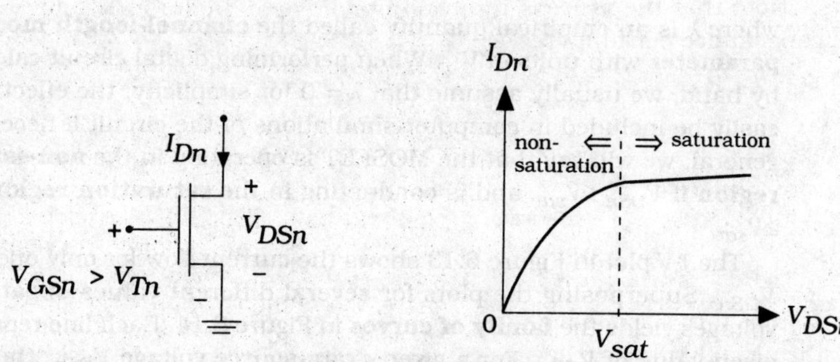

Figure 6.13 *I-V* characteristics as a function of V_{DSn}

$$I_{Dn} = \frac{\beta_n}{2}[2(V_{GSn} - V_{Tn})V_{DSn} - V_{DSn}^2] \tag{6.29}$$

which describes a parabola. The peak occurs at the point where

$$\frac{\partial I_{Dn}}{\partial V_{DSn}} = 0 \tag{6.30}$$

Evaluating the derivative and equating the result to 0 gives

$$\frac{\partial}{\partial V_{DSn}}[2(V_{GSn} - V_{Tn})V_{DSn} - V_{DSn}^2] = 2(V_{GSn} - V_{Tn}) - 2V_{DSn} = 0 \tag{6.31}$$

The solution to this equation defines a special value of V_{DSn} called the **saturation voltage**

$$\begin{aligned} V_{sat} &= V_{DSn}\big|_{\text{peak current}} \\ &= V_{GSn} - V_{Tn} \end{aligned} \tag{6.32}$$

that is shown in the plot. For larger drain-source voltages that satisfy $V_{DSn} \geq V_{sat}$, the current is approximately independent of V_{DSn} and is given by

$$I_{Dn} = \frac{\beta_n}{2}(V_{GSn} - V_{Tn})^2 \tag{6.33}$$

This is identical to that given in equation (6.20) and is called the **saturation current** since it is the largest value of I_{Dn} that can flow for a given value of V_{GSn}. A more detailed analysis shows that the saturation current does increase slightly for $V_{DSn} \geq V_{sat}$. This is often modeled by the equation

$$I_{Dn} = \frac{\beta_n}{2}(V_{GSn} - V_{Tn})^2[1 + \lambda(V_{DSn} - V_{sat})] \tag{6.34}$$

where λ is an empirical quantity called the **channel-length modulation parameter** with units of V^{-1}. When performing digital circuit calculations by hand, we usually assume that $\lambda = 0$ for simplicity; the effect of λ can easily be included in computer simulations of the circuit if necessary. In general, we will say that the MOSFET is operating in the **non-saturation region** if $V_{DSn} \leq V_{sat}$, and is conducting in the **saturation region** if $V_{DSn} \geq V_{sat}$.

The *I-V* plot in Figure 6.13 shows the current flow for only one value of V_{GSn}. Superposing the plots for several different values of gate-source voltages yields the **family of curves** in Figure 6.14. Each line represents a given value of V_{GSn}. For a given drain-source voltage V_{DSn}, the current increases with V_{GSn} as indicated. The separation between the non-saturated and saturated operational regions is given by the saturation current

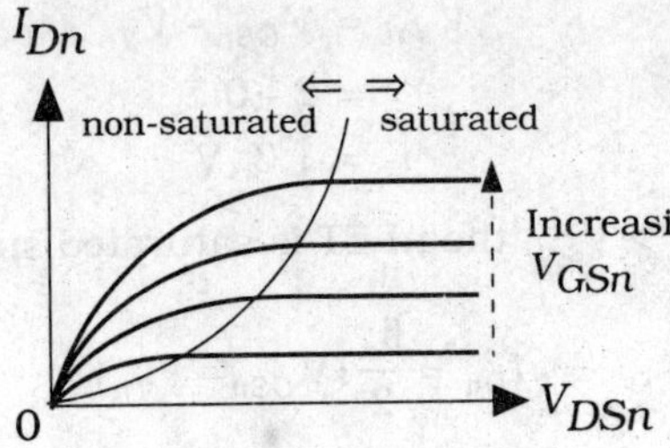

Figure 6.14 nFET family of curves

$$I_{Dn} = \frac{\beta_n}{2} V_{sat}^2 \tag{6.35}$$

where $V_{sat} = (V_{GSn} - V_{Tn})$ depends upon the value of the gate-source voltage. This set of equations allows us to find the drain current I_{Dn} once we know the voltages.

Example 6.2

Consider an n-channel MOSFET with the following characteristics:

$t_{ox} = 10$ nm, $\mu_n = 520$ cm^2/V-s, $(W/L) = 8$, $V_{Tn} = +0.70$ V

This information allows us to find the device equations. We will start by finding the oxide capacitance using

$$C_{ox} = \frac{\varepsilon_{ox}}{t_{ox}} = \frac{(3.9)(8.854\times10^{-14})}{10\times10^{-7}} = 3.45\times10^{-7}\ \text{F/cm}^2 \tag{6.36}$$

The process transconductance is found from

$$k'_n = \mu_n C_{ox} = (520)(3.45\times10^{-7}) = 1.79\times10^{-4}\ \text{A/V}^2 \tag{6.37}$$

or, $k'_n = 179\ \mu\text{A/V}^2$. The device transconductance may now be calculated from

$$\beta'_n = k'_n\left(\frac{W}{L}\right) = 179(8) = 1.435\ \text{mA/V}^2 \tag{6.38}$$

Let us now calculate the drain current for different voltage combinations.

Suppose that we apply voltages of $V_{GSn} = 2$ V and $V_{DSn} = 2$ V to the nFET. The first task is to determine the state of conduction, i.e., is the transistor operating in the saturated or non-saturated region? Once this is known, we can use the appropriate equation. The saturation voltage is

$$\begin{aligned} V_{sat} &= V_{GSn} - V_{Tn} \\ &= 2 - 0.7 \\ &= 1.3 \text{ V} \end{aligned} \tag{6.39}$$

Since $V_{DS} = 2\text{ V} > V_{sat}$, the nFET is saturated such that

$$\begin{aligned} I_{Dn} &= \frac{\beta_n}{2}(V_{GSn} - V_{Tn})^2 \\ &= \left(\frac{1.435}{2}\right)(2 - 0.7)^2 \\ &= 1.213 \text{ mA} \end{aligned} \tag{6.40}$$

Now let us lower the drain-source voltage to $V_{DSn} = 1.2$ V while maintaining $V_{GSn} = 2$ V. The saturation voltage is still given by

$$V_{sat} = V_{GSn} - V_{Tn} = 1.3 \text{ V} \tag{6.41}$$

but now $V_{DSn} = 1.2\text{ V} < V_{sat}$, which says that the transistor is non-saturated. The current is then computed from

$$\begin{aligned} I_{Dn} &= \frac{\beta_n}{2}[2(V_{GSn} - V_{Tn})V_{DSn} - V_{DSn}^2] \\ &= \left(\frac{1.435}{2}\right)[2(1.3)(1.2) - (1.2)^2] \\ &= 1.21 \text{ mA} \end{aligned} \tag{6.42}$$

This set of calculations illustrates the general current characteristics of a MOSFET.

6.2.1 SPICE Level 1 Equations

Channel-length modulation effects are easily included in SPICE simulations, but tend to be somewhat cumbersome for hand calculations that use the equation set above. An alternate set of MOSFET equations that follows SPICE LEVEL 1 models is to write the non-saturation current which is valid for $V_{DSn} \le V_{sat}$ in the form

$$I_{Dn} = \frac{\beta_n}{2}[2(V_{GSn} - V_{Tn})V_{DSn} - V_{DSn}^2](1 + \lambda V_{DSn}) \tag{6.43}$$

This provides a continuous transition to the saturation current

$$I_{Dn} = \frac{\beta_n}{2}(V_{GSn} - V_{Tn})^2(1 + \lambda V_{DSn}) \tag{6.44}$$

that is valid for $V_{DSn} \ge V_{sat}$. This is not consistent with a physical analysis since channel-length modulation occurs only in a saturated device. However, it makes circuit analysis easier. These forms are quite common in analog CMOS design. However, channel-length modulation effects do not affect hand calculations of digital circuits enough to justify the increased algebraic complexity, so they are rarely used in hand calculations here.

6.2.2 Body-Bias Effects

Up to this point we have ignored the presence of the p-type substrate. In reality, the MOSFET is a four-terminal device with the substrate being the **bulk** (B) terminal of the device. **Body-bias effects** occur when a voltage V_{SBn} exists between the source and bulk terminals of a nFET as in Figure 6.15. The body-bias V_{SBn} voltage increases the threshold voltage of the device such that

$$V_{Tn} = V_{TOn} + \gamma(\sqrt{2|\phi_F| + V_{SBn}} - \sqrt{2|\phi_F|}) \tag{6.45}$$

where γ is the body-bias coefficient with units of $V^{1/2}$ and $2|\phi_F|$ is the bulk Fermi potential term from equation (6.14). The term V_{TOn} is the zero body-bias threshold voltage

$$V_{TOn} = V_{Tn}|_{V_{SBn}=0} \tag{6.46}$$

and is the value quoted in a set of processing specifications. The body-bias coefficient can be estimated by

$$\gamma = \frac{\sqrt{2q\varepsilon_{Si}N_a}}{C_{ox}} \tag{6.47}$$

where $q = 1.6 \times 10^{-19}$ C is the fundamental charge unit, $\varepsilon_{Si} = 11.8\varepsilon_o$ is the permittivity of silicon, and N_a is the acceptor doping in the p-type substrate. The value of γ is usually quoted in the process specification. Note that thin oxides decrease the value of γ.

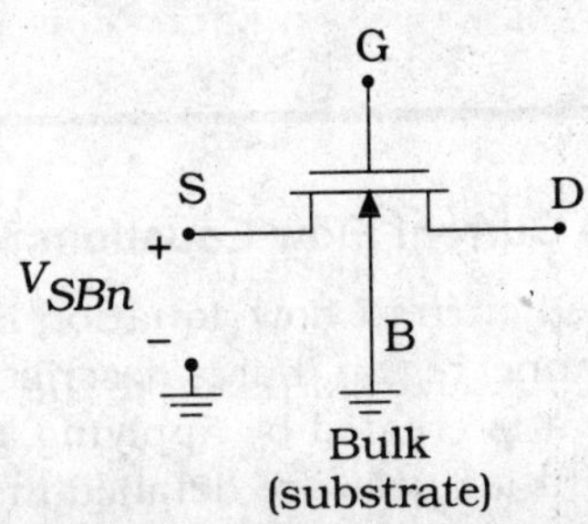

Figure 6.15 Bulk electrode and body-bias voltage

Example 6.3

Consider an nFET where $V_{T0n} = 0.7$ V, $\gamma = 0.08$ $V^{1/2}$, and $2|\phi_F| = 0.58$ V. The threshold voltage depends on the body-bias voltage V_{SBn} according to

$$V_{Tn} = 0.70 + 0.08(\sqrt{0.58 + V_{SBn}} - \sqrt{0.58}) \tag{6.48}$$

Some values can be computed as follows:

V_{SBn} (V)	V_{Tn} (V)
0	0.70
1	0.74
2	0.77
3	0.79

The function is plotted in Figure 6.16, which illustrates the characteristic square root dependence.

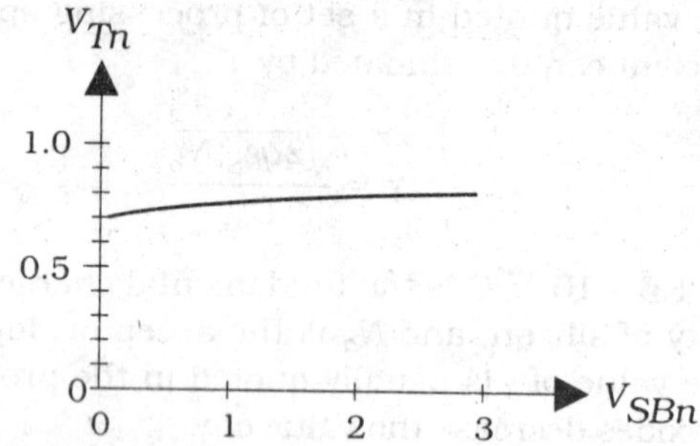

Figure 6.16 Body-bias effect

6.2.3 Derivation of the Current Flow Equations[5]

The non-saturated current flow equation is obtained by analyzing the physics of the channel region that is described of the electron charge density Q_e C/cm^2 that is created by applying a gate-source voltage $V_{GSn} > V_{Tn}$. The important features are detailed in Figure 6.17. Physically, the

5 This section may be skipped without loss of continuity. The reader may jump to Section 6.3 where the main discussion is resumed.

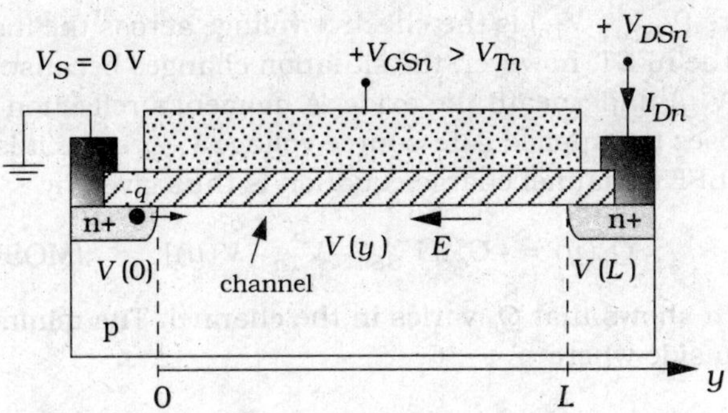

Figure 6.17 Channel voltage in a MOSFET

drain-source voltage V_{DSn} applied across the device induces an electric field E that points from the drain to the source (remember that, by definition, the drain is the side at the higher voltage). Since electrons have a negative charge $-q$, they experience a force in a direction opposite to that of the electric field. The electrons thus move *from* the source and flow through the channel *to* the drain; this is the origin of the electrode names. In electronics, we usually deal with **conventional current** which moves in the direction of positive charge; current flows in a direction that is opposite to the direction of electron motion. Applying this to the nFET shows that the current flows from the drain to the source as shown.

Now that the qualitative aspects of the physics have been discussed, let us analyze the situation in greater depth. From electromagnetic theory we know that electric fields are conservative. This means that there exists an electrostatic potential (or voltage) $V(y)$ such that

$$E(y) = -\frac{dV}{dy} \tag{6.49}$$

where y is a coordinate that is defined as shown in the drawing. $V(y)$ is called the **channel voltage** and is due to the applied drain-source voltage V_{DSn}. At the ends of the channel, it has the known values of

$$\begin{aligned} V(0) &= 0 \\ V(L) &= V_{DSn} \end{aligned} \tag{6.50}$$

which act as boundary conditions on the problem and indicate that $V(y)$ decreases from the drain to the source. The existence of the channel voltage alters the charge in the channel and makes Q_e a function of the coordinate y. To understand this, recall the electron charge density in a simple MOS structure (not a FET) is given by

$$Q_e = -C_{ox}(V_{GSn} - V_{Tn}) \qquad \text{(MOS value)} \tag{6.51}$$

where $(V_{GSn} - V_{Tn})$ is the effective voltage across the insulating oxide layer. For the nFET, however, the situation changes because of the channel voltage $V(y)$ underneath the oxide. A moment's reflection will verify that $V(y)$ opposes the applied gate-source voltage V_{GSn} since it is a positive number. The nFET channel charge equation is thus given by

$$\mathcal{Q}_e(y) = -C_{ox}[V_{GSn} - V_{Tn} - V(y)] \qquad \text{(MOSFET)} \tag{6.52}$$

which shows that $\mathcal{Q}_e$ varies in the channel. The minimum value is on the drain side where

$$\mathcal{Q}_e(L) = -C_{ox}[V_{GSn} - V_{Tn} - V_{DSn}] \tag{6.53}$$

while the maximum charge density is found at the source with

$$\mathcal{Q}_e(0) = -C_{ox}[V_{GSn} - V_{Tn}] \tag{6.54}$$

The functional dependence $\mathcal{Q}_e(y)$ is significant because it means that the charge density is nonuniform. This in turn implies that the *I-V* relationship will be non-linear.

The equation for I_{Dn} can be obtained by applying the above observations to the channel geometry illustrated in Figure 6.18. To handle the varying charge density, let us start with the differential channel segment that has a length dy as shown. The current I_{Dn} flows through this segment and causes a voltage drop

$$dV = I_{Dn} dR \tag{6.55}$$

where dR is the differential resistance

$$dR = \frac{dy}{\sigma_n A_n} \tag{6.56}$$

of the segment. In this equation, σ_n is the conductivity and A_n is the cross-sectional area. Since the conductivity of an n-type region is given by

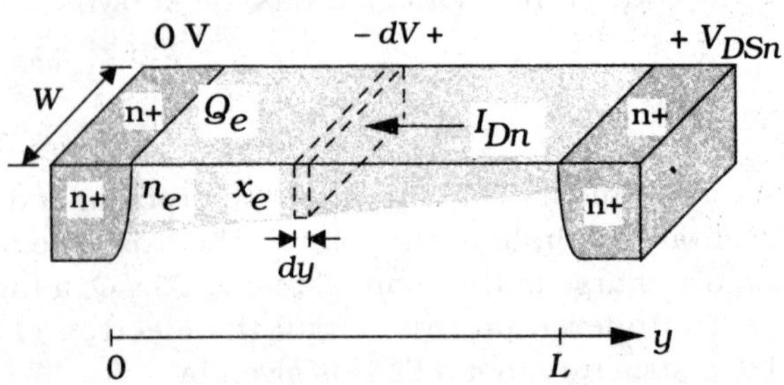

Figure 6.18 Channel geometry

$\sigma_n = q\mu n$, we may rewrite the denominator in the form

$$\sigma_n A_n = q\mu_n n_e W x_e \tag{6.57}$$

where n_e is the electron density in units of cm^{-3} and x_e is the thickness of the channel at that point. The channel charge density is equivalent to

$$Q_e = -qn_e x_e \tag{6.58}$$

This can be seen by noting the units of Q_e are C/cm^2 and that the given quantities combine on physical grounds; the negative sign is due to the fact that Q_e is defined as a negative number. Substituting this into resistance equation then yields

$$dV = -\frac{I_{Dn}dy}{\mu_n W Q_e} = \frac{I_{Dn}dy}{\mu_n W C_{ox}(V_{GSn} - V_{Tn} - V)} \tag{6.59}$$

using the expression for Q_e from equation (6.52). This can be rearranged and integrated to read

$$I_{Dn}\int_0^L dy = \mu_n W C_{ox}\int_0^{V_{DSn}} [(V_{GSn} - V_{Tn}) - V]dV \tag{6.60}$$

The limits of integration have been chosen as $y = 0$ to $y = L$ to include the entire channel. The voltage integral on the right-hand side uses the equivalent channel voltages at these points, i.e., $V(0) = 0$ V and $V(L) = V_{DSn}$ Assuming that the term $(V_{GSn}-V_{Tn})$ on the right side is independent of the channel voltage V gives

$$I_{Dn}L = \mu_n W C_{ox}[(V_{GSn} - V_{Tn})V_{DSn} - V_{DSn}^2] \tag{6.61}$$

so that

$$I_{Dn} = \mu_n C_{ox}\left(\frac{W}{L}\right)[(V_{GSn} - V_{Tn})V_{DSn} - V_{DSn}^2] \tag{6.62}$$

This is the same as the non-saturated current expression given earlier in equation (6.29).

One interesting point concerning the channel arises when we extend the analysis to the saturation voltage $V_{sat} = (V_{GSn} - V_{Tn})$. Equation (6.53) gives the channel charge at the drain side. Substituting the saturation voltage $V_{DSn} = V_{sat}$ gives

$$Q_e(L) = -C_{ox}[V_{GSn} - V_{Tn} - V_{sat}] = 0 \tag{6.63}$$

i.e., the charge density appears to fall to 0 when at the saturation voltage. A more detailed analysis shows that the charge does not really fall to zero, but is in fact small. This corresponds to a phenomenon known as **channel pinch-off** in the FET. Formally, it is the border between saturation and non-saturation regions of operation. For $V_{DSn} > V_{sat}$, the pinch-off of the charge limits the current flow (hence the term **saturation**) and the pinch-off effect itself decreases the effective length of the channel (hence the **channel-length modulation** factor λ).

6.3 The FET RC Model

The equations of current flow above illustrate that the nFET exhibits **non-linear** *I-V* characteristics. This property makes it difficult to analyze electrical circuits that use FETs because the circuit equations themselves become non-linear; hand calculations thus become quite tedious. The solution, of course, is to use a CAD tool such as SPICE to perform the difficult analyses. But this does not solve the problem that VLSI designers face: they must **create** circuits that have the proper electrical characteristics. This pinpoints the difference between **analysis** and **design**: analysis deals with studying a new network that has resulted from the design process. Designers are true problem solvers in that they use existing knowledge as a basis for building new systems.

There are two approaches to dealing with the problem of messy transistor equations. The first is to let circuit specialists deal with the issues introduced by the non-linear devices. Skilled electronic designers are indispensable in the chip design process. VLSI system design, on the other hand, is based on logic and digital architectures; engineers working at the systems level also need to understand FET circuitry. This provides the basis of the second approach: create a simplified **linear model** of the device that is useful at the logic and system level. By its very nature, the model will ignore most of the details of the current flow. It will, however, be much simpler to use for tracing signal flows in complex networks at the system level. If we can work at least some of the important transistor characteristics into the model, then it can be used to provide a basis for the first design phase. Simplified linear models also allow us to develop techniques that compare various algorithms for choosing the most efficient VLSI approach.

The linear model that will be used in our treatment is shown in Figure 6.19. This simplifies the nFET to a resistor R_n, two capacitors (C_S and C_D), and an assert-high logic-controlled switch. The values of the linear components depend on the aspect ratio $(W/L)_n$ of the nFET in a manner that will be developed in the next two subsections.

6.3.1 Drain-Source FET Resistance

Field-effect transistors are inherently non-linear, so we must be careful about the concept of using a linear resistor with a fixed value of R_n to model the current flow through an nFET.

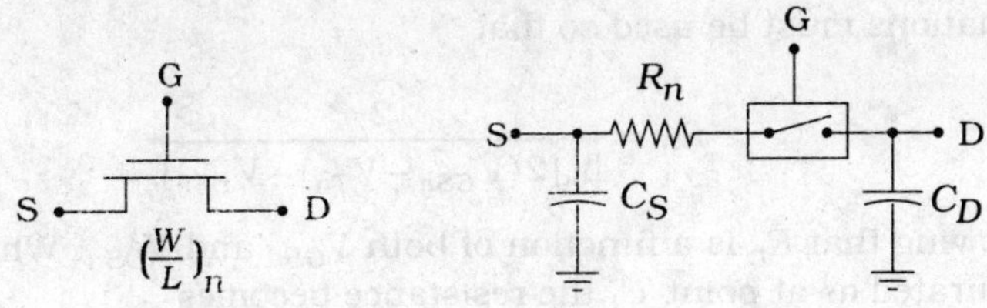

(a) nFET Symbol (b) Linear model for nFET

Figure 6.19 RC model of an nFET

Consider the situation shown in Figure 6.20. In Figure 6.20(a), the gate-source voltage is assumed to be set with a value $V_{GSn} > V_{Tn}$, to make the nFET active. The current I_{Dn} is then a function of the drain-source voltage V_{DSn} as plotted in Figure 6.20(b). The drain-source resistance at any point on the curve is then given by

$$R_n = \frac{V_{DSn}}{I_{Dn}} \tag{6.64}$$

The non-linear effects are due to the fact that I_{Dn} varies with V_{DSn}, which makes R_n itself a function of V_{DSn}.

The effects of this dependence can be seen by writing the resistance equations for the three points labeled 'a', 'b', and 'c' shown in the drawing. For small values of V_{DSn} (point 'a'), the current is approximated by

$$I_{Dn} \approx \beta_n (V_{GSn} - V_{Tn}) V_{DSn} \tag{6.65}$$

by ignoring the squared term V_{DSn}^2 in the non-saturated current flow equation (6.29). The resistance is then

$$R_n \approx \frac{1}{\beta_n (V_{GSn} - V_{Tn})} \tag{6.66}$$

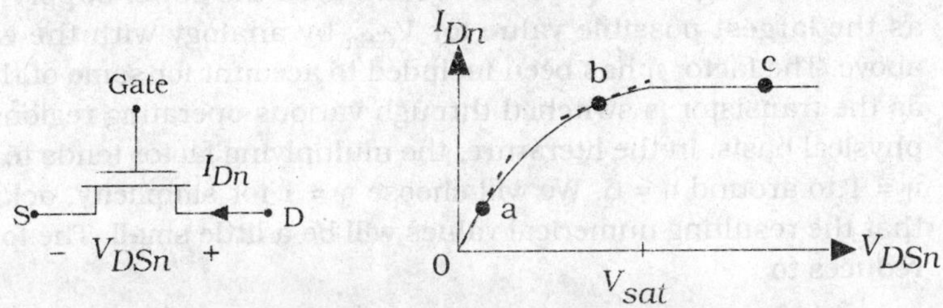

Figure 6.20 Determining the nFET resistance.

so that R_n varies with V_{GSn}. At point 'b', the full non-saturated current equations must be used so that

$$R_n = \frac{2}{\beta_n[2(V_{GSn} - V_{Tn}) - V_{DSn}]} \tag{6.67}$$

showing that R_n is a function of both V_{GSn} and V_{DSn}. When the device is saturated as at point 'c', the resistance becomes

$$R_n = \frac{2V_{DSn}}{\beta_n(V_{GSn} - V_{Tn})^2} \tag{6.68}$$

by using equation (6.20) which ignores channel-length modulation. Once again, the resistance varies with both V_{GSn} and V_{DSn}.

These equations illustrate that it is not possible to define a constant value for R_n and still maintain the correct current-flow behavior. Note, however, that in all cases, R_n is inversely proportional to β_n, i.e.,

$$R_n \propto \frac{1}{\beta_n} \tag{6.69}$$

This is simply a statement that a device with a large β_n conducts more current than one with a small β_n. Using the definition

$$\beta_n = k'_n\left(\frac{W}{L}\right)_n \tag{6.70}$$

shows that the important parameter is the device aspect ratio $(W/L)_n$. Qualitatively, increasing the width W of the nFET decreases the resistance.

With this in mind, we will introduce a simple equation for modeling the resistance as a function of the aspect ratio (or, width) of the transistor by writing

$$R_n = \frac{\eta}{\beta_n(V_{DD} - V_{Tn})} \tag{6.71}$$

In constructing this equation, we have used the power supply voltage V_{DD} as the largest possible value for V_{GSn} by analogy with the expressions above. The factor η has been included to account for some of the variation as the transistor is switched through various operating regions; it has no physical basis. In the literature, the multiplying factor tends to range from $\eta = 1$ to around $\eta = 6$. We will choose $\eta = 1$ for simplicity, acknowledging that the resulting numerical values will be a little small. The formula then reduces to

$$R_n = \frac{1}{\beta_n(V_{DD} - V_{Tn})} \ \Omega \tag{6.72}$$

which is the final form. The unit of the resistance R_n is ohms, which is consistent with the units established by the denominator.

Example 6.4

Consider an nFET that has a channel width $W = 8$ μm, a channel length of $L = 0.5$ μm, and is made in a process where $k'_n = 180$ μA/V^2, $V_{Tn} = 0.70$ V, and $V_{DD} = 3.3$ V. The linearized drain-source resistance is computed as

$$R_n = \frac{1}{\beta_n(V_{DD} - V_{Tn})} \tag{6.73}$$

so that substituting the values gives

$$R_n = \frac{1}{(180\times10^{-6})\left(\frac{8}{0.5}\right)(3.3-0.7)} = 133.5\ \Omega \tag{6.74}$$

If we shrink the channel width to $W = 5$ μm while keeping all other quantities the same, the resistance increases to

$$R_n = 133.5\left(\frac{8}{5}\right) = 213.6\ \Omega \tag{6.75}$$

where we have simply scaled the value by noting that R_n is inversely proportional to the channel width. It is important to remember that these values are not actual values for the nFET resistance, but are used only for simplified modeling.

6.3.2 FET Capacitances

A MOSFET has several parasitic capacitances that must be included in the simplified switching model. As we will see in later developments, the maximum switching speed of a CMOS circuit is determined by the capacitances.

MOS Capacitances

The metal-oxide-semiconductor layering scheme is intrinsically a capacitor, so let us analyze its value first. Figure 6.21(a) shows the circuit model. If we look into the gate terminal of the FET, we see the **gate capacitance** C_G that is due to the MOS structure. Since this is the region that has a gate oxide thickness of t_{ox}, it is described by the oxide capacitance per unit area C_{ox}. Denoting the area of the gate region by A_G gives us

$$C_G = C_{ox}A_G \tag{6.76}$$

in farads, which is taken to be the capacitance between the gate terminal and ground. For the simple geometry shown in Figure 6.21(b) the gate

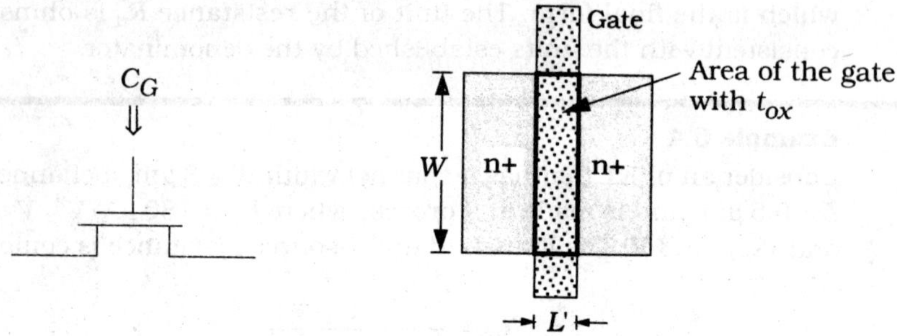

(a) CIrcuit perspective (b) Physical origin

Figure 6.21 Gate capacitance in a FET

area is $A_G = WL'$ where W is the channel width and L' is the drawn channel length. L' is just the channel length that is defined by the extent of the gate region when viewed from the top of the layout drawing. Thus,

$$C_G = C_{ox} WL' \tag{6.77}$$

gives the important result that the gate capacitance is proportional to the width W of the channel.

We also describe the MOS contributions using the gate-source capacitance C_{GS} and the gate-drain capacitance C_{GD} shown in Figure 6.22. These two parasitics are complicated because their values change with the voltages due to the changing shape cf the channel region. When we have $C = C(V)$, the capacitance is said to be **non-linear**. In VLSI system design, we will usually employ a circuit simulation program such as SPICE to handle the detailed calculations. For our purposes, we will simply estimate the values by writing

$$C_{GS} \approx \frac{1}{2} C_G \approx C_{GD} \tag{6.78}$$

In other words, we will just divide the gate capacitance by 2 and split it equally between C_{GS} and C_{GD}. Although this isn't extremely accurate, it

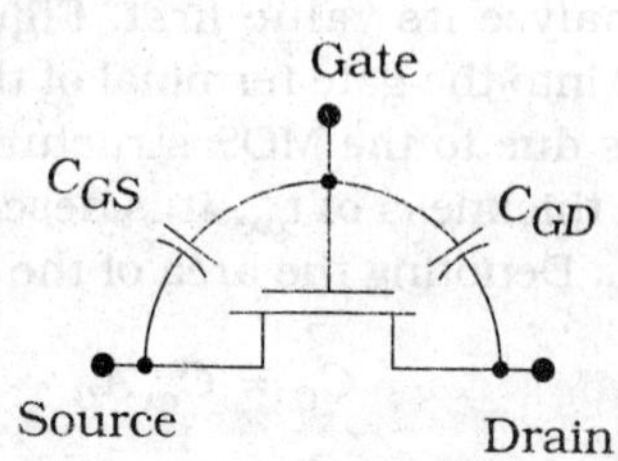

Figure 6.22 Gate-source and gate-drain capacitance

allows us to focus on the large-scale characteristics. Proper use of a CAD tool suite will provide final verification.

Example 6.5

Consider a FET with an oxide capacitance of $C_{ox} = 3.45 \times 10^{-7}$ F/cm^2 and a gate with dimensions $W = 8$ μm and $L' = 0.5$ μm. The gate capacitance formula gives

$$C_G = (3.45\times10^{-7})(8\times10^{-4})(0.5\times10^{-4}) \tag{6.79}$$

While this is a simple calculation, let us reduce it even further by noting that $C_{ox} = 3.45 \times 10^{-7} = 3.45$ fF/μm^2 where we recall that 1 fF = 10^{-15} F. Then

$$C_G = 3.45(8)(0.5) = 13.8 \text{ fF} \tag{6.80}$$

The gate-source and gate-drain contributions are then estimated by

$$C_{GS} \approx \frac{1}{2}C_G = 6.9 \text{ fF} = C_{GS} \tag{6.81}$$

These are typical orders of magnitude for FET capacitances. It is important to keep in mind that we are always dealing with device capacitances that are on the order of a few fF.

Junction Capacitance

Semiconductor physics reveals that a pn junction automatically exhibits capacitance due to the opposite polarity charges involved. This is called **junction** or **depletion** capacitance and is found at every drain or source region of a FET. Figure 6.23 illustrates the presence of the pn junctions and the associated capacitances C_{SB} (source-bulk) and C_{DB} (drain-bulk).

We usually characterize this capacitance by introducing a parameter C_j with units of F/cm^2 such that the total capacitance is

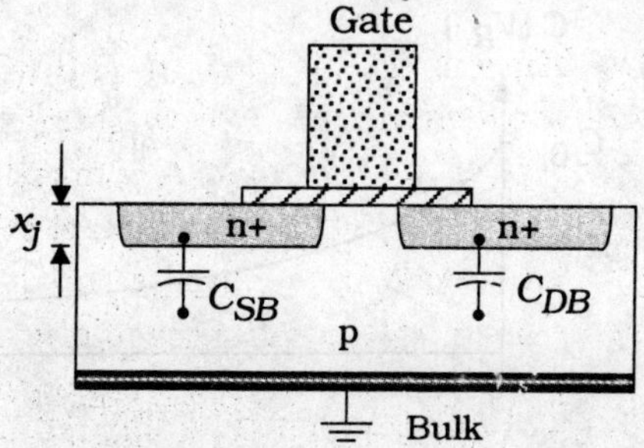

Figure 6.23 Junction capacitances in a MOSFET

$$C_0 = C_j A_{pn} \text{ F} \tag{6.82}$$

where A_{pn} is the area of the junction in units of cm^2. The value of C_j is determined by the processing, and varies with doping levels.

There are two complications in applying this formula to the nFET. The first is that this capacitance also varies with the voltage. With a reverse-bias voltage of V_R applied, this is usually modeled by an equation of the form

$$C = \frac{C_0}{\left(1 + \frac{V_R}{\phi_o}\right)^{m_j}} \tag{6.83}$$

where C_0 is the **zero-bias capacitance** (with $V_R = 0$), ϕ_o is the **built-in potential** of the junction, and m_j is called the **grading coefficient** of the junction. Both ϕ_o and m_j are determined by the doping characteristics. A special case is that of an **abrupt** or **step** junction where the doping changes from a constant acceptor density N_a to a constant donor density N_d. In this case, $m_j = 1/2$ and the built-in voltage is computed from

$$\phi_o = \left(\frac{kT}{q}\right) \ln\left[\frac{N_d N_a}{n_i^2}\right] \tag{6.84}$$

Another simple model is the **linearly graded junction** where the doping transition is a linear function of position. This gives a grading coefficient of $m_j = 1/3$, while the built-in potential ϕ_o can be calculated if the details of the doping are known. For our purposes, we will always assume that C_j, ϕ_o, and m_j are known parameters. In general, the maximum value of the capacitance is $C = C_0$ when $V_R = 0$; increasing the reverse voltage across the junction causes C to decrease as illustrated in Figure 6.24. We will use the zero-bias values as estimates in hand calculations, and turn to the use of CAD tools when more accurate values are needed.

The second complication that we need to consider in calculating the pn junction capacitance is the geometry of the pn junctions. The cross-sec-

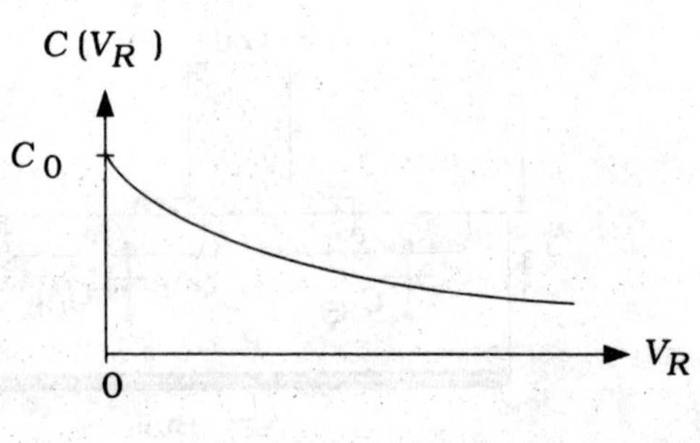

Figure 6.24 Junction capacitance variation with reverse voltage

tional view shown in Figure 6.23 shows that the n+ regions are "embedded" a depth x_j (called the **junction depth**) within the p-substrate. When computing the area A_{pn} of the pn junction, we must be careful to include both the bottom and the side contributions. Figure 6.25 illustrates the geometry. The top view of the FET in Figure 6.25(a) defines the channel width W of the transistor, and the extent X (away from the gate) of the n+ region. The 3-dimensional aspects of the pn junction area calculation are illustrated in Figure 6.25(b). Since the n+ region may be visualized as an "open box" structure, it is possible to decompose the boundaries into the bottom and sidewall sections shown. The area of the bottom region is easily seen to be

$$A_{bot} = XW \tag{6.85}$$

which is equal to the area of the n+ region seen in the top view. Then, denoting the zero-bias junction capacitance per unit area of this region by C_j with unit of F/cm^2, the capacitance due to the bottom section is

$$C_{bot} = C_j XW \tag{6.86}$$

To compute the sidewall capacitance C_{sw}, we note that the total sidewall area is obtained by adding the four contributions. Each sidewall section has a height equal to the junction depth x_j. Sidewall sections 1 and 2 have areas of $(W \times x_j)$, while Sidewall sections 3 and 4 have areas of $(X \times x_j)$. Adding the terms gives

$$\begin{aligned} A_{sw} &= 2(W \times x_j) + 2(X \times x_j) \\ &= x_j P_{sw} \end{aligned} \tag{6.87}$$

where P_{sw} is the **sidewall perimeter** in units of cm such that

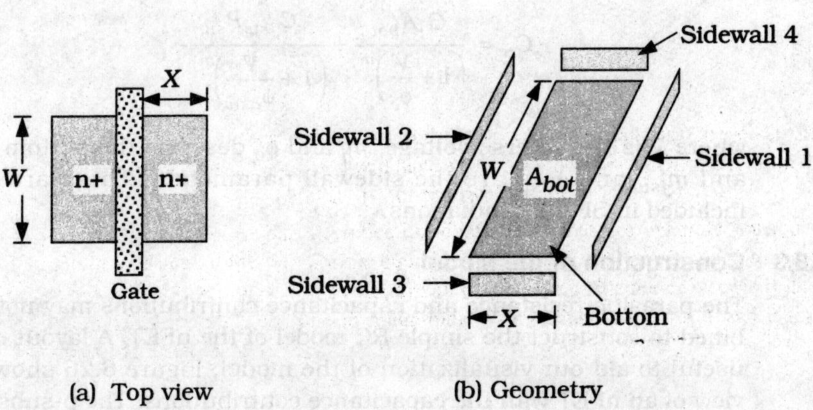

Figure 6.25 Calculation of the FET junction capacitance

$$P_{sw} = 2(W + X) \tag{6.88}$$

for the rectangular geometry shown in the example. The sidewall capacitance is obtained by multiplying by the junction capacitance per unit area. This is usually modified to the form

$$C_{sw} = C_{jsw} P_{sw} \text{ farads} \tag{6.89}$$

where

$$C_{jsw} = C_j x_j \text{ F/cm} \tag{6.90}$$

is the **sidewall capacitance per unit perimeter**. This is convenient to use because the perimeter P_{sw} can be found directly from the layout drawing. In practice, C_{jsw} is specified as a processing parameter while C_j automatically refers to the bottom capacitance.

These formulas ignore the gate overlap L_o of the n+ regions underneath the gate. For hand calculations, these should be included by changing

$$X \rightarrow (X + L_o) \tag{6.91}$$

everywhere. In a SPICE simulation, the drawn values of L and W are used to describe the circuit and gate overlap (and other) correction factors are included through the modeling information.

The total zero-bias capacitance of the n+ region is given by adding the bottom and sidewall contributions:

$$\begin{aligned} C_n &= C_{bot} + C_{sw} \\ &= C_j A_{bot} + C_{jsw} P_{sw} \end{aligned} \tag{6.92}$$

which can be used to compute both C_{SB} and C_{DB}. It is worthwhile to note that the non-linear characteristics of the bottom and sidewall junctions are usually distinct. This gives a non-linear variation of the form

$$C_n = \frac{C_j A_{bot}}{\left(1 + \dfrac{V}{\phi_o}\right)^{m_j}} + \frac{C_{jsw} P_{sw}}{\left(1 + \dfrac{V}{\phi_{osw}}\right)^{m_{jsw}}} \tag{6.93}$$

where V is the reverse voltage, m_j and ϕ_o describe the bottom junction, and m_{jsw} and ϕ_{osw} are the sidewall parameters. These are routinely included in SPICE simulations.

6.3.3 Construction of the Model

The parasitic resistance and capacitance contributions may now be combined to construct the simple RC model of the nFET. A layout drawing is useful to aid our visualization of the model. Figure 6.26 shows the top view of an nFET with the capacitance contributions. The p-substrate surrounding the transistor is at ground potential. A signal entering from

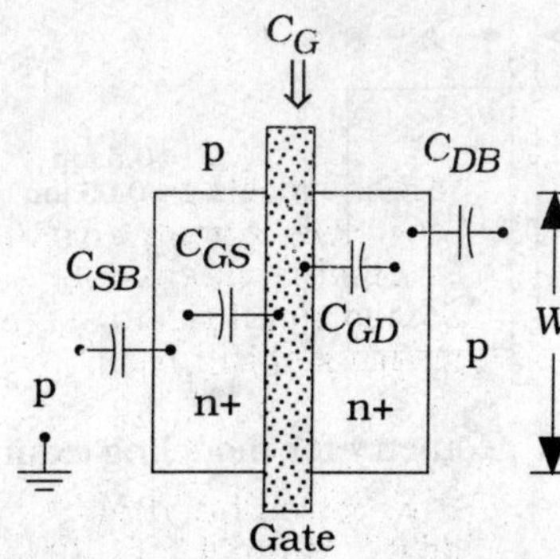

Figure 6.26 Physical visualization of FET capacitances

either side sees both an MOS term (C_{GS} or C_{GD}) and a junction parasitic (C_{SB} or C_{DB}).

The physical layout forms the basis for the schematic-level circuit in Figure 6.27(a), where the capacitors are divided into source and drain components. The simplest approach is to write

$$
\begin{aligned}
C_S &= C_{GS} + C_{SB} \\
C_D &= C_{GD} + C_{DB}
\end{aligned}
\tag{6.94}
$$

which approximates the total capacitance by summing all contributions that touch a given node. Moreover, we will use zero-bias values for simplicity in all hand calculations. It is important to note that the resistance R_n is inversely proportional to the aspect ratio $(W/L)_n$, while the capacitances increase with the channel width W.

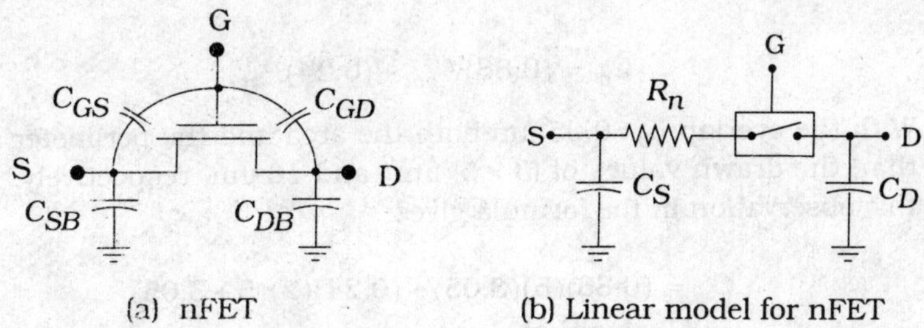

Figure 6.27 Final construction of the nFET RC model

Example 6.6

Let us create a switch model for the nFET shown in Figure 6.28; the measurements are given in units of microns (μm). First, since we are given

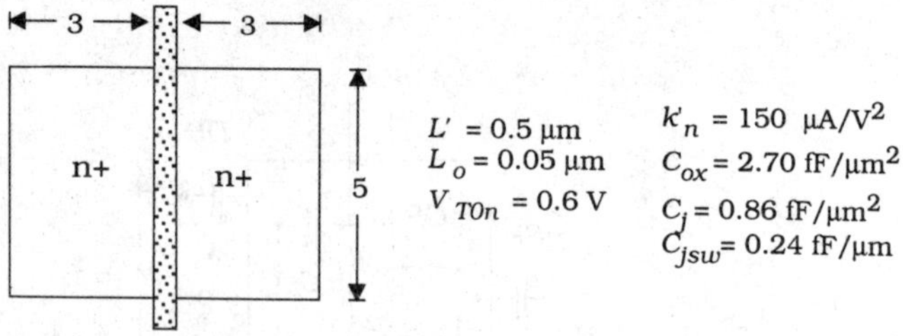

Figure 6.28 FET geometry for modeling example

that the overlap distance is $L_o = 0.05$ µm, the electrical channel length is $L = 0.5-2(0.05) = 0.4$ µm. The channel width is shown as $W = 5$ µm. Assuming a power supply voltage of $V_{DD} = 3.3$ V, the linear resistance is given by

$$R_n = \frac{1}{\left(\frac{5}{0.4}\right)(150\times10^{-6})(3.3-0.6)} = 197.5\ \Omega \tag{6.95}$$

If the sheet resistance of the n+ regions were known, the parasitic resistance could be found and added to this value.

The gate capacitance is

$$C_G = (2.7)(5)(0.5) = 6.75 \text{ fF} \tag{6.96}$$

so that

$$C_{GS} = C_{GD} = 3.375 \text{ fF} \tag{6.97}$$

by taking one-half of the gate value. The junction capacitance for either side is

$$C_n = (0.86)A_{bot} + (0.24)P_{sw} \tag{6.98}$$

With the overlap L_o =0.05 µm both the area and the perimeter are larger than the drawn values of (3 × 5) µm^2 and 16 µm, respectively. Including this observation in the formula gives

$$\begin{aligned} C_n &= (0.86)(5)(3.05) + (0.24)(2)(5+3.05) \\ &= 16.98 \text{ fF} \end{aligned} \tag{6.99}$$

The final drain and source capacitances are then

$$C_D = C_S = 16.98 + 3.375 = 20.36 \text{ fF} \tag{6.100}$$

which completes the calculations.

This simple model provides a reasonable basis for design estimates. To use it in a circuit problem, we just substitute the model for the transistor and then apply standard linear circuit techniques. Since it ignores the inherent non-linearities of the FET, the analysis will have limited accuracy. Increased precision is obtained from computer simulations that are performed after the initial design has produced a candidate circuit. Simplified device modeling is an important part of the VLSI design process as it allows us to create basic networks very quickly. However, these must always be checked and "fine-tuned" using CAD tools.

6.4 pFET Characteristics

A p-channel MOSFET is the electrical complement of an nFET. This was seen in Chapter 2 where the nFET was modeled as an assert-high switch while the pFET behaved as an assert-low switch. The complementary characteristics are even more evident at the device level. Suppose that we start with an nFET and want to modify it to form a pFET. All that needs to be done to the structure is

- Change all n-type regions to p-type regions
- Change all p-type regions to n-type regions

and the resulting device will in fact be a pFET. This shown in Figure 6.29. We have chosen a p-type substrate for both devices, which then makes it necessary to include the n-well region to embed the pFET in. Both devices are assumed to have the same oxide thickness of t_{ox} so that

$$C_{ox} = \frac{\varepsilon_{ox}}{t_{ox}} \tag{6.101}$$

describes both nFETs and pFETs. This means that the basic mechanism of the field effect is identical to that discussed for the nFET. However, since the polarities of the regions have been reversed, both the direction of the electric fields and the polarities of the charges will be opposite.

The structural details of a pFET are provided in Figure 6.30. The channel length L is defined as the distance between the edges of the source

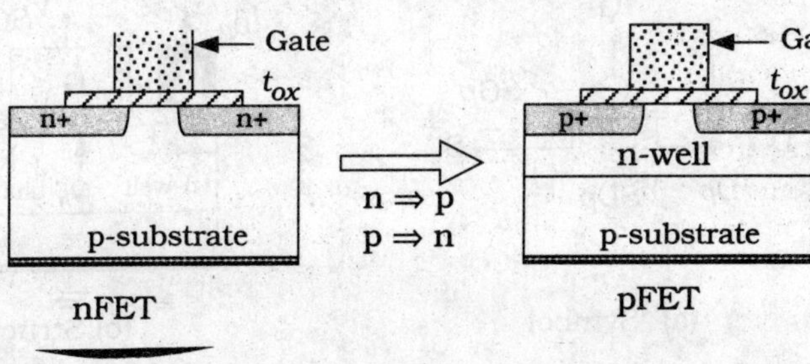

Figure 6.29 Transforming an nFET to a pFET

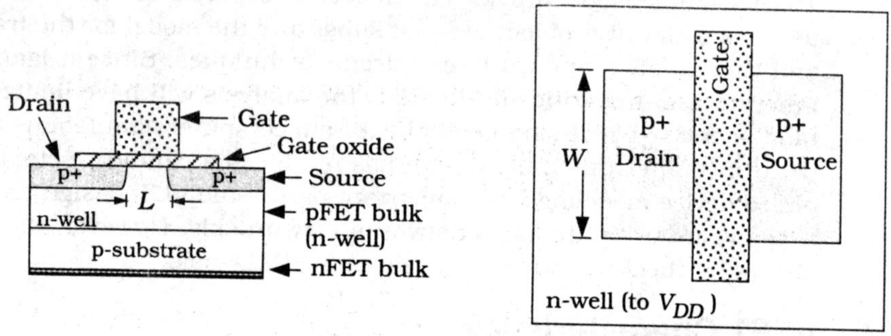

(a) Side view (b) Top view

Figure 6.30 Structural details of a pFET

and drain p+ regions as shown in the side view of Figure 6.30(a), while the channel width W is defined by the extent of the p+ regions as in the top view of Figure 6.30(b); these feature sizes are the same as those used to define the nFET. The presence of the n-well is shown in both drawings, and is an important region of a pFET since it acts as the bulk electrode for the device. Electrically, the n-well is tied to the positive power supply voltage V_{DD} which acts to insure that the voltage is well defined. As with the nFET, the naming of the source and drain terminals requires that we know the relative voltage levels. The pFET, however, uses definitions that are exactly opposite to those used for an nFET. This means that the p+ side at the higher voltage is the source, while the remaining side (at a lower voltage) is the drain.

A p-channel MOSFET uses positively charged holes for current flow. The pFET current I_{Dp} and the device voltages are defined in Figure 6.31 where it has been assumed that the right side of the device is the source in both drawings. First note that the schematic symbol in Figure 6.31(a)

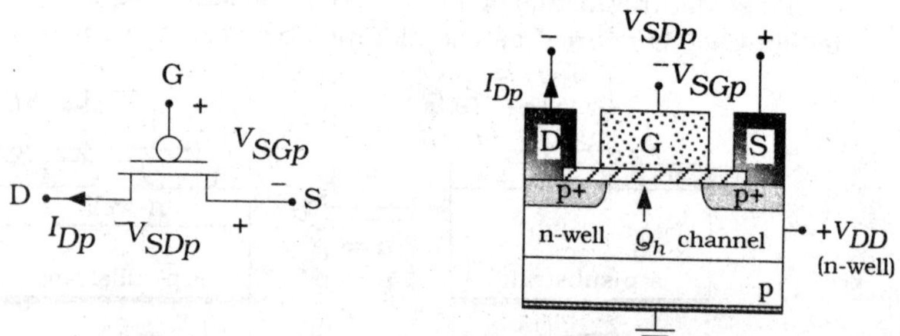

(a) Symbol (b) Structure

Figure 6.31 Current and voltages in a pFET

shows the current I_{Dp} as flowing out of the drain electrode. This is because positive charge moves from the source to the drain, which gives current in that direction. The pFET voltages are referenced to the source and are denoted by V_{SGp} (the source-gate voltage) and V_{SDp} (the source-drain voltage); note that these are opposite in polarity from the analogous nFET quantities V_{GSn} and V_{DSn}. The structural view in Figure 6.31(b) includes the fact that the n-well layer is electrically connected to the power supply voltage V_{DD}.

The conduction through the pFET is governed by the source-gate voltage V_{SGp}. The MOS structure consisting of the gate, the oxide, and the n-well layers is characterized by a pFET threshold voltage V_{Tp}. By convention, V_{Tp} is a negative number with typical values of V_{Tp} = -0.5 V to about V_{Tp} = -1.0 V. From the physical viewpoint, the value of V_{SGp} determines whether the gate is sufficiently negative with respect to the source to create a layer of holes under the gate oxide and thus establish a positive hole charge density of $\mathcal{Q}_h$ C/cm^2 that can be used as a channel between the source and drain. This is summarized by the statements

$$\begin{aligned} \mathcal{Q}_h &= 0 \text{ for } (V_{SGp} < |V_{Tp}|) \\ \mathcal{Q}_h &\text{ exists for } (V_{SGp} > |V_{Tp}|) \end{aligned} \tag{6.102}$$

where we have used the absolute value $|V_{Tp}|$ of the threshold voltage. The first line corresponds to the situation where the gate voltage is not sufficiently negative to induce the formation of a hole conduction layer in the n-well, while the second case is where V_{SGp} is large enough to insure that the gate voltage can attract the holes and form the channel. The role of the source-drain voltage V_{SDp} is to move the charge from the source to the drain if the channel exists.

The pFET threshold voltage can be computed from

$$V_{Tp} = -\frac{1}{C_{ox}}\sqrt{2q\varepsilon_{Si}N_d(2\phi_{Fp})}\ -2\phi_{Fp} + V_{FBp} \mp \frac{qD_I}{C_{ox}} \tag{6.103}$$

where N_d is the donor doping in the n-well,

$$2\phi_{Fp} = 2\left(\frac{kT}{q}\right)\ln\left(\frac{N_d}{n_i}\right) \tag{6.104}$$

is the surface potential needed to create the hole layer in the pFET, V_{FBp} is flatband voltage for the pFET MOS structure, and the last term represents the threshold adjustment ion implant step that has an ion dose of D_I. The minus sign '-' is used if donors are implanted, while the plus sign '+' corresponds to the case where acceptors are used.

The conduction modes for a pFET are summarized in Figure 6.32. The cutoff condition is portrayed in Figure 6.32(a). In this situation, V_{SGp} is less than $|V_{Tp}|$ so that $\mathcal{Q}_h = 0$ and no channel exists. This gives $I_{Dp} = 0$

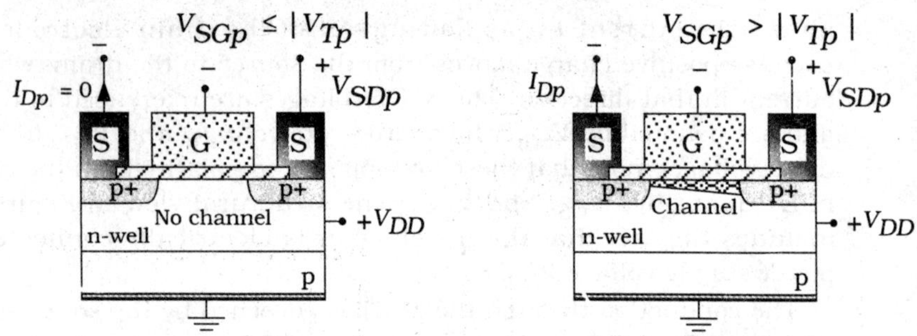

(a) Cutoff (b) Active bias

Figure 6.32 Conduction modes of a pFET

which can be modeled as an open switch. Active operation is shown in Figure 6.32(b) and is defined by $V_{SGp} \geq |V_{Tp}|$. The hole conduction layer forms and gives rise to the channel as shown. Since the electric field points from the right side to the left side, the positively charged holes originate from the source (right) and flow to the drain (left). The pFET current I_{Dp} thus flows out of the drain electrode as shown.

The current-voltage characteristics of a pFET can described using the same approach as that introduced for nFETs. In Figure 6.33, the source-drain voltage V_{SDp} is specified to be V_{DD} (the power supply value) while the source-gate voltage V_{SGp} is increased. For $V_{SGp} \leq |V_{Tp}|$, the device is in cutoff with $I_{Dp} = 0$ since no channel exists. When V_{SGp} is elevated above $|V_{Tp}|$, the charge layer Q_h is formed and the device is active. The current can be approximated by the square-law expression

$$I_{Dp} = \frac{\beta_p}{2}(V_{SGp} - |V_{Tp}|)^2 \tag{6.105}$$

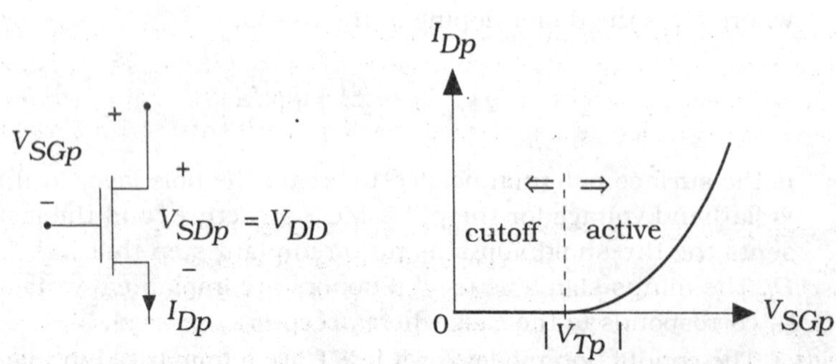

Figure 6.33 Gate-controlled pFET current-voltage characteristics

where

$$\beta_p = k'_p\left(\frac{W}{L}\right)_p \tag{6.106}$$

is the pFET device transconductance with units of A/V^2. The quantity $(W/L)_p$ is the pFET aspect ratio, and k'_p is the pFET process transconductance parameter

$$k'_p = \mu_p C_{ox} \tag{6.107}$$

with units of A/V^2. In this equation, μ_p is the hole mobility. These definitions are identical to the nFET parameters except that μ_p must be used to describe the motion of holes in silicon. A typical value for the surface hole mobility in silicon at room temperature is μ_p = 220 cm^2/V-sec; this is noticeably lower than the electron value (around 550 cm^2/V-sec) quoted earlier. A typical ratio is

$$r = \frac{\mu_n}{\mu_p} \approx 2 - 3 \tag{6.108}$$

Note that the important multiplying factors in the FET currents are transconductance parameters

$$\beta_n = k'_n\left(\frac{W}{L}\right)_n \qquad \beta_p = k'_p\left(\frac{W}{L}\right)_p \tag{6.109}$$

The difference between k'_p and k'_n can lead to some unique design choices for $(W/L)_n$ and $(W/L)_p$ when nFETs and pFETs are used in the same circuit.

Figure 6.34 shows the more general case where V_{SGp} is held constant while V_{SDp} is increased. Each value of V_{SGp} gives a distinct plot of I_{Dp} vs. V_{SDp} which results in the family of curves shown. The saturation voltage

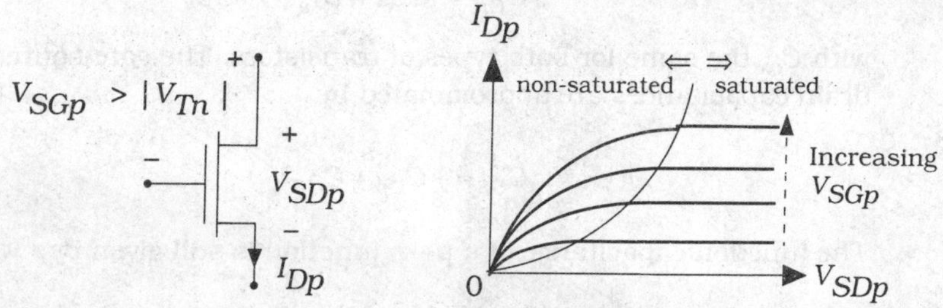

Figure 6.34 pFET *I-V* family of curves

for a pFET is defined by

$$V_{sat} = V_{SGp} - |V_{Tp}| \tag{6.110}$$

such that non-saturated conduction occurs for $V_{SDp} \leq V_{sat}$ and is described by

$$I_{Dp} = \frac{\beta_p}{2}[2(V_{SGp} - |V_{Tp}|)V_{SDp} - V_{SDp}^2] \tag{6.111}$$

while saturation occurs for $V_{SDp} \geq V_{sat}$ with

$$I_{Dp} = \frac{\beta_p}{2}(V_{SGp} - |V_{Tp}|)^2 \tag{6.112}$$

Saturated conduction was portrayed previously in Figure 6.33; a FET can be recognized as being saturated if the voltage between the source and drain is large (compared to V_{sat}).

6.4.1 pFET Parasitics

The parasitic resistance and capacitances of the pFET are calculated in the same manner as for the nFET. A linearized pFET resistance can be introduced as

$$R_p = \frac{1}{\beta_p(V_{DD} - |V_{Tp}|)} \tag{6.113}$$

which illustrates the dependence

$$R_p \propto \frac{1}{\beta_p} = \frac{1}{k_p'(W/L)_p} \tag{6.114}$$

Large aspect ratios thus give small resistances that allow for larger current flows.

The capacitances are computed using the same equations as for nFETs. For example, the input gate capacitance is given by

$$C_{Gp} = C_{ox}(WL)_p \tag{6.115}$$

with C_{ox} the same for both types of transistors. The gate-source and gate-drain capacitances are approximated by

$$C_{GS} \approx \frac{1}{2}C_{Gp} \approx C_{GD} \tag{6.116}$$

The junction capacitance of a p+-n junction is still given by

$$C_p = C_j A_{bot} + C_{jsw} P \tag{6.117}$$

but it is important to remember that the numerical values of C_j and C_{jsw}

are different for nFETs and pFETs because of differences in doping. Linear RC modeling of a pFET is identical to that shown in Figure 6.27 for the nFET, except that pFET values and an assert-low switch are used.

6.5 Modeling of Small MOSFETs

The equations presented in this chapter are simplified models that are useful for initial design estimates. They are reasonably accurate in **long-channel** MOSFETs where L is larger than about 20–30 μm; these are still found in discrete (separate individual) devices. Modern IC technology has reduced the channel length of production-line VLSI transistors to $L = 0.13$ μm, and this value is still shrinking. The physics of submicron sized devices is quite complicated. It is not possible to find closed form expressions that accurately describe these transistors. At the circuit design level, we turn instead to two levels of modeling: scaling theory and computer models.

6.5.1 Scaling Theory

Scaling theory deals with the "incredible shrinking transistor" and directs us toward the behavior of a device when its dimensions are reduced in a structured manner.

Consider a transistor that has a channel width W and a channel length L. We wish to find out how the main electrical characteristics change when both dimensions are reduced by a scaling factor $s > 1$ such that the new (scaled) transistor has sizes

$$\tilde{W} = \frac{W}{s} \qquad \tilde{L} = \frac{L}{s} \tag{6.118}$$

We note that the original transistor has a gate area of $A = WL$ while the scaled FET occupies

$$\tilde{A} = \frac{A}{s^2} \tag{6.119}$$

For example, $s = 2$ implies that the scaled device occupies only 25% of the area of the original. This provides ample motivation for continuing to improve the lithographic process.

Now let us consider the device transconductance. Since both W and L are scaled by the same factor, the aspect ratio is invariant:

$$\left(\frac{W}{L}\right) = \left(\frac{\tilde{W}}{\tilde{L}}\right) \tag{6.120}$$

The oxide capacitance is given by

$$C_{ox} = \frac{\varepsilon_{ox}}{t_{ox}} \tag{6.121}$$

where t_{ox} is the thickness of the gate oxide. If the new FET has a thinner oxide that is decreased as

$$\tilde{t}_{ox} = \frac{t_{ox}}{s} \tag{6.122}$$

then the scaled device has

$$\tilde{C}_{ox} = \frac{\varepsilon_{ox}}{\left(\frac{t_{ox}}{s}\right)} = sC_{ox} \tag{6.123}$$

i.e., it is increased by a factor of s. Since the process transconductance is given by $k' = \mu C_{ox}$, the device transconductance $\beta = k'(W/L)$ is increased in the scaled device to

$$\tilde{\beta} = sk'\left(\frac{W}{L}\right) = s\beta \tag{6.124}$$

Note, however, that the ability to scale L and W by s does not imply that the oxide thickness can be reduced by the same factor, so one must be careful when applying this relation. If it does hold, then the FET resistance

$$R = \frac{1}{\beta(V_{DD} - V_T)} \tag{6.125}$$

has a scaled value

$$\tilde{R} = \frac{1}{s\beta(V_{DD} - V_T)} \tag{6.126}$$

If we do not alter the voltages applied to the reduced-size FET, then the resistance is decreased according to

$$\tilde{R} = \frac{R}{s} \tag{6.127}$$

On the other hand, if we can scale the voltages in the small device to new values of

$$\tilde{V}_{DD} = \frac{V_{DD}}{s}, \qquad \tilde{V}_T = \frac{V_T}{s} \tag{6.128}$$

then the resistance of the scaled FET would be unchanged with

$$\tilde{R} = R \tag{6.129}$$

This provides the basis of **voltage scaling** where we reduce the voltages as the device dimensions decrease.

To see the effects of scaling the voltage, consider a scaled MOSFET with reduced voltages of

$$\tilde{V}_{GS} = \frac{V_{DS}}{s}, \quad \tilde{V}_{GS} = \frac{V_{GS}}{s} \tag{6.130}$$

such that the non-saturated current of the original device is given by

$$I_D = \frac{\beta}{2}[2(V_{GS} - V_T)V_{DS} - V_{DS}^2] \tag{6.131}$$

Applying the scaling formulas gives the current in the scaled FET as

$$\tilde{I}_D = \frac{s\beta}{2}\left[2\left(\frac{V_{GS}}{s} - \frac{V_T}{s}\right)\frac{V_{DS}}{s} - \frac{V_{DS}^2}{s^2}\right] = \frac{I_D}{s} \tag{6.132}$$

The power dissipation of the transistor is

$$\tilde{P} = \tilde{V}_{DS}\tilde{I}_D = \frac{V_{DS}I_D}{s^2} \tag{6.133}$$

i.e., it is reduced by a factor of $1/s^2$. This is a motivating factor for reducing the power supply voltage as the size of FETs decreases.

The actual value of the power supply voltage V_{DD} is a system-level decision that is often used to reduce the power dissipation of the circuit. The value of the threshold voltage V_T is controlled in the processing. Although some changes in the operating voltages can be made, the reduction is usually different from the geometric scaling specified by s. This does, however, provide a general set of guidelines on what to expect.

6.5.2 Small-Device Effects

As the size of MOSFETs decreased in the 1980's and 1990's, the natural approach was to provide corrections to the current flow equations that would account for newly observed effects. Many new types of phenomena were discovered and studied, and much of the jargon and terminology remains in use today.

The most important geometrical parameter in a VLSI FET is the channel length L. Since the aspect ratio (W/L) determines the maximum current flow through the transistor, reducing L allows us to simultaneously reduce W while still maintaining the same aspect ratio. In the next chapter we will demonstrate that the aspect ratio is the primary circuit design parameter. The scaled circuit thus consumes less area, but still maintains some of the important circuit characteristics.

When the channel length is reduced below about 20 μm, it is found that the threshold voltage decreases from its long-channel value $V_{T,long}$. This is called a **geometrical short-channel effect** (SCE) and is expressed by an equation of the form

$$V_T = V_{T,long} - (\Delta V_T)_{SCE} \tag{6.134}$$

where $(\Delta V_T)_{SCE}$ increases with decreasing L. The reduction can be calculated by accounting for the charge more accurately than is done in long-channel derivation. The **narrow-width effect** (NWE) is also a geometrical correction that increases the threshold voltage as W decreases. This can be expressed in the form

$$V_T = V_{T,long} + (\Delta V_T)_{NWE} \tag{6.135}$$

and is due to fringing electric fields that are ignored in the long-channel analysis. Minimum-size devices may exhibit both SCEs and NWEs.

Reducing L also causes a change in the channel conduction characteristics. Consider a FET with a drain-source voltage V_{DS} applied. The channel electric field may be estimated by

$$E \approx \frac{V_{DS}}{L} \tag{6.136}$$

which shows that E increases as L decreases. The velocity of a charged particle in silicon is observed to follow the dependence illustrated in Figure 6.35. For small values of E, the velocity increases linearly as expressed by

$$v = \mu E \tag{6.137}$$

which defines the mobility μ used in the derivation of the FET current. However, as the electric field intensity is increased, $v(E)$ enters the nonlinear region and the mobility is no longer a constant. The value eventually hits the **saturation velocity** v_s, which is about 10^7 cm/sec for electrons in silicon at room temperature. The simplified equations are therefore no longer valid, and must be modified. Modern short-channel FETs routinely enter the velocity saturation region.

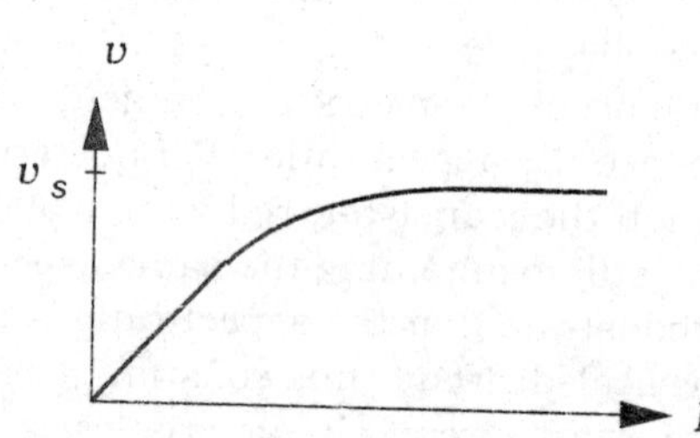

Figure 6.35 Velocity-field relation for charged particles in silicon

The velocity is also useful for estimating the transit time τ_t, which is the time required for a charge to traverse the channel. This is viewed as the fundamental limit on how fast a FET can switch. To obtain a simple expression, we use the velocity v to write

$$\tau_t = \frac{v}{L} \tag{6.138}$$

In the linear region of the v-E plot,

$$\tau_t = \frac{\mu E}{L} = \frac{\mu V_{DS}}{L^2} \tag{6.139}$$

This gives another reason for decreasing the channel length: the value of τ_t is decreased, indicating faster switching. Once the particle velocity saturates, the transit time goes to the constant

$$\tau_t = \frac{v_s}{L} \tag{6.140}$$

so the effect is muted somewhat. Another effect of short channels is that fewer electrons are involved in the current flow. Many assumptions involving the statistically derived charge concentrations start to become invalid.

Many other effects have been observed in small-geometry MOSFETs, and research continues as the device dimensions continue to shrink and new transistors are proposed and developed. The interested reader is directed toward the current literature for more details.

6.5.3 SPICE Modeling

Over the years we have learned that it is not possible to derive closed-form equations that accurately describe modern transistors. Luckily, the development of sophisticated CAD tools allows us to perform accurate simulations at the device and circuit level. Device simulators are beyond the scope of this book. Circuit simulation, on the other hand, is a routine procedure in the design of VLSI circuits. The design flow is to first create the logic circuit using FET switching theory, then estimate electrical characteristics using simplified equations. The circuit is then simulated, and the results are used to refine the electrical design. The most widely used circuit simulation engine is SPICE.[6] This program was conceived and written at the University of California, Berkeley, to aid in the design of integrated circuits. Since it is considered a standard in the industry, we will center our discussion around it. Several implementations of SPICE are available, but they are all similar in operation.

---

6 This is an acronym for Simulation Program with Integrated Circuit Emphasis.

MOSFETs in SPICE are entered into a circuit listing using an element statement of the form

Mname ND NG NS NB model_name L=length W=width <AS, PS, AD, PD >

where

- Mname is the name of the FET, such as M1 or Mn_out
- ND, NG, NS, NB are the node numbers (or names, if allowed) of the drain, gate, source, and bulk, respectively.
- model_name is the name of the .model line that provide the process parameter listing.
- AS, PS, AD, PD are the (optional) area and perimeter of the drain (AD, PD) and source (AS, PS) for the device. Areas are in units of m^2, and perimeters must be specified in units of m.

The important electrical parameters are included in the .model line which has the form

.model < listing >

where < listing > is a list of numerical values. There are many different MOSFET models in use. They are distinguished using the

Level = N

statement in < listing > where the value of N defines the equation set. The original SPICE allowed Level = 1, 2, 3 where Level 1 is based on a modified form of the equation set derived in Section 6.2.3. The Level 2 model, also called the bulk-charge equations, is more accurate, while Level 3 is an empirical model. Levels 1 and 2 lose accuracy when applied to modern submicron devices, but are often used for initial estimates because they allow very quick simulations.

We usually enter drawn values for all dimensions in the element statement. For example, the transistor in Example 6.6 would be described by

MExa6_6 10 20 30 0 nFET L=0.4U W=5U AD=15P PD=16U AS=15P PS=16U

where P is the pico scaler, and U is the micro scaler. The difference between the drawn values and the effective (electrical) values is computed from information supplied in the

.MODEL nFET <parameters>

listing supplied for the process. This makes the translation from layout to simulation file much easier.

In modern CMOS, the BSIM model set provides the most accurate SPICE simulations.[7] Unfortunately, the parameter set itself is somewhat cryptic and the values do not always have a direct relationship to the simple analytic expressions. A detailed treatment of the BSIM model can be found in Reference [2]. In VLSI design, we generally interpret the model as

7 BSIM stands for Berkeley Submicron IGFET Model, where IGFET stands for Insulated Gate FET. In everyday usage, IGFET and MOSFET are used interchangeably.

a given set of parameters that can be used in a CAD tool suite. Extracting the netlist from a layout allows an electrical simulation to be run.

6.6 References for Further Reading

[1] R. Jacob Baker, Harry Li, and David E. Boyce, **CMOS Circuit Design, Layout and, Simulation,** IEEE Press, Piscataway, NJ, 1998.

[2] Yuhua Cheng and Chenming Hu, **MOSFET Modeling and BSIM3 User's Guide**, Kluwer Academic Press, Norwell, MA, 1999.

[3] Richard S. Muller and Theodore I. Kamins, **Device Electronics for Integrated Circuits**, 2nd ed., John Wiley & Sons, New York, 1992.

[4] Robert F. Pierret, **Semiconductor Device Fundamentals**, Addison-Wesley, Reading, MA, 1996.

[5] Ben G. Streetman and Sanjay Banerjee, **Solid State Electronic Devices**, 5th ed., Prentice Hall, Upper Saddle River, NJ, 1999.

[6] Jasprit Singh, **Semiconductor Devices**, John Wiley & Sons, New York, 2001.

[7] S. M. Sze, **Semiconductor Devices**, 2nd ed., Wiley-Interscience, New York, 1981.

[8] John P. Uyemura, **CMOS Logic Circuit Design**, Kluwer Academic Publishers, Norwell, MA, 1999.

[9] Edward S. Yang, **Microelectronic Devices**, McGraw-Hill, New York, 1988.

6.7 Problems

[6.1] A CMOS process produces gate oxides with a thickness of t_{ox} = 100 Å. The FET carrier mobility values are given as μ_n = 550 cm^2/V-sec and μ_p = 210 cm^2/V-sec.

(a) Calculate the oxide capacitance per unit area in units of fF/μm^2.

(b) Find the process transconductance values for nFETs and pFETs. Place your answer in units of μA/V^2.

[6.2] An nFET with W = 10 μm and L = 0.35 μm is built in a process where k'_n = 110 μA/V^2 and V_{Tn} = 0.70 V. Assume V_{SBn} = 0 V.

(a) Find the current if the voltages are set to V_{GSn} = 2 V, V_{DSn} = 1.0 V.

(b) Find the current if the voltages are set to V_{GSn} = 2 V, V_{DSn} = 2 V.

[6.3] An nFET has a device transconductance of β_n = 2.3 mA/V^2 and a threshold voltage of 0.76 V. Assume V_{SBn} = 0 V.

(a) Find the current if the voltages are set to V_{GSn} = 1 V, V_{DSn} = 2.5 V.

(b) Find the current if the voltages are set to V_{GSn} = 2 V, V_{DSn} = 2.5 V.

(c) Find the current if the voltages are set to V_{GSn} = 3 V, V_{DSn} = 2.5 V.

[6.4] Consider a pFET that has a gate oxide thickness of t_{ox} = 60 Å. The hole mobility is measured to be 220 cm^2/V-sec, and the aspect ratio is $(W/L) = (12/1)$. Assume that V_{DD} = 3.3 V and $|V_{Tp}|$ = 0.7 V.

(a) Calculate the process transconductance k'_p in units of mA/V^2.

(b) Find the device transconductance β_p and the resistance R_p.

[6.5] An nFET has a gate oxide with a thickness of t_{ox} = 120 Å. The p-type bulk region is doped with boron at a density of $N_a = 8 \times 10^{14}$ cm^{-3}. It is given that V_{TOn} = 0.55 V and $(W/L) = 10$.

(a) Calculate the body bias coefficient γ.

(b) What is the device threshold voltage if a body bias voltage of V_{SBn} = 2 V is applied?

(c) The electron mobility is μ_n = 540 cm^2/V-sec. Calculate the drain current with bias voltages of V_{GSn} = 3 V, V_{DSn} = 3 V, and V_{SBn} = 3 V applied to the device.

[6.6] Construct the RC switch model for the FET layout in Figure P6.1. Assume a power supply voltage of 3 V and that the dimensions are in units of microns.

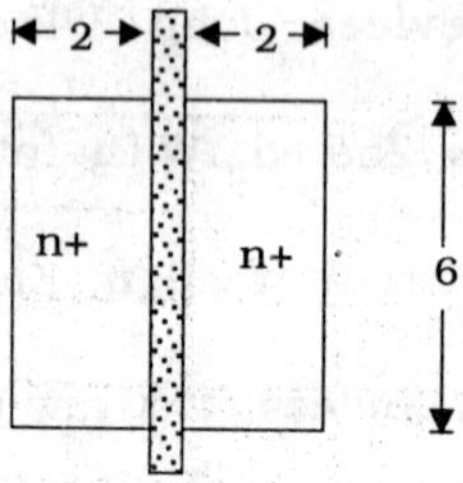

$L' = 0.5\ \mu m$
$L_o = 0.05\ \mu m$
$V_{TOn} = 0.6$ V

$k'_n = 150\ \mu A/V^2$
$C_{ox} = 2.70$ fF/µm^2
$C_j = 0.86$ fF/µm^2
$C_{jsw} = 0.24$ fF/µm

Figure P6.1 Transistor layout geometry for Problem 6.6

[6.7] Write a SPICE description of the nFET in Figure P6.1. Use your listing to obtain the family of I_D versus V_{DS} curves.

[6.8] Consider the FET geometry shown in Figure P6.1 where the sheet resistance of the n+ regions is 30 Ω, and the poly gate has a sheet resistance of 26 Ω. Compute the parasitic resistances R_{n+} and R_{poly} associated with these parameters by determining the appropriate geometry that applies for each. How would these parasitics affect the device operation?

[6.9] An nFET with W = 20 µm and L = 0.5 µm is built in a process where k'_n = 120 µA/V^2 and V_{Tn} = 0.65 V. The voltages are set to a value of $V_{GSn} = V_{DSn} = V_{DD}$ = 5 V.

(a) Is the transistor saturated or non-saturated?

(b) Calculate the drain-source resistance using the proper equation for the transistor.

(c) Compare your value in (b) with that found using equation (6.71) with a value of $\eta = 1$.

[6.10] An nFET with L = 0.5 µm is built in a process where k'_n = 100 µA/V^2 and V_{Tn} = 0.70 V. The gate-source voltage is set to a value of $V_{GSn} = V_{DD}$ = 3.3 V. Calculate the required channel width to obtain a resistance of R_n = 950 Ω using equation (6.71) with for a value of $\eta = 1$.

Electronic Analysis of CMOS Logic Gates

7

In the previous chapter we examined the electrical characteristics of MOSFETs. This sets the foundation for analyzing the behavior of transistors in CMOS logic circuits in this chapter. The treatment centers on the important areas of switching speed and layout design, and provides the foundation for much of modern chip design.

7.1 DC Characteristics of the CMOS Inverter

The CMOS inverter gives the basis for calculating the electrical characteristics of logic gates. Consider the circuit shown in Figure 7.1. The input voltage V_{in} determines the conduction states of the two FETs Mn and Mp. This produces the output voltage V_{out} of the gate. Two types of calculations are needed to characterize a digital logic circuit. A **DC analysis** determines V_{out} for a given value of V_{in}. In this type of calculation, it is assumed that V_{in} is changed very slowly, and that V_{out} is allowed to stabilize before a measurement is made. A DC analysis provides a direct mapping of the input to the output, which in turn tells us the voltage ranges

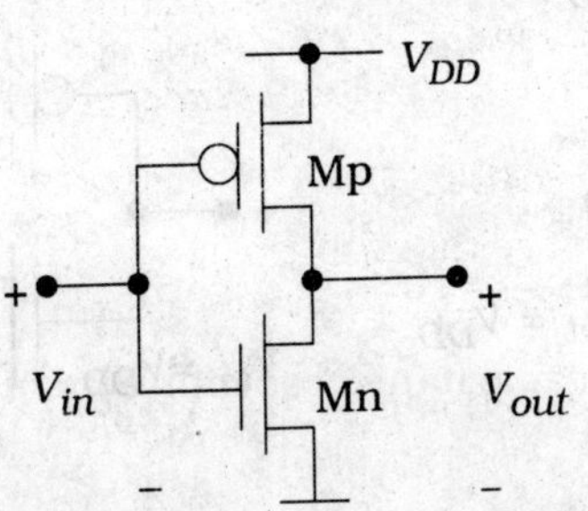

pFET: $V_{Tp} < 0$

$$\beta_p = k'_p \left(\frac{W}{L}\right)_p$$

nFET: $V_{Tn} > 0$

$$\beta_n = k'_n \left(\frac{W}{L}\right)_n$$

Figure 7.1 The CMOS inverter circuit

that define Boolean logic 0 and logic 1 values. The second type of characterization is called a **transient analysis**. In this case, the input voltage is an explicit function of time $V_{in}(t)$ corresponding to a changing logic value. The response of the circuit is contained in $V_{out}(t)$. The delay between a change in the input and the corresponding change at the output is the fundamental limiting factor for high-speed design. In this section we will concentrate on the DC analysis. The transient response is analyzed in the next section.

The DC characteristics of the inverter are portrayed in the **voltage transfer characteristic (VTC)**, which is a plot of V_{out} as a function of V_{in}. This is obtained by varying the input voltage V_{in} in the range from 0 V to V_{DD} and finding the output voltage V_{out}. The end point values are easily found with the aid of the circuits in Figure 7.2. If V_{in} is equal to 0 V as in Figure 7.2(a), Mn is off while Mp is on. Since the pFET is on, it connects the output to the power supply and gives $V_{out} = V_{DD}$. This defines the **output high voltage** of the circuit as

$$V_{OH} = V_{DD} \tag{7.1}$$

i.e., the highest output voltage is the value of the power supply V_{DD}. The opposite case with $V_{in} = V_{DD}$ is illustrated in Figure 7.2(b). This turns on Mn while Mp is in cutoff. The output node is then connected to 0 V (ground) through the nFET, defining the **output low voltage**

$$V_{OL} = 0 \text{ V} \tag{7.2}$$

The **logic swing** at the output is

$$\begin{aligned} V_L &= V_{OH} - V_{OL} \\ &= V_{DD} \end{aligned} \tag{7.3}$$

Since this is equal to the full value of the power supply, this is called a **full-rail output**.

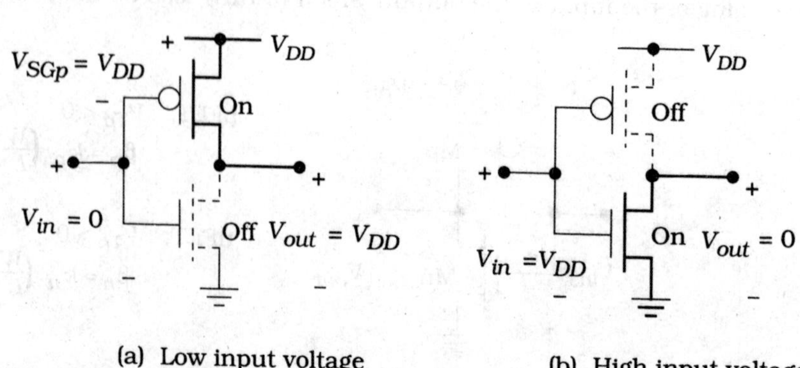

Figure 7.2 V_{OH} and V_{OL} for the inverter circuit

The VTC for the circuit is obtained by starting with an input voltage of V_{in} = 0 V and then increasing it up to a value of $V_{in} = V_{DD}$. This results in the plot shown in Figure 7.3. The details can be understood by writing the device voltages in terms of the input and output voltages:

$$V_{GSn} = V_{in}$$
$$V_{SGp} = V_{DD} - V_{in} \tag{7.4}$$

Mn is in cutoff so long as $V_{in} \leq V_{Tn}$. Since the output voltage is high with a value $V_{out} = V_{DD}$, any input voltage in the range labeled as "0" can be interpreted as a logic 0 input. Increasing V_{in} causes a downward transition in the VTC. This is because the input voltage turns the nFET on while the pFET is still conducting. Note, however, that increasing V_{in} decreases V_{SGp}, so the pFET becomes a less efficient conductor and the output voltage falls. Mp goes into cutoff when

$$V_{in} = V_{DD} - |V_{Tp}| \tag{7.5}$$

For V_{in} greater than this value, V_{out} = 0 V since only the nFET is active. This shows that there is a range of input voltages that act as logic 1 input values as indicated by the "1" on the VTC.

The logic 0 and 1 voltage ranges are defined by the changing slope of the VTC. Point 'a' in the drawing is where the slope has a value of -1, and defines the **input low voltage** V_{IL}. By definition, a logic 0 input voltage is defined by

$$0 \leq V_{in} \leq V_{IL} \tag{7.6}$$

The second -1 slope point is labeled as 'b' and defines the **input high voltage** V_{IH}. This is used to define a logic 1 input voltage as

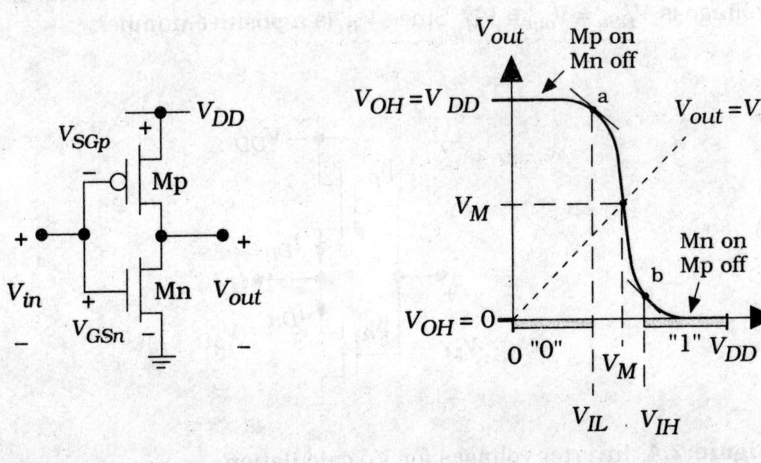

Figure 7.3 Voltage transfer curve for the NOT gate

$$V_{IH} \leq V_{in} \leq V_{DD} \tag{7.7}$$

The **voltage noise margins** are

$$\begin{aligned} VNM_H &= V_{OH} - V_{IH} \\ VNM_L &= V_{IL} - V_{OL} \end{aligned} \tag{7.8}$$

for high and low states, respectively. The noise margins give a quantitative measure of how stable the inputs are with respect to coupled electromagnetic signal interference.

While it is possible to calculate the exact voltages that define logic 0 and 1 input voltages, it is simpler to introduce the midpoint voltage V_M shown in the VTC. This is defined as the point where the VTC intersects the unity gain line that is defined by $V_{out} = V_{in} = V_M$. A value of $V_{in} = V_M$ is in the transition region and does not represent a Boolean quantity. However, for V_{in} less than V_M the input voltage is toward the logic 0 values while $V_{in} > V_M$ indicates that the input is on the logic 1 side. Knowing the value of V_M thus tells us the center point for input transitions.

To calculate the midpoint voltage we set $V_{in} = V_{out} = V_M$ as shown in Figure 7.4. Equating the drain currents of the FETs gives

$$I_{Dn} = I_{Dp} \tag{7.9}$$

but we need to find the operating region (saturation or non-saturation) of each FET before we can use the expression. Consider first the nFET and recall that the saturation voltage is given by

$$\begin{aligned} V_{sat} &= V_{GSn} - V_{Tn} \\ &= V_M - V_{Tn} \end{aligned} \tag{7.10}$$

where we have used $V_{in} = V_{GSn} = V_M$ in the second line. The drain-source voltage is $V_{DSn} = V_{out} = V_M$. Since V_{Tn} is a positive number,

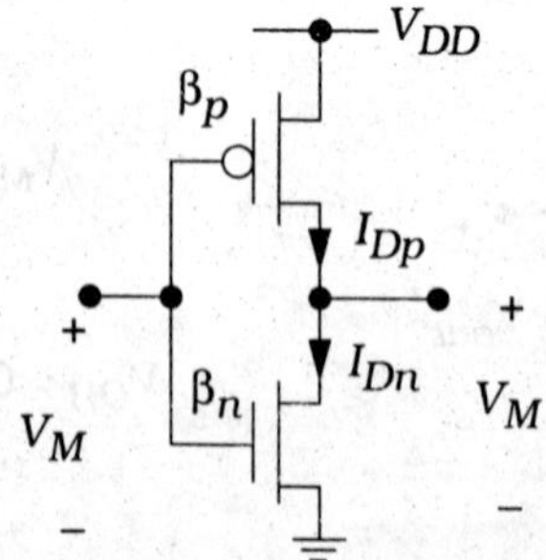

Figure 7.4 Inverter voltages for V_M calculation

$$V_{DSn} > V_{sat} = V_M - V_{Tn} \tag{7.11}$$

which says that Mn must be saturated. The same arguments can be applied to the pFET Mp since $V_{SGp} = V_{SDp}$. Using the saturation current equations from Chapter 6 gives

$$\frac{\beta_n}{2}(V_M - V_{Tn})^2 = \frac{\beta_p}{2}(V_{DD} - V_M - |V_{Tp}|)^2 \tag{7.12}$$

Dividing by β_p and taking the square root gives

$$\sqrt{\frac{\beta_n}{\beta_p}}(V_M - V_{Tn}) = V_{DD} - V_M - |V_{Tp}| \tag{7.13}$$

Simple algebra then gives the midpoint voltage as

$$V_M = \frac{V_{DD} - |V_{Tp}| + \sqrt{\frac{\beta_n}{\beta_p}}V_{Tn}}{1 + \sqrt{\frac{\beta_n}{\beta_p}}} \tag{7.14}$$

This equation shows that V_M is set by the nFET-to-pFET ratio

$$\frac{\beta_n}{\beta_p} = \frac{k'_n\left(\frac{W}{L}\right)_n}{k'_p\left(\frac{W}{L}\right)_p} \tag{7.15}$$

Since k'_n and k'_p are set in the processing, the ratio of the FET sizes establishes the switching point. It is important to remember that nFETs and pFETs have different mobility factors with a typical ratio of

$$\frac{k'_n}{k'_p} \approx 2 \text{ to } 3 \tag{7.16}$$

depending upon the details of the processing. This fact has a significant effect on the choices we make in both the sizing of individual transistors, and the types of circuits that are used in advanced VLSI designs. Note that, since C_{ox} is approximately the same for both FET types.

$$\frac{k'_n}{k'_p} = \frac{\mu_n}{\mu_p} = r \tag{7.17}$$

where r is the mobility ratio introduced in Chapter 5.

A **symmetrical inverter** VTC is one that has equal "0" and "1" input voltage ranges. This can be achieved by choosing

$$V_M = \frac{1}{2}V_{DD} \tag{7.18}$$

in equation (7.12). Rearranging gives us the design equation

$$\frac{\beta_n}{\beta_p} = \left(\frac{\frac{1}{2}V_{DD} - |V_{Tp}|}{\frac{1}{2}V_{DD} - V_{Tn}}\right)^2 \tag{7.19}$$

This allows us to compute the transistor sizes for this particular choice of V_M. Note that if $V_{Tn} = |V_{Tp}|$, then a symmetric design requires that

$$\beta_n = \beta_p \tag{7.20}$$

i.e., the device transconductance values of the two FETs are equal. It is important to remember that β is proportional to the aspect ratio (W/L) of a MOSFET, and that (W/L) is the actual design variable.

Example 7.1

Consider a CMOS process with the following parameters

$$\begin{aligned} k'_n &= 140\ \mu A/V^2 \qquad & V_{Tn} &= +0.70\ \text{V} \\ k'_p &= 60\ \mu A/V^2 \qquad & V_{Tp} &= -0.70\ \text{V} \end{aligned} \tag{7.21}$$

with $V_{DD} = 3.0$ V.

Consider the case where $\beta_n = \beta_p$. We can verify that this is a symmetrical design by calculating

$$V_M = \frac{3 - 0.7 + \sqrt{1}(0.7)}{1 + \sqrt{1}} = 1.5\ \text{V} \tag{7.22}$$

so that V_M is one-half the value of the power supply voltage. To achieve this design, we must choose the device aspect ratios such that

$$\frac{\beta_n}{\beta_p} = \frac{k'_n\left(\frac{W}{L}\right)_n}{k'_p\left(\frac{W}{L}\right)_p} = 1 \tag{7.23}$$

where we recall that the process transconductance parameters k' are given by $k' = \mu_n C_{ox}$, and are set by the processing. For the present case, we rearrange the expression to read

$$\left(\frac{W}{L}\right)_p = \frac{k'_n}{k'_p}\left(\frac{W}{L}\right)_n \tag{7.24}$$

so that

$$\left(\frac{W}{L}\right)_p = \left(\frac{140}{60}\right)\left(\frac{W}{L}\right)_n = 2.33\left(\frac{W}{L}\right)_n \tag{7.25}$$

This shows that the pFET must be about 2.33 times larger than the nFET.

Let us now examine the case where the nFET and the pFET have the same aspect ratio: $(W/L)_n = (W/L)_p$. With the values provided in the problem statement,

$$\frac{\beta_n}{\beta_p} = \frac{k'_n}{k'_p} = 2.33 \tag{7.26}$$

so that the midpoint voltage is given by

$$V_M = \frac{3 - 0.7 + \sqrt{2.33}\ (0.7)}{1 + \sqrt{2.33}} = 1.33 \text{ V} \tag{7.27}$$

This choice shifts V_M to a value that is smaller than $(V_{DD}/2)$.

Figure 7.5 illustrates the difference in the layout between an inverter that uses the two design styles. The channel length is the same for both transistors in the inverter, leaving the channel widths W_p and W_n as the design variables. In Figure 7.5(a), the pFET has a width of about $W_p \approx 2\ W_n$ which gives V_M of about $(V_{DD}/2)$. Equal size transistors are used in the layout of Figure 7.5(b), so that the circuit has $V_M < (V_{DD}/2)$. It is important to remember that we are only dealing with the DC characteristics at the moment. As we will see in the next section, the switching properties of the two designs are also affected by the aspect ratios.

The derivation and examples above illustrate the importance of the FET aspect ratios in the DC behavior of the logic gate. At the physical level, the relative device sizes contained in the ratio (β_n/β_p) determine the switching points. In general, increasing (β_n/β_p) decreases the value of the midpoint voltage V_M. This dependence is illustrated in the plot of Figure

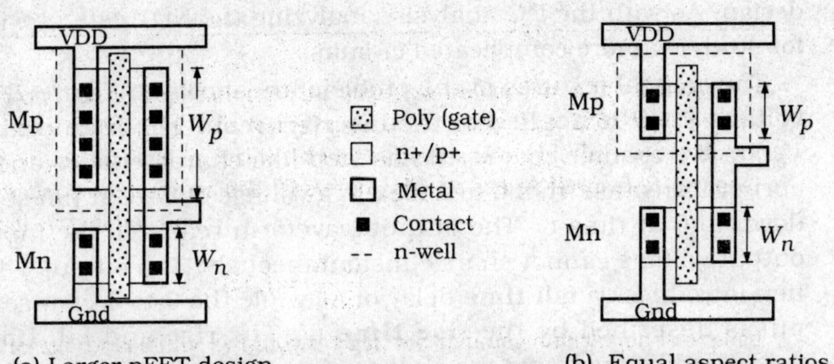

Figure 7.5 Comparison of the layouts for Example 7.1

7.6. With the parameters shown, a symmetrical design with $\beta_n = \beta_p$ gives $V_M = (V_{DD}/2) = 1.5$ V. Increasing the ratio to $(\beta_n/\beta_p) = 1.5$ gives $V_M \approx 1.42$ V, while $(\beta_n/\beta_p) = 2.5$ decreases the midpoint voltage to $V_M \approx 1.31$ V. It is also possible to use a ratio of $(\beta_n/\beta_p) < 1$, which shifts the VTC toward the right, i.e., $V_M > (V_{DD}/2)$. However, this is rarely used since the pFET aspect ratios get quite large.

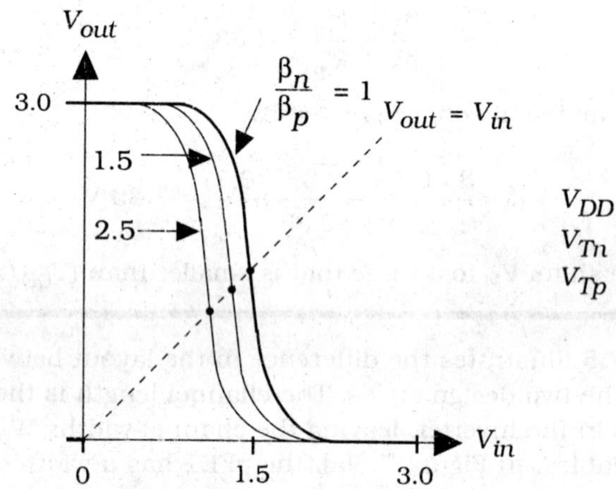

Figure 7.6 Dependence of V_M on the device ratio

7.2 Inverter Switching Characteristics

High-speed digital system design is based on the ability to perform calculations very quickly. This requires that logic gates introduce a minimum amount of time delay when the inputs change. Designing fast logic circuits is one of the more challenging (but critical) aspects of VLSI physical design. As with the DC analysis, analyzing the NOT gate provides a basis for studying more complicated circuits.

The general features of the problem are shown in Figure 7.7. An input voltage $V_{in}(t)$ is applied to the inverter, resulting in an output voltage $V_{out}(t)$. We assume that $V_{in}(t)$ has step-like characteristics and makes an abrupt transition from 0 to 1 (i.e., to a voltage of V_{DD}) at time t_1, and back down to 0 at time t_2. The output waveform reacts to the input, but the output voltage cannot change instantaneously. The output 1-to-0 transition introduces a **fall time** delay of t_f, while the 0-to-1 change at the output is described by the **rise time** t_r. The rise and fall times can be calculated by analyzing the electronic transitions of the circuits.

The rise and fall time delays are due to the parasitic resistance and

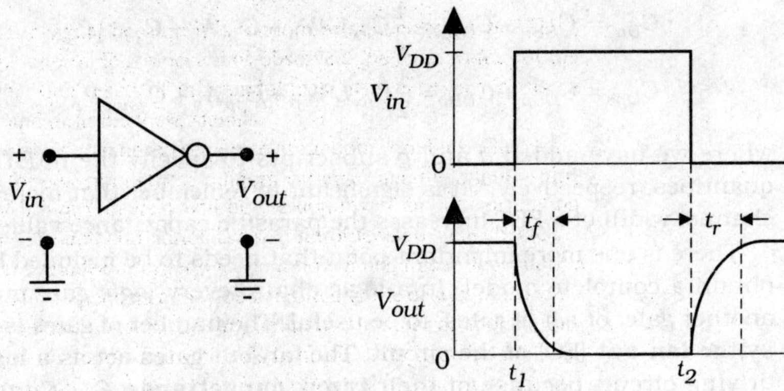

Figure 7.7 General switching waveforms

capacitances of the transistors. Consider the NOT circuit shown in Figure 7.8(a). Both FETs can be replaced by their switch equivalents, which results in the simplified RC model in Figure 7.8(b). It is worth recalling that the actual values of the components depend upon the device dimensions. Once we specify the aspect ratios $(W/L)_n$ and $(W/L)_p$, we can calculate R_n and R_p using

$$R_n = \frac{1}{\beta_n(V_{DD} - V_{Tn})}$$
$$R_p = \frac{1}{\beta_p(V_{DD} - |V_{Tp}|)} \qquad (7.28)$$

Knowing the layout dimensions of each FET allows us to find the capacitances C_{Dn} and C_{Dp} at the output node. The formulas are given by

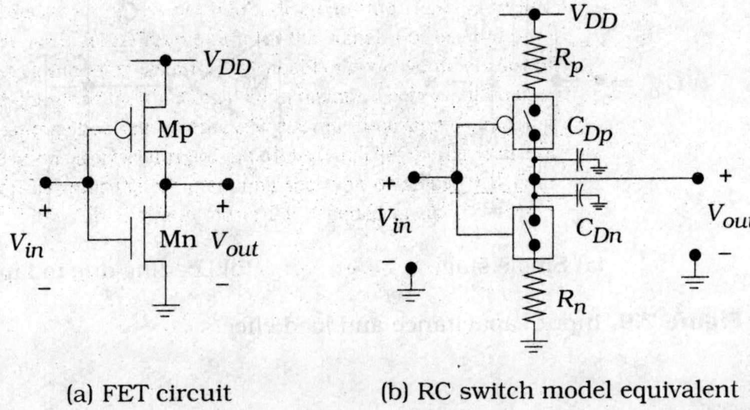

(a) FET circuit (b) RC switch model equivalent

Figure 7.8 RC switch model equivalent for the CMOS inverter

$$
\begin{aligned}
C_{Dn} &= C_{GSn} + C_{DBn} = \frac{1}{2}C_{ox}L'W_n + C_{jn}A_n + C_{jswn}P_n \\
C_{Dp} &= C_{GSp} + C_{DBp} = \frac{1}{2}C_{ox}L'W_p + C_{jp}A_p + C_{jswp}P_p
\end{aligned}
\tag{7.29}
$$

where we have added n and p subscripts to specify the nFET or pFET quantities, respectively.[1] It is significant to remember that increasing the channel width of a FET increases the parasitic capacitance values.

There is one more important point that needs to be included before we obtain a complete model. In a logic chain, every logic gate must drive another gate, or set of gates, to be useful. The number of gates is specified by the **fan-out** (FO) of the circuit. The fan-out gates act as a **load** to the driving circuit because of their **input capacitance** C_{in}. Consider the inverter shown in Figure 7.9(a). The input capacitance of the inverter is just the sum of the FET capacitances

$$C_{in} = C_{Gp} + C_{Gn} \tag{7.30}$$

Figure 7.8(b) shows the effect of input capacitance for a fan-out of FO = 3. The input capacitance to each gate acts as an **external load capacitance** C_L to the driving gate. In this example, it is easily seen that

$$C_L = 3C_{in} \tag{7.31}$$

is the value of the load presented to the NOT gate.

We may now calculate the switching times of the inverter. Figure 7.10 illustrates the general problem. A CMOS NOT gate is used to drive an external load capacitance C_L as in Figure 7.10(a). This gives the complete

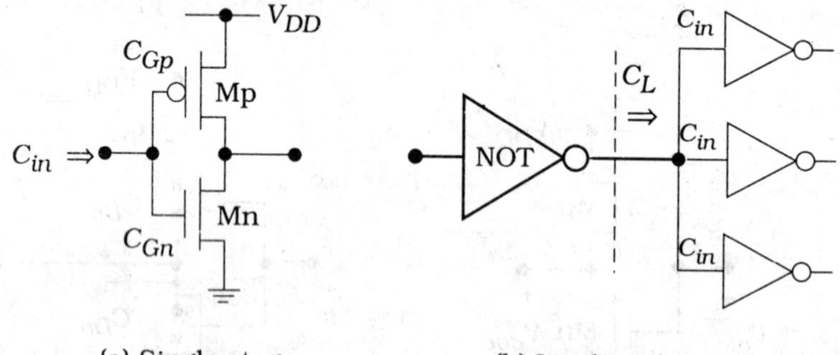

(a) Single stage (b) Loading due to fan-out

Figure 7.9 Input capacitance and load effects

---

[1] Note that the source capacitances C_{Sp} and C_{Sn} do not enter the problem as they are at the power supply and ground, respectively, and have constant voltages.

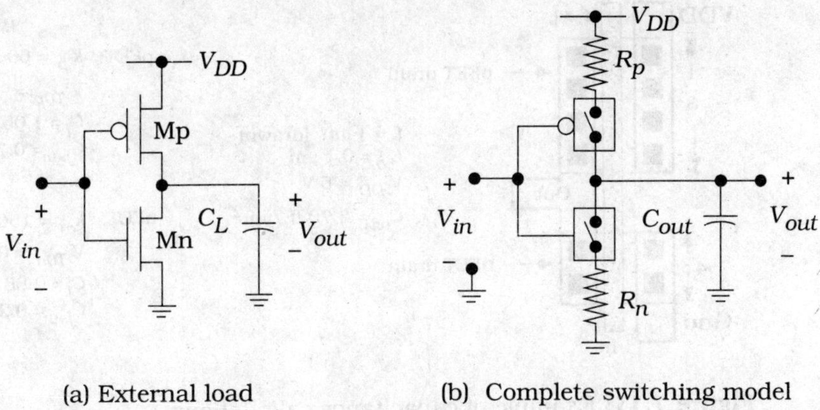

Figure 7.10 Evolution of the inverter switching model

switching model shown in Figure 7.10(b) where the total output capacitance is defined as

$$C_{out} = C_{FET} + C_L \tag{7.32}$$

The FET capacitances shown earlier in Figure 7.8 have been merged into the single term

$$C_{FET} = C_{Dn} + C_{Dp} \tag{7.33}$$

and are the parasitic internal contributions that cannot be eliminated. These add with C_L since all elements are in parallel. The total output capacitance C_{out} is the load that the gate must drive; the numerical value varies with the load.

Example 7.2

Let us apply this analysis to find the capacitances in the NOT gate shown in Figure 7.11. It is assumed that all dimensions have units of microns (µm).

First we will find the gate capacitances using

$$\begin{aligned} C_{Gp} &= (2.70)(1)(8) = 21.6 \text{ fF} \\ C_{Gn} &= (2.70)(1)(4) = 10.8 \text{ fF} \end{aligned} \tag{7.34}$$

Next, note that the overlap distance L_o is specified as 0.1 µm, which should be included in the area and perimeter factors in the junction capacitances. For the pFET, the p+ capacitance is

$$C_p = C_j A_{bot} + C_{jsw} P_{sw} \tag{7.35}$$

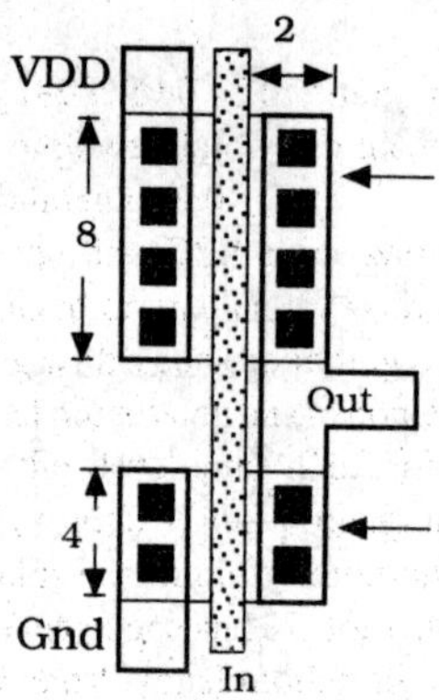

$L = 1\ \mu m$ (drawn)
$L_o = 0.1\ \mu m$
$V_{DD} = 5$ V
$C_{ox} = 2.70\ fF/\mu m^2$

pFET: $k'_p = 60\ \mu A/V^2$
$V_{T0p} = -0.7$ V
$C_j = 1.05\ fF/\mu m^2$
$C_{jsw} = 0.32\ fF/\mu m$

nFET: $k'_n = 150\ \mu A/V^2$
$V_{T0n} = 0.6$ V
$C_j = 0.86\ fF/\mu m^2$
$C_{jsw} = 0.24\ fF/\mu m$

Figure 7.11 Example of capacitance calculations

so

$$C_p = (1.05)(8)(2.1) + (0.32)2(8 + 2.1) = 24.10 \text{ fF} \tag{7.36}$$

The total capacitance at the pFET drain is therefore given by

$$C_{Dp} = \frac{21.6}{2} + 24.10 = 34.9 \text{ fF} \tag{7.37}$$

The nFET drain is analyzed using the same approach. The n+ junction capacitance is

$$C_n = (0.86)(4)(2.1) + (0.24)(2)(4 + 2.1) = 10.15 \text{ fF} \tag{7.38}$$

so that

$$C_{Dn} = \frac{10.8}{2} + 10.15 = 15.55 \text{ fF} \tag{7.39}$$

is the total capacitance at the drain of the nFET. Adding gives

$$\begin{aligned} C_{FET} &= C_{Dp} + C_{Dn} \\ &= 34.9 + 15.55 \\ &= 50.45 \text{ fF} \end{aligned} \tag{7.40}$$

as the total internal FET capacitance. The total capacitance at the output is

$$C_{out} = 50.45 + C_L \tag{7.41}$$

in fF, where C_L is the external load (also in fF).

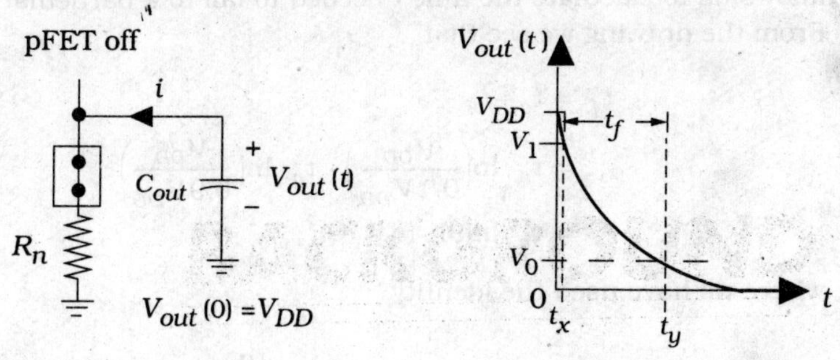

(a) Discharge circuit (b) Output waveform

Figure 7.12 Discharge circuit for the fall time calculation

7.2.1 Fall Time Calculation

Let us start by calculating the output fall time t_f. We will shift the time origin such that V_{in} changes from 0 to V_{DD} at time $t = 0$. The initial condition at the output is $V_{out}(0) = V_{DD}$. When the input is switched, the nFET goes active while the pFET is driven into cutoff. In terms of the switch models, the nFET switch is closed and the pFET switch is open. This gives us the simplified discharge circuit shown in Figure 7.12(a). The capacitor C_{out} is initially charged to a voltage V_{DD}, and is allowed to discharge to 0 V through the nFET resistance R_n. The current leaving the capacitor is

$$i = -C_{out}\frac{dV_{out}}{dt} = \frac{V_{out}}{R_n} \tag{7.42}$$

which gives the differential equation for the discharge event. Solving with the initial condition $V_{out}(0) = V_{DD}$ results in the well-known form

$$V_{out}(t) = V_{DD}e^{-t/\tau_n} \tag{7.43}$$

where

$$\tau_n = R_n C_{out} \tag{7.44}$$

is the nFET **time constant** with units of seconds. The function is plotted in Figure 7.12(b).

The fall time is traditionally defined to be the time interval from $V_1 = 0.9\ V_{DD}$ to $V_0 = 0.1\ V_{DD}$, which are respectively known as the 90% and the 10% voltages as referenced to the full rail swing of V_{DD}. Rearranging the solution to the form

$$t = \tau_n \ln\left(\frac{V_{DD}}{V_{out}}\right) \tag{7.45}$$

allows us to calculate the time t needed to fall to a particular voltage V_{out}. From the drawing we see that

$$\begin{aligned} t_f &= t_y - t_x \\ &= \tau_n \ln\left(\frac{V_{DD}}{0.1V_{DD}}\right) - \tau_n \ln\left(\frac{V_{DD}}{0.9V_{DD}}\right) \\ &= \tau_n \ln(9) \end{aligned} \tag{7.46}$$

where we have used the identity

$$\ln(a) - \ln(b) = \ln\left(\frac{a}{b}\right) \tag{7.47}$$

in the last step. Approximating $\ln(9) \approx 2.2$ gives the final result

$$t_f \approx 2.2\tau_n \tag{7.48}$$

as the fall time for the circuit. The output fall time in a generic digital logic gate is usually called the output **high-to-low time** t_{HL} and is identical to the value computed here:

$$t_{HL} = t_f \tag{7.49}$$

The two symbols will be used interchangeably in the discussion.

7.2.2 The Rise Time

The rise time calculation follows in the same manner. Initially, the input voltage is at $V_{in} = V_{DD}$ and is switched to $V_{in} = 0$ V; we time shift this event to occur at $t = 0$ for simplicity. This turns on the pFET while simultaneously driving the nFET into cutoff, so that the simplified charging circuit of Figure 7.13(a) is valid. The output voltage at $t = 0$ is given by $V_{out}(0) = 0$ V.

The charging current is given by

$$i = C_{out}\frac{dV_{out}}{dt} = \frac{V_{DD} - V_{out}}{R_p} \tag{7.50}$$

Solving and applying the initial condition gives the exponential form

$$V_{out}(t) = V_{DD}[1 - e^{-t/\tau_p}] \tag{7.51}$$

where the pFET time constant is defined by

$$\tau_p = R_p C_{out} \tag{7.52}$$

Figure 7.13(b) shows the output voltage as a function of time. The rise time is taken between 10% and 90% points such that

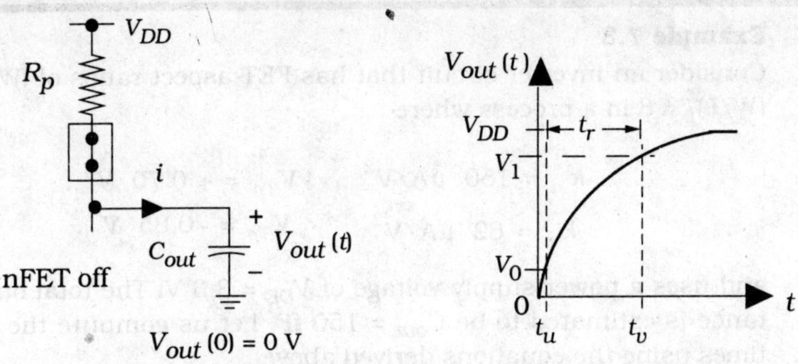

(a) Charge circuit (b) Output waveform

Figure 7.13 Rise time calculation

$$t_r = t_v - t_u \tag{7.53}$$

A little algebra yields the expression

$$t_r = \ln(9)\tau_p \approx 2.2\tau_p \tag{7.54}$$

for the rise time t_r. This has the same form as the fall time t_f because of the symmetry of the charge and discharge circuits. The rise time is identical to the output **low-to-high time** t_{LH}; the symbols will be used interchangeably.

The low-to-high time t_{LH} and the high-to-low time t_{HL} represent the shortest amount of time needed for the output to change from a logic 0 to logic 1 voltage, or from a logic 1 to a logic 0 voltage, respectively. Let us assume that the input is a square wave with a period of T sec such that the voltage is 0 for $(T/2)$ and V_{DD} for a $(T/2)$ time interval.[2] We then define the **maximum signal frequency** as

$$f_{max} = \frac{1}{t_{HL} + t_{LH}} = \frac{1}{t_r + t_f} \tag{7.55}$$

since this is the largest frequency that can be applied to the gate and still allow the output to settle to a definable state.[3] If the signal frequency exceeds f_{max}, the output voltage of the gate will not have sufficient time to stabilize to the correct value.

2 This defines what is known as a 50% duty cycle.

3 This definition assumes that t_{HL} and t_{LH} have the same order of magnitude to be useful.

Example 7.3

Consider an inverter circuit that has FET aspect ratios of $(W/L)_n = 6$ and $(W/L)_p = 8$ in a process where

$$\begin{aligned} k'_n &= 150\ \mu A/V^2 & V_{Tn} &= +0.70\ V \\ k'_p &= 62\ \mu A/V^2 & V_{Tp} &= -0.85\ V \end{aligned} \tag{7.56}$$

and uses a power supply voltage of $V_{DD} = 3.3$ V. The total output capacitance is estimated to be $C_{out} = 150$ fF. Let us compute the rise and fall times using the equations derived above.

Consider first the fall time. The pFET resistance is given by

$$\begin{aligned} R_p &= \frac{1}{\beta_p(V_{DD} - |V_{Tp}|)} \\ &= \frac{1}{(62\times10^{-6})(8)(3.3-0.85)} \\ &= 822.9\ \Omega \end{aligned} \tag{7.57}$$

The time constant for the charging event is computed using the RC product R_pC_{out} to find

$$\tau_p = (822.9)(150\times10^{-15}) = 123.43\ \text{ps} \tag{7.58}$$

where 1 ps (picosecond) is 10^{-12} sec. The rise time is

$$t_r = 2.2\tau_p = 271.55\ \text{ps} \tag{7.59}$$

The fall time is calculated in a similar manner. First, we find the nFET resistance

$$\begin{aligned} R_n &= \frac{1}{\beta_n(V_{DD} - V_{Tn})} \\ &= \frac{1}{(150\times10^{-6})(6)(3.3-0.70)} \\ &= 427.35\ \Omega \end{aligned} \tag{7.60}$$

so that the discharge time constant is

$$\tau_p = (427.35)(150\times10^{-15}) = 64.1\ \text{ps} \tag{7.61}$$

The fall time is

$$t_f = 2.2\tau_n = 141.0\ \text{ps} \tag{7.62}$$

Combining these results, the maximum signal frequency is

$$f_{max} = \frac{1}{t_r + t_f} = \frac{1}{(271.55 + 141.0)\times 10^{-12}} = 2.42 \quad \text{GHz} \tag{7.63}$$

where 1 GHz = 10^9 Hz. Although this is a very high frequency, it is important to remember that this refers only to a single inverter.

7.2.3 The Propagation Delay

The propagation delay time t_p is often used to estimate the "reaction" delay time from input to output. When we use step-like input voltages, the propagation delay is defined by the simple average of the two time intervals shown in Figure 7.14 by

$$t_p = \frac{(t_{pf} + t_{pr})}{2} \tag{7.64}$$

In this expression, t_{pf} is the output fall time from the maximum level to the "50%" voltage line, i.e., from V_{DD} to $(V_{DD}/2)$; t_{pr} is the propagation rise time from 0 V to $(V_{DD}/2)$. Using the exponential equations for V_{out} we obtain

$$\begin{aligned} t_{pf} &= \ln(2)\tau_n \\ t_{pr} &= \ln(2)\tau_p \end{aligned} \tag{7.65}$$

Approximating ln(2) ≈ 0.693 then gives

$$t_p \approx 0.35(\tau_n + \tau_p) \tag{7.66}$$

The propagation delay time is a useful estimate of the basic delay, but does not provide detailed information on the rise and fall times as individual quantities. Propagation delays are commonly used in basic logic simulation programs.

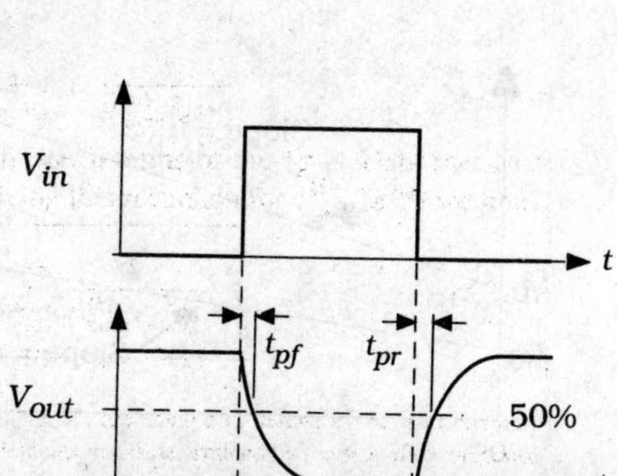

Figure 7.14 Propagation time definitions

7.2.4 General Analysis

The rise and fall time equations provide the basis for high-speed CMOS design. We can manipulate them to show us how to design single logic gates and then characterize the behavior of the gates when used in logic cascades.

To see the important factors, recall that the total output capacitance consists of two terms such that

$$C_{out} = C_{FET} + C_L \tag{7.67}$$

C_{FET} represents the parasitic capacitances of the transistors, while C_L is the external load. The layout geometry establishes the value of C_{FET}, but the load capacitance C_L varies with the application. Substituting this expression into the rise and fall time equations gives

$$\begin{aligned} t_r &\approx 2.2R_p(C_{FET} + C_L) \\ t_f &\approx 2.2R_n(C_{FET} + C_L) \end{aligned} \tag{7.68}$$

which can be cast into the forms

$$\begin{aligned} t_r &= t_{r0} + \alpha_p C_L \\ t_f &= t_{f0} + \alpha_n C_L \end{aligned} \tag{7.69}$$

These show that the rise and fall times are linear functions of the load capacitance C_L. The general behavior of both quantities is shown in Figure 7.15. Under zero-load conditions ($C_L = 0$), the inverter drives its own capacitances such that

$$\begin{aligned} t_r &= t_{r0} \approx 2.2R_p C_{FET} \\ t_f &= t_{f0} \approx 2.2R_n C_{FET} \end{aligned} \tag{7.70}$$

are determined solely from the inverter parameters. When an external

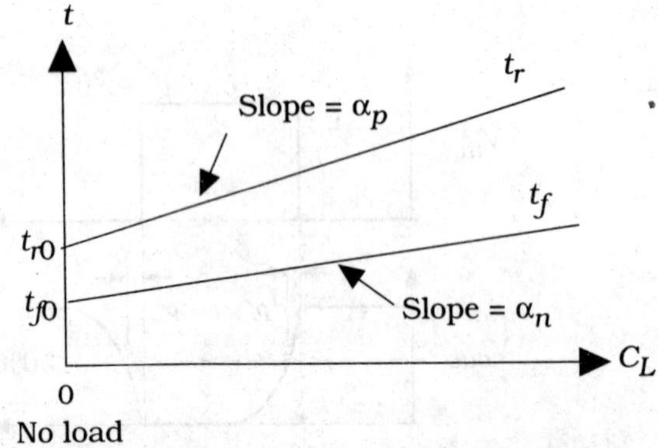

Figure 7.15 General behavior of the rise and fall times

load C_L is added, the switching times increase in a linear fashion. Large capacitive loads may cause problems because of longer delays. The dependence is described by the slope values

$$\alpha_p = 2.2R_p = \frac{2.2}{\beta_p(V_{DD} - |V_{Tp}|)} \tag{7.71}$$

and

$$\alpha_n = 2.2R_n = \frac{2.2}{\beta_n(V_{DD} - V_{Tn})} \tag{7.72}$$

Note that these are inversely proportional to the aspect ratios since

$$\beta_p = k_p'\left(\frac{W}{L}\right)_p, \qquad \beta_n = k_n'\left(\frac{W}{L}\right)_n \tag{7.73}$$

For a given load capacitance C_L, t_r and t_f can be reduced by using large FETs. However, increasing the aspect ratio of a transistor implies that it will consume more area on the chip, which in turn decreases the number of devices that can be placed on the die area allocated for the circuit. Designing for speed thus decreases the integration density of the circuit. This is called the **speed versus area trade-off** which says that

Fast circuits consume more area than slow circuits

Chip designers regularly face the problem of minimizing the switching delays without requiring excessive amounts of silicon "real estate," which is slang for chip area.

Example 7.4

Let us use the results of Example 7.3 to find the general delay equations for the case where the internal FET capacitance is $C_{FET} = 80$ fF.

The rise time t_r is controlled by the pFET that has a resistance of R_p = 822.9 Ω. The slope is given by

$$\alpha_p = 2.2R_p = 1,810.4\Omega \tag{7.74}$$

while

$$\begin{aligned} t_{r0} &\approx 2.2R_pC_{FET} \\ &= 2.2(822.9)(80\times10^{-15}) \\ &= 144.9 \text{ ps} \end{aligned} \tag{7.75}$$

The rise time can thus be written in the form

$$\begin{aligned} t_r &= t_{r0} + \alpha_p C_L \\ &= 144.9 + 1.810C_L \text{ ps} \end{aligned} \tag{7.76}$$

which requires that C_L be in units of fF.

For the fall time equation, we calculate

$$\alpha_n = 2.2(427.35) = 940.2\Omega \tag{7.77}$$

and

$$t_{f0} \approx 2.2(940.2)(80\times10^{-15}) = 165.5 \text{ ps} \tag{7.78}$$

yielding

$$t_f = 165.5 + 0.940 C_L \text{ ps} \tag{7.79}$$

as the general expression.

As an example of using these equations, suppose that the load is specified as C_L = 150 fF. We compute

$$\begin{aligned} t_r &= 144.9 + 1.810(150) = 416.4 \text{ ps} \\ t_f &= 165.5 + 0.940(150) = 306.5 \text{ ps} \end{aligned} \tag{7.80}$$

for the rise and fall times at the output. This corresponds to a maximum switching frequency for the gate of $f_{max} \approx 1.38$ GHz.

The relative values of $(W/L)_n$ and $(W/L)_p$ determine the shape of the output waveform. For example, if we design the circuit such that

$$R_p = R_n \tag{7.81}$$

then the output waveform is symmetrical with

$$t_r = t_f \tag{7.82}$$

To equalize the resistances we must design the circuit such that

$$\beta_p(V_{DD} - |V_{Tp}|) = \beta_n(V_{DD} - V_{Tn}) \tag{7.83}$$

is satisfied. If $V_{Tn} = |V_{Tp}|$, then the requirement reduces to

$$\beta_p = \beta_n \tag{7.84}$$

which gives the DC midpoint voltage at $V_M = (V_{DD}/2)$. This illustrates the fact that the nFET/pFET ratio (β_n/β_p) determines the DC midpoint voltage, while the individual values of β_n and β_p establish the switching times t_f and t_r, respectively.

7.2.5 Summary of the Inverter Circuit

It is worth taking the time to summarize the results of our study to this point. The electrical characteristics of an isolated CMOS inverter are established by two sets of parameters:

- The processing variables, such as k' and V_T values, and parasitic capacitances,

and,

- The transistor aspect ratios $(W/L)_n$ and $(W/L)_p$.

VLSI designers do not have any control over the processing parameters, as they are set by the details of the manufacturing sequence. Device sizing thus becomes the critical issue in high-speed circuit design.

System design is accomplished by using cascades of logic gates to perform the necessary binary operations. In electrical terms, the logic flow path establishes the load capacitance C_L seen by each gate. The choice of aspect ratios is the key to achieving the desired transient response of a chain of gates.

7.3 Power Dissipation

An important characteristic of CMOS integrated circuits is the power dissipated by a particular design technique. The general problem is shown in Figure 7.16. The current I_{DD} flowing from the power supply to ground gives a dissipated power of

$$P = V_{DD} I_{DD} \tag{7.85}$$

Since the value of the voltage supply V_{DD} is assumed to be a constant, we can find the value of P by studying the nature of the current flow. We usually divide the currents into DC and dynamic (or switching) contributions, so let us write

$$P = P_{DC} + P_{dyn} \tag{7.86}$$

where P_{DC} is the DC term and P_{dyn} is due to dynamic switching events.

The DC contribution can be calculated by examining the voltage transfer curve reproduced in Figure 7.17(a). When the input voltage V_{in} is stable at a low logic 0 value, the nFET Mn is off; as seen earlier in Figure 7.2, there is no direct current flow path between V_{DD} and ground. Ideally, the

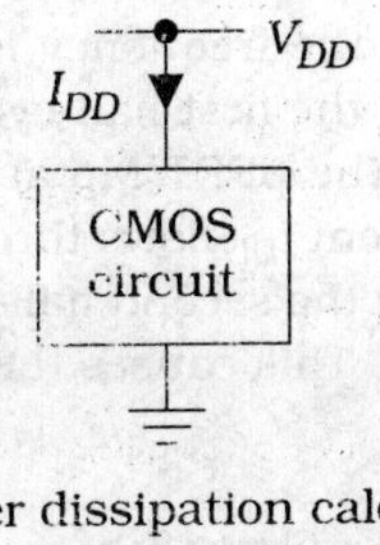

Figure 7.16 Origin of power dissipation calculation

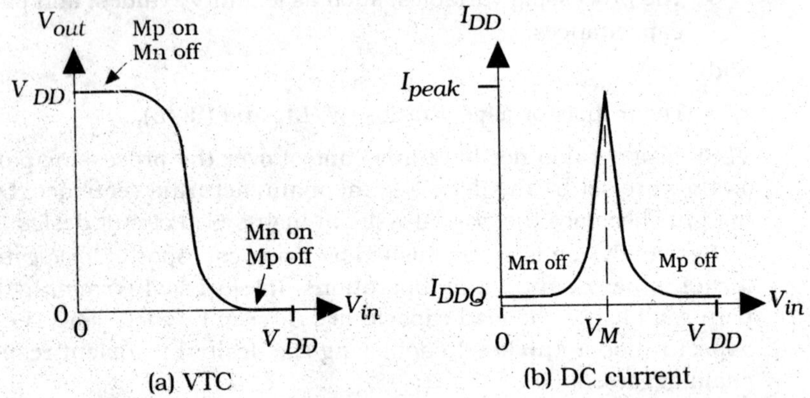

Figure 7.17 DC current flow

DC current flow for this case would be $I_{DD} = 0$, but in a realistic circuit, small **leakage currents** exist.[4] The value is denoted as I_{DDQ} and is called the **quiescent** leakage current. When V_{in} is switched, the current flow reaches a peak value I_{peak} at V_M as shown in Figure 7.17(b). However, when the input reaches a logic 1 voltage, then the pFET Mp turns off, once again preventing a direct current flow path. If we assume that the inputs are in stable 0 or 1 states as in an idle system, the DC power dissipation is given by

$$P_{DC} = V_{DD} I_{DDQ} \tag{7.87}$$

The leakage current I_{DDQ} is usually quite small, with a typical value on the order of a picoampere per gate. The value of P_{DC} is thus quite small. This consideration was a major factor in the move to CMOS in the mid-1990's.

To find the dynamic power dissipation P_{dyn}, we use a square-wave input voltage $V_{in}(t)$ as shown in Figure 7.18(a). The waveform has a period T corresponding to a switching frequency of

$$f = \frac{1}{T} \tag{7.88}$$

with units of Hertz; the frequency is the number of cycles completed in one second. During the first half-cycle, the input voltage is at a value $V_{in} = 0$. This turns on the pFET Mp as shown in Figure 7.18(b). Since the nFET is off, the current i_{DD} flows through Mp and charges C_{out} to a voltage of $V_{out} = V_{DD}$. During the second half-cycle, the input voltage is high, turning on the nFET Mn. This causes the discharge event illustrated in Figure

4 These are discussed in more detail in Chapter 9.

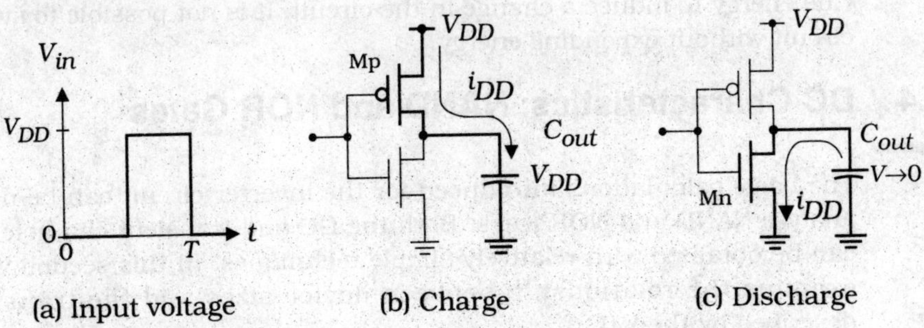

Figure 7.18 Circuit for finding the transient power dissipation

7.18(c) where V_{out} decays to 0 V. The dynamic power P_{dyn} arises from the observation that a complete cycle effectively creates a path for current to flow from the power supply to ground: during the charge event, current flows to the capacitor C_{out} while the discharge path to ground completes the circuit.

To calculate P_{dyn}, we note that the charging event leaves C_{out} with a voltage of $V_{out} = V_{DD}$. This corresponds to a stored electric charge on the capacitor of

$$Q_e = C_{out} V_{DD} \tag{7.89}$$

which has units of coulombs. When the capacitor is discharged through the nFET, the same amount of charge is lost. The average power dissipated over a single cycle with a period T is

$$P_{av} = V_{DD} I_{DD} = V_{DD}\left(\frac{Q_e}{T}\right) \tag{7.90}$$

Substituting for Q_e gives

$$P_{sw} = C_{out} V_{DD}^2 f \tag{7.91}$$

as the switching power. Combining the DC and dynamic power terms gives the total power as

$$P = V_{DD} I_{DDQ} + C_{out} V_{DD}^2 f \tag{7.92}$$

which will usually be dominated by the dynamic term. This illustrates an extremely important point:

- The dynamic power dissipation is proportional to the signal frequency

In other words, a fast circuit dissipates more power than a slow circuit. If we double the switching speed, then the dynamic power dissipation doubles. These are simply statements of the physical law that we must provide energy to induce a change in the circuit. It is not possible to switch a circuit without expending energy.

7.4 DC Characteristics: NAND and NOR Gates

The basic calculations introduced for the inverter circuit can be used to analyze NAND and NOR gates. Both the DC and transient characteristics can be obtained with relatively simple techniques. In this section we will examine the relationship between device sizes and the transitions described by the VTC.

7.4.1 NAND Analysis

Let us start with the NAND2 gate illustrated in Figure 7.19. We will analyze the case where like-polarity FETs have the same aspect ratio. This means that both pFETs are described by β_p and both nFETs have the same β_n. Since the pFETs are in parallel while the nFETs are in series, the circuit behaves quite differently from the simple inverter.

The presence of two independent inputs implies that more than one VTC curve is needed to describe the circuit. Suppose that we look for transitions where V_{out} is initially high at V_{DD} and then falls to 0 V when inputs are changed. Figure 7.20(a) summarizes the possible starting points that can lead to this situation. In case (i), both V_A and V_B are at 0 V and then switched to the bottom line condition where $V_A = V_B = V_{DD}$ such that $V_{out} = 0$ V. Since both inputs are increased at the same time, this describes the case for simultaneous input switching. The other two possibilities (ii) and (iii) describe cases where only a single input is changed. For example, in (ii) V_A is changed from 0 V to V_{DD} while V_B is held constant at V_{DD}. These three possibilities lead to the three distinct transitions shown in the plot of Figure 7.20(b). This shows that the simultaneous switching case is "pushed to the right" compared to the single-switched input cases.

It is instructive to calculate the value of the midpoint voltage V_M for the

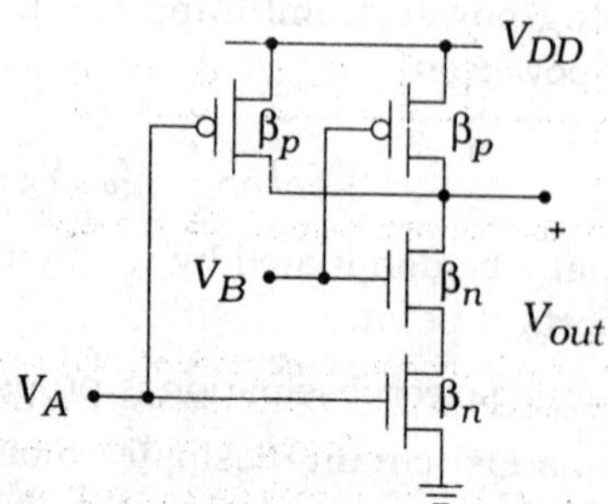

Figure 7.19 NAND2 logic circuit

	V_A	V_B	V_{out}
(i)	0	0	V_{DD}
(ii)	0	V_{DD}	V_{DD}
(iii)	V_{DD}	0	V_{DD}
	V_{DD}	V_{DD}	0

(a) Transition table

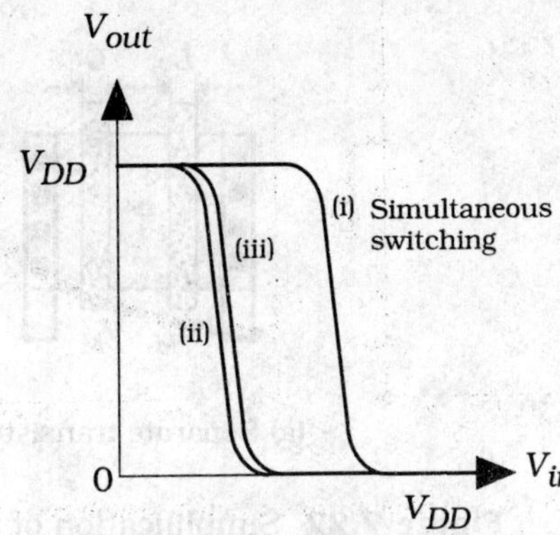

(b) VTC family

Figure 7.20 NAND2 VTC analysis

case of simultaneous switching using layout drawings. The circuit problem is illustrated in Figure 7.21, where W_n and W_p are the nFET and pFET channel widths, respectively. All transistors are assumed to have the same channel length L. Now then, for this case both input voltages V_A and V_B are equal to V_M. On the layout plot, both gates are thus at the same potential and can be connected to simplify the calculations.

Consider the nFETs first. In Figure 7.22(a), the layout is shown in its original form with two separate series-connected transistors. Let us "merge" the two gates together into one to obtain the patterning shown in Figure 7.22(b). If we ignore the n+ region that separates the two gates, then the structure can be approximated as a single nFET with an aspect ratio of $(W_n/2L)$ as shown. Since the original nFETs each had a device transconductance of β_n, the single equivalent transistor is described by the value $(\beta_n/2)$.

The pFETs can be combined in a similar manner. The original parallel-connected transistors are illustrated in Figure 7.23(a). Owing to the paral-

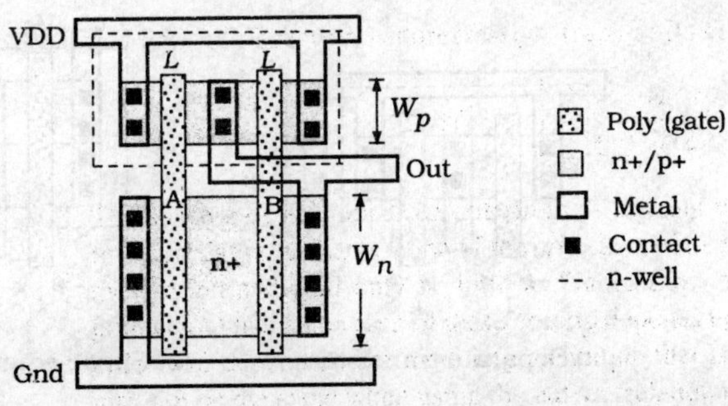

Figure 7.21 Layout of NAND2 for V_M calculation

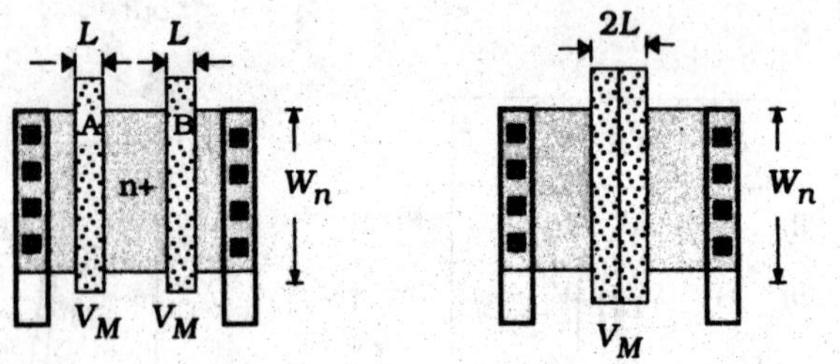

(a) Separate transistors (b) Single equivalent FET

Figure 7.22 Simplification of the series-connected nFETs

lel wiring, the left and right sides are electrically the same point, so that the two may be simplified into the single gate structure shown in Figure 7.23(b). In this case, the two combine to act as a single pFET with an aspect ratio of $(2W_p/L)$. If the original devices each have β_p, then the equivalent structure acts as a pFET with $2\beta_p$.

Let us now use these results to find V_M for the case of simultaneous switching. Replacing the transistor pairs by their single-FET equivalents gives the inverter circuit in Figure 7.24, where the nFET and pFET transconductances are $(\beta_n/2)$ and $2\beta_p$, respectively. The calculation then proceeds in the same manner as for the "normal" NOT gate. Both transistors are saturated, so equating currents gives

$$\frac{(\beta_n/2)}{2}(V_M - V_{Tn})^2 = \frac{(2\beta_p)}{2}(V_{DD} - V_M - |V_{Tp}|)^2 \tag{7.93}$$

Taking square roots of both sides and solving for the midpoint voltage results in the expression

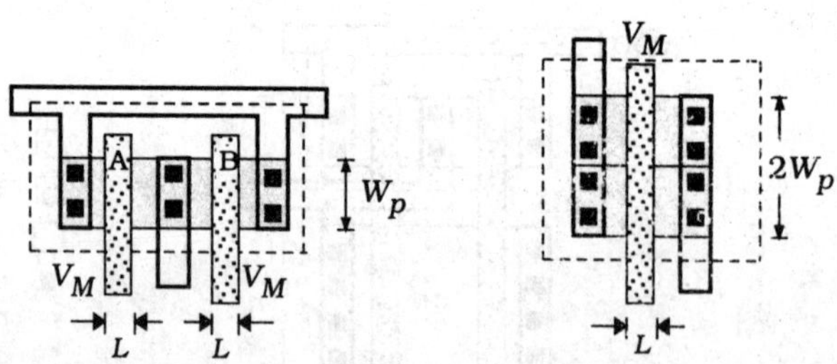

(a) Separate transistors (b) Single equivalent FET

Figure 7.23 Simplification of parallel-connected pFETs

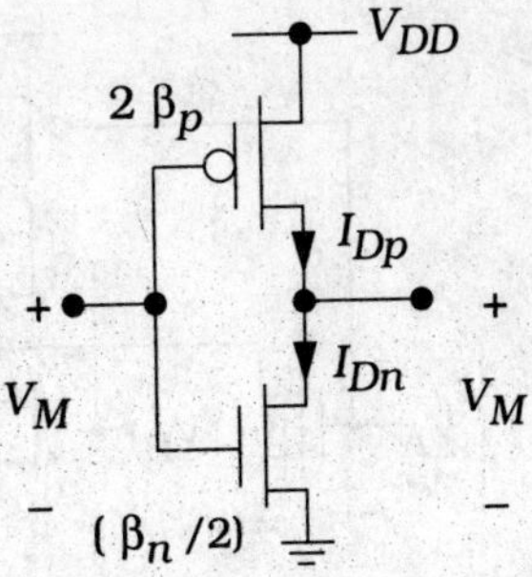

Figure 7.24 Simplified V_M circuit for the NAND2 gate

$$V_M = \frac{V_{DD} - |V_{Tp}| + \frac{1}{2}\sqrt{\frac{\beta_n}{\beta_p}}V_{Tn}}{1 + \frac{1}{2}\sqrt{\frac{\beta_n}{\beta_p}}} \tag{7.94}$$

This has the same form as the NOT gate in equation (7.14), except that the square root term is multiplied by a factor of (1/2). This reduces the denominator, which is why the VTC curve is shifted toward the right. If we apply the same reasoning to an *N*-input NAND gate, the simultaneous switching point is found to be

$$V_M = \frac{V_{DD} - |V_{Tp}| + \frac{1}{N}\sqrt{\frac{\beta_n}{\beta_p}}V_{Tn}}{1 + \frac{1}{N}\sqrt{\frac{\beta_n}{\beta_p}}} \tag{7.95}$$

The right shift is due to the series-connected nFETs, since their resistances add.

7.4.2 NOR Gate

The NOR2 gate can be analyzed using the same techniques. We assume that the nFETs have the same β_n and that both pFETs are described by β_p as shown in the basic circuit of Figure 7.25. To construct VTC, note that $V_{out} = V_{DD}$ requires that $V_A = V_B = 0$ V. If either input (or both) are switched to logic 1 values, then the output will fall to $V_{out} = 0$ V. The three combinations are listed in the function table of Figure 7.26(a). As with the NAND2 gate, there are three distinct transitions shown in the VTC family of Figure 7.26(b). Case (i) describes the simultaneous switching event where both V_A and V_B are increased from 0 V toward V_{DD}. This case is the leftmost plot in the VTC family, exactly opposite to that found for the NAND2. Single-input switching cases (ii) and (iii) are distinct, but are close to each other.

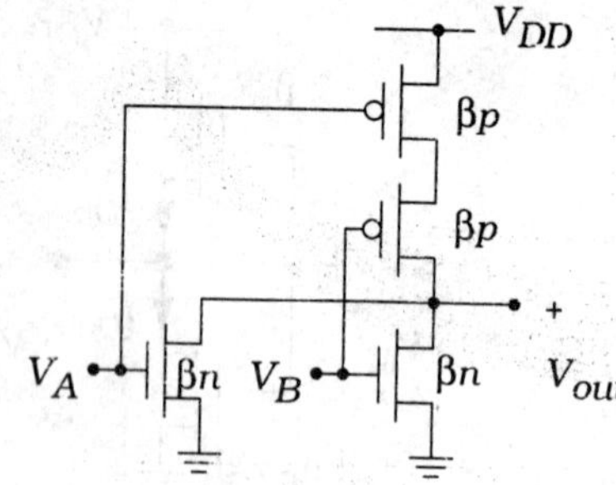

Figure 7.25 NOR2 circuit

The techniques of combining series and parallel transistors may be used to compute V_M for the simultaneous switching case. Since the nFETs are in parallel, they may be combined to a single equivalent nFET with a transconductance of $2\beta_n$. The series-connected pFETs act as a single pFET with $(\beta_p/2)$ which gives rise to the simplified equivalent circuit in Figure 7.27. Equating the saturation currents using the effective transconductance values gives us

$$\frac{(2\beta_n)}{2}(V_M - V_{Tn})^2 = \frac{(\beta_p/2)}{2}(V_{DD} - V_M - |V_{Tp}|)^2 \tag{7.96}$$

This may be solved to give

$$V_M = \frac{V_{DD} - |V_{Tp}| + 2\sqrt{\frac{\beta_n}{\beta_p}}V_{Tn}}{1 + 2\sqrt{\frac{\beta_n}{\beta_p}}} \tag{7.97}$$

	V_A	V_B	V_{out}
	0	0	V_{DD}
(iii)	0	V_{DD}	0
(ii)	V_{DD}	0	0
(i)	V_{DD}	V_{DD}	0

(a) Transition table

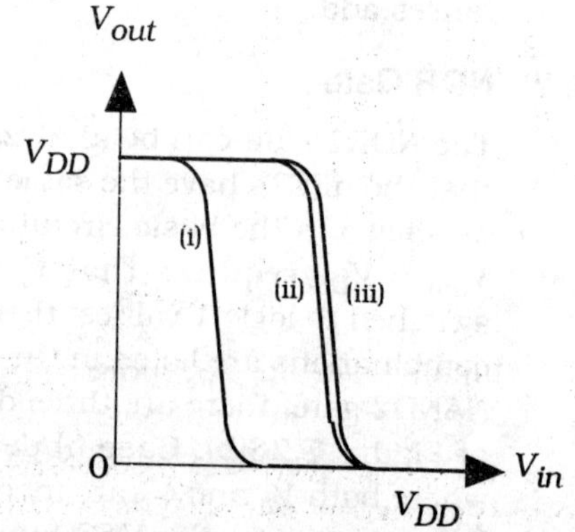

(b) VTC family

Figure 7.26 NOR2 VTC construction

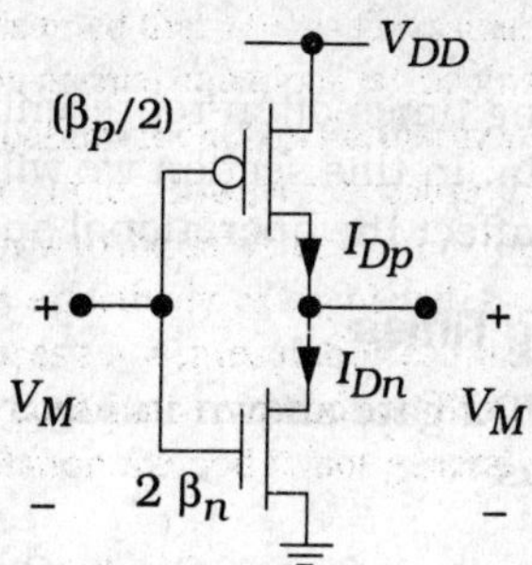

Figure 7.27 NOR2 V_M calculation for simultaneous switching

Comparing this with the NOT and NAND expressions shows that the only difference is the factor of 2 multiplying the square root term. This increases the denominator, which decreases the value of V_M from that of an inverter with a device ratio of (β_n/β_p). The midpoint voltage for an N-input NOR gate is

$$V_M = \frac{V_{DD} - |V_{Tp}| + N\sqrt{\frac{\beta_n}{\beta_p}}V_{Tn}}{1 + N\sqrt{\frac{\beta_n}{\beta_p}}} \tag{7.98}$$

It is worthwhile noting that the NAND and NOR gates tend have opposite behaviors with respect to the reference NOT gate VTC.

As a final comment, we note that both the NAND and NOR gates exhibit low DC power dissipation values of

$$P_{DC} = V_{DD}I_{DDQ} \tag{7.99}$$

since there is no direct current flow path from the power supply to ground when the inputs are stable logic 0 or logic 1 values. The low power characteristic of the gates is due to the use of complementary pairs and series-parallel structuring of the transistor arrays. Dynamic power is still present in the general form

$$P_{sw} = C_{out}V_{DD}^2 f_{gate} \tag{7.100}$$

which shows the dependence on gate switching frequency f_{gate}. Since it takes more than a single input to switch the gate, f_{gate} is different from the basic switching frequency used for the inverter. This is discussed in more detail later.

7.5 NAND and NOR Transient Response

Transient switching times often represent the limiting factor in designing a digital logic chain. In this section we will examine how the FET topology and device sizing affect the operational speed of the gate.

7.5.1 NAND2 Switching Times

Consider the NAND2 gate shown in Figure 7.28. The total output capacitance is denoted as

$$C_{out} = C_{FET} + C_L \tag{7.101}$$

where C_L is the external load and

$$C_{FET} = C_{Dn} + 2C_{Dp} \tag{7.102}$$

represents the parasitic internal FET capacitances. Note that there are two contributions of C_{Dp} since two pFETs are connected to the output node. The drawing identifies the transistors by their resistance values

$$R_p = \frac{1}{\beta_p(V_{DD} - |V_{Tp}|)}, \qquad R_n = \frac{1}{\beta_n(V_{DD} - V_{Tn})} \tag{7.103}$$

The transient calculations are based on finding RC time constants for the charge time (t_r or t_{LH}) and fall time (t_f or t_{HL}) for the transitions. The procedure is complicated by the presence of two inputs. We will concentrate on estimating the worst-case values of the switching times.

Let us consider the rise time t_r first. The output voltage is initially at a value V_{out} (0) = 0 V and is then charged to V_{DD}. If only one pFET is conducting, we obtain the simplified charging circuit shown in Figure 7.29(a) where C_{out} charges through a pFET resistance R_p. Since this looks like the charging circuit for a simple inverter, we can write

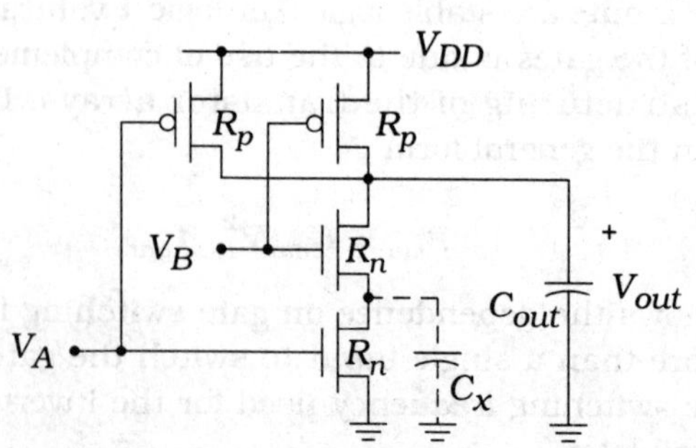

Figure 7.28 NAND2 circuit for transient calculations

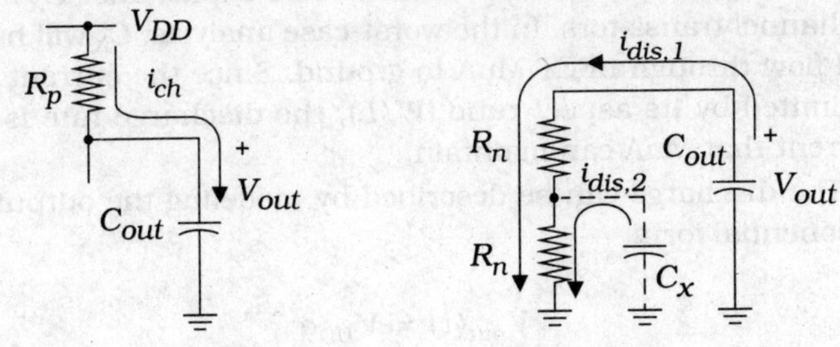

Figure 7.29 NAND2 subcircuits for estimating rise and fall times

$$V_{out}(t) = V_{DD}[1 - e^{-t/\tau_p}] \tag{7.104}$$

where

$$\tau_p = R_p C_{out} \tag{7.105}$$

is the time constant. The rise time is thus given by

$$t_r \approx 2.2\tau_p \tag{7.106}$$

This is considered to be a "worst-case" situation since only one pFET is charging C_{out}. Note that this can be cast into the linear form

$$t_r = t_0 + \alpha_0 C_L \tag{7.107}$$

where

$$t_0 = 2.2 R_p C_{FET} \tag{7.108}$$

is the zero-load value, and

$$\alpha_0 = 2.2 R_p \tag{7.109}$$

is the slope of t_r as a function of the load capacitance C_L. If both pFETs are conducting, then the equivalent resistance is lowered to $(R_p/2)$ since the two are in parallel; this would be the "best-case" event, i.e., the one with the shortest charging time. Design is usually based on worst-case analysis since we want to insure that the circuit operates under all conditions.

The situation is more complicated when we analyze the fall time t_f where C_{out} discharges through the series-connected nFET chain. RC modeling of each device leads to the "ladder" network shown in Figure 7.29(b). While the main item of interest is discharging C_{out}, the situation is complicated by the presence of the inter-FET capacitance C_X between the two n-channel transistors. In the worst-case analysis, C_X will have charge that will flow through nFET MnA to ground. Since the current through a FET is limited by its aspect ratio (W/L), the discharge rate is limited by the current that MnA can maintain.

The discharge can be described by modeling the output voltage in the exponential form

$$V_{out}(t) = V_{DD}e^{-t/\tau_n} \tag{7.110}$$

such that the time constant is given by the **Elmore formula** as

$$\tau_n = C_{out}(R_n + R_n) + C_X R_n \tag{7.111}$$

This estimates the time constant as the superposition of time constants

$$\tau_n = \tau_{n1} + \tau_{n2} \tag{7.112}$$

where

$$\tau_{n1} = C_{out}(R_n + R_n) \tag{7.113}$$

is the time constant for C_{out} discharging through two nFETs, each with a resistance R_n; this is shown by the current $i_{dis,1}$ in the drawing. The other term

$$\tau_{n2} = C_X R_n \tag{7.114}$$

is the time constant for C_X discharging through one nFET with a resistance R_n. This is corresponds to the discharge current $i_{dis,2}$. The fall time t_f is then given by

$$t_f \approx 2.2\tau_n \tag{7.115}$$

Substituting the time constant expression transforms this into

$$t_f \approx 2.2[(C_{FET} + C_L)(2R_n) + C_X R_n] \tag{7.116}$$

Grouping terms results in the linear expression

$$t_f = t_1 + \alpha_1 C_L \tag{7.117}$$

with a zero-load delay of

$$t_1 = 2.2R_n(2C_{FET} + C_X) \tag{7.118}$$

and a slope of

$$\alpha_1 = 4.4R_n \tag{7.119}$$

where the multiplier is from (2×2.2). Although we are able to write t_f as a linear function of C_L, both the zero-load delay and the slope are affected by the series-connected nFETs in the discharge circuitry.

The Elmore formulation of time constants for RC ladder-type networks illustrates that series-connected FETs lead to longer delays in CMOS circuits. To understand this comment, let us rewrite equation (7.111) as

$$\tau_n = R_n(2C_{out} + C_X) \tag{7.120}$$

In this form, we can interpret the time constant as R_n multiplying an effective capacitance with a value

$$C_{eff} = 2C_{out} + C_X \tag{7.121}$$

which is larger than twice the output capacitance. Alternately, we may write

$$\tau_n = C_{out}(2R_n) + C_X R_n \tag{7.122}$$

which clearly shows the effect of the series-connected FETs in the term $2R_n$ and the increase due to the parasitic capacitance C_X. Regardless of the interpretation one chooses, it is important to remember that series-connected FET chains can lead to excessive logic delays.

7.5.2 NOR2 Switching Times

The analysis of the NOR2 transients proceeds in the same manner. Figure 7.30 shows the circuit with FET resistances and the capacitances. The output capacitance for any gate is given by the general form

$$C_{out} = C_{FET} + C_L \tag{7.123}$$

For the NOR2 circuit, the internal capacitance can be broken down into components as

$$C_{FET} = 2C_{Dn} + C_{Dp} \tag{7.124}$$

since there are two nFETs connected to the output node but only one

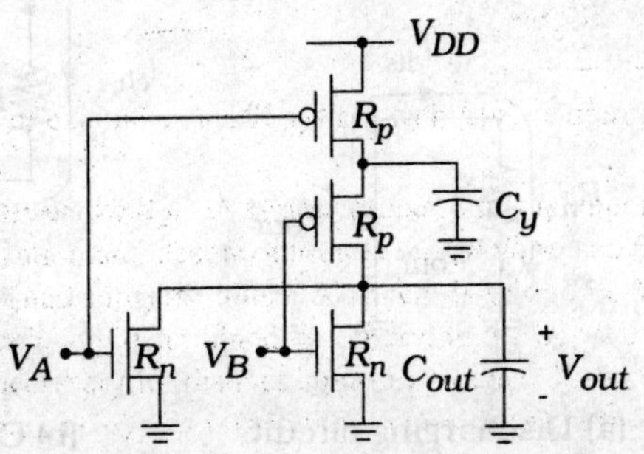

Figure 7.30 NOR2 circuit for switching time calculations

pFET. The inter-FET capacitance C_y represents the parasitic contributions between the two pFETs.

Figure 7.31 shows the subcircuits for the output transients. The fall time t_f may be computed using the worst-case circuit in Figure 7.31(a) where only one nFET acts to discharge the output capacitance. We thus write the output voltage as

$$V_{out}(t) = V_{DD}e^{-t/\tau_n} \tag{7.125}$$

with

$$\tau_n = R_n C_{out} \tag{7.126}$$

as the time constant. The fall time is then given by

$$t_f \approx 2.2\tau_n \tag{7.127}$$

which is identical to that for a simple inverter. Expanding C_{out} gives the linear dependence

$$t_f = t_1 + \alpha_1 C_L \tag{7.128}$$

where the zero-load delay is

$$t_1 = 2.2R_n C_{FET} \tag{7.129}$$

and the slope is

$$\alpha_1 = 2.2R_n \tag{7.130}$$

These results are similar to the NOT gate, but it is important to remember that C_{FET} is larger for the NOR2 gate.

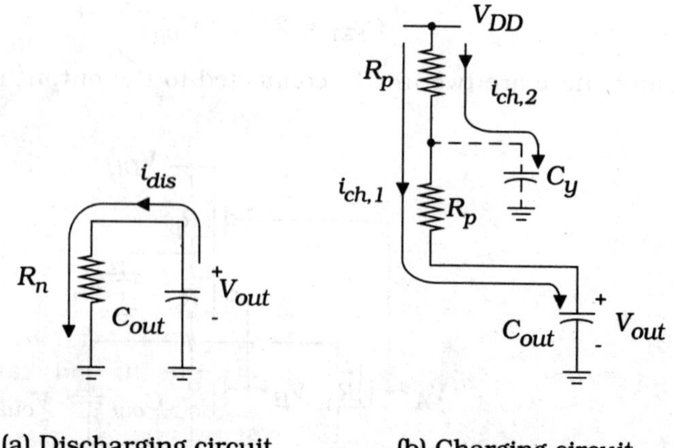

(a) Discharging circuit (b) Charging circuit

Figure 7.31 Subcircuits for the NOR2 transient calculations

The charging circuit for finding the rise time t_r is shown in Figure 7.31(b). We will write the output voltage in the exponential form

$$V_{out}(t) = V_{DD}[1 - e^{-t/\tau_p}] \tag{7.131}$$

However, since C_y will be charged during this event, we must use the Elmore formula to find the time constant. The two paths are shown as $i_{ch,1}$ and $i_{ch,2}$ in the drawing. The primary charge path due to $i_{ch,1}$ is described by a time constant

$$\tau_1 = C_{out}(R_p + R_p) \tag{7.132}$$

while that associated with $i_{ch,2}$ is

$$\tau_2 = C_y R_p \tag{7.133}$$

Superposing gives the total effective time constant in the form

$$\begin{aligned} \tau_p &= \tau_1 + \tau_2 \\ &= C_{out}(2R_p) + C_y R_p \end{aligned} \tag{7.134}$$

such that the rise time is

$$t_r = 2.2\tau_p \tag{7.135}$$

Since the series-connected pFETs introduce a large time constant, the rise time may be quite large compared to the fall time. Substituting for C_{out} gives the linear equation

$$t_r = t_0 + \alpha_0 C_L \tag{7.136}$$

where

$$t_0 = 2.2R_p(2C_{FET} + C_y) \tag{7.137}$$

and

$$\alpha_0 = 4.4R_p \tag{7.138}$$

characterize the dependence of t_r on C_L. As with the NAND2 gate, the presence of series-connected FETs slows down the associated switching time.

7.5.3 Summary

The analyses above illustrate that the NAND and NOR gates exhibit complementary characteristics at both the DC and transient levels. This arises because they are constructed using complementary series-parallel transistor arrangements.

While the DC characteristics are important, most design effort is directed toward minimizing delays through logic chains. The study above allows us to make some general statements about NAND and NOR gates as compared to the simpler NOT circuit. First, we have seen that the rise time can be written in the form

$$t_r = t_0 + \alpha_0 C_L \tag{7.139}$$

while the fall time has the same structure with

$$t_f = t_1 + \alpha_1 C_L \tag{7.140}$$

The constants (t_0 and α_0 for the rise time, and t_1 and α_1 for the fall time) depend upon the parasitic transistor resistances and capacitances. The constants are the smallest for a NOT gate, so we often use it as a reference. This, of course, is because the inverter consists of only two FETs. In general, adding complementary transistor pairs increases the delay times because C_{FET} is increased. The number of inputs to a logic gate is called the **fan-in** (FI). Since every input is connected to a complementary pair, we can state that

- Switching delays increase with the fan-in.

This says, for example, a NAND3 gate will be slower than a NAND2 gate if the two use the same size transistors. Of course, the actual delay depends upon the value of the load capacitance C_L such that

- Switching delays increase with the external load.

Since logic functions are implemented using cascades of gates, the effect of this dependence varies with the circuit.

Let us summarize the results of the NAND and NOR analysis. As with the inverter, the electrical characteristics of these gates are set by

- The processing variables and
- The aspect ratios $(W/L)_n$ and $(W/L)_p$ of every FET

Furthermore, series transistors introduced us to the problem of parasitic capacitance between the two devices. This factor leads us to make one additional statement

- The details of the layout geometries affect the transient response of the logic gate.

We thus conclude that the physical layout and structure of the circuitry is a critical factor in designing high-speed logic networks.

7.6 Analysis of Complex Logic Gates

The analysis techniques developed for the NAND and NOT circuits may b extended to analyze complex CMOS logic gates with AOI and OAI struc turing. The most important problem is the transient delay associated wit series-connected FETs.

Consider the complex logic gate shown in Figure 7.32. This implements the logic function

$$f = \overline{x \cdot (y+z)} \tag{7.141}$$

with series-parallel FET arrays. The aspect ratio values shown in the drawing are the critical parameters that affect the rise and fall times. The fall time is governed by the nFETs. If we assume that they are all the same size with

$$\left(\frac{W}{L}\right)_{nx} = \left(\frac{W}{L}\right)_{ny} = \left(\frac{W}{L}\right)_{nz} \tag{7.142}$$

then the nFET resistance R_n can be used to describe each one. The worst-case fall time will occur when $x = 1$, but only one of the ORed inputs y or z is 1. This results in a 2-FET series pair that must handle the discharge of the output capacitor

$$C_{out} = C_{FET} + C_L \tag{7.143}$$

With the capacitance C_n in the chain, the time constant is

$$\tau_n = R_n C_n + 2R_n C_{out} \tag{7.144}$$

which gives a fall time of

$$\begin{aligned} t_f &= 2.2\tau_n \\ &= 2.2R_n[C_n + 2(C_{FET} + C_L)] \\ &= t_1 + \alpha_1 C_L \end{aligned} \tag{7.145}$$

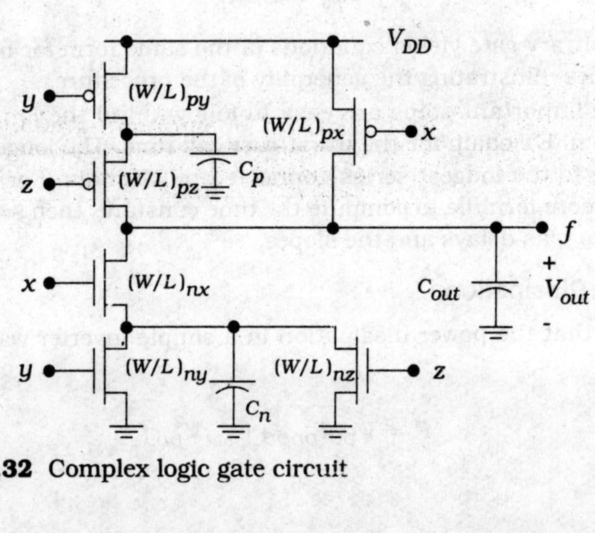

Figure 7.32 Complex logic gate circuit

where

$$t_1 = 2.2R_n(C_n + 2C_{FET}) \quad (7.146)$$

is the zero-load time, and

$$\alpha_1 = 2.2R_n \quad (7.147)$$

is the slope.

The rise time t_r is determined by the pFETs. If these are chosen with equal aspect ratios

$$\left(\frac{W}{L}\right)_{px} = \left(\frac{W}{L}\right)_{py} = \left(\frac{W}{L}\right)_{pz} \quad (7.148)$$

then we can use the same R_p for each device. The limiting series chain is with the y and z input p-channel transistors; the x-input pFET provides the fast switching, and could be decreased to half-size without affecting the results. The series chain gives a time constant of

$$\tau_p = R_pC_p + 2R_pC_{out} \quad (7.149)$$

where C_p is the parasitic capacitance between the pFETs. The worst-case rise time is thus of the form

$$t_r = t_0 + \alpha_0 C_L \quad (7.150)$$

where the zero-load delay is

$$t_0 = 2.2R_p(C_p + 2C_{FET}) \quad (7.151)$$

and the slope is

$$\alpha_0 = 2.2R_p \quad (7.152)$$

An arbitrary gate yields equations of the same form for both the rise and fall times, illustrating the generality of the procedure.

The important steps are easy to follow. Find the longest series-connected nFET chain for the worst-case fall time. The longest rise time will be due to the longest series-connected pFET chain. For both cases, use the Elmore formula to compute the time constant, then separate terms for the zero bias delays and the slopes.

7.6.1 Power Dissipation

Recall that the power dissipation in a simple inverter was written in the form

$$P = V_{DD}I_{DDQ} + C_{out}V_{DD}^2 f \quad (7.153)$$

When we analyze a general static CMOS logic gate, the DC term is still small, but the dynamic switching power P_{dyn} becomes important in high-speed, high-density designs.

To model the dynamic power dissipation of an arbitrary gate we recall that P_{dyn} originates from an output switching event. First, the output capacitor C_{out} is charged from 0 V to V_{DD}, corresponding to an output logic $0 \rightarrow 1$ transition. Then, C_{out} discharges to give a $1 \rightarrow 0$ transition, completing the cycle. To model the number of transitions that take place over a switching period T we introduce the **activity coefficient** a that represents the probability that an output $0 \rightarrow 1$ transition takes place during one period. The dynamic power is then modified to read

$$P_{dyn} = aC_{out}V_{DD}^2 f \tag{7.154}$$

For a network that consists of N gates, the total dynamic power is more generally written in the form

$$P_{dyn} = \sum_{i=1}^{N} a_i C_i V_i V_{DD} f \tag{7.155}$$

where, for the i-th gate, a_i is the activity coefficient and C_i is the node capacitance that charges to a maximum value of V_i.

Activity coefficients can be determined from truth tables. Figure 7.33 provides the truth tables for the NOR2 and NAND2 functions. We will assume that each input combination has equal probability of occurring. Let us analyze the NOR2 transitions first. Since the activity factor a_{NOR2} is the probability that the gate makes a $0 \rightarrow 1$ transition, it can be calculated by

$$a = p_0 p_1 \tag{7.156}$$

where p_0 is the probability that the output is initially at 0, and p_1 the probability that it makes a transition to 1. The truth table shows us that $p_0 = (3/4)$ and $p_1 = (1/4)$, so

A	B	$\overline{A+B}$	$\overline{A \cdot B}$
0	0	1	1
0	1	0	1
1	0	0	1
1	1	0	0

Figure 7.33 Truth tables for determining activity coefficients

$$a_{NOR2} = \left(\frac{3}{4}\right)\left(\frac{1}{4}\right) = \frac{3}{16} \tag{7.157}$$

The NAND2 gate can be analyzed in the same manner. For this gate, the truth table shows that $p_0 = (1/4)$ and $p_1 = (3/4)$ so

$$a_{NAND2} = \left(\frac{3}{4}\right)\left(\frac{1}{4}\right) = \frac{3}{16} \tag{7.158}$$

has the same value as the NOR2 gate. If we look at 3-input gates, the truth tables give

$$a_{NOR3} = \frac{7}{64} = a_{NAND3} \tag{7.159}$$

Similarly, we can calculate

$$a_{XNOR2} = \frac{1}{4} = a_{XOR2} \tag{7.160}$$

since $p_0 = (1/4) = p_1$. The technique can be applied to an arbitrary gate.

The limit on this simple treatment is that, in practice, we rarely have input combinations that occur with equal probability. More advanced techniques have been developed to handle these situations. The interested reader is directed to Reference [2] for an excellent discussion of the details. Reference [8] is a very thorough analysis of power dissipation and low-power design.

7.7 Gate Design for Transient Performance

High-speed circuits are limited by the switching time of individual gates. Logic formation determines the series and parallel connections of the transistors. The aspect ratios are the critical design parameters for both the DC and transient switching times. Once these are specified and the transistors are created in the layout, all of the parasitics are set.

The DC switching characteristics are often considered less important than the switching speed. It is common to design a gate to have the desired transient times, and then check the DC VTC to insure that it is acceptable. This approach is based on the fact that the individual nFET and pFET aspect ratios determine the switching response, while the DC transition point is a result of the ratio of the nFET to pFET values. For example, the value of β_n/β_p gives V_M for an inverter, while t_r depends primarily on β_p and t_f is established by β_n.

The design philosophy used to select aspect ratios varies with the situation. A straightforward approach is to use the inverter as a reference and then attempt to design other gates that have approximately the same switching times. Since the NOT gate is the simplest, it can be built using relatively small transistors. We will use the device transconductance

$$\beta = k'\left(\frac{W}{L}\right) \tag{7.161}$$

as being equivalent to the aspect ratio.

Figure 7.34(a) shows an inverter with device sizes specified by β_p and β_n, which we will assume are known. These set the rise and fall times t_r and t_f for the circuit, which serve as the reference switching times. Since both transistors drive the same capacitance, the difference is in the resistance values

$$R_p = \frac{1}{\beta_p(V_{DD} - |V_{Tp}|)}, \qquad R_n = \frac{1}{\beta_n(V_{DD} - V_{Tn})} \tag{7.162}$$

Recall that a symmetrical inverter has

$$\beta_n = \beta_p \tag{7.163}$$

and requires the device sizes to be related by

$$\left(\frac{W}{L}\right)_p = r\left(\frac{W}{L}\right)_n \tag{7.164}$$

where

$$r = \frac{k'_n}{k'_p} \tag{7.165}$$

is the process transconductance ratio. A nonsymmetrical design that uses equal size transistors such that $\beta_n > \beta_p$ is also commonly used as a reference.

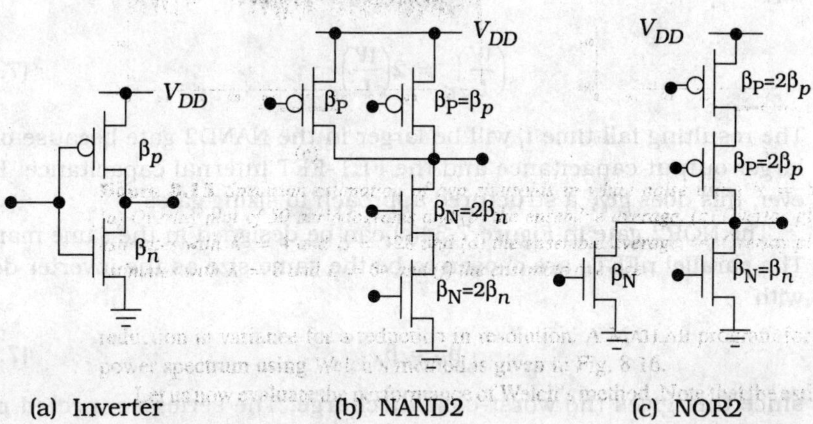

Figure 7.34 Relative FET sizing

Let us use these values to find the device sizes β_P and β_N for the NAND2 gate in Figure 7.34(b) with the philosophy that we want to achieve similar rise and fall times. Consider first the parallel pFETs. Since the worst-case situation is where only one transistor contributes to the rise time, we may select the same size as the inverter:

$$\beta_P = \beta_p \tag{7.166}$$

The actual rise time t_r will be longer than that of the inverter because C_{out} is larger. The series-connected nFET chain has to be modeled as two series-connected resistors between the output and ground, with a total value of

$$R = R_N + R_N \tag{7.167}$$

where

$$R_N = \frac{1}{\beta_N(V_{DD} - V_{Tn})} \tag{7.168}$$

Using the inverter as a reference, we set

$$R = R_n = 2R_N \tag{7.169}$$

Substituting,

$$\frac{1}{\beta_n(V_{DD} - V_{Tn})} = \frac{2}{\beta_N(V_{DD} - V_{Tn})} \tag{7.170}$$

which has the solution

$$\beta_N = 2\beta_n \tag{7.171}$$

i.e., the series-connected nFETs are twice as large as the inverter transistor:

$$\left(\frac{W}{L}\right)_N = 2\left(\frac{W}{L}\right)_n \tag{7.172}$$

The resulting fall time t_f will be larger in the NAND2 gate because of the larger output capacitance and the FET-FET internal capacitance. However, this does give a structured approach to sizing gates.

The NOR2 gate in Figure 7.34(c) can be designed in the same manner. The parallel nFETs are chosen to be the same size as the inverter device with

$$\beta_N = \beta_n \tag{7.173}$$

since this gives the worst-case discharge. The series-connected pFET resistances add to a total of $2R_P$. Equating this to the inverter resistance

R_p gives

$$\frac{1}{\beta_p(V_{DD}-|V_{Tp}|)} = \frac{2}{\beta_P(V_{DD}-|V_{Tp}|)} \tag{7.174}$$

so that

$$\beta_P = 2\beta_p \tag{7.175}$$

indicating that the pFETs are twice as large as the inverter transistors:

$$\left(\frac{W}{L}\right)_P = 2\left(\frac{W}{L}\right)_p \tag{7.176}$$

The main problem is that pFETs are intrinsically slow, so that the value of $(W/L)_p$ may be large to begin with.

This technique can be extended to larger chains. For n series-connected FETs, the size must be n times larger than the inverter value. The NAND3 gate in Figure 7.35(a) would thus be designed with

$$\beta_N = 3\beta_n, \qquad \beta_P = \beta_p \tag{7.177}$$

such that

$$\left(\frac{W}{L}\right)_N = 3\left(\frac{W}{L}\right)_n, \qquad \left(\frac{W}{L}\right)_P = \left(\frac{W}{L}\right)_p \tag{7.178}$$

while the NOR3 gate in Figure 7.35(b) would have

$$\beta_N = \beta_n, \qquad \beta_P = 3\beta_p \tag{7.179}$$

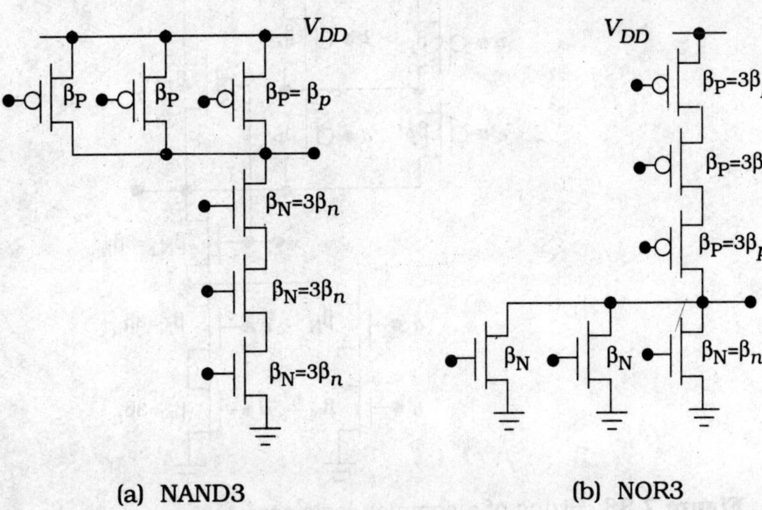

Figure 7.35 Sizing for 3-input gates

with

$$\left(\frac{W}{L}\right)_N = \left(\frac{W}{L}\right)_n \quad , \qquad \left(\frac{W}{L}\right)_P = 3\left(\frac{W}{L}\right)_p \tag{7.180}$$

Since the reference values β_n and β_p are arbitrary, the sizes can be adjusted as needed to accommodate reasonable values. Also note that if we select a symmetric inverter design with $\beta_n = \beta_p$, then the resulting gates will also be approximately symmetric.

Complex logic gates can be designed in the same manner. Consider the gate in Figure 7.36 that has an output of

$$f = \overline{(a \cdot b + c \cdot d) \cdot x} \tag{7.181}$$

using series-parallel structuring. Consider the nFET array first. Any discharge event will have current flow through a minimum of three series-connected nFETs. The device sizes would all be the same with the value

$$\beta_N = 3\beta_n = \beta_{N1} \tag{7.182}$$

The pFET array is a little different. The worst-case charge path is through two series-connected transistors on the left side of the circuit. The sizes would be

$$\beta_P = 2\beta_p \tag{7.183}$$

for the pFETs in the inputs *a*, *b*, *c*, and *d*. The *x*-input pFET is alone, so

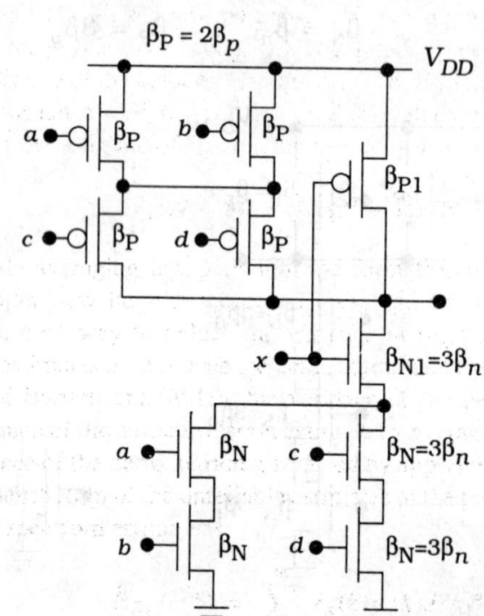

Figure 7.36 Sizing of a complex logic gate

that we can select its size as being the same as for an inverter:

$$\beta_{P1} = \beta_p \tag{7.184}$$

Alternately, the choice

$$\beta_{P1} = \beta_P = 2\beta_p \tag{7.185}$$

may lead to simpler layout since only a single size pFET would be used. Note that the two options for β_{P1} result in different input capacitances for the *x*-input.

Although this approach provides a nice structured methodology, it leads to large transistors. The designer must decide whether the real estate consumption is worth the added speed. This becomes more complicated as the number of FETs increases since the FET-to-FET parasitic capacitance terms in the Elmore time constant formula will also increase. In practice, we may just select a standard cell that meets the area allocation and then find the overall speed of the logic cascade. If the design is not fast enough, we can apply some of the techniques in the next chapter to find a better design.

7.8 Transmission Gates and Pass Transistors

Transmission gates consist of an nFET/pFET pair wired in parallel as shown in Figure 7.37(a). The RC switching model shown in Figure 7.37(b) consists of a TG resistance R_{TG} and capacitances that account for the parasitic contributions of both FETs. Even though the FETs are in parallel, one usually dominates the conduction process at any given time. For example, a logic 0 transmission is controlled by the nFET. Owing to this, a reasonable approximation for the linear resistance is

$$R_{TG} = \max(R_n, R_p) \tag{7.186}$$

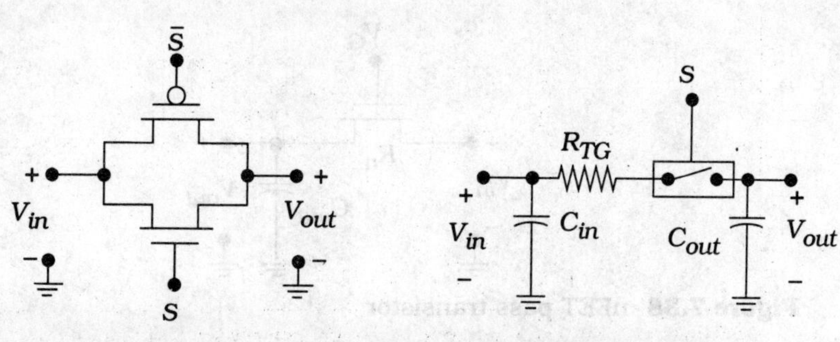

(a) Circuit (b) RC model

Figure 7.37 Transmission gate modeling

i.e., we use the larger of the two values. The capacitances are obtained by adding the contributions. For example, assuming that the left side is at a lower voltage than the right side,

$$C_{in} = C_{S,n} + C_{D,p} \tag{7.187}$$

since the left side of the nFET is the source, while the same node is the drain of the pFET.[5] We note the trade-off in selecting the aspect ratios for the two transistors: large values of (W/L) decrease the resistance, but a large W implies large capacitances. This has made TGs less and less attractive during the evolution of high-density VLSI.

An important electrical feature of the transmission gate (and the pass FETs discussed below) is that there are no direct signal connections to the power supply V_{DD} or ground. Static logic circuits are able to provide full rail outputs $V_{OH} = V_{DD}$ and $V_{OL} = 0$ V by using a power supply rail. Since the TG is not used in this manner, the driving circuit (the one preceding the transmission gate) is responsible for providing the input signal voltages. However, the TG appears to be an RC parasitic to the driving gate, so the response is slower than if the TG were absent. Additional buffer circuits are thus needed to maintain the speed.

Pass transistors are single FETs that pass the **signal** between the drain and source terminals instead of a fixed power supply value. "Pass FETs" can be used in place of transmission gates in most circuits. They require less area and wiring, but cannot pass the entire voltage range. When choosing between the two polarities, nFETs are preferred for this application since the larger electron mobility implies faster switching than could be obtained with pFETs of the same size.

The basic nFET pass circuit is shown in Figure 7.38. The switch is controlled by the gate voltage V_G. If $V_G = 0$, then the transistor is off and there is no connection between the input and output. Placing a high voltage of $V_G = V_{DD}$ drives the nFET active, and current can flow. For the case of a

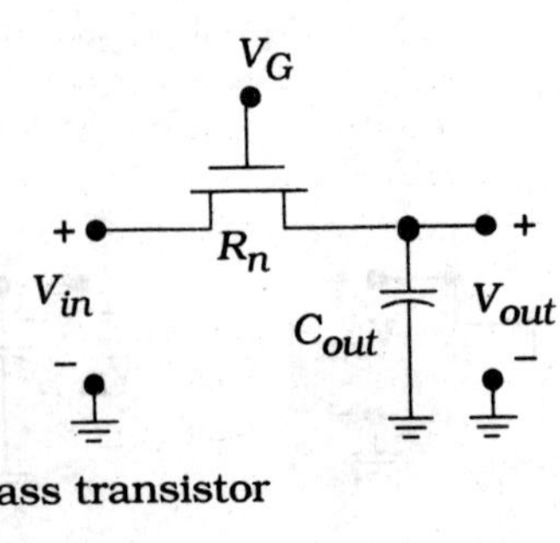

Figure 7.38 nFET pass transistor

5 Remember that the drain and source are determined by the relative voltages.

logic 1 transfer, we use an input voltage of $V_{in} = V_{DD}$. Assuming an initial condition of $V_{out}(t = 0) = 0$, the analysis gives[6]

$$V_{out}(t) = V_{max}\left(\frac{t/2\tau_n}{1 + t/2\tau_n}\right) \tag{7.188}$$

where

$$V_{max} = V_{DD} - V_{Tn} \tag{7.189}$$

is the maximum voltage transferred through an nFET as seen by taking the limit

$$\lim_{t \to \infty} V_{out}(t) = V_{max} \tag{7.190}$$

This clearly exhibits the threshold drop problem. The time constant is defined by

$$\tau_n = R_n C_{out} \tag{7.191}$$

but does not have the same interpretation as when it appears in an exponential. The rise time needed for the output voltage to rise from 0 V to a value of 0.9 V_{max} is calculated as

$$t_r = 18\tau_n \tag{7.192}$$

These results show that the logic 1 transfer event is slow and suffers from the threshold loss problem.

A logic 0 transfer is analyzed by placing $V_{in} = 0$ V. With the initial condition $V_{out}(0) = V_{max}$, the analysis gives

$$V_{out}(t) = V_{max}\left(\frac{2e^{-(t/\tau_n)}}{1 + e^{-(t/\tau_n)}}\right) \tag{7.193}$$

where the time constant has the same definition. This exponential function has the limit

$$\lim_{t \to \infty} V_{out}(t) = 0 \tag{7.194}$$

which shows that an nFET can pass a logic 0 without any problems. The fall time needed for the output to change from V_{max} to the 10% voltage $0.1\,V_{max}$ is

$$t_f = \ln(19)\tau_n \approx 2.94\tau_n \tag{7.195}$$

6 See Reference [10] for the details of the derivation.

Comparing the rise and fall times shows that

$$t_r \approx 6t_f \tag{7.196}$$

so the rise time is the limiting factor. The plot in Figure 7.39 is an example of the shapes of the input versus the output waveforms for an nFET pass transistor.

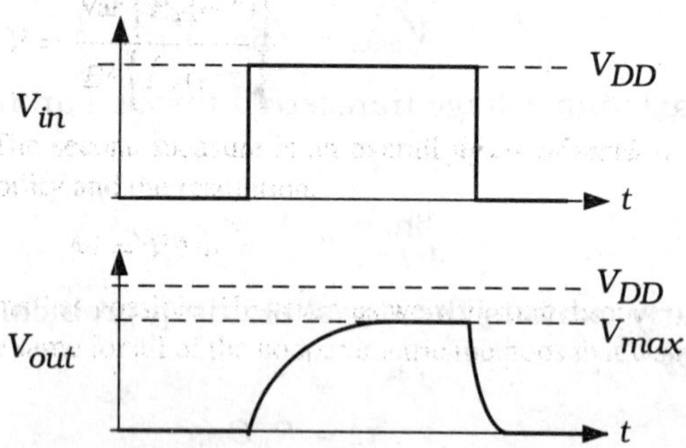

Figure 7.39 Voltage waveforms for a nFET pass transistor

If we use a pFET as a pass transistor, we find complementary results. The maximum voltage through the FET is V_{DD}, and the output charges quite rapidly with a rise time of

$$t_r = 2.94\tau_p \tag{7.197}$$

where

$$\tau_p = R_p C_{out} \tag{7.198}$$

The pFET is thus able to pass a strong logic 1 voltage. When a logic 0 is applied at the input, however, the output discharges to a level

$$V_{min} = |V_{Tp}| \tag{7.199}$$

with a fall time of

$$t_f = 18\tau_p \tag{7.200}$$

The discharge is thus the limiting factor. These results are expected due to the complementary behavior of nFETs and pFETs.

The analysis shows that pass transistors cannot be accurately modeled as simple RC circuits, since the threshold losses and the asymmetrical rise and fall times would be ignored. Regardless of this fact, however, it is common practice to model a pass FET using R_n or R_p in hand calculations during the initial design phase. This allows for quick modeling estimates and is a valuable approximation technique. More precise calculations can be obtained using a computer simulation.

7.9 Comments on SPICE Simulations

The analyses performed in this chapter provide the theoretical basis for designing CMOS logic gates. They allow one to estimate the behavior of a circuit and illustrate the dependence of the overall performance on individual device parameters.

Analytic treatments are intrinsically limited by the accuracy of the device models. In the case of MOSFETs, the square law model is only a low-order approximation to the true behavior. Another level of estimation was introduced with the assumption of step-like input voltage waveforms. We have also ignored the voltage dependence of the capacitances to simplify the analysis. In chip design, the operation of a circuit must be verified by computer simulations. These are not foolproof, as convergence problems and computational noise can affect the results. However, they do provide reasonable verification once the designer becomes familiar with the problem areas. In this section we will examine a few important features of SPICE simulations.

A SPICE netlist for a circuit is obtained from the extraction routine in the layout editor. Each element is represented by a separate line in the listing, and the elements are wired according to the layout. To run a simulation, we must add power supply values, input voltages, and modeling information. As an example, suppose that we extract the netlist from an inverter layout and obtain the following listing:

```
M1 15 17 20 20 NFET W=5U L=0.5U
M2 15 17 12 12 PFET W=10U L=0.5U
```

This identifies the two transistors using arbitrary device and node numbering. In the listing, M1 is an nFET while M2 is a pFET. Since the MOSFET node order is Drain-Gate-Source-Bulk, the input to the inverter is the common gate node 17, while the inverter output is taken from the drain node 15. Node 20 must be grounded, while node 12 is the power supply. Some of the more powerful extractors would also provide drain and source dimensions for the junction capacitance calculations in the form

```
M1 15 17 20 20 NFET W=5U L=0.5U AD=12.5P PD=15U AS=20P PS=18U
M2 15 17 12 12 PFET W=10U L=0.5U AD=25P PD=25U AS=40P PS=36U
```

If the extractor does not find the area and perimeter of the drain and source, it must be added by hand.

To run a full simulation, we will add elements to give the following listing:

```
NOT SIMULATION
VDD 12 0 5V
```

```
M1 15 17 20 20 NMOS W=5U L=0.5U AD=12.5P PD=15U AS=20P PS=18U
M2 15 17 12 12 PMOS W=10U L=0.5U AD=25P PD=25U AS=40P PS=36U
RGND 20 0 1U
CLOAD 15 0 100F
.MODEL NFET NMOS <parameter listing ... >
.MODEL PFET PMOS <parameter listing ... >
...
```

where the first line is the name of the circuit and CLOAD has been selected as a 100 fF external load capacitor. RGND is a 1 μΩ resistor to pull node 20 to ground; alternately, we could renumber the netlist or the layout editor may allow it to be defined in the layout before the extraction.[7]

The input voltage at node 17 allows us to model more realistic waveforms. One useful SPICE construct is the PULSE waveform shown in Figure 7.40. It is specified by a statement of the form

```
VIN 17 0 PULSE(V1 V2 TD TR TF PW PER)
```

where V1 and V2 are the start and final voltages, TD is the time delay before the transition starts, TR is the rise time, TF is the fall time, PER and is the period before the waveform repeats itself. This allows us to calculate low-to-high and high-to-low transition times that are more accurate than those found using step-like inputs. Another useful waveform is the exponential source EXP that is specified by a listing of the form

```
VIN_EXP 17 0 EXP (V1 V2 TD1 TAU1 TD2 TAU2)
```

where TD1 and TAU1 are the time delay and time constant for the V1-to-V2 transition, while TD2 and TAU2 are for the opposite case. In both cases, the time values need to be carefully chosen to represent a simulation that provides information on the transient response by displaying the changes

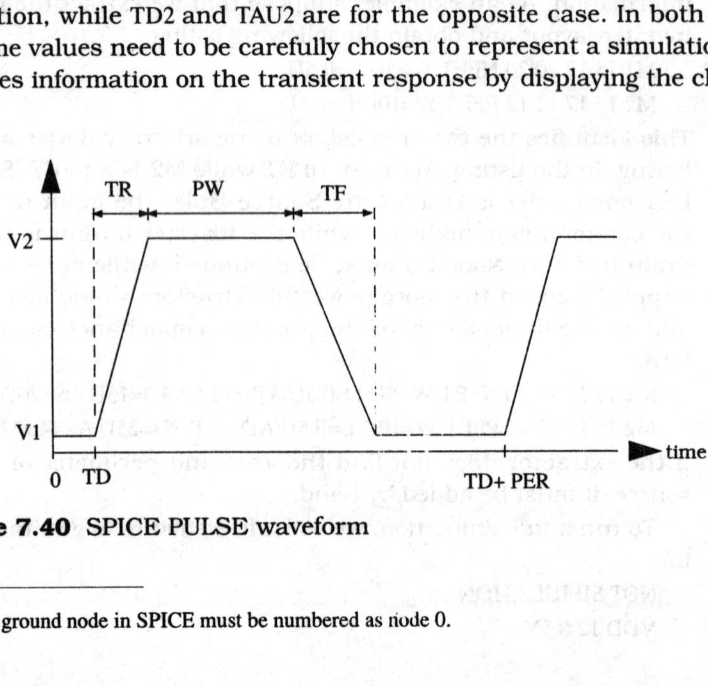

Figure 7.40 SPICE PULSE waveform

7 Recall that the ground node in SPICE must be numbered as node 0.

in the output as smooth functions of time. These can be estimated from the RC model.

The voltage transfer curve is obtained by a DC sweep initiated by the dot command

```
.DC VIN 0 VDD VSTEP
```

which starts at VIN = 0 and increments by VSTEP to a final value of VDD. The transient response is calculated by

```
.TRAN TSTEP TSTOP
```

This starts at time 0 and increments by time units of TSTEP until the time TSTOP is reached. These two commands provide the most critical operating characteristics of the circuit discussed in this chapter.

The same techniques can be applied to modeling any CMOS circuit. One fine point that sometimes causes confusion is where a common active (n+ or p+) region is shared by adjacent gates. The designation of drain or source is arbitrary, and the total area and perimeter can be split between the two FETs as desired. Care must be taken to insure that the total area and the total perimeter length specified for the two transistors do not exceed the actual layout.

Example 7.5

Consider the two FETs in Figure 7.41. The shared region has a total area of (10)(8) = 80, and a total perimeter of 2(10+8) = 36. M1 uses this as a source region while M2 declares it to be a drain. The split could be listed by writing

```
M1 ... AS=40P PS=18U
M2 ... AD=40P PD=18U
```

which is an equal division. Another choice would be

```
M1 ... AS=10P PS=4.5U
M2 ... AD=70P PD=31.5U
```

which would work equally well.

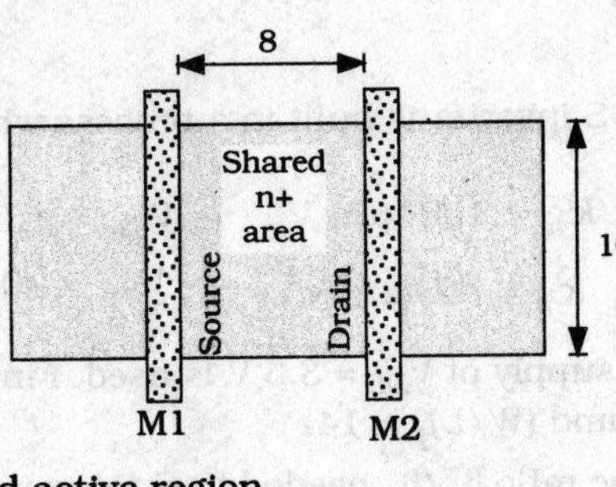

Figure 7.41 Shared active region

More tips and tricks of SPICE modeling of CMOS circuits can be found in the references. As with learning any code, experience is the best teacher.

7.10 References for Further Study

[1] R. Jacob Baker, Harry W. Li, and David E. Boyce, **CMOS Circuit Design, Layout, and Simulation**, IEEE Press, Piscataway, NJ, 1988.

[2] Abdellatif Bellaouar and Mohamed I. Elmasry, **Low-Power Digital VLSI Design**, Kluwer Academic Publishers, Norwell, MA, 1995.

[3] Yuhua Cheng and Chemning Hu, **MOSFET Modeling & BSIM3 User's Guide**, Kluwer Academic Publishers, Norwell, MA, 1999.

[4] Tor A. Fjeldly, Trond Ytterdal, and Michael Shur, **Introduction to Device Modeling and Circuit Simulation**, John Wiley & Sons, New York, 1998.

[5] Ken Martin, **Digital Integrated Circuit Design**, Oxford University Press, New York, 2000.

[6] Jan Rabaey, **Digital Integrated Circuits**, Prentice Hall, Upper Saddle River, NJ, 1996.

[7] Michael Reed and Ron Rohrer, **Applied Introductory Circuit Analysis**, Prentice Hall, Upper Saddle River, NJ, 1999.

[8] Kaushik Roy and Sharat C. Prasad, **Low-Power CMOS VLSI Circuit Design**, Wiley-Interscience, New York, 2000.

[9] Michael John Sebastian Smith, **Application-Specific Integrated Circuits**, Addison-Wesley, Reading, MA, 1997.

[10] John P. Uyemura, **CMOS Logic Circuit Design**, Kluwer Academic Publishers, Norwell, MA, 1999.

[11] Andrei Vladimirescu, **The SPICE Book**, John Wiley & Sons, New York, 1994.

[12] Gary K. Yeap, **Practical Low Power Digital VLSI Design**, Kluwer Academic Publishers, Norwell, MA, 1998.

7.11 Problems

[7.1] A CMOS inverter is built in a process where

$$k'_n = 100\ \mu A/V^2 \qquad V_{Tn} = +0.70\ V$$
$$k'_p = 42\ \mu A/V^2 \qquad V_{Tp} = -0.80\ V \tag{7.201}$$

and a power supply of $V_{DD} = 3.3$ V is used. Find the midpoint voltage V_M if $(W/L)_n = 10$ and $(W/L)_p = 14$.

[7.2] Find the ratio β_n/β_p needed to obtain an inverter midpoint voltage of $V_M = 1.3$ V with a power supply of 3 V. Assume that $V_{Tn} = 0.6$ V and $V_{Tp} =$ -0.82 V. What would be the relative device sizes if $k'_n = 110\ \mu A/V^2$ and the mobility values are related by $\mu_n = 2.2\ \mu_p$?

[7.3] An inverter uses FETs with $\beta_n = 2.1$ mA/V^2 and $\beta_p = 1.8$ mA/V^2. The threshold voltages are given as $V_{Tn} = 0.60$ V and $V_{Tp} = -0.70$ V and the power supply has a value of $V_{DD} = 5$ V. The parasitic FET capacitance at the output node is estimated to be $C_{FET} = 74$ fF.

(a) Find the midpoint voltage V_M.

(b) Find the values of R_n and R_n.

(c) Calculate the rise and fall times at the output when $C_L = 0$.

(d) Calculate the rise and fall times when an external load of value C_L = 115 fF is connected to the output.

(e) Plot t_r and t_f as functions of C_L.

[7.4] Find the midpoint voltage for the inverter layout shown in Figure 7.11.

[7.5] Consider the NOT gate shown in Figure 7.11 when an external load of $C_L = 80$ fF is connected to the output. Note that the electrical channel length is $L = 0.8$ μm.

(a) Find the input capacitance of the circuit.

(b) Find the values of R_n and R_p.

(c) Calculate rise and fall times for the inverter.

[7.6] Simulate the circuit in Figure 7.11 using SPICE. Perform both a DC and a transient simulation assuming an external load of $C_L = 100$ fF.

[7.7] A CMOS NAND2 is designed using identical nFETs with a value of $\beta_n = 2\beta_p$; the pFETs are the same size. The power supply is chosen to be $V_{DD} = 5$ V, and the device threshold voltages are given as $V_{Tn} = 0.60$ V and $V_{Tp} = -0.70$ V.

(a) Find the midpoint voltage V_M for the case of simultaneous switching.

(b) What would be the midpoint voltage for an inverter made with the same β-specification?

[7.8] A CMOS NOR2 gate is designed using nFETs with a value of β_n. The pFETs are both described by $\beta_p = 2.2\beta_n$. Find the value of V_M for the case of simultaneous switching if $V_{DD} = 3.3$ V, $V_{Tn} = 0.65$ V, and $V_{Tp} = -0.80$ V.

[7.9] A NAND3 gate uses identical nFETs with an aspect ratio of 4. The nFET process transconductance is 120 μA/V^2, and the threshold voltage is 0.55 V. A power supply of 5 V is chosen for the circuit.

Find the value of the pFET β_p needed to create a gate where the case of simultaneous switching gives a midpoint voltage of $V_M = 2.4$ V. Assume that $V_{Tp} = -0.90$ V and $r = 2.4$.

[7.10] Consider the nFET chain shown in Figure P7.1. This represents a portion of a NAND3 gate. The output capacitance has a value of $C_{out} = 130$ fF, while the internal values are $C_1 = 36$ fF and $C_2 = 36$ fF. The transistors are identical with $\beta_n = 2.0$ mA/V^2 in a process where $V_{DD} = 3.3$ V and $V_{Tn} = 0.70$ V.

(a) Find the discharge time constant for $C_{out} = 130$ fF using the Elmore formula for a ladder RC network.

(b) Find the time constant if we ignore C_1 and C_2. What is the percentage error introduced if we do not include the internal capacitors?

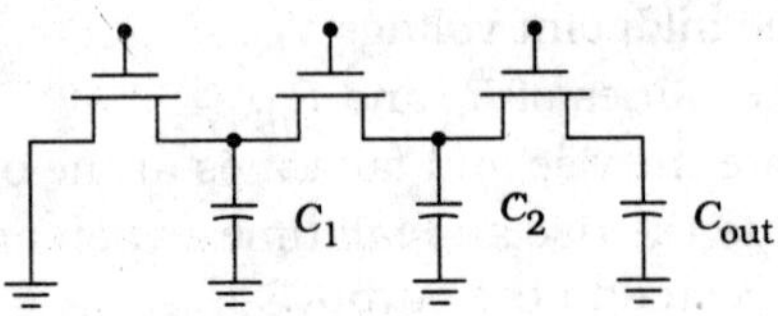

Figure P7.1

[7.11] Consider a complex CMOS logic gate that implements the function

$$F = \overline{a \cdot b + c \cdot d \cdot e} \tag{7.202}$$

(a) Design the logic circuit.

(b) An inverter with $\beta_n = \beta_p$ is used as a sizing reference. Find the device sizes in the gate if we choose to equalize the nFET and pFET resistances.

[7.12] A CMOS logic gate that implements the function

$$F = \overline{x \cdot (y + z) + x \cdot w} \tag{7.203}$$

is needed in a control network.

(a) Design the logic circuit.

An inverter with $\beta_n = \beta_p$ is used as a sizing reference.

(b) Find the device sizes in the gate if we choose to equalize the nFET and pFET resistances.

(c) Suppose instead that we use transistors that are the same size as the inverter values. Identify the worst-case nFET and pFET paths that will slow down the response.

[7.13] An OAI function of the form

$$f = \overline{(a + b) \cdot (b + c) \cdot d} \tag{7.204}$$

is built using series-parallel CMOS structuring.

(a) Design the circuit.

(b) An inverter with $\beta_n = 1.5\ \beta_p$ is used as a sizing reference. Find the transistor sizes needed to equalize the path resistances in both the nFET and pFET chain.

(c) Expand the function into AOI form, and then apply the same sizing philosophy. Which design (the AOI or the OAI) requires the smallest total transistor area?

[7.14] The nFET in Figure P7.2 has $\beta_n = 1.50$ mA/V^2 and is used as a pass transistor as shown. The process uses $V_{DD} = 5.0$ V and $V_{Tn} = 0.5$ V. A logic 1 voltage $V_{in} = V_{DD}$ is applied to the input side, while the output node has a total capacitance of C_{out} = 84 fF. The output capacitor is initially uncharged.

(a) Find the time constant for the logic 1 charging event.

(b) Calculate the rise time in units of picoseconds.

(c) The input is switched to V_{in} = 0 V. Calculate the fall time.

(d) Simulate the pulse response using SPICE to produce the input and output waveforms.

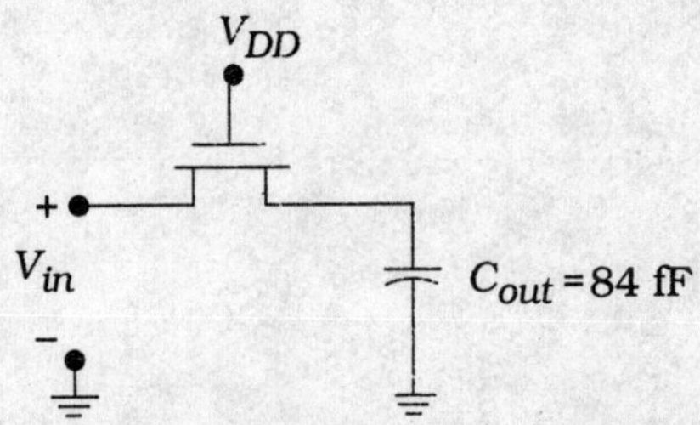

Figure P7.2

[7.15] The pFET pass transistor in Figure P7.3 has an aspect ratio of 8 in a process where k'_p = 60 μA/V^2, V_{DD} = 3.3 V, and V_{Tp} = -0.8 V. At time t=0, the output capacitor is charged to a voltage of V_{DD} while the input is switched to V_{in} = 0 V.

(a) Find the fall time at the output node.

(b) The input is switched back to V_{DD}. Find the rise time needed to drive the output voltage back up to its high value.

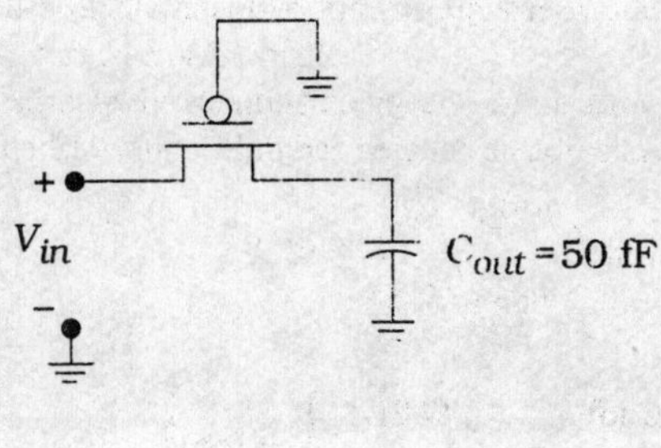

Figure P7.3

Designing High-Speed CMOS Logic Networks

8

Modern CMOS technology is capable of fabricating MOSFETs with channel lengths smaller than 0.1 μm. The channel width W of a FET establishes the aspect ratio (W/L) that is the critical parameter in determining the electrical characteristics of a logic circuit.

Systems designers must take a global view where the logic and architectural features are the first order of business, and the circuits are chosen to implement the necessary functions. In VLSI, however, the ability to meet system timing targets is intimately related to the switching speed of the logic circuits. If the timing specifications cannot be met by the circuitry, then we may be forced to modify the logic.

In this chapter we will initiate our study of high-speed system design and learn techniques to select transistor sizes. These methods are useful for designing both library collections and custom designs. The techniques presented in this chapter are an integral part of high-speed VLSI design, and are heavily oriented toward electronics. Owing to the specialized nature of the material, some readers may prefer to skip this and the following chapter in a first reading, and refer back to them as needed.

8.1 Gate Delays

In the previous chapter we found that the output switching times of the CMOS logic gate in Figure 8.1 are described by the linear expressions

$$\begin{aligned} t_r &= t_{r0} + \alpha_p C_L \\ t_f &= t_{f0} + \alpha_n C_L \end{aligned} \tag{8.1}$$

where C_L is the external load capacitance. Given the layout geometry and processing parameters, the equation set allows us to analyze the switching performance of an arbitrary gate. VLSI designers are faced with the

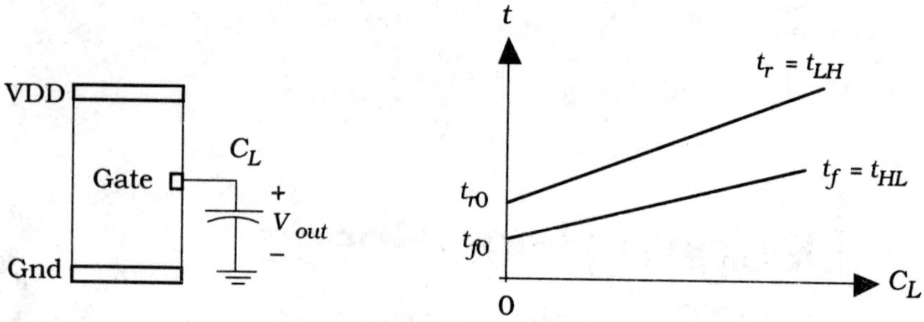

Figure 8.1 Output switching times

opposite problem. It is their responsibility to choose the logic cascades and then specify aspect ratios for every transistor. The system timing specifications must be met while working within a limited real estate allocation. This provides motivation for developing a structured approach to estimating logic delays in CMOS gates.

Let us examine an approach that uses the minimum-size MOSFET as a basis. The layout geometry is shown in Figure 8.2(a). The drawn aspect ratio (W/L) and the active dimension X are determined by the design rules. Once these are known, we can define the parasitics for the device and use them for reference. Let us denote **unit** FET parameters with the subscript 'u' such that the transistor resistance is

$$R_u = \frac{1}{k'\left(\frac{W}{L}\right)_u (V_{DD} - V_T)} \tag{8.2}$$

while

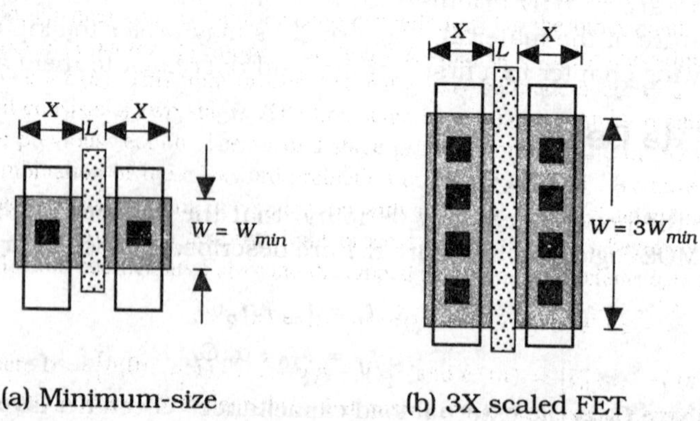

Figure 8.2 Unit transistor reference

$$
\begin{aligned}
C_{Gu} &= C_{ox}(WL)_u \\
C_{Du} &= (C_{GD} + C_{DB})_u \\
C_{Su} &= (C_{GS} + C_{SB})_u
\end{aligned}
\tag{8.3}
$$

give the capacitance values. These are assumed to be known parameters in the analysis. To create a design methodology, we will specify that all transistor sizes are integer multiples of the minimum width $W_{min} = W_u$. An example is the $m = 3$ FET shown in Figure 8.2(b). In general, this gives

$$\left(\frac{W}{L}\right)_m = m\left(\frac{W}{L}\right)_u \tag{8.4}$$

with $m = 1, 2, 3, \ldots$ as the size specifier. The resistance and gate capacitance of the m-sized FET are written in terms of the unit transistor as

$$
\begin{aligned}
R_m &= \frac{R_u}{m} \\
C_{Gm} &= mC_{Gu}
\end{aligned}
\tag{8.5}
$$

We will scale the FET so that X is the same as for the unit FET. For arbitrary m, this implies that the drain and source capacitances scale approximately as

$$
\begin{aligned}
C_{Dm} &\approx mC_{Du} \\
C_{Sm} &\approx mC_{Su}
\end{aligned}
\tag{8.6}
$$

These will be used as equalities in our treatment. Combining with the resistance formula gives the result

$$R_m C_m = R_u C_u = \text{constant} \tag{8.7}$$

which is very useful in scaling theory.

Now suppose that we design an inverter using the minimum-size geometry for both the nFET and the pFET. This results in the layout shown in Figure 8.3(a); note that $\beta_n > \beta_p$ for this design. The rise time for this circuit is controlled by the pFET and can be expressed as

$$t_{ru} = t_{r0} + \alpha_{pu} C_L \tag{8.8}$$

The fall time

$$t_{fu} = t_{f0} + \alpha_{nu} C_L \tag{8.9}$$

is governed by the nFET parameters. Since $R_p > R_n$, $t_{r0} > t_{f0}$, and $\alpha_{pu} > \alpha_{nu}$ so, for a given load C_L, $t_{ru} > t_{fu}$. The midpoint voltage is

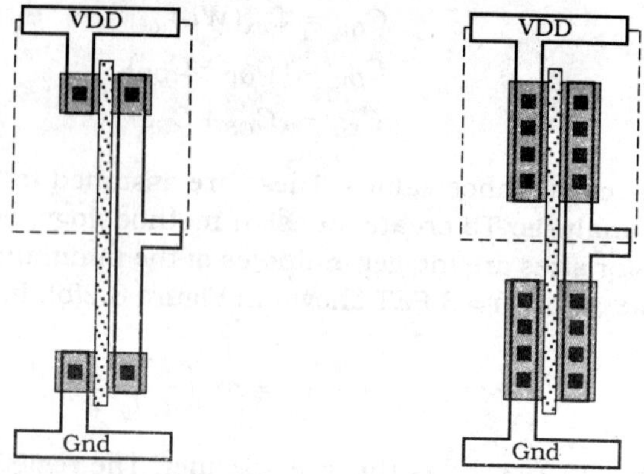

Figure 8.3 Inverter designs using scaled transistors

$$V_M = \frac{V_{DD} - |V_{Tp}| + \sqrt{r}V_{Tn}}{1 + \sqrt{r}} \tag{8.10}$$

where $r = (\mu_n / \mu_p)$ is the mobility ratio. The input capacitance is a minimum value for a complementary pair

$$C_{in} = 2C_u = C_{min} \tag{8.11}$$

since both transistors are minimum-size devices.

If we scale the FETs by a factor $m = 3$, then we arrive at the layout in Figure 8.3(b). This does not change the midpoint voltage, but does alter the switching times. To find the response of the new circuit, first note that the zero-load times t_{r0} and t_{f0} are (approximately) constants as demonstrated by equation (8.7). The slope parameter α decreases as $(1/m)$ because of the decrease in resistance by the same factor. Thus,

$$\begin{aligned} t_{r3} &= t_{r0} + \frac{\alpha_{pu}}{3}C_L \\ t_{f3} &= t_{f0} + \frac{\alpha_{nu}}{3}C_L \end{aligned} \tag{8.12}$$

describes the scaled circuit. The input capacitance for this gate is

$$C_{in} = 3C_{min} \tag{8.13}$$

Consider next the NAND2 gate in Figure 8.4(a) that uses minimum-size transistors. The switching equations must be modified for this circuit. First, recall that the zero-load times t_{r0} and t_{f0} are proportional to the product of C_{FET} and the resistance. In the inverter, two FETs contribute to the capacitance. Since there are now three FETs that touch the output node, we introduce a factor of (3/2) multiplying the internal capacitance.[1] The resistances scale in a different manner. The pFET resistance R_p is the same as that for an inverter, while the nFET resistance R_n between the output node and ground is doubled because of the series connection; this increases both t_{f0} and α_{nu} by a factor of 2. Including these multipliers in the equation gives

$$\begin{aligned} t_r &= \left(\frac{3}{2}\right)t_{r0} + \alpha_{pu}C_L \qquad \text{(Unit NAND2)} \\ t_f &= 3t_{f0} + 2\alpha_{nu}C_L \end{aligned} \tag{8.14}$$

This ignores the capacitance between the series-connected nFETs, but does illustrate the trends. The input capacitance is

$$C_{in} = C_{min} \tag{8.15}$$

since an nFET/pFET pair consists of minimum-size devices.

If we scale the transistors with $m = 3$, as in Figure 8.4(b), then the equations must be modified. Both α factors are reduced by $(1/m)$ because of the decrease in resistance. The decrease in resistance counteracts the increase in C_{FET}, so that the zero-load terms are unchanged. Thus,

$$t_r = \left(\frac{3}{2}\right)t_{r0} + \frac{\alpha_{pu}}{3}C_L \tag{8.16}$$

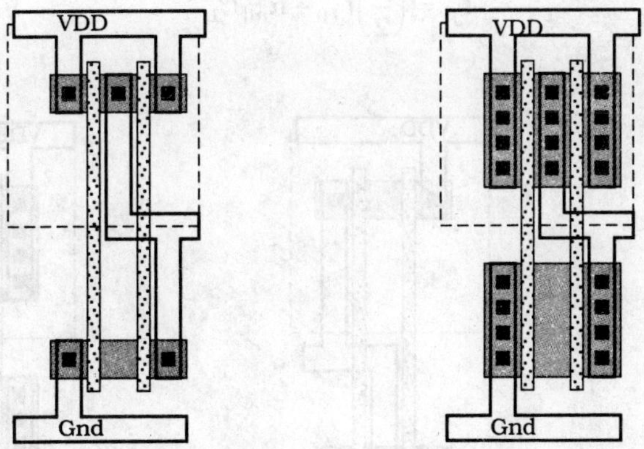

(a) Unit transistors (b) $m = 3$ circuit

Figure 8.4 NAND2 gate scaling

---

1 This assumes that the nFET and pFET capacitances are equal, which is not true even if they are the same size.

and

$$t_f = 3t_{f0} + \frac{2\alpha_{nu}}{3}C_L \tag{8.17}$$

provides the scaled response times. The input capacitance is

$$C_{in} = 3C_{min} \tag{8.18}$$

If N is the fan-in (number of inputs), then we may extrapolate the analysis to write

$$\begin{aligned} t_r &= \left(\frac{N+1}{2}\right)t_{r0} + \frac{\alpha_{pu}}{m}C_L \\ t_f &= (N+1)t_{f0} + \frac{N\alpha_{nu}}{m}C_L \end{aligned} \quad \text{(NAND-}N\text{)} \tag{8.19}$$

for an N-input NAND gate that uses m-sized FETs. In this case

$$C_{in} = mC_{min} \tag{8.20}$$

gives the input capacitance.

A NOR2 gate can be analyzed using the same techniques. The unit-transistor layout in Figure 8.5(a) has switching times that can be approximated by

$$\begin{aligned} t_r &= 3t_{r0} + 2\alpha_{pu}C_L \\ t_f &= \left(\frac{3}{2}\right)t_{f0} + \alpha_{nu}C_L \end{aligned} \quad \text{(Unit NOR2)} \tag{8.21}$$

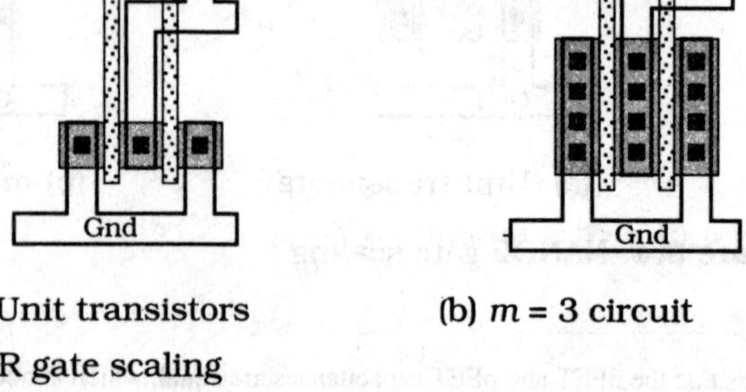

(a) Unit transistors (b) $m = 3$ circuit

Figure 8.5 NOR gate scaling

The $m = 3$ scaled circuit in Figure 8.5(b) modifies the expressions to

$$t_r = 3t_{r0} + \frac{2\alpha_{pu}}{3}C_L$$
$$t_f = \left(\frac{3}{2}\right)t_{f0} + \frac{\alpha_{nu}}{3}C_L \tag{8.22}$$

because of the decrease in the slope parameters α. For N inputs and general scaling factor m, these may be extended to

$$t_r = (N+1)t_{r0} + \frac{N\alpha_{pu}}{m}C_L \quad \text{(NOR-}N\text{)}$$
$$t_f = \left(\frac{N+1}{2}\right)t_{f0} + \frac{\alpha_{nu}}{m}C_L \tag{8.23}$$

for an N-input NOR gate. Also,

$$C_{in} = mC_{min} \tag{8.24}$$

gives the input capacitance.

These equations clearly demonstrate the dependence of the switching times and input capacitance on

- Number of inputs N (fan-in)
- Transistor scaling factor m

The input capacitance is important because it is a measure of how much a gate loads the stage that is driving it.

This technique of gate design provides a structured approach for estimating delays. For a logic chain with M stages, we may approximate the total delay through the chain by summing the individual delays:

$$t_d = \sum_{i=1}^{M} t_i \tag{8.25}$$

The individual contributions depend upon the gate type (i.e., NOT, NAND, etc.) and its size, in addition to the size and type of the next gate in the chain. We also need to be aware of the difference between rise and fall times.

As an example, consider the logic chain in Figure 8.6 where the input is originally at 0 and then makes a transition to a 1. The stages are scaled with increasing values of m, and the output is a capacitor with a value of $C = 4\ C_{min}$. The total delay is

$$t_d = t_{NOT}\big|_{m=1} + t_{NAND2}\big|_{m=2} + t_{NOR2}\big|_{m=3} \tag{8.26}$$

where the first and third terms represent fall times, while the second term

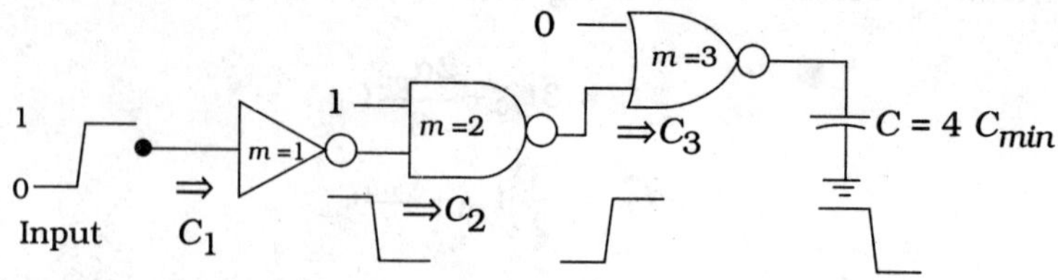

Figure 8.6 Delay time example

is a rise time. Applying the equations above gives the terms as

$$
\begin{aligned}
t_{NOT}\big|_{m=1} &= t_{f0} + \alpha_{nu} 2C_{min} \\
t_{NAND2}\big|_{m=2} &= \left(\frac{3}{2}\right)t_{r0} + \frac{\alpha_{pu}}{2}3C_{min} \\
t_{NOR2}\big|_{m=3} &= \left(\frac{3}{2}\right)t_{f0} + \frac{\alpha_{nu}}{3}4C_{min}
\end{aligned}
\tag{8.27}
$$

so that the total chain delay is

$$
\begin{aligned}
t_d &= \left(\frac{5}{2}\right)t_{f0} + \left(\frac{10}{3}\right)\alpha_{nu}C_{min} + \left(\frac{3}{2}\right)t_{r0} + \left(\frac{3}{2}\right)\alpha_{pu}C_{min} \\
&= \frac{1}{2}(5t_{f0} + 3t_{r0}) + \left[\left(\frac{10}{3}\right)\alpha_{nu} + \left(\frac{3}{2}\right)\alpha_{pu}\right]C_{min}
\end{aligned}
\tag{8.28}
$$

It is important to note that the expression for t_d will change if different inputs are applied. Overall, the technique allows us to estimate delays through logic cascades in a uniform manner.

Although the analysis has been performed using minimum-size transistors for both the nFET and pFET, it is straightforward to modify the analysis for a symmetrical design with $\beta_n = \beta_p$. In this case, the inverter rise and fall times are equal and given by

$$t_s = t_0 + \alpha C_L \tag{8.29}$$

for a circuit with $W_n = W_{min}$ and $W_p = rW_{min}$. The input capacitance is increased to

$$
\begin{aligned}
C_{in} &= C_u(1 + r) \\
&= C_{inv}
\end{aligned}
\tag{8.30}
$$

which now becomes the reference. Scaling the transistors in the NOT gate by m gives

$$t_s = t_0 + \frac{\alpha}{m}C_L \tag{8.31}$$

as discussed in the previous chapter. The analysis of multi-input gates such as the NAND and NOR circuits proceeds in the same manner. Note that if m is used to scale both nFETs and pFETs equally, the rise and fall times will be unequal for gates with $N > 1$. Equalization of the switching times can be achieved only if the two FET types are different sizes. If the parallel-connected FETs are increased by m then the series-connected transistors must be increased by a factor mN to obtain a symmetrical design.

Other approaches have been developed to estimate the delay through a logic chain. One simple technique is to use the minimum-size inverter as a basis, and then build up NAND and NOR gates for increasing numbers of inputs N. If the switching delay is plotted as a function of load capacitance C_L, one obtains a trend such as that shown in Figure 8.7. By definition, an inverter is described by the $N = 1$ plot and gives the basis for writing a delay time of

$$t_d = (A + Bn)\tau_{min} \tag{8.32}$$

where A and B are dimensionless constants,

$$\tau_{min} = R_{min} C_{min} \tag{8.33}$$

is the time constant for the minimum size inverter, and

$$n = \frac{C_L}{C_{min}} \tag{8.34}$$

is the number of minimum load factors being driven by the stage. These are taken to be empirically measured quantities, i.e., curve fitting parameters. Alternately, they may be generated by a circuit simulation. If the fan-in is increased to $N = 2$ (for either a NAND2 or a NOR2 gate), then the worst-case delay time has a large zero-load value and a steeper slope. The same comment holds as we increase to $N = 3$. An empirical fit is obtained

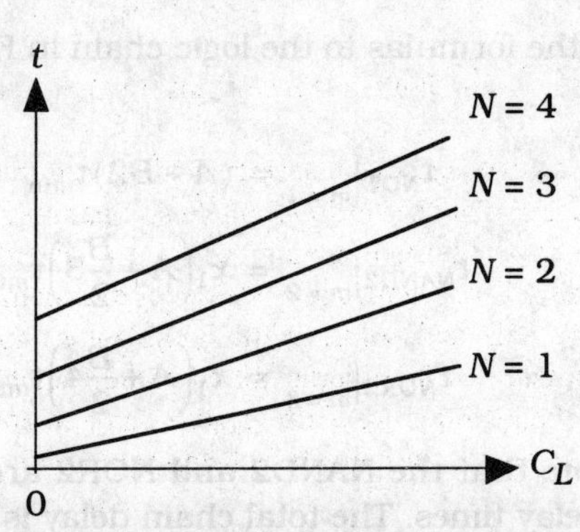

Figure 8.7 Delay times as a function of fan-in N

by multiplying t_d by a factor x_1 that accounts for the increases in the form

$$t_{d,N} = (x_1)^{(N-1)}(A+Bn)\tau_{min} \tag{8.35}$$

For example, if the increase from $N = 1$ to $N = 2$ is 17% per input, this means that $x_1 = 1.17$ and

$$t_{d,N} = (1.17)^{(N-1)}(A+Bn)\tau_{min} \tag{8.36}$$

In practice, an average value of many comparisons would be used. If the transistors are scaled by a factor $m = 1, 2, \ldots$, then we would modify the expression to

$$t_{d,N}^{m} = (x_1)^{(N-1)}\left(A+\frac{B}{m}n\right)\tau_{min} \tag{8.37}$$

to account for the increased drive strength. Also, for a complex N-input logic gate, the delay would be even larger since the internal circuit capacitances will increase and slow down charging or discharge events. In this case, we multiply by another empirical parameter $x_2 > 1$ to obtain

$$t_{d,N}^{m} = x_2(x_1)^{(N-1)}\left(A+\frac{B}{m}n\right)\tau_{min} \tag{8.38}$$

In practice, one would expect around a 5 to 20% increase due to additional FET parasitics.

While this approach is approximate in nature, it does reflect the physical fact that the switching times increase with the fan-in. If we apply the delay estimate to gates in a uniform manner, then it allows us to compare the delay through various cascaded arrangements. The actual numerical values are not accurate, but we would expect the *relative* values to have merit.

Example 8.1

Let us apply the formulas to the logic chain in Figure 8.6. The three terms are

$$\begin{aligned} t_{NOT}\big|_{m=1} &= (A+B2)\tau_{min} \\ t_{NAND2}\big|_{m=2} &= x_1\left(A+\frac{B}{2}3\right)\tau_{min} \\ t_{NOR2}\big|_{m=3} &= x_1\left(A+\frac{B}{2}4\right)\tau_{min} \end{aligned} \tag{8.39}$$

where we note that the NAND2 and NOR2 are treated as having equal worst-case delay times. The total chain delay is

$$\frac{t_d}{\tau_{min}} = [x_1 + 1]A + \left[\left(\frac{7}{2}\right)x_1 + 2\right]B \tag{8.40}$$

If $x_1 = 1.17$, then

$$\frac{t_d}{\tau_{min}} = 2.17A + 6.1B \tag{8.41}$$

is the delay compared to a single inverter.

As we will see in later chapters, the ability to estimate the delay through a logic chain is an important skill for high-speed design. In realistic digital systems design, one can usually find distinctly different equations or algorithms that produce the same result. Each, however, will use a different type of logic cascade. Techniques such as these provide a basis for deciding on the design that will be the fastest.

8.2 Driving Large Capacitive Loads

Many of the important points in high-speed design can be obtained from studying the characteristic delay through inverter circuits. The analyses form the basis for several well-known design techniques that can be extended to include arbitrary gates.

Consider the NOT gate in Figure 8.8 where the circuit drives the external load capacitance C_L; the internal parasitic capacitance C_{FET} due to the transistors is not shown explicitly in the drawing. The electrical characteristics are determined by the values of β_n and β_p. A symmetric design with $\beta_n = \beta_p = \beta$ will be used for simplicity. Since $\beta = k'(W/L)$, this means that the aspect ratios are related by

$$\left(\frac{W}{L}\right)_p = r\left(\frac{W}{L}\right)_n \tag{8.42}$$

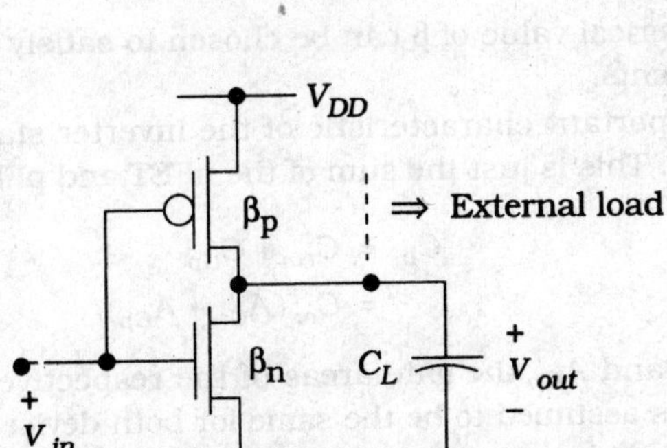

Figure 8.8 CMOS inverter circuit

where r is the mobility ratio

$$r = \frac{\mu_n}{\mu_p} = \frac{k'_n}{k'_p} > 1 \tag{8.43}$$

Assuming equal magnitude threshold voltages $V_{Tn} = |V_{Tp}| = V_T$ gives equal FET resistances of

$$R_n = R_p = R = \frac{1}{\beta(V_{DD} - V_T)} \tag{8.44}$$

This design yields a VTC with a midpoint voltage of $V_M = (V_{DD}/2)$ and equal rise and fall times. For a 0-to-1 transition at the output, the voltage $V_{out}(t)$ across C_L is of the form

$$V_{out}(t) = V_{DD}[1 - e^{-t/\tau}] \tag{8.45}$$

while a 1-to-0 change is described by

$$V_{out}(t) = V_{DD}e^{-t/\tau} \tag{8.46}$$

In both expressions, the time constant is given by the product

$$\tau = RC_{out} = R(C_{FET} + C_L) \tag{8.47}$$

The generic switching time delay $t_s = t_r = t_f$ is then given in the form

$$t_s = t_0 + \alpha C_L \tag{8.48}$$

where t_0 is the zero-load delay and α is the slope of the t_s vs. C_L plot. The value of t_0 is almost invariant to changes in the circuit, while α is proportional to the resistance R:

$$\alpha \propto R = \frac{1}{\beta(V_{DD} - V_T)} \tag{8.49}$$

The numerical value of β can be chosen to satisfy the transient response requirements.

An important characteristic of the inverter stage is its input capacitance C_{in}. This is just the sum of the nFET and pFET gate capacitances

$$\begin{aligned} C_{in} &= C_{Gn} + C_{Gp} \\ &= C_{ox}(A_{Gn} + A_{Gp}) \end{aligned} \tag{8.50}$$

with A_{Gn} and A_{Gp} the gate areas of the respective devices. The channel length L is assumed to be the same for both devices. If we ignore the gate overlap L_o and approximate $L = L'$ then

$$\begin{aligned} C_{in} &= C_{ox}L(W_n + W_p) \\ &= (1+r)(C_{ox}LW_n) \\ &= (1+r)C_{Gn} \end{aligned} \tag{8.51}$$

where we have used equation (8.42) in the second line.

Now suppose that we use the inverter to drive an identical gate as shown in Figure 8.9. In this case, the load C_{L1} seen by gate 1 is

$$C_{L1} = C_{in} \tag{8.52}$$

Since the load capacitance is the same as the gate's own input capacitance, we call this a **unit load** value. The switching time is given by

$$t_{s1} = t_0 + \alpha C_{in} \tag{8.53}$$

which is a convenient reference for analyzing the performance of the gate when it is used to drive other loads.

If the load capacitance is increased to a very large value $C_L >> C_{in}$, then the switching times increase proportionately. To keep t_s small, we may decrease α by using larger transistors to decrease the resistance. Increasing the value of β compensates for the larger load and demonstrates the speed-versus-area trade-off. Suppose that the aspect ratios are increased by the scaling factor $S > 1$. The new device transconductance is

$$\beta' = S\beta \tag{8.54}$$

so that the resistance is reduced to

$$R' = \frac{R}{S} \tag{8.55}$$

The slope is also decreased to a new value of

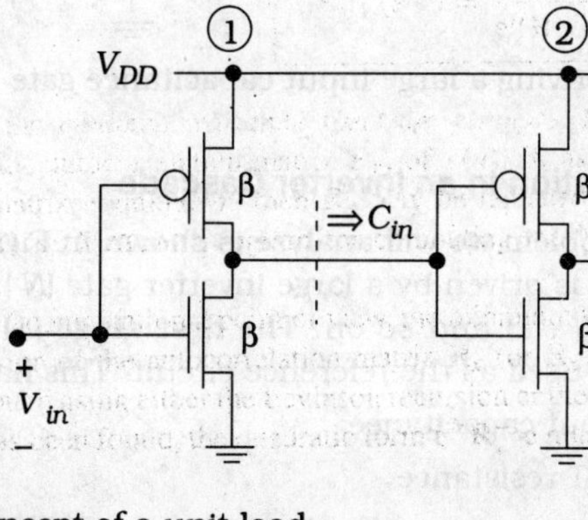

Figure 8.9 Concept of a unit load

$$\alpha' = \frac{\alpha}{S} \tag{8.56}$$

These combine to give the switching time equation for the new inverter as

$$t_s = t_0 + \left(\frac{\alpha}{S}\right) C_L \tag{8.57}$$

The compensation factor $(1/S)$ allows us to drive larger values of C_L. If the load has a value $C_L = SC_{in}$, then the switching time is the same as for a unit load. Scaling the transistor also affects the input capacitance since

$$W_n' = SW_n \tag{8.58}$$

increases the gate area. The new value is given by

$$C_{in}' = SC_{in} \tag{8.59}$$

i.e., it increases by the same scaling factor S. This introduces another problem that is illustrated in Figure 8.10. The large input capacitance acts as the load $C_{L,d} = C_{in}$ to the driving circuit. To compensate for this effect, we must increase the size of the transistor in the driving gate, which increases its value of $C_{in,d}$. This in turn makes it more difficult to drive. Clearly, a methodology would be useful in solving this problem.

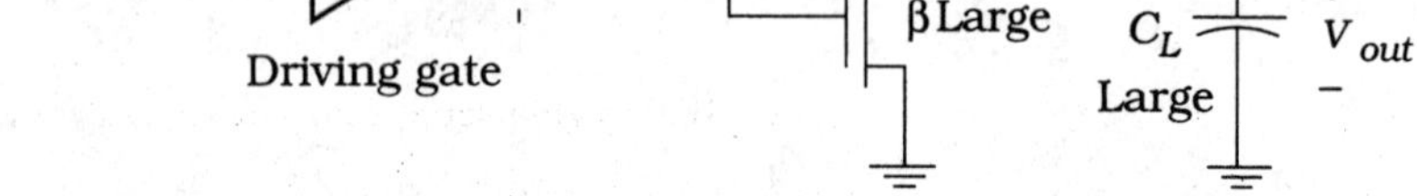

Figure 8.10 Driving a large input capacitance gate

8.2.1 Delay Minimization in an Inverter Cascade

The general problem we will analyze is shown in Figure 8.11. A large load capacitance C_L is driven by a large inverter gate (N), which is driven by a smaller gate (N - 1) and so on. The first stage (1) is a "standard size" inverter that is used as the reference circuit. This has known parameters

C_1 = input capacitance

R_1 = FET resistance

β_1 = device transconductance

The stages are monotonically scaled such that 1 is the smallest and N is the largest:

$$\beta_1 < \beta_2 < \beta_3 < \ldots < \beta_{N-1} < \beta_N \tag{8.60}$$

The sizes of the NOT symbols have been adjusted to show the relative sizing. The simplest scaling is obtained by increasing the size of the transistors by a factor $S > 1$ from one stage to the next. This means that

$$\begin{aligned} \beta_2 &= S\beta_1 \\ \beta_3 &= S\beta_2 \end{aligned} \tag{8.61}$$

and so on. The general expression

$$\beta_{j+1} = S\beta_j \tag{8.62}$$

relates the j-th and $(j+1)$-st stages.

The main problem is as follows. A signal is placed at the input to inverter 1. We want to find the number of states N and the scaling factor S that will minimize the time needed for the signal to reach the load C_L. This can be solved by first studying the characteristics of a typical stage, and then applying the results to the chain.

First note that using β_1 as the reference value in the scaling implies that

$$\begin{aligned} \beta_2 &= S\beta_1 \\ \beta_3 &= S\beta_2 = S^2\beta_1 \\ \beta_4 &= S\beta_3 = S^3\beta_1 \end{aligned} \tag{8.63}$$

or, in general,

$$\beta_j = S^{(j-1)}\beta_1 \tag{8.64}$$

for $j = 2$ to N. The input capacitance scales with β_j, so that

$$C_j = S^{(j-1)}C_1 \tag{8.65}$$

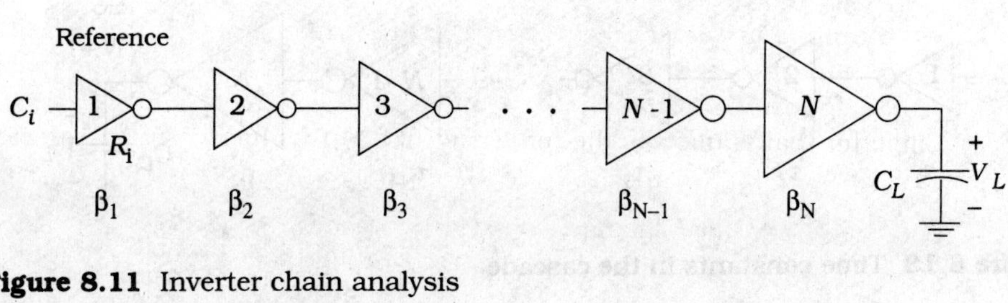

Figure 8.11 Inverter chain analysis

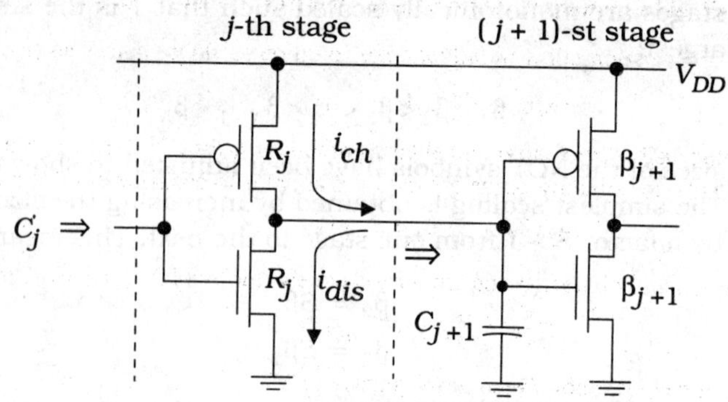

Figure 8.12 Characteristics of a typical stage in the chain

is the value into the j-th stage. The FET resistance scales as $(1/\beta_j)$, leading us to write

$$R_j = \frac{R_1}{S^{(j-1)}} \tag{8.66}$$

as the j-th stage resistance.

Now let us calculate the behavior of a typical stage. Figure 8.12 shows the j-th and $(j+1)$-st stages in the chain. The charging current i_{ch} and discharging current i_{dis} for the j-th stage are shown in the drawing. If we make the simplifying assumption that the load capacitance $C_{j+1} >> C_{FET,j}$ then the time constant for the j-th stage is

$$\tau_j = R_j C_{j+1} \tag{8.67}$$

for $j = 1$ to N. This result can be used to analyze the total delay through the chain. Figure 8.13 shows the time constants for each stage. The total time constant for the chain can be computed by just summing each term to give

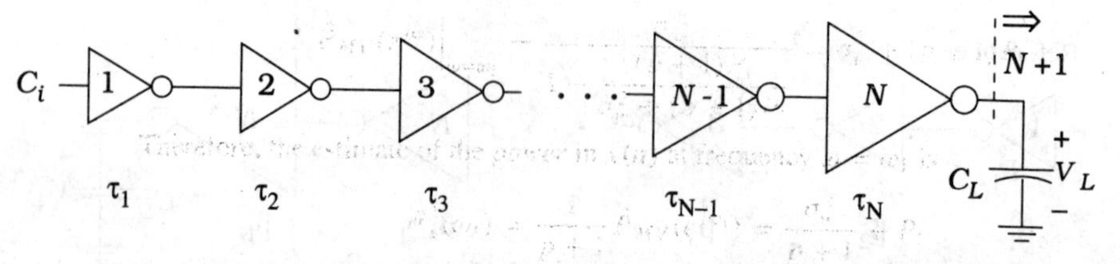

Figure 8.13 Time constants in the cascade

$$\begin{aligned}\tau_d &= \tau_1 + \tau_2 + \tau_3 + \ldots + \tau_{N-1} + \tau_N \\ &= R_1C_2 + R_2C_3 + R_3C_4 + \ldots + R_{N-1}C_N + R_NC_L\end{aligned} \tag{8.68}$$

where we have used the N-stage load of C_L by assigning it the symbol

$$\begin{aligned}C_L &= C_{N+1} \\ &= S^N C_1\end{aligned} \tag{8.69}$$

The second step is added to maintain consistency with the numbering scheme. Substituting the scaling relations in equations (8.65) and (8.66), this reduces to

$$\tau_d = R_1SC_1 + \frac{R_1}{S}S^2C_1 + \frac{R_1}{S^2}S^3C_1 + \ldots + \frac{R_1}{S^{N-2}}S^{N-1}C_1 + \frac{R_1}{S^{N-1}}S^NC_1 \tag{8.70}$$

Simplifying gives

$$\begin{aligned}\tau_d &= SR_1C_1 + SR_1C_1 + SR_1C_1 + \ldots + SR_1C_1 + SR_1C_1 \\ &= N(SR_1C_1)\end{aligned} \tag{8.71}$$

since every term is identical. The total delay is thus given by

$$\tau_d = NS\tau_r \tag{8.72}$$

where $\tau_r = R_1C_1$ is a reference time constant. This is a very important result to remember; qualitatively, it tells us to equalize the signal delay through each stage.

Now let us turn to the problem of minimizing the delay. The unknowns are N and S, so we need two equations. One is equation (8.72) for the total delay. The other can be obtained from equation (8.69) which is a boundary condition for the problem. Starting with the form

$$C_L = S^N C_1 \tag{8.73}$$

we divide by C_1 and take the natural logarithm of both sides to arrive at the form

$$\ln(S^N) = \ln\left(\frac{C_L}{C_1}\right) = N\ln(S) \tag{8.74}$$

This allows us to write

$$N = \frac{\ln\left(\frac{C_L}{C_1}\right)}{\ln(S)} \tag{8.75}$$

as the second equation. Substituting into equation (8.72) then gives the delay time constant in the form

$$\tau_d = \tau_r \ln\left(\frac{C_L}{C_1}\right)\left[\frac{S}{\ln(S)}\right] \tag{8.76}$$

which is a function only of S. To minimize τ_d we apply the derivative condition

$$\frac{\partial \tau_d}{\partial S} = \frac{\partial}{\partial S}\left[\frac{S}{\ln(S)}\right] = 0 \tag{8.77}$$

Differentiating,

$$\frac{1}{\ln(S)} - \frac{S}{S[\ln(S)]^2} = 0 \tag{8.78}$$

or,

$$\ln(S) = 1 \tag{8.79}$$

This is a very interesting expression since the solution is

$$S = e \tag{8.80}$$

i.e., the Euler $e = 2.71...$ is the scaling factor for a minimum delay chain. The number of stages for this design is

$$N = \frac{\ln\left(\frac{C_L}{C_1}\right)}{\ln(S)} = \ln\left(\frac{C_L}{C_1}\right) \tag{8.81}$$

The total delay through the chain is

$$\tau_d = e \ln\left(\frac{C_L}{C_1}\right)\tau_r \tag{8.82}$$

which is the minimum value, and completes the solution to the problem.

Example 8.2

To see how the results may be applied, suppose that we want to drive a load capacitor of value $C_L = 10$ pF (where 1 pF = 10^{-12} F). The input stage is defined with $C_1 = 20$ fF $= 20 \times 10^{-15}$ F and has $\beta_1 = 200\ \mu A/V^2$. The number of stages N needed to minimize the delay is calculated as

$$N = \ln\left(\frac{10\times10^{-12}}{20\times10^{-15}}\right) = \ln(500) \tag{8.83}$$

Since $\ln(500) \approx 6.21$, we will select $N = 6$ to obtain a non-inverting chain. The results gave us a scaling factor of $S = e$ if the N equation is exact. However, since we have rounded N to a (useful) integer value, the scaling factor is more correctly given by rearranging equation (8.73) to the form

$$S = \left(\frac{C_L}{C_1}\right)^{1/N} \tag{8.84}$$

For our example, this gives

$$S = (500)^{1/6} = 2.82 \tag{8.85}$$

which is slightly larger than the ideal value.

The design consists of 6 inverters with device transconductances of

$$\begin{aligned} \beta_2 &= (2.82)\beta_1 \\ \beta_3 &= (2.82^2)\beta_1 = (8)\beta_1 \\ \beta_4 &= (2.82^3)\beta_1 = (22)\beta_1 \\ \beta_5 &= (2.82^4)\beta_1 = (63)\beta_1 \\ \beta_6 &= (2.82^5)\beta_1 = (178)\beta_1 \end{aligned} \tag{8.86}$$

where we have rounded to the nearest integer. Note how rapidly the FET sizes increase when approaching the output stage.

The idealized calculation above tends to underestimate the scaling factor S because the analysis ignored the presence of the parasitic FET capacitance. In practice, $S > e$ and the value depends on the processing. To see the origin of the increase, let us redo the calculation with the parasitic transistor capacitances included.

Figure 8.14 shows the j-th stage circuit with the parasitic FET capacitance C_{Fj} included at the output. The time constant for this stage is now given by the time constant

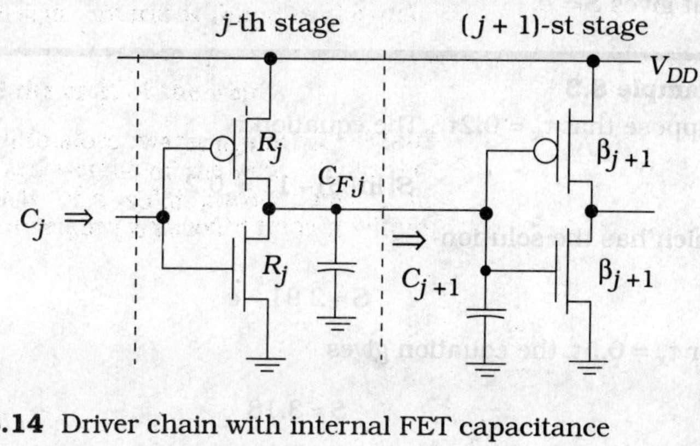

Figure 8.14 Driver chain with internal FET capacitance

$$\tau_j = R_j(C_{F,j} + C_{j+1}) \tag{8.87}$$

since the transistors must drive both $C_{F,j}$ and C_{j+1}. Parasitic FET capacitance is proportional to the width of the transistor, so that the scaling relation is

$$C_{F,j} = S^{(j-1)}C_{F,1} \tag{8.88}$$

where $C_{F,1}$ is the capacitance of the stage 1 FETs. With this, the delay time constant for the entire chain is

$$\tau_d = R_1(C_{F,1} + C_2) + R_2(C_{F,2} + C_3) + \ldots + R_N(C_{F,N} + C_L) \tag{8.89}$$

Using the scaling relations shows that each stage has a parasitic term of $R_1C_{F,1}$ so that the total delay is

$$\tau_d = NR_1C_{F,1} + N(SR_1C_1) \tag{8.90}$$

Using equation (8.75) for N gives the form

$$\tau_d = \left[\frac{\tau_x}{\ln(S)} + \tau_r\left(\frac{S}{\ln(S)}\right)\right]\ln\left(\frac{C_L}{C_1}\right) \tag{8.91}$$

where

$$\tau_x = R_1C_{F,1} \tag{8.92}$$

Differentiating with respect to S and setting the result to 0 gives the minimization condition in the form

$$S[\ln(S) - 1] = \frac{\tau_x}{\tau_r} \tag{8.93}$$

which is a transcendental equation whose solution depends on the ratio of τ_x to τ_r. Note that for $\tau_x = 0$ this degenerates into the simpler equation that gives $S = e$.

Example 8.3

Suppose that $\tau_x = 0.2\tau_r$. The equation is

$$S[\ln(S) - 1] = 0.2 \tag{8.94}$$

which has the solution

$$S \approx 2.91 > e \tag{8.95}$$

For $\tau_x = 0.5\tau_r$ the equation gives

$$S \approx 3.18 \tag{8.96}$$

Finally, for $\tau_x = \tau_r$ we find

$$S \approx 3.59 \tag{8.97}$$

This illustrates the dependence of the scaling factor on the parasitics.

It is important to remember that the algorithm minimizes the time delay from the input to the output, and often specifies transistor sizes that are too large to be practical. This is especially true if we increase the scaling factor to account for the parasitics while attempting to design around a very large output capacitance.

8.3 Logical Effort

The scaling of logic cascades has been a mainstay technique since the beginnings of digital MOS/VLSI circuits. It serves as a guide for designing fast logic chains, and provides many qualitative features that are applied in everyday circuits.

Sutherland et al. have reformulated the ideas contained in the scaling analysis and used them to develop a generalized technique called **Logical Effort**. Logical Effort characterizes gates and how they interact in logic cascades, and provides techniques to minimize the delay. It allows the theory to be extended to include standard logic gates such as NAND and NOR, in addition to complex logic gate circuits. In this section we will examine the basics of the approach to learn how it can be used to design high-speed chains. The interested reader is directed to Reference [8] for a complete and well-written treatment of this useful technique.

8.3.1 Basic Definitions

The starting point is to define an inverter as a reference gate. The simplest approach is to use a symmetric NOT gate where $\beta_n = \beta_p$ and the device aspect ratios are related by

$$\left(\frac{W}{L}\right)_p = r\left(\frac{W}{L}\right)_n \tag{8.98}$$

The important difference between the two FETs is the value of $r > 1$. Figure 8.15 shows the reference circuit for a 1X design. The relative values of the aspect ratios (1 and r) are included next to the transistors. The circuit can be applied to any value of $(W/L)_n$ that defines the reference circuit, but the 1X reference is the smallest sizing in the logic chain. Larger devices are obtained by scaling the circuit. For example, a 4X NOT gate would have nFET and pFET sizes of 4 and $4r$, respectively.

The **logical effort** ***g*** of a gate is defined by the ratio of capacitance to that of the reference gate:

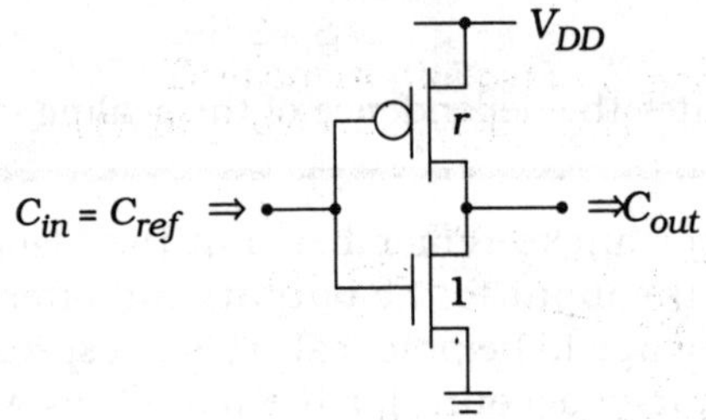

Figure 8.15 Reference inverter for logical effort

$$g = \frac{C_{in}}{C_{ref}} \tag{8.99}$$

Note that the parameter g has the same name as the technique; to distinguish between the two, we will treat the technique as a proper noun and use capital letters: **Logical Effort**. For the 1X inverter,

$$C_{in} = C_{ox}(A_{Gn} + A_{Gp}) \tag{8.100}$$

where A_{Gn} and A_{Gp} are the areas of the respective gates

$$A_{Gn} = W_n L \quad \text{and} \quad A_{Gp} = W_p L \tag{8.101}$$

with L the common channel length. Since $W_p = r\,W_n$,

$$\begin{aligned} C_{in} &= C_{ox} L W_n (1 + r) \\ &= C_{Gn}(1 + r) \\ &= C_{ref} \end{aligned} \tag{8.102}$$

defines the reference input capacitance C_{ref}. Then, by definition, the logical effort of the 1X inverter is

$$g_{NOT} = \frac{C_{ref}}{C_{ref}} = 1 \tag{8.103}$$

The value of $g_{NOT} = 1$ provides the basis for comparing the performance of other gates. Note that the nFET gate capacitance C_{Gn} is the base unit of input capacitance.

The **electrical effort** h is defined by the capacitance ratio

$$h = \frac{C_{out}}{C_{in}} \tag{8.104}$$

where C_{out} is the *external* load capacitance seen at the output. One word of caution in the notation: in the context of Logical Effort, C_{out} is the same as C_L used in the rest of the book. The notation has been changed in this section to allow a smoother transition for those who want to pursue deeper studies in the technique. The electrical effort is the ratio of electrical drive strength that is required to drive C_{out} relative to that needed to drive its own input capacitance C_{in}.

The absolute **delay time** d_{abs} through the inverter is written in the form

$$d_{abs} = \kappa R_{ref}(C_{p,\,ref} + C_{out}) \quad \text{sec} \tag{8.105}$$

using the circuit drawn in Figure 8.16. The reference FET resistance R_{ref} is the same for both transistors since the design is symmetric. The total capacitance at the output node consists of the external value C_{out} and the internal parasitic capacitance $C_{p,ref}$ (i.e., the FET capacitance C_{FET} in our notation). The factor κ is the scaling multiplier; to obtain correlation with the analysis in Chapter 6, we would choose $\kappa = \ln(9) \approx 2.2$.

Now consider an inverter that is scaled by a factor $S > 1$. The relative transistor sizes are increased to S and rS for the nFET and pFET, respectively. The FET resistance decreases to

$$R = \frac{R_{ref}}{S} \tag{8.106}$$

and the parasitic capacitance increases to

$$C_p = SC_{p,\,ref} \tag{8.107}$$

The delay for the scaled gate is then

$$\begin{aligned} d_{abs} &= \kappa R(C_p + C_{out}) \\ &= \kappa \frac{R_{ref}}{S}(SC_{p,\,ref} + C_{out}) \end{aligned} \tag{8.108}$$

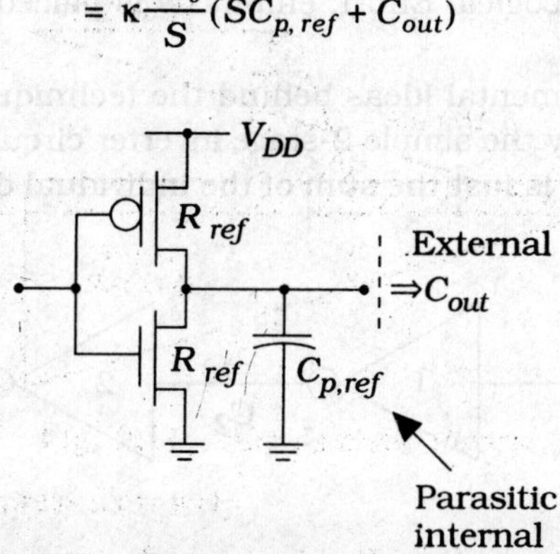

Figure 8.16 Delay circuit for a 1X inverter

Now note that the input capacitance for the scaled gate is

$$C_{in} = SC_{ref} \tag{8.109}$$

Distributing the terms then gives

$$\begin{aligned} d_{abs} &= \kappa\frac{R_{ref}}{S}SC_{p,\,ref} + \kappa\frac{R_{ref}}{S}C_{out} \\ &= \kappa R_{ref}C_{p,\,ref} + \kappa\frac{R_{ref}}{S}\left(\frac{C_{out}}{C_{ref}}\right)C_{ref} \\ &= \kappa R_{ref}C_{p,\,ref} + \kappa R_{ref}C_{ref}\left(\frac{C_{out}}{C_{in}}\right) \end{aligned} \tag{8.110}$$

Defining the reference time constant

$$\tau = \kappa R_{ref}C_{ref} \tag{8.111}$$

allows us to factor the delay into the form

$$d_{abs} = \tau(h + p) \tag{8.112}$$

where h is the electrical effort and

$$p = \frac{\tau_{par}}{\tau} = \frac{R_{ref}C_{p,\,ref}}{R_{ref}C_{ref}} \tag{8.113}$$

is the delay term associated with the parasitic capacitance. The **normalized delay**

$$d = \frac{d_{abs}}{\tau} = h + p \tag{8.114}$$

is unitless, and provides the important information about the gate. In the technique of Logical Effort, emphasis is placed on finding d for different paths.

The fundamental ideas behind the technique of Logical Effort can be understood by the simple 2-stage inverter circuit in Figure 8.17. The total **path delay** D is just the sum of the individual delays as expressed by

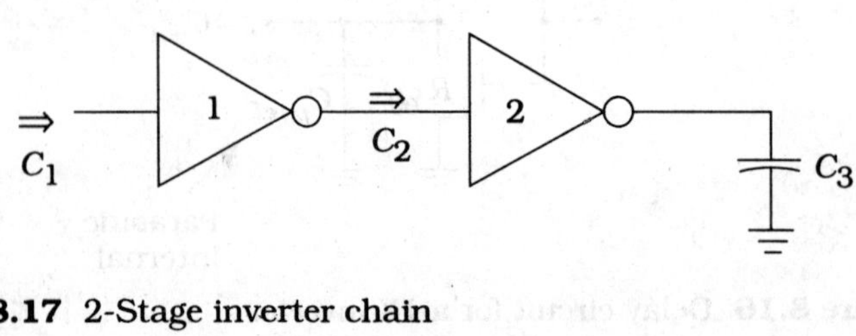

Figure 8.17 2-Stage inverter chain

$$D = d_1 + d_2 \\ = (h_1 + p_1) + (h_2 + p_2) \tag{8.115}$$

where

$$h_1 = \frac{C_2}{C_1}, \quad h_1 = \frac{C_3}{C_2} \tag{8.116}$$

are the individual electrical effort values. The **path electrical effort** H is defined as the ratio

$$H = \frac{C_{last}}{C_{first}} \tag{8.117}$$

and can be expressed as the product

$$H = h_1 h_2 \tag{8.118}$$

as seen from

$$H = \left(\frac{C_2}{C_1}\right)\left(\frac{C_3}{C_2}\right) = \frac{C_3}{C_1} \tag{8.119}$$

The product form is a general property of H. Using

$$h_2 = \frac{H}{h_1} \tag{8.120}$$

the path delay equation becomes

$$D = (h_1 + p_1) + \left(\frac{H}{h_1} + p_2\right) \tag{8.121}$$

The primary goal of Logical Effort techniques is to minimize the delay time though logic chains. For the present case, this condition can be found by calculating the derivative

$$\frac{\partial D}{\partial h_1} = \frac{\partial}{\partial h_1}\left[(h_1 + p_1) + \left(\frac{H}{h_1} + p_2\right)\right] \tag{8.122}$$

The parasitic terms p_1 and p_2 are constants to the differentiation so

$$\frac{\partial D}{\partial h_1} = 1 - \frac{H}{h_1^2} = 0 \tag{8.123}$$

Using $H = h_1 h_2$, the equation shows that the path delay is minimized if

$$h_1 = h_2 \tag{8.124}$$

Since the delay through an inverter is proportional to h, this is equivalent to saying that the path delay is minimized by equalizing the delay through each stage. This, of course, is the same conclusion we arrived at in the more rigorous analysis.

8.3.2 Generalization

The real power of the Logical Effort technique is that it can be generalized to include arbitrary CMOS logic gates. The calculations allow one to estimate delays through logic cascades and provide scaling relationships for minimum-delay designs.

The first step toward generalizing the technique is to develop expressions for the logical effort parameter g of basic CMOS gates. All calculations are referenced to the 1X reference inverter with an input capacitance C_{ref} and transistor resistance R_{ref}. The simplest designs are those that maintain a symmetrical design, i.e., $R_n = R_p = R_{ref}$. This requires us to adjust the sizes of series-connected transistors.

Figure 8.18(a) shows a symmetric 1X NAND2 gate. The pFET sizes are still r, since the worst-case path from the output to the power supply is the same as an inverter. The nFETs, however, must be twice as large as

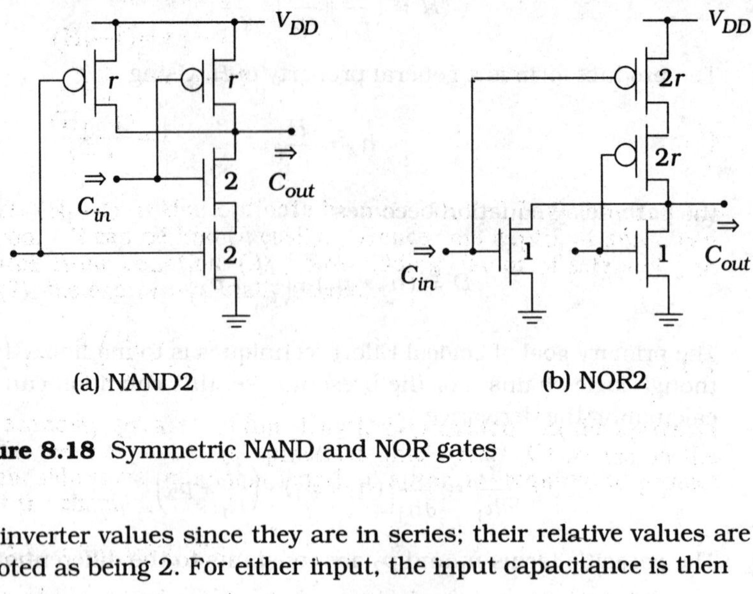

Figure 8.18 Symmetric NAND and NOR gates

the inverter values since they are in series; their relative values are thus denoted as being 2. For either input, the input capacitance is then

$$C_{in} = C_{Gn}(2 + r) \tag{8.125}$$

so that the logical effort for the NAND2 gate is

$$g_{NAND2} = \frac{C_{Gn}(2 + r)}{C_{ref}} = \frac{2 + r}{1 + r} \tag{8.126}$$

This is sufficient to characterize the gate for the delay calculation.

The 1X NOR2 circuit in Figure 8.18(b) is analyzed in the same manner. The parallel-connected nFETs have a relative size of 1 while the pFETs are chosen to have sizes of $2r$ to make R_p the same as R_{ref}. The input capacitance is then

$$C_{in} = C_{Gn}(1+2r) \tag{8.127}$$

so that the logical effort of the gate is

$$g_{NOR2} = \frac{C_{Gn}(1+2r)}{C_{ref}} = \frac{1+2r}{1+r} \tag{8.128}$$

Note that the numerical values of g depend upon the ratio r.

These results may be generalized to larger fan-in gates. An n-input NAND gate will have n-parallel pFETs with size r and n-series nFETs that have a sizing n. The capacitance seen at an input is

$$C_{in} = C_{Gn}(n+r) \tag{8.129}$$

so that the logical effort is

$$g_{NAND} = \frac{n+r}{1+r} \tag{8.130}$$

An n-input NOR gate has a logical effort of

$$g_{NOR} = \frac{1+nr}{1+r} \tag{8.131}$$

as can be verified using the same approach. It is easily seen that any basic CMOS gate can be characterized for a value of logical effort g.

The delay through a general gate is expressed as

$$d = gh + p \tag{8.132}$$

The primary effect of the logical effort parameter g is to modify the first term to account for the difference in drive characteristics among various gates. For a logic cascade with N stages, each gate will be characterized by a delay

$$d_i = g_i h_i + p_i \tag{8.133}$$

for $i = 1$ to N. The **total path delay** D is the sum

$$D = \sum_{i=1}^{N} d_i = \sum_{i=1}^{N} (g_i h_i + p_i) \tag{8.134}$$

The **path logical effort** G is just the product of the individual factors

$$G = \prod_{i=1}^{N} g_i = g_1 g_2 ... g_N \tag{8.135}$$

and the **path electrical effort** H is defined in a similar manner by

$$H = \prod_{i=1}^{N} h_i = h_1 h_2 ... h_N \tag{8.136}$$

These combine to give the **path effort** F

$$\begin{aligned} F &= GH \\ &= (g_1 h_1)(g_2 h_2)(g_3 h_3)...(g_N h_N) \\ &= f_1 f_2 \cdots f_N \end{aligned} \tag{8.137}$$

A minimum delay through the cascade is achieved if

$$g_i h_i = \text{constant} = \hat{f} \tag{8.138}$$

for every i. This is consistent with our conclusions for the simple 2-stage inverter chain. The optimum path effort is thus

$$F = \hat{f}^N \tag{8.139}$$

so that the fastest design is where each stage has

$$gh = \hat{f} = F^{1/N} \tag{8.140}$$

This is the main equation of Logical Effort. The composition of an N-stage logic chain allows us to find the value of F. Each stage can be sized to accommodate the optimum electrical effort value

$$h_i = \frac{\hat{f}}{g_i} \tag{8.141}$$

The optimized path delay is then

$$\hat{D} = NF^{1/N} + P \tag{8.142}$$

where

$$P = \sum_{i=1}^{N} p_i \tag{8.143}$$

is the sum of the parasitic delays. In general, p_{ref} for an inverter is the smallest, with multiple-input gates exhibiting larger parasitic delay times. One simple estimate is to write

$$p = np_{ref} \tag{8.144}$$

as the parasitic delay for an n-input gate.

Example 8.4

Let us analyze the logic cascade in Figure 8.19 using the technique of Logical Effort. We will assume values of C_4 = 500 fF and C_1 = 20 fF. First, the path logical effort is given by

$$\begin{aligned} G &= g_{NOT}g_{NOR2}g_{NAND2} \\ &= (1)\left(\frac{1+2r}{1+r}\right)\left(\frac{2+r}{1+r}\right) \end{aligned} \tag{8.145}$$

Assuming a value of $r = 2.5$, we compute

$$G = (1)\left(\frac{6}{3.5}\right)\left(\frac{4.5}{3.5}\right) = 2.2 \tag{8.146}$$

The path electrical effort is

$$H = \frac{C_4}{C_1} = \frac{500}{20} = 25 \tag{8.147}$$

so that the path effort is

$$F = GH = 55 \tag{8.148}$$

The optimum stage effort is

$$\hat{f} = F^{1/N} = (55)^{1/3} = 3.8 \tag{8.149}$$

which gives a total path delay of

$$\begin{aligned} \hat{D} &= 3(3.8) + P \\ &= 11.41 + P \end{aligned} \tag{8.150}$$

where

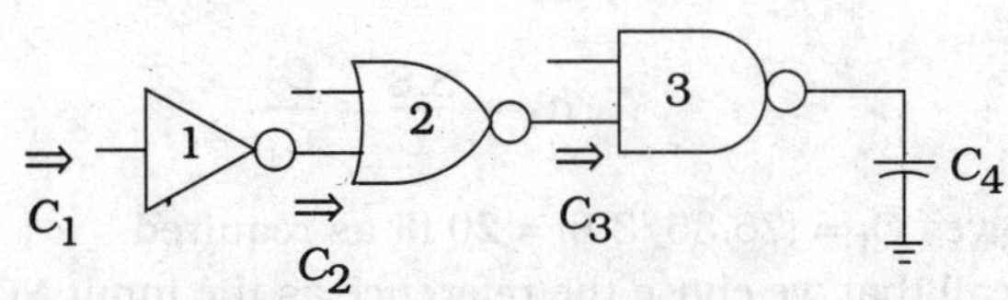

Figure 8.19 Logic cascade for Example 8.4

$$P = (p_{NOT} + p_{NOR2} + p_{NAND2}) \tag{8.151}$$

is the parasitic delay term that is determined by the process specifications.

The sizing equations are obtained from the analysis using the optimized quantities. Starting from the NAND2 gate at the output with $g_{NAND2} = (4.5/3.5) = 1.29$ we have

$$h_3 = \frac{3.8}{1.29} = 2.95 = \frac{C_4}{C_3} \tag{8.152}$$

so that

$$C_3 = \frac{500}{2.95} = 169.5\text{ fF} \tag{8.153}$$

Since C_3 in the input capacitance into a NAND2 gate, we may use equation (8.125) to write a scaled gate as

$$\begin{aligned} C_3 &= S_3 C_{Gn}(2+r) \\ &= S_3(4.5C_{Gn}) \end{aligned} \tag{8.154}$$

where S_3 is the scaling factor.

The NOR2 gate is analyzed in the same manner. Since $g_{NOR2} = 1.71$, we have

$$h_2 = \frac{3.8}{1.71} = 2.22 = \frac{C_3}{C_2} \tag{8.155}$$

Thus,

$$C_2 = \frac{169.5}{2.22} = 76.35\text{ fF} \tag{8.156}$$

The input capacitance into the NOR2 gate is

$$\begin{aligned} C_2 &= S_2 C_{Gn}(1+2r) \\ &= S_2(6C_{Gn}) \end{aligned} \tag{8.157}$$

The input NOT gate is defined to have a logical effort of 1 so

$$h_1 = \frac{3.8}{1} = \frac{C_2}{C_1} \tag{8.158}$$

This gives $C_1 = (76.35/3.8) = 20$ fF as required.

Recall that we chose the reference as the input NOT gate with $C_1 = C_{ref} = 2.5\ C_{Gn}$. The NOR and NAND gates then scale as

$$S_2 = \frac{76.35}{(6)(3.5C_{Gn})} = \frac{3.64}{C_{Gn}}$$
$$S_3 = \frac{169.5}{(4.5)(3.5C_{Gn})} = \frac{10.76}{C_{Gn}} \tag{8.159}$$

to achieve the minimum delay. These scaling values are referenced to a capacitance of

$$C_{Gn} = \frac{20}{3.5} = 5.71 \text{ fF} \tag{8.160}$$

where

$$C_{Gn} = C_{ox} W_n L \tag{8.161}$$

gives the reference nFET channel width W_n.

Another approach is to choose a minimum size 1X inverter as the reference. If, for example, C_{ref} = 8 fF for a 1X gate, then the scale factors are S_1 = 2.5 (for the NOT gate), S_2 = 1.59, and S_3 = 4.71. Usually the reference can be chosen for convenience.

8.3.3 Optimizing the Number of Stages

A well-known characteristic of CMOS logic cascades is the fact that one can often insert inverters into a logic chain and decrease the total delay time. While this may play against simple intuition developed in introductory logic design courses, it is based in the fact that distributing out the drive strength among several stages is more important than counting the number of logic symbols. Logical Effort shows this feature using the path delay *D*.

First, note that the logical effort of an inverter is g_{NOT} = 1. Since

$$G = g_1 g_2 \cdots g_N \tag{8.162}$$

multiplying by additional factors of g_{NOT} does not change the numerical value of the path effort

$$F = GH \tag{8.163}$$

Delay time minimization is expressed by

$$\begin{aligned} \hat{f} &= F^{1/N} \\ &= (GH)^{1/N} \end{aligned} \tag{8.164}$$

such that the total path delay is

$$\hat{D} = NF^{1/N} + P \tag{8.165}$$

In general, $F^{1/N}$ decreases with increasing N. Thus, it may be possible to obtain a smaller path delay by inserting the inverters. Note, however, that the increased parasitic delay in P due to the extra inverters will offset some of the performance.

Example 8.5

To see the dependence, suppose that $F = 200$. For $N = 3$,

$$3(200)^{1/3} = 17.54 \tag{8.166}$$

For $N = 4$,

$$4(200)^{1/4} = 15.04 \tag{8.167}$$

and $N = 5$ gives

$$5(200)^{1/5} = 14.43 \tag{8.168}$$

However, if we try $N = 10$, then the term increases

$$10(200)^{1/10} = 16.99 \tag{8.169}$$

so that we have passed the optimum number of stages.

An analysis of the problem shows that the optimum number of stages for a given F is obtained by solving the transcendental equation [8]

$$F^{1/N}[1 - \ln(F^{1/N})] + p_{ref} = 0 \tag{8.170}$$

This can be rewritten into a simpler looking form by defining

$$\rho = F^{1/N} \tag{8.171}$$

so that

$$\rho[1 - \ln(\rho)] + p_{ref} = 0 \tag{8.172}$$

A moment's reflection confirms that this has the same form as equation (8.93) that was derived from circuit considerations, thus demonstrating the equivalence of the two approaches. The power of Logical Effort is that it is not restricted to inverters.

For small values of p_{ref}, the approximate solutions are

$$\rho \approx 0.71 p_{ref} + 2.82 \tag{8.173}$$

which is useful for estimating the optimum value of N during an initial design phase.

8.3.4 Logical Area

The real estate area is important, particularly in scaled designs. An estimate of the circuit requirements can be obtained using Logical Effort quantities by simply summing the gate areas of each FET by calculating the logical area (*LA*) for the *i*-th gate using

$$LA_i = W_i \times L \tag{8.174}$$

where L is the channel length and W_i is determined by the sizing. For example, the logical area of a 1X NOT gate with $L = 1$ unit is

$$LA_{NOT} = 1+r \tag{8.175}$$

which accounts for the pFET and nFET sizes. If this is scaled by a factor $S > 1$, then the logical area increases to

$$LA_{NOT} = S(1+r) \tag{8.176}$$

Similarly, a scaled NOR2 gate has

$$LA_{NOR2} = S(1+2r) \tag{8.177}$$

while

$$LA_{NAND2} = S(2+r) \tag{8.178}$$

applies to a NAND2 gate. For a network with M gates, the total logical area is

$$LA = \sum_{i=1}^{M} LA_i \tag{8.179}$$

This allows a simple metric for comparing area requirements of different designs. Note, however, that since it ignores drain and source spacings, interconnect wiring, well, etc., it is only a rough estimate.

8.3.5 Branching

The technique of Logical Effort applies to a well-defined path. When a logic gate drives two or more gates, the data path splits and we must account for presence of the gates that are not in the main path, but contribute capacitance. This situation is portrayed in the logic diagram of Figure 8.20, where the main path of interest from In to Out has been highlighted. Tracing the circuit shows two branching points. In both cases, the NOR2 gates add capacitance to the NAND2 loads and cannot be ignored.

These effects are handled by introducing the **branching effort** *b* at every branch point such that

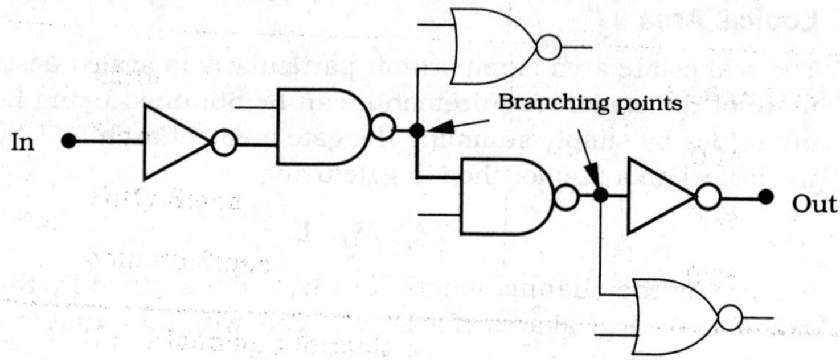

Figure 8.20 Branching

$$b = \frac{C_T}{C_{path}} \qquad (8.180)$$

where C_{path} is the capacitance in the main logic path, and

$$C_T = C_{path} + C_{off} \qquad (8.181)$$

represents the total capacitance seen at the node. In this equation, C_{off} includes all capacitance contributions that are off of the main path. The branching effort has the property that $b > 1$ and accounts for the additional loading. The **path branching effort** is given by the product

$$B = \prod_i b_i \qquad (8.182)$$

where b_i are the individual branching efforts.

Example 8.6

Consider the logic network in Figure 8.20. At the first branch point, a NAND2 gate drives another on-path NAND2, and an off-path NOR2 gate. Assuming unit gate sizes, the branching effort b_1 for this point is

$$\begin{aligned} b_1 &= \frac{C_{NAND2} + C_{NOR2}}{C_{NAND2}} \\ &= \frac{(2+r)+(1+2r)}{(2+r)} \\ &= \frac{3(1+r)}{(2+r)} \end{aligned} \qquad (8.183)$$

The second branch point in the drawing is described by

$$b_2 = \frac{C_{NOT} + C_{NOR2}}{C_{NOT}} \qquad (8.184)$$

or

$$b_2 = \frac{(1+r)+(1+2r)}{(1+r)} = \frac{(2+3r)}{(1+r)} \tag{8.185}$$

The path branching effort is then

$$B = \frac{3(1+r)}{(2+r)}\frac{(2+3r)}{(1+r)} = \frac{3(2+3r)}{(2+r)} \tag{8.186}$$

for the selected path from In to Out.

Once the path branching effort has been calculated, we modify the path effort F to read

$$F = GHB \tag{8.187}$$

and the calculation proceeds in the same manner as for the simpler case without branching. This allows us to extend Logical Effort to arbitrary logic configurations and analyze every path for relative delay.

8.3.6 Summary

This short discussion of Logical Effort illustrates the usefulness of the technique. It is particularly valuable in advanced systems design where we have the choice of several algorithms that lead to the same result. Logical Effort allows us to compare the performance of the different circuits to see which is better for our design. These considerations will be discussed in later chapters of the book.

8.4 BiCMOS Drivers[2]

BiCMOS is a modified CMOS technology that includes bipolar junction transistors as circuit elements. In digital design, BiCMOS stages are used to drive high-capacitance lines more efficiently than MOSFET-only circuits. BiCMOS processing is more expensive than standard CMOS, and bipolar transistors have an intrinsic voltage drop that cannot be avoided making them undesirable for low-voltage applications.

8.4.1 Bipolar Junction Transistor Characteristics

A bipolar junction transistor (BJT) is a 3-terminal element that obtains its electrical characteristics from the properties of pn junctions. There are two types of BJTs, npn and pnp. The current flowing through an npn

2 This section can be skipped without loss of continuity in the discussion.

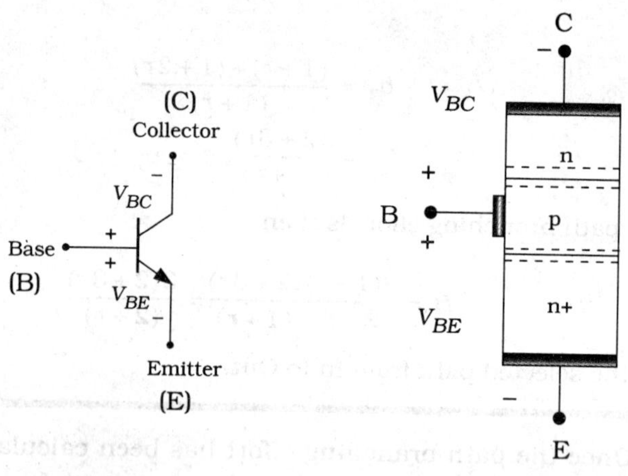

(a) Symbol (b) Structure

Figure 8.21 Symbol and structure of an npn BJT

transistor is due mostly to electrons, while that through a pnp device is due to holes. Since electrons are faster than holes, we concentrate on using npn devices in high-speed BiCMOS circuits.

The circuit symbol for an npn BJT is shown in Figure 8.21(a). The device has three terminals that are called the base (B), the emitter (E) with the arrowhead, and the collector (C). A simplified "prototype" structure of the npn BJT is shown in Figure 8.21(b); this illustrates the npn layer that gives the device its name. The drawing shows that the npn transistor can be viewed as two back-to-back pn junction diodes, one between the base and emitter terminals and the other between the base and collector electrodes. Current flow through the BJT is controlled by two voltages, the base-emitter voltage V_{BE} and the base-collector voltage V_{BC}, that bias the two pn junctions. They are defined to be positive values when the "+" polarity is applied to the p-type base layer. A positive voltage indicates a forward bias on the junction that allows current flow, while a negative voltage is a reverse bias.

The operation of the bipolar transistor is complicated by the fact that the voltages can be either positive or negative (reversed polarity). Consider the situation shown in Figure 8.22(a). The currents I_C, I_B, and I_E are determined by the voltages, but each combination of polarities gives a different mode of operation. These are summarized by the plot shown in Figure 8.22(b) that indicates the polarities of V_{BE} and V_{BC} by quadrants. **Forward-active bias** is defined by $V_{BE} > 0$ and $V_{BC} < 0$, i.e., the base-emitter junction is forward biased and the base-collector junction is reverse biased. This mode of operation allows for amplification and controlled current flow, and is used for analog circuits. The opposite case

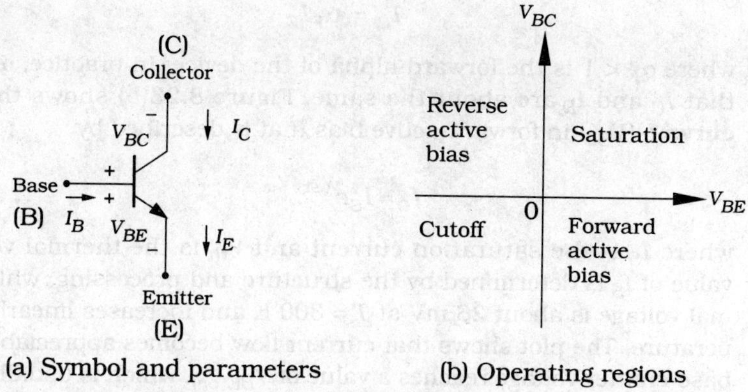

Figure 8.22 Operating regions of the bipolar junction transistor

where $V_{BE} < 0$ and $V_{BC} > 0$ is called **reverse-active bias**, and is used only in a few special cases. If both junctions are forward biased with $V_{BE} > 0$ and $V_{BC} > 0$, the device is said to be in **saturation**. In this case, large currents can flow through the device but the transistor does not control the values. It is important to remember that saturation in a BJT has no relation to a saturated FET. The final case is where both junctions are reverse biased with $V_{BE} < 0$ and $V_{BC} < 0$. Only small leakage currents flow and the BJT is said to be in **cutoff**. This can be modeled as an open switch.

Bipolar transistors are faster than MOSFETs but are more complicated to build into an integrated circuit. Let us examine forward-active bias to understand why a bipolar circuit can provide faster switching. Figure 8.23(a) shows the device with this bias. The collector and emitter currents are related by

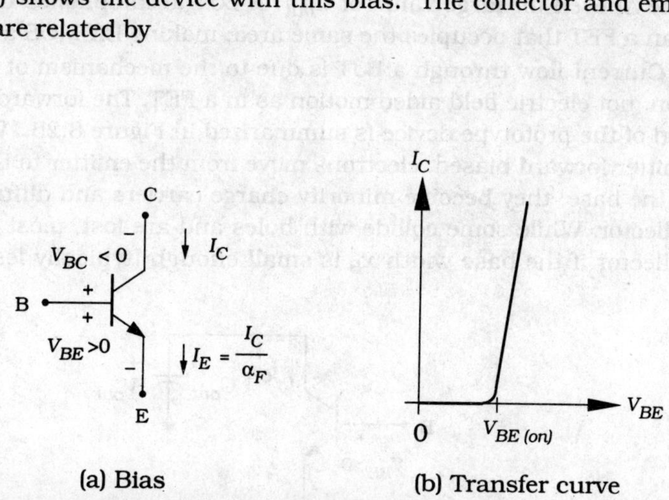

Figure 8.23 Forward-active bias in a BJT

$$I_C = \alpha_F I_E \tag{8.188}$$

where $\alpha_F < 1$ is the forward-alpha of the device; in practice, $\alpha_F \approx 0.99$ so that I_C and I_E are about the same. Figure 8.23(b) shows the **transfer curve** $I_C(V_{BE})$ in forward-active bias that is described by

$$I_C \approx I_S e^{V_{BE}/V_{th}} \tag{8.189}$$

where I_S is the saturation current and V_{th} is the thermal voltage. The value of I_S is determined by the structure and processing, while the thermal voltage is about 26 mV at $T = 300$ K and increases linearly with temperature. The plot shows that current flow becomes appreciable when the base-emitter voltage reaches a value of $V_{BE(on)}$, which is usually estimated to be about 0.5 V to 0.7 V. Once this is achieved, the current increases exponentially with increasing V_{BE}.

Consider the simple circuit shown in Figure 8.24. With the BJT in forward-active bias, the current flow out of the capacitor is

$$I_C = -C_{out}\frac{dV_{out}}{dt} = I_S e^{V_{BE}/V_{th}} \tag{8.190}$$

We can estimate the discharge time by

$$\Delta t \approx \frac{(\Delta V_{out})}{I_C C_{out}} \tag{8.191}$$

where ΔV_{out} is the change in voltage. The values of I_C can be large, easily reaching tens to hundreds of milliamperes, which reduces the discharge time Δt even for large value of C_{out}. A BJT accomplishes the task faster than a FET that occupies the same area, making BiCMOS attractive.

Current flow through a BJT is due to the mechanism of particle diffusion, not electric field aided motion as in a FET. The forward active operation of the prototype device is summarized in Figure 8.25. With the base-emitter forward biased, electrons move from the emitter to the base. Once in the base, they become minority charge carriers and diffuse toward the collector. While some collide with holes and are lost, most will reach the collector if the base width x_B is small enough (typically less than about

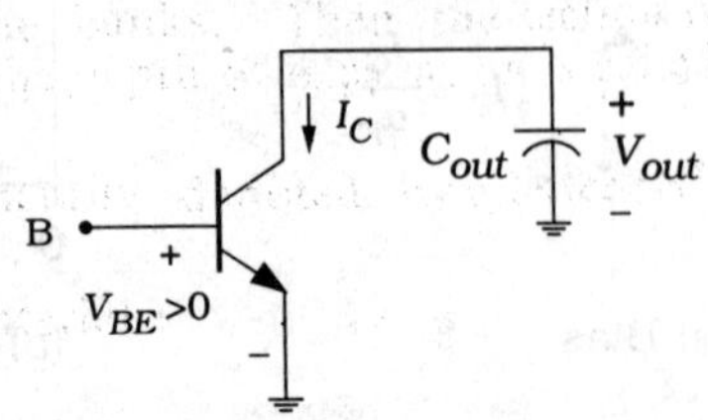

Figure 8.24 Discharge of a capacitor using a BJT

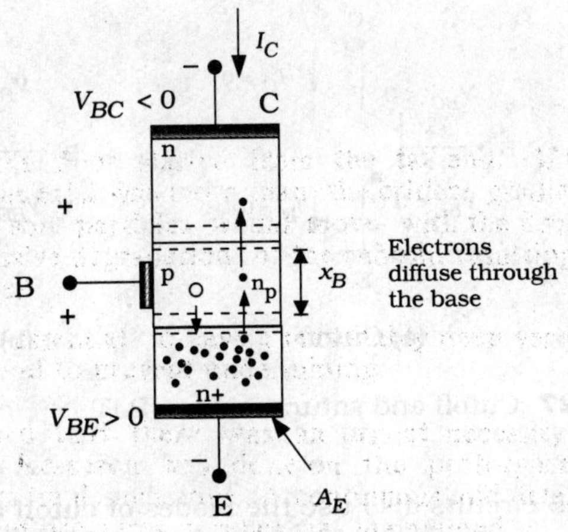

Figure 8.25 Forward-bias operation

0.5 μm). This establishes current flow from collector to the emitter. An analysis shows that the saturation current is given by

$$I_S = qA_E \frac{D_n n_i^2}{x_B N_{aB}} \tag{8.192}$$

where A_E [cm^2] is the emitter area, D_n [cm^2/sec] is the electron diffusion coefficient in the base and is a measure of the diffusive motion, q is the electron charge, and N_{aB} [cm^{-3}] is the acceptor doping in the base. A typical value for the saturation current is I_S = 0.1 pA = 10^{-13} A. While this is quite small, the exponential dependence of the current on V_{BE} gives large values of I_C. The cross-sectional view of an integrated bipolar transistor is shown in Figure 8.26. The prototype structure can be seen in the center region underneath the emitter n+ region. Since specialized layers are required to create the device, the processing of a BiCMOS chip is more expensive than a basic CMOS design.

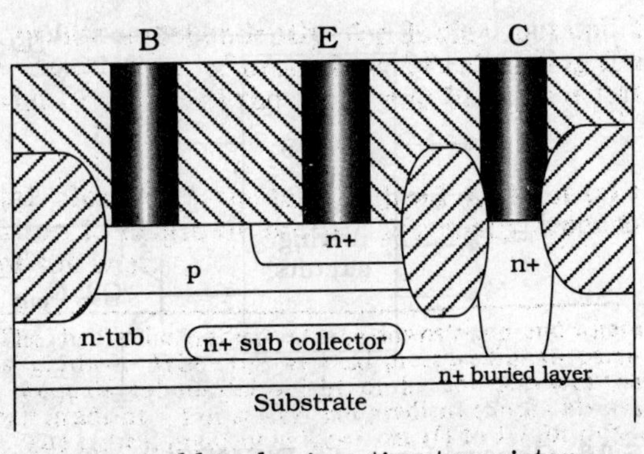

Figure 8.26 An integrated bipolar junction transistor

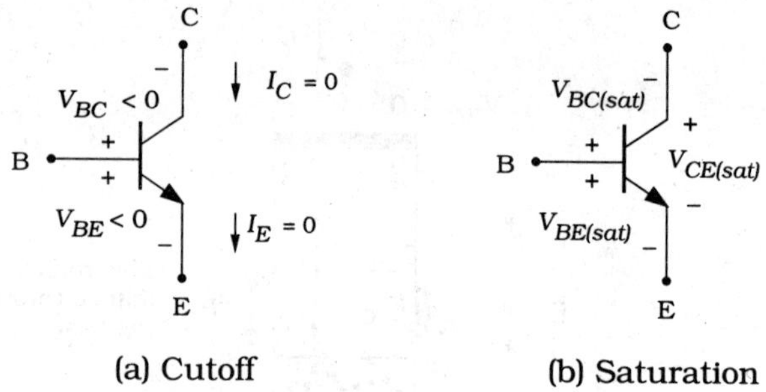

Figure 8.27 Cutoff and saturation in a BJT

BiCMOS circuits also use the modes of cutoff and saturation, which are summarized in Figure 8.27. In cutoff, both junctions are reverse-biased and both I_C and I_E are approximately 0 as in Figure 8.27(a). The device is saturated when both junctions are forward biased; this case is shown in Figure 8.27(b). In this case, the values of the currents are determined by the circuits that are connected to the transistor. The junction voltages take on constant values of $V_{BE(sat)}$ and $V_{BC(sat)}$ with typical values of around 0.8 V and 0.7 V, respectively. The collector-emitter voltage is thus about $V_{CE(sat)} \approx 0.1$ V by using Kirchhoff's law.

8.4.2 Driver Circuits

BiCMOS circuits employ CMOS logic circuits that are connected to a bipolar output driver stage. A general structure is shown in Figure 8.28. The CMOS network is used to provide logic operations and drive the output bipolar transistors Q1 and Q2. Only one BJT is active at a time. Transistor Q1 provides the high output voltage while Q1 discharges the output capacitance and gives the low output state.

The inverting circuit in Figure 8.29 gives an example of the operational

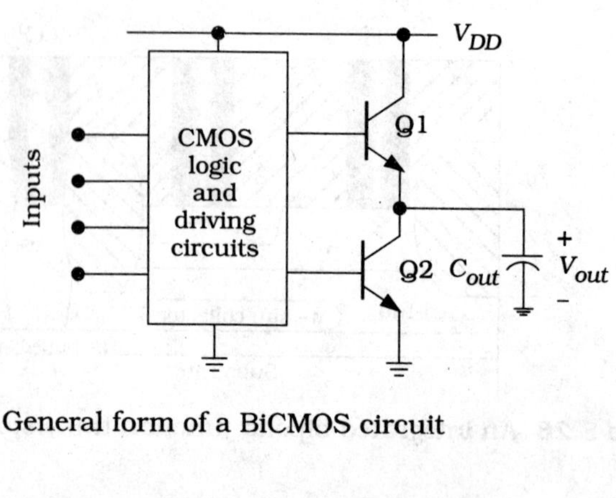

Figure 8.28 General form of a BiCMOS circuit

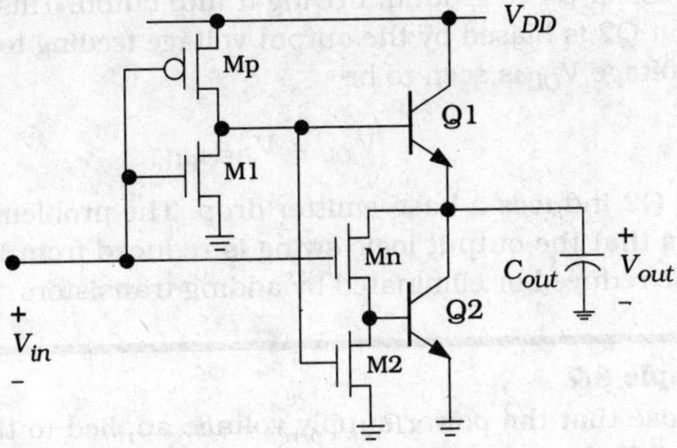

Figure 8.29 An inverting BiCMOS driver circuit

details. The NOT logic operation is performed by FETs Mp and Mn, even though they are separated from each other. The other two FETs M1 and M2 are used to provide paths to remove charge from the base terminals of Q1 and Q2, respectively. This speeds up the switching of the circuit, enhancing its use as an output driver.

Let us examine the DC operation of the circuit. Consider first the case where the input voltage is at a value of $V_{in} = 0$ V. This turns Mp on, while M1 and Mn are off. Since Mp and M1 form an inverter, the base of Q1 is high at a voltage of V_{DD}, and it goes active; the same voltage turns on M2, which grounds the base of Q2 and drives it into cutoff. The output high voltage V_{OH} for this case can be calculated from the subcircuit shown in Figure 8.30(a). Noting that Q1 will eventually enter saturation, we have

$$V_{OH} = V_{DD} - V_{BE(sat)} \tag{8.193}$$

since the voltage is dropped a value of $V_{BE(sat)}$ from the base to the output. The subcircuit for the case where $V_{in} = V_{DD}$ is shown in Figure 8.30(b). Now we see that Mp is off while M1 and Mn are on. M1 connects

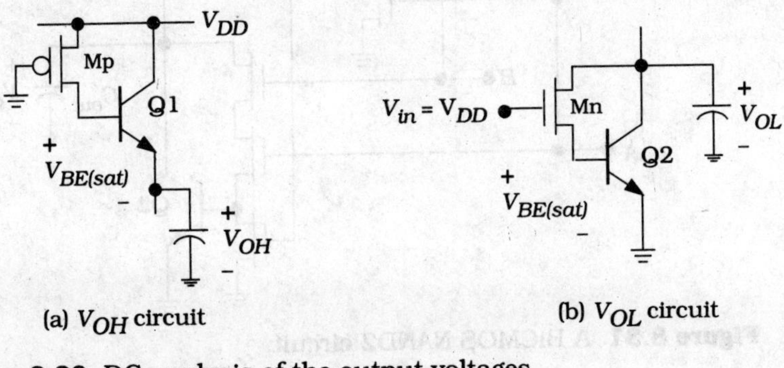

(a) V_{OH} circuit

(b) V_{OL} circuit

Figure 8.30 DC analysis of the output voltages

the base of Q1 to ground, driving it into cutoff. This in turn shuts off M2 so that Q2 is biased by the output voltage feeding to the base. The output low voltage V_{OL} is seen to be

$$V_{OL} = V_{BE(sat)} \tag{8.194}$$

since Q2 induces a base-emitter drop. The problem with this configuration is that the output logic swing is reduced from V_{DD} by $2V_{BE(sat)}$. This can be reduced or eliminated by adding transistors.

Example 8.7

Suppose that the power supply voltage applied to the BiCMOS circuit is $V_{DD} = 5$ V. Assuming that $V_{BE(sat)} = 0.8$ V,

$$\begin{aligned} V_{OH} &= 5 - 0.8 = 4.3 \text{ V} \\ V_{OL} &= 0.8 \text{ V} \end{aligned} \tag{8.195}$$

which implies a logic swing of 3.4 V at the output. This can be improved by redesigning the output stage.

The CMOS circuitry can be modified to provide logic functions. A NAND2 gate based on this design is shown in Figure 8.31. A careful examination of the circuit shows that the logic is formed by the parallel pFETs driving Q1, and the series nFETs between the collector and base of Q2. The other FETs are used as pull-down devices to turn off the output transistors. Other logic functions can be designed using this as a basis. In

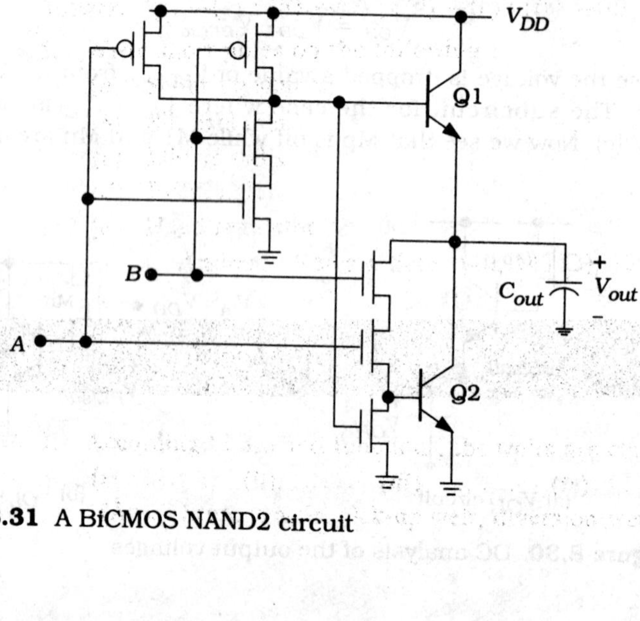

Figure 8.31 A BiCMOS NAND2 circuit

general, the upper output transistor uses a standard-design CMOS circuit as a driver. The nFET section is replicated and placed in between the collector and base of the lower output transistor; adding a pull-down nFET to the base completes the design.

It is apparent that BiCMOS circuits are more complicated than their CMOS equivalents. If we write the total output capacitance as

$$C_{out} = C_{transistors} + C_L \tag{8.196}$$

where C_L is the external load, we see that the parasitic transistor capacitance $C_{transistor}$ will be larger in a BiCMOS circuit due to the additional devices present. This leads to an important conclusion: BiCMOS is only effective for large values of C_L. A typical plot of time delay t_d versus C_L is shown in Figure 8.32. Due to the higher parasitic device capacitance, the CMOS and BiCMOS behaviors cross at a value $C_L = C_X$. For $C_L < C_X$, a standard CMOS design provides faster switching than a BiCMOS circuit. The speed increase is seen only for loads where C_L is much larger than C_X. This restricts the application of BiCMOS circuits to applications such as driving long data buses. Moreover, the cost and problem of V_{BE} drops are important factors in using the technology in digital VLSI.

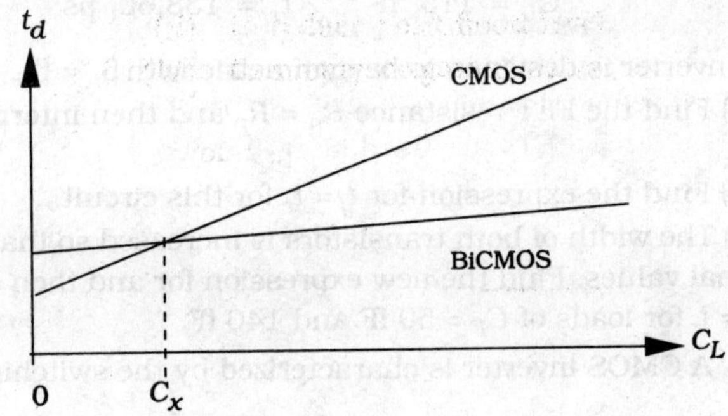

Figure 8.32 Gate delay versus external load capacitance

8.5 Books for Further Reading

[1] R. Jacob Baker, Harry W. Li, and David E. Boyce, **CMOS Circuit Design, Layout, and Simulation**, IEEE Press, Piscataway, NJ, 1998.

[2] Abdellatif Bellaouar and Mohamed I. Elmasry, **Low-Power Digital VLSI Design**, Kluwer Academic Publishers, Norwell, MA, 1995.

[3] Kerry Bernstein, et. al, **High Speed CMOS Design Styles**, Kluwer Academic Publishers, Norwell, MA, 1998.

[4] Ken Martin, **Digital Integrated Circuits**, Oxford University Press, New York, 2000.

[5] Robert F. Pierret, **Semiconductor Device Fundamentals**, Addison Wesley, Reading, MA, 1996.

[6] Jan M. Rabaey, **Digital Integrated Circuits**, Prentice Hall, Upper Saddle River, NJ, 1996.

[7] Jasprit Singh, **Semiconductor Devices**, John Wiley & Sons, New York, 2001.

[8] Ivan P. Sutherland, Bob Sproull, and David Harris, **Logical Effort**, Morgan-Kauffman Publishers, Inc., San Francisco, 1999.

[9] John P. Uyemura, **CMOS Logic Circuit Design**, Kluwer Academic Publishers, Norwell, MA, 1999.

[10] Neil H. E. Weste and Kamran Eshraghian, **Principles of CMOS VLSI Design**, 2nd ed., Addison-Wesley, 1993.

[11] Edward S. Yang, **Microelectronic Devices**, McGraw-Hill, New York, 1988.

8.6 Problems

[8.1] A CMOS inverter circuit has the following characteristics:

$$\begin{aligned} C_L &= 100 \text{ fF} \qquad t_r = 123.75 \text{ ps} \\ C_L &= 115 \text{ fF} \qquad t_r = 138.60 \text{ ps} \end{aligned} \tag{8.197}$$

The inverter is designed to be symmetric with $\beta_n = \beta_p$, and $V_{Tn} = |V_{Tp}|$.

(a) Find the FET resistance $R_n = R_p$ and then internal FET capacitance C_{FET}.

(b) Find the expression for $t_f = t_r$ for this circuit.

(c) The width of both transistors is increased so that they are 3.2× their original values. Find the new expression for and then calculate the values of $t_f = t_r$ for loads of C_F = 50 fF and 140 fF.

[8.2] A CMOS inverter is characterized by the switching times

$$\begin{aligned} t_r &= 430 + 3.68 C_L \text{ ps} \\ t_f &= 300 + 2.56 C_L \text{ ps} \end{aligned} \tag{8.198}$$

with the external load capacitance C_L in units of fF.

(a) Plot the rise and fall times for the range C_L = 0 to C_L = 200 fF.

(b) A three-inverter cascade is built using identical circuits. Find the worst-case delay through the chain if the output capacitance to each NOT gate is C_L = 45 fF.

[8.3] Consider the logic chain shown in Figure P8.1. The input at A is switched from a 1 to a 0. Find an expression for the delay time through the chain using the procedure developed for the network shown in Figure 8.6.

[8.4] A CMOS process is characterized by C_{ox} = 8 fF/µm², r = 2.6, and L = 0.4 µm. The magnitudes of the nFET and pFET threshold voltages are

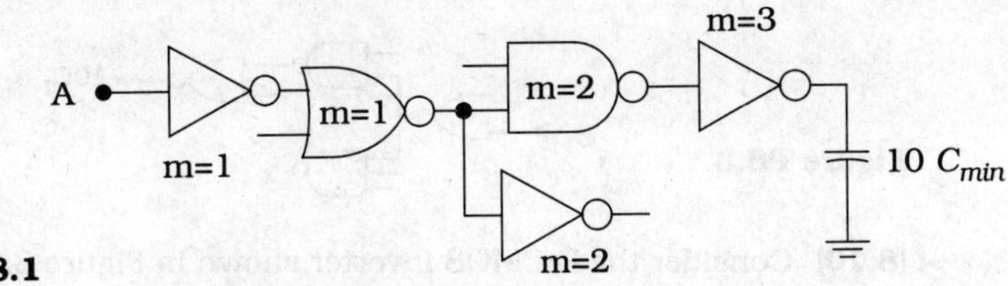

Figure P8.1

equal. A symmetrical inverter is designed using an nFET with a channel width of 2.2 μm. This acts as the input stage to a driver chain that has a load of C_L = 38 pF at the end. The design stipulates that the chain must produce an inverted signal with minimum delay from the input stage to the load.

(a) Calculate the input capacitance C_{in} of the inverter in units of fF.

(b) Apply idealized scaling to find the number of stages needed in the chain.

(c) It is known that an nFET with a channel width of W = 1 μm has a resistance of $R_n \approx 1725\ \Omega$. Given this, can you find the total delay time through the chain? If not, what other information is needed?

[8.5] Design a driver chain that will drive a load capacitance of C_L= 40 pF if the initial stage has an input capacitance of C_{in} = 50 fF. Use ideal scaling to determine the number of stages and the relative sizes.

[8.6] An interconnect line is described by a capacitance per unit length of c' = 0.86 pF/cm. The line itself runs over a significant portion of the chip and has a total length of 272 μm. A "standard" inverter has an input capacitance of 52 fF and uses symmetrical devices with $\beta_n = \beta_p$. The mobility ratio is r = 2.8 for the process. This is used as the first stage in a driver chain for the interconnect.

Use the idealized theory to design the driver chain with the constraint that the output must be non-inverting.

[8.7] Solve equation (8.93) for the case $\tau_x = 0.72\ \tau_r$.

[8.8] Consider the logic cascade shown in Figure P8.2. Use Logical Effort to find the relative size of each stage needed to minimize delay through the chain. Assume symmetric gates with r = 2.5.

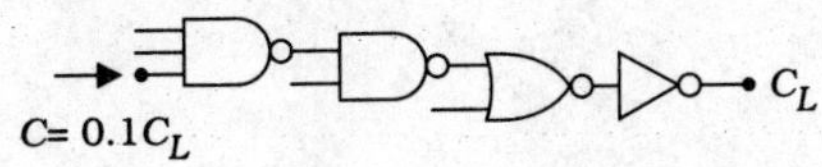

Figure P8.2

[8.9] The logic chain in Figure P8.3 is constructed in a process with r = 2.5. Determine the optimum sizing for each stage for the "highlighted" path indicated using the technique of Logical Effort.

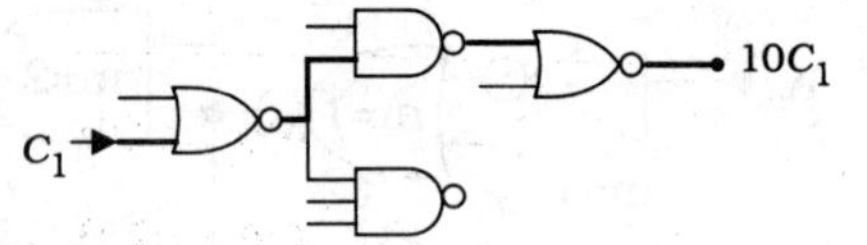

Figure P8.3

[8.10] Consider the BiCMOS inverter shown in Figure 8.29. Suppose that we replace the bottom BJT Q2 with a large nFET in its place, but leave Q1 in as the pull-up driver.

Draw the resulting circuit including only the CMOS driver circuitry needed for Q1. What is the logic swing for this design?

[8.11] Construct a BiCMOS NOR2 circuit using the circuit in Figure 8.29 as a basis.

[8.12] Design a digital BiCMOS circuit that implements the function

$$f = \overline{a + b \cdot c} \tag{8.199}$$

[8.13] Can you design a BiCMOS circuit that has $V_{OH} = V_{DD}$ and $V_{OL} = 0$ V by keeping the basic structure discussed, but modifying the output circuit? Hint: remember that a standard CMOS design has these values.

Advanced Techniques in CMOS Logic Circuits

9

A wide variety of CMOS circuit design styles have been published that are useful in the design of high-speed VLSI networks. All are based on simple logic gates, but operate in distinct ways. Most advanced techniques have been developed to overcome one or more problems that have arisen as VLSI applications have increased over the years. Some are very general, while others are used only for special cases. In this chapter we will unleash a sampling of the modern CMOS circuit techniques that are used in VLSI. This will provide a basis for applications in later chapters.

9.1 Mirror Circuits

Mirror circuits are based on series-parallel logic gates, but are usually faster and have a more uniform layout. The basic idea of a mirror is seen from the XOR truth table in Figure 9.1. Output 0's imply that an nFET chain is conducting to ground, while an output 1 means that a pFET group provides support from the power supply. The important aspect of this observation is that there are equal numbers of input combinations that produce 0's and 1's.

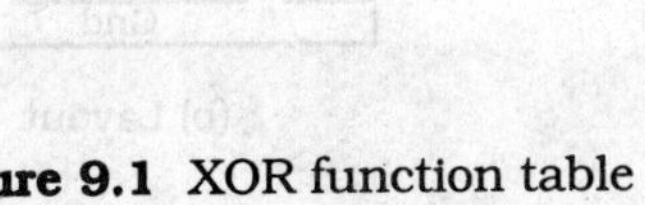

a	b	$a \oplus b$	On devices
0	0	0	← nFET
0	1	1	← pFET
1	0	1	← pFET
1	1	0	← nFET

Figure 9.1 XOR function table

A mirror circuit uses the same transistor topology for the nFETs and pFETs. Applying this to the XOR function yields the circuit in Figure 9.2(a). The input combinations are shown for each branch. The "mirror" effect can be understood by placing a mirror along the output line, facing either up or down. The mirror image seen in the mirror will be the other side of the circuit. A layout for an XOR cell is shown in Figure 9.2(b); the pFETs are larger than the nFETs to compensate for the lower process transconductance (k') values.

The advantages of the mirror circuit are more symmetric layouts and shorter rise and fall times. The latter comment can be understood using the RC switch model in Figure 9.3. Every path between the output and a power supply rail consists of two resistors and a parasitic inter-FET capacitor. The Elmore time constant is of the form

$$\tau_x = C_{out}(2R_x) + C_x R_x \tag{9.1}$$

where the subscript x is either n or p, depending upon the branch. Approximating the output voltage as being exponential gives the rise and fall time expressions

$$t_r \approx 2.2\tau_p \\ t_f \approx 2.2\tau_n \tag{9.2}$$

While the form is the same as for an AOI network, the rise time will be smaller because the parasitic capacitance C_p is smaller. This is due to the fact that the mirror circuit has only two pFETs contributing the C_p, while an AOI network has four transistors at that node.

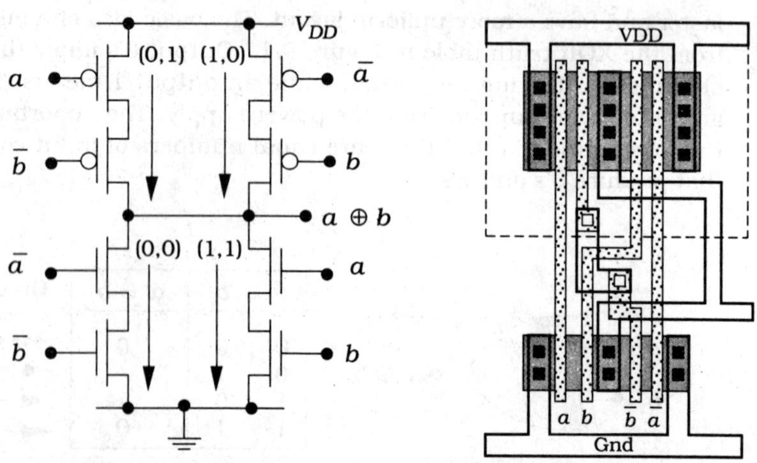

(a) Circuit (b) Layout

Figure 9.2 XOR mirror circuit

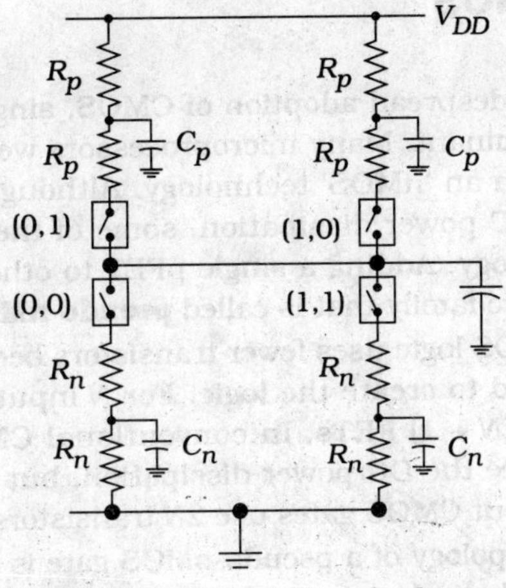

Figure 9.3 Switch model for transient calculations

The idea is easily used to create the XNOR circuit in Figure 9.4. It has the same basic features of the XOR. The relationship

$$\overline{a \oplus b} = \overline{\overline{a} \cdot b + a \cdot \overline{b}} \tag{9.3}$$

shows that only the inputs a and $\overline{a}$ need to be switched. Other mirror circuits will be introduced later in the text in the context of specific applications such as adder circuits.

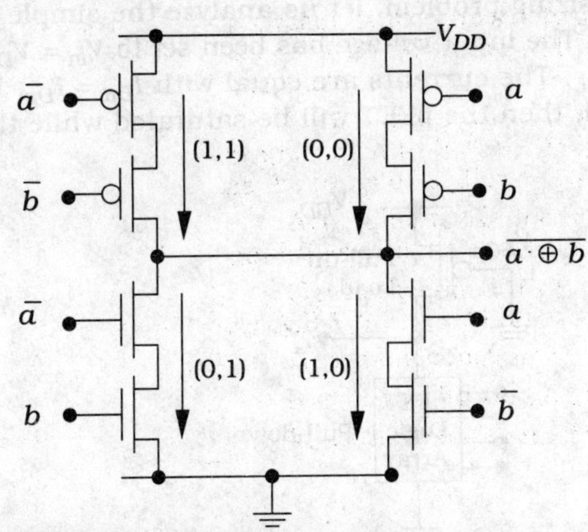

a	b	$\overline{a \oplus b}$
0	0	1
0	1	0
1	0	0
1	1	1

Figure 9.4 Exclusive-NOR (XNOR) mirror circuit

9.2 Pseudo-nMOS

Prior to the widespread adoption of CMOS, single-FET polarity logic circuits were dominant. Many microprocessors were designed using nFET-only circuits in an 'nMOS' technology. Although nMOS was abandoned due to high DC power dissipation, some of the main ideas are used in CMOS technology. Adding a single pFET to otherwise nFET-only circuit produces a logic family that is called **pseudo-nMOS**.

Pseudo-nMOS logic uses fewer transistors because only the nFET logic block is needed to create the logic. For *N* inputs, a pseudo-nMOS logic gate requires $(N + 1)$ FETs. In conventional CMOS, the pFET group is added to reduce the DC power dissipation, but the logic is superfluous. Standard *N*-input CMOS gates use $2N$ transistors.

The basic topology of a pseudo-nMOS gate is drawn in Figure 9.5. The single pFET is biased active since the grounded gate gives $V_{SGp} = V_{DD}$. It acts as a **pull-up** device that tries to pull the output *f* to the power supply voltage V_{DD}. Logic is performed by the nFET array that is designed using the same techniques we have seen. The array acts as a large switch between the output *f* and ground. If the switch is open, the pFET pulls up the output to a voltage $V_{OH} = V_{DD}$. If the nFET switch is closed, then the array acts as a **pull-down** device that tries to pull *f* down to ground. However, since the pFET is always biased on, V_{OL} can never achieve the ideal value of 0 V. It is tempting to use pseudo-nMOS circuits to reduce the FET count and area. However, this logic family is more complicated because the relative sizes of the transistors set the numerical value of V_{OL} and care must be taken to insure that V_{OL} is small enough to be an electronic logic 0 voltage.

To illustrate the sizing problem, let us analyze the simple inverter shown in Figure 9.6. The input voltage has been set to $V_{in} = V_{DD}$ so the output voltage is V_{OL}. The currents are equal with $I_{Dn} = I_{Dp}$. If V_{OL} is assumed to be small, then the pFET will be saturated while the nFET

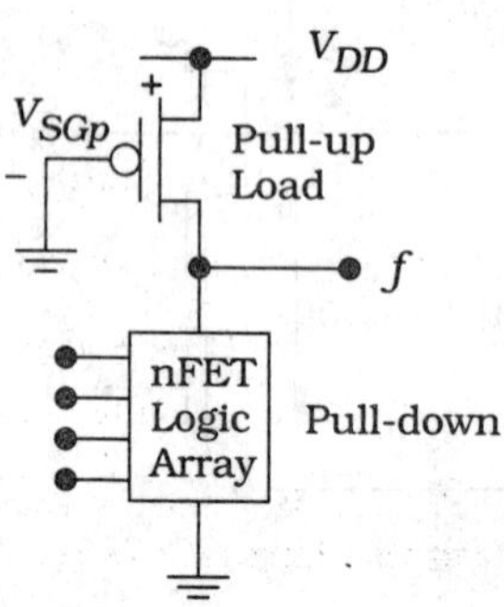

Figure 9.5 General structure of a pseudo-nMOS logic gate

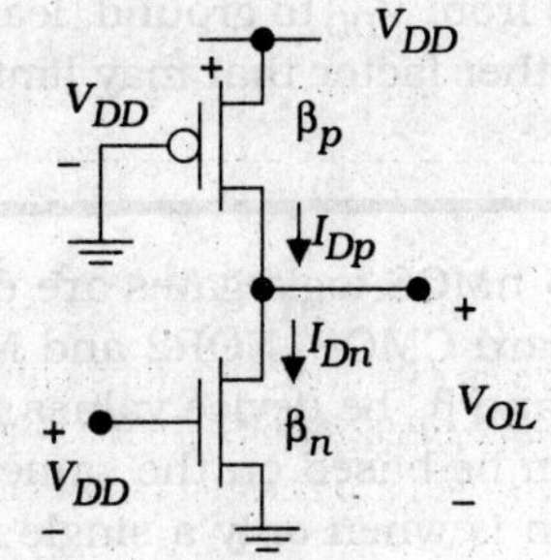

Figure 9.6 Pseudo-nMOS inverter

operates in the non-saturation region. The KCL equation thus assumes the form

$$\frac{\beta_n}{2}[2(V_{DD}-V_{Tn})V_{OL}-V_{OL}^2] = \frac{\beta_p}{2}(V_{DD}-|V_{Tp}|)^2 \tag{9.4}$$

which is a quadratic equation for V_{OL}. Solving gives the physical root

$$V_{OL} = (V_{DD}-V_{Tn}) - \sqrt{(V_{DD}-V_{Tn})^2 - \frac{\beta_p}{\beta_n}(V_{DD}-|V_{Tp}|)^2} \tag{9.5}$$

The value of V_{OL} thus depends on the ratio $(\beta_n/\beta_p) > 1$. Increasing the device ratio decreases the output low voltage. Because of this characteristic, pseudo-nMOS is a type of **ratioed** logic where the relative device sizes set V_{OL} or V_{OH}.

Example 9.1

Consider a CMOS process with $V_{DD} = 5$ V, $V_{Tn} = +0.7$ V, $V_{Tp} = -0.8$ V, $k'_n = 150\ \mu A/V^2$, and $k'_p = 68\ \mu A/V^2$. A pseudo-nMOS inverter sized with $(W/L)_n = 4$ and $(W/L)_p = 6$ gives an inverter with an output-low voltage of

$$V_{OL} = 4.3 - \sqrt{(4.3)^2 - \frac{408}{600}(4.2)^2} = 1.75\ \text{V} \tag{9.6}$$

which is too large since it would not be interpreted as a logic 0 by a circuit of the same type. If we increase the nFET size to $(W/L)_n = 8$ and decrease the pFET to $(W/L)_p = 2$, the calculation gives

$$V_{OL} = 4.3 - \sqrt{(4.3)^2 - \frac{136}{1200}(4.2)^2} = 0.24\ \text{V} \tag{9.7}$$

which is acceptable since it is below the voltage $V_{in} = V_{Tn}$ that turns the nFET on. This illustrates that the choice of aspect ratios is critical to this design style. It is important to note that when $V_{in} = V_{DD}$, a current flow path is established from V_{DD} to ground, leading to a large DC power dissipation. This is another factor that may limit the use of pseudo-nMOS circuits.

General pseudo-nMOS logic gates are designed using the same nFET arrays as in standard CMOS. NOR2 and NAND2 examples are shown in Figure 9.7. Let β_n and β_p be device values for an inverter. The NOR2 gate in Figure 9.7(a) can be based on the same β-values since the worst-case pull-down situation is when only a single nFET is active. This argument can be extended to an *N*-input NOR gate. The NAND2 gate in Figure 9.7(b) is complicated by the series nFETs. To obtain the same pull-down characteristics of the inverter, the logic transistors must be increased to $2\beta_n$ to provide the same total nFET resistance from the output to ground. This is a general problem with pseudo-nMOS logic gates that require series logic FETs.

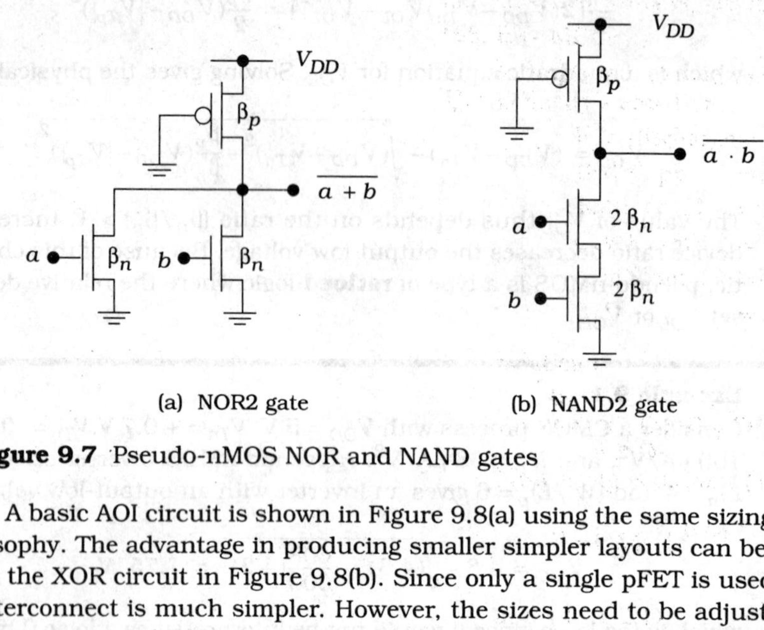

Figure 9.7 Pseudo-nMOS NOR and NAND gates

A basic AOI circuit is shown in Figure 9.8(a) using the same sizing philosophy. The advantage in producing smaller simpler layouts can be seen by the XOR circuit in Figure 9.8(b). Since only a single pFET is used, the interconnect is much simpler. However, the sizes need to be adjusted to insure proper electrical coupling to the next stage. The problems associated with pseudo-nMOS limit its usage to situations where the layout problems are critical, or to some special switching situations where it yields simpler circuitry.

9.3 Tri-State Circuits

A tri-state circuit produces the usual 0 and 1 voltages, but also has a third **high-impedance** *Z* (or Hi-Z) state that is the same as an open cir-

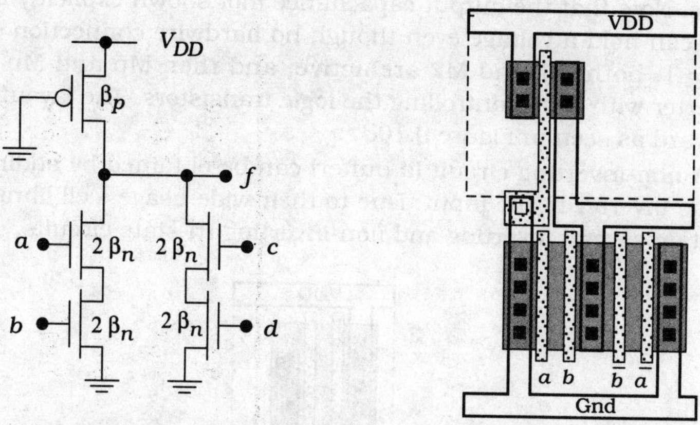

(a) General circuit (b) Layout example

Figure 9.8 AOI gate in pseudo-nMOS logic

cuit. Tri-state circuits are useful for isolating circuits from common bus lines.

The symbol for a tri-state inverter is shown in Figure 9.9(a). The enable signal *En* controls the operation. With *En* = 0, the output is "tri-stated" which means that *f* = *Z*. Normal operation occurs with *En* = 1. A CMOS circuit is shown in Figure 9.9(b). FETs M1 and M2 are the tri-stating devices. The $\overline{En}$ signal is applied to the pFET M1, while *En* controls M2. With *En* = 0, both M1 and M2 are off, and the output is isolated from both the power supply and ground. This is the circuit condition of the Hi-*Z*

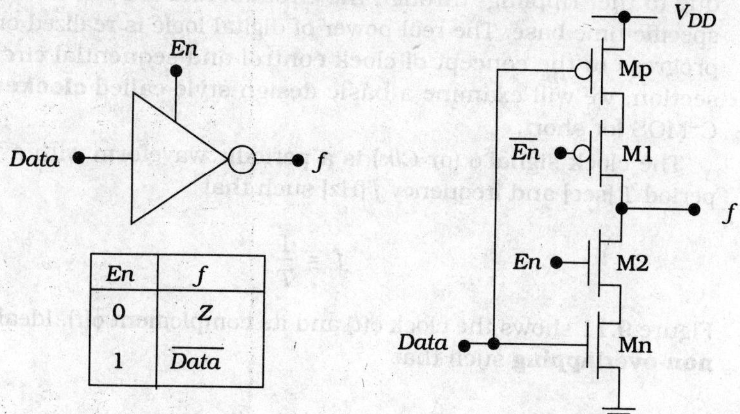

En	*f*
0	*Z*
1	$\overline{Data}$

(a) Symbol and operation (b) CMOS circuit

Figure 9.9 Tri-state inverter

state. Note that the output capacitance (not shown explicitly in the drawing) can hold a voltage even though no hardwire connection exists. With $En = 1$, both M1 and M2 are active, and then Mp and Mn act like an inverter with *Data* controlling the logic transistors. The layout is straightforward as seen in Figure 9.10.

A non-inverting circuit (a buffer) can be obtained by adding a regular static inverter to the input. Due to their wide usage, cell libraries usually contain several inverting and non-inverting tri-state circuits.

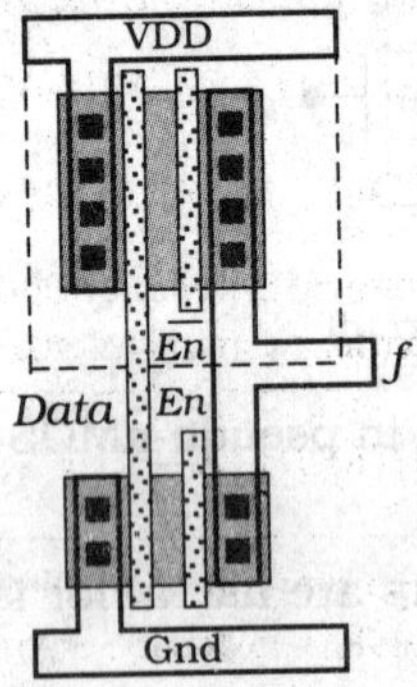

Figure 9.10 Tri-state layout

9.4 Clocked CMOS

Up to this point, all of the circuits we have examined have been completely **static** in nature. The output of a static logic gate is valid so long as the input values are valid and the circuit has stabilized. Logic delays are due to the "rippling" through the circuits, and are not referenced to any specific time base. The real power of digital logic is realized only when we progress to the concept of clock control and sequential circuits. In this section, we will examine a basic design style called **clocked CMOS**, or C^2MOS for short.

The clock signal ϕ (or *Clk*) is a periodic waveform with a well-defined period T [sec] and frequency f [Hz] such that

$$f = \frac{1}{T} \tag{9.8}$$

Figure 9.11 shows the clock $\phi(t)$ and its complement $\bar{\phi}(t)$. Ideally, these are **non-overlapping** such that

$$\phi(t) \cdot \bar{\phi}(t) = 0 \tag{9.9}$$

for all times t. However, if $\phi(t)$ is defined to have a minimum value of 0 V and a maximum of V_{DD}, then

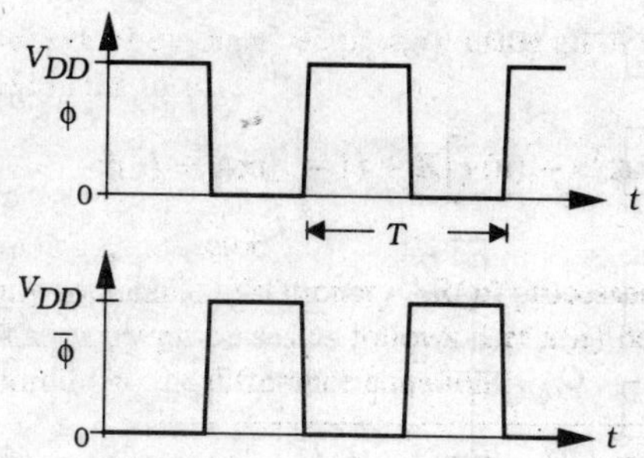

Figure 9.11 Clocking signals

$$\bar{\phi}(t) = V_{DD} - \phi(t) \tag{9.10}$$

so that the clocks overlap slightly during a transition. It may be advantageous to create a set of clocks that are truly non-overlapping for all times.

The general structure of a C^2MOS gate is shown in Figure 9.12. It is composed of a static logic circuit with tri-state output network (made up of FETs M1 and M2) that is controlled by ϕ and $\bar{\phi}$. The operation of the circuit can be understood using the clocking waveform shown. When $\phi = 1$, both M1 and M2 are active. Since both the pFET and nFET logic blocks are connected to the output node, the circuit degenerates to a standard static logic gate. The output $f(a, b, c)$ is valid during this time, establishing the voltage V_{out} on the output capacitance C_{out}. When the clock changes to a value of $\phi = 0$, both M1 and M2 are in cutoff, so the output is in a high-impedance state Hi-Z. During this time interval, the FET logic arrays are not connected to the output, so the inputs have no effect. Instead, the output voltage is held on C_{out} until the clock returns to a value of $\phi = 1$.

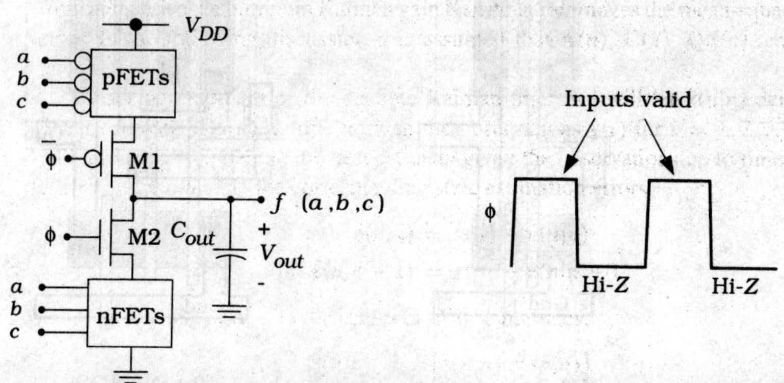

Figure 9.12 Structure of a C^2MOS gate

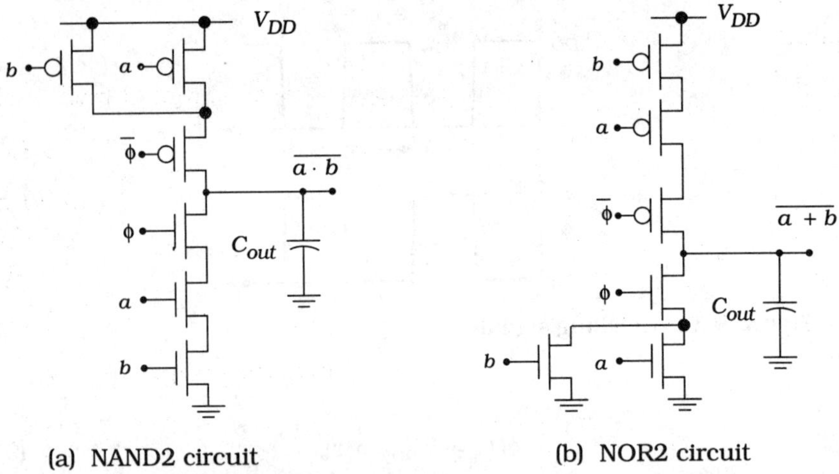

Figure 9.13 Example of clocked-CMOS logic gates

The transistor arrays are designed using the same techniques as for standard logic gates. The circuits for a NAND2 and a NOR2 are shown in Figure 9.13, subdrawings (a) and (b), respectively. Layout is similar to the tri-state circuit with the clock replacing the enable signal. The layouts in Figure 9.14 provide one approach to placing and connecting the transistors. Note that the presence of the series-connected clocking FETs automatically lengthens both the rise and fall times of the circuit.

Clocked CMOS is useful because we can synchronize the data flow

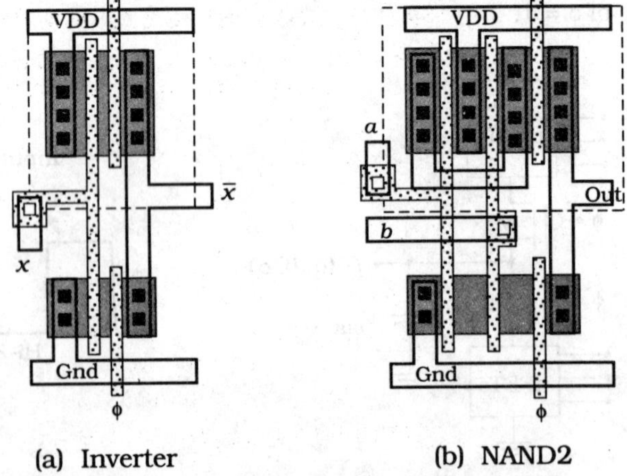

Figure 9.14 Layout examples of C^2MOS circuits

through a logic cascade by controlling the internal operation of the gate. Every cycle of ϕ allows a new group of data bits to enter the network. One drawback is that the output node cannot hold the charge on V_{out} very long due to a phenomenon called **charge leakage**. This places a lower limit on the allowable clock frequency.

The basics of charge leakage are shown in Figure 9.15(a). Even though the transistors are in cutoff, it is not possible to block all current flow using a FET. If a voltage is applied to the drain or source, a small **leakage current** flows into, or out of, the device. There are many contributions to the leakage current. One is due to the required bulk connections that are shown in the drawing. The pFET bulk is the nWell region, which is connected to the power supply V_{DD}. Since the pFET source is a p+ region, this creates a pn junction (a diode) that admits a small leakage current i_p flowing on to the node. The nFET has the same problem, with i_n flowing from the output to the p-substrate. Denoting the current off of the capacitor by i_{out}, we may sum the contributions to obtain

$$\begin{aligned} i_{out} &= i_n - i_p \\ &= -C_{out}\frac{dV}{dt} \end{aligned} \tag{9.11}$$

where we have used the capacitor I-V relation in the second line; note the presence of a minus sign to indicate that i_{out} flows out of the positive terminal.

To see the effects of the leakage currents, suppose that we have an initial voltage $V(t = 0) = V_1$ stored on the capacitor. If $i_n > i_p$, then $i_{out} = I_L$ is a positive number, indicating current flow off of the capacitor. Rewriting the equation as

$$I_L = -C_{out}\frac{dV}{dt} \tag{9.12}$$

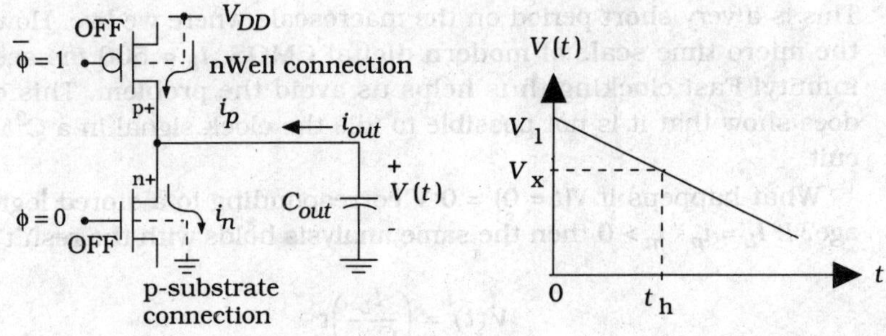

(a) Bulk leakage currents (b) Logic 1 voltage decay

Figure 9.15 Charge leakage problem

we may rearrange it to read

$$\int_{V_1}^{V(t)} dV = -\int_0^t \left(\frac{I_L}{C_{out}}\right) dt \tag{9.13}$$

Assuming that I_L is a constant, the equation may be integrated to yield

$$V(t) = V_1 - \left(\frac{I_L}{C_{out}}\right) t \tag{9.14}$$

which is a linear decay of the voltage with time. This is plotted in Figure 9.15(b). As the voltage decreases, it eventually reaches a minimum logic 1 value that is shown as V_x in the plot. If V falls below this value, it will incorrectly be interpreted as a logic 0 voltage. The **hold time** t_h corresponds to the maximum time that the logic 1 voltage can be stored. By definition, this occurs when

$$V(t_h) = V_1 - \left(\frac{I_L}{C_{out}}\right) t_h = V_x \tag{9.15}$$

Rearranging,

$$t_h = \left(\frac{C_{out}}{I_L}\right)(V_1 - V_x) \tag{9.16}$$

gives the hold time for this case. An order of magnitude estimate of the hold time can be obtained by estimating the capacitance as 50 fF, the leakage current as 0.1 pA, and the voltage change as 1 V. These values give

$$t_h = \left(\frac{50 \times 10^{-15}}{10^{-13}}\right)(1) = 0.5 \text{ sec} \tag{9.17}$$

This is a very short period on the macroscale where we live. However, on the micro time scale of modern digital CMOS, t_h = 500 ms seems like infinity! Fast clocking thus helps us avoid the problem. This estimate does show that it is not possible to idle the clock signal in a C^2MOS circuit.

What happens if $V(t = 0) = 0$ V corresponding to a stored logic 0 voltage? If $I_L = i_p - i_n > 0$ then the same analysis holds with the result that

$$V(t) = \left(\frac{I_L}{C_{out}}\right) t \tag{9.18}$$

i.e., the charging current I_C increases the voltage in time. This means that the logic 0 voltage may drift, so that we again require a minimum clock frequency.

In submicron devices, the charge leakage problem is exacerbated by the existence of another FET leakage current called the **subthreshold current** I_{sub}. This is a drain-source current that flows even though the gate voltage is less than V_T. A simple estimate for the subthreshold current is

$$I = I_{D0}\left(\frac{W}{L}\right)e^{-(V_{GS}-V_T)/(nV_{th})} \tag{9.19}$$

where I_{D0} varies with V_{DS}, V_{th} is the thermal voltage $(kT/q) \approx 26$ mV at 300 K, and n is a parameter that varies with capacitance. A conservative value of I_{D0} is around 10^{-9} A, which noticeably reduces the hold time. With the previous values of capacitance and voltage and $V_{GS} = 0$, the hold time estimate is

$$t_h = \left(\frac{50\times10^{-15}}{10^{-9}}\right)(1) = 50\ \mu\text{s} \tag{9.20}$$

for leakage through a unity aspect ratio FET. In addition, other contributions to the leakage current originate from the physical structure and the materials used to create the silicon circuit. It would not be unreasonable to find a total charge leakage current of $I_L = 0.1\ \mu\text{A} = 10^{-7}$ A in a submicron device. With this level of leakage, the hold time is reduced to

$$t_h = \left(\frac{50\times10^{-15}}{10^{-7}}\right)(1) = 0.5\ \mu\text{sec} \tag{9.21}$$

This clearly indicates that charge storage on a capacitive node is a limited-time event, and places important constraints on our logic circuits.

Although we have been approximating the leakage currents as having a constant value for simplicity, a deeper analysis shows that they are voltage-dependent functions. The general differential equation is of the form

$$I_L(V) = -C_{out}(V)\,\frac{dV}{dt} \tag{9.22}$$

where we have noted that the output capacitance C_{out} also depends on voltage. If we know the explicit functions for $I_L(V)$ and $C_{out}(V)$, then

$$\int_0^t dt = \int_{V_x}^{V(t)} \frac{C_{out}(V)}{I_L(V)}\,dV = t \tag{9.23}$$

can be integrated to give $V(t)$. A more practical approach is to use a numerical solution. The dependence of the quantities on V results in a non-linear decay, such as the example illustrated in Figure 9.16. The hold time is still defined in the same manner. At the circuit design level, charge

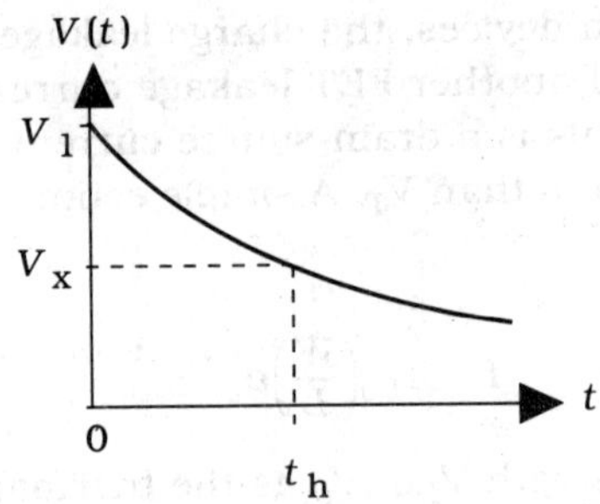

Figure 9.16 General voltage decay

leakage information is usually obtained from circuit simulations.

Charge leakage occurs whenever we attempt to hold charge on a node capacitance using a MOSFET in cutoff. Many of the advanced circuits in the remainder of this chapter have this characteristic, and it is important to remember to check for the problem. Simple SPICE models of MOSFETs do not accurately account for leakage currents. The best results to date are obtained using the BSIM equations.

Motivation for Future Research

While charge leakage is an important problem in dynamic circuits, this discussion highlights the problem of achieving an "open switch" using a MOSFET. As the dimensions shrink, the drain-to-source leakage current increases and the device looks less and less like the idealized switch that was used to design CMOS logic networks. This is one of the most critical problems in digital submicron VLSI. Device researchers are continually looking at the problem. In terms of silicon technology, two main approaches are prevalent. One technique is to reduce the leakage by refining the fabrication process using different materials and variations in the FET structures. Over the years, this has resulted in better devices that have "manageable" leakage current levels that circuit designers must work around.

The other approach is to develop new types of transistors to replace the standard MOSFET. Novel devices with improved characteristics have been proposed and built, and many promising structures have appeared in the literature. However, device research tends to be initially concerned with creating a single transistor, not a high-density VLSI chip. Manufacturing problems often limit the usage of the device in these applications. Another problem is that circuit and logic designers must learn the characteristics of a device before they can develop digital design methodologies. A technique that works with standard MOSFETs probably won't be the best choice for circuits based on transistors that have different I-V characteristics, if it works at all.

Shrinking the size of a MOSFET is often taken as natural evolution of the processing technology. The development of submicron sized FETs had a marked effect on circuit design techniques. Introducing new switching devices would affect all levels of the VLSI design hierarchy, and much research would have to be completed before high-density designs could be implemented. VLSI designers must be continually aware of changes in the field.

9.5 Dynamic CMOS Logic Circuits

A **dynamic logic gate** uses clocking and charge storage properties of MOSFETs to implement logic operations. The clock provides a synchronized data flow which makes the technique useful in designing sequential networks. The characterizing feature of a dynamic logic gate is that the result of a calculation is valid only for a short period of time. While this makes the circuits more difficult to design and use, they require fewer transistors and may be faster than static cascades.

Dynamic circuits are based on the circuit illustrated in Figure 9.17. The clock ϕ drives a complementary pair of transistors Mn and Mp; these control the operation of the circuit and provide synchronization. Logic is implemented using an nFET array between the output node and ground. The output voltage V_{out} is taken across the output capacitor C_{out}.

The clocking signal ϕ defines two distinct modes of operation during every cycle. When $\phi = 0$ the circuit is in **precharge** with Mp on and Mn off. This establishes a conducting path between V_{DD} and the output, allowing C_{out} to charge to a voltage of $V_{out} = V_{DD}$. Mp is often called the precharge FET. Since the bottom of the nFET logic block is not connected to ground during precharge, the inputs have no effect.

A clock transition to $\phi = 1$ drives the circuit into the **evaluation** mode where Mp is off and Mn is on. The inputs are valid and control the switching in the nFET logic array; Mn is usually called the evaluate transistor. If

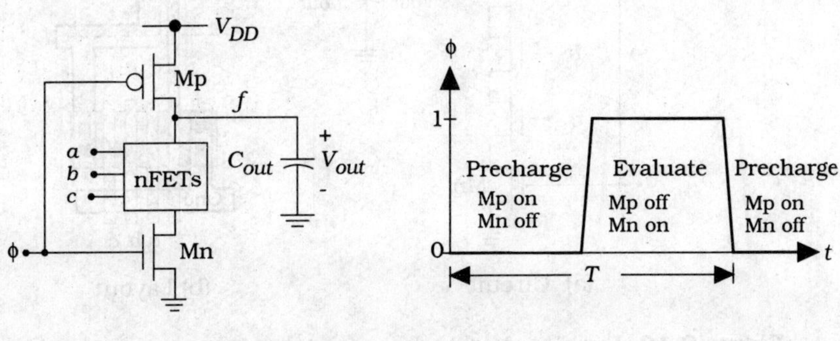

Figure 9.17 Basic dynamic logic gate

the logic block acts like a closed switch, then C_{out} can discharge through the logic array and Mn; this gives the final result of $V_{out} = 0$ V, corresponding to a logic $f = 0$. If the inputs cause the block to behave like an open switch from top to bottom, the charge on C_{out} is held and $V_{out} = V_{DD}$; logically, this is an output of $f = 1$. Charge leakage eventually drops the output to $V_{out} \rightarrow 0$ V, which would be an incorrect logic value. The hold time is determined by the circuitry. In general, this consideration places a minimum frequency stipulation on the clock.

A dynamic NAND3 circuit is shown in Figure 9.18(a). Logic formation is achieved using the three series-connected FETs. The output

$$f = \overline{a \cdot b \cdot c} \tag{9.24}$$

is valid only during the evaluation period when $\phi = 1$. Layout is straightforward as shown by the example in Figure 9.18(b). Since the evaluation nFET Mn is in series with the logic block, C_{out} must discharge through four transistors. Increasing the sizes of the nFETs will reduce the fall time.

As mentioned above, charge leakage reduces the voltages held on the output node when $f = 1$. A detailed analysis of the circuit shows that another problem called **charge sharing** can occur when the clock makes the transition to $\phi \rightarrow 1$. It has the effect of reducing the output voltage even before charge leakage effects become noticeable.

The origin of the charge sharing problem is the parasitic node capacitance C_1 and C_2 between FETs as shown in Figure 9.19. The clock has been set at $\phi = 1$ so that Mp is off, isolating the output node from the power supply. The initial voltage on C_{out} at the start of the evaluation

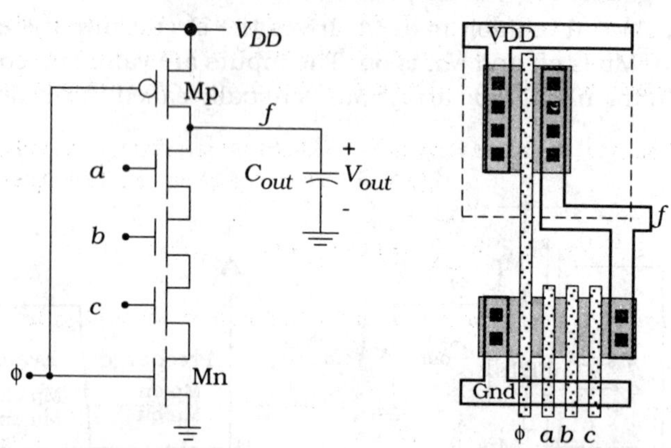

(a) Circuit (b) Layout

Figure 9.18 Dynamic logic gate example

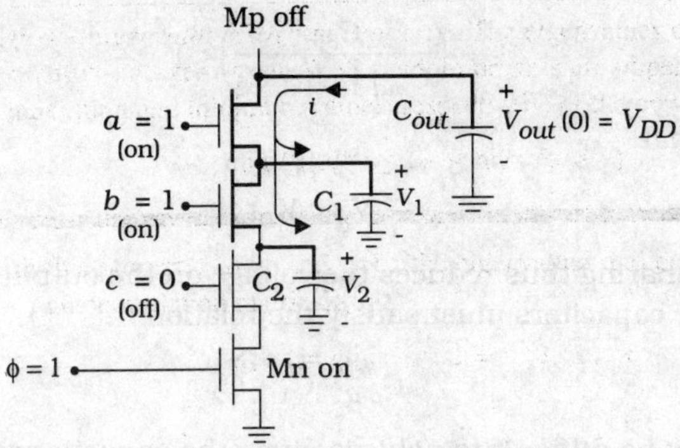

Figure 9.19 Charge sharing circuit

interval is $V_{out} = V_{DD}$ as shown. Assuming that the capacitor voltages V_1 and V_2 are both 0 V at this time, the total charge on the circuit is

$$Q = C_{out} V_{DD} \tag{9.25}$$

The worst-case charge sharing condition for this circuit is when the inputs are at $(a, b, c) = (1, 1, 0)$. With $c = 0$, there is no discharge path to ground, so that the output voltage should remain high. However, since the a- and b-input FETs are on, C_{out} is electrically connected to C_1 and C_2 as indicated by the darkened lines. The current i flows because V_{out} is initially larger than V_1 or V_2. This corresponds to the transfer of charge from C_{out} to both C_1 and C_2. Using the relationship $Q = CV$ shows that V_{out} decreases while V_1 and V_2 increase. The current flow ceases when the voltages are equal with a final value

$$V_{out} = V_2 = V_1 = V_f \tag{9.26}$$

The total charge on the circuit is then distributed according to

$$\begin{aligned} Q &= C_{out} V_f + C_1 V_f + C_2 V_f \\ &= (C_{out} + C_1 + C_2) V_f \end{aligned} \tag{9.27}$$

Applying the principle of conservation of charge, this must be equal to the initial charge in the system:

$$Q = (C_{out} + C_1 + C_2) V_f = C_{out} V_{DD} \tag{9.28}$$

Solving for the final voltage gives

$$V_f = \left(\frac{C_{out}}{C_{out} + C_1 + C_2} \right) V_{DD} \tag{9.29}$$

Since

$$\left(\frac{C_{out}}{C_{out} + C_1 + C_2}\right) < 1 \tag{9.30}$$

we see that

$$V_f < V_{DD} \tag{9.31}$$

Charge sharing thus reduces the voltage on the output node. To keep V_{out} high, the capacitors must satisfy the relation

$$C_{out} \gg C_1 + C_2 \tag{9.32}$$

This may be difficult to achieve since the capacitance values are determined by the layout dimensions. After charge sharing takes place, the node is still subject to charge leakage, which continues to drop the voltage with time.

9.5.1 Domino Logic

Domino logic is a CMOS logic style obtained by adding a static inverter to the output of the basic dynamic gate circuit. The resulting structure is shown in Figure 9.20. The precharge and evaluate events still occur, but now it is the capacitor C_X between the dynamic stage and the inverter that is affected. A clock value of $\phi = 0$ defines the precharge. During this time, C_X is charged to a voltage $V_X = V_{DD}$ which forces the output voltage to $V_{out} = 0$ V. Inputs are valid during the evaluation interval when $\phi = 1$. If C_X holds its charge, V_X remains high and $V_{out} = 0$ V indicates a logic 0 output. If C_X discharges, then $V_X \rightarrow 0$ V and $V_{out} \rightarrow V_{DD}$. This corresponds to a logic 1 output.

Domino logic gates are **non-inverting** because of the output inverter. Two examples of this characteristic are shown in Figure 9.21. The AND gate in Figure 9.21(a) is easily understood: if $a = b = 1$, then the internal node discharges to 0 V, forcing the output to a logic 1 (V_{DD}). Similarly, the OR gate in Figure (9.21b) gives a 1 output if either $a = 1$ or $b = 1$. This

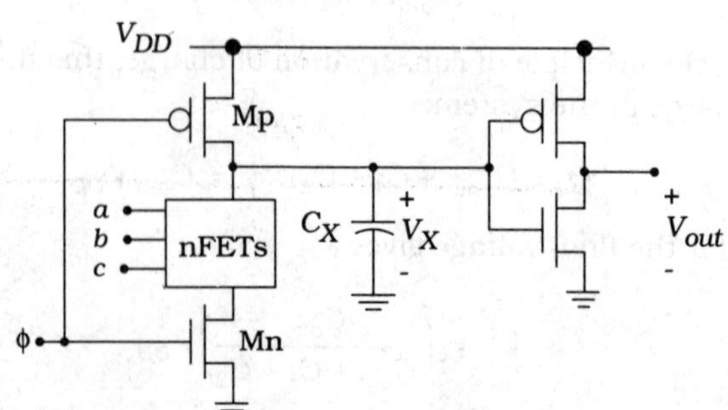

Figure 9.20 Domino logic stage

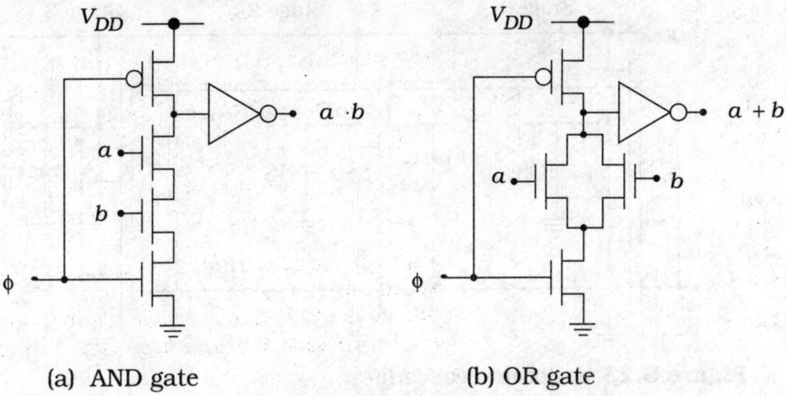

Figure 9.21 Non-inverting domino logic gates

makes logic design using only domino gates somewhat tricky since the NOT operation is required for a complete set of logic operations.[1] While one can add inverters, it is found that this causes the possibility of introducing a hardware glitch into the circuit, and is usually avoided. Inverters are used only at the beginning or the end of a domino chain. An example of a domino layout is shown in Figure 9.22 for an AND3 gate. This is just a dynamic NAND3 circuit cascaded into a static inverter, so the layout preserves the features of general dynamic logic.

Domino logic derives its name from the manner in which a cascade operates. A 3-stage network is shown in Figure 9.23. Every stage is con-

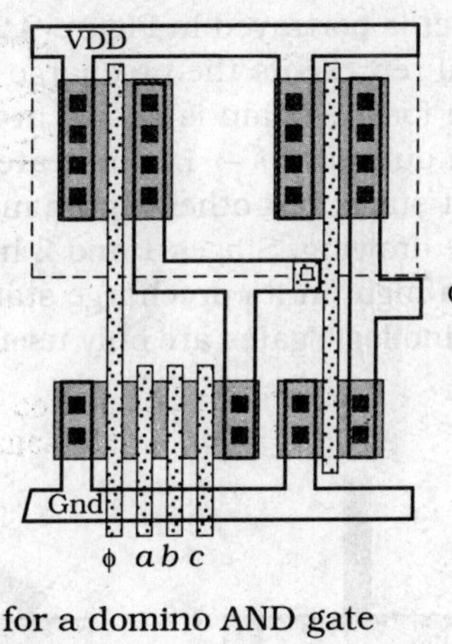

Figure 9.22 Layout for a domino AND gate

[1] A complete set of logic operations is one that is capable of producing any logic combination. Without the NOT operator, functions such as the XOR and XNOR are not possible.

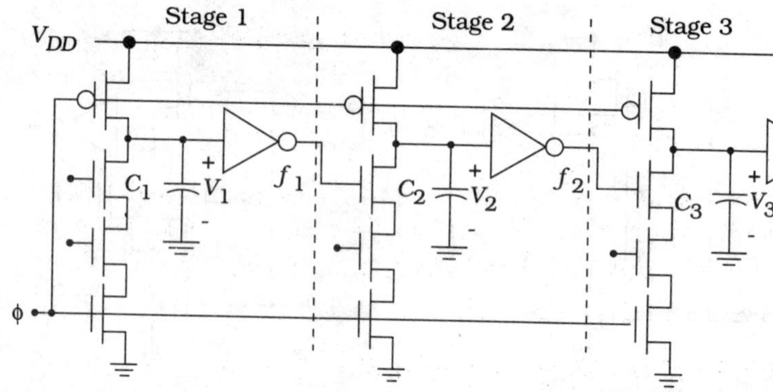

Figure 9.23 A domino cascade

trolled by the same clock phase ϕ. During a precharge event with $\phi = 0$, capacitors C_1, C_2, and C_3 are simultaneously charged to V_{DD}. This causes the outputs f_1, f_2, and f_3 to all be 0. When $\phi = 1$, the entire chain undergoes evaluation. In a domino cascade, this is like a "domino chain reaction" that must start at the first stage and then propagate stage by stage to the output. To understand this comment, suppose that we monitor the second stage output f_2 and see it switch from its precharge value $f_2 = 0$ to $f_2 = 1$ during the evaluation interval. The only way this could have happened is if C_2 discharged, but this requires that $f_1 = 1$ to turn on the nFET in the discharge chain. Applying the same logic to the first stage, f_1 can switch to 1 only if C_1 has discharged. Extending this argument, we see that $f_3 \rightarrow 1$ occurs only if both Stage 1 and Stage 2 have made the same transition.

The domino effect is portrayed in Figure 9.24 to help visualize the process. Figure 9.24(a) represents the precharge event by dominos standing on end. Evaluation for the chain is shown in Figure 9.24(b). A discharge event that gives an output of $f \rightarrow 1$ is indicated by a falling domino. This can topple the next stage, but other inputs may keep the discharge from taking place. In the drawing, Stages 1 and 2 have undergone a discharge, but Stage 3 remain high (in its precharge state). Note that the operation indicates that domino logic gates are only useful in cascades.

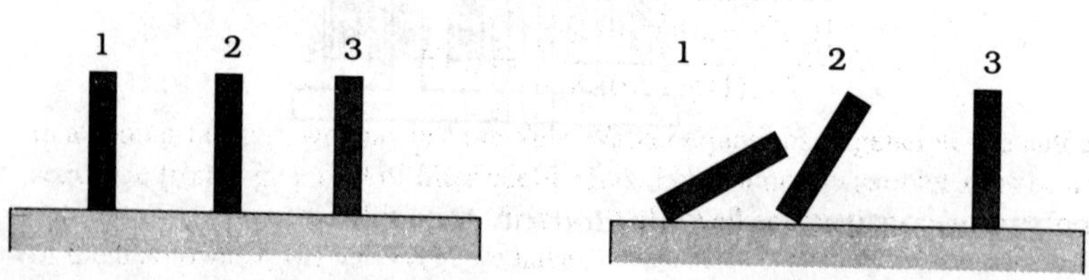

(a) Precharge (b) Evaluate

Figure 9.24 Visualization of the domino effect

The domino cascade must have an evaluation interval that is long enough to allow every stage time to discharge. This means that charge sharing and charge leakage processes that reduce the interval voltage V_X may be limiting factors. **Charge-keeper** circuits have been developed to combat this problem. Two are shown in Figure 9.25. In Figure 9.25(a), a pFET MK is biased active to allow a small current to replenish charge on C_X. The aspect ratio of the charge-keeper FET must be small so that it does not interfere with a discharge event in any significant manner; this is called a 'weak' device. Another approach is shown in Figure 9.25(b). An inverter controls the gate of the weak pFET. If an internal discharge of C_X does occur, then the output voltage V_{out} increases. Feeding this through the inverter shuts the pFET off and allows the discharge to continue.

An interesting extension of the basic domino circuit is that of **Multiple-Output Domino Logic** (**MODL**). This type of circuit allows two or more outputs from a single logic gate, making it quite unique. The structure of a 2-output MODL stage is shown in Figure 9.26. The logic array has been split into two separate blocks denoted as F and G, which creates an additional output node. Adding an inverter and a precharge transistor results in the two outputs

$$
\begin{aligned}
f_1 &= G \\
f_2 &= F \cdot G
\end{aligned}
\tag{9.33}
$$

This is easily understood by studying the logic network. If the G-logic block acts like a closed switch, then it produces an output of $f_1 = G$. If this occurs, then it is possible for the second logic block F to induce a discharge by also acting as a closed switch. This dependence produces the ANDing relation between the two outputs. While this is quite restrictive, the nesting of the AND operation does appear in several important computational algorithms such as the carry look-ahead adder.

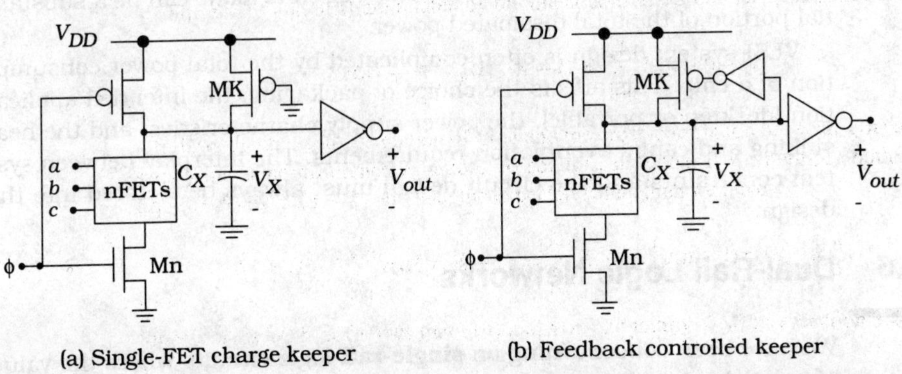

Figure 9.25 Charge-keeper circuits

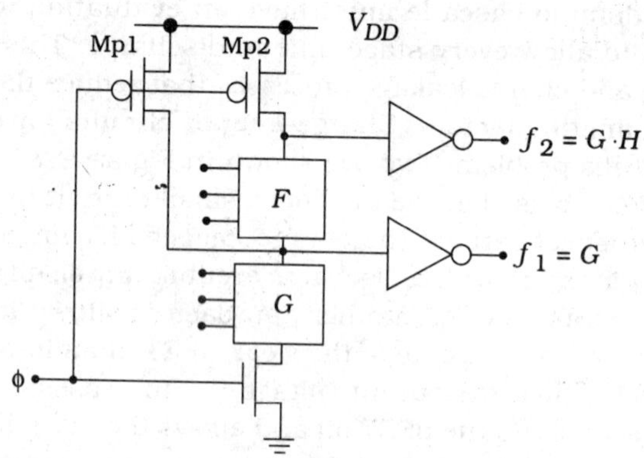

Figure 9.26 Structure of a MODL circuit

9.5.2 Power Dissipation of Dynamic Logic Circuits

CMOS dynamic logic circuits can be designed to provide very fast switching with modest real estate consumption. They have been successfully used in several well-known chips and are the basis of DRAMs and other important computer components. Unfortunately, they can be quite power hungry which may limit their usage.

In a dynamic circuit, the clock ϕ defines the precharge and evaluate operations in every cycle. Since charge cannot be held on a capacitive node, every precharge cycle will pull current from the voltage source, adding to the overall power dissipation of the circuit. The clock circuits themselves require dynamic power to drive the FETs. In the standard configuration, every stage presents a capacitance of

$$C_L = C_{Gp} + C_{Gn} \tag{9.34}$$

to the clock drivers corresponding to the precharge and evaluate transistors. The power consumption of the clock circuits alone can be a substantial portion of the total dissipated power.

VLSI system design is often complicated by the total power consumption of a chip. This affects the choice of packaging, the intended application (desktop or portable), the power supply characteristics, and the heat sinking and cabinet ventilation requirements. The interplay between system constraints and the circuit design must always be factored into the design.

9.6 Dual-Rail Logic Networks

We have been concentrating on **single-rail** logic circuits where the value of a variable is either a 0 or a 1 only. In **dual-rail** networks, both the variable x and its complement $\overline{x}$ are used to form the difference

$$f_x = (x - \bar{x}) \tag{9.35}$$

Using the quantity f_x provides an increase in the switching speed. This can be seen by calculating the time derivative as

$$\frac{df_x}{dt} = \left(\frac{dx}{dt} - \frac{d\bar{x}}{dt}\right) \tag{9.36}$$

and noting that

$$\frac{d\bar{x}}{dt} \approx -\left|\frac{dx}{dt}\right| \tag{9.37}$$

since x increases while $\bar{x}$ decreases, and vice versa. Thus

$$\frac{df_x}{dt} \approx 2\left|\frac{dx}{dt}\right| \tag{9.38}$$

so that the rate of change of f_x is approximately twice that of a single variable. Translated into logic terms, this means that the switching speed is almost twice as fast as can be obtained in a single-rail circuit.

The complicating factor in dual-rail circuits is the increase in circuit complexity and wiring overhead. Every input and output is now a doublet consisting of the variable and its complement. The circuits are correspondingly more complicated, and can be tricky to deal with. However, the speed advantage makes them worth studying. Some even provide structured and compact layout schemes.

9.6.1 CVSL

Most dual-rail CMOS circuits are loosely based around **differential cascode voltage switch logic**, which goes under the acronyms **DCVS logic** or **differential CVSL**; we will adopt the latter one here. CVSL provides for dual-rail logic gates that have latching characteristics built into the circuit itself. The output results f and $\bar{f}$ are held until the inputs induce a change.

The basic structure of a CVSL logic gate is shown in Figure 9.27. The input set consists of the variables (a, b, c) and their complements ($\bar{a}$, $\bar{b}$, $\bar{c}$) that are routed into an nFET 'logic tree' network. The logic tree is modeled as a pair of complementary switches Sw1 and Sw2 such that one is closed while the other is open as determined by the inputs. The state of the switches establishes the outputs. For example, if Sw1 is closed then $f = 0$. The opposite side ($\bar{f}$) is forced to the complementary state ($\bar{f} = 1$) by the action of the pFET latch.

The latch is controlled by the left and right source-gate voltages V_l and V_r shown in the drawing. Suppose that Sw2 is closed, forcing $\bar{f} = 0$ on the right side. In this case,

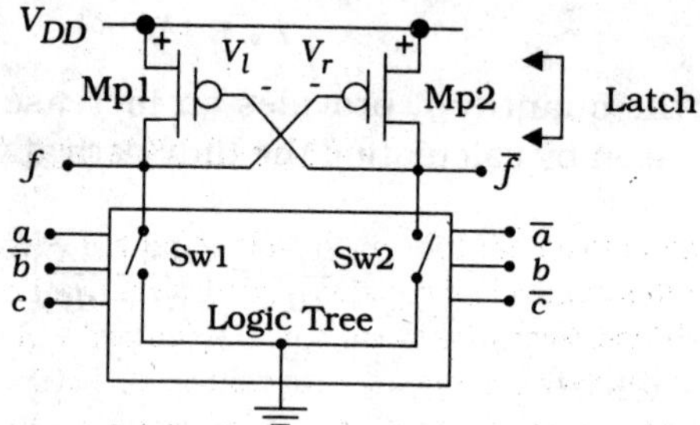

Figure 9.27 Structure of a CVSL logic gate

$$V_l = V_{DD} \tag{9.39}$$

which turns on Mp1. With Mp1 conducting, the left output node sees a path to the power supply, giving V_{DD} there; this is the $f = 1$ state. The ability to set the latch using a pull down on one side helps make the stage react quickly.

Several techniques have been published for designing the logic network. A straightforward approach is to use separate circuits for the left and right sides. Figure 9.28(a) is an AND/NAND circuit that has inputs of (a, b) on the right and $(\bar{a}, \bar{b})$ on the left; it is important to remember that dual-rail logic gates require pairs of complementary inputs and outputs. The formation of the NAND operation on the right side uses series-nFETs which is identical to nFET logic in standard CMOS. To obtain the left circuit, we simply use the DeMorgan identity

$$\overline{a \cdot b} = \bar{a} + \bar{b} \tag{9.40}$$

which, from our study of bubble pushing, indicates parallel nFETs with complemented inputs. An OR/NOR circuit is drawn in Figure 9.28(b). The logic formation follows the same approach as for the AND/NAND circuit.

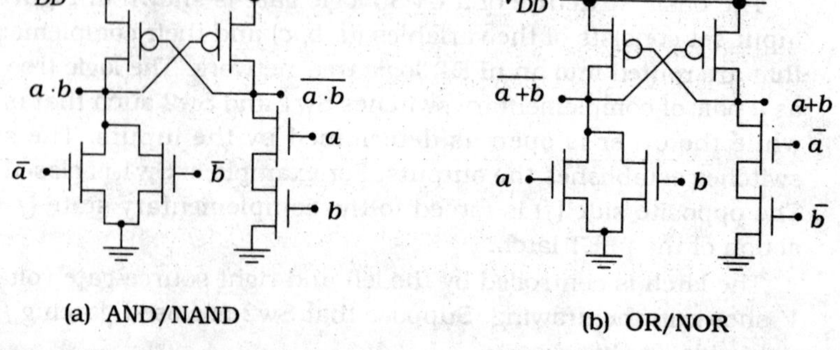

Figure 9.28 CVSL gate examples

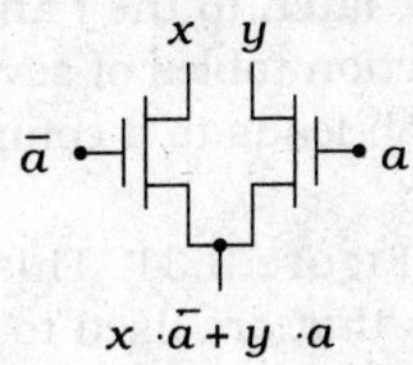

(a) Simple nFET pair

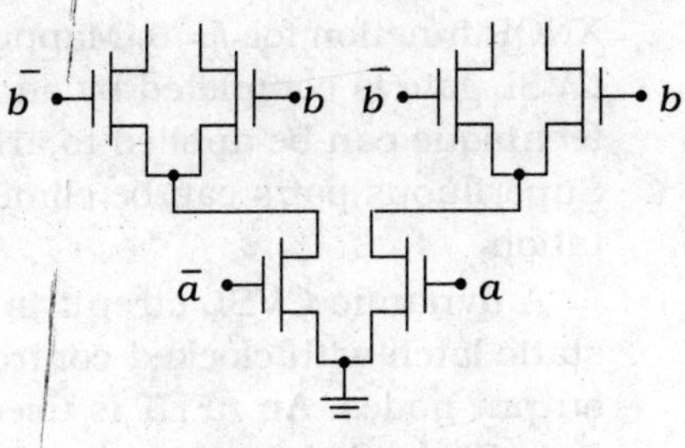

(b) Stacked pairs

Figure 9.29 nFET logic pairs

A more important observation is that the OR/NOR and AND/NAND gates are identical in form; only the locations of the inputs are different. This symmetry is due to the fact that OR and AND are logical duals.

Logic trees provide a more structured approach to designing the switching network. These are based on pairs of nFETs that are driven by complementary inputs as shown in Figure 9.29(a). With x and y applied to the top of the pair, the pair acts like a 2:1 MUX with a (bottom) output of

$$x \cdot \bar{a} + y \cdot a \tag{9.41}$$

Qualitatively, this says that x is transmitted if $a = 0$, while the output is y if $a = 1$. The pair $(\bar{a}, a)$ thus corresponds to an input pattern of (0, 1) which is the same way that input combinations are listed in a function table. If $x = y$, then the output is always x and the FETs can be eliminated. A 2-level stack of nFET pairs is shown in Figure 9.29(b). The b-input pairs on the upper row correspond to the input sequence (01) (01), while the bottom pair (a-inputs) has the sequence (01). This provides a one-to-one mapping from a 2-input function table to the nFET arrays.

An example is the gate in Figure 9.30. The output f of the truth table has the sequence (1001) indicating the XOR function for $f = 1$, and the

f	1 0 0 1
b	0 1 0 1
a	0 0 1 1

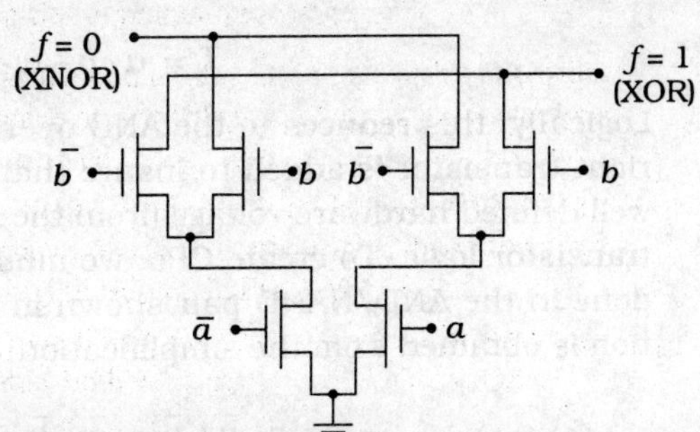

Figure 9.30 Example of a logic tree using nFET pairs

XNOR function for $f = 0$. Mapping the table gives the logic tree shown. The CVSL gate is completed by adding a pFET latch to the f and $\bar{f}$ lines. This technique can be applied to arbitrary function tables of several variables. Superfluous pairs can be eliminated, which leads to a compact representation.

A dynamic CVSL circuit is shown in Figure 9.31. This replaces the static latch with clocked-controlled pFETs that are used to precharge the output nodes. An nFET is used at the bottom of the tree for the evaluation. Simplified notation has been used in the schematic. Each '– +' box corresponds to an nFET pair with the variable applied to the '+' side, and the complement to the '-' side. Two reductions have been made translating the function table to the logic tree. This is because the left entries for f have the sequence 00 11, which allows both c-level pairs to be eliminated.

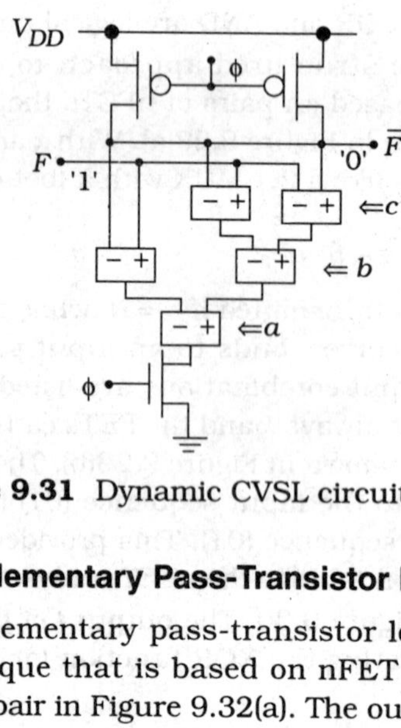

F	0	0	1	1	0	1	1	0
c	0	1	0	1	0	1	0	1
b	0	0	1	1	0	0	1	1
a	0	0	0	0	1	1	1	1

Figure 9.31 Dynamic CVSL circuit with 3-level logic tree

9.6.2 Complementary Pass-Transistor Logic

Complementary pass-transistor logic (**CPL**) is an interesting dual-rail technique that is based on nFET logic equations. Let us examine the nFET pair in Figure 9.32(a). The output is given by

$$f = a \cdot b + \bar{a} \cdot a \tag{9.42}$$

Logically, this reduces to the AND operation $f = a \cdot b$ since $\bar{a} \cdot a = 0$. The right transistor is added to insure that the output $f = 0$ when $a = 0$ is a well defined hardware voltage (from the input a). This is the basis of pass-transistor logic. To create CPL, we must add the NAND function. This is done in the AND/NAND pair shown in Figure 9.32(b). The NAND operation is obtained from the simplification

$$a \cdot \bar{b} + \bar{a} = \bar{a} + \bar{b} = \overline{a \cdot b} \tag{9.43}$$

Since nFETs suffer from threshold losses, static output inverters have

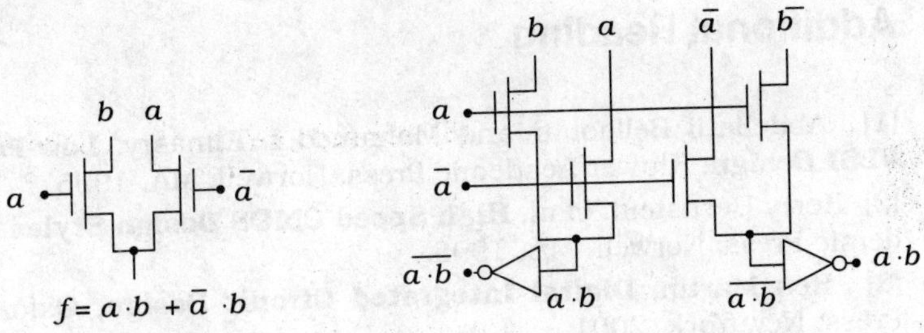

(a) AND gate (b) AND/NAND array

Figure 9.32 CPL AND/NAND circuit

been added to restore the voltages to full-rail values. These are not necessary until the full power supply is required, but they also help to speed up the circuit.

A unique feature of CPL is that several 2-input gates can be created by using the same transistor topology with different input sequences. Figure 9.33(a) shows an OR/NOR array. Comparing this with the AND/NAND shows that we have simply switched a and $\bar{a}$ on the FET inputs. An XOR/XNOR pair is shown in Figure 9.33(b). This is achieved by changing the top (drain) inputs. CPL also allows for 3-input logic gates with similar properties.

CPL is an interesting approach because it provides compact logic gates and the cell layout is reusable. The main drawbacks are the threshold loss and the fact that an input variable may have to drive more than one FET terminal. Similar approaches designed to overcome these problems have been proposed in the literature, but all result in more complex circuits.

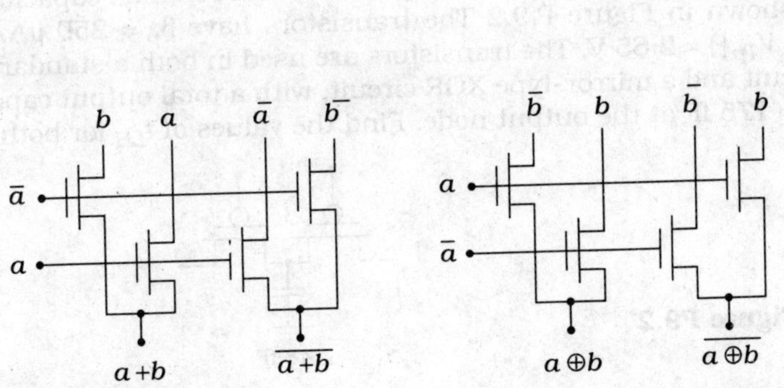

(a) OR/NOR (b) X OR/XNOR

Figure 9.33 2-input CPL arrays

9.7 Additional Reading

[1] Abdellatif Bellaouar and Mohamed I. Elmasry, **Low-Power Digital VLSI Design**, Kluwer Academic Press, Norwell, MA, 1995.

[2] Kerry Bernstein, et al, **High Speed CMOS Design Styles**, Kluwer Academic Press, Norwell, MA, 1998.

[3] Ken Martin, **Digital Integrated Circuit Design**, Oxford University Press, New York, 2001.

[4] Jan Rabaey, **Digital Integrated Circuits**, Prentice Hall, Upper Saddle River, NJ, 1996.

[5] John P. Uyemura, **CMOS Logic Circuit Design**, Kluwer Academic Press, Norwell, MA, 1999.

[6] Neil H. E. Weste and Kamran Eshraghian, **Principles of CMOS VLSI Design**, 2nd ed., Addison-Wesley, Reading, MA, 1993.

9.8 Problems

[9.1] One of your colleagues decides to use a mirror circuit to implement the 2-input function described in the truth table of Figure P9.1.

(a) Does the function have the correct symmetry required to build a mirror circuit? If so, construct the logic gate.

(b) Is the mirror circuit an intelligent design for this situation? Explain.

a	b	f
0	0	0
0	1	1
1	0	0
1	1	1

Figure P9.1

[9.2] Two series-connected pFETs have a common capacitance of 48 fF as shown in Figure P.9.2 The transistors have β_p = 250 μA/V^2 and (V_{DD} - $|V_{Tp}|$) = 2.65 V. The transistors are used in both a standard AOI XOR circuit and a mirror-type XOR circuit, with a total output capacitance of C_{out} = 175 fF at the output node. Find the values of t_{LH} for both designs.

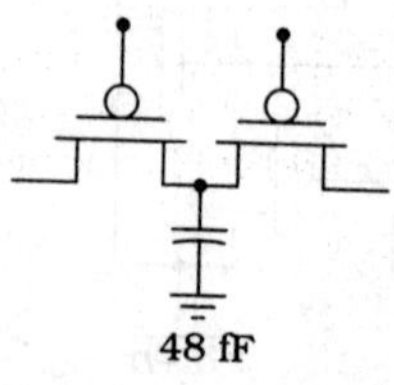

Figure P9.2

[9.3] Consider a CMOS process that is characterized by V_{DD} = 5 V, V_{Tn} = 0.7 V, V_{Tp} = -0.85V, k'_n = 120 μA/V^2, and k'_p = 55 μA/V^2. A pseudo-nMOS inverter is designed using an nFET aspect ratio of 4.

(a) Find the pFET aspect ratio needed to achieve V_{OL} = 0.3 V.

(b) Suppose instead that we select a pFET aspect ratio of $(W/L)_p = 3$. Find V_{OL} for this case.

[9.4] Consider the process described in Problem 9.3. Design a NAND2 gate and a NAND3 gate that both have $V_{OL} = 0.4$ V. The pFET is specified to have an aspect ratio of 2. Then compare the transistor area of the two gates.

[9.5] Draw the pseudo-nMOS circuits that provide the following logic operations.
(a) $f = \overline{a \cdot b + c}$; (b) $h = \overline{(a+b+c) \cdot x + y \cdot z}$; (c) $F = \overline{a + (c \cdot [x + (y \cdot z)])}$

[9.6] Consider the dual expressions

$$g = \overline{x \cdot y + z \cdot w} \qquad G = \overline{(x+y) \cdot (z+w)} \tag{9.44}$$

Which form (AOI or OAI) would provide the best performance when built using pseudo-nMOS design?

[9.7] Design a tri-state circuit that is in a high-impedance state when the control signal $T = 1$, and acts as a non-inverting buffer when $T = 0$.

[9.8] Design a clocked CMOS circuit that implements the function

$$f = \overline{a \cdot (b+c) + x \cdot y} \tag{9.45}$$

[9.9] The output node of a C^2MOS circuit is tri-stated with a clock signal of $\phi = 0$. The output capacitance at the node is $C_{out} = 76$ fF. The leakage currents are estimated to be $i_n = 0.46$ μA and $i_p = 127$ nA. The output voltage must be maintained above a value of 2.4 volts to be interpreted as a logic 1 stage by the next stage.

(a) Find the hold time at the output node if $V_{DD} = 5$ V.

(b) Find the hold time at the output node if $V_{DD} = 3.3$ V.

[9.10] Consider a charge leakage equation in the form

$$I_L(V) = -C_{out}\frac{dV}{dt} \tag{9.46}$$

where C_{out} is a constant, but the leakage current is described by

$$I_L(V) = B\frac{V}{V_0} \tag{9.47}$$

where B and V_0 are constants.

(a) Solve the differential equation for $V(t)$ using $V(0) = V_0$.

(b) Find an expression for the hold time t_h if the minimum logic 1 voltage is $V_x = 0.4\ V_0$.

[9.11] Draw the circuit diagram for a dynamic logic gate that has an output of

$$f = \overline{a \cdot b + c \cdot a} \tag{9.48}$$

using the smallest number of transistors.

[9.12] Draw the circuit diagram for a dynamic logic gate that has an output of

$$F = \overline{a \cdot (b + c + d)} \tag{9.49}$$

[9.13] The output voltage stored on the 100 fF capacitor in Figure P9.3 has an initial value of 5 V when $A = B = 0$. Find the value of V_{out} if the signals are changed to $A = 0$, $B = 1$.

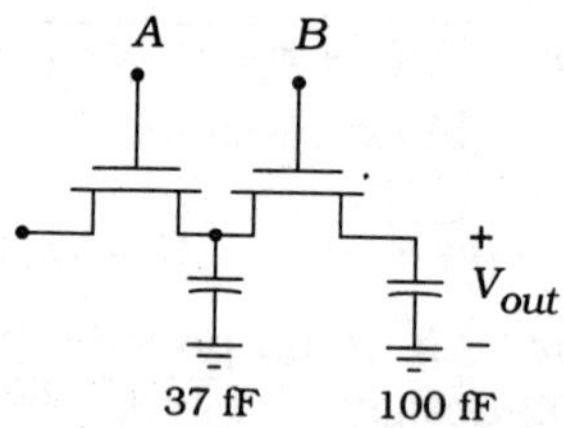

Figure P9.3

[9.14] Four nFETs are used as pass transistor as shown in Figure P9.4. The input voltage is set to $V_{in} = V_{DD} = 5$ V, and it is given that $V_{Tn} = 0.75$V.

(a) For the first case, suppose that the signals are initially at $(A, B, C, D) = (1, 1, 0, 0)$ and are then switched to $(A, B, C, D) = (0, 1, 1, 1)$. Find the final value of V_{out}.

(b) Suppose instead that signals are initially at $(A, B, C, D) = (1, 1, 1, 0)$ and are then switched to $(A, B, C, D) = (0, 0, 1, 1)$. Find the final value of V_{out}.

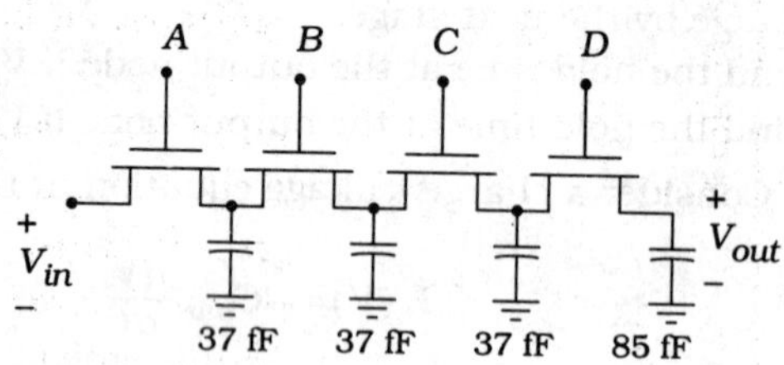

Figure P9.4

[9.15] Construct an MODL circuit that provides the two outputs

$$F = a \cdot b \qquad G = (a \cdot b) \cdot (c + d) \tag{9.50}$$

[9.16] Find the CVSL gate for the function table in Figure P9.5 by constructing an nFET logic tree.

f	1	1	0	1	0	0	1	1
c	0	1	0	1	0	1	0	1
b	0	0	1	1	0	0	1	1
a	0	0	0	0	1	1	1	1

Figure P9.5

Part 3

The Design of VLSI Systems

System Specifications Using Verilog® HDL

10

Hardware description languages (HDLs) are an ideal vehicle for hierarchical design. A system can be specified from the highest abstract architectural level down to primitive logic gates and switches.

Two HDLs dominate the field: VHDL (VHSIC HDL)[1] and Verilog® HDL. VHDL started as a government effort to unify projects from different contractors, while Verilog was the result of private development. Both are now standardized and widely used in industry, so either one could be presented here. Verilog was chosen because of its popularity in VLSI design. Compared to VHDL, it is a relatively loose and free-flowing language, and most chip designers feel that it adheres to their way of thinking. Verilog is structured after the C programming language and uses similar procedures and constructs. We should note, however, that C or C++ themselves can be used as an HDL [9], and several companies develop their own language.

This chapter introduces the basic concepts of the Verilog language. If you are familiar with VHDL from another course, you will find that learning Verilog is straightforward. If this is your first trek into an HDL, don't worry; the road is smooth and the ideas are easy to master.

10.1 Basic Concepts

A hardware description language allows us to specify the components that make up a digital system using words and symbols instead of having to use a pictorial representation like a block or logic diagram. Every component is defined by its input and output ports, the logic function it per-

1 VHSIC is a DoD acronym for Very High-Speed Integrated Circuits; DoD is an acronym for the Department of Defense.

forms, and timing characteristics such as delays and clocking. An entire digital system can be described in text format using a prescribed set of rules and **keywords** (reserved words). The file is then processed with the language compiler, and the output can be analyzed for proper operations. This can be applied to simple logic gates or to an entire microprocessor design. Logic verification using an HDL is usually considered mandatory to validate the design.

A typical design hierarchy is portrayed in Figure 10.1. At the highest level is a **behavioral** description that describes the system in terms of its architectural features. This is generally quite abstract in that it does not contain any details on how to implement the design. Once the behavioral model is simulated and refined, the design moves down to the **register-transfer level (RTL)**. An RTL description of a digital network concentrates on how the data moves about the system from unit to unit, and the main operations. State machines and sequential circuits can be introduced at this level. Timing windows are checked and rechecked, and validation of the design is again a primary objective.

The next level in the design process is called **synthesis**. In fully automated design, the RTL description is sent through a synthesis tool that produces a netlist of the hardware components needed to actually build the system. One of the more popular synthesis tools is Synopsis®. The success or failure of the synthesis process often depends upon the skill of the code writer. Not all HDL constructs can be synthesized, with a typical estimate hovering somewhere around 50%.

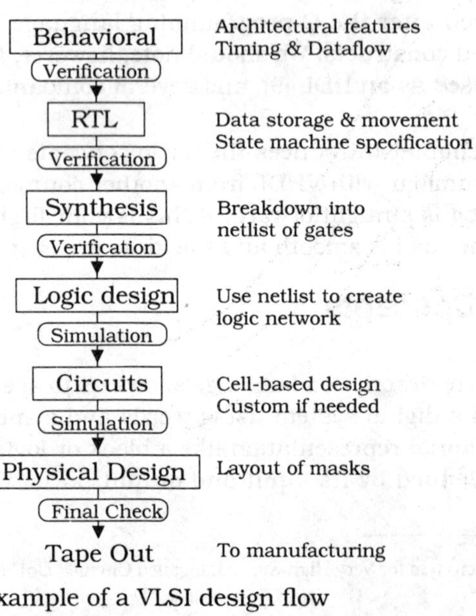

Figure 10.1 Example of a VLSI design flow

After the synthesis step, the netlist is used to design the logic network. Verification at this level consists of simulations to insure that the logic is correct. Once the logic is validated, the cell library can be used to design the circuits. Components are wired together, and both the electrical characteristics and the logic are verified using simulation. The cell instances and wirings are translated into silicon patterns in the physical design phase. After verifying the layout, the design is (at last!) complete and sent to manufacturing for the first silicon test chip.

Verilog HDL provides for descriptions of a digital system at all of the levels listed above. Every level is related to every other level, and the hierarchical design philosophy is linked by the different types of code. Each level has its own coding style using certain sets of commands and constructs. Verilog even provides for switch modeling of MOSFETs, although it is not as robust and sensitive to the CMOS processing variables as a circuit simulator such as SPICE. Verilog-A is an extension of an intrinsically digital language to the analog world.

The concept that links the various levels is that of a **module**. A Verilog module is the description of a unit that performs some function. It may be as simple as a basic FET switch, or as complex as a 64-bit ALU. Instantiations of simple modules are used to create more complex modules. The hierarchical structure is analogous to that used in the design of cells in a layout editor that was discussed earlier in the book.

Our treatment of Verilog will start at the digital logic level where simple gates are used to build more complex logic units. Once the structure of the language is understood, higher levels of abstraction are introduced.

10.2 Structural Gate-Level Modeling

Structural modeling describes a digital logic network in terms of the components that make up the system. Gate-level modeling is based on using primitive logic gates and specifying how they are wired together. It is the easiest to learn since it parallels the ideas developed in elementary logic.

Verilog is built using certain keywords that are understood by the compiler. Included in the group are primitives (such as logic gates), signal types, and commands. In our listings, Verilog code will use a sans serif font of this type, and will be indented from the main text. Keywords will be **boldface** using the same font. At the structural modeling level, the keywords are often primitive logic operations (gates) which results in a very readable coding style. A straightforward approach to learning Verilog is to study how a logic network is translated into a Verilog description using a line-by-line analysis. This will illustrate the ideas and syntax in a direct manner.

10.2.1 Verilog by Example

Consider the 4-input AOI circuit shown in Figure 10.2. The logic is constructed using primitive AND and NOR gates that take the inputs a, b, c, d and produce an output of

$$f = \text{NOT}(a \cdot b + c \cdot d) \tag{10.1}$$

Let us examine the listing for the Verilog module that describes the network by its internal structure. We will then study the details to learn how the module was constructed.

```
module AOI4 (f, a, b, c, d) ;
    input a, b, c, d ;
    output f ;
    wire w1, w2 ;
    and G1 (w1, a, b ) ;
    and G2 (w2, c, d ) ;
    nor G3 (f, w1, w2) ;
endmodule
```

A first reading of the listing exhibits the structure and syntax of a Verilog module. The keyword **module** defines the start of the listing for a network that has the name AOI4. The last line of the listing **endmodule** indicates that the description of the module is complete. The names of output and input "identifiers" are then listed in parentheses, with the output f first and then the inputs a, b, c, d. Semicolons are used as delimiters in Verilog; their usage should be memorized.

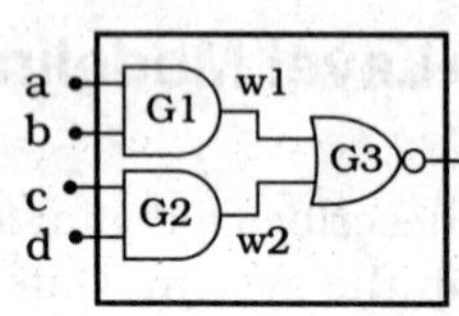

Figure 10.2 AOI module example

The next group of lines are the port keywords **input** and **output** that identify the input and output variables. The **wire** keyword identifies w1 and w2 as internal values that are needed to describe the network, but are not **input** or **output** ports. A **wire** declaration is a datatype called a net. A net value is determined by the output of the driving gate. In this case, w1 and w2 are the outputs of AND2 gates, which are in turn determined by the input values.

The structure of the logic is specified by the next three lines. These are instances of primitive AND and NOT gates that are part of the Verilog language. A gate instance has the form

```
gate_name instance_name (out, in_1, in_2, in_3, ... ) ;
```

where instance_name is an optional specifier that is used to correlate gates to their listing. In our example, we have named the gates G1, G2, and G3, so these appear in the listing. The compiler will interpret the code in the same manner if these are left out.

A structural listing provides a unique one-to-one correspondence with the components of a logic network. Suppose that we start with the following module description and then construct the logic diagram from it.

```
module Example (s_out, c_out , in_0, in_1 ) ;
    input in_0, in_1 ;
    output s_out, c_out ;
    xor (s_out, in_0, in_1) ;
    and (c_out ,in_0, in_1 ) ;
endmodule
```

This results in the internal details shown in Figure 10.3. This was drawn by starting with the input ports for in_0 and in_1, adding the gates (**xor** and **and**) with the specified wiring, and then pulling the outputs (s_out, and c_out) from the central region of the module. The logic equations are

$$s\_out = (in\_0) \oplus (in\_1)$$

$$c\_out = (in\_0 \cdot in\_1)$$

which is recognized as the sum and carry-out of a half-adder. These examples illustrate the fact that a Verilog structural description is equivalent to the information contained in a standard logic diagram.

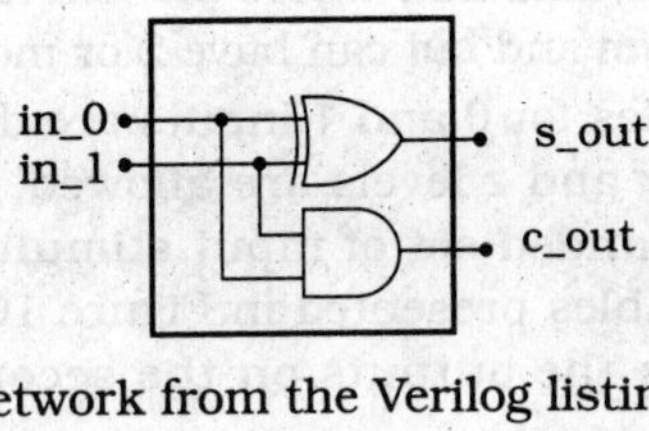

Figure 10.3 Logic network from the Verilog listing

Before proceeding further, let us examine some of the basics of writing Verilog descriptions.

Identifiers

Identifiers are names of modules, variables, and other objects that we can reference in the design. Examples of identifiers used so far include AOI4, a, b, in_0 , and s_out. Identifiers consist of upper- and lowercase letters, digits 0 through 9, the underscore character (_), and the dollar sign ($). The first character must be a letter or the underscore in normal usage. An identifier must be a single group of characters. For example, input_control_A is a single object, but input control A is not allowed as a single identifier.

It is important to point out that the Verilog language is **case sensitive**.

One must be careful to not mix upper- and lowercase letters, as they will mean different things. For example, in_0, In_0, and IN_0 are all distinct and are not interchangeable. Listings are insensitive to white space, so you may insert as many spaces or blank lines to help readability.

Value Set

The value set refers to the specific values that a binary variable can have. Verilog provides four levels for the values needed to describe hardware: 0, 1, x, and z. The 0 and 1 levels are the usual binary values. A 0 is either a logic 0 or a FALSE statement, while a 1 indicates either a logic 1 or a TRUE statement. The context determines which interpretation is valid. An x represents an unknown value, and z is the high-impedance (Hi-Z) value. The unknown value x is important as there are many situations where there is insufficient information. For example, when we first power up a circuit, the outputs of logic gates are unknown; we must wait for an input set to establish a value.

In addition to the four levels, 0 and 1 values can be subdivided into eight "strengths." These are used to model various physical phenomena that degrade the signals that contend for control of a line. Strengths will be discussed in more detail later.

Gate Primitives

Primitive logic function keywords provide the basis for structural modeling at this level. The important operations in Verilog are **and**, **nand**, **or**, **nor**, **xor**, **xnor**, **not**, and **buf**, where **buf** is a non-inverting drive buffer. All gates except for **not** and **buf** can have 2 or more inputs.

The truth tables for 0 and 1 inputs are defined in the usual manner. However, since x and z levels are allowed, we must define how a gate reacts to an expanded set of input stimuli. The **buf** and **not** gates are defined by the tables presented in Figure 10.4. The input values on the top row produce the outputs on the second row, making these self-explanatory.

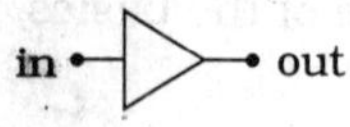

in	0	1	x	z
out	0	1	x	x

(a) **buf** primitive

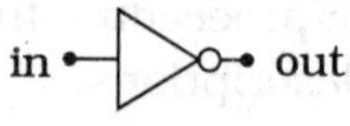

in	0	1	x	z
out	1	0	x	x

(b) **not** primitive

Figure 10.4 Function maps for **buf** and **not** gates

Figure 10.5 provides the truth table for the multiple-input gates **and**, **nand**, **or**, **nor**, **xor**, and **xnor**. The tables themselves are for two inputs and must be extrapolated for 3 or more inputs. The format of the tables are standard in Verilog, and have the structure of a Karnaugh map. The top row gives the values for one input, while the left column is the other. The output value out for each possibility is read from the matrix contained within the box by aligning a row with a column. The 4 × 4 sub-matrix in

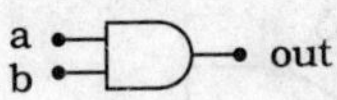

and

	0	1	x	z
0	0	0	0	0
1	0	1	x	x
x	0	x	x	x
z	0	x	x	x

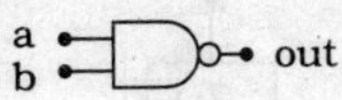

nand

	0	1	x	z
0	1	1	1	1
1	1	0	x	x
x	1	x	x	x
z	1	x	x	x

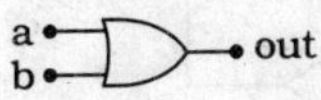

or

	0	1	x	z
0	0	1	x	x
1	1	1	1	1
x	x	1	x	x
z	x	1	x	x

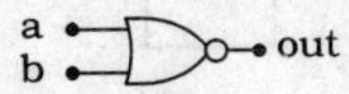

nor

	0	1	x	z
0	1	0	x	x
1	0	0	0	0
x	x	0	x	x
z	x	0	x	x

xor

	0	1	x	z
0	0	1	x	x
1	1	0	x	x
x	x	x	x	x
z	x	x	x	x

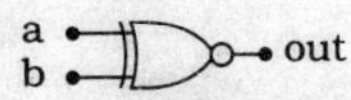

xnor

	0	1	x	z
0	1	0	x	x
1	0	1	x	x
x	x	x	x	x
z	x	x	x	x

Figure 10.5 Multiple-input gate maps

the upper left-hand corner is easily recognized as the standard K-map for 0 and 1 inputs.

Tri-state primitives are **bufif0**, **bufif1**, **notif0**, and **notif1**. The names help remember the operation. The **bufif0** gate is a buffer if the control is 0; if the control is 1, then it is tri-stated with a Hi-Z output. Similarly, **notif1** acts as a **not** if the control is 1, while a control of 0 gives a Hi-Z output. Tri-state gates have one input, but can have more than one output corresponding to their usages as drivers. To describe them we use the form

tristate_name instance_name (out_0, out_1, out_2, ... , input, control);

where instance_name is the optional name of the instance. The logic maps for these primitives are summarized in Figure 10.6. An example of a tri-state circuit is the 2:1 MUX shown in Figure 10.7. The logic for this network is

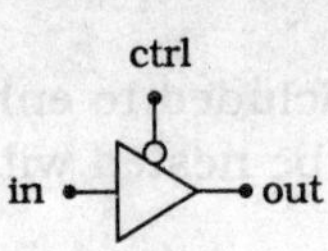

bufif0

in \ ctrl	0	1	x	z
0	0	z	L	L
1	1	z	H	H
x	x	z	x	x
z	x	z	x	x

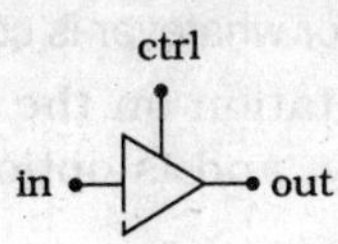

bufif1

in \ ctrl	0	1	x	z
0	z	0	L	L
1	z	1	H	H
x	z	x	x	x
z	z	x	x	x

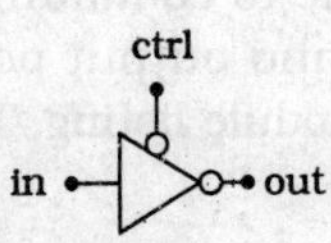

notif0

in \ ctrl	0	1	x	z
0	1	z	H	H
1	0	z	L	L
x	x	z	x	x
z	x	z	x	x

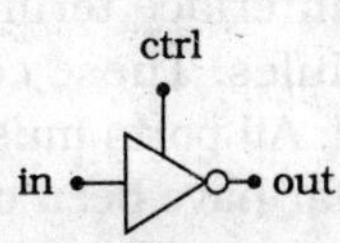

notif1

in \ ctrl	0	1	x	z
0	z	1	H	H
1	z	0	L	L
x	z	x	x	x
z	z	x	x	x

Figure 10.6 Maps for tri-state primitives

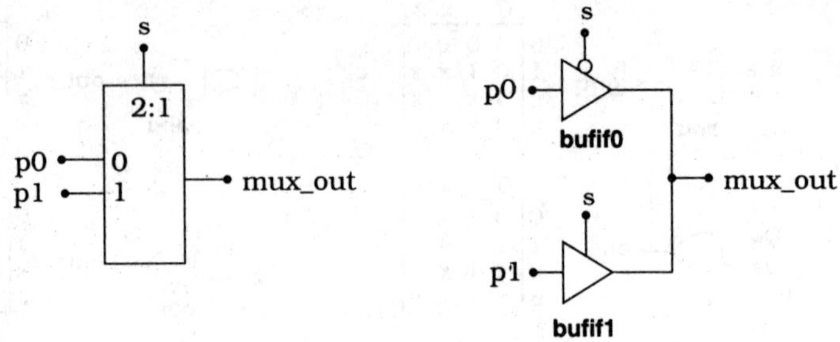

Figure 10.7 2:1 MUX using tri-state primitives

$$out = p0 \cdot \overline{s} + p1 \cdot s \qquad (10.2)$$

and is described by the Verilog listing

```
module 2_1_mux (out, p0, p1, s) ;
   input p0, p1, s ;
   output out ;
   bufif0 (mux_out , p0 , s) ;
   bufif1 (mux_out , p1 , s) ;
endmodule
```

Other primitives will be introduced later. These include MOSFET switches and other useful components.

Comment Lines

Comments are useful for documenting code. In the statement

```
xor (s_out, in_0, in_1) ; // This line produces s_out
```

everything to the right of the // is ignored by the compiler. If the comment extends over two or more lines, then we use /* to denote the start of the comment on the first line, and */ for the end on the last line, as in

```
/* If we have a long comment that we want to insert
   then we may extend it into multiple lines
   or whatever is convenient */
```

The indentation on the second line has been included to enhance the readability, and is optional. Comments cannot be nested within other comments.

Ports

Ports are interface terminals that allow a module to communicate with other modules. These correspond to the input and output points on a library cell. All ports must be declared within a module listing. The examples thus far have been of the form

```
input in_0, in_1 ;
output s_out, c_out ;
```

A bidirectional port is declared with the syntax

```
inout IO_0, IO_1 ;
```

where the identifiers IO_0 and IO_1 can be used as either inputs or outputs to the module.

Consider next the NOR-based SR latch in Figure 10.8. A Verilog module description for this circuit can be written in the form

```
module sr_latch (q, q_bar, s, r) ;
   input s, r ;
   output q, q_bar ;
   reg q, q_bar ;
   nor (q_bar, s, q), (q, r, q_bar) ;
endmodule
```

Two new features have been introduced. The first is the register (**reg**) datatype specification. A register datatype is one whose value is held until it is overwritten by another value. In the current usage, this allows the values of q and q_bar to be held for communication to another port in a different module. Note that q and q_bar are specified as both **reg** and **output** ports. A Verilog **reg** datatype should not be interpreted as a hardware register, such as a D-type flip-flop. Instead, just think of them as lines that can hold their values without any external driver. A **reg** quantity is classified as a type of net specification.

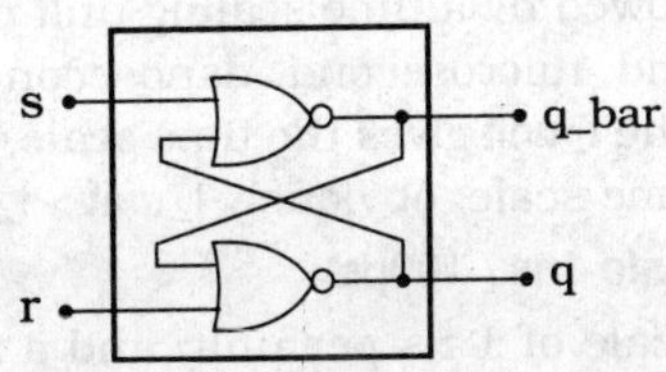

Figure 10.8 SR latch

The second new aspect is the multiple instancing of the **nor** gate primitive using one line. Since the input and output listings are distinct, each is grouped inside a set of parentheses, and a comma is used as a delimiter between the two. The single line thus represents two independent gates. This technique can be extended to multiple gates. Including instance names with each helps decipher the code.

Gate Delays

A hardware description language must use modeling that allows the simulation to include time delays. Verilog provides several techniques for introducing delays at the gate level.

The logic delay through a gate is sometimes modeled using a single delay time (propagation delay) from the input to the output. Delays are specified in instantiations using the pound sign (#) as in

```
nand #(prop_delay) G1 (output, in_a, in_b) ;
```

where prop_delay is the value of the delay. If the rise and fall times are known separately, they can be used by writing

```
nand #( t_rise, t_fall ) G1 (output, in_a, in_b) ;
```

The turn off delay can also be included as in

```
nand #(t_rise, t_fall , t_off ) G1 (output, in_a, in_b) ;
```

The number of values in the #(listing) determines the manner in which Verilog interprets the information. A single entry implies a propagation delay, two entries imply t_r and t_f values, while three entries add the turn off time.

Numerical values of gate delay values are specified as integer values of an internal time step unit. For example,

```
and #(4, 2) A1 (out, A_in, B_in) ;
```

assigns t_rise = 4 units and t_fall = 2 units. Relative units are sufficient for a broad class of simulations, so it is not necessary to use absolute time values (i.e., seconds).

If numerical values are desired, then one uses a compiler directive of the form

```
'timescale t_unit / t_precision
```

in the listing. In this expression, t_unit and t_precision can have values of 1, 10, or 100 followed by a time scaling unit of s, ms, us, ns, ps, or fs for second, millisecond, microsecond, nanosecond, picosecond, or femtosecond, respectively. The t_unit gives the time scale, while t_precision gives the resolution of the time scale; obviously t_unit > t_precision. For example

```
'timescale 1ns / 100ps
```

gives a time scale of 1 ns per unit, and a resolution of the time scale as 100 ps. If a gate instance is written as

```
xor #(10) (out, A_0, A_1) ;
```

is used, the absolute delay through the gate is 10 × t_unit = 10 ns. If we change the time scale to

```
'timescale 10ns / 1ns
```

the absolute delay is 10 × 10ns = 100 ns. The value of t_precision = 1 ns determines the resolution; for example, if one specifies a time delay of 10.748 ns, the value would be rounded to 11 ns.

Gate delays allow us to monitor the response of a network in a dynamic environment. Let us simulate the module shown in Figure 10.9 for the inputs a, b, and c shown in the waveform. The Verilog listing below introduces the concept of a stimulus module that provides the signals.

```
// This module has gate delays
module DelayEx (out, a, b, c) ;
    input a, b, c ;
```

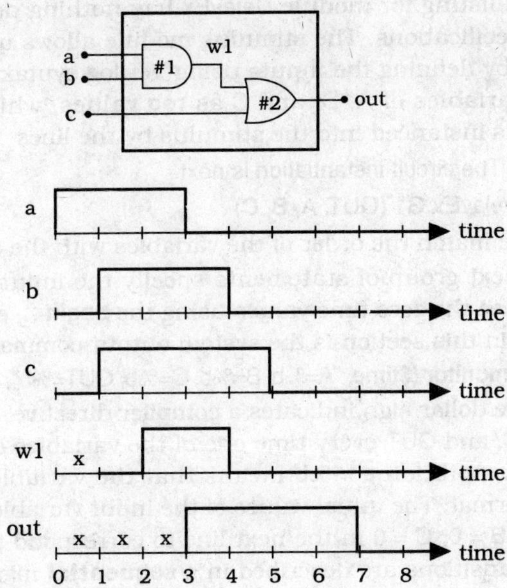

Figure 10.9 Gate delay example

```
    output out ;
    wire w1 ;
    and #1 (w1, a, b) ;
    or  #2 (out, w1, c) ;
endmodule
// The stimulus module provides the input signals
module stimulus ;
    reg A, B, C ; // Hold input values
    wire OUT ; // This is a driven output value
// The circuit instantiation is next
    DelayEx G1 (OUT, A, B, C) ;
initial
    begin
    $monitor ($time, "A=%b, B=%b, C=%b, OUT=%b", A, B, C, Out) ;
        A=1 ; B=0 ; C=0 ;
    #1 B=1 ; C=1 ;
    #2 A=0 ;
    #1 B=0 ;
    #1 C=0 ;
    #3 $finish ;
    end
endmodule
```

The first listing for module DelayEx has nothing new in it except for the delay specifications. The stimulus module allows us to "test" the module DelayEx by defining the inputs using Verilog syntax. For the stimulus, we define variables of A, B, and C as **reg** values, while OUT is a **wire**. The module is instanced into the stimulus by the lines

```
// The circuit instantiation is next
DelayEx G1 (OUT, A, B, C) ;
```

where we match the order of the variables with the defining module.

The next group of statements specify the inputs. The **initial** directive establishes the zero time values using the **begin** ... **end** structure. Embedded within this section is the system output command

```
$monitor ($time, "A=%b, B=%b, C=%b, OUT=%b", A, B, C, Out) ;
```

where the dollar sign indicates a compiler directive. This provides outputs of A, B, C, and OUT every time one of the variables changes. As explained later, the notation a = %b means that the variable a is to be shown in binary format. The initial values of the input variables are assigned values of A = 1, B = 0, C = 0 in the next line to correspond to the waveforms. The signal transitions are described in a **sequential** manner by statements of the form

```
#1 B=1 ; C=1 ;
#2 A=0 ;
#1 B=0 ;
#1 C=0 ;
```

These must be executed in order. The stimulus at **#1** means that at time 0 +1 = 1, both B and C are logic 1 values. The next line at **#2** assigns the value A = 0 at 2 time units after the first line; for this example, the absolute time is 1+2 = 3 units. The next line resets B to 0 at 3+1 = 4 units, and the final line resets C to 0 at 4+1 = 5 time units. It is easily verified that this describes the input waveforms. The final directive #3 **$finish** indicates that the simulation is completed at time 5+3 = 8 time units. Finally, **end** closes the **begin** procedure. Simulating this yields the waveform for out shown in the drawing.

This example provides an idea of how to build a testbench for Verilog code. Once the network is defined, different stimulus modules can be written to test the logic. The concept is illustrated schematically in Figure 10.10. The stimulus module is usually separate from the logic module so that the inputs can be changed without affecting the logic. The Verilog work environment allows the two to be linked during the simulation. The details vary with the compiler implementation, so it is important to read the documentation. Logic verification is one of the most important aspects of high-level VLSI design.

Number Specifications

In the delay example, input stimuli were defined via statements such as

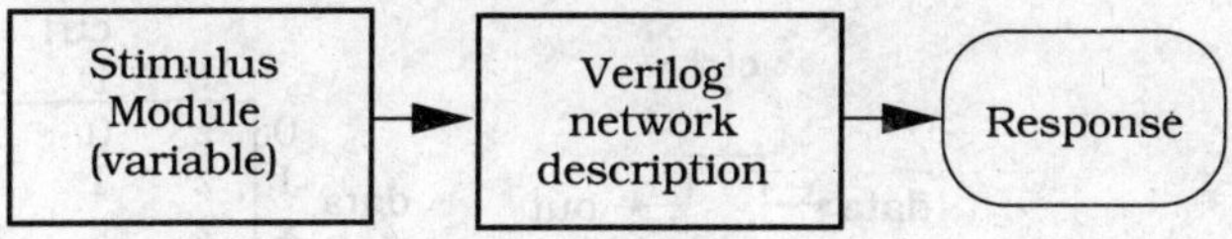

Figure 10.10 Testbench concept

```
A = 1 ; B = 0 ; C = 0 ;
```

These are interpreted as default binary values. Values can also be specified in base-*r* for radix values of 2 (binary, b), 8 (octal, o), 10 (decimal, d), and 16 (hexadecimal, h) using a format of

```
<size> '<base designator> <value>
```

with <size> a decimal number indicating the number of bits in the number. Some examples are

```
1'b0        // 1-bit binary number with a value of 0
4'b1011     // 4-bit binary word with a value of 1011
16'h1a36    // 16-bit number with a value of hexadecimal 1a36
```
3'd4 // 3-bit number with a decimal value of $4 = 100_2$

Values can be declared in a listing. For example, the code

```
reg reset ;
initial
    begin
    reset = 1'b1 ;  // initialize reset to a value of 1
    #10 reset = 1'b0 ; // reset to 0 after 10 time units
    end
```

allows us to specify the value of reset as required.

10.3 Switch-Level Modeling

Verilog allows switch-level modeling that is based on the behavior of MOSFETs. Although circuit-level simulators (such as SPICE) are much more accurate for performing critical electrical calculations, Verilog coding is useful for verifying logic flow through networks that consist of both transistors and logic gates. More importantly, switch-level models have a direct one-to-one correspondence with CMOS circuits and logic gates as discussed in Chapter 2. The ability to construct Verilog descriptions of complex system-level designs all the way down to basic CMOS circuits demonstrates the power of hierarchical design.

The switch primitives are named **nmos** and **pmos**, and behave in the same manner as the transistors with the same names. Figure 10.11 summarizes the behavior of both. Verilog syntax for these primitives is in the

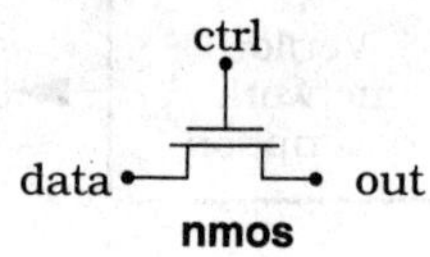

data \ ctrl	0	1	x	z
0	z	0	L	L
1	z	1	H	H
x	z	x	x	x
z	z	z	z	z

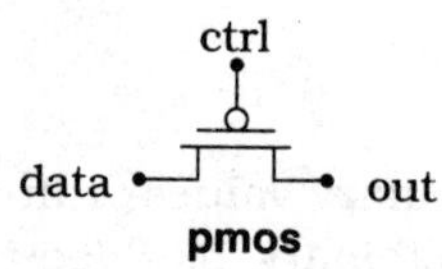

data \ ctrl	0	1	x	z
0	0	z	L	L
1	1	z	H	H
x	x	z	x	x
z	z	z	z	z

Figure 10.11 Switch-level primitives

form

```
nmos name (out, data, ctrl) ;
pmos name (out, data, ctrl) ;
```

where name is the optional instance identifier. For ctrl values (applied to the gate) that are 0 and 1, the behavior is identical to FETs. The **nmos** switches are open for ctrl = 0 and closed for ctrl = 1, while **pmos** switches are closed for ctrl = 0 and open for ctrl = 1. An open switch induces a high-impedance state with out = z. The tables also list two new entries, L and H, for the value of out when ctrl is x or z. The (low) symbol L stands for 0 or z, while the (high) symbol H represents 1 or z. The basis of this ambiguity is non-trivial. It is related to the physical concept that the output node can store charge, so that out may be related to an earlier value.

MOS switches can be used to describe CMOS logic gates. The simple NOT circuit in Figure 10.12 has the Verilog description

```
// CMOS inverter switch network
module fet_not (out, in) ;
    input input ;
    output output ;
    supply1 vdd ;
    supply0 gnd ;
    pmos p1 (vdd, output, input) ;
    nmos n1 (gnd, output, input) ;
endmodule
```

The circuit and listing has been used to introduce two new Verilog keywords **supply1** and **supply0** that define the power supply vdd and ground gnd connections. These represent the strongest logic 1 and logic 0 drivers, respectively. The Verilog module treats these as the data into the FETs, while the gate input is the switch ctrl.

The same constructs can be used to model arbitrary CMOS logic gates.

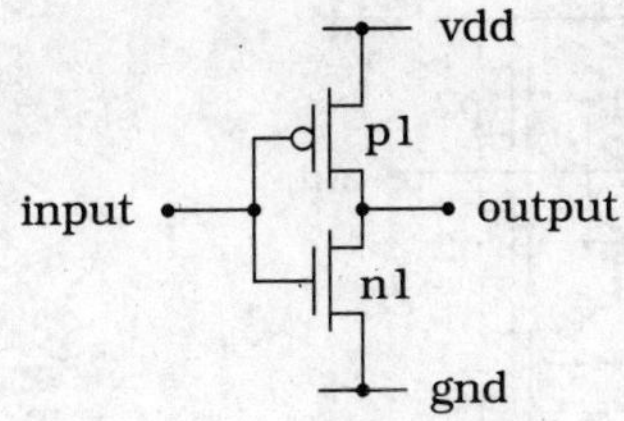

Figure 10.12 CMOS inverter using Verilog switches

The NAND2 and NOR2 switching networks in Figure 10.13 are described by the module

```
// CMOS logic gates
module fet_nand2 (out, in_a, in_b) ;
    input in_a, in_b ;
    output out ;
    wire wn ; // This wire connects the series nmos switches
    supply1 vdd ;
    supply0 gnd ;
    pmos p1 (vdd, out, in_a) ;
    pmos p2 (vdd, out, in_b) ;
    nmos n1 (gnd, wn, in_a ) ;
    nmos n2 (wn, out, in_b ) ;
endmodule
```

for the NAND gate, and

```
module fet_nor2 (out, in_a , in_b ) ;
    input in_a, in_b ;
    output out ;
    wire wp ; // This connects the series pmos switches
    supply1 vdd ;
    supply0 gnd ;
    pmos p1 (vdd, wp, in_a) ;
    pmos p2 (wp , out, in_b) ;
    nmos n1 (gnd, out, in_a ) ;
    nmos n2 (gnd, out, in_b ) ;
endmodule
```

for the NOR gate. These can be verified using a line-by-line comparison.

Another useful set of primitives includes pull-up and pull-down components that have the keywords **pullup** and **pulldown**. These can be modeled as resistors that are connected to **supply1** and **supply0** as shown in Figure 10.14(a) and are described by

```
pullup (out_1) ; // This gives a high output
```

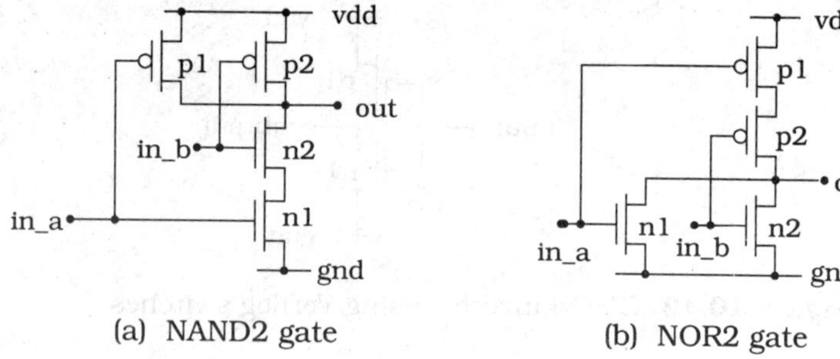

(a) NAND2 gate (b) NOR2 gate

Figure 10.13 Logic gate construction

```
pulldown (out_0) ; // This gives a low output
```

in a Verilog listing. The output strengths are called **pull1** and **pull0**, and are weaker than the **supply1** and **supply0** levels. Pull primitives are used in various ways to model circuits. For example, a **pullup** can be used as a load device as in the nMOS NOR3 gate drawn in Figure 10.14(b). The Verilog description is

```
module fet_nor2 (out, in_a, in_b, in_c ) ;
   input in_a, in_b ;
   output out ;
   supply0 gnd ;
   nmos na (gnd, out, in_a) ,
        nb (gnd, out, in_b) ,
        nc (gnd, out, in_c) ;
   pullup (out) ;
endmodule
```

Note that **pullup** and **pulldown** require only one identifier. This is because only a single wire is provided out of each "device" equivalent.

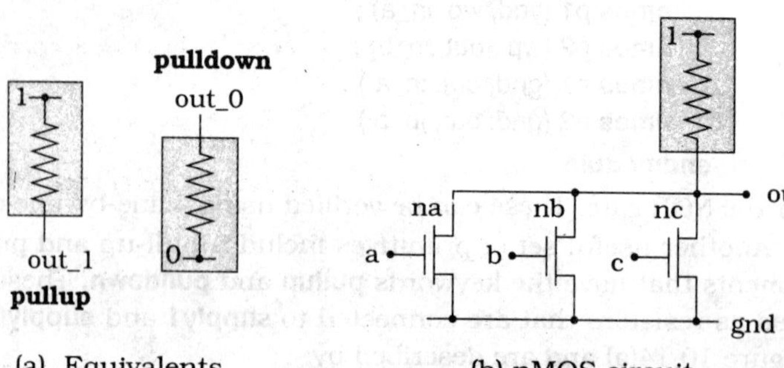

(a) Equivalents (b) nMOS circuit

Figure 10.14 Pull-up and pull-down primitives

The cmos Primitive

Verilog models CMOS transmission gates using the **cmos** keyword.[2] The symbol and function table are shown in Figure 10.15. To instance the TG, we use the syntax

```
cmos tg1 (out, data , n_ctrl , p_ctrl) ;
```

with data being the input. In most cases, n_ctrl and p_ctrl are complementary signals. However, the table lists the most general case where the two are separate. In practice, this may occur because of an inverter delay when generating one signal from the other.

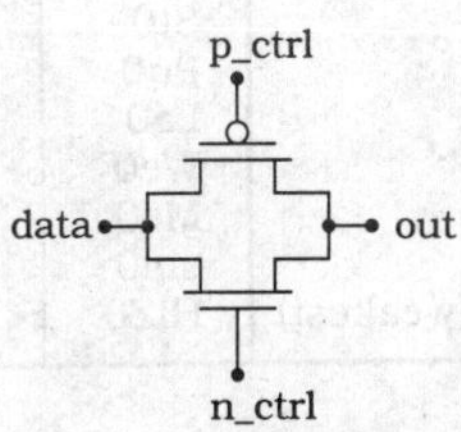

cmos

n_ctrl	p_ctrl	data 0	data 1	data x	data z
0	0	0	1	x	z
0	1	z	z	z	z
0	x	L	H	x	z
0	z	L	H	x	z
1	0	0	1	x	z
1	1	0	1	x	z
1	x	0	1	x	z
1	z	0	z	x	z
x	0	0	1	x	z
x	1	L	H	x	z
x	x	L	H	x	z
x	z	L	H	x	z
z	0	0	1	x	z
z	1	L	H	x	z
z	x	L	H	x	z
z	z	L	H	x	z

Figure 10.15 Verilog **cmos** transmission gate.

Delay Times

The syntax for time delays is identical to that used for logic gates. Delays are specified in time units using the pound sign designator # (times). The number of entries in (times) determines their meaning. One entry is the propagation delay, two entries mean (t_rise, t_fall), and three entries imply (t_rise , t_fall, t_off). Some examples are

```
nmos #(2) n1 (out, data, ctrl) ;
pmos #(3, 4) p1 (out_p, data_in, p_ctrl) ;
cmos #(2, 3, 3 ) TG1 (output, input, n_sig, p_sig ) ;
```

These are not always related to physical load-dependent values, so care must be exercised when specifying device delays.

Strength Levels

In addition to the strengths 0, 1, x, and z, variables are allowed to take on

2 Note that we will use lower case boldface letters with a sans serif font to distinguish the keyword from the CMOS technology.

different strength levels. These are used in cases where two or more signals contend for control of a net, or to describe a physical loss of voltage. Figure 10.16 summarizes the ranges for both logic 1 and logic 0 values. When there is a contention by various signals, the stronger one dominates. The strengths are useful for modeling voltage changes, such as threshold losses through pass transistors. The strengths can be specified as needed, or we can introduce resistive switches that have signal altering characteristics included in their definitions.

Logic 1			**Logic 0**	
Strength Level	**Name**	**Type**	**Name**	**Strength Level**
supply1	Su1	drive (strongest)	Su0	supply0
strong1	St1	drive	St0	strong0
pull1	Pu1	drive	Pu0	pull0
large1	La1	storage	La0	large0
weak1	We1	drive	We0	weak0
medium1	Me1	storage	Me0	medium0
small1	sm1	storage	sm0	small0
high-z1	HiZ1	high-Z (weakest)	HiZ0	high-z0

Figure 10.16 Strength levels in Verilog

Resistive (rmos) Switches

Realistic MOSFETs have drain-source resistance that can modify the signal strength passing through them. Some of the effects can be included by using resistive MOS switches which are gate-controlled in the same manner as regular switches, but the devices alter the output strength. The FET equivalent primitives are **rnmos**, **rpmos**, and **rcmos**. The instancing syntax is the same as for non-resistive (ideal) switches. For example,

```
rnmos #(1, 2, 2 ) fet_1 (output, input, gate_ctrl) ;
```

specifies a resistive nFET. The main difference is that input-output strength relations are defined by the list in Figure 10.17. This is useful for including physical effects such as threshold voltage losses through nFET pass transistors. While a SPICE simulation at the electronics level is much more accurate, these are useful for modeling the switching behavior in non-critical paths.

10.4 Design Hierarchies

The concept of primitive, modules, and instancing provides the basis for hierarchical design in Verilog. Up to this point we have learned how to

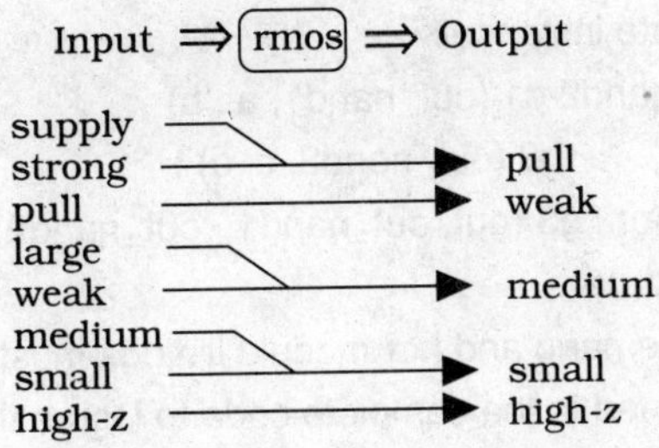

Figure 10.17 Resistive (**rmos**) input-output strength map

write Verilog code at the gate-level and the switch-level. These two levels can be used separately, or intermixed within a single module. We will use these two modeling levels as a vehicle for learning the fundamentals of hierarchical design.

Let us start with a simple example. Suppose that we have constructed the switch-level models for the NAND2 and NOR2 gates using the circuits illustrated in Figure 10.13. These are described by the Verilog modules that were named fet_nand2 and fet_nor2, respectively. Our objective is to create an AND4 gate module using these two gates for instances. Figure 10.18 shows the logic diagram; the formation of the AND4 operation is easily verified using bubble pushing. Let us construct a Verilog module for the gate by instancing the switch-level modules.

```
module fet_and2 (out, a, b, c, d) ;
   input a, b, c, d ;
   output out ;
   wire out_nor, out_nand1, out_nand2 ;
```

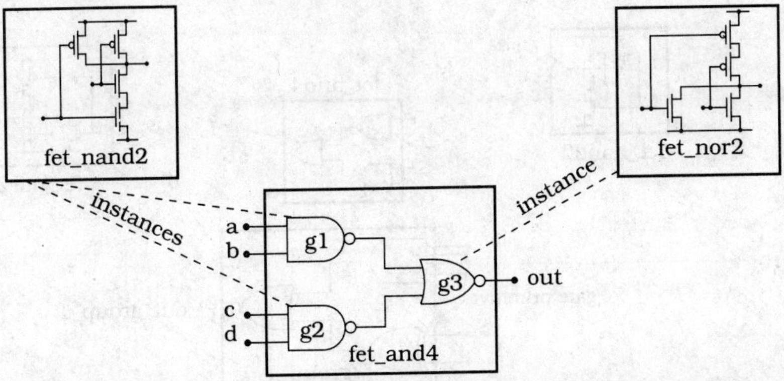

Figure 10.18 Creating an AND4 gate module

```
    // Gate instances
    fet_nand2 g1 (out_nand1, a , b) ,
              g2 (out_nand2, c, d) ;
    fet_nor2 g3 (out, out_nand1 , out_nand2) ;
endmodule
    /* The nand and nor module listings must be
    included in the complete code to insure that they
    are defined for instancing */
```

This illustrates the instancing procedure, where it is assumed that the modules fet_nand2 and fet_nor2 have been defined using the previously written modules. Now suppose that we want to build a more complex network using the fet_and4 module. The new module, which we will call group_1, can be constructed using any entries that have been defined. Figure 10.19 illustrates how the cell can be built using instances (dashed lines) of switch-level modules and the fet_and4 module, combined with the Verilog primitive XOR gate. The basic features of the module are summarized by the general form

```
module group_1 (out_group_1, . . .) ;
    ... // input and wire declarations
    output out_group_1 ;
        // Gate instances
    fet_and4 ( . . . ) ;
    fet_nor2 ( . . . ) ;
    xor ( . . . ) ;
endmodule
```

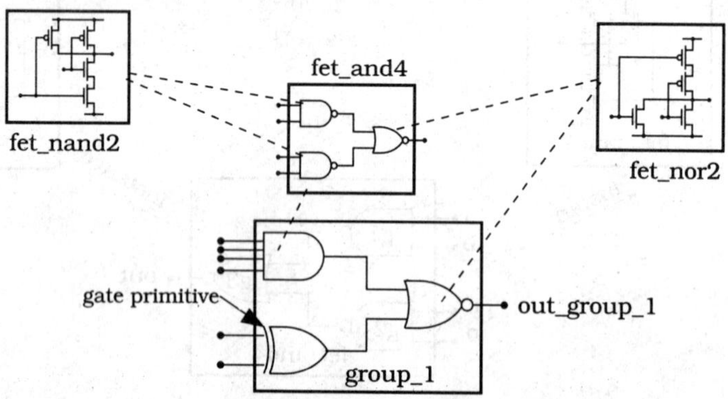

Figure 10.19 Building the next level of hierarchy

which shows the mixing of the levels and primitives (the **xor**). Of course, the new group_1 module can itself be instanced in the next higher level, and so on. This type of procedure allows us to design VLSI switching and logic networks in a structured manner that can be documented and traced. Verification is simplified as errors can often be located more easily by their occurrences in certain modules. And, there is a one-to-one correspondence between the HDL description and the usage of a cell library in the physical design phase.

Let us now consider the problems that we encounter at the VLSI system level. It is not possible to follow every bit as it moves through a complex system, so we must move to a higher level of modeling. This increases the abstraction of the viewpoint and coding necessary to reflect the architectural features. Suppose that we need to include a 32-bit adder in our design. At the architectural level, the important characteristics of a module would be the function it performs and delay and timing aspects, since these are critical for interfacing it to other modules. In terms of the block shown in Figure 10.20, we would concentrate on specifics such as the word size (32 bits), the inputs (*a* and *b*) and output (*s*), and any control signals that are used (to indicate, for example, signed or unsigned addition). The internal details of the module are not very interesting at this level; one does not need to know *how* the circuits produce the results to use the unit in a design. Of course, the circuits are important if we want to actually build the adder.

Modern VLSI system design starts at the top architectural level and works downward to the physical level, since we must first insure that the design is valid before worrying about polygons on silicon. As mentioned earlier, this is called **top-down** design. It intrinsically assumes that we can build the needed units in silicon and interface them together to meet the system specifications. Experience is the best guide for projecting the limits of the silicon area and speed and relating this to the architecture. As chip complexity increases, this becomes more difficult. Luckily, both silicon technology and CAD tools improve every year.

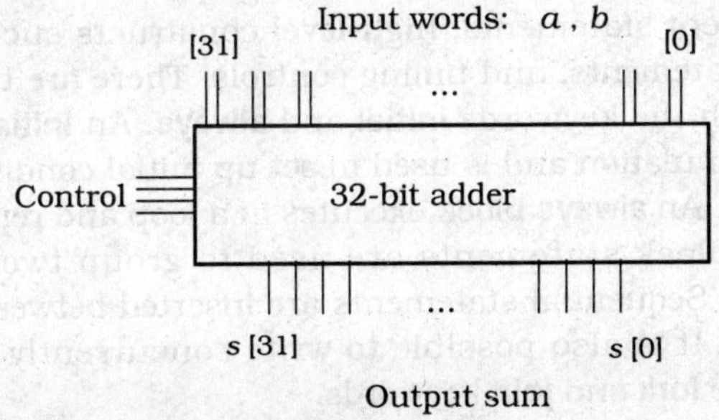

Figure 10.20 Functional 32-bit adder block

HDLs provide a powerful vehicle for system-level design by introducing different levels of abstraction. The highest Verilog level is called **behavioral** modeling. As implied by its name, it concentrates on describing the general *behavior* of units to characterize how they will work when embedded in a larger system. Timing is often the most critical feature in a behavioral model. The internal details of a unit are not specified, nor do they affect the modeling; it is assumed that the specifications are a result of physically realizable internal circuitry.

The next level of abstraction down is usually termed Register-Transfer Level (**RTL**) modeling. RTL concentrates on specifying the movement of data among hardware sections. The name itself arises from the fact that synchronous digital systems rely very heavily on the use of clock-controlled storage registers. Data transfers take place at specific times dictated by the clocking. An RTL specification is viewed as being the link between purely abstract modeling and hardware design. RTL code is often the input to the synthesis stage of design (see Figure 10.1) that produces gate netlists.

The remaining section of this chapter is an introduction to high-level behavioral modeling in Verilog. The treatment covers the basics of behavioral and RTL coding with short examples to clarify the structure and concepts. Advanced constructs and coding techniques are introduced for specific applications in later chapters.

10.5 Behavioral and RTL Modeling

Verilog behavioral modeling is based on specifying a group of concurrent procedures that characterize a block. Emphasis is on an accurate representation of the architecture, with most of the implementation details ignored. This feature makes the coding style quite abstract.

The basis for behavioral modeling is the construction of **procedural blocks**. As implied by its name, a procedural block is a listing of statements that describe how a set of operations are performed. Many of these resemble constructs in the C programming language, and they introduce a new level of abstraction to the design process. Procedural blocks contain assignment statements, high-level constructs such as loops and conditional statements, and timing controls. There are two types of block that start with the keywords **initial** and **always**. An **initial** block executes once in the simulation and is used to set up initial conditions and step-by-step dataflow. An **always** block executes in a loop and repeats during the simulation. Block statements are used to group two or more statements together. Sequential statements are inserted between the keywords **begin** and **end**. It is also possible to write concurrently executed statements using the **fork** and **join** keywords.

Let us start by writing a module for a clock variable clk. We will assume a clock period of 10 time units so that the variable must change every 5 time units as illustrated in Figure10.21.

```
module clock ;
reg clk ;
// The next statement starts the clock with a value of 0 at t = 0
    initial
        clk = 1'b0 ;
// When there is only one statement in the block, no grouping is required
    always
        #5 clk = ~ clk ;
    initial
        #500 $finish ; // End of the simulation
endmodule
```

The cyclic action is obtained using the NOT operator ~ in the statement

```
# 5 clk = ~ clk ;
```

Since this falls within an **always** statement the command is executed in a loop until the simulation ends at 500 time units.

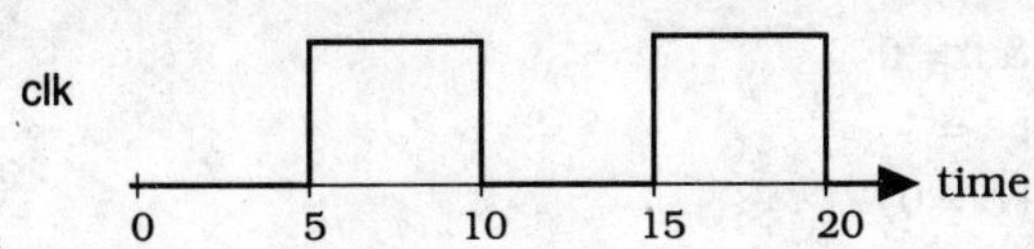

Figure 10.21 Clocking waveform clk

Operators

The Verilog operators such as ~ are summarized in Figure 10.22 for future reference. Note that some symbols such as & are used differently depending upon the context. We will study a few to understand how they work.

Consider first the behavior of the reduction or unary operators (i.e., operations on a single number.) Suppose we assign the binary values a = 1101 and b = 0000. Bit-wise negation gives

~ a = 0010

~ b = 1111

as it operates on each bit independently. A logical negation evaluates to

! a = 0

! b = 1

The logical operator !A gives the logical inverse of A. If A contains all zeros, then it is false (0). If it is non-zero, then it is true (1); !A gives the inverse of the value of A. Reduction operators operate on each bit of the number

```
Arithmetic
+     addition
-     subtraction
*     multiplication
/     division
%     modulus

Shift Operations
>>    shift right
<<    shift left

Relational and Logical
>     greater than
>=    greater than or equal to
<     less than
<=    less than or equal to
!     negation
&&    logical and
||    logical or
==    identity
!=    logical inequality
===   case equality
!==   case inequality

Bit-wise
~     not
&     and
~&    nand
|     or
~|    nor
^     xor
~^    xnor
^~    xnor

Reduction (unary)
&     reduction and
|     reduction or
~&    reduction nand
~|    reduction nor
^     reduction xor
^~    reduction xnor
~^    reduction xnor
```

Figure 10.22 Verilog operators

and result in a single bit true (1) or false (0) value. For example, with a and b defined as previously stated,

& a = 0

& b = 0

| a = 1

| b = 0

^ a = 1

^ b = 0

The symbol '|' used for the OR is called a **pipe**.

The next group are the binary operators that have two operands. These are used in both bit-wise and logical contexts. With a = 1010 and b = 0011, the bit-wise application of the operators acts in a bit-by-bit manner:

a & b = 0010

a | b = 1011

a ^ b = 1001

In a logical context, the answer is a single true (non-zero) or false (all zeros) number.

a && b = 1

a || b = 1

a && c = 0

where c = 0000.

The equality operators are =, ==, and ===. The assignment operator = is used to copy the value from the right side of an expression to the left side as in

a = 4'b1010

The equality operator == is used in

```
a == b
```

to express "a is equal to b." Identity is written as

```
c === d
```

which says that "c is identical to d."

Timing Controls

Timing controls statements dictate the times when actions take place. There are three types of timing controls that are used in a procedural block. A simple delay is specified using **#** <time> as the clock example. An edge-triggered control is of the form @ (signal). In the statement

```
@(posedge clk) reg_1 = reg 2 ;
```

the **posedge** keyword is used to induce the assignment when the clock clk is rising from a 0 to a 1, or from x or z to 1. The positive edge of the clock is shown in Figure 10.23. Similarly, a negative-edge triggered event can be described by a statement of the form

```
@(negedge clk) output = a_in ;
```

A negative edge transition is from a 1 to a 0, or from x or z to 0. Edge-triggered statements can include the possibility of several signals changing by using the **or** keyword. Level-triggered events are modeled using the **wait** keyword.

```
wait (clk) q_out = d_in ;
```

effects the transfer when clk is a 1. In general, the **wait** directive executes when the expression is logically TRUE (i.e., non-zero).

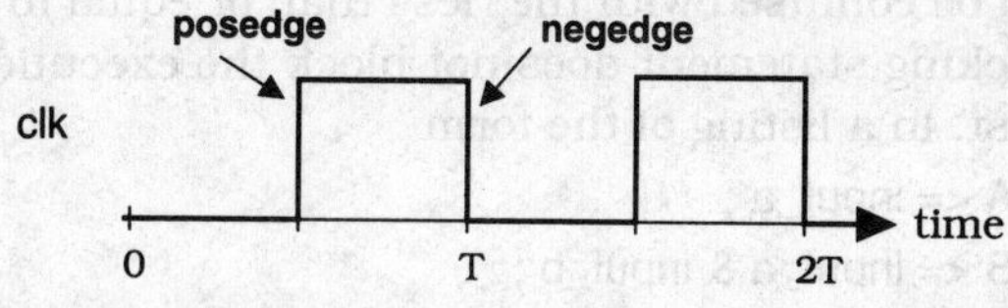

Figure 10.23 Clock edges

Procedural Assignments

A procedural assignment is used to change or update the values of reg and other variables. They are usually divided into **blocking** and **non-blocking** assignments.

Blocking assignments are executed in the order that they are listed, and allow for straightforward sequential and parallel blocks. The assignment operator " = " is used in these statements. Consider the simple code listing

```
reg a, b, c, reg_1, reg_2 ;
initial
```

```
begin
    a = 1 ;
    b = 0 ;
    c = 1 ;
    # 10 reg_1 = 1'b0 ;
    # 5 c = reg_1 ;
    # 5 reg_2 = b ;
end
```

The sequence of the statements is important. The assignments for a, b, and c are all performed at time 0. After 10 time units, reg_1 is assigned the value of binary 1. Then 5 time units after this event (at 15 time units) c is assigned the value of reg_1. Finally, at 20 time units, reg_2 = b is executed. The events are summarized in Figure 10.24.

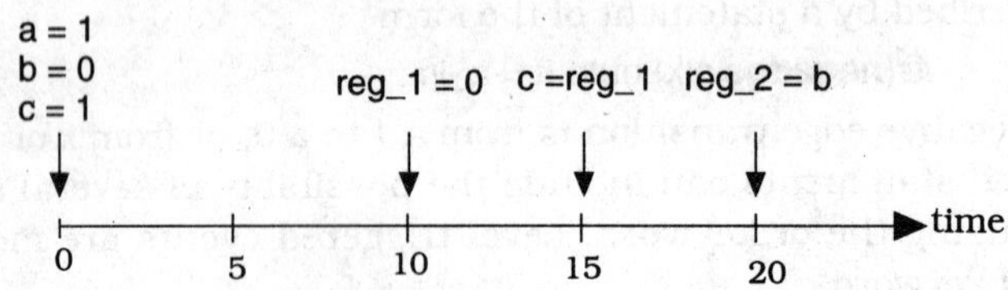

Figure 10.24 Timing for non-blocking assignment example

Non-blocking assignments contain the <= assignment operator, which is not to be confused with the "less than or equal to" relational operator. A non-blocking statement does not block the execution of other statements in the list. In a listing of the form

```
A <= input_a ;
B <= input_a & input_b ;
```

the right-hand sides are evaluated first according to any timing control statements; the values are then transferred to the left-hand side variables. Non-blocking assignments require careful consideration of related quantities. Consider the coding below.

```
initial
    begin
        in_a = 1 ;
        in_b = 0 ;
    end
always
    # 10 clk = ~ clk ;
always @ (posedge clk)
```

```
    begin
        in_a <= in_b ;
        in_b <= in_a ;
    end
```

This has the effect of switching the values of in_a and in_b from their initial assignments on every rising clock edge.

Conditional Statements

Condition-dependent statements are very powerful constructs in behavioral modeling. They exist at the highest level of abstraction and describe events using C-like statement groups.

The **if / else if** constructs allow different outcomes depending upon current conditions. Let us examine the module listing

```
module if_else_example ( ctrl , alu_op, clk) ;
input [ 1:0 ] alu_op ;
input clk ;
output ctrl_a ;
reg ctrl_a ;
always @ ( posedge clk )
  begin
    if (alu_op == 0)
        ctrl = 0 ;
    else if (alu_op == 1)
        ctrl = 1 ;
    else
        $display (" Signal ctrl is greater than 1") ;
    end
endmodule
```

This detects the value of alu_op on the positive edge of the clock clk and gives three possible outcomes for ctrl. The last **else** line takes care of the unlisted values of alu_op = 2, 3. The **if /else** constructs can be nested, with the **else** being associated with the nearest **if**.

A case statement is another powerful construct. It has the form

```
case (condition)
```

and stipulates outcomes that depend on the value of condition. This is seen in the simple 2:1 multiplexor description below.

```
module simple_mux (mux_out, p0, p1, select ) ;
input p0, p1 ;
input select ;
output mux_out ;
```

```
always @ (select or p0 or p1)
   case ( select )
   1'b0 : mux_out = p0 ;
   1'b1 : mux_out = p1 ;
   endcase
endmodule
```

The **case** listing accounts for the two possible values of select, and dictates which input p0 or p1 is sent to mux_out whenever there is a change in the variables listed in the **always @** statement.

Another type of coding is obtained using looping statements. The **repeat** loop executes a set of statements a specific number of times. For example, suppose that the variable counter has a value of 10. Then,

```
repeat (counter)
   begin
      ...
   end
```

performs the procedures listed in between the **begin / end** statements in a loop manner a total of counter = 10 times. A related keyword is **while** with the syntax

```
while ( condition )
   begin
      ...
   end
```

This executes the **begin / end** block so long as condition is true (non-zero). If condition is initially false (zero), then the entire block is ignored. Verilog also has the **for** construct with the syntax

```
for ( condition ) ...
```

that allows the sensing of condition expressions to execute the statement block.

A **forever** loop does exactly what its name implies: it executes for the entire length of the simulation. The clock generation module below illustrates this construct.

```
module clk_1 ;
reg clk ;
initial
   begin
      clk = 0 ;
      forever
         begin
         # 5 clk = 1 ;
```

```
        # 5 clk = 0 ;
        end
    end
endmodule
```

A few other conditional constructs are available, but these illustrate the most-used coding styles.

Dataflow Modeling and RTL

Dataflow modeling describes a system by how the data moves and is processed. As with general behavioral modeling, a dataflow description is a high-level abstraction that does not provide structural details. Although the definition tends to vary, register-transfer level (RTL) modeling is usually interpreted as a combination of dataflow and behavioral coding styles. It uses high-level constructs that can be used as an input into a synthesis tool, which is then used to generate a gate netlist. Not all behavioral keywords and statements can be synthesized, so RTL centers around a restricted set. Mastery of RTL usually revolves around learning how to write synthesizable code, which is far beyond the scope of this book.

We will be content with introducing the **assign** keyword. Continuous assignments define relationships and values. For example, statements of the form

```
assign a = ~ b & c ;
assign out_1 = ( a | b ) & (c | d ) ;
```

can be used to define combinational logic operations. A useful conditional statement is

assign output = (*something*) ? < *true condition* > : < *false condition* >

In this case, *something* represents a variable or a statement. The value of output depends if *something* is true or false. The description of a 2:1 MUX can be written as

```
module mux_2 (out, p0, p1, select ) ;
input p0, p1 ;
input select ;
output mux_out ;
assign out = (select ) ? p1 : p0 ;
endmodule
```

Statements of this type can be nested. We will see more examples of **assign** and other dataflow constructs in later chapters.

10.6 References

Verilog is a rich and powerful language with many intricate details. All of the books listed below provide more in-depth treatments than can be included in this text. Reference [3] is a textbook, while [2] and [8] are written with a textbook flavor, but with fewer examples and details. Reference [7] is designed for a rapid introduction to Verilog. Books [9] and [12] are comprehensive references.

[1] Mark Gordon Arnold, **Verilog Digital Computer Design**, Prentice-Hall PTR, Upper Saddle River, NJ, 1999.

[2] J. Bhasker, **A Verilog HDL® Primer**, Star Galaxy Press, Allentown, PA, 1997.

[3] Michael D. Ciletti, **Modeling, Synthesis and Rapid Prototyping with the Verilog HDL**, Prentice-Hall, Upper Saddle River, NJ, 1999.

[4] Ken Coffman, **Real World FPGA Design with Verilog**, Prentice-Hall PTR, Upper Saddle River, NJ, 2000.

[5] Dan Fitzpatrick and Ira Miller, **Analog Behavioral Modeling with the Verilog-A Language**, Kluwer Academic Press, Norwell, MA, 1999.

[6] Pran Kurup and Taher Abasi, **Logic Synthesis Using Synopsys®**, 2nd ed., Kluwer Academic Publishers, Norwell, MA, 1997.

[7] James M. Lee, **Verilog Quickstart!**, Kluwer Academic Publishers, Norwell, MA, 1998.

[8] Samir Palnitkar, **Verilog® HDL**, SunSoft Press (Prentice-Hall), Mountain View, CA, 1996.

[9] Vivek Sagdeo, **The Complete Verilog Book**, Kluwer Academic Publishers, Norwell, MA, 1998.

[10] Bruce Shrive and Bennett Smith, **The Anatomy of a High-Performance Microprocessor**, IEEE Computer Society Press, Los Alamitos, CA, 1998.

[11] David R. Smith and Paul D. Franzon, **Verilog Styles for Synthesis of Digital Systems**, Prentice-Hall, Upper Saddle River, NJ, 2000.

[12] Donald E. Thomas and Philip R. Moorby, **The Verilog® Hardware Description Language**, 4th ed., Kluwer Academic Press, Norwell, MA, 1998.

[13] Bob Zeidman, **Verilog Designer's Library**, Prentice-Hall PTR, Upper Saddle River, NJ, 1999.

10.7 Problems

[10.1] Write the gate-level structural description for the module illustrated in Figure P10.1.

[10.2] Consider the logic network illustrated in Figure P10.2. Construct a Verilog module that describes the circuit.

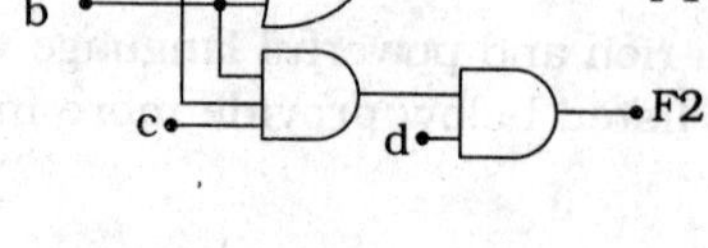

Figure P10.1

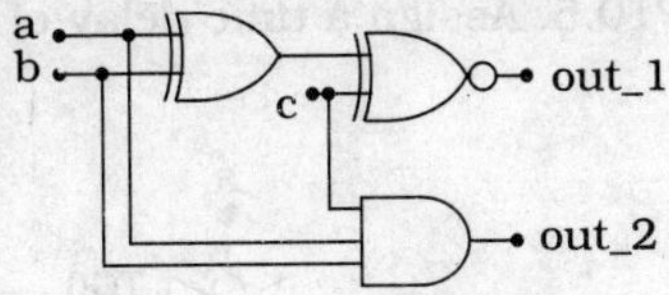

Figure P10.2

[10.3] Write a Verilog description of the NAND latch in Figure P10.3. Include a time delay of 2 units for each NAND gate.

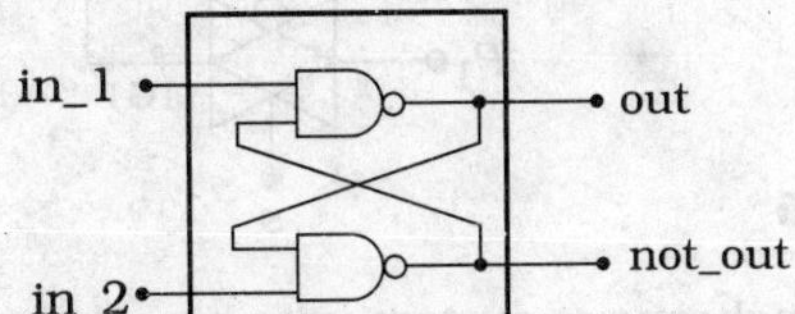

Figure P10.3

[10.4] Construct the Verilog module for the logic network shown in Figure P10.4. Assume that the NOT gates have a time delay of 1 unit, while the AND2 gates have a delay of 2 units.

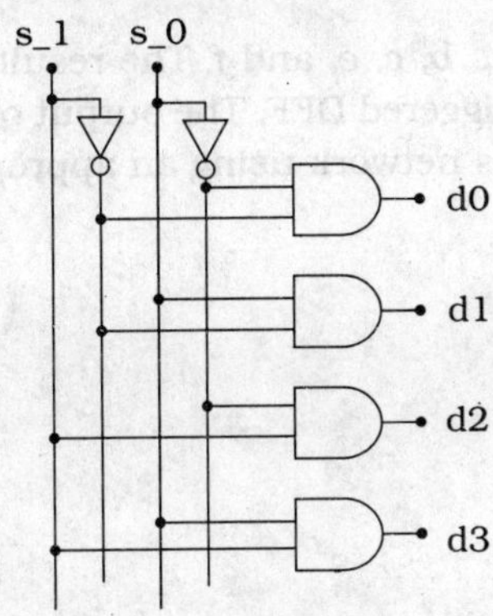

Figure P10.4

[10.5] Construct the circuit diagram for a CMOS logic gate that implements the function

$$f = \overline{a \cdot (b + c) + b \cdot d} \tag{10.3}$$

Then write a Verilog description of the circuit using the **nmos** and **pmos** primitives.

[10.6] Construct a pseudo-nMOS logic gate for the function

$$F = \overline{a \cdot b \cdot c + a \cdot (d + e)} \tag{10.4}$$

Then use the **nmos** and **pullup** primitives to write a Verilog description of the circuit.

[10.7] Use the **cmos** primitive to write a Verilog module listing for the 2:1 MUX in Figure P10.5. Assign a time delay of 2 units to each transmission gate.

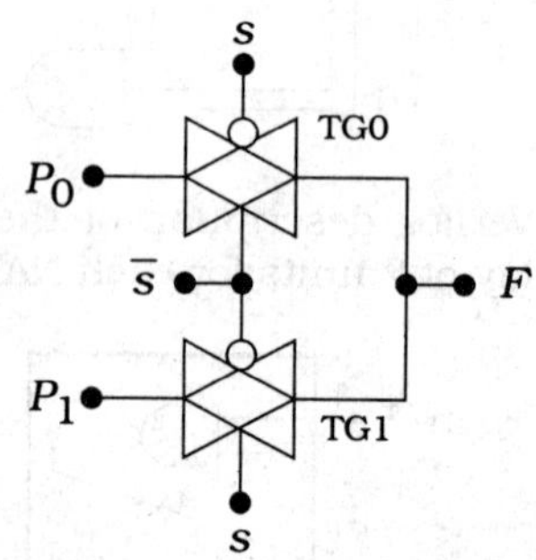

Figure P10.5

[10.8] A synchronous system uses complementary clock phases ϕ and $\overline{\phi}$ to control data flow. Construct a Verilog module that provides two non-overlapping clocking signals clk and clk_bar with a period of 40 time units.

[10.9] A combinational logic network produces the function

$$A = \bar{a} \cdot b + c \cdot e + a \cdot (\bar{c} + f) \tag{10.5}$$

from the inputs a, b, c, e, and f. The result A is connected to the input of a positive-edge triggered DFF. The output of the DFF is q. Write the Verilog description of this network using an appropriate set of primitives.

General VLSI System Components 11

VLSI systems design revolves around a library of component functions. Primitive entries include FETs and basic logic gates, but higher level functions are needed to build the system hierarchy. In this chapter we will study a few examples of system components that are used to build large-scale systems and are commonly found in a VLSI cell library. The list of cells presented here is not comprehensive; additional components will be introduced in later chapters. Instead, the approach is intended to emphasize the connection between a high-level architectural specification and the resulting circuits and silicon implementation.

11.1 Multiplexors

Multiplexors are indispensable in modern digital design. A MUX consists of n input lines and one output f. The main function of the component is to use an m-bit select word to connect one of the inputs to the output. To cover every input, we must choose m such that $n = 2^m$. An alternate way to specify this is to introduce the base-2 logarithm such that $\log_2(2) = 1$. Then, using

$$m = \log_2(2^m) \tag{11.1}$$

we can say that

$$m = \log_2(n) \tag{11.2}$$

gives the number of select lines.

The simplest example is a 2-to-1 multiplexor. There are several ways to describe the component. A behavioral description using the **case** statement was presented in the previous chapter, and is repeated here for reference. The input lines are designated as p0 and p0, and the select bit is denoted by the identifier select.

```
module simple_mux (mux_out, p0, p1, select ) ;
input p0, p1 ;
input select ;
output mux_out ;
always @ (select )
    case ( select )
    1'b0 : mux_out = p0 ;
    1'b1 : mux_out = p1 ;
    endcase
endmodule
```

Given this description, several different logic and circuit implementations can be obtained. A gate-level NAND implementation is illustrated in Figure 11.1. Applying DeMorgan's theorem allows us to bubble-push to obtain the SOP (AND-OR) form

$$f = p_0 \cdot \bar{s} + p_1 \cdot s \tag{11.3}$$

Since a NAND2 gate requires four FETs, the entire network with drivers can be implemented in 16 transistors.[1]

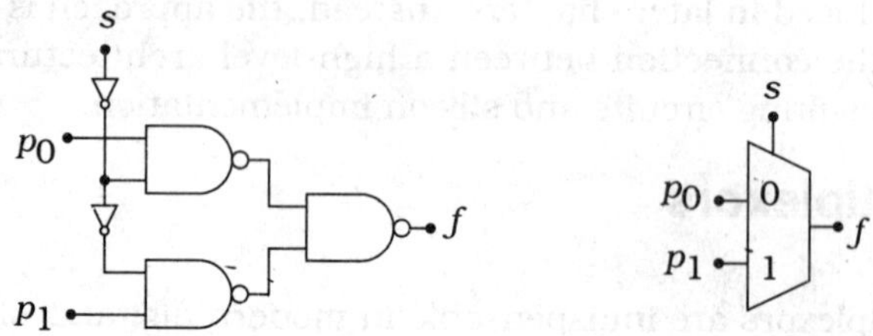

Figure 11.1 Gate-level NAND 2:1 multiplexor

The transmission-gate circuit in Figure 11.2(a) could also be a candidate in a CMOS technology. This circuit uses four FETs for the path logic (two for each TG). If we include a buffering NOT pair for the select bit, then the total FET count is increased to 8. The main problem with this circuit is that the TGs have parasitic resistance and capacitance that slow down the response. Figure 11.2(b) shows the same configuration using only nFET switches. With the select drivers, the FET count is reduced to 6. However, we have added an inverting driver at the output to compen-

[1] Note that this network can also be described using a structural listing.

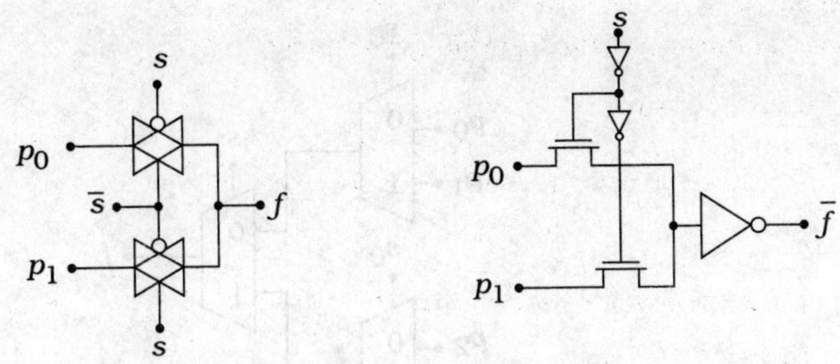

(a) TG circuit (b) Pass transistors

Figure 11.2 Multiplexor using switch logic

sate for the fact that the nFET passes only a voltage range of 0 V to V_{max} where

$$V_{max} = V_{DD} - V_{Tn} \tag{11.4}$$

with V_{Tn} the threshold voltage. The inverter helps to restore the output to the full-rail range [0, V_{DD}]. Although the FET count is the same as for the TG circuit, the layout wiring will be easier. When translating the high-level description to silicon, considerations such as these become important factors.

Larger multiplexors can be designed using primitive gates or by instancing 2:1 devices. Consider a 4:1 MUX described by

```
module bigger_mux (out_4, p0, p1, p2, p3, s0, s1) ;
input p0 , p1, p2, p3 ;
input s0, s1 ;
output out_4 ;
assign out_4 = s1 ? (s0 ? p3 : p2 ) : (s0 ? p1 : p0) ;
endmodule
```

Since this is a high-level abstraction, no details about the internal structure are given. However, the **assign** statement can be interpreted as three 2:1 separate multiplexors with the first ? : using s1, and the second and third occurrences based on s0. This implies the structure illustrated in Figure 11.3. The select bit s0 is used to select (p0, p2) or (p1, p3) in the first stage devices. The final selection is achieved with s1 determining the actual output *f* which is the same as out_4 in the Verilog listing.

Another implementation is the gate-level construction shown in Figure 11.4. Using this as a guide, the equivalent Verilog structural description would be

```
module gate_mux_4 (out_gate, p0, p1, p2, p3, s0, s1) ;
input p0, p1, p2, p3 ;
input s0, s1 ;
```

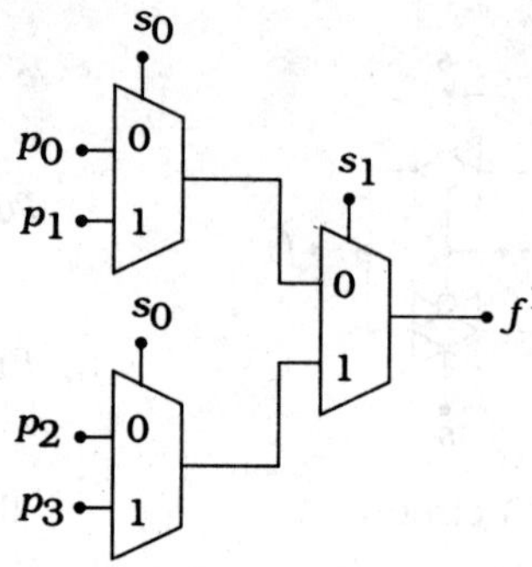

Figure 11.3 A 4:1 MUX using instanced 2:1 devices

```
wire w1, w2, w3, w4 ;
output out_gate ;
nand (w1, p_0, ~s1, ~s0) ,
     (w2, p_1, ~s1, s0) ,
     (w3, p_2, s1, ~s0) ,
     (w4, p_3, s1, s0) ,
     (out_gate, w1, w2, w3, w4) ;
endmodule
```

The NOT gates have been modeled using ~ operators, but they could have been instanced with primitive **not** gates with the same result. In standard logic, this is equivalent to the SOP expression

$$f = p_0 \cdot \overline{s_1} \cdot \overline{s_0} + p_1 \cdot \overline{s_1} \cdot s_0 + p_2 \cdot s_1 \cdot \overline{s_0} + p_3 \cdot s_1 \cdot s_0 \tag{11.5}$$

obtained from applying basic logic.

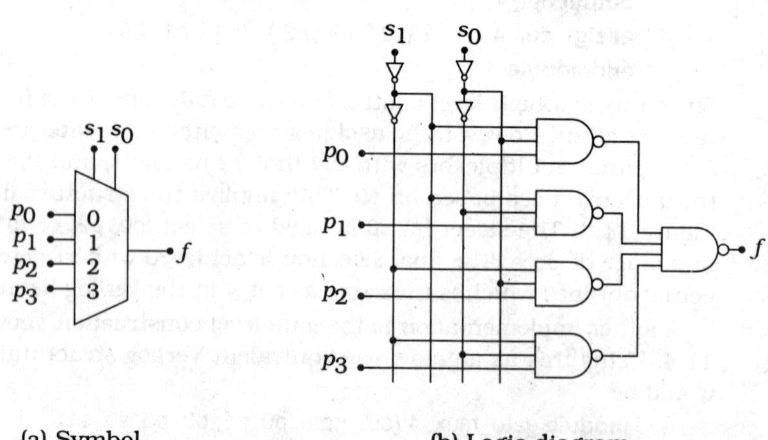

Figure 11.4 Gate-level 4:1 MUX

Yet another network is the pass-FET array shown in Figure 11.5. This uses the ANDing properties of nFETs to implement the logic expression directly. The structural description of this circuit is given by

```
module tg_mux_4 (f, p0, p1, p2, p3, s0, s1) ;
input p0, p1, p2, p3 ;
input s0, s1 ;
wire w0, w1, w2, w3, w_o , w_x ;
output f ;
nmos (p0, w0, ~ s1), (w0, w_o, ~ s0) ;
nmos (p1, w1, ~ s1), (w1, w_o, s0) ;
nmos (p2, w2, s1), (w2, w_o, ~ s0) ;
nmos (p3, w3, s1), (w3, w_o, s0) ;
not (w_x, w_o), (f , w_x ) ;
endmodule
```

where the **nmos** instances have been grouped to make them easier to trace. The wires between the FETs in the *i*-th line are labeled wi for i = 0, 1, 2, 3 and w_o is the common output wire for the four paths. The wire between the inverters is w_x. A simple layout strategy for the cell is shown in Figure 11.6. The one-to-one correspondence is obvious, but the packing density could be improved by moving the FETs and rerouting.

The split-array nMOS/pMOS circuit shown in Figure 11.7 provides the function using similar reasoning. Every input sees a path to the output through both an nFET and a pFET chain. Since pFETs can pass logic 1 voltages while nFETs pass logic 0 voltages, the output has a full-rail swing from 0 V to V_{DD}. Output restoring buffers are not required in this case, but the circuit will be much faster if they are added. This arrangement uses the ideas of complementary pairs that are wired in a manner similar to TGs. However, since the nFETs and pFETs circuits are sepa-

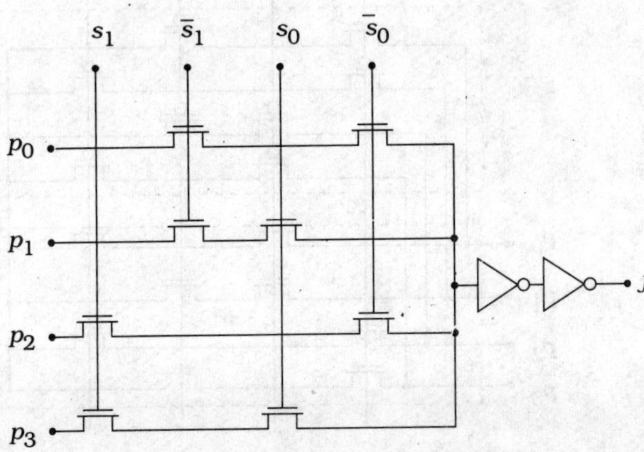

Figure 11.5 4:1 MUX using nFET pass transistors

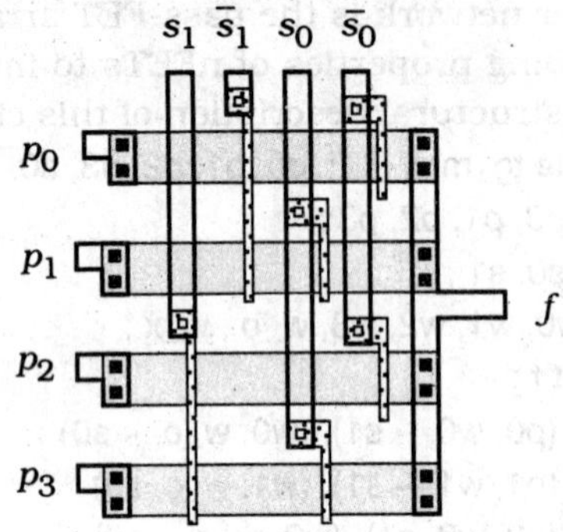

Figure 11.6 Simple 4:1 pass-FET MUX layout

rate, the interconnect wiring will be simpler than if TGs were used for every switch. These examples illustrate the variations that are possible in translating a high-level HDL construct to basic logic circuits.

Architectural specifications treat *n*-bit words in about the same manner as single-bit entities. Suppose that we have two 8-bit words

$$a = a_7a_6a_5a_4a_3a_2a_1a_0$$
$$b = b_7b_6b_5b_4b_3b_2b_1b_0 \tag{11.6}$$

that we want to use as inputs to a 2:1 MUX. The output

$$f = f_7f_6f_5f_4f_3f_2f_1f_0 \tag{11.7}$$

is determined by the select bit *s* such that

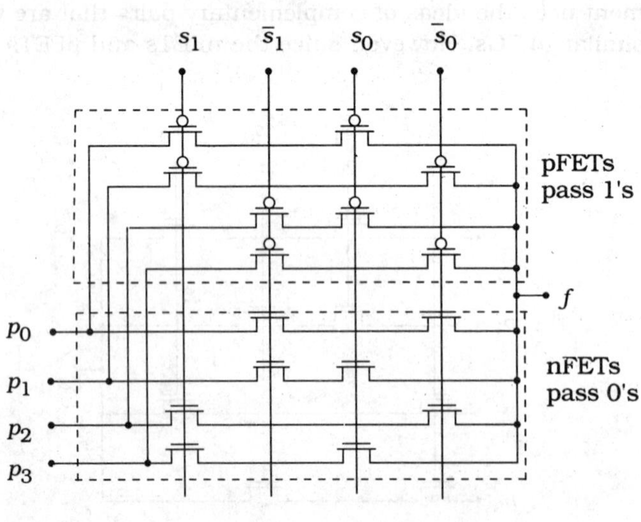

Figure 11.7 Split-array 4:1 MUX for full-rail output

$$f_i = a_i \cdot \bar{s} + b_i \cdot s \tag{11.8}$$

for $i = 0, ..., 7$. This of course implies that we should use 8 identical 2:1 MUXes that are all controlled by the same select bit s.

At the system level, we tend to treat a and b as single objects. The MUX symbol in Figure 11.8(a) identifies the width of the word using the slash notation (/) across the lines. A Verilog description can be written using 8-bit vectors as specified by the [7 : 0] in the listing

```
module mux_2-1_8b (f, a, b, s ) ;
input [7 : 0] a, b ;
input s ;
output [7 : 0] f ;
   assign f = s ? b : a ;
endmodule
```

The extension to larger word sizes is accommodated by simply resizing the vectors. The bit-level implementation is more complicated, since we must take n parallel 2:1 MUX units as in Figure 11.8(b). At the physical level, every 1-bit MUX consumes area and is characterized by a set of delay times. To build the 8-bit network, we can tile eight identical cells as shown in Figure 11.9. Since silicon circuits (and all logic gates) are designed at the bit level, the layout area and wiring may become the limiting factor.

This discussion is intended to illustrate an important point. Given a high-level architectural description of a digital network, several choices can be made for the circuitry. Each choice leads to a distinct physical design with its own layout and switching characteristics. In a critical timing path the TG circuits may not be fast enough. In a different situation, a limited real estate allocation may make area more important. VLSI system design is not just writing good code or pushing polygons. Every level of the hierarchy has some effect on every other level. In a top-down design approach, we often find that the behavioral specifications can be imple-

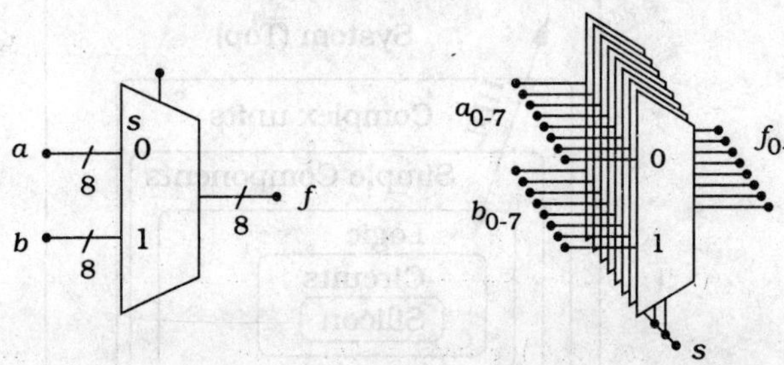

(a) Symbol (b) Bit-level realization

Figure 11.8 A vector 2:1 MUX

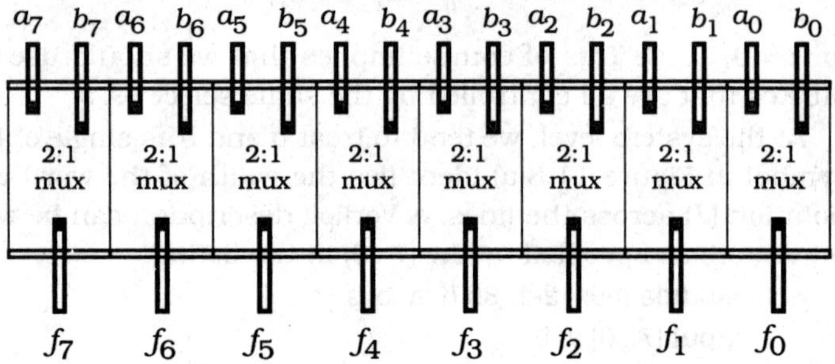

Figure 11.9 Single-bit cell tiling for an 8-bit 2:1 MUX

mented in several ways. Some are better than the others, but the choice depends upon the metrics as they are applied throughout the design cycle.

This theme will be maintained throughout the chapter. Several logic components will be analyzed at different levels to illustrate the connection between a high-level architectural description and the circuit or silicon network. While the components are important in their own right, the critical system-level features emerge only after the components are used to construct more complex logic units. The hierarchy is illustrated by the nesting diagram in Figure 11.10. In top-down design, we start with a set of specifications and design the high-level architectural model in an HDL. The progression downward to silicon eventually results in a silicon device, but the interaction among the various levels is linked in a critical manner. No system can operate faster than each subsequent level allows, while the silicon real estate budget is the bottom line limit. And, of course, we must be able to manufacture the chip and sell it at a price that users are willing to pay while still maintaining a profit margin!

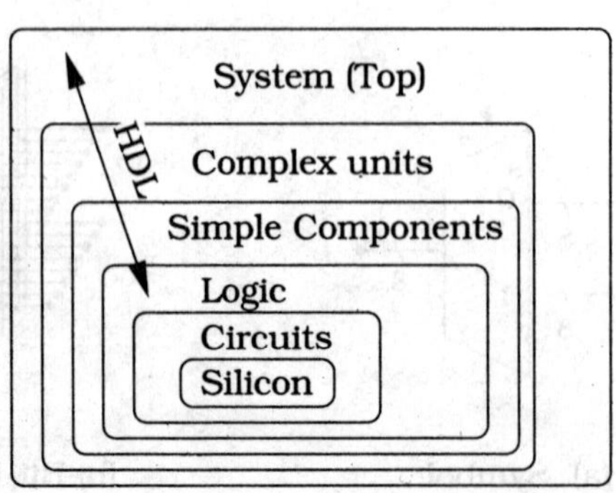

Figure 11.10 System-level hierarchy

11.2 Binary Decoders

A binary n/m row decoder accepts an n-bit control word and activates one of the m-output lines while the other $(m - 1)$ lines are not affected. An **active-high** decoder sets a 1 on the selected line and keeps the others at 0. An **active-low** decoder is just the opposite, with the selected line reset to 0 while the remaining lines are at 1.

A 2/4 active-high decoder symbol and function table is shown in Figure 11.11(a). The 2-bit select word $s_1 s_0$ activates the line corresponding to its specified decimal value 0, 1, 2, 3. The function table yields the equations

$$
\begin{aligned}
d_0 &= \overline{s_1} \cdot \overline{s_0} = \overline{s_1 + s_0} \\
d_1 &= \overline{s_1} \cdot s_0 = \overline{s_1 + \overline{s_0}} \\
d_2 &= s_1 \cdot \overline{s_0} = \overline{\overline{s_1} + s_0} \\
d_3 &= s_1 \cdot s_0 = \overline{\overline{s_1} + \overline{s_0}}
\end{aligned}
\tag{11.9}
$$

A straightforward NOR-gate implementation is shown in Figure 11.11(b). This gives the basis for the structural description

```
module decode_4 (d0, d1, d2, d3, s0, s1) ;
input s0, s1 ;
output d0, d1, d2, d3 ;
nor   (d3, ~s0, ~s1) ,
      (d2, ~s0, s1) ,
```

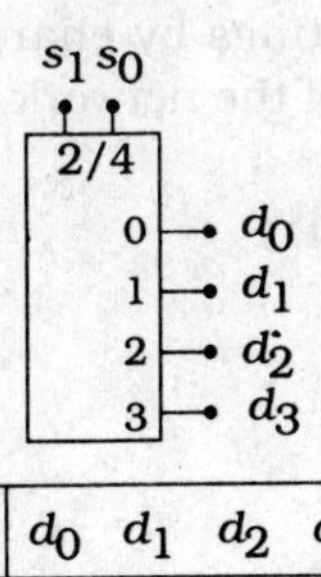

s_1	s_0	d_0	d_1	d_2	d_3
0	0	1	0	0	0
0	1	0	1	0	0
1	0	0	0	1	0
1	1	0	0	0	1

(a) Symbol and table

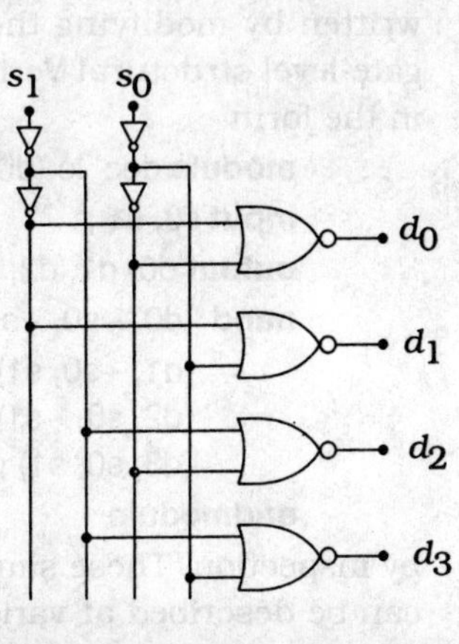

(b) NOR2 implementation

Figure 11.11 An active-high 2/4 decoder

```
            (d1, s0, ~s1) ,
            (d0, s0, s1) ;
endmodule
```

where we have absorbed the NOT drivers into the notation using the ~ operator.

An equivalent architectural description using **case** keywords can be written as

```
module dec_4 (d0, d1, d2, d3, sel ) ;
input [1 : 0] sel ;
output d0, d1, d2, d3 ;
case (sel)
    0 : d0 = 1, d1 = 0 , d2 = 0, d3 = 0 ;
    1 : d0 = 0, d1 = 1 , d2 = 0, d3 = 0 ;
    2 : d0 = 0, d1 = 0 , d2 = 1, d3 = 0 ;
    3 : d0 = 0, d1 = 0 , d2 = 0, d3 = 1 ;
endmodule
```

which explicitly lists each possibility depending on the decimal value of sel. Another approach would be to use **assign** procedures. This represents an abstract high-level description of the operation that contains no structural information. While one can understand the operation of the unit, it must be translated into a lower level description before it can be built. This provides another example of equivalent hierarchical views.

An active-low decoder is shown in Figure 11.12. In this case, the selected output is driven low while the others remain at logic 1 values. The design is achieved by simply replacing the NOR2 gates with NAND2 gates and complementing the inputs to each gate. The HDL code can be written by modifying the active-high listings by changing the logic. The gate-level structural Verilog description of the network can be constructed in the form

```
module dec_lo (d0, d1, d2, d3, s0, s1) ;
input s0, s1 ;
output d0, d1, d2, d3 ;
nand (d0, ~s0, ~s1) ,
     (d1, ~s0, s1) ,
     (d2, s0, ~s1) ,
     (d3, s0, s1) ;
endmodule
```

by inspection. These simple examples clearly show how large components can be described at various levels. The one that is most used in practice depends upon the problem and its level in the design hierarchy. In general, no single solution will be optimal for all situations.

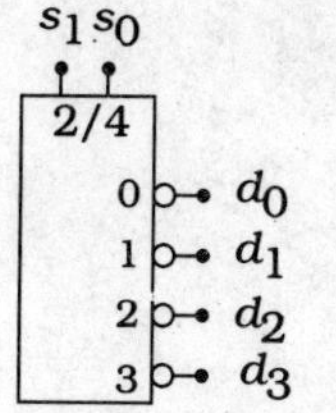

s_1	s_0	d_0	d_1	d_2	d_3
0	0	0	1	1	1
0	1	1	0	1	1
1	0	1	1	0	1
1	1	1	1	1	0

(a) Symbol and table

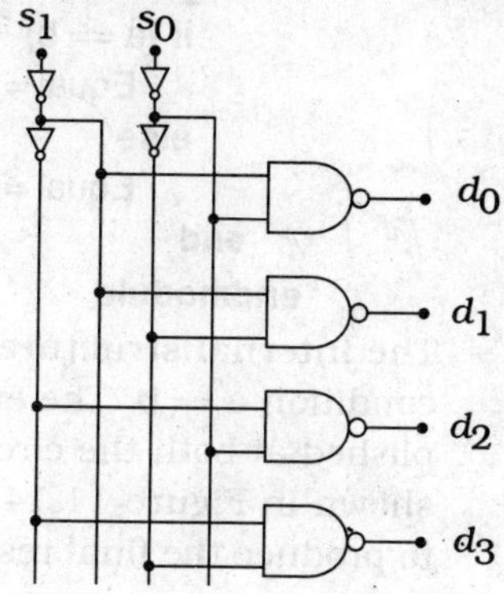

(b) NAND2 implementation

Figure 11.12 Active low 2/4 decoder

11.3 Equality Detectors and Comparators

An equality detector compares two n-bit words and produces an output that is 1 if the inputs are equal on a bit-by-bit basis. A simple 4-bit circuit is shown in Figure 11.13. This uses the equality (XNOR) relation

$$\overline{a_i \oplus b_i} = 1 \tag{11.10}$$

iff $a_i = b_i$ as a means to compare the inputs. If every XNOR produces a 1, then the output AND gate gives *Equal* = 1; otherwise, *Equal* = 0.

A Verilog listing for the operation is

```
module equality (Equal , a, b ) ;
input [3 : 0] a, b ;
output Equal ;
```

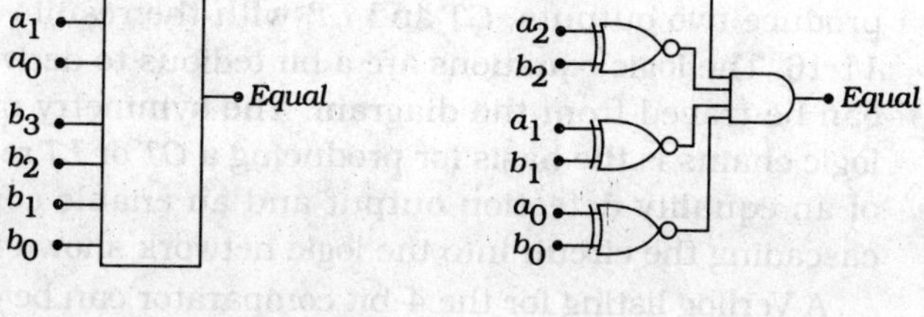

Figure 11.13 A 4-bit equality detector

```
always @ (a or b)
    begin
        if (a == b)
            Equal = 1 ;
        else
            Equal = 0 ;
    end
endmodule
```

The internal structure of the circuitry is hidden in the logical equality condition a == b. The extension to an arbitrary word size is easily accomplished at both the circuit and HDL level. An example is the 8-bit version shown in Figure 11.14. This uses two 4-bit circuits with ANDed outputs to produce the final result.

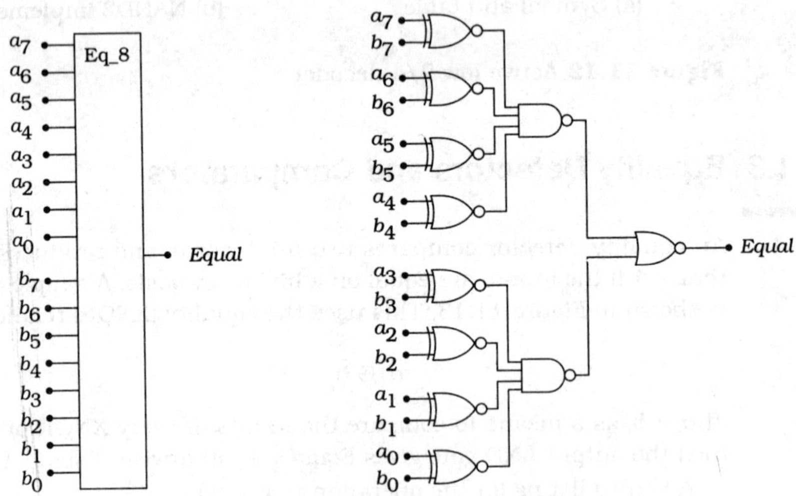

Figure 11.14 8-bit equality detector

Magnitude comparator circuits are used to compare two words a and b and determine if $a > b$ or $a < b$ is true; the equality condition $a = b$ may also be detected by the logic. The logic for a 4-bit magnitude comparator is shown in Figure 11.15. The input words are used on a bit-wise basis to produce two outputs, *GT* and *LT*, with the results summarized in Figure 11.16. The logic equations are a bit tedious to derive, but the signal paths can be traced from the diagram. The symmetry of the upper and lower logic chains is the basis for producing a *GT* or *LT* result. Optional features of an equality detection output and an enable control can be added by cascading the circuit into the logic network shown in Figure 11.17.

A Verilog listing for the 4-bit comparator can be constructed as follows.

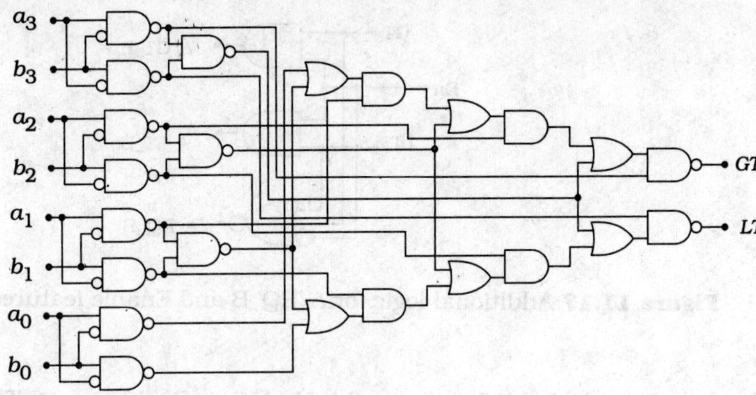

Figure 11.15 4-bit magnitude comparator logic

```
module comp_4 (GT, LT, a, b ) ;
input [ 3 : 0 ] a; b ;
output GT, LT ;
always @ (a or b)
    begin
        if (a > b)
            GT = 1 , LT = 0 ;
        elseif (a < b)
            GT = 0 , LT = 1 ;
        else
            GT = 0 , LT = 0 ;
    end
endmodule
```

The high-level description masks the internal structure completely, making it appropriate for architectural simulations. However, the logic and circuit implementations can be quite complicated .

The hierarchical design technique allows us to build an 8-bit comparator using two 4-bit circuits (Comp 4) and an interfacing network. The main circuit is shown in Figure 11.18. The lower Comp 4 block accepts the lower 4 bits of each word, while the upper block uses bits 4–7. The

Condition	GT	LT
$a > b$	1	0
$a < b$	0	1
$a = b$	0	0

Figure 11.16 Comparator output summary

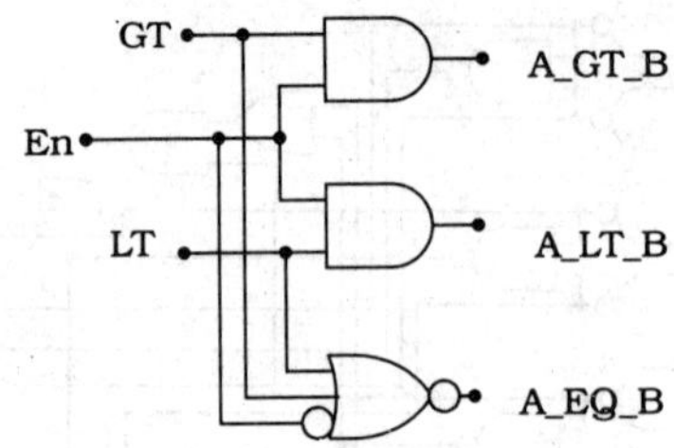

Figure 11.17 Additional logic for A_EQ_B and Enable features

interface block labeled Comp 8 includes an Enable input. The logic diagram for the interfacing network is shown in Figure 11.19. The upper inputs are the *GT* Comp 4 outputs, while the lower inputs are *LT* values from the 4-bit comparison circuits. These are then compared to produce the outputs. Including the AND gate and the NAND-NOT cascades (at the A_GT_B and A_LT_B outputs) allows us to generate the equality signal A_EQ_B that will be 1 if the words are equal. Note that A_GT_B and A_LT_B are both zero when A_EQ_B = 1.

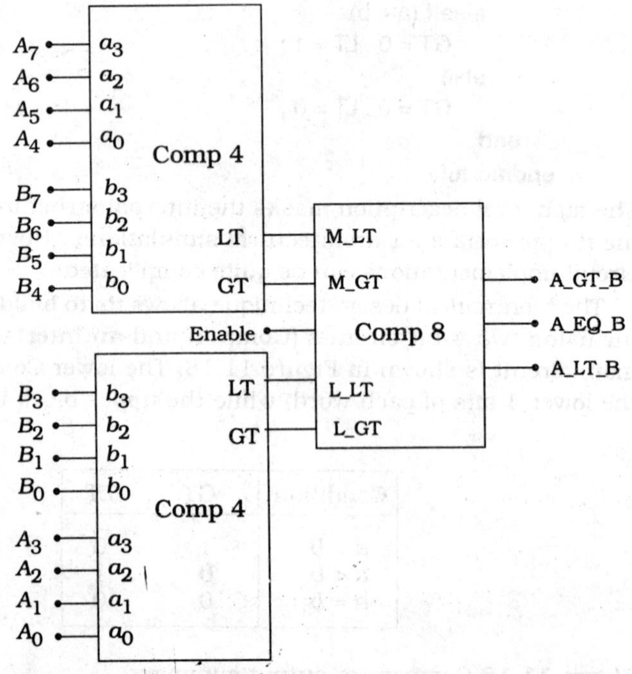

Figure 11.18 8-bit comparator system

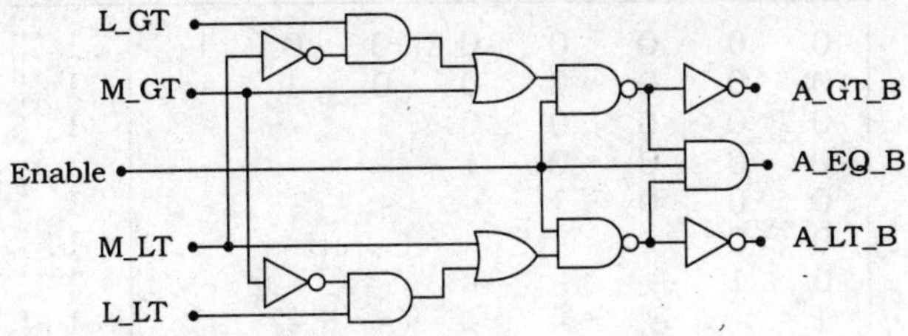

Figure 11.19 Comp 8 logic diagram

11.4 Priority Encoder

A priority encoder examines the input bits of an *n*-bit word and produces an output that indicates the position of the highest priority logic 1 bit. Consider an 8-bit word

$$d = d_7 d_6 d_5 d_4 d_3 d_2 d_1 d_0 \tag{11.11}$$

and let us assign the highest priority to bit d_7, the second highest priority to d_6, and so on. The operation of a priority encoder is to detect the presence of 1's in d; if two or more bits are at a logic 1 value, then the input with the highest priority takes precedence. If we use d as the input to an 8-bit priority encoder, then the output word

$$Q = Q_3 Q_2 Q_1 Q_0 \tag{11.12}$$

is coded to indicate the highest priority bit. A function table for this scheme is provided in Figure 11.20. The bit Q_3 is equal to 1 if any input bit is a 1. The 3-bit word $Q_2 Q_1 Q_0$ is encoded to indicate the highest priority input bit. There is no formal logic symbol, so we will use the simple box shown in Figure 11.21 when the device is used in a system design.

The logic for the network is drawn in two parts. The first section in Figure 11.22 shows the input buffers and complement generators for each bit. The output logic for the $Q2$ and $Q3$ is simple and is given by the expressions

$$\begin{aligned} Q2 &= (d_0 + d_1 + d_2 + d_3) \cdot \overline{(d_4 + d_5 + d_6 + d_7)} \\ Q3 &= (d_0 + d_1 + d_2 + d_3) + \overline{(d_4 + d_5 + d_6 + d_7)} \end{aligned} \tag{11.13}$$

as can be verified from the schematic. The $Q0$ and $Q1$ encoders use the buffered and complemented inputs as shown in the circuits of Figure 11.23. The logic equation for the $Q0$ circuit is

$$Q0 = \overline{d_7} \cdot [d_6 + \overline{d_5} \cdot (d_4 + \overline{d_3} \cdot [d_d + \overline{d_1} \cdot d_0])] \tag{11.14}$$

d_7	d_6	d_5	d_4	d_3	d_2	d_1	d_0	Q_3	Q_2	Q_1	Q_0
0	0	0	0	0	0	0	1	1	0	0	0
0	0	0	0	0	0	1	-	1	0	0	1
0	0	0	0	0	1	-	-	1	0	1	0
0	0	0	0	1	-	-	-	1	0	1	1
0	0	0	1	-	-	-	-	1	1	0	0
0	0	1	-	-	-	-	-	1	1	0	1
0	1	-	-	-	-	-	-	1	1	1	0
1	-	-	-	-	-	-	-	1	1	1	1
0	0	0	0	0	0	0	0	0	0	0	0

d_7 has highest priority
d_0 has lowest priority

$Q_3 = 1$ when $d_i = 1$
for any $i = 0, \ldots, 7$

Figure 11.20 Function table for an 8-bit priority encoder

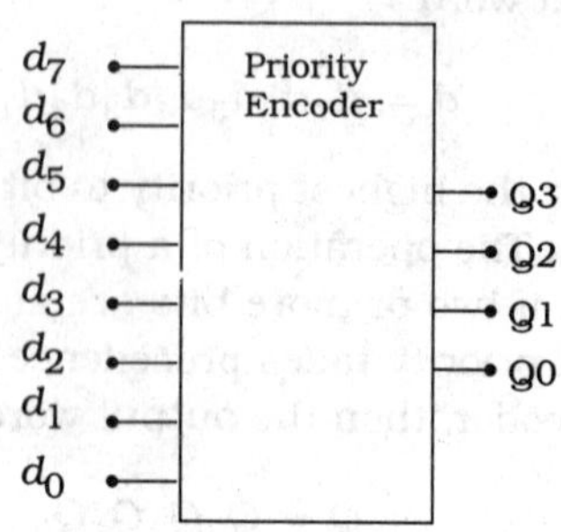

Figure 11.21 Symbol for priority encoder

while

$$Q1 = \overline{d_7} \cdot \overline{d_6} \cdot [d_5 + \overline{d_4} + \overline{d_3} \cdot \overline{d_2} \cdot (d_1 + d_0)] \tag{11.15}$$

gives the $Q1$ bit.

Even though the internal details of the circuit are complicated, the behavioral description is concerned only with the overall functional behavior. One implementation for the module is

```
module priority_8 (Q, Q3, d ) ;
input [ 7: 0 ] d ;
output Q3 ;
output [ 2: 0 ] Q ;
always @ (d)
   begin
      Q3 = 1 ;
      if ( A[7] ) Q = 7 ;
```

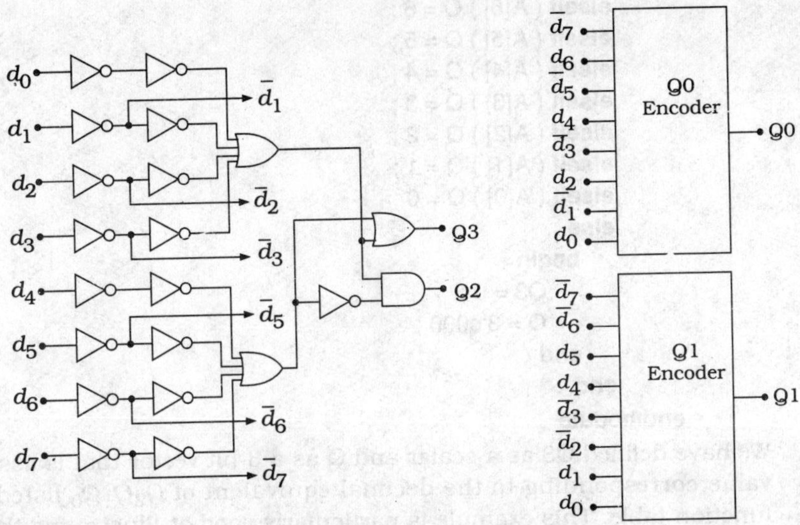

Figure 11.22 Logic diagram for the priority encoder

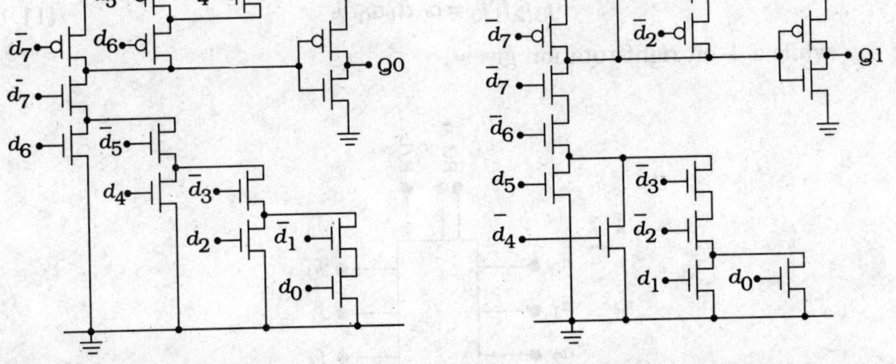

(a) $Q0$ circuit

(b) $Q1$ circuit

Figure 11.23 $Q0$ and $Q1$ circuits for the 8-bit priority encoder

```
        elseif ( A[6] ) Q = 6 ;
        elseif ( A[5] ) Q = 5 ;
        elseif ( A[4] ) Q = 4 ;
        elseif ( A[3] ) Q = 3 ;
        elseif ( A[2] ) Q = 2 ;
        elseif ( A[1] ) Q = 1 ;
        elseif ( A[0] ) Q = 0 ;
        else
            begin
                Q3 = 0 ;
                Q = 3'b000 ;
            end
    end
endmodule
```

We have defined Q3 as a scalar and Q as a 3-bit vector that is assigned a value corresponding to the decimal equivalent of $Q_2Q_1Q_0$ listed in the function table. This example is particularly good at illustrating the separation of a high-level versus a low-level description. The translation of the HDL to the circuit diagram is not a simple problem. Moreover, other equivalent circuits and logic algorithms can be constructed, each with different area and switching properties.

11.5 Shift and Rotation Operations

Shift and rotation units are useful in many different networks. Consider a 4-bit word $a_3a_2a_1a_0$ as the input into the general rotation unit shown in Figure 11.24. The output is a rotated word $f_3f_2f_1f_0$. An n-bit rotation is specified using the control word RO_n, while the L/R bit defines a left or right movement. For example, a 1-bit left rotation yields an output of

$$f_3f_2f_1f_0 = a_2a_1a_0a_3 \tag{11.16}$$

while a 1-bit right rotation gives

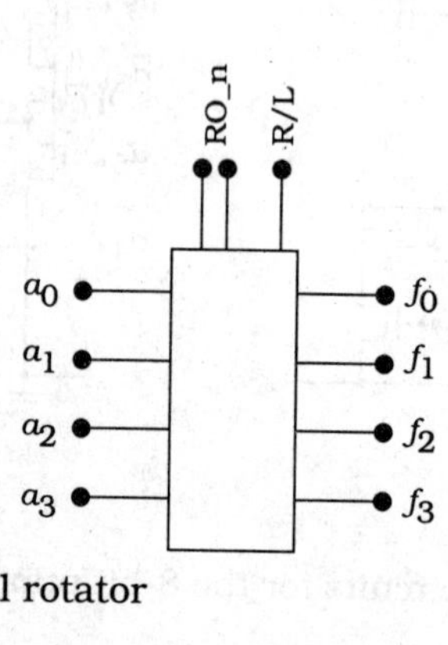

Figure 11.24 General rotator

$$f_3f_2f_1f_0 = a_0a_3a_2a_1 \quad (11.17)$$

A rotation exhibits wrap-around behavior where a bit that is pushed out of the word is added to the other side. A shift operation forces a 0 into the empty space. If we modify the unit to give a 1-bit shift left operation, then an input of $a_3a_2a_1a_0$ produces an output of

$$f_3f_2f_1f_0 = a_2a_1a_00 \quad (11.18)$$

with a similar behavior for a shift right operation.

Verilog provides bit-wise shift operators of

```
<< // This is a shift left operation
>> // This is a shift right operation
```

that can be used to specify vector shifts; both fill slots with 0s. These are shown in the example code

```
reg [7:0] a ;
reg [7:0] new_1 ;
reg [3:0] new_2 ;
reg [3:0] b ;
    new_1 = a >> b ; // This shifts the 7-bit word a by b-bits to the right
    new_2 = a << b ; // This shifts a by b-bits to the left
...
```

A rotation can be specified in a number of different ways. The simplest is a bit-by-bit assignment as in the clocked-behavior unit that is described by the listing

```
...
reg [3:0] ;
always @ (posedge clk)
    begin // This is a bit-by-bit rotate left
        a[0] <= a[3] ;
        a[1] <= a[0] ;
        a[2] <= a[1] ;
        a[3] <= a[2] ;
    end
...
```

More general rotations can be included by adding control bits.

There are different ways to implement rotations and shifts in VLSI circuits. While standard FF-based designs can be used, simpler networks based on FET switching properties yield highly regular designs. Consider the rotation of a 4-bit word. The switching network in Figure 11.25 uses four control bits Ror_0, Ror_1, Ror_2, Ror_3 to specify an n-bit rotation. The signals are created using combinational logic; only one of these is a 1 at any given time. Signal routing of the input bits $a_3a_2a_1a_0$ is accomplished by using nFETs as switches that give the desired connection to the output lines $f_3f_2f_1f_0$. A left-rotate array may be created by rewiring

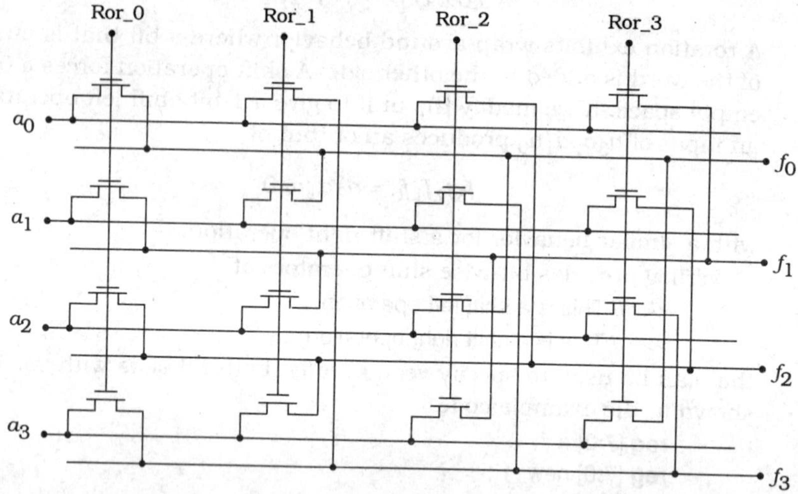

Figure 11.25 A 4-bit rotate-right network

the FETs to the configuration shown in Figure 11.26. The routing of both networks can be verified by simply tracing the input-output paths for each column of control FETs. The two may be combined into a single array that uses a control word Ro_n to specify the number of bits, and a separate control bit for left/right shifting.

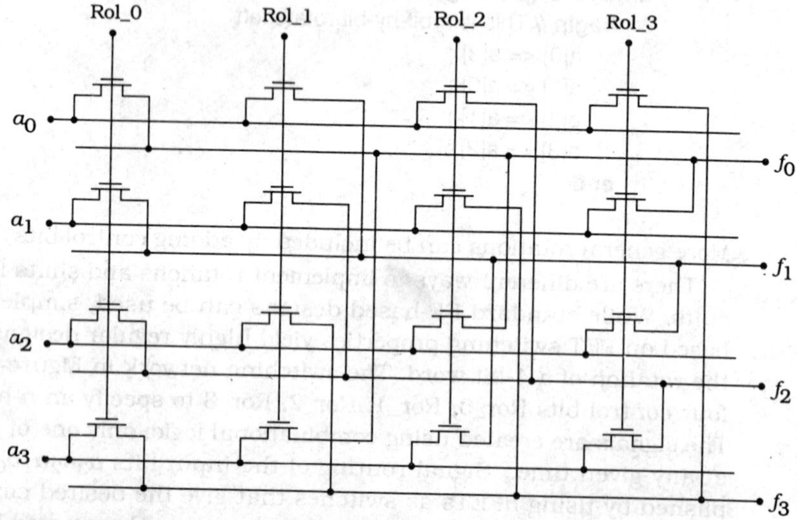

Figure 11.26 Left-rotate switching array

A related component is the **barrel shifter** that is specified as having an $m \times n$ architecture, with m the number of bits in the input word and n the number of output bits. Common situations are where we have $m = 2n$ and $m = n$. Figure 11.27 shows an 8×4 unit. The decimal value of the control word *shift* defines the output $f_3 f_2 f_1 f_0$ in terms of the input word $a_7 a_6 a_5 a_4 a_3 a_2 a_1 a_0$ as summarized in the table. This unit can be built using an nFET array with the result shown in Figure 11.28. As with the rotator designs, each column of transistors is controlled by a single-bit signal Sh_n where n = 0, 1, 2, 3, 4 in this design. These are produced by a combinational logic network from the 3-bit control word *shift*. Only one of the column signals is 1 at a time, so that the input-to-output paths are defined by the column of transistors that is active. A 4×4 network with wrap-around is just a rotator. Barrel shifters are useful in ALUs (arithmetic and logic units) for bit manipulations. The overall structure of the circuit itself provides a basis for designing integrated cross-bar switching networks and signal routers for applications in telecommunications and parallel processing.

The nFET array is extremely regular, which makes it relatively easy to layout at the CMOS physical design level. Library cells can be as simple as individual FETs, or as complicated as $p \times q$ sub-units that are used to create larger arrays. The main drawbacks of the nFET-only design are the threshold voltage drop problem (and the associated weak-1 transmission) and parasitic-limited switching times. Drivers can be added to speed up the circuits and restore the voltage swings. Alternately, transmission gates can be used as a 1-for-1 replacement of the FETs. Although the pFET area consumption is small, the routing complexity increases considerably.

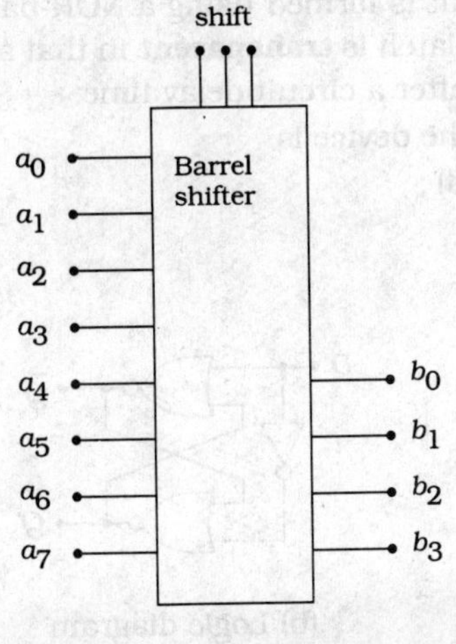

shift	$b_0 b_1 b_2 b_3$
0	$a_0 a_1 a_2 a_3$
1	$a_1 a_2 a_3 a_4$
2	$a_2 a_3 a_4 a_5$
3	$a_3 a_4 a_5 a_6$
4	$a_4 a_5 a_6 a_7$

Figure 11.27 An 8×4 barrel shifter

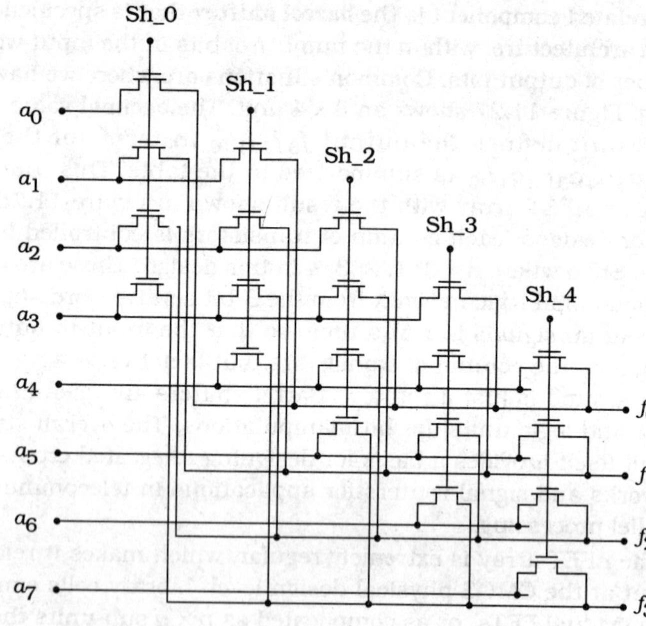

Figure 11.28 FET-array barrel shifter

11.6 Latches

A latch is a device that can receive and hold an input bit. A simple D-latch forms the basis for many designs. The symbol for the D-latch is shown in Figure 11.29(a), and a logic diagram is provided in Figure 11.29(b). By inspection we see that the circuit is formed using a NOR-based SR latch with complemented inputs. The latch is **transparent** in that a change in D is seen at the outputs Q and $\overline{Q}$ after a circuit delay time.

A behavioral description for the device is

```
module d_latch (q, q_bar, d) ;
input d ;
output q, q_bar ;
```

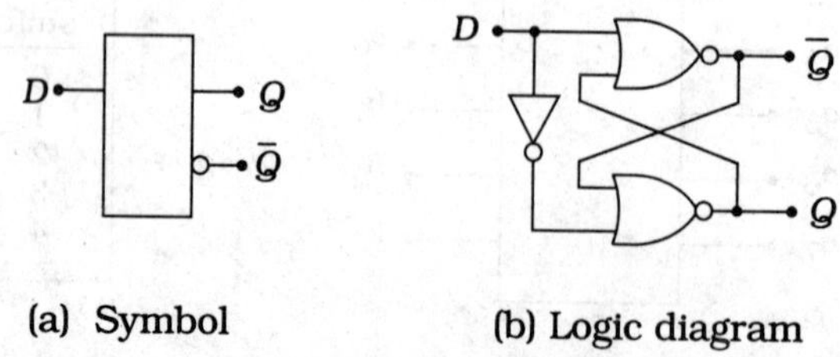

Figure 11.29 D-latch

```
reg q, q_bar ;
always @ (d)
   begin
      # (t_d) q = d ;
      # (t_d) q_bar = ~d ;
   end
endmodule
```

We have used t_d for the time delay. The declaration models the action of the cross-coupled NOR circuit that can hold the input state. The equivalent structural description is

```
module d_latch_gates (q, q_bar, d ) ;
input d ;
output q, q_bar ;
wire not_d ;
not (not_d, d) ;
nor # (t_nor) g1 (q_bar, q, d),
    # (t_nor) g2 (q, q_bar, not_d);
endmodule
```

This provides a gate-by-gate guide for the device at the circuit and physical design level

Combinational Logic Designs

The CMOS circuit can be constructed using either the logic diagram or the structural description. A direct translation is shown in Figure 11.30. At the physical level, this can be created by instancing two NOR2 cells and one NOT cell, and then adding the interconnect wiring. Alternately, a custom layout would probably consume less area.

An enable control *En* can be added to the basic D-latch by routing the inputs through AND gates as illustrated in Figure 11.31. An enable bit of $En = 0$ blocks the inputs by forcing 0's to the AND outputs, which places the SR latch into a hold state. If $En = 1$, then the values of D and $\overline{D}$ are

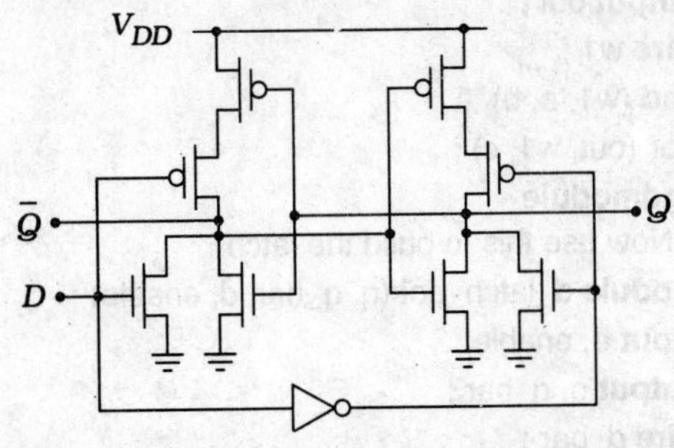

Figure 11.30 CMOS circuit for a D-latch

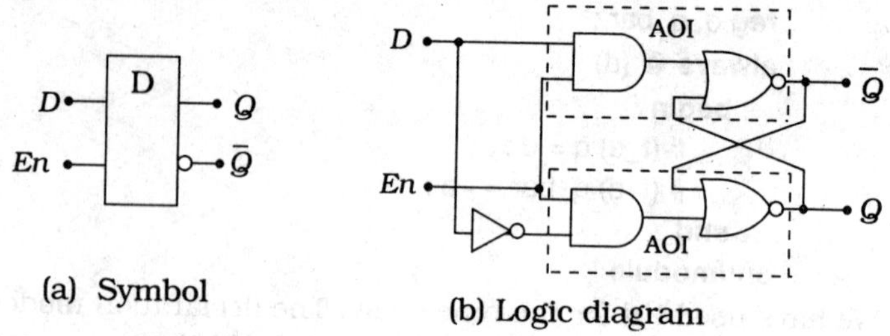

Figure 11.31 Gated D-latch with *Enable* control

admitted to the NOR circuits. To include this control in the behavioral description, we rewrite the code as

```
module d_latch (q, q_bar, d , enable) ;
input d, enable ;
output q, q_bar ;
reg q, q_bar ;
always @ (d and enable)
    begin
        # ( t_d ) q = d ;
        # ( t_d ) q_bar = ~ d ;
    end
endmodule
```

A condition of enable = 1 is required to change the state.

The structural description could be based on a brute-force listing of the circuit. However, since CMOS allows for complex logic gates as primitives, it makes more sense to describe the complex gate and then instance it in the listing.

```
// First define the AOI module
module aoi_2_1 (out, a, b, c) ;
input a, b, c ;
output out ;
wire w1 ;
and (w1, a, b) ;
nor (out, w1, c) ;
endmodule
// Now use this to build the latch
module d_latch_aoi (q, q_bar, d, enable) ;
input d, enable ;
output q, q_bar ;
wire d_bar ;
not (d_bar, d) ;
```

```
aoi_2_1 (q_bar, d, enable, q) ;
aoi_2_1 (q, enable, d_bar, q_bar) ;
endmodule
```

The module name aoi_2_1 is interpreted as an AOI gate with 2 inputs to the AND, and 1 input to the OR. Although the notation is not standard, it is widely used in practice. Note that the order of the inputs must be preserved when the module is instanced. The CMOS circuit diagram for this implementation is shown in Figure 11.32, and consists of one NOT gate and two AOI circuits.

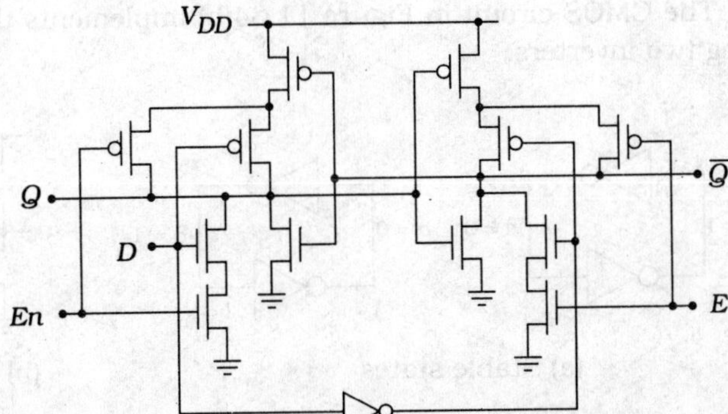

Figure 11.32 AOI CMOS gate for D-latch with *Enable*

CMOS VLSI Latch

Many static D-latches CMOS VLSI are constructed from inverters and TGs or pass FETs. This design is based on the characteristics of the simplest static storage configuration called a **bistable circuit**. A bistable circuit is one that can store (or, hold) either a logic 0 or a logic 1 indefinitely (or, at least as long as power is applied.)

A basic bistable circuit consists of two inverters as shown in Figure 11.33(a). The gates are wired such that the output of one is the input to the other, forming a closed loop. Any closed loop with an even number of inverters gives a bistable circuit. If we use three inverters as in Figure

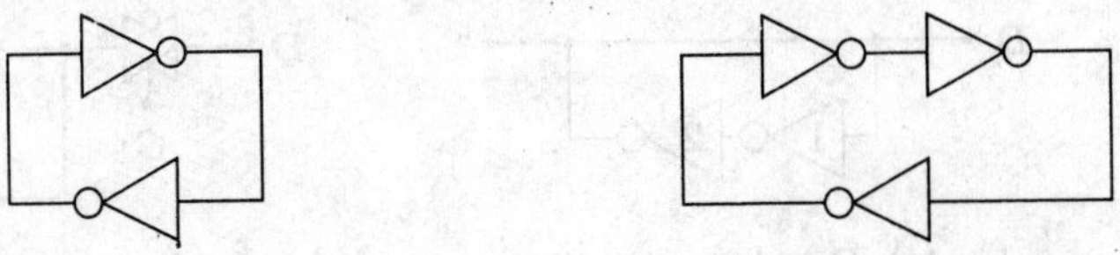

(a) Bistable circuit (b) Ring oscillator

Figure 11.33 Closed-loop inverter configurations

11.33(b), the resulting circuit is unstable and cannot hold a bit value. A closed loop with an odd number of inverters is often called a **ring oscillator** as the signal at any point oscillates in time.

The storage mechanism of the bistable circuit is illustrated in Figure 11.34(a). If the left side is at a value $a = 1$, then tracing the signal path through the upper inverter shows that the right side has a value of $\overline{a} = 0$. If we continue and trace the signal to the left through the lower inverter, we obtain the starting point of $a = 1$. This shows that the state $a = 1$ is stable in that it can be maintained by the circuit itself. The same arguments can be applied to the case where $a = 0$, as shown in the same figure. The CMOS circuit in Figure 11.34(b) implements the bistable circuit using two inverters.

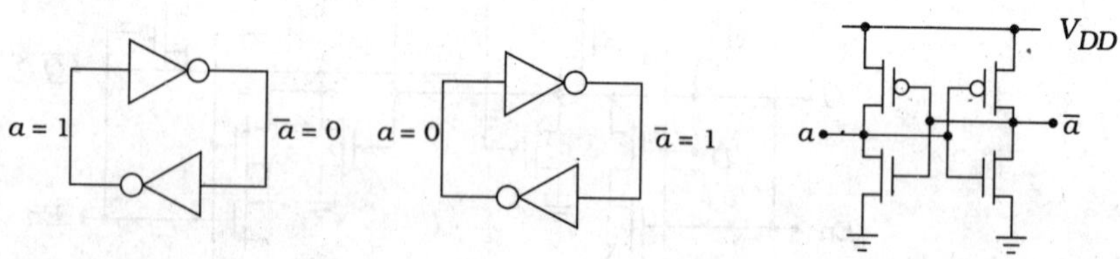

(a) Stable states (b) CMOS circuit

Figure 11.34 Operation of a bistable circuit

To create a D-latch, we must provide an entry node for the input bit. One simple idea is the **receiver circuit** shown in Figure 11.35(a). The value of D is held by the bistable circuit formed by the inverter pair. The circuit helps the line resist changes in D, making it useful as an input stage to a receiver module that is driven by an external line. The presence of the **feedback loop** from the output of Inverter 2 to the input of Inverter 1 provides the desired latching, but complicates the design. Inverter 1 needs to detect changes, but Inverter 2 cannot be so strong that it prohibits a change in state. In general, Inverter 1 can use relatively large FETs,

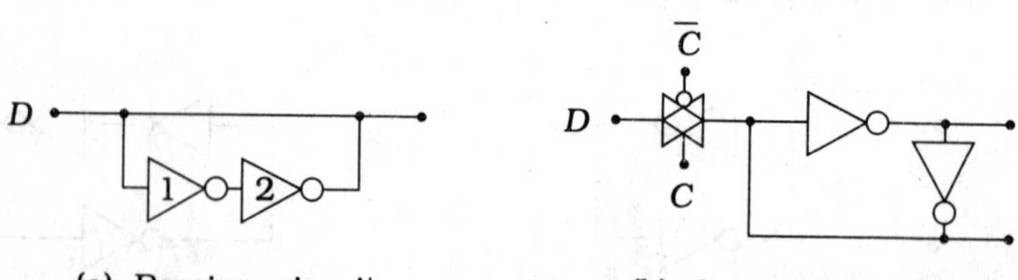

(a) Receiver circuit (b) Controlled loading

Figure 11.35 Adding an input node to the bistable circuit

but Inverter 2 is purposely made weaker by using small transistors.

Adding a transmission gate at the input as shown in Figure 11.35(b) gives us the ability to control the loading. When $C = 0$ (so $\overline{C} = 1$), the TG acts as an open switch and the circuit holds the values of Q and $\overline{Q}$ at the output. When the control bit is set to $C = 1$ ($\overline{C} = 0$), the TG conducts and allows the input bit D to be transferred to the latching circuit. During this time the latch is transparent and the outputs go to $Q = D$ and $\overline{Q} = \overline{D}$. If C is reset to 0, then the state is held. The control bit C is thus equivalent to the enable signal *En* used previously. The inverter design constraints still apply.

While these circuits are easy to build in CMOS, they are relatively slow devices. This is because the bistable circuit tries to hold onto the stored values and resists changes as discussed for Figure 11.34(a) . If we force a change of the stored voltage, the feedback that exists from the output back to the input fights the transition. A solution to this problem is to add another switch that breaks the feedback loop during which a value is stored. The TG circuit shown in Figure 11.36(a) accomplishes this by using an oppositely phased switch between the inverters. In many cases, chip designers prefer to use nFETs in place of TGs because of the simpler wiring. This is shown in Figure 11.36(b); note that the input FET is controlled by C, while the feedback transistor is switched by the complement $\overline{C}$.

The operation of the nFET latch is summarized in Figure 11.37. A load of the input data bit D occurs when $C = 1$. As shown in Figure 11.37(a), this turns on the input FET and opens the feedback loop. The input thus sees a simple NOT chain, and loads very quickly. A control bit of $C = 0$ defines the hold state, and is illustrated in Figure 11.37(b). This turns off the input FET but the feedback loop is established to hold the values of Q and $\overline{Q}$. Since the logic equation for a TG is identical to that for an nFET, this description also applies to the TG-based circuit.

C^2MOS circuits provide the basis for another D-latch design style. The latch in Figure 11.38(a) uses a C^2MOS inverter as the input stage with the

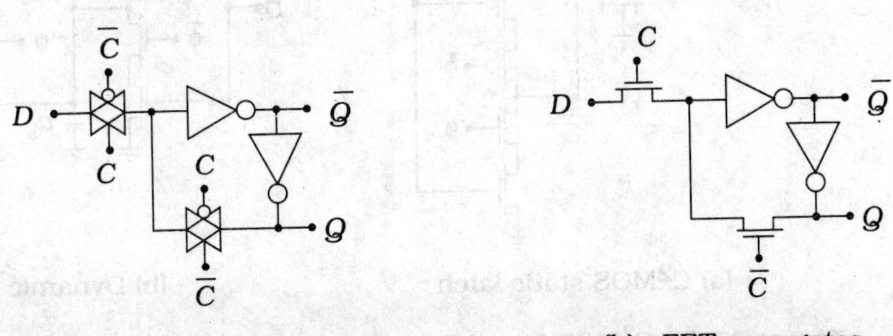

(a) CMOS TG version (b) nFET pass gates

Figure 11.36 D-latch using oppositely phased switches

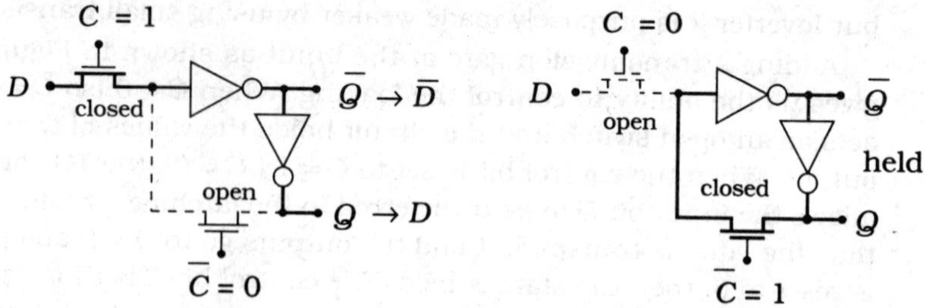

(a) Load with $C = 1$ (b) Hold with $C = 0$

Figure 11.37 Operation of the D-latch

clock ϕ controlling the loading. When $\phi = 0$, D is admitted into the circuit where it passes through the static inverter. This gives $Q = D$ after the characteristic delay time. When the clock makes a transition to $\phi = 1$, the first stage is driven into a Hi-Z state, while the feedback loop is closed by the oppositely phased inverter. This holds the output until the next clock cycle.

A variation that uses a pair of cascaded C^2MOS inverters is shown in Figure 11.38(b). This operation of this network is quite different in that it is a true dynamic circuit, i.e., it uses charge storage on capacitive nodes. The parasitic capacitor C_s between the two stages acts as the storage device for this circuit. With $\phi = 0$, D is admitted and the charge corresponding to $\overline{D}$ is stored on C_s. When the clock changes to $\phi = 1$, the out-

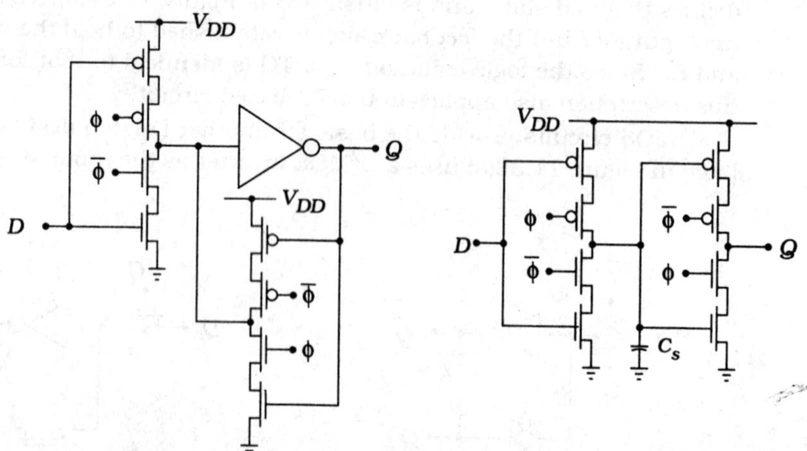

(a) C^2MOS static latch (b) Dynamic latch

Figure 11.38 C^2MOS-based D-latch circuits

put of the first stage is in a Hi-Z state which holds the charge on C_S. This clock phase activates the second stage which has an output of Q. The value of Q will be that of $\overline{D}$ delayed by the rise or fall time of the circuit. It is important to note that charge leakage limits the time that C_S can hold the state. Although introduced here as a latch, this circuit is often used as a time-delay element for synchronous networks.

11.7 D Flip-Flop

A flip-flop differs from a latch in that it is non-transparent. The D-type flip-flop (DFF) is the most commonly used flip-flop in CMOS circuits. The basic DFF design is a master-slave configuration obtained by cascading two oppositely phased D-latches as in Figure 11.39. The clock signal ϕ controls the operation and provides synchronization. The master latch allows an input of D when $\phi = 0$ and M1 acts as a closed switch. During this time, nFETs M2 and M3 are open circuits. When the clock makes a transition to $\phi \rightarrow 1$, switches M2 and M3 close and effect the transfer of the bit to the slave. The input to the master is blocked since M1 is open with $\phi = 1$. The master-slave circuit acts as a **positive edge-triggered** device since the value of D during the positive clock edge defines the value of the bit that is transferred to the slave and available as Q.[2] Note that once the bit is latched into the slave at time t_1, it does not appear at the output until

$$t_1 + t_{FET} + t_{NOT} \tag{11.19}$$

where t_{label} is the rise or fall time of the specified element. The symbol for a positive edge-triggered DFF is shown in Figure 11.40(a). The 'triangle' denotes an edge-triggering input. Adding a bubble produces the symbol

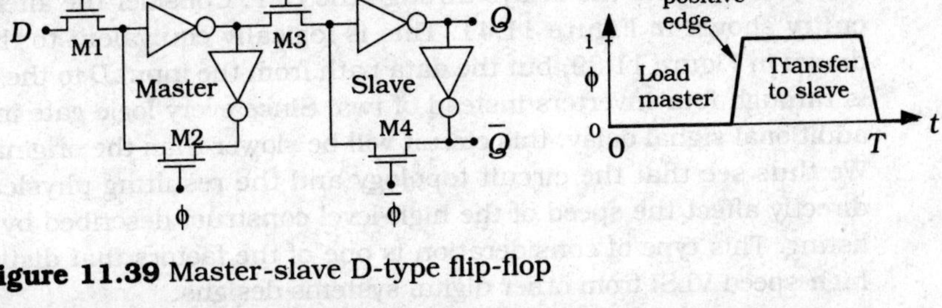

Figure 11.39 Master-slave D-type flip-flop

2 In the strictest sense, there is a difference between a master-slave FF and a true positive edge-triggered circuit. However, in VLSI design, the terminology is used interchangeably.

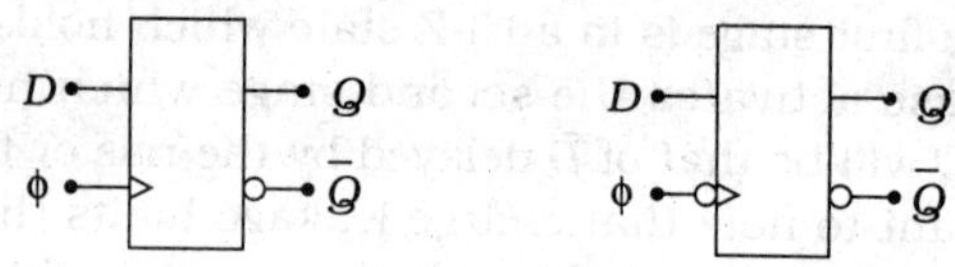

(a) Positive edge-triggered DFF

(b) Negative edge-triggered DFF

Figure 11.40 Edge-triggered DFF symbols

for a **negative edge-triggered** DFF in Figure 11.40(b). At the circuit level, all that needs to be done is to interchange the ϕ and $\overline{\phi}$ signals.

A Verilog behavioral description of a positive edge-triggered DFF can be written in the following manner.

```
module positive_dff (q, q_bar, d, clk) ;
input d, clk ;
output q, q_bar ;
reg q, q_bar ;
always @ (posedge clk)
  begin
     q = d ;
     q_bar = ~ d ;
  end
endmodule
```

In a realistic application, a set of delay times would be needed. A negative edge-triggered module is obtained by modifying the **always** statement to

```
always @ (negedge clk)
```

This changes the stimulus in an obvious manner. A structural description is straightforward using the **nmos** and **not** primitive elements to model the circuit in Figure 11.39, and is left as an exercise.

An important point to reiterate here is that the design of the CMOS circuitry determines the delays through the DFF. Consider the alternate circuitry shown in Figure 11.41. This is logically equivalent to the circuit drawn in Figure 11.39, but the data path from the input D to the output Q is through four inverters instead of two. Since every logic gate introduces additional signal delay, this circuit will be slower than the original design. We thus see that the circuit topology and the resulting physical design directly affect the speed of the high-level construct described by the HDL listing. This type of consideration is one of the factors that distinguishes high-speed VLSI from other digital systems designs.

It is possible to add direct **clear** and **set** capabilities to the circuit by changing the gate functions. One approach is to use NAND2 logic. Consider the case in Figure 11.42 where one input is a control bit s and the

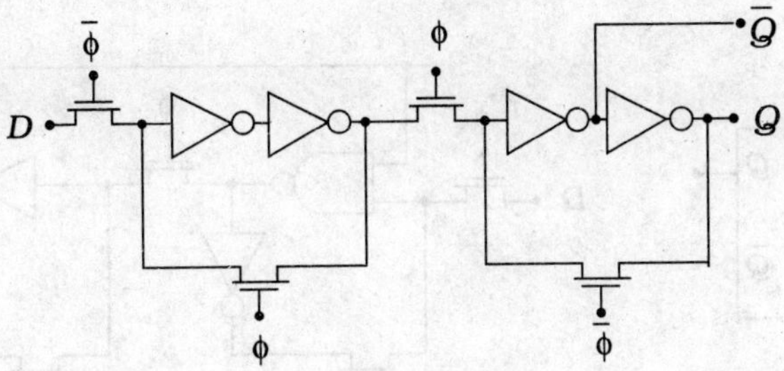

Figure 11.41 Alternate circuitry for the master-slave DFF

other is a data value *in*. When $s = 0$, the output is 1 regardless of the value of *in*. If $s = 1$, then $out = \overline{in}$ as shown. Replacing selected inverters with NAND2 gates yields DFFs with assert-low clear or set inputs, or both. Figure 11.43(a) provides the ability to clear (to 0) the contents of the latch using the *Clear* control. With *Clear* = 1, the NAND gates act as inverters and the circuit behaves as a normal DFF. When *Clear* = 0, the NAND in the slave forces an output of $Q = 0$. The output of the master is forced to a logic 1, which inverts to an output $Q = 0$. A Verilog listing that describes this device is

```
module dff_clear (q, d, clear, clk) ;
input d, clk, clear ;
output q ;
reg q ;
always @ (posedge clk )
        q = d ;
always @ ( clear )
    if ( clear )
        assign q = 0 ;
    else
        deassign q ;
endmodule
```

This uses the **deassign** statement that is executed if clear is 0. This returns the value of q to its value established in the q = d line.

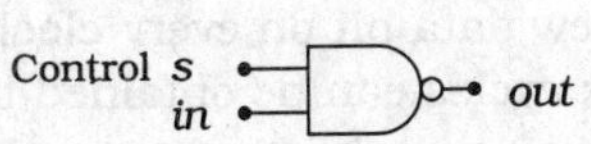

s	in	out
0	0	1
0	1	1
1	0	*in*
1	1	$\overline{in}$

Figure 11.42 NAND2 used as a control element

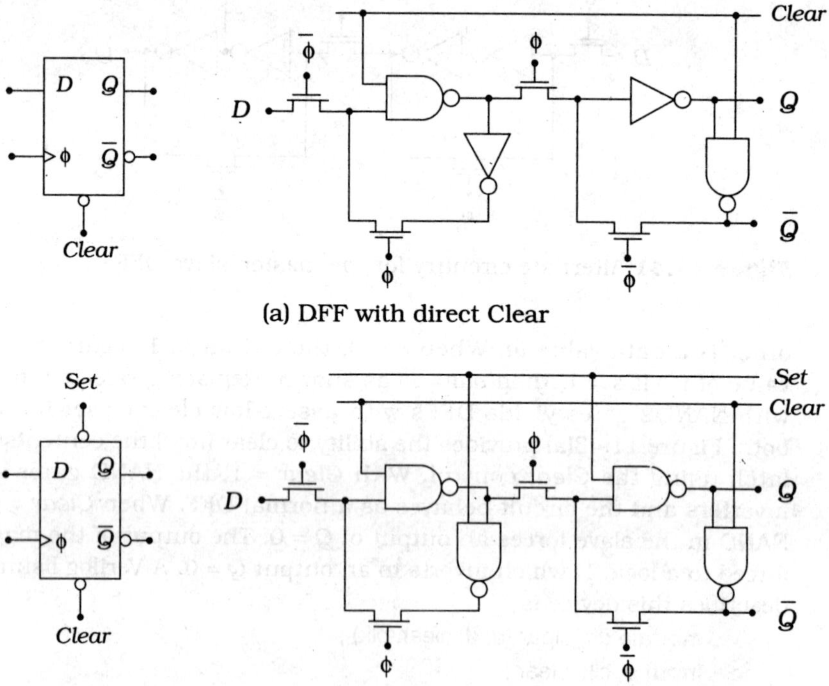

(a) DFF with direct Clear

(b) DFF with direct Clear and Set

Figure 11.43 DFF circuits with assert-low Clear and Clear/Set controls

A DFF with both clear and set controls is shown in Figure 11.43(b). The set capabilities are achieved by substituting NAND2 gates for the other two inverters. A condition of *Set* = 1 gives normal operation, while *Set* = 0 forces the slave to an output of $Q = 1$. This is reinforced by the output of the master. Note that *Clear* and *Set* cannot both be 0 at the same time as this forces $Q = \overline{Q}$. It is common to find several variations of DFFs in a cell library including features such as input buffers, clock buffers, and inputs with combinational logic gates. For example, a toggle flip-flop (TFF) that changes states on every rising clock edge can be created by adding a feedback loop from $\overline{Q}$ to *D* as shown in Figure 11.44. The assert-low *Set* logic modification allows the initial value to be established as a 1.

A basic DFF loads a new data bit on every clock edge. Storage over an arbitrary number of clock cycles can be obtained by adding a control signal and associated logic to the circuit. A simple classical solution is shown in Figure 11.45(a). The control signal *Load* controls a 2:1 MUX. If the control signal is asserted with *Load* = 1, the MUX allows a new data value *D* to enter the DFF. A control bit with a value *Load* = 0 takes the

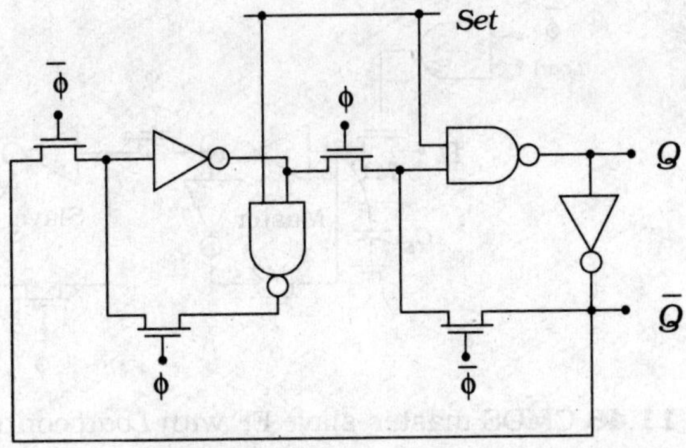

Figure 11.44 DFF modified to a TFF circuit using feedback

output Q and redirects it back to the input. This type of circuit is usually integrated as a single element with the simplified symbol shown in Figure 11.45(b). A simple Verilog description that includes the *Load* control is

```
module dff_load (q, q_bar, d, load, clk) ;
input d, clk, load ;
output q, q_bar ;
reg q, q_bar ;
always @ (posedge clk )
  begin
     if ( load ) q = d ;
     q_bar = ~ d ;
  end
endmodule
```

Another approach to designing a controlled-loading DFF is shown in Figure 11.46. This is the basic CMOS master-slave arrangement with a modified control signal

$$\bar{\phi} \cdot Load \tag{11.20}$$

applied to the input FET M1. This allows loading only when both *Load*

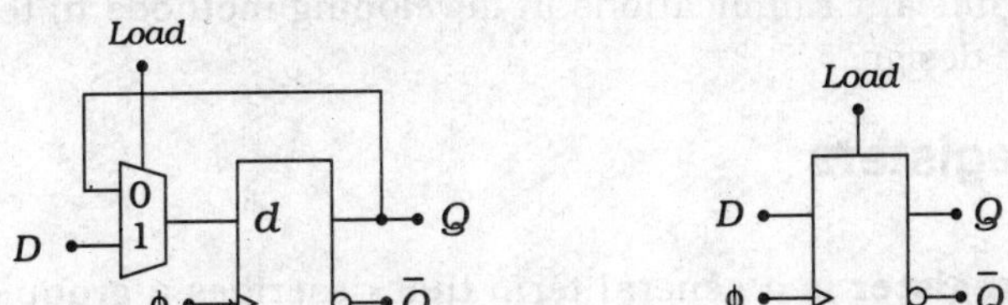

(a) Wiring diagram (b) Symbol

Figure 11.45 D-type flip-flop with *Load* control

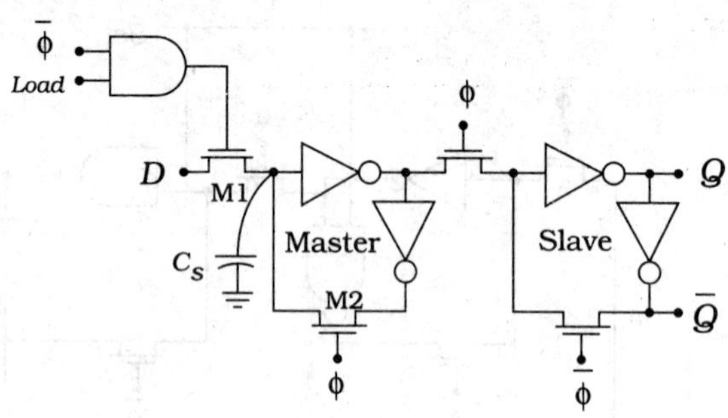

Figure 11.46 CMOS master-slave FF with *Load* control

and $\overline{\phi}$ are 1. The operation of the circuit is shown in Figure 11.47. A load condition is shown in Figure 11.47(a), and is identical to the operation in the original circuit. The value of D establishes the voltage V_{in} at the input to the master circuit of either 0 V or V_{max} corresponding to a logic 0 or a logic 1, respectively. Note that this voltage establishes the charge state of the storage capacitor C_s shown in the schematic.

There are two possible conditions for a data hold state when *Load* = 0. When $\phi = 1$, the master latch holds the value of the data bit using a closed feedback loop, and transfers it to the slave and the outputs. This is shown in Figure 11.47(b). If, on the other hand, $\phi = 0$ and *Load* = 0 at the same time, then the circuit switches are in the states shown in Figure 11.47(c). The master feedback loop is open. Storage is achieved by holding the charge on C_s, making this a quasi-dynamic circuit that is subject to charge leakage effects. Note that the slave feedback loop is closed during this time. This establishes the voltage V_a. When the clock returns to $\phi = 0$, the circuit of Figure 11.47(b) is again valid. This has two effects. First, the master feedback loop closes and establishes static hold capabilities. Second, the slave voltage V_a reinforces the value of the voltage on C_s. It is important to remember that the charge leakage may place a lower limit on the usable clock frequency. In particular, idling the clock may cause the circuit to exhibit long delays when the clock is restarted. This may have significant ramifications in developing methods to test a chip that uses the design.

11.8 Registers

A **register** is a general term that describes a group of circuits that are used to store a word as a unit entity. A 1-bit register is simply a single flip-flop, whereas an n-bit register loads and holds an n-bit word. Registers are important components in VLSI design. They allow us to design sequential circuits and state machines that form the basis of modern dig-

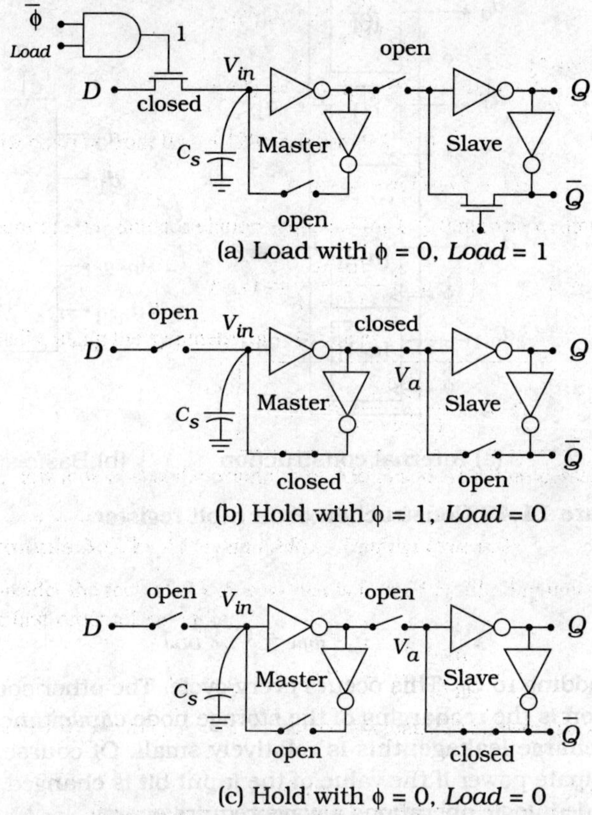

Figure 11.47 Operation of the CMOS DFF with load control

ital systems. The transition from single-bit logic to word-handling units is a critical step in the design hierarchy.

An *n*-bit positive edge-triggered register can be built by paralleling *n* single-bit DFFs as shown in Figure 11.48(a). The register symbol shown in Figure 11.48(b) masks the details of the individual circuits, but provides the basic operation: *n*-bit input words are loaded into the register on a rising clock edge. The outputs are valid after a circuit-induced delay time. Hold capabilities can be added as discussed in the previous section.

Since DFF's are intrinsically clock-controlled, they affect the dynamic power dissipation of the network. The DFF operation described previously in Figure 11.47 shows two sources of dynamic power. One is due to the four clocking FETs needed to control the operation of the master-slave arrangement. These load the clock drivers and increase the dynamic power

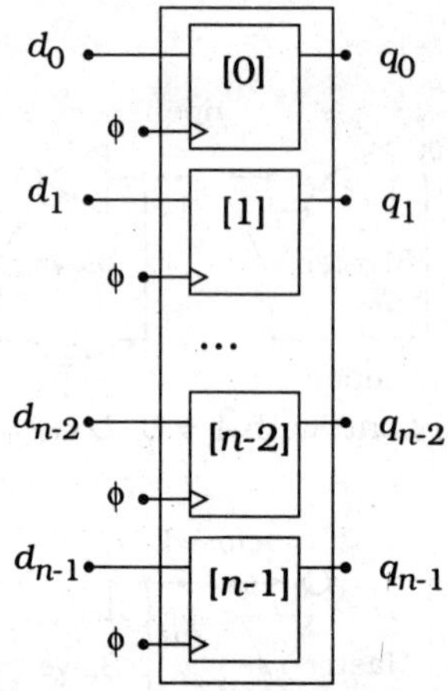

(a) Internal construction

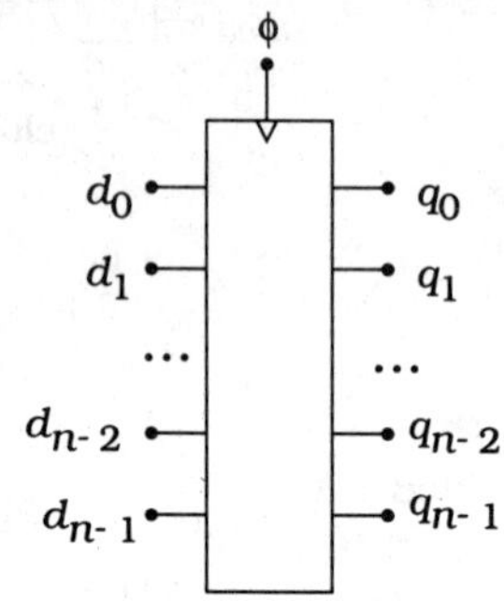

(b) Basic symbol

Figure 11.48 Construction of an n-bit register

$$P_{dyn} = C_L V_{DD}^2 f \tag{11.21}$$

by adding to C_L. This occurs every cycle. The other source of power dissipation is the recharging of the storage node capacitance C_s to compensate for charge leakage; this is relatively small. Of course, the inverters also dissipate power if the value of the input bit is changed, but this cannot be avoided; logic operations always require energy.

A purely static multi-output port solution is shown in Figure 11.49. This circuit is the basic two-inverter storage circuit with access transistors and an output driver. The write-enable signal *WE* controls the loading, while the two read-enable signals (*RE_a* and *RE_b*) are used to demultiplex the output to *Qa* or *Qb*. Since this uses single-ended inputs, the design of the inverters is important. Inverter 1 should be sensitive to the input and have reasonable drive strength for quick response. Rela-

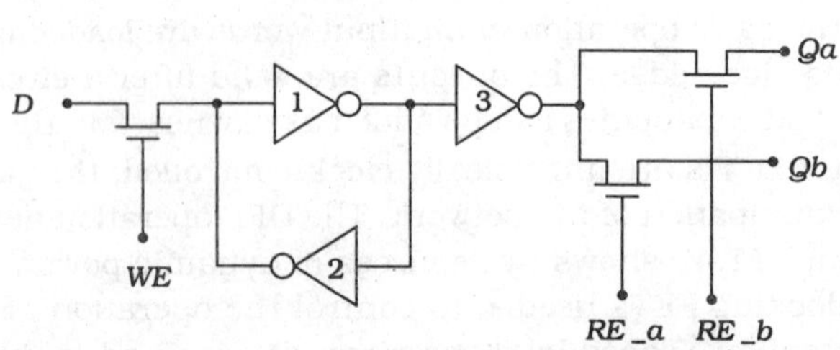

Figure 11.49 One-bit static multiport register circuit

tively large transistors would be appropriate for this and inverter 3. Inverter 2 provides the feedback that latches the data bit. Since it will resist changes at the input, it is usually designed to be much weaker with small transistors. These considerations are identical to those discussed in the context of the latch in Figure 11.35(a). Clocking is not included in this primitive circuit, and must be added to the input if needed. A simple solution is to add a clock-controlled nFET in series with the input transistor, or create a composite control signal $WE \cdot \phi$ for the existing circuit.

An n-bit register using this circuit is shown in Figure 11.50. This provides one input and two outputs for every bit. Additional ports can be added if necessary. Since there is no clock power dissipation associated with this design, it is useful in situations where the contents may be held for a long period of time. In particular, it can be used to build a **register file**, which is a collection of word-size storage registers.

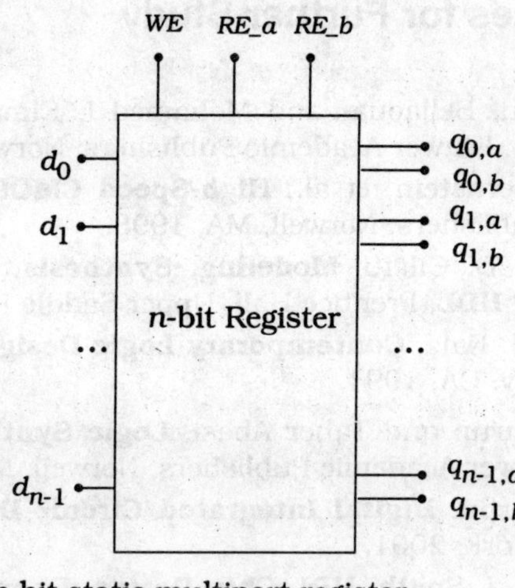

Figure 11.50 An n-bit static multiport register

11.9 The Role of Synthesis

The logic components in this chapter were chosen to illustrate techniques of translating a high-level HDL description down to a logic or circuit schematic. In each case, the large-scale function could be implemented using different logic networks or CMOS design styles. Since every realization of a given function will result in different values for the switching speed and silicon real estate, the role of the designer is complicated by the need to make the correct choice. CAD toolsuites allow different designs to be built and characterized for comparison. This is particularly important for designing critical datapaths or complex logic networks.

Architectural details reside at the top of the design hierarchy. High-level abstract descriptions are used to define and verify the operation of the system. Once the model is validated at this level, it moves down to a descriptive level that serves as the basis for logic synthesis. Logic synthesis tools are designed to translate HDL code into a logic network that is made up of predefined primitives stored in the library. The output is then used to build the circuitry so long as there is a one-to-one association between the logic primitives and the CMOS cells.

Synthesis is one of the most important aspects of design automation. It transfers much of the tedious work to the machine. However, the synthesized solutions are not always the "best" in terms of area or speed. In critical units, it may be necessary to custom design the logic, the circuits, or both, to achieve the necessary characteristics. The examples in this chapter illustrate the complexity of this process, and the interplay among the various levels.

11.10 References for Further Study

[1] Abdellatif Bellaouar and Mohamed I. Elmasry, **Low-Power Digital VLSI Design**, Kluwer Academic Publishers, Norwell, MA, 1995.

[2] Kerry Bernstein, et al., **High-Speed CMOS Design Styles**, Kluwer Academic Publishers, Norwell, MA, 1998.

[3] Michael D. Ciletti, **Modeling, Synthesis, and Rapid Prototyping with Verilog HDL**, Prentice Hall, Upper Saddle River, NJ, 1999.

[4] Randy H. Katz, **Contemporary Logic Design**, Benjamin/Cummings, Redwood City, CA, 1994.

[5] Pran Kurup and Taher Abasi, **Logic Synthesis Using Synopsys®**, 2nd ed., Kluwer Academic Publishers, Norwell, MA, 1997.

[6] Ken Martin, **Digital Integrated Circuit Design**, Oxford University Press, New York, 2001.

[7] Douglas J. Smith, **HDL Chip Design**, Doone Publications, Madison, AL, 1996.

[8] Michael J.S. Smith, **Application-Specific Integrated Circuits**, Addison-Wesley Longman, Reading, MA, 1997.

[9] John P. Uyemura, **A First Course in Digital Systems Design**, Brooks/Cole Publishers, Monterey, CA, 2000.

[10] John P. Uyemura, **CMOS Logic Circuit Design**, Kluwer Academic Publishers, Norwell, MA, 1999.

[11] Neil H. E. Weste and Kamran Eshraghian, **Principles of CMOS VLSI Design**, 2nd ed., Addison-Wesley, Reading, MA, 1993.

11.11 Problems

[11.1] Consider the problem of building a 4:1 MUX.

(a) Design a 4:1 MUX using transmission gates.

(b) Use the Verilog **cmos** primitive to write a structural description of your circuit.

(c) Construct an RC equivalent circuit for the MUX assuming that the resistance of a TG is R and each side of the switch has a capacitance C. Then use the Elmore formula to find the delay time constant of the worst-case path through the multiplexor.

[11.2] Construct a Verilog module listing for a 16:1 MUX that is based on the **assign** statement. Use a 4-bit select word $s_3s_2s_1s_0$ to map the selected input p_i (i = 0, ..., 15) to the output.

[11.3] Consider an 8:1 multiplexor that is constructed using smaller MUXes as primitives.

(a) Construct an 8:1 multiplexor using 4:1 and 2:1 MUX units.

(b) Select a logic circuit to implement the design.

(c) Assume that the gates are built using static CMOS circuits. Apply the technique of Logical Effort to designing the gates if the output of the 8:1 MUX is to drive a capacitor C_{out} that is $10C_{inv}$, where C_{inv} is the capacitance of a unit inverter.

[11.4] Design a 2/4 active-high decoder using only transmission gates in the main logic paths. Then construct the Verilog description of your network with the **cmos** primitive.

[11.5] Design a 2/4 active-low decoder using NOR gates. Then

(a) Construct the Verilog structural listing.

(b) Modify your Verilog code to include an input enable control. Then build the new circuit.

[11.6] Design a 4:1 MUX using standard dynamic or domino CMOS logic.

[11.7] Design a 4-bit left/right rotation unit that specifies the number of bits in the rotation by the 2-bit word Ro_n, and the direction of rotation by a single bit R/L such that $R/L = 1$ denotes a right rotation and $R/L = 0$ is a left rotation. Base your design on the FET arrays shown in Figures 11.25 and 11.26.

[11.8] Write a structural Verilog description of the 8-bit equality detector circuit shown in Figure 11.14.

[11.9] Consider the 4-bit shift register shown in Figure P11.1. The data stream D consists of sequential bits d_0, d_1, d_2, and d_3. The timing is set such that the first bit d_0 enters Stage 0 on the first clock edge. On the next rising edge, d_1 enters Stage 0, while d_0 moves to stage 1, and so on.

(a) Write a Verilog description of the shift register using DFF modules as primitives. You may use one from the book or write one of your own.

(b) Select a CMOS design technique for the DFFs and use it to construct the circuit.

(c) Now write a Verilog description of the shift register using **nmos** and **pmos** primitives.

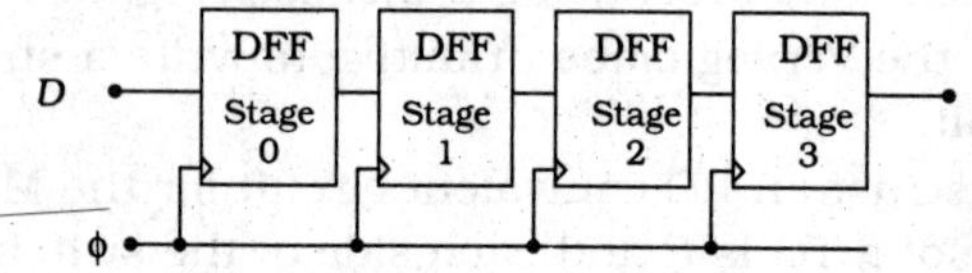

Figure P11.1

[11.10] Construct a Verilog module for an 8-bit register that loads words on rising clock edges iff the control bit *En* is 1. You may use any level of description.

[11.11] Write a structural description of the register circuit shown in Figure 11.49.

Arithmetic Circuits in CMOS VLSI

12

Arithmetic functions such as addition and multiplication have a special significance in VLSI designs. Many applications require these basic operations, but good silicon implementations have been a challenge since the early days of digital chip building. In this chapter we will examine binary adders in detail, and extend the discussion to include multipliers.

12.1 Bit Adder Circuits

Consider two binary digits x and y. The binary sum is denoted by $x + y$ such that

$$\begin{aligned} 0+0 &= 0 \\ 0+1 &= 1 \\ 1+0 &= 1 \\ 1+1 &= 10 \end{aligned} \tag{12.1}$$

where the result in the last line is a binary 10 (i.e., 2 in base-10). This simple example illustrates the problem with addition. If we take two base-r numbers with digits 0, 1, ... , (r-1), then the sum of two numbers can be out of the range of the digit set itself. This, of course, is the origin of the concept of a carry-out. In the binary sum 1+1, the result 10 is viewed as a 0 with a 1 shifted to the left to give a "carry-out of 1."

A **half-adder** circuit has 2 inputs (x and y) and 2 outputs (the sum s and the carry-out c) and is described by the table provided in Figure 12.1. The outputs are given by the basic equations

$$\begin{aligned} s &= x \oplus y \\ c &= x \cdot y \end{aligned} \tag{12.2}$$

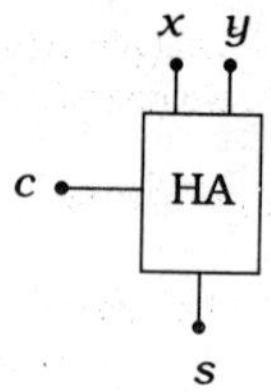

x	y	s	c
0	0	0	0
0	1	1	0
1	0	1	0
1	1	0	1

Figure 12.1 Half-adder symbol and operation

which are taken directly from the table. A high-level Verilog behavioral description of the cell can be written as

```
module half_adder (sum, c_out, x, y ) ;
input x, y ;
output sum, c _out ;
assign { c _out , sum } = x + y ;
endmodule
```

This defines x and y as single-bit quantities, and then uses the concatenation operator { } to obtain a 2-bit result. This operator "connects" binary segments in the order listed to create a single result. Alternately, we may construct the gate-level network shown in Figure 12.2. This is described by the structural model

```
module half_adder_gate (sum, c_out, x, y ) ;
input x, y ;
output sum, c _out ;
and (c_out, x, y ) ;
xor (sum, x, y ) ;
endmodule
```

using primitive gate instances. Two more possibilities are shown in Figure 12.3. The half-adder in Figure 12.3(a) uses NAND2 gates, while the alternate in Figure 12.3(b) is a NOR-based design. Preference might be given to the NAND design since it avoids series pFET chains, but a half-adder is simple enough so that the difference is not a major factor.

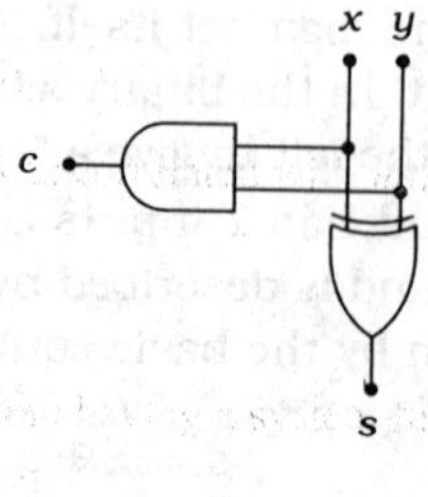

Figure 12.2 Half-adder logic diagram

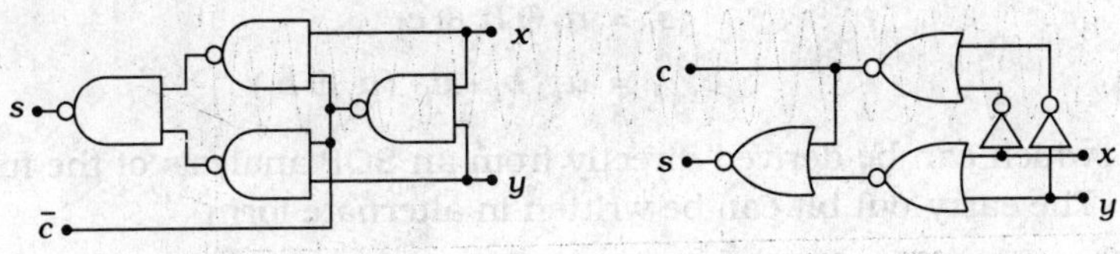

(a) NAND2 logic (b) NOR-based network

Figure 12.3 Alternate half-adder logic networks

A more complicated problem is adding n-bit binary words. Consider two 4-bit numbers $a = a_3a_2a_1a_0$ and $b = b_3b_2b_1b_0$. Adding gives

$$\begin{array}{r} a_3a_2a_1a_0 \\ +\ b_3b_2b_1b_0 \\ \hline c_4\ s_3s_2s_1s_0 \end{array} \tag{12.3}$$

where $s = s_3s_2s_1s_0$ is the 4-bit result and c_4 is the carry-out bit. To design an adder for the binary words, we break the problem down to bit level adders on a column-by-column basis. In the standard carry algorithm, each of the i-th columns (i = 0, 1, 2, 3) operates according to the **full-adder** equation

$$\begin{array}{r} c_i \\ a_i \\ +\ \ b_i \\ \hline c_{i+1}\ \ s_i \end{array} \tag{12.4}$$

where c_i is the carry-in bit from the (i -1)-st column, and c_{i+1} is the carry-out bit for the column. The operation is described by the full-adder table given in Figure 12.4 along with a simple schematic symbol. The most common expressions for the network are

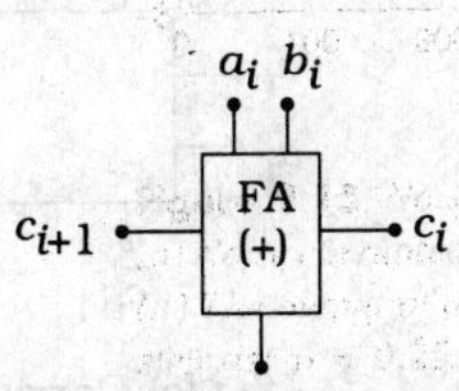

a_i	b_i	c_i	s_i	c_{i+1}
0	0	0	0	0
0	1	0	1	0
1	0	0	1	0
1	1	0	0	1
0	0	1	1	0
0	1	1	0	1
1	0	1	0	1
1	1	1	1	1

Figure 12.4 Full-adder symbol and function table

$$\begin{aligned} s_i &= a_i \oplus b_i \oplus c_i \\ c_{i+1} &= a_i \cdot b_i + c_i \cdot (a_i \oplus b_i) \end{aligned} \tag{12.5}$$

which can be derived directly from an SOP analysis of the function table. The carry-out bit can be written in alternate form

$$c_{i+1} = a_i \cdot b_i + c_i \cdot (a_i + b_i) \tag{12.6}$$

if desired.

A particularly compact circuit implementation is obtained using dual-rail complementary pass-transistor logic (CPL). The basic building block is the CPL 2-input array that has the generic form illustrated in Figure 12.5(a). The sum circuit is shown in Figure 12.5(b); the output of the first XOR/XNOR gate is the pair

$$a_i \oplus b_i \quad \text{and} \quad \overline{(a_i \oplus b_i)} \tag{12.7}$$

The second gate produces the sum by

$$s_n = \overline{(a_i \oplus b_i)} \cdot c_i + (a_i \oplus b_i) \cdot \bar{c}_i \tag{12.8}$$

The carry circuit in Figure 12.5(c) employs the 2-input array as a combinational logic element. For example, the top array on the left-hand side produces outputs of

$$\begin{aligned} &\bar{a}_i \cdot b_i + \bar{b}_i \cdot \bar{c}_i \\ &\bar{b}_i \cdot c_i + a_i \cdot b_i \end{aligned} \tag{12.9}$$

while the upper right circuit gives

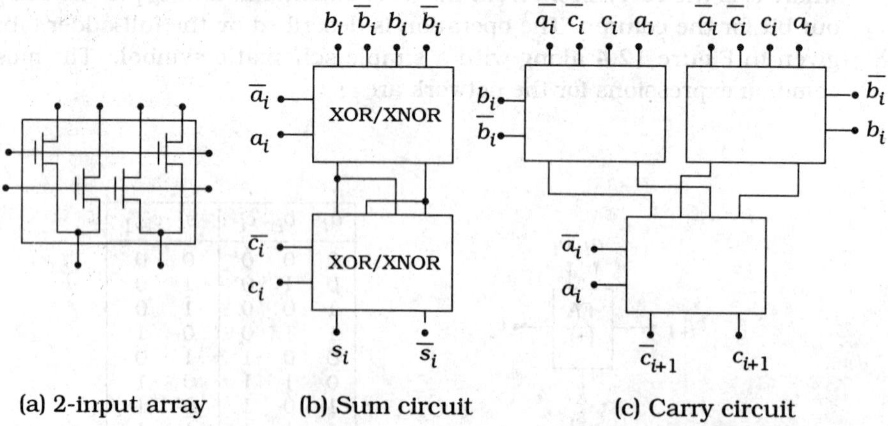

(a) 2-input array (b) Sum circuit (c) Carry circuit

Figure 12.5 CPL full-adder design

$$\overline{a}_i \cdot \overline{b}_i + b_i \cdot \overline{c}_i$$
$$b_i \cdot c_i + a_i \cdot \overline{b}_i \tag{12.10}$$

The last gate uses these to produce c_{i+1} and $\overline{c}_{i+1}$. Although this is a simple-looking solution, it must be remembered that CPL is a dual-rail technique that requires complementary variable pairs such as $(a_i, \overline{a}_i)$ at every stage. Also, because the threshold voltage loss drops the value of a logic 1 voltage as it passes through an nFET, restoring buffers or latching circuits are needed at the output. CPL is thus a somewhat specialized solution to implementing the CMOS full-adder. We note in passing that a CPL half-adder is easy to construct since it requires only the XOR/XNOR and AND/NAND functions.

A behavioral description of the full-adder is obtained by a simple modification of the half-adder model to the form

```
module full_adder (sum, c_out, a , b , c_in) ;
input a, b, c_in ;
output sum, c _out ;
assign { c _out, sum } = a + b + c_in ;
endmodule
```

All variables are scalars (single bits) and the concatenation operator creates the two outputs. Structural modeling can be based on the gate-level network shown in Figure 12.6(a). This is a straightforward one-to-one translation of the equation set. At this level, the module takes the form

```
module full_adder_gate (sum, c_out, a, b, c_in ) ;
input a, b, c_in ;
output sum, c _out ;
wire w1, w2, w3 ;
xor (w1, a, b) ,
     (sum, w1, c_in) ;
and (w2, a, b) ,
     (w3 , w1, c_in) ;
or (c_out, w2, w3) ;
endmodule
```

where we have used slightly different variable identifiers to make the code more readable. A full-adder can also be built from two HA modules as shown in Figure 12.6(b). Using instances of the module defined by the listing

```
module half_adder_gate (sum, c_out, x, y) ;
...
```

gives

```
module full_adder_HA (sum, c_out , a, b, c_in) ;
input a, b, c_in ;
output sum, c _out ;
wire wa, wb, wc ;
```

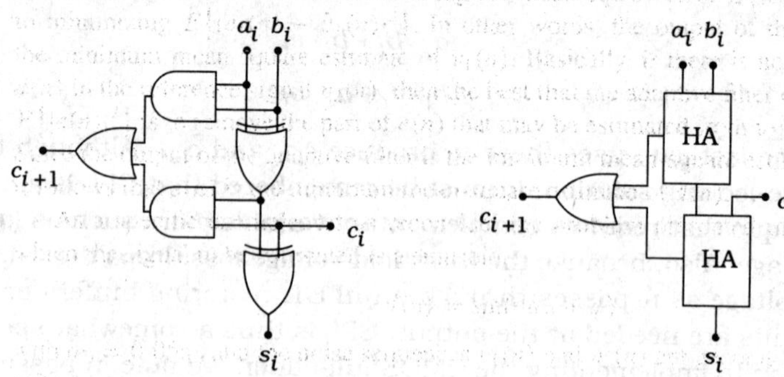

(a) Gate-level logic (b) HA-based design

Figure 12.6 Full-adder logic networks

```
half_adder_gate (wa, wb, a, b) ;
half_adder_gate (sum, wc, wa, c_in) ;
or (c_out, wb, wc) ;
endmodule
```

as the description.

Owing to the importance of the full adder, several implementations have been developed over the years. An AOI algorithm for static CMOS logic circuits can be obtained by writing the carry-out bit using equation (12.6). This allows us to write

$$\bar{s}_i = (a_i + b_i + c_i) \cdot \bar{c}_{i+1} + (a_i \cdot b_i \cdot c_i) \tag{12.11}$$

so that both c_{i+1} and $\bar{s}_i$ are in SOP form. Moreover, $\bar{s}_i$ uses $\bar{c}_{i+1}$ so that we can design an AOI gate for $\bar{c}_{i+1}$ and use the output to feed another AOI gate for $\bar{s}_i$. Figure 12.7 shows the construction of the two OAOI networks.

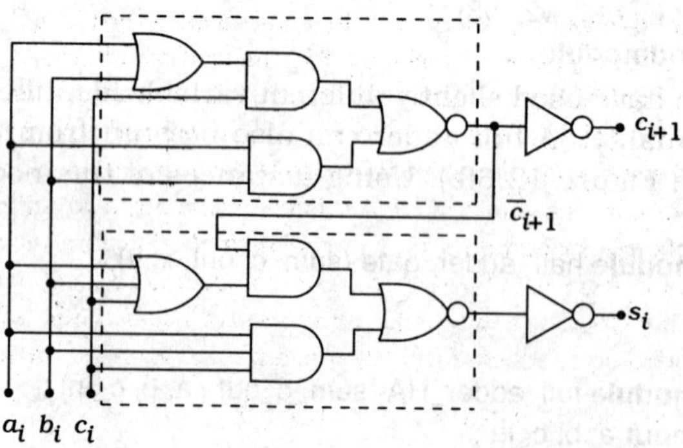

Figure 12.7 AOI full-adder logic

The upper circuit produces $\bar{c}_{i+1}$ and the lower one gives s_i after inversion. It is a straightforward exercise to design both OAOI circuits using series-parallel CMOS gates. Note, however, that equation (12.11) contains four OR operations, which indicates that the bottom AOI gate will have 4 series-connected pFETs. This may induce an unacceptably long delay in a word adder arrangement.

To find a more efficient circuit, consider the nFET array for the carry-out circuit as implied by the AOI logic diagram. Using standard construction gives the nFET circuit in Figure 12.8(a). We see that there are two main pull-down paths corresponding to the terms

$$\begin{gathered} a_i \cdot b_i \\ c_i \cdot (a_i + b_i) \end{gathered} \tag{12.12}$$

If either of these evaluates to 1, then the output is pulled to 0 (ground). The AND term is important when a_i and b_i are both 1; the OR term gives a pull down if either $a_i = 1$ or $b_i = 1$ while $c_i = 1$. This leads us to construct a pFET mirror circuit to yield the total gate shown in Figure 12.8(b). The series pFETs give a pull up to 1 (V_{DD}) if $a_i = b_i = 0$, which is the opposite of the nFET pull-down condition. If only one is 0, then the output is determined by the value of c_i.

To complete the building of a mirror CMOS full-adder, let us write the sum bit $\bar{s}_i$ as a simple OR gate such that

$$\bar{s}_i = A + B \tag{12.13}$$

where

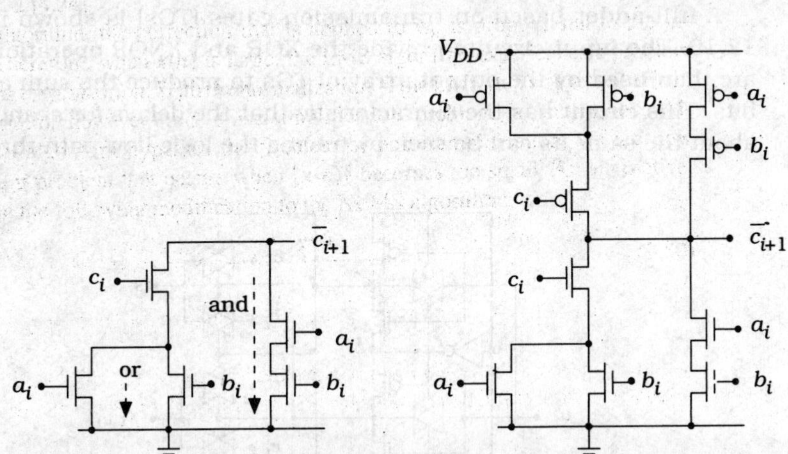

(a) Standard nFET logic (b) Mirror circuit

Figure 12.8 Evolution of carry-out circuit

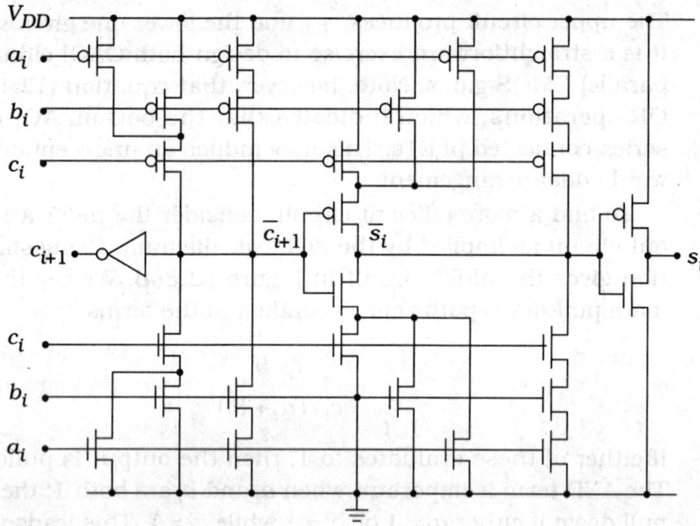

Figure 12.9 Mirror AOI CMOS full-adder

$$A = (a_i + b_i + c_i) \cdot \overline{c}_{i+1}$$
$$B = (a_i \cdot b_i \cdot c_i) \quad (12.14)$$

This has the same characteristics as the carry-out circuit, and allows us to construct the complete full adder shown in Figure 12.9. This is faster than a series-parallel realization, and facilitates the layout because of its mirrored FET arrays.

A full-adder based on transmission gates (TGs) is shown in Figure 12.10. The input circuits provide the XOR and XNOR operations which are then used by the output array of TGs to produce the sum and carry bits. This circuit has the characteristic that the delays for s_i and c_{i+1} are about the same as can be seen by tracing the logic flow path through the

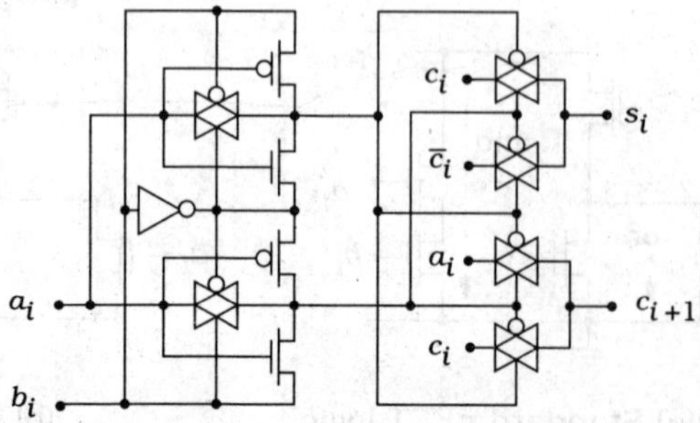

Figure 12.10 Transmission-gate full-adder circuit

upper and lower sections. If the input bits are applied simultaneously, then both the sum and carry-out bits will be valid at about the same time. This is distinctly different from the AOI circuit in which the carry-out bit is produced first and then used in calculating the sum.

12.2 Ripple-Carry Adders

Now that we have a basis for adding single bits, let us extend the problem to adding binary words. In general, adding two n-bit words yields an n-bit sum and a carry-out bit c_n that can either be used as the carry-in to another higher order adder, or act as an overflow flag. A general symbol is shown in Figure 12.11. We will use $n = 4$ in our initial discussions.

Ripple-carry adders are based on the addition equation

$$\begin{array}{r} c_3\ c_2\ c_1\ c_0 \\ +\ a_3 a_2 a_1 a_0 \\ +\ b_3 b_2 b_1 b_0 \\ \hline c_4\ s_3 s_2 s_1 s_0 \end{array} \tag{12.15}$$

where c_i represents the carry-in bit from the previous column. We will keep the 0-th carry-in bit c_0 for generality. Note that by including the carry-in word this is really adding three binary words. An n-bit ripple-carry adder requires n full-adders with the carry-out bit c_{i+1} used as in the carry-in bit to the next column. This is shown in Figure 12.12 for the case of 4-bit words.

A high-level model can be constructed using Verilog vectors as illustrated by the following code.

```
module four_bit_adder (sum, c_4, a, b, c_0) ;
input [ 3 : 0 ] a, b ;
input c_0 ;
output [ 3 : 0 ] sum ;
output c _4 ;
assign { c _4, sum } = a + b + c_0 ;
endmodule
```

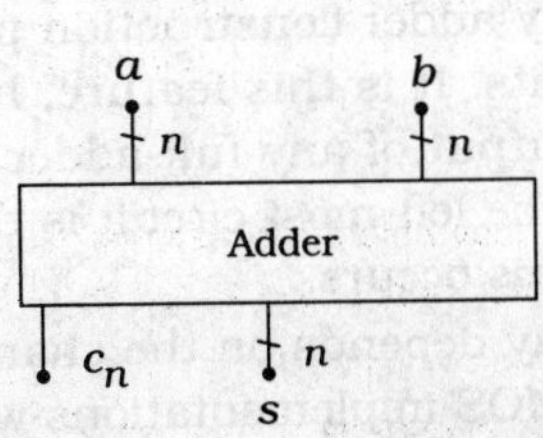

Figure 12.11 An n-bit adder

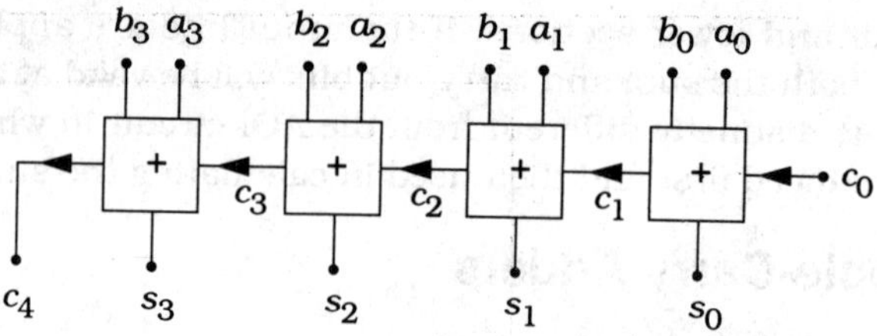

Figure 12.12 A 4-bit ripple-carry adder

This uses concatenation to create a 5-bit output that contains both sum and the carry-out bit c_4. Another approach to modeling this is to use four full-adder modules that are wired together as shown in the drawing:

```
module FA_modules (sum, c_4, a, b, c_0) ;
input [ 3 :0 ] a, b ;
input c_0 ;
output [ 3 : 0 ] sum ;
output c _4 ;
wire c _1, c_2, c_3 ;
/ * The single-bit FA modules instanced below have the syntax
    full_adder (sum, c_out, a, b, c_in) */
full_adder fa0 (sum[0], c_1, a[0], b[0], c_0 ) ;
full_adder fa1 (sum[1], c_2, a[1], b[1], c_1 ) ;
full_adder fa2 (sum[2], c_3, a[2], b[2], c_2 ) ;
full_adder fa3 (sum[3], c_4, a[3], b[3], c_3 ) ;
endmodule
```

This example uses the notation

sum[i], a[i], and b[i]

to define the i-th bit of a vector. This is straightforward to understand. If we define a quantity such as

```
input [7: 0] Q ;
```

with Q = 10001110, then Q[0] = 0, Q[1] = 1, Q[2] = 1, and so on. As always, it is assumed that instanced modules such as

```
full_adder (sum, c_out, a, b, c_in) ;
```

are defined elsewhere in the listing. They may be written at any level from a behavioral description down to a gate-level structural listing.

The ripple-carry adder construction provides for easy connections of neighboring circuits. It is this feature, however, that makes the design slow. Since the output of any full-adder is not valid until the incoming carry bit is valid, the left-most circuit is the last to react. The word result is not valid until this occurs.

The overall delay depends on the characteristics of the full-adder circuits. Different CMOS implementations will produce different worst-case delay paths. For our purposes, let us assume that the AOI mirror CMOS

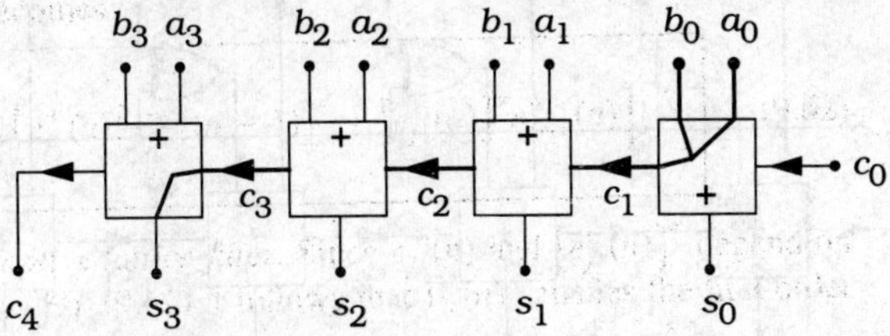

Figure 12.13 Worst-case delay through the 4-bit ripple adder

full-adder in Figure 12.9 is used in the 4-bit network. Since the carry-out is required to calculate the sum, the carry delay from c_i to c_{i+1} is minimized. Figure 12.13 shows the longest delay path for the adder where the carry bits are transferred through every stage; it is assumed that all inputs are valid at the same time. Let us start by summing the individual delays to get the total delay t_{4b} as

$$t_{4b} = t_{d3} + t_{d2} + t_{d1} + t_{d0} \tag{12.16}$$

where t_{di} is the worst-case delay through the i-th stage. The contributions for each stage can be evaluated. For the 0-th bit, $t_{d0} = t_d\,(a_0, b_0 \rightarrow c_1)$, which is the time for the inputs to produce the carry-out bit. The delay through sections 1 and 2 is the same, and is from the carry-in to the carry-out: $t_{d1} = t_{d2} = t_d\,(c_{in} \rightarrow c_{out})$. Finally, the delay in the last stage 3 in this design is the time needed to produce the output sum bit s_3, which we write as $t_{d3} = t_d\,(c_{in} \rightarrow s_3)$. Thus, the total delay is

$$t_{4b} = t_d(c_{in} \rightarrow s_3) + 2t_d(c_{in} \rightarrow c_{out}) + t_d(a_0, b_0 \rightarrow c_1) \tag{12.17}$$

If we extend this to an n-bit ripple-carry adder, then the worst-case delay is

$$t_{n-bit} = t_d(c_{in} \rightarrow s_{n-1}) + (n-2)t_d(c_{in} \rightarrow c_{out}) + t_d(a_0, b_0 \rightarrow c_1) \tag{12.18}$$

which shows that the delay is of order n. Symbolically, we express this as

$$\text{delay} \sim O(n) \tag{12.19}$$

The ripple structure is therefore not a good choice for large word sizes.

Before progressing into more advanced adder designs, let us recall that a 2's complement subtractor can be built by adding XOR gates and an *add_sub* control bit as shown in Figure 12.14. When *add_sub* = 0, the XORs pass the b_i bits and the output is the sum $(a+b)$. A control bit of *add_sub* = 1 changes the XORs into inverters, and the complemented values $\bar{b}_i$ enter the full adders; *add_sub* = 1 also acts as a carry-in of $c_0 = 1$. These operations combine to give the 2's complement algorithm for the

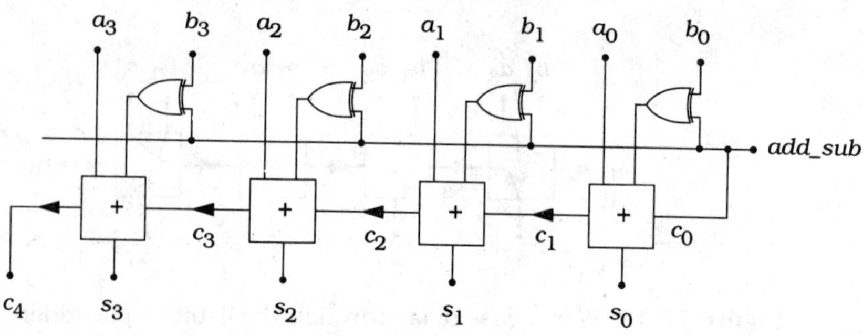

Figure 12.14 4-bit adder-subtractor circuit

difference $(a - b)$. This technique is also applicable in a limited manner to other adder networks.

12.3 Carry Look-Ahead Adders

Carry look-ahead (CLA) adders are designed to overcome the latency introduced by the rippling effect of the carry bits. The CLA algorithm is based on the origin of the carry-out bit in the equation

$$c_{i+1} = a_i \cdot b_i + c_i \cdot (a_i \oplus b_i) \tag{12.20}$$

for the cases that give $c_{i+1} = 1$. Since either term may cause this output, we treat the two separately. First, if $a_i \cdot b_i = 1$, then $c_{i+1} = 1$. We call

$$g_i = a_i \cdot b_i \tag{12.21}$$

the **generate** term, since the inputs are viewed as "generating" the carry-out bit. Note that if $g_i = 1$, then we must have $a_i = b_i = 1$. The second term represents the case where an input carry $c_i = 1$ may be "propagated" through the full-adder. This will happen if the **propagate** term

$$p_i = a_i \oplus b_i \tag{12.22}$$

is equal to 1; if $p_i = 1$ then $g_i = 0$ since the XOR operation produces a 1 iff the inputs are not equal. Figure 12.15 shows the behavior of the generate and propagate terms. With these definitions, the equation for the carry-out bit is

$$c_{i+1} = g_i + p_i \cdot c_i \tag{12.23}$$

The main idea of the CLA is to first calculate the values of p_i and g_i for every bit, then use them to find the carry bits c_{i+1}. Once these are found, the sum bits are given by

$$s_i = p_i \oplus c_i \tag{12.24}$$

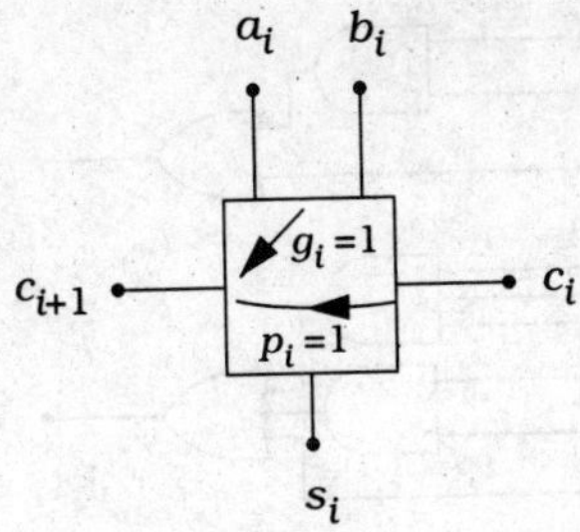

	g_i	p_i
	$a_i \; b_i$	$a_i \oplus b_i$
$a_i = b_i = 0$	0	0
$a_i = b_i = 1$	1	0
$a_i \neq b_i$	0	1

$c_{i+1} = a_i \; b_i + c_i \cdot (a_i \oplus b_i)$

Figure 12.15 Basis of the carry look-ahead algorithm

for every *i*. This avoids the need to ripple the carry bits serially down the chain.

Let us analyze the 4-bit CLA equations. With c_0 assumed known, we have

$$c_1 = g_0 + p_0 \cdot c_0 \tag{12.25}$$

The expressions for c_2, c_3, and c_4 have the same form with

$$\begin{aligned} c_2 &= g_1 + p_1 \cdot c_1 \\ c_3 &= g_2 + p_2 \cdot c_2 \\ c_4 &= g_3 + p_3 \cdot c_3 \end{aligned} \tag{12.26}$$

These can be expressed using primitive generate and propagate terms by noting that c_i can be substituted into c_{i+1} in succession. The first reduction is obtained by substituting c_1 into the c_2 equation to arrive at

$$c_2 = g_1 + p_1 \cdot (g_0 + p_0 \cdot c_0) \tag{12.27}$$

Expanding,

$$c_2 = g_1 + p_1 \cdot g_0 + p_1 \cdot p_0 \; c_0 \tag{12.28}$$

Similarly, substituting c_2 into c_3 gives

$$\begin{aligned} c_3 &= g_2 + p_2 \cdot (g_1 + p_1 \cdot g_0 + p_1 \cdot p_0 \cdot c_0) \\ &= g_2 + p_2 \cdot g_1 + p_2 \cdot p_1 \cdot g_0 + p_2 \cdot p_1 \cdot p_0 \cdot c_0 \end{aligned} \tag{12.29}$$

Finally, the carry-out bit is

$$\begin{aligned} c_4 &= g_3 + p_3 \cdot (g_2 + p_2 \cdot g_1 + p_2 \cdot p_1 \cdot g_0 + p_2 \cdot p_1 \cdot p_0 \cdot c_0) \\ &= g_3 + p_3 \cdot g_2 + p_3 \cdot p_2 \cdot g_1 + p_3 \cdot p_2 \cdot p_1 \cdot g_0 + p_3 \cdot p_2 \cdot p_1 \cdot p_0 \cdot c_0 \end{aligned} \tag{12.30}$$

These equations show that every carry bit can be found from the generate and propagate terms. Moreover, the algorithm yields nested SOP expressions. The logic diagram for the 4-bit network is shown in Figure 12.16

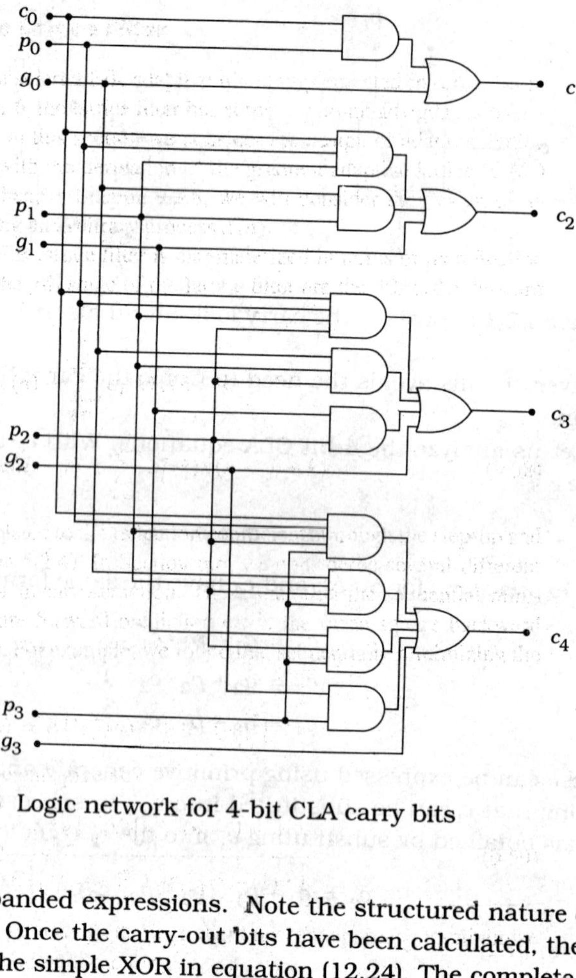

Figure 12.16 Logic network for 4-bit CLA carry bits

using the expanded expressions. Note the structured nature of the gate arrangement. Once the carry-out bits have been calculated, the sums are found using the simple XOR in equation (12.24). The complete adder circuit is shown in Figure 12.17 where the "CLA Network" box represents the carry bit logic in Figure 12.16. This illustrates a marked departure from the ripple-carry design.

The high-level abstract Verilog description of a 4-bit adder can be used to describe any adder, including the CLA-based design. However, we can rewrite the behavioral code to better illustrate the internal algorithm in an explicit manner. The **assign**-based RTL module below illustrates this idea.

```
module CLA_4b (sum, c_4, a, b, c_0 ) ;
input [ 3 : 0 ] a, b ;
input c_0 ;
output [ 3 : 0 ] sum ;
output c _4 ;
wire p0, p1, p2, p3, g0, g1, g2, g3 ;
wire c1, c2, c3, c4 ;
```

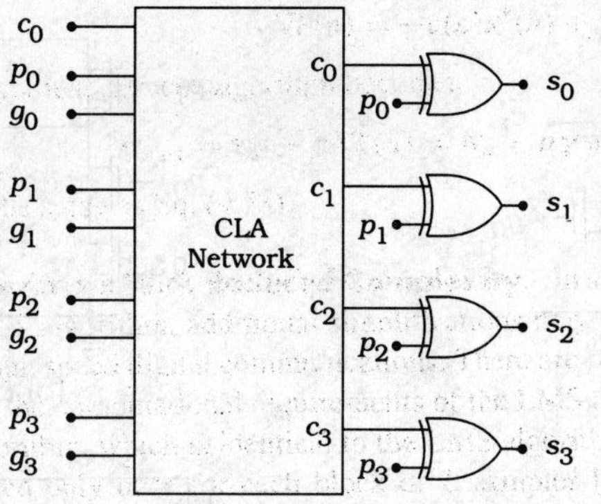

Figure 12.17 Sum calculation using the CLA network

```
assign
        p0 = a[0] ^ b[0] ,
        p1 = a[1] ^ b[1] ,
        p2 = a[2] ^ b[2] ,
        p3 = a[3] ^ b[3] ,
        g0 = a[0] & b[0] ,
        g1 = a[1] & b[1] ,
        g2 = a[2] & b[2] ,
        g3 = a[3] & b[3] ;
assign
        c1 = g0 | ( p0 & c_0 ) ,
        c2 = g1 | ( p1 & g0 ) | ( p1 & p0 & c_0 ) ,
        c3 = g2 | ( p2 & g1 ) | ( p2 & p1 & g0 ) | ( p2 & p1 & p0 & c_0 ) ,
        c4 = g3 | ( p3 & g2) | ( p3 & p2 & g1 ) | ( p3 & p2 & p1 & g0 )
             | ( p3 & p2 & p1 & p0 & c_0 ) ;
assign
        sum [0] = p0 ^ c_0 ,
        sum [1] = p1 ^ c1 ,
        sum [2] = p2 ^ c2 ,
        sum [3] = p3 ^ c3 ,
         c_4 = c4 ;
endmodule
```

Adding delay times on each statement provides the final information needed for a simulation. The repetitive nature of the CLA equations can be implemented in a more efficient coding style by using the Verilog **for** procedure.

To translate the CLA algorithms into circuits, we use the logic construction techniques developed in Chapter 2 to create the nFET arrays shown in Figure 12.18. Note that each carry-out circuit $\bar{c}_i$ forms the basis

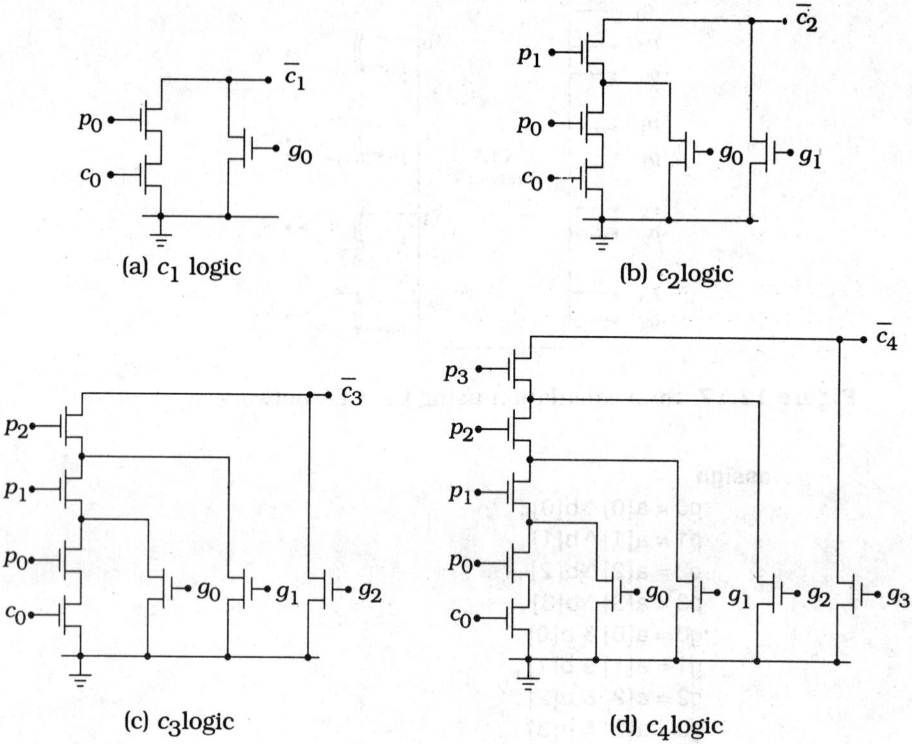

Figure 12.18 nFET logic arrays for the CLA terms

for the next higher term $\overline{c}_{i+1}$. This is due to the nesting property of the algorithm.

Once the nFET logic is designed, it can be used in a variety of circuits. Figure 12.19 shows three possibilities. The structure in Figure 12.19(a) represents standard complementary structuring where we create a pFET array using bubble pushing to obtain the series-parallel pFET array. The static pseudo-nMOS approach in Figure 12.19(b) could be used, but we would have to be concerned about device ratios to insure that the output low voltage V_{OL} is sufficiently small without using excessively large nFETs. This is avoided if we opt for a dynamic logic family as in Figure 12.19(c). This, however, introduces timing problems and gives outputs that are only valid for a short period of time. Obviously, the selection of the circuit family involves considering many factors.

Let us examine the possibility of using full complementary static circuits. The $\overline{c}_1$ circuit in Figure 12.20(a) shows the complete nFET/pFET arrays with series-parallel structuring. This, however, has a form similar to the carry-out circuit analyzed earlier in Figure 12.8, so that we may create the mirror-equivalent logic gate shown in Figure 12.20(b). The pro-

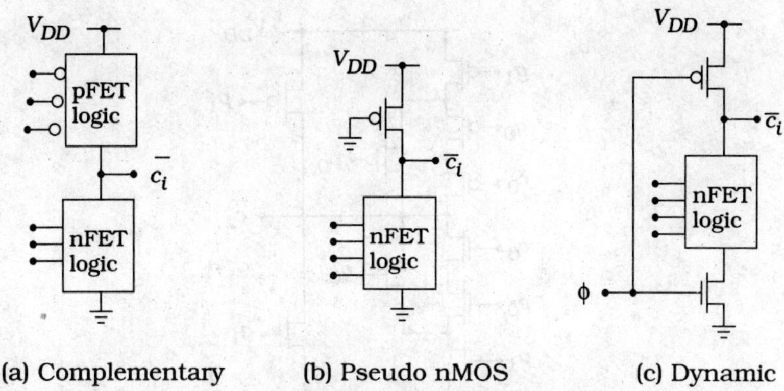

Figure 12.19 Possible uses of the nFET logic arrays in Figure 12.18

cess can be continued for the remaining bits. For example, Figure 12.21 shows the mirror circuit for $\overline{c_2}$. Note the symmetry in the arrays. This feature allows for a more structured layout at the physical design level.

Another approach is to use multiple-output domino logic (MODL) as a basis. This is possible because the nesting of the carry bits from one bit to the next gives the ANDing relationship needed to implement MODL. To see this analytically, recall that we had

$$
\begin{aligned}
c_1 &= g_0 + p_0 \cdot c_0 \\
c_2 &= g_1 + p_1 \cdot c_1 \\
&= g_1 + p_1 \cdot (g_0 + p_0 \cdot c_0)
\end{aligned}
\tag{12.31}
$$

We can use c_1 as one output, and c_2 as another output with the two related by the AND operation. Since this type of relationship is valid for c_3 and c_4, only a single MODL gate is needed to produce all four carry bits.

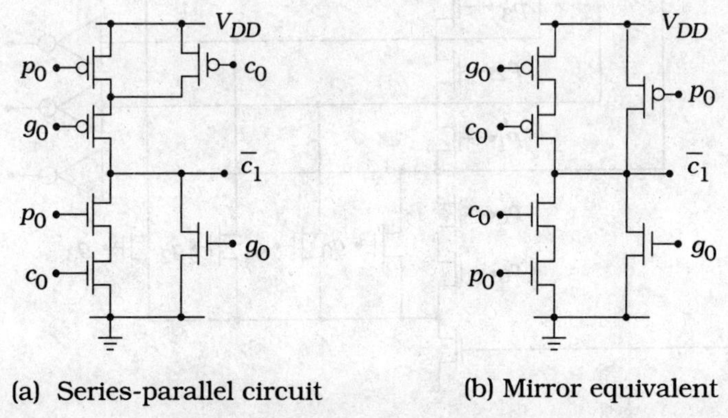

Figure 12.20 Static CLA mirror circuit

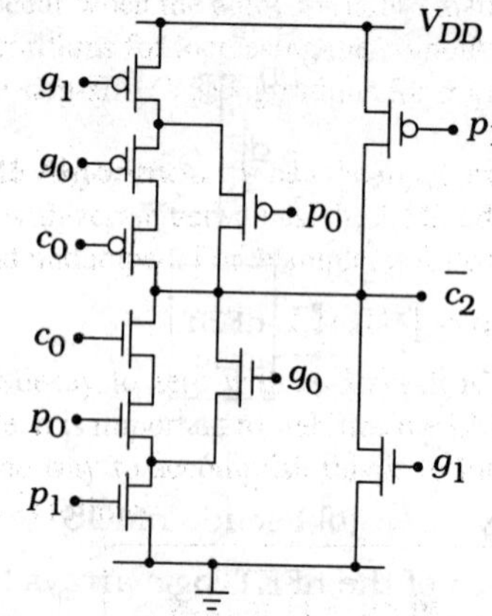

Figure 12.21 Static mirror circuit for $\bar{c}_2$

Moreover, MODL is a non-inverting logic family with the inverters built into the structure. Figure 12.22 shows the 4-bit MODL carry circuit where the logic array provides separate outputs for each carry bit. A precharge pFET has been added for every internal node. When the circuit is in precharge with the clock at $\phi = 0$, all of the outputs are driven to 0. The logic network accepts the inputs during the evaluation phase ($\phi = 0$). If the c_1 network allows for a discharge of the internal node and produces $c_1 = 1$, then one of the conditions

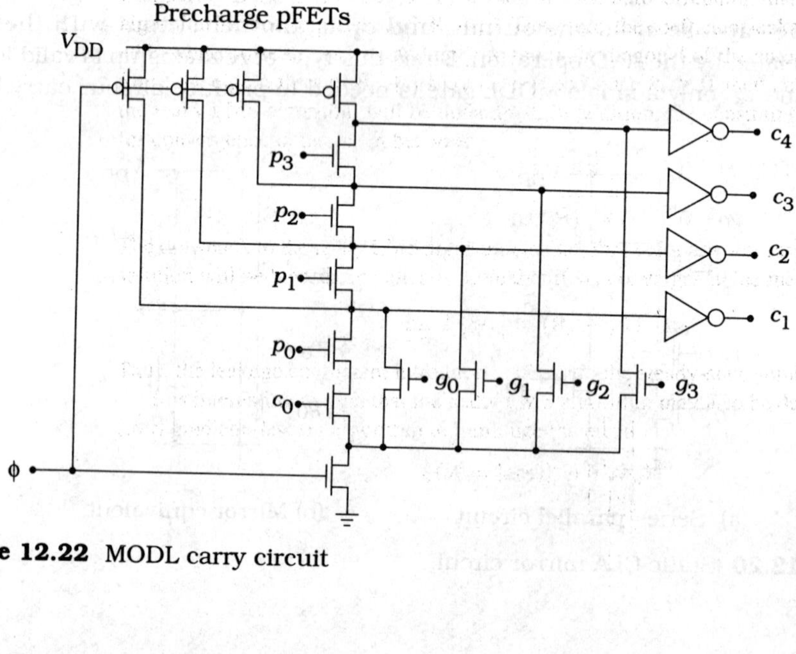

Figure 12.22 MODL carry circuit

$$
\begin{aligned}
g_0 &= 1, \text{ or} \\
p_0 \cdot c_0 &= 1
\end{aligned}
\tag{12.32}
$$

holds. If $p_0 \cdot c_0 = 1$ then a value of $p_1 = 1$ will drive c_2 to 1. Alternately, the carry-out may be generated with $g_1 = 1$. This type of interdependence continues upward in the logic network to produce c_3 and c_4. The ability to use a single logic gate to produce the four carry-out bits is very attractive. The layout will have the complexity of a single c_4 mirror gate with fewer transistors. However, it must be remembered that MODL is a dynamic circuit technique, and is subject to the usual limitations: clocking is mandatory, the output is subject to charge leakage and charge sharing, and the series-connected nFET chains will give long discharge times unless large FETs are used.

12.3.1 Manchester Carry Chains

The Manchester carry scheme is a particularly elegant approach to dealing with CLA bits. It is based on building a switch-logic network for the basic equation

$$
c_{i+1} = g_i + p_i \cdot c_i \tag{12.33}
$$

that can be cascaded to feed to successively stages.

Consider a full adder with inputs a_i, b_i, and c_i. We will use the generate and propagate expressions

$$
\begin{aligned}
g_i &= a_i \cdot b_i \\
p_i &= a_i \oplus b_i
\end{aligned}
\tag{12.34}
$$

to introduce the **carry-kill** bit k_i such that

$$
\begin{aligned}
k_i &= \overline{a_i + b_i} \\
&= \bar{a}_i \cdot \bar{b}_i
\end{aligned}
\tag{12.35}
$$

This term gets its name from the fact that if $k_i = 1$, then $p_i = 0$ and $g_i = 0$, so that $c_{i+1} = 0$; $k_i = 1$ thus "kills" the carry-out bit. This can be verified from the table in Figure 12:23 that shows the values of p_i, g_i, and k_i for all possible inputs. Note that for a given input set (a_i, b_i), only one of the three quantities is a logic 1

a_i	b_i	p_i	g_i	k_i
0	0	0	0	1
0	1	1	0	0
1	0	1	0	0
1	1	0	1	0

Figure 12.23 Propagate, generate, and carry-kill values

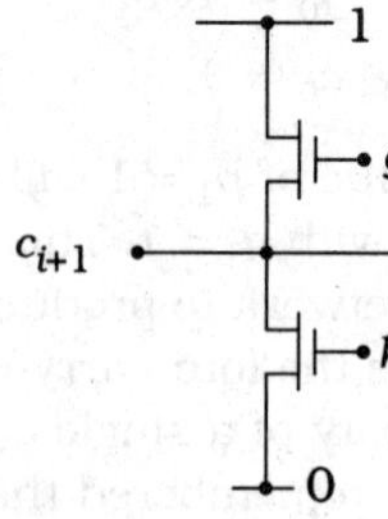

Figure 12.24 Switching network for the carry-out equation

The Manchester carry scheme is based on this behavior. Since only one of the three quantities p_i, g_i, and k_i can be a 1, we can construct the switch-level circuit using FETs shown in Figure 12.24. The topology has been chosen such that only one FET is a closed switch at a time. The operation can be understood by examining each possibility. First, if we have $(a_i, b_i) = (0, 0)$, then $k_i = 1$ and $c_{i+1} = 0$. If $a_i \neq b_i$, then $p_i = 1$ and the input bit c_i is propagated through the circuit to give $c_{i+1} = c_i$. Finally, an input of $(a_i, b_i) = (1, 1)$ indicates that a carry-out has been generated by a term $g_i = 1$, so $c_{i+1} = 1$. At the circuit level, it is important to note that using only nFETs induces a threshold voltage drop on a logic 1 transmission through the transistor.

Several different Manchester carry circuits can be built. Two are shown in Figure 12.25. The static logic gate in Figure 12.25(a) uses $\overline{c_i}$ as an input. First, suppose that $p_i = 0$. This opens M1 and blocks the input $\overline{c_i}$ from propagating through, but also turns on nFET M3. If $g_i = 0$, pFET M4 is on and pulls the output to $\overline{c_{i+1}} = 1$. If $g_i = 1$, then both nFETs M2 and M3 are on while M4 is off, giving an output of $\overline{c_{i+1}} = 0$. The case where $p_i = 1$ is more complicated. The generate term g_i must be 0, so pFET M4 is on while the nFET chain acts as an open circuit since M3 is off. The output is

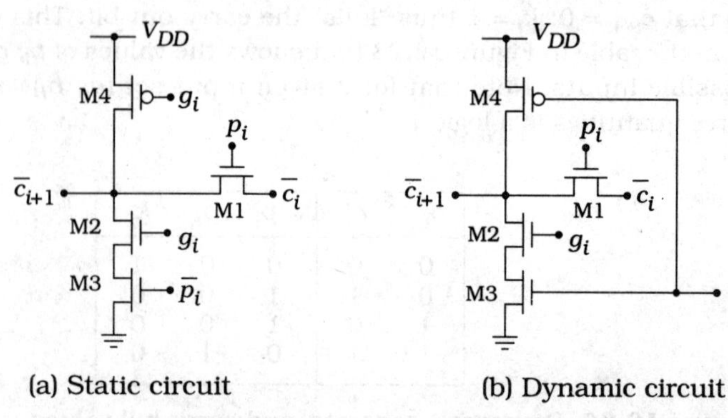

(a) Static circuit (b) Dynamic circuit

Figure 12.25 Manchester circuit styles

then controlled by $\bar{c}_i$. If $\bar{c}_i = 1$, then this is transmitted to the output and supported by the pFET connection to the power supply so that $\bar{c}_{i+1} = 1$. However, if $\bar{c}_i = 0$, then the circuit reduces to a pseudo-nMOS inverter made up of M4 and M1, with $p_i = 1$ at the input.[1] To obtain a low output of $\bar{c}_{i+1} = 0$ we must choose the nFET/pFET size ratio to be large enough to give a low output voltage.

A dynamic circuit is shown in Figure 12.25(b). The logic is similar to the static design except that the evaluation nFET M3 replaces a logic transistor. During the precharge ($\phi = 0$), the output node is brought to a logic 1 voltage. Evaluation takes place when the clock switches to $\phi = 1$. A carry propagation occurs if $p_i = 1$, while the node discharges to 0 if $g_i = 1$. This circuit can be used to build the Manchester carry chain shown in Figure 12.26. Every stage undergoes precharge when $\phi = 0$. The carry bits are available during the evaluation time with the longest time delay for c_4.

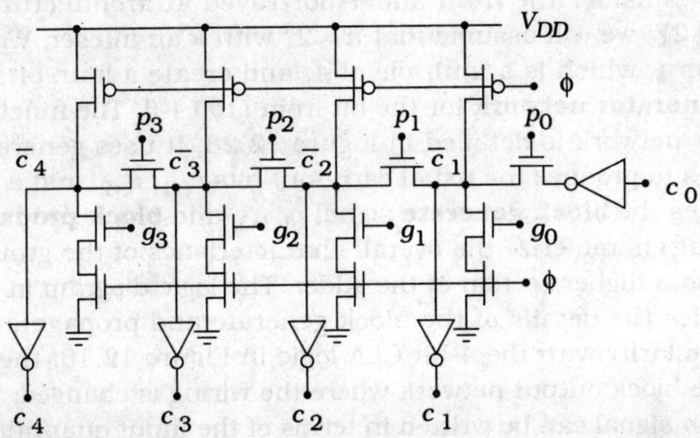

Figure 12.26 Dynamic Manchester carry chain

12.3.2 Extension to Wide Adders

The carry look-ahead equations can be extended to adders wider than 4-bits, but one must be careful of hardware delays due to the increased gate count through the longest delay path. For example, if we use a brute-force approach for an 8-bit design, then the carry-out bit c_8 would have a term of the form

$$p_7 \cdot p_6 \cdot p_5 \cdot p_4 \cdot p_3 \cdot p_2 \cdot p_1 \cdot p_0 \cdot c_0 \tag{12.36}$$

that would have to be dealt with. Various techniques have been published to obtain more efficient CLA networks for wide adders. Consider the addi-

1 This can be seen by grounding the input $\bar{c}_i$ and covering up the M2-M3 transistors.

tion of two n-bit words. Work by von Neumann and others has shown that the longest carry chain has an average length of[2]

$$\log_2(n) \tag{12.37}$$

For example, the average carry chain in an 8-bit adder is

$$\log_2(8) = \log_2(2^3) = 3 \tag{12.38}$$

while a 32-bit adder has an average length

$$\log_2(32) = \log_2(2^5) = 5 \tag{12.39}$$

This implies that the length of the carry circuits does not have to span the entire length of the word, but can be broken up into smaller segments. Multilevel CLA networks are based on this philosophy.

Consider the n-bit adder portrayed at architectural level in Figure 12.27; we will assume that $n = 2^k$ with k an integer. We select a bit position i, which is a multiple of 4, and create a four-bit **lookahead carry generator network** for the bit from i to $i + 3$. The function of the generator network is detailed in Figure 12.28. It uses generate and propagate bits to produce the usual carry-out bits c_{i+1}, c_{i+2}, and c_{i+3}, but also calculates the **block generate** signal $g_{[i,i+3]}$ and **block propagate** signal $p_{[i,i+3]}$ that characterize the overall characteristics of the group and can be fed into a higher section of the adder. The logic diagram in Figure 12.29 provides the details of the block generate and propagate signals. Note the similarity with the 4-bit CLA logic in Figure 12.16; the difference lies in the block output network where the wiring is changed. The block generation signal can be written in terms of the input quantities as

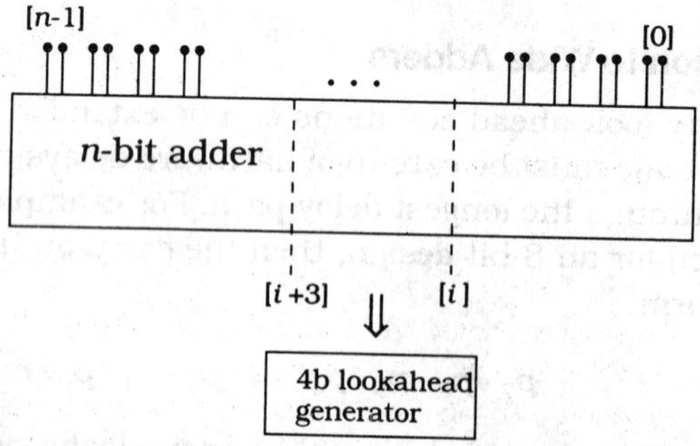

Figure 12.27 An n-bit adder network

---

2 See Reference [4].

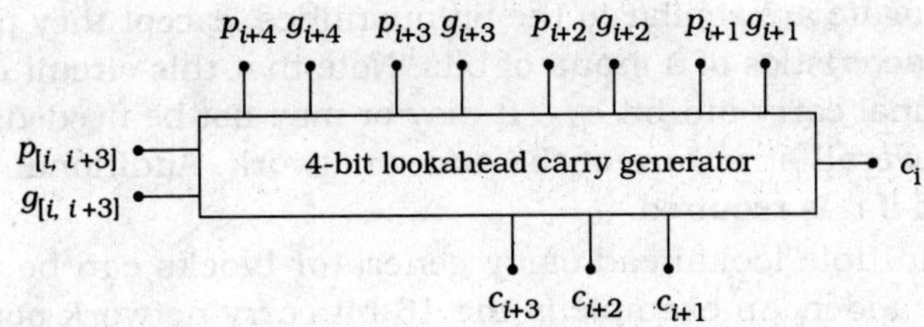

Figure 12.28 4-bit lookahead carry generator signals

$$g_{[i,\,i+3]} = g_{i+3} + p_{i+3} \cdot g_{i+2} + p_{i+3} \cdot p_{i+2} \cdot g_{i+1} + p_{i+3} \cdot p_{i+2} \cdot p_{i+1} \cdot g_i \tag{12.40}$$

and is taken out of the gate labeled **or1** in the diagram. The block propagation is

$$p_{[i,\,i+3]} = p_{i+3} \cdot p_{i+2} \cdot p_{i+1} \cdot p_i \tag{12.41}$$

which is the output of gate **and1** in the diagram. The block generate and

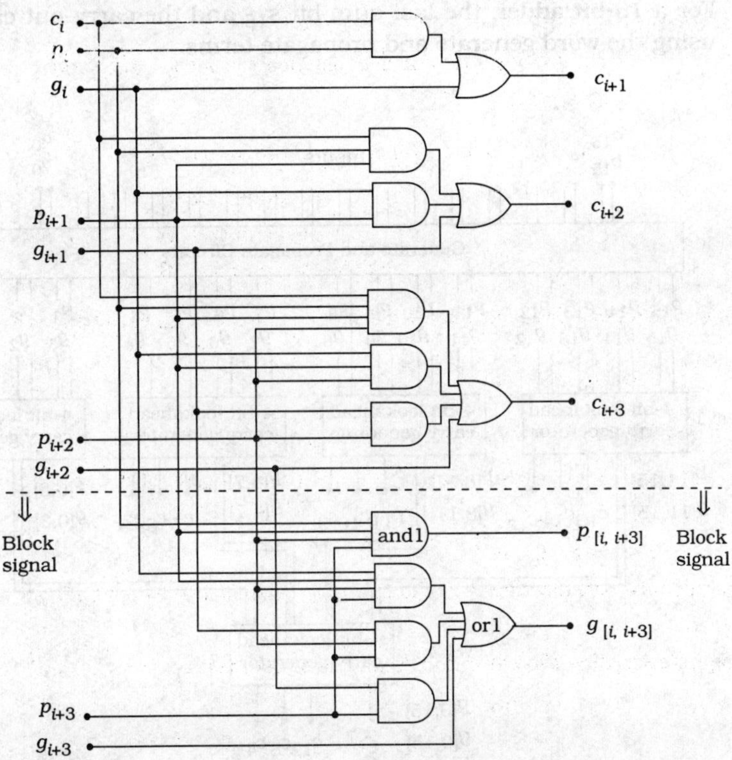

Figure 12.29 Block lookahead generator logic

propagate are similar to the bit quantities, except they provide the overall characteristics of a group of bits. Note that this circuit does not compute the final carry-out bit c_{i+4}. It may or may not be needed, depending upon the overall structure of the adder network. Additional logic can be provided if it is required.

Multiple lookahead carry generator blocks can be used to design a wide adder. An example is the 16-bit carry network portrayed in Figure 12.30. The inputs $a_{15} \ldots a_0$ and $b_{15} \ldots b_0$ are fed into the generate and propagate network that produces the values $(p_{15}, g_{15}), \ldots, (p_0, g_0)$ for use in the CLA blocks. The CLA subsystem is usually described in levels. At Level 1, four 4-bit lookahead carry generator networks are used to provide the carry-out bits c_{i+3}, c_{i+2}, c_{i+1}, and the block generate and propagate terms $g_{[i,i+3]}$ and $p_{[i,i+3]}$ for $i = 0, 4, 8, 12$. The block terms are then sent to the single Level 2 4-bit lookahead carry network. The Level 2 block produces carry-out bits c_4, c_8, c_{12}, and the word generate and propagate terms $g_{[0,15]}$ and $p_{[0,15]}$. At this point, all of the carry bits except c_{15} have been calculated for use in the sum equation

$$s_i = p_i \oplus c_i \tag{12.42}$$

For a 16-bit adder, the last sum bit s_{15} and the carry-out can be found using the word generate and propagate terms.

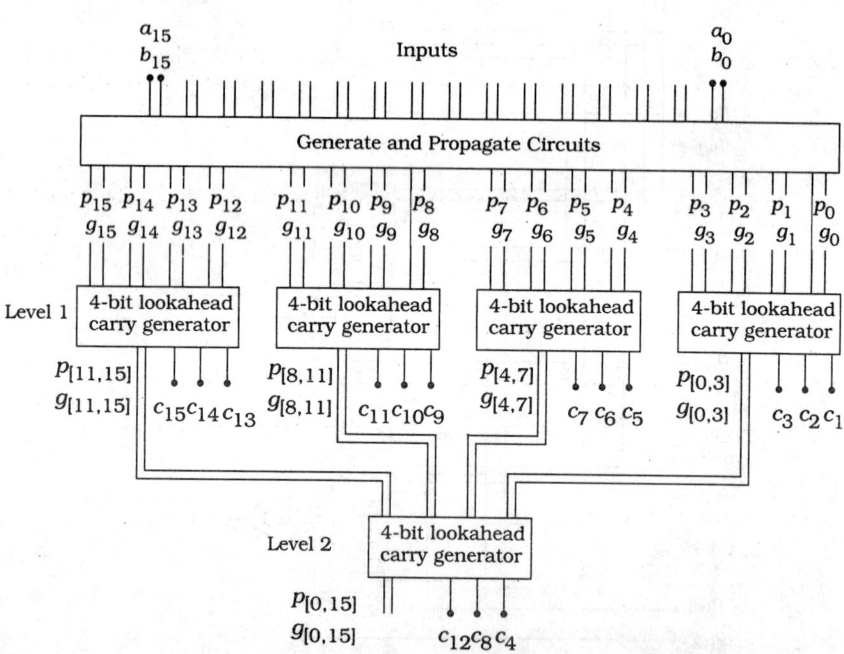

Figure 12.30 Multilevel CLA block scheme for a 16-bit adder

A 64-bit adder can be obtained by adding another level of lookahead carry blocks to the 16-bit network. The scheme is shown in Figure 12.31. Four 16-bit blocks are used to produce four sets of group generate and propagate terms. These are then fed into the Level 3 block that provides the final carry-out bits. It is important to note that each block produces carry-out bits for use in the sum calculations. The carry-out bits are available at times that vary with the level where the circuitry is. Level 1 bits are available first, Level 2 bits second, and Level 3 bits are the final ones out of the network. There is no *a priori* reason for using 4-bit lookahead carry generator circuits; smaller or larger widths are acceptable.

We have examined the basic concepts involved in CLA structures here. The interested reader is directed to Reference [2] for a more detailed discussion.

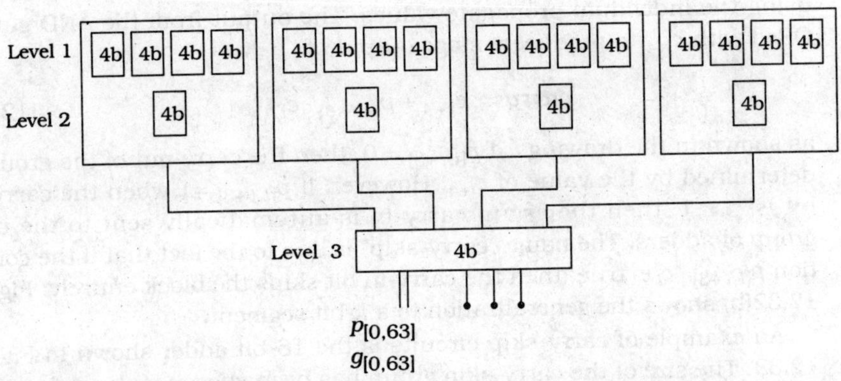

Figure 12.31 64-bit CLA adder architecture

12.4 Other High-Speed Adders

Several alternate approaches to designing fast word adders have been published in the literature. All have the objective of decreasing the computation time, and each has different trade-offs. This section examines a few of these designs illustrating the variations that one has in translating from a high-level architectural description down to the circuit level.

12.4.1 Carry-Skip Circuits

A **carry-skip adder** is designed to speed up a wide adder by aiding the propagation of a carry bit around a portion of the entire adder. The idea is illustrated in Figure 12.32(a) for the case of a 4-bit adder. The carry-in bit is designated as c_i, and the adder itself produces a carry-out bit of c_{i+4}. The carry-skip circuitry consists of two logic gates. The AND gate accepts the carry-in bit and compares it to the group propagate signal

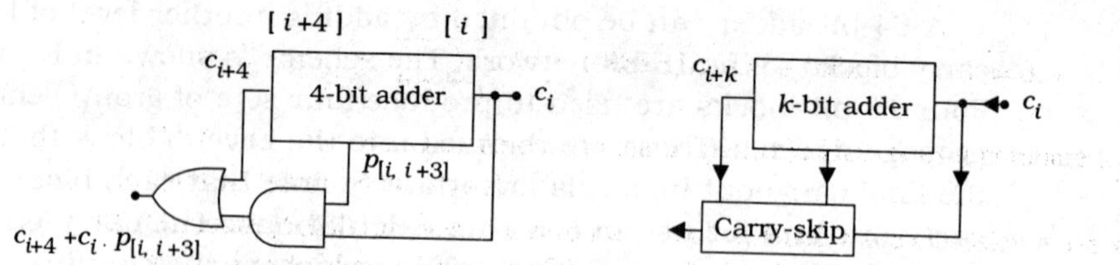

(a) Carry-skip logic (b) Generalization

Figure 12.32 Carry-skip circuitry

$$p_{[i, i+3]} = p_{i+3} \cdot p_{i+2} \cdot p_{i+1} \cdot p_i \tag{12.43}$$

using the individual propagate values. The output from the AND gate is ORed with c_{i+4} to produce a stage output of

$$carry = c_{i+4} + p_{[i, i+3]} \cdot c_i \tag{12.44}$$

as shown in the drawing. If $p_{[i,i+3]} = 0$, then the carry-out of the group is determined by the value of c_{i+4}. However, if $p_{[i,i+3]} = 1$ when the carry-in bit is $c_i = 1$, then the group carry-in is automatically sent to the next group of adders. The name "carry-skip" is due to the fact that if the condition $p_{[i,i+3]} \cdot c_i$ is true, then the carry-in bit skips the block entirely. Figure 12.32(b) shows the generalization to a k-bit segment.

An example of carry-skip circuits is the 16-bit adder shown in Figure 12.33. The size of the carry-skip group has been chosen as $k = 4$ for every segment. The worst-case delay through this circuit is when $c_0 = 0$ and the 0-th bit adder produces a carry-out bit of $c_1 = 1$. If ripple adders are used, then the worst-case situation is where this bit emerges as $c_4 = 1$, and then skips the next segment groups [7,4] and [11,8] and enters the final block, where it ripples through to the output as $c_{16} = 1$.

The size k of a carry-skip block affects the overall speed of the scheme. It has been shown that the optimal block size for an n-bit adder that minimizes the delay can be estimated as

$$k = \sqrt{\frac{n}{2}} \tag{12.45}$$

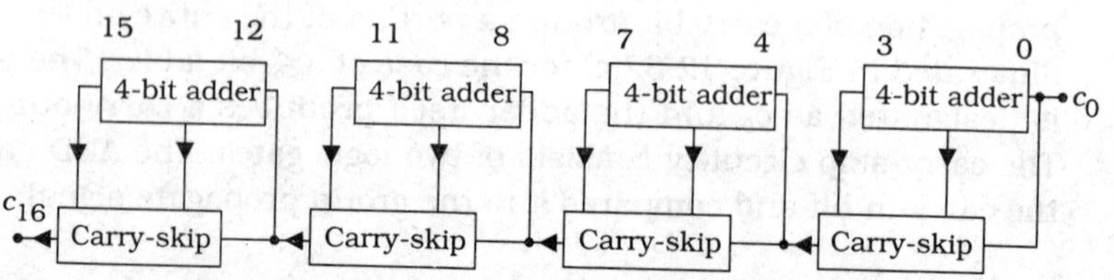

Figure 12.33 A 16-bit adder using carry-skip circuits

For $n = 16$, the block size would be $k \approx 3$. Alternately, a variable k-value can be used. The carry-skip circuits can be nested to create multilevel networks. Figure 12.34 shows an example of a 2-level carry-skip adder.

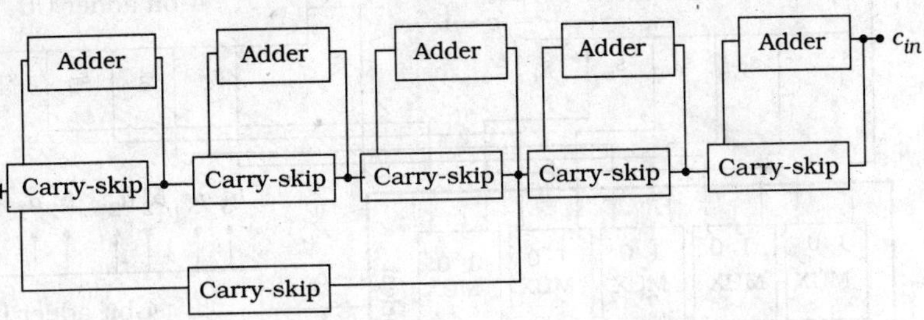

Figure 12.34 A 2-level carry-skip adder

12.4.2 Carry-Select Adders

Carry-select adders use multiple narrow adders to create fast wide adders. Consider the addition of two n-bit numbers with $a = a_{n-1} \ldots a_0$ and $b = b_{n-1} \ldots b_0$. At the bit level, the adder delay increases from the least significant 0-th position upward, with the $(n-1)$-th requiring the most complex logic. A carry-select adder breaks the addition problem into smaller groups. For example, we can split the n-bit problem into two $(n/2)$-bit sections, then give special attention to the higher order group that adds the word segments $a_{n-1} \ldots a_{n/2}$ and $b_{n-1} \ldots b_{n/2}$. The carry delay will then center around the carry-out bit $c_{n/2}$ produced by the sum of lower order word segments $a_{(n/2)-1} \ldots a_0$ and $b_{(n/2)-1} \ldots b_0$. We know that there are only two possibilities for the carry bit:

$$c_{n/2} = 0 \text{ or } c_{n/2} = 1 \tag{12.46}$$

A carry-select adder provides two separate adders for the upper words, one for each possibility. A MUX is then used to select the valid result.

As a concrete example, consider an 8-bit adder that is split into two 4-bit groups. The lower-order bits $a_3a_2a_1a_0$ and $b_3b_2b_1b_0$ are fed into the 4-bit adder L to produce the sum bits $s_3s_2s_1s_0$ and a carry-out bit c_4 as shown in Figure 12.35. The higher order bits $a_7a_6a_5a_4$ and $b_7b_6b_5b_4$ are used as inputs to two 4-bit adders. Adder U0 calculates the sum with a carry-in of $c = 0$, while U1 does the same only it has a carry-in value of $c = 1$. Both sets of results are used as inputs to an array of 2:1 MUXes. The carry bit c_4 from the adder L is used as the MUX select signal. If $c_4 = 0$, then the results of U0 are sent to the output, while a value of $c_4 = 1$ selects the results of U1 for $s_7s_6s_5s_4$. The carry-out bit c_8 is also selected by the MUX array.

This design speeds up addition of the word by allowing the upper and lower portions of the sum to be calculated simultaneously. The price paid is that it requires an additional word adder, a set of multiplexors, and the

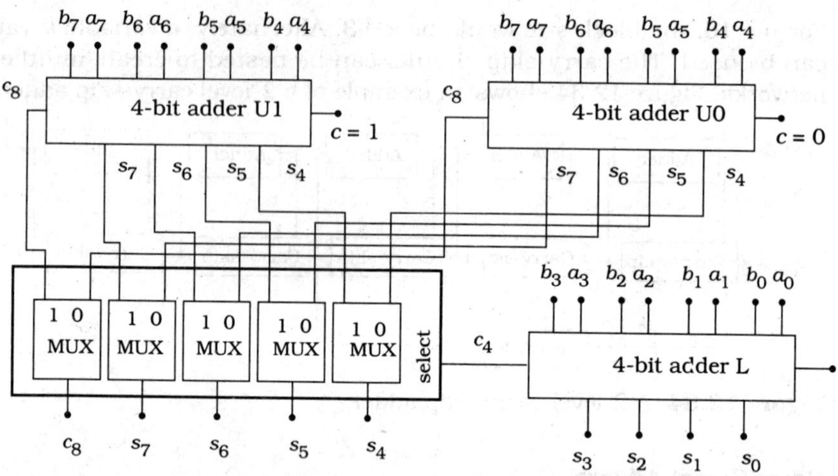

Figure 12.35 8-bit carry-select adder

associated interconnect wiring. The design becomes viable if speed is more important than area consumption. Carry-select adders can be created using multiple levels, but the hardware costs increase accordingly.

12.4.3 Carry-Save Adders

Carry-save adders are based on the idea that a full-adder really has three inputs and produces two outputs as shown in Figure 12.36(a). While we usually associate the third input with a carry-in, it could equally well be used as a "regular" value. In Figure 12.36(b), the FA is used as a 3-2 reduction network where it starts with bits from 3 words, adds them, and then has an output that is 2-bits wide. We can build an n-bit carry-save adder by using n separate adders as in Figure 12.37. The name "carry-save" arises from the fact that we save the carry-out word instead of using it immediately to calculate a final sum.

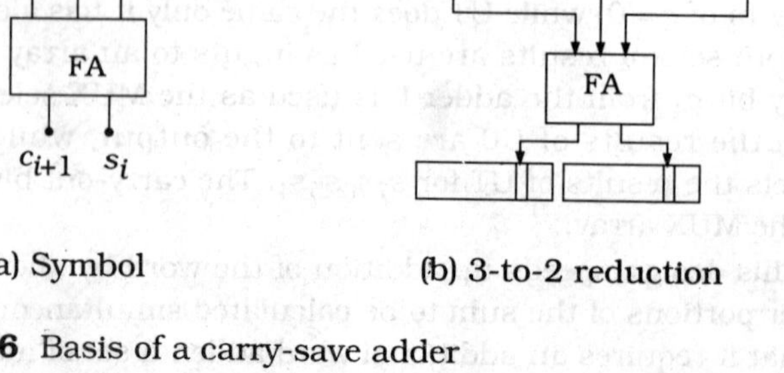

(a) Symbol (b) 3-to-2 reduction

Figure 12.36 Basis of a carry-save adder

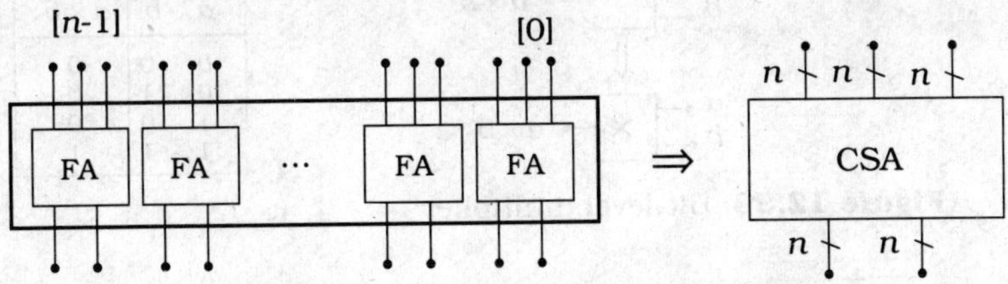

Figure 12.37 Creation of an n-bit carry-save adder

Carry-save adders (CSAs) are useful in situations where we need to add more than two numbers. Since the design automatically avoids the delay in the carry-out bits, a CSA chain may be faster than using standard adders or cycling with a clocked synchronous network.

An example is the 7-to-2 reduction scheme shown in Figure 12.38. This starts with 7 n-bit words a, b, ..., g and uses five CSA units to reduce it down to two words. If we want a final sum, then a normal CPA (carry-propagate adder) can be used at the bottom of the chain to add the two values together.

12.5 Multipliers

Binary multiplication is based on the basic operations

$$0 \times 0 = 0 \ , \ \ 0 \times 1 = 0 \ , \ \ 1 \times 0 = 0 \ , \ \ 1 \times 1 = 1 \qquad (12.47)$$

If we multiply two bits a and b, then we see that the logical AND operation

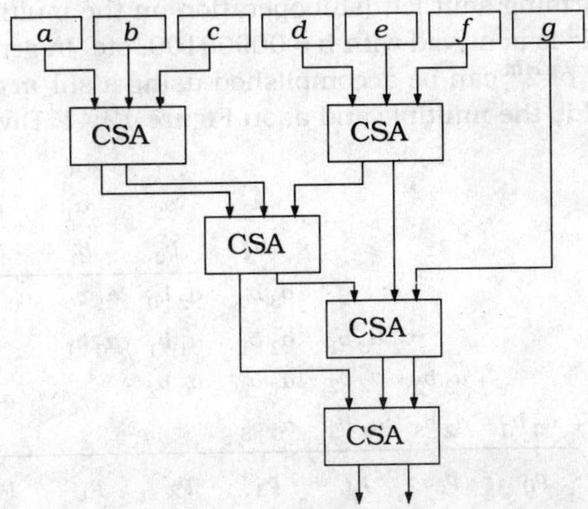

Figure 12.38 A 7-to-2 reduction using carry-save adders

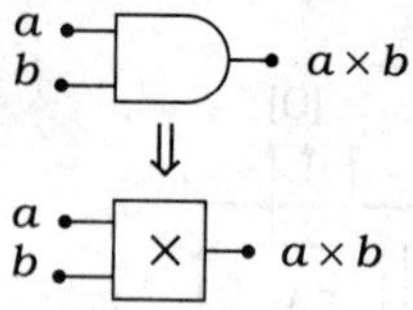

a	b	$a \times b$
0	0	0
0	1	0
1	0	0
1	1	1

Figure 12.39 Bit-level multiplier

produces the same result as summarized by the symbols in Figure 12.39. Bit-level multiplication is thus a trivial operation.

The complexity arises when we multiply n-bit words. Let us specify a word length of $n = 4$ to illustrate the main ideas. With input values given by $a = a_3a_2a_1a_0$ and $b = b_3b_2b_1b_0$, the product $a \times b$ is given by the 8-bit $(2n)$ result

$$p = p_7p_6p_5p_4p_3p_2p_1p_0 \tag{12.48}$$

as shown in Figure 12.40. Each bit b_i multiplies the multiplicand a on a bit-by-bit basis. The product term from least significant bit b_0 is aligned to the multiplicand, while the next term (due to b_1) is shifted one column left. The array builds until every bit of the multiplier is used. The product bits p_i are obtained by summing each of the i-th columns, accounting for a carry from the $(i-1)$-th column. A simple expression is

$$p_i = \sum_{i=j+k} a_j b_k + c_{i-1} \tag{12.49}$$

where $c_{i-1} = 0$ for $(i - 1) \le 0$.

A special case worth remembering is multiplication by 2. For an 8-bit word, this corresponds to an input of $b = 00000010$, which is equivalent to performing shift left (shl) operation on the multiplicand. Multiplication by $4 = 2^2$ is achieved with $b = 00000100$, etc. In general, multiplication by a factor of 2^m can be accomplished using a shl_m operation on a register that holds the multiplicand as in Figure 12.41. Division by 2^k is obtained

				a_3	a_2	a_1	a_0	multiplicand
				$\times\ b_3$	b_2	b_1	b_0	multiplier
				a_3b_0	a_2b_0	a_1b_0	a_0b_0	
			$+\ a_3b_1$	a_2b_1	a_1b_1	a_0b_1		
		$+\ a_3b_2$	a_2b_2	a_1b_2	a_0b_2			
	$+\ a_3b_3$	a_2b_3	a_1b_3	a_0b_3				
p_7	p_6	p_5	p_4	p_3	p_2	p_1	p_0	product

Figure 12.40 Multiplication of two 4-bit words

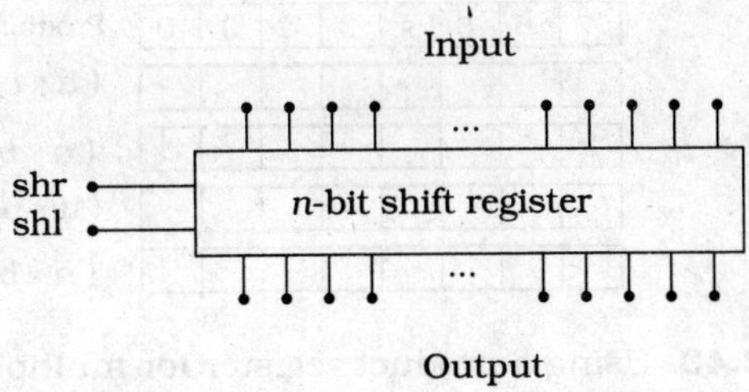

Figure 12.41 Shift register for multiplication or division by a factor of 2

with a shift right (shr_*k*) command. Bit overflow should be monitored to insure that the results are valid.

A high-level Verilog description of the 4 × 4-bit multiplier can be written as

```
module mult_4 (product, a , b) ;
input [ 3 : 0 ] a, b ;
output [ 7 : 0 ] product ;
assign #(t_delay) product = a * b ;
endmodule
```

for use in initial architectural simulations. The word size can be adjusted as needed using the vector specification. However, the critical time delay parameter t_delay depends upon the implementation, and accurate values are necessary to establish a sound design.

The details of the multiplication procedure yield specific techniques for calculating the product with binary switching networks. One way to view the process is that the bit b_i multiplies the entire word a; each term $(a \times b_i)$ has a base-10 weighting of 2^i. This is shown in Figure 12.42. Starting from the first line $(a \times b_0) \times 2^0$ and working down, each factor of 2^i represents a shift in position. A simple view of this process is to use a product register as shown in Figure 12.43. Each term $(a \times b_i)$ occupies a shifted position such that the product is obtained by adding the terms in each column.

$$
\begin{array}{cccccccccl}
 & & & & a_3 & a_2 & a_1 & a_0 & & \text{multiplicand} \\
 & & & \times & b_3 & b_2 & b_1 & b_0 & & \text{multiplier} \\
\hline
 & & & & (a_3 & a_2 & a_1 & a_0) \times b_0 & & (a \times b_0)\,2^0 \\
 & & & (a_3 & a_2 & a_1 & a_0) \times b_1 & & & (a \times b_1)\,2^1 \\
 & & (a_3 & a_2 & a_1 & a_0) \times b_2 & & & & (a \times b_2)\,2^2 \\
+ & (a_3 & a_2 & a_1 & a_0) \times b_3 & & & & & (a \times b_3)\,2^3 \\
\hline
p_7 & p_6 & p_5 & p_4 & p_3 & p_2 & p_1 & p_0 & & \text{product}
\end{array}
$$

Figure 12.42 Alternate view of multiplication process

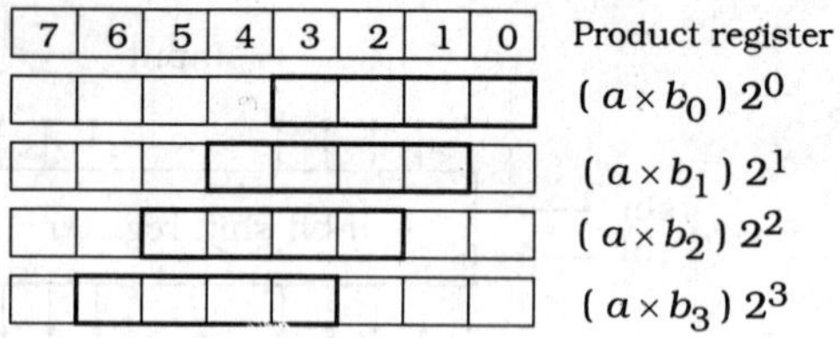

Figure 12.43 Using a product register for multiplication

A practical implementation is based on the sequence illustrated in Figure 12.44. The left side of the register allows for parallel loading of a 4-bit word. The product is created by successive addition and shift-right operations as shown. Note that the carry-out bits are tracked by shifting them into the left register. For the multiplication of two n-bit words, the algo-

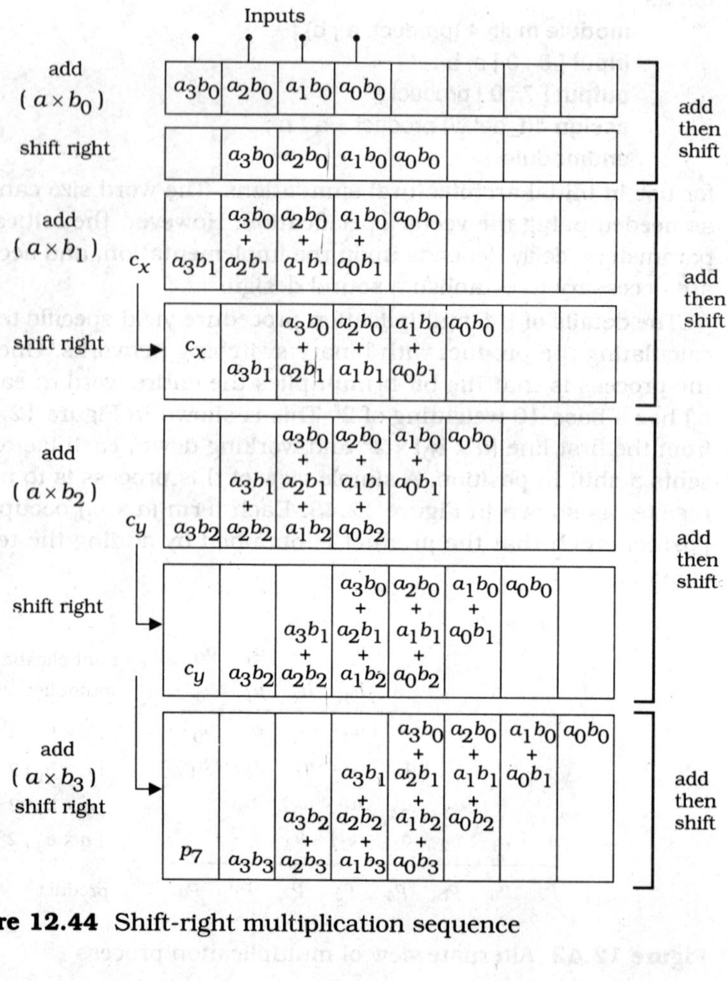

Figure 12.44 Shift-right multiplication sequence

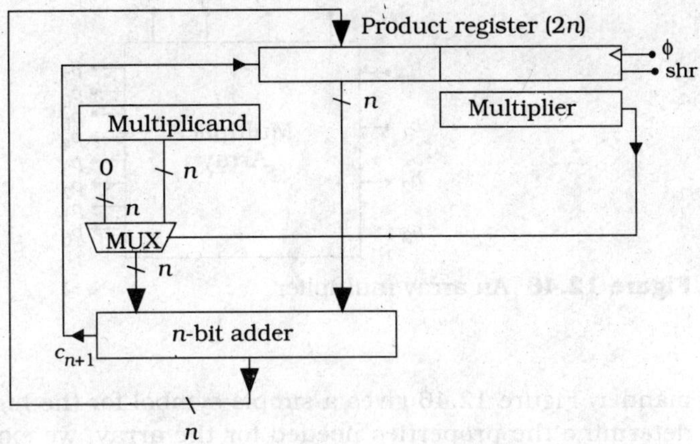

Figure 12.45 Register-based multiplier network

rithm for the product can be expressed as[3]

$$p_{i+1} = (p_i + a2^n b_i)2^{-1} \tag{12.50}$$

with $p_n = p$ the final answer such that $p_0 = 0$. The factor $(p_i + a\, b_i\, 2^n)$ gives the addition while 2^{-1} accounts for a right shift. The factor of 2^n multiplying a is used to compensate for the 2^{-n} introduced by the right shift at the end of the calculation. The algorithm can be used to create the register-based hardwire multiplier network shown in Figure 12.45, which can be built using standard VLSI cells and sequential design. Note that the multiplier bits b_i are used to control a 2:1 multiplexor. If $b_i = 0$, an n-bit zero word is sent to the adder, while $b_i = 1$ directs the multiplicand a to the input. The Booth algorithm can be added to the network, as can several other improvements. Also, we note in passing that a left-shift algorithm can be derived, leading to a different hardware implementation.

This type of design can be coded in Verilog using **assign** statements and shift operators. Accurate delay times are needed to accurately simulate the system. These in turn depend upon the characteristics of the VLSI cells in the library. The complexity of multiplier units is reflected in the length of the HDL code and the time needed to design an efficient network.

12.5.1 Array Multipliers

An array multiplier accepts the multiplier and multiplicand and uses an array of cells to calculate the bit products $a_j \cdot b_k$ individually in a parallel

3 See Reference [2].

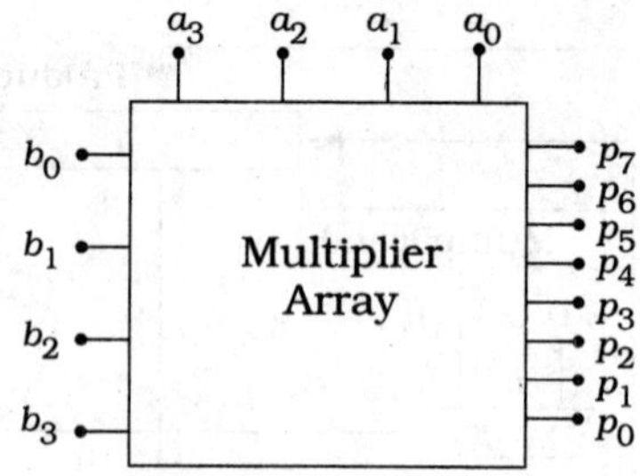

Figure 12.46 An array multiplier

manner. Figure 12.46 gives a simple symbol for the high-level view. To determine the properties needed for the array, we expand the view to show the structure of the multiplication procedure in Figure 12.47. Each block requires that we first calculate the bit product $a_j \cdot b_k$ and then add it to other contributions in column $i = (j + k)$. This produces the sum

$$p_i = \sum_{i=j+k} a_j b_k + c_{i-1} \tag{12.51}$$

for each product bit. An equivalent description of the operation is obtained by writing the base-10 values

$$A = \sum_{j=0}^{n-1} a_j 2^j \qquad B = \sum_{k=0}^{n-1} b_k 2^k \tag{12.52}$$

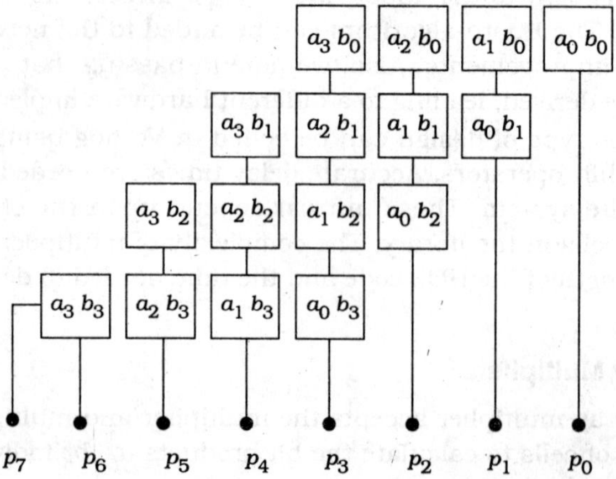

Figure 12.47 Modularized view of the multiplication sequence

and then forming the product

$$P = AB$$

$$= \left(\sum_{j=0}^{n-1} a_j 2^j\right)\left(\sum_{k=0}^{n-1} b_k 2^k\right) \tag{12.53}$$

$$= \sum_{i=0}^{n-1}\sum_{j=0}^{n-1} a_j b_k 2^{j+k}$$

Then we see that the terms $a_j \cdot b_k$ provide the bit value and 2^{j+k} the weighting.

The general structure of a 4 × 4 array is shown in Figure 12.48. This scheme calculates the bit products $a_j \cdot b_k$ using AND gates. The product bits are formed using adders in each column. The adders are arranged in a carry-save chain as can be seen by noting that the carry-out bits are fed to the next available adder in the column to the left. The array multiplier accepts all of the input bits simultaneously. The longest delay in the calculation of the product bits depends on the speed of the adders. The

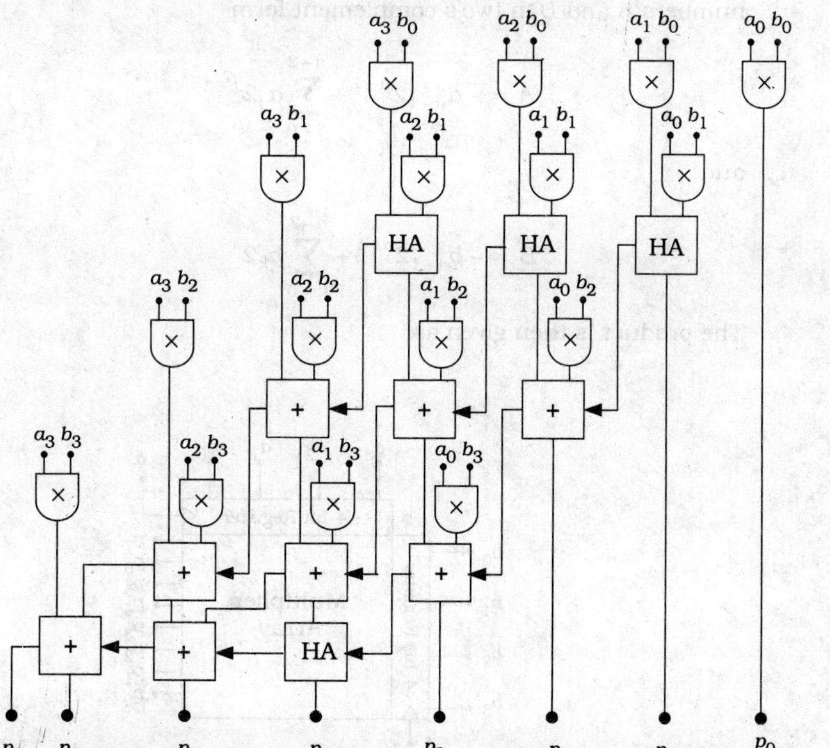

Figure 12.48 Details for a 4× 4 array multiplier

carry-chain in p_7 that originates from the carry bits from the p_1 column and propagates through the p_2 - p_6 quantities would be an obvious problem. Input registers may be added to synchronize the dataflow as shown in Figure 12.49. An output register may also be used if necessary. In general, an array multiplier for n-bit words requires n $(n$-2) full-adders, n half-adders, and n^2 AND gates. The gate count allows an estimate of the required area based on the library entries.

For layout purposes, it is useful to see if the cells can be arranged to give a more rectangular overall shape. An initial plan is obtained by using a regular interconnect pattern for the input bits, and then placing the units themselves in the order of the dataflow. The array structure starts to evolve as illustrated by the first-cut patterning in Figure 12.50. The actual placement can be adjusted to accommodate interconnect wiring and the different cell sizes.

12.5.2 Other Multipliers

Many multiplier algorithms and circuits have been published in the literature. The Baugh-Wooley multiplier is based on two's complement numbers for use in signed arithmetic. For this case, we write the input numbers A and B in two's complement form

$$A = -a_{n-1}2^{n-1} + \sum_{j=0}^{n-2} a_j 2^j \tag{12.54}$$

and

$$B = -b_{n-1}2^{n-1} + \sum_{k=0}^{n-2} b_k 2^k \tag{12.55}$$

The product is then given as

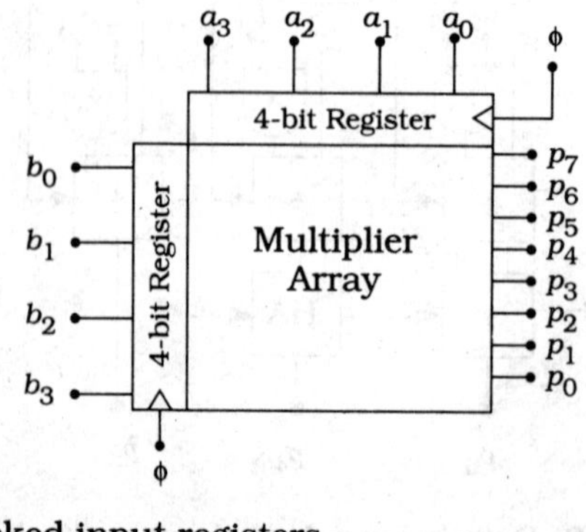

Figure 12.49 Clocked input registers

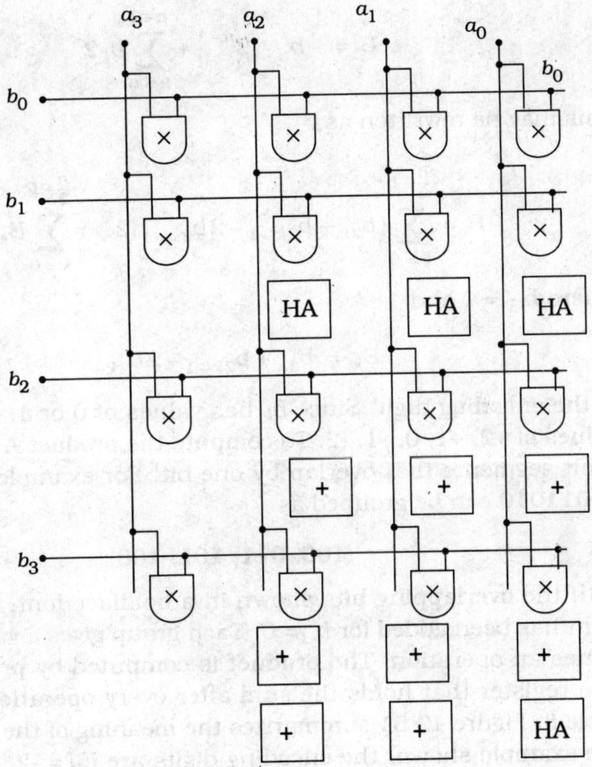

Figure 12.50 Initial cell placement for the array

$$P = a_{n-1}b_{n-1}2^{2(n-1)} + \sum_{j=0}^{n-2}\sum_{k=0}^{n-2} b_k a_j 2^{j+k} + -a_{n-1}\sum_{k=0}^{n-2} b_k 2^{k+n-1} - b_{n-1}\sum_{j=0}^{n-2} a_j 2^{j+n-1} \qquad (12.56)$$

which can be implemented using adders and subtractors. This can be transformed to an adder-only scheme by using bit complements [1].

The Booth algorithm, familiar from studies in basic computer architecture [5], examines the multiplier word B and searches for 0's since these have no effect on the sum. This may be used to encode groups of bits in B to produce a control digit that specifies the operation to be performed on the multiplicand A. To see the basis of the technique, we start with the two's complement form

$$B = -b_{n-1}2^{n-1} + \sum_{k=0}^{n-2} b_k 2^k \tag{12.57}$$

This may be rewritten as

$$B = \sum_{k=0}^{\frac{n}{2}-1} [b_{2k} + b_{2k-1} - 2b_{2k+1}]2^k = \sum_{k=0}^{\frac{n}{2}-1} E_k 2^k \tag{12.58}$$

where $b_{-1} = 0$ and

$$E_k = b_{2k} + b_{2k-1} - 2b_{2k+1} \tag{12.59}$$

is the encoding digit. Since b_k has values of 0 or 1, E_k can have decimal values of +2, +1, 0, -1, -2. To compute the product $A \times B$, we divide B into 3-bit segments that overlap by one bit. For example, the 8-bit word B = 1**0**01**1**01**0** can be grouped as

$$\mathbf{1}00, \mathbf{0}11, \mathbf{1}01, \mathbf{1}00 \tag{12.60}$$

with the overlapping bits shown in a boldface font. The last zero on the right has been added for b_{-1}= 0. Each group gives a value of E_k that determines an operation. The product is computed by providing a dual-word size register that holds the sum after every operation is completed. The table in Figure 12.51 summarizes the meaning of the encoded values. For the example shown, the encoding digits are E_k = -2. +2. -1, -2. The VLSI circuit can thus be constructed using relatively simple logic along with standard adder cells, making it attractive for multiplying large words. Another adder circuit called the Wallace tree can be used to improve the network by using carry-save adders for the sum.

b_{2k+1}	b_{2k}	b_{2k-1}	E_k	Effect on sum
0	0	0	0	add 0
0	0	1	+ 1	add *A*
0	1	0	+ 1	add *A*
0	1	1	+ 2	shift A left, add
1	0	0	- 2	take two's (A), shift left, add
1	0	1	- 1	add two's (*A*)
1	1	0	- 1	add two's (*A*)
1	1	1	0	add 0

Figure 12.51 Summary of Booth encoded digit operations

12.6 Summary

Arithmetic circuits are created by using binary algorithms to suggest structures that fit well into the VLSI principles of regular layout, repetition of cells, and fast circuits. In this chapter we have examined some of the more important issues associated with issues of implementation. Only the basics have been presented. High-radix algorithms, floating-point numbers, and a host of other topics await the interested reader who is willing to pursue deeper studies.

Arithmetic circuits will continue to be of primary importance as microprocessors and other VLSI circuits evolve to even higher levels of performance. This represents a fascinating field for future research endeavors.

12.7 References

[1] Abdellatif Bellaouar and Mohamed I. Elmasry, **Low-Power Digital VLSI Design**, Kluwer Academic Publishers, Norwell, MA, 1995.

[2] James M. Feldman and Charles T. Retter, **Computer Architecture**, McGraw-Hill, New York, 1994.

[3] Ken Martin, **Digital Integrated Circuit Design**, Oxford University Press, New York, 2000.

[4] Behrooz Parhami, **Computer Arithmetic**, Oxford University Press, New York, 2000. A comprehensive, in-depth treatment of the subject.

[5] David A. Patterson and John L. Hennessy, **Computer Organization & Design**, 2nd ed., Morgan-Kaufmann Publishers, San Francisco, 1998.

[6] Jan M. Rabaey, **Digital Integrated Circuits**, Prentice Hall, Upper Saddle River, NJ, 1996.

[7] Bruce Shriver and Bennett Smith, **The Anatomy of a High-Performance Microprocessor**, IEEE Computer Society Press, Los Alamitos, CA, 1998.

[8] William Stallings, **Computer Organization and Architecture**, 4th ed., Prentice Hall, Upper Saddle River, NJ, 1996.

[9] John P. Uyemura, **CMOS Logic Circuit Design**, Kluwer Academic Publishers, Norwell, MA, 1999.

[10] Neil H.E. Weste and Kamran Eshraghian, **Principles of CMOS VLSI Design**, 2nd ed., Addison-Wesley, Reading, MA, 1993.

[11] Wayne Wolf, **Modern VLSI Design**, 2nd ed., Prentice Hall PTR, Upper Saddle River, NJ, 1998.

12.8 Problems

[12.1] Design a half-adder that has inputs *a* and *b* using pseudo-nMOS. Then construct the gate-level Verilog description using **nmos** and any other primitives that are needed.

[12.2] Consider the CMOS dual-rail CPL logic family.

(a) Draw the circuit diagram for a half-adder circuit using the 2-input array in Figure 12.5(a) as a basis.

(b) Write a Verilog module description for a 2-input array using **nmos** primitives. Then instance the module to create the half-adder model.

(c) Use the 2-input array module in part (b) to model the CPL full-adder.

[12.3] Draw the circuits for p_i and g_i needed for a 4-bit CLA in each of the following CMOS technologies:

(a) Static CMOS; (b) Domino CMOS; and (c) TG logic.

[12.4] Construct the CMOS circuits for the CLA bits $\overline{c}_2$ and $\overline{c}_3$ using series-parallel nFET-pFET structuring. Identify the longest delay path in each.

[12.5] Construct the static mirror circuits for the CLA bits $\overline{c}_3$ and $\overline{c}_4$ using Figures 12.20 and 12.21 as a guide.

[12.6] Consider the static Manchester carry circuit shown in Figure 12.25(a). Examine the FET sizing problem for a carry-propagate event if $V_{DD} = 3$ V, $r = 2.5$, $k'_n = 150\ \mu A/V^2$, and $V_{Tn} = |V_{Tp}| = 0.7$ V.

[12.7] Consider the dynamic Manchester carry chain in Figure 12.26.

(a) Draw the RC equivalent circuit for the carry chain starting at c_0 (the output of the inverter) to $\overline{c}_4$. Assume that each transistor has a resistance R, and that the output node of every gate has a capacitance C_{out}.

(b) The chain is precharged when $\phi = 0$ and undergoes evaluation when ϕ switches to 1. What is the value of $\overline{c}_4$ directly at the start of the evaluation interval?

(c) How will charge leakage affect the operation of the chain?

[12.8] Consider 64-bit and 128-bit adders. What is the average length of a carry chain for each?

[12.9] Design a 16-bit carry-select adder using 4-bit adder blocks.

[12.10] Construct a 2 × 2 array multiplier circuit with latching inputs. Then write a Verilog description for your design.

[12.11] Consider the 4 × 4 array multiplier in Figure 12.48. Can this be used as a building block to create an 8 × 8 array multiplier? If so, detail the problems and modifications that need to be made.

[12.12] Provide the basic design for an 8 × 8 array multiplier. How many adders, full adders, etc., are required to build the circuit?

[12.13] Determine the Booth encoded digits E_k for the following words.

(a) A = 10110011

(b) A = 01101101

(c) A = 01010010

Memories and Programmable Logic

13

Memories are indispensable in modern digital systems. They provide for short- and long-term storage of binary variables and words. The VLSI aspects of CMOS memories are interesting because they are designed using a cell library and exhibit repetitive layout geometries. This chapter discusses the design of semiconductor memory arrays and concludes with an introduction to more general programmable logic structures.

13.1 The Static RAM

The acronym **RAM** stands for **random-access memory**, and implies a memory array that allows access to any bit (or group of bits) as needed. In practice, however, the meaning of "RAM" has evolved to imply a memory with both read and write capabilities to distinguish it from a read-only memory (ROM) array.

Static random-access memory **(SRAM)** cells use a simple bistable circuit to hold a data bit. A static RAM cell can hold the stored data bit so long as the power is applied to the circuit. SRAMs have three operational modes. When the cell is in a **hold** state, the value of the bit is stored in the cell for future usage. During a **write** operation, a logic 0 or 1 is fed to the cell for storage. The value of the stored bit is transmitted to the outside world during a **read** operation.

Figure 13.1 shows the general circuit scheme. A pair of cross-coupled inverters provides the storage, while two **access transistors** MAL and MAR provide read and write operations. The access transistors are controlled by the **word line** signal WL that defines the operational modes. When $WL = 0$, both access FETs are off and the cell is isolated. This defines the hold condition. To perform a read or write operation, the word line is brought up to a value of $WL = 1$. This turns on the access transis-

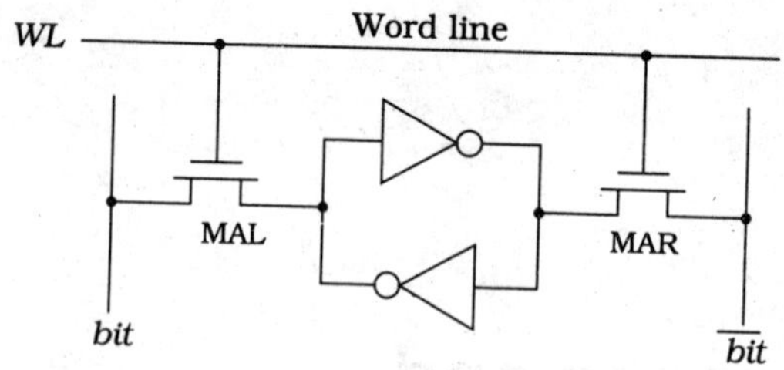

Figure 13.1 General SRAM cell

tors connecting the dual-rail data lines *bit* and $\overline{bit}$ to the outside circuitry; these are often called the **bit** and **bit-bar** lines, respectively. A write operation is performed by placing voltages on the *bit* and $\overline{bit}$ lines, which then act as inputs. Dual-rail logic helps increase the writing speed. For a read operation, the bit and bit-bar lines act as outputs and are fed into a **sense amplifier** that determines the stored state. The distinction between read and write operations is obtained by circuitry outside the cell array.

Two types of CMOS cells are dominant in practice. The circuit in Figure 13.2(a) is called the 6-transistor (6T) design and uses standard CMOS inverters. The 4-transistor (4T) uses resistors as load devices in an nMOS circuit as in Figure 13.2(b). The resistors are made using an undoped poly layer that resides above the silicon (transistor) level. This can yield a smaller cell area and allow higher packing density, but requires that an additional polysilicon layer and masking step be added to the process. The electrical characteristics of the two are quite different since the 4T cell uses a very large (typically greater than about 1 GΩ) passive pull-up resistor. We will concentrate on the dominant 6T design here.

The basic circuit level design issues revolve around choosing the values of the transistor aspect ratios to insure that the cell can hold a state while still allowing it to be changed during a write operation without excessive delay. Figure 13.3 shows the main parameters. A symmetric design is assumed such that β_A is the device transconductance of both

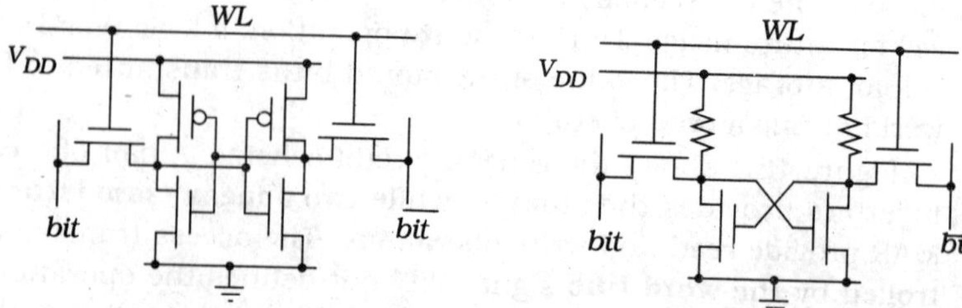

(a) 6T cell

(b) 4T cell with poly resistors

Figure 13.2 CMOS SRAM circuits

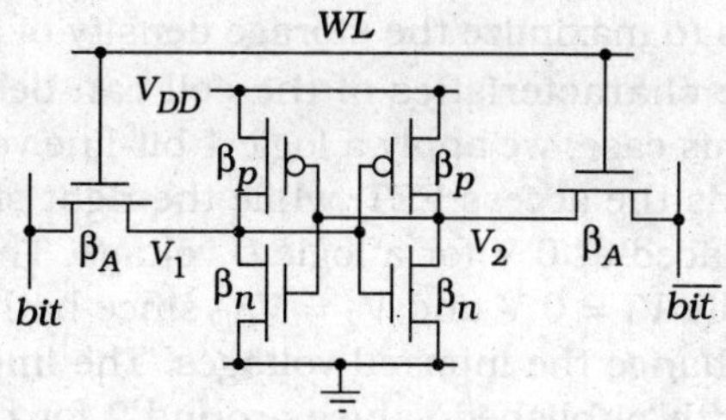

Figure 13.3 6T SRAM cell design parameters

access FETs, while the storage cell itself uses nFETs and pFETs with sizes described by β_n and β_p, respectively.

Stability of the hold state depends upon the functionality of the cross-coupled inverter cell. The inverter ratio (β_n/β_p) establishes the midpoint voltage V_M of each NOT gate, which in turn sets the characteristics of the feedback. This is usually described by a curve known as a **butterfly plot** that is obtained by forcing an input on one of the internal nodes and plotting the response on the other side, then performing the same operation to the other side. The superposed plots give the butterfly shape as in Figure 13.4. The **static noise margin**, labeled as SNM in the drawing, is the separation between the curves along a 45° slope and has units of volts. Its value indicates the level of immunity that the cell has to unwanted voltage changes due to coupled electromagnetic signals that are collectively called **noise**. A reasonable noise margin is needed for robust storage. The 6T cell design gives higher SNM values than the resistor-load 4T design, making it more attractive in noisy high-density environments.[1] Although the values of β_n and β_p can be adjusted to create different butterfly characteristics, the storage FETs are commonly chosen to have the smallest possible

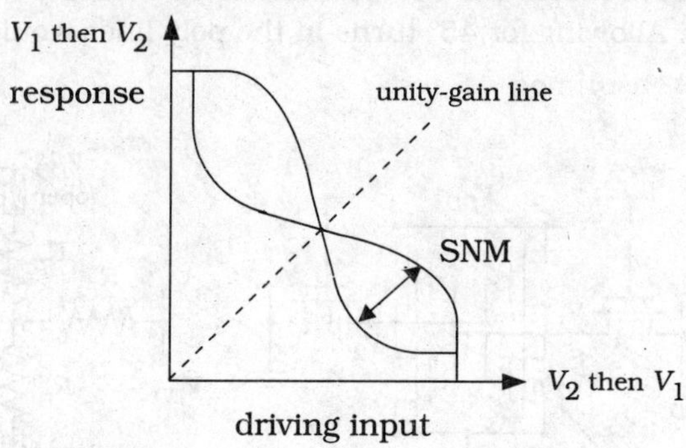

Figure 13.4 Butterfly plot

[1] The problem of electrical noise is discussed in Chapter 14 in the context of interconnect analysis.

aspect ratios to maximize the storage density of an SRAM array.

The write characteristics of the cell can be understood from Figure 13.5(a). In this case, we apply a logic 1 bit-line voltage of V_{DD} to the left bit line that feeds the access FET, while the right side (bit-bar line) is simultaneously placed at 0 V for a logic 0 voltage. The worst-case condition is where initially $V_1 = 0$ V and $V_2 = V_{DD}$ since both the bit and bit-bar voltages must change the internal voltages. The important design parameter is (β_A / β_n) with published values around 2 for the 6T cell. The reasoning behind this statement can be seen from the resistor model of the circuit shown in Figure 13.5(b). The input voltage V_{DD} is responsible for increasing V_1 to a logic 1 level. However, the nFET switch (at the bottom of R_n) is closed and pulls V_1 to 0 V and the feedback loop with the other inverter tries to hold this value. Selecting $\beta_A > \beta_n$ implies that $R_A < R_n$, which allows the access FET to be more effective in increasing V_1 to the level needed to switch the stored state. If cell area is the overriding factor, then (β_A / β_n) may be chosen to have a value closer to 1. Note that since both FETs are n-channel devices, the design ratio reduces to the ratio of aspect ratios

$$\frac{(W/L)_{nA}}{(W/L)_n} \tag{13.1}$$

in the layout; in the literature, this is sometimes called simply *the* β-ratio.

SRAM cell layout is driven by the desire to simultaneously minimize the cell area while providing port locations that will allow for high-density arrays. Figure 13.6 shows an approach to cell design that uses perpendicular lines in Metal1 and Metal2 to form the power supplies (VDD and VSS) and the bit, bit_bar lines. The storage cell is contained in the central part of the cell. The n+ regions of nFETs are extended beyond the inverter circuits to form the access transistors with the word lines running vertically. Allowing for 45° turns in the poly lines would help reduce the area.

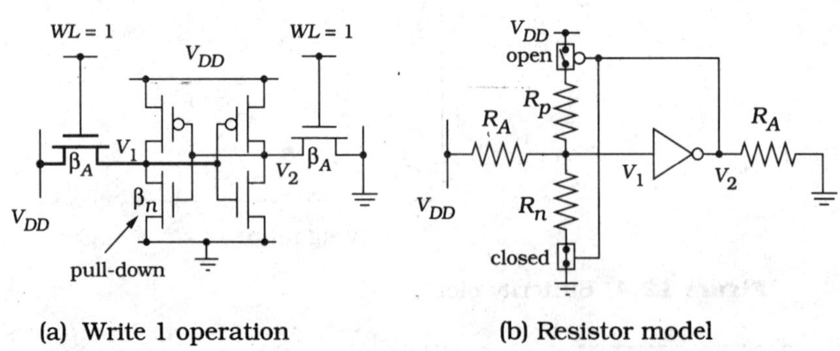

(a) Write 1 operation (b) Resistor model

Figure 13.5 Writing to an SRAM

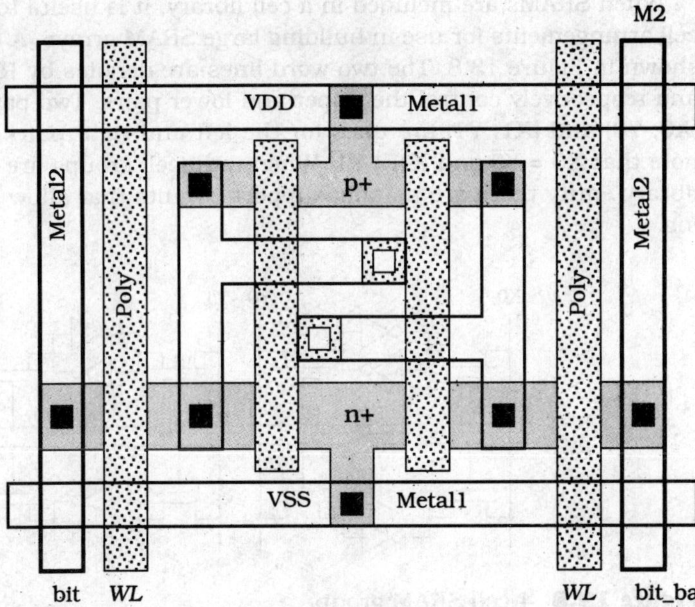

Figure 13.6 Example of a basic SRAM cell layout

Multiple-port SRAM cells provide cell access to more than one pair of bit/bit-bar lines. A 2-port cell is shown in Figure 13.7. The word line *WL*_1 controls the read/write operations for the *bit*_1 lines, while *WL*_2 provides the same control for the *bit*_2 lines. Additional logic must be added to avoid conflicts between the two ports. Multiport memories can simplify system wiring and layout, since different logic sections can share a memory block. At the system level, however, a method for tracking the contents of the memory and a priority access scheme must be developed to insure correct operation.

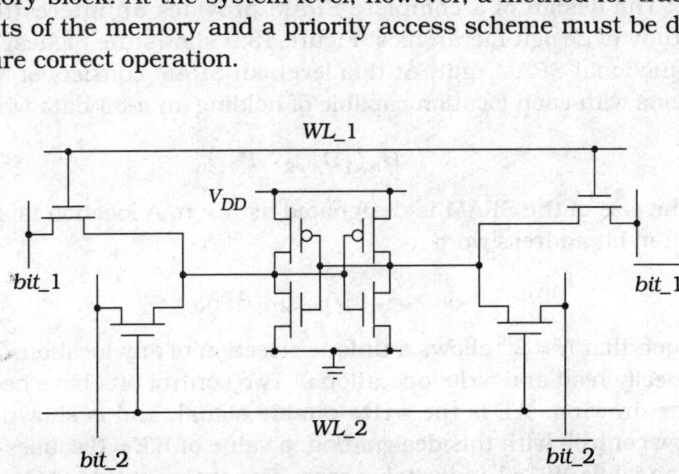

Figure 13.7 A 2-port CMOS SRAM cell

When SRAMs are included in a cell library, it is useful to create multicell arrangements for use in building large SRAM arrays. A 4-cell group is shown in Figure 13.8. The two word lines are denotes by RW0 and RW1, and respectively control the upper and lower pairs. Two pairs of bit lines (X0, Y0) and (X1, Y1) are used for the left and right pairs, respectively; note that $X0 = \overline{Y0}$ and $X1 = \overline{Y1}$. When multicell groups are included as a library entity there are usually support circuits that allow easy interfacing.

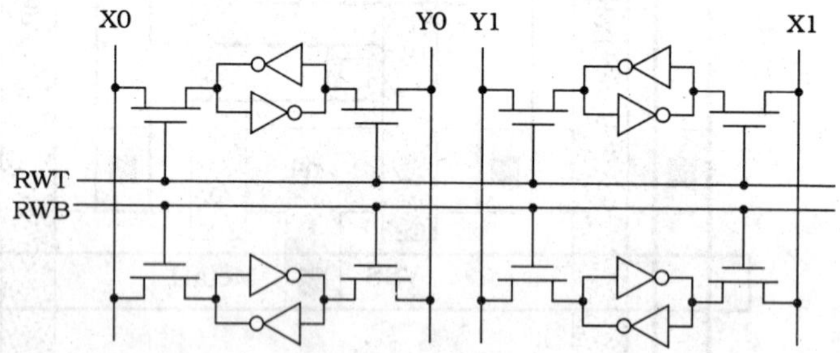

Figure 13.8 4-cell SRAM group

13.2 SRAM Arrays

Static RAM arrays are created by replicating the basic storage cell and adding the necessary peripheral circuitry. The objective is to obtain the highest storage density for a given cell layout; short access times are also important in the majority of applications.

The design of a complete SRAM provides an interesting and useful study of design hierarchies. Figure 13.9 shows the highest level view of a functional SRAM unit. At this level, an SRAM consists of N storage locations with each location capable of holding an n-bit data word

$$D_{n-1}D_{n-2}\ldots D_1 D_0 \tag{13.2}$$

The size of the SRAM is designated as $N \times n$. A location is specified using an m-bit address word

$$A_{m-1}A_{m-2}\ldots A_1 A_0 \tag{13.3}$$

such that $N = 2^m$ allows a unique selection of any location. This is used to specify read and write operations. Two control bits have been included in the drawing. WE is the **write-enable** signal, and is shown as an assert-low control; with this designation, a value of $WE = 0$ causes a write operation while $WE = 1$ indicates a read. The entire unit is under the control of

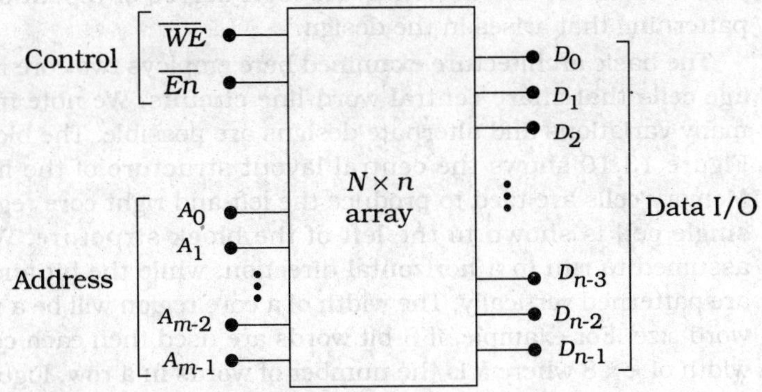

Figure 13.9 High-level view of an SRAM

the assert-low enable signal *En*. When $En = 1$, the read and write circuitry is disabled and the memory is in a hold state. A value of $En = 0$ is needed to activate the read/write operations. At the chip level, *En* would be renamed the **chip select** *CS* or **chip enable** *CE*.

Example 13.1

A 128K × 8 SRAM chip holds 128K 8-bit words for a total of 1 Mb of total storage. The address word must have a width of

$$\begin{aligned} m &= \log_2(128K) \\ &= 17 \end{aligned} \tag{13.4}$$

to select every 8-bit word location.

Verilog does not provide primitives for 2-dimensional memories. However, the **reg** data type can be used to write statements that describe SRAMs at the system level. An example is the 2KB storage unit sram_1 in the code segment

```
...
reg [7 : 0 ] sram_1 [0 : 2047]
...
```

This defines 8-bit words (i.e., 1 byte) using **reg** [7 : 0] that are identified by sram_1 with addresses from 0 to 2047. This can be modified for any word or memory size. The simplicity of the Verilog high-level description masks the complexity of the internal structure of the unit sram_1. To see the physical implementation of the memory, we will start with an architectural view and then progress downward and study some of the circuitry. This can be used to write Verilog models at lower hierarchical levels, which are useful in the verification of the architecture. One key point to note during the discussion is the high degree of repetition and regular patterning that arises in the design.

The basic architecture examined here employs two core regions of storage cells that share central word-line circuits. We note in passing that many variations and alternate designs are possible. The block diagram in Figure 13.10 shows the central layout structure of the memory array. Memory cells are tiled to produce the left and right core regions shown. A single cell is shown to the left of the block structure. Word lines are assumed to run in a horizontal direction, while the bit and bit-bar lines are patterned vertically. The width of a core region will be a multiple of the word size. For example, if 8-bit words are used then each core will have a width of $k \times 8$ where k is the number of words in a row. Figure 13.11 is an enlarged view of a core section that illustrates the structural detail of the cells. The structure can be used for both the left and right cores of the present example. The regular patterning in the schematic is maintained at the physical silicon level. Cell layout is based on finding a pattern that allows for proper placement and wiring with a high packing density.

The outputs of a centrally located active-high row decoder provide the word line signals to the storage cells. The address word specifies a particular row, which is then driven high. The access transistors of the selected row cells are turned on, permitting the read/write operations to take place. The location of the circuitry allows a single decoder to be used for both the left and right memory cores. A library-based static decoder circuit can be instanced directly into the design. The row decoder outputs are fed into row driver circuits that are used to drive the word lines of the arrays. Drivers are needed because of the large capacitive load presented by the long interconnects and the access transistors connected to each

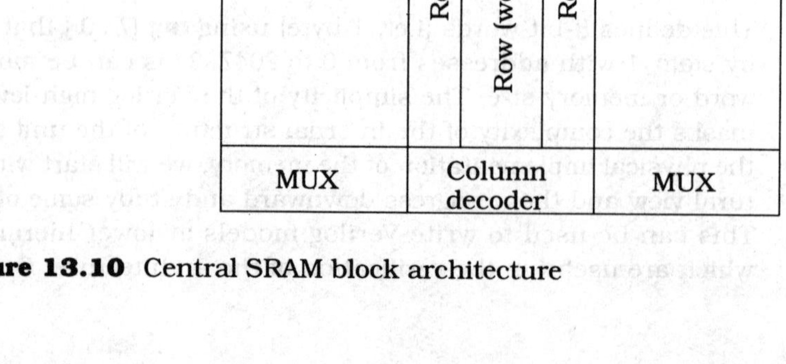

Figure 13.10 Central SRAM block architecture

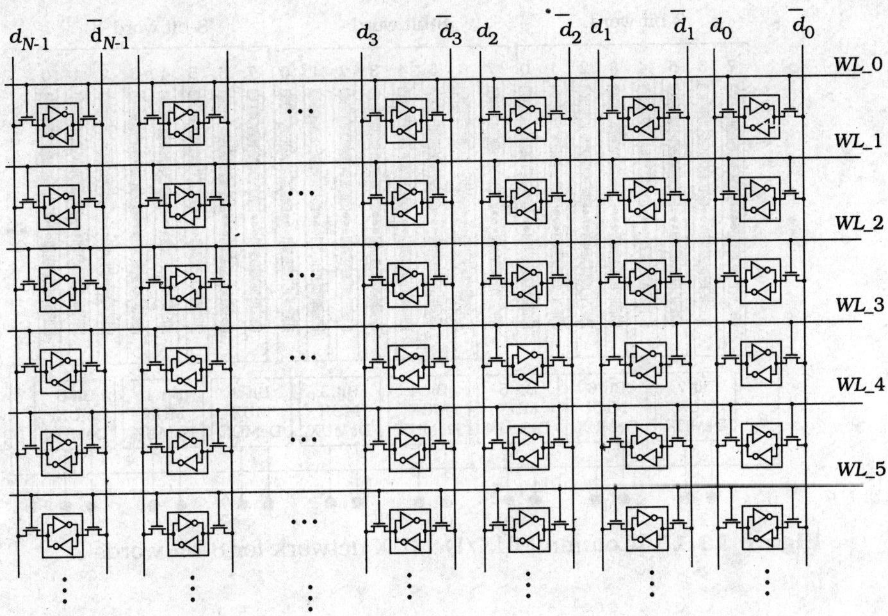

Figure 13.11 Cell arrangement in a core region

word line. A basic row driver design is shown in Figure 13.12. The output from the decoder is designated as *Dec_out*. The first pFET acts as a pull-up device, while the second pFET reinforces the output using feedback around the NAND2 gate. A sized inverter chain is used to provide the drive capability for the word line.

It is seen from the array in Figure 13.11 that the input/output bit and bit-bar data lines of the cells form the columns of the memory matrix. The data flow is thus visualized to be vertical for both read and write operations. Once a word line is driven high by the row decoder, every cell in the row is accessible. To choose a particular *k*-bit word in the row, we must add the group of column decoder circuits that select a particular set of *k* columns in the matrix. The MUX sections shown in Figure 13.10 are con-

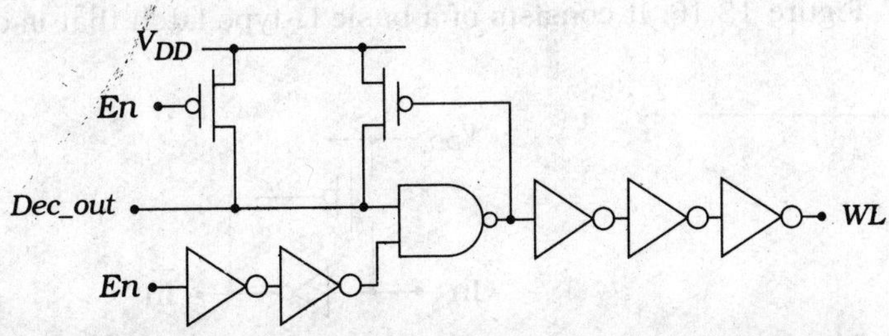

Figure 13.12 Row driver circuit

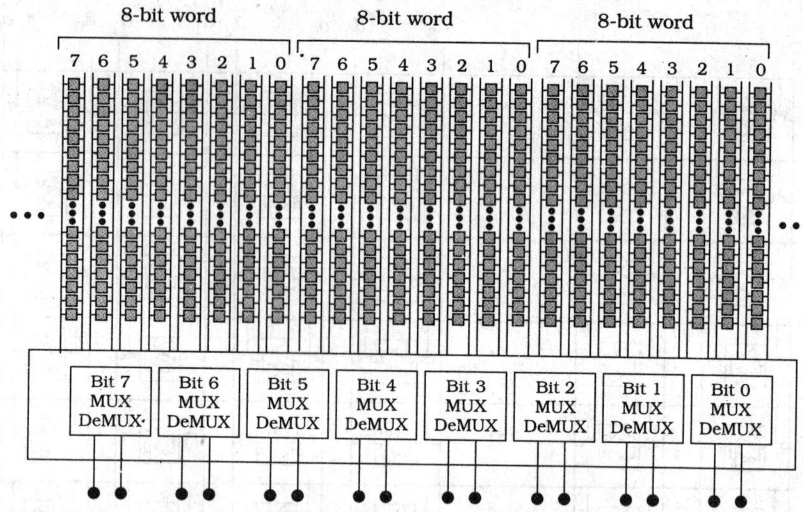

Figure 13.13 Column MUX/DeMUX network for 8-bit words

trolled by the column decoder to steer the selected (bit, bit-bar) groups. The overall structure of the column-select network is shown in Figure 13.13 for the case of 8-bit words. Each MUX/DeMUX block is connected to the appropriate data line of each word; for example, the bit_0 and bit-bar_0 lines of every word are wired to the Bit 0 MUX/DeMUX block. A read operation requires an output from the cells, so the circuits act as multiplexors. For a write operation, the DeMUX mode must be used to steer a data word into the proper columns. Column drivers are used to feed the MUXes; a simple feedback-oriented logic 1 driver design is shown in Figure 13.14. When *In* =1, the output is a 0 which turns on the pull-up pFET; the pFET is wired to help maintain the high input voltage.

To clarify the addressing scheme, let us examine the simplest case where the *m*-bit address word $A = A_{m-1} \ldots A_2A_1A_0$ is divided into row and column groups with *x* rows and *y* columns such that $x + y = m$. In the block diagram of Figure 13.15, the address is fed into an address latching register that allows it to stabilize. An address latching circuit is shown in Figure 13.16. It consists of a basic D-type latch that is controlled by the

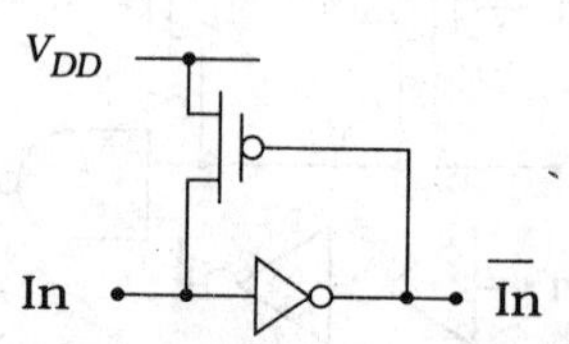

Figure 13.14 Logic 1 column driver

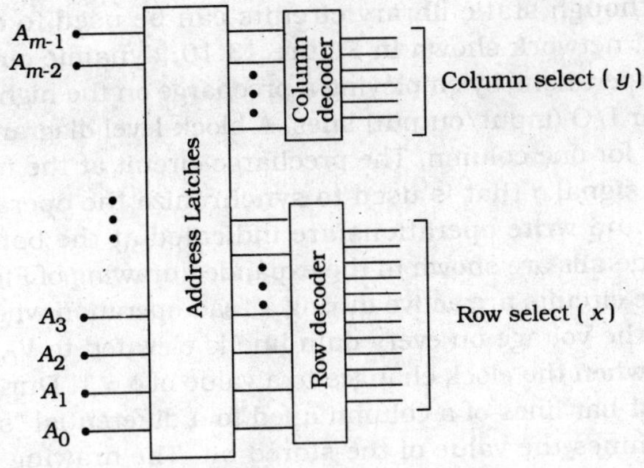

Figure 13.15 Basic addressing scheme

enable signal E that is derived from En and other control signals to synchronize the system. This circuit latches onto an input address bit *Add_in* when E makes a transition from 0 to 1. The outputs A and $\bar{A}$ are then divided into column and row segments and used as inputs to the decoder networks. This allows us access to any group of words in the cell matrix.

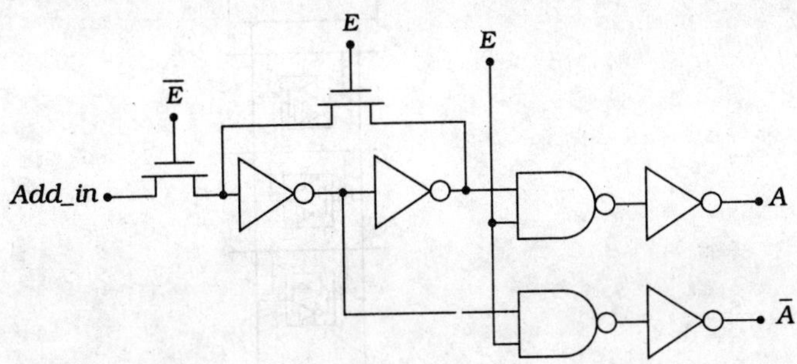

Figure 13.16 Address latch circuit

Example 13.2

The 128K × 8 SRAM chip requires a 17-bit address word. If we use a dual core arrangement with one word per word line, then we need 64K word lines. If we expand each word line to 64 bits (= 8 words) then the number of word lines is reduced to 8K. The 17-bit address $A_{16} \ldots A_0$ can thus be divided into a 4-bit column address group of $A_{16}A_{15}A_{14}A_{13}$ and a 13-bit row address group of $A_{12} \ldots A_0$. Other array sizes divide the address word proportionately.

Although static library circuits can be used to construct the entire SRAM network shown in Figure 13.10, dynamic circuits provide faster read operations by employing a precharge on the high-capacitance bit and bit-bar I/O (input/output) lines. A block-level diagram is shown in Figure 13.17 for one column. The precharge circuit at the top is controlled by a clock signal ϕ that is used to synchronize the operation and data flow. Read and write operations are indicated at the bottom of the column. More details are shown in the expanded drawing of Figure 13.18. The precharge circuits are active during a read operation when $\phi = 0$; during this time, the voltage on every data line is elevated to V_{DD}. Evaluation takes place when the clock changes to a value of $\phi = 1$. During this time, the bit and bit-bar lines of a column a fed to a differential "sense" amplifier that determines the value of the stored bit. The drawing also shows the column MUX circuits. Each word is selected by a control signal; *Col_0* is used as an example in the figure. When *Col_0* = 1, the nFETs are active and the entire group of bit and bit-bar lines are connected to the read/write circuit blocks. A separate column select signal is used for every word group.

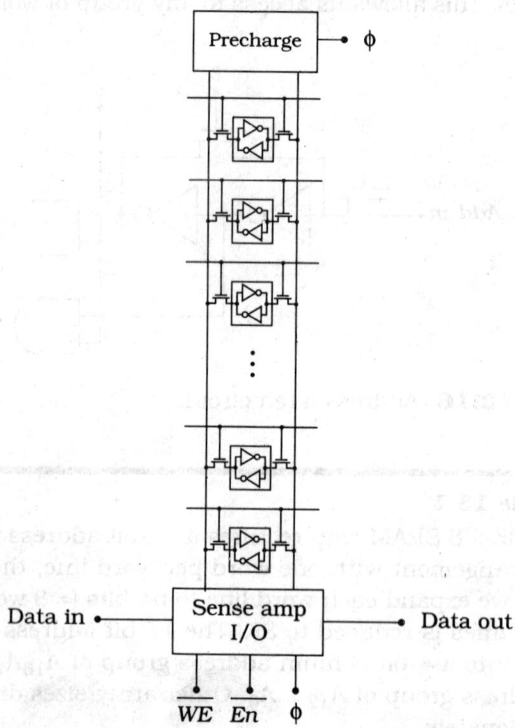

Figure 13.17 Precharge and I/O circuits for a single column

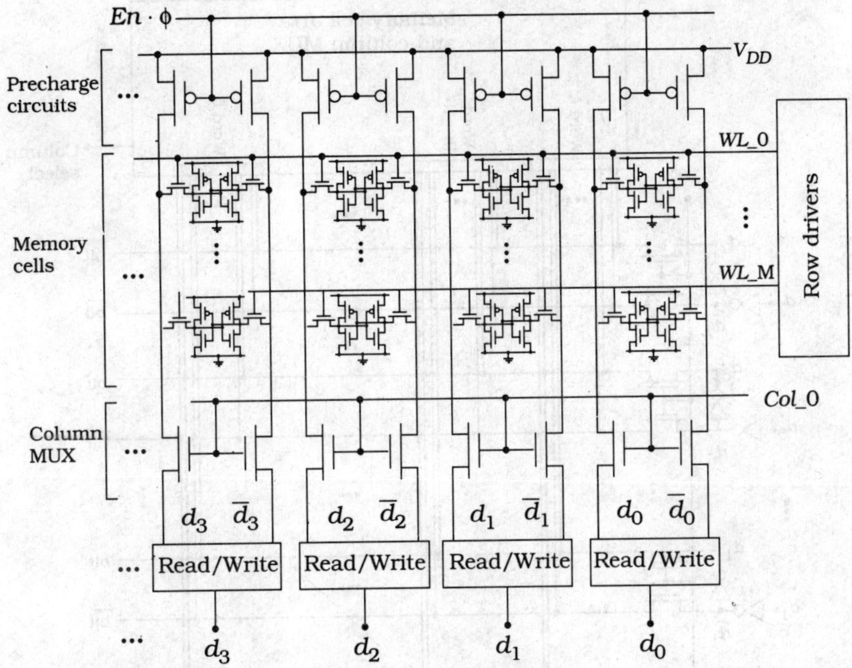

Figure 13.18 Expanded view of column circuitry

The read/write circuitry performs several functions including

- Directing the data flow into the array during a write operation, or out of the array during a read operation.
- Connecting the read and write circuits to the bit and bit-bar lines of every column.
- Providing amplifiers to detect and amplify the outputs during a read operation.

An example of the write circuitry is shown in Figure 13.19 for an 8-bit word design. The input bits d_7, d_6, ..., d_1, d_0 are inverted and buffered to provide complementary pairs (d_i, $\overline{d_i}$). When the write enable control bit has a value $WE = 1$, the nFETs act as closed switches connecting the data pairs to the bit and bit-bar columns. As shown in the schematic, every bit pair is fed to the appropriate locations that define the 8-bit word column groups. The column multiplexor circuits (not shown explicitly in the drawing) determine which column receives the input word.

Additional circuitry is required to detect the stored bit values during a read operation. The block-level circuit for one bit shown in Figure 13.20 is based on the use of differential amplifiers (denoted by triangular symbols) with + and – inputs. An identical circuit is required for each output bit. A

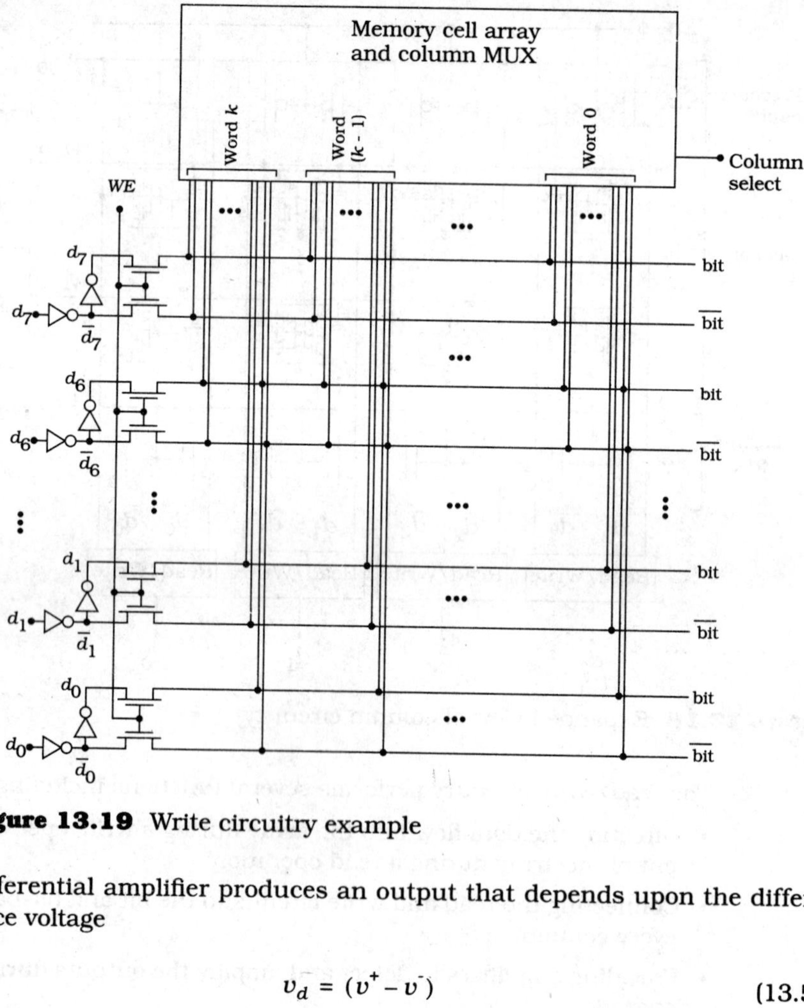

Figure 13.19 Write circuitry example

differential amplifier produces an output that depends upon the difference voltage

$$v_d = (v^+ - v^-) \tag{13.5}$$

between the input voltages v^+ and v^-. The output voltage of the amp is

$$v_{out} = Av_d = A(v^+ - v^-) \tag{13.6}$$

where $A > 1$ is the voltage gain of the amplifier. When used in an SRAM, the inputs are the bit and bit-bar signals from the storage cells. The circuit shown in the drawing uses a two-level sensing scheme. The first level consists of a pair of differential amplifiers that are fed oppositely phased inputs. The outputs are then combined to a single differential amplifier that outputs the result to a data latch. To make these compatible with the dynamic column precharge circuitry, the sense amps themselves are controlled by the clock signal ϕ.

The transistor-level details for a differential amplifier are shown in Fig-

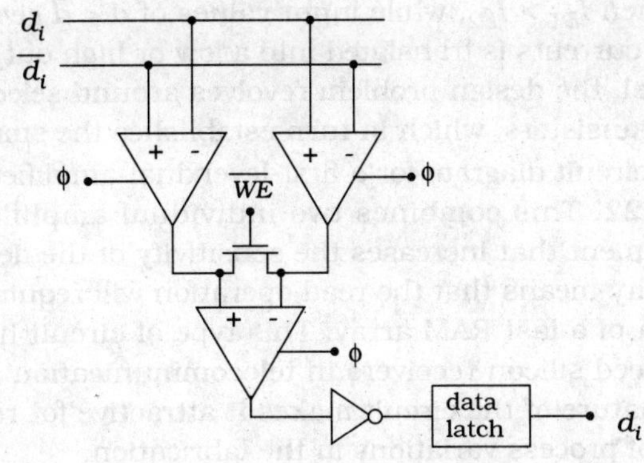

Figure 13.20 Example of a sensing scheme for the read operation

ure 13.21(a). This is a standard design that is based on two input nFETs Mn1 and Mn2 that accept complementary inputs d and $\overline{d}$. The pFETs Mp1 and Mp2 are used as **active load** devices and act like non-linear pull-up resistors. The difference signal (d-$\overline{d}$) due to the bit and bit-bar voltages controls the currents I_{D1} and I_{D2} flowing in the nFETs. When the voltage associated with d is large, I_{D1} increases; similarly, increasing the $\overline{d}$ voltage increases I_{D2}. The total currents are limited by the current through the clock-controlled nFET Mn such that

$$I_{SS} = I_{D1} + I_{D2} \tag{13.7}$$

which is valid when $\phi = 1$ in this design. Analyzing the circuit yields the current flow characteristics illustrated in Figure 13.21(b), which portrays the currents as a function of $(d - \overline{d})$.

When the inputs are the same with $d = \overline{d}$, $(d - \overline{d}) = 0$ and $I_{D1} = I_{D2}$. During an SRAM read operation, one voltage will be higher than the other. If

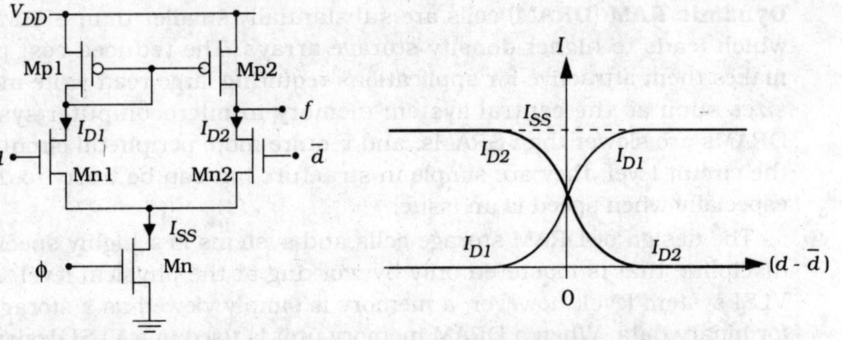

(a) Circuit diagram (b) Current flow

Figure 13.21 Single-ended differential amplifier

$d > \overline{d}$ then $I_{D1} > I_{D2}$, while input values of $d < \overline{d}$ give $I_{D2} > I_{D1}$. The difference in currents is translated into a low or high output voltage. At the circuit level, the design problem revolves around selecting the aspect ratios of the transistors, which in turn establishes the small signal gain.

The circuit diagram for a first-level dual-amplifier pair is shown in Figure 13.22. This combines two individual amplifiers in a cross-driven arrangement that increases the sensitivity of the detection circuit. A high sensitivity means that the read operation will require less time, leading to the idea of a fast RAM array. This type of circuit has also been used for high-speed silicon receivers in telecommunication applications. The balanced nature of the circuit makes it attractive for reducing noise and the effects of process variations in the fabrication.

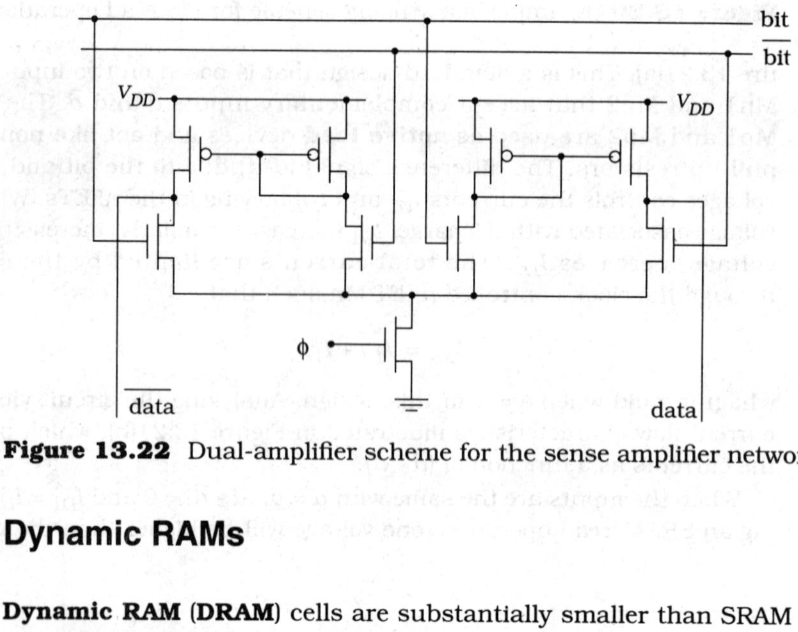

Figure 13.22 Dual-amplifier scheme for the sense amplifier network

13.3 Dynamic RAMs

Dynamic RAM (DRAM) cells are substantially smaller than SRAM cells, which leads to higher density storage arrays. The reduced cost per bit makes them attractive for applications requiring large read-write memory sizes such as the central system memory in microcomputer systems. DRAMs are slower than SRAMs, and require more peripheral circuitry. At the circuit level, they are simple in structure but can be tricky to design, especially when speed is an issue.

The design of DRAM storage cells and systems is a highly specialized discipline that is mastered only by working at the physical level. At the VLSI system level, however, a memory is simply viewed as a storage unit for binary data. When a DRAM memory unit is used in a VLSI design, it is usually instanced from a library entry that has been designed by a specialized group.[2] Owing to this observation, our discussion of DRAMs will

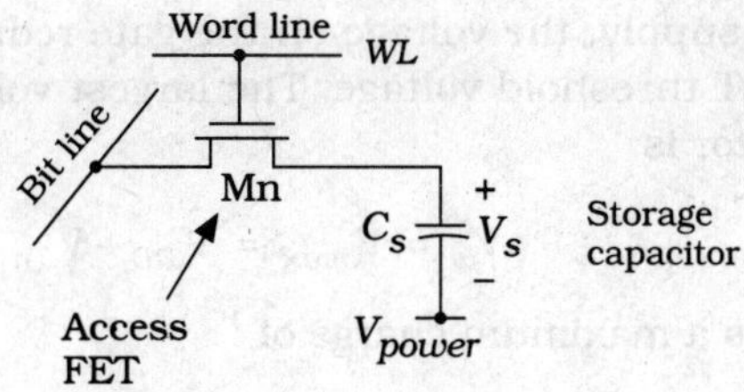

Figure 13.23 1T DRAM cell

be limited to understanding the basics to see the operation and trade-offs.

A 1-transistor (1T) DRAM cell is shown in Figure 13.23. It consists of a single access nFET Mn and a storage capacitor C_s. The cell is controlled by the word line signal *WL* and a single bit line provides the I/O path to the cell. The bottom of the capacitor is connected to one of the power supply rails, and is denoted as V_{power} in the drawing; either V_{DD} or V_{SS} may be used. The storage mechanism is based on the concept of temporary charge retention on the capacitor. A voltage V_s across the capacitor corresponds to a stored charge Q_s of

$$Q_s = C_s V_s \tag{13.8}$$

With $V_s = 0$ V, $Q_s = 0$ and the charge state is a logic 0. Conversely, a large value of V_s gives a large Q_s, which is defined to be a logic 1 charge state.

The write operation is shown in Figure 13.24(a) for the case where $V_{power} = V_{SS} = 0$ V. Applying V_{DD} to the nFET gate turns on the access transistor and allows access to the storage capacitor. The input data voltage V_d controls the current to/from C_s. A logic 0 data voltage $V_d = 0$ V results in a voltage $V_s = 0$ V across the capacitor, corresponding to a charge state of $Q_s = 0$. If we apply a logic 1 data voltage $V_d = V_{DD}$ equal to

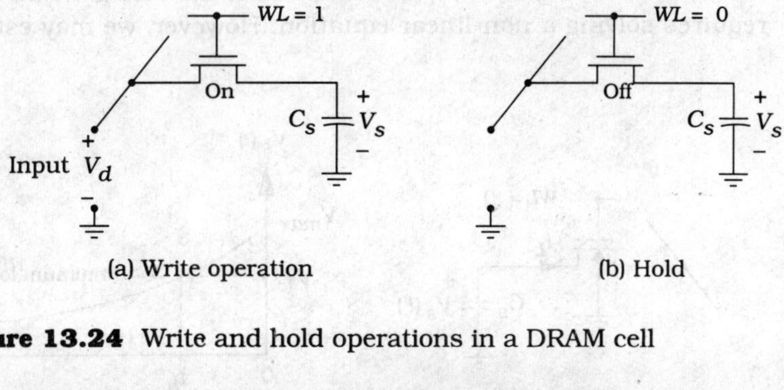

(a) Write operation (b) Hold

Figure 13.24 Write and hold operations in a DRAM cell

---

2 SRAMs are often viewed in the same manner.

the power supply, the voltage on the gate reduces the transmitted signal by an nFET threshold voltage. The largest voltage that can be passed to the capacitor is

$$V_s = V_{max} = V_{DD} - V_{Tn} \tag{13.9}$$

which gives a maximum charge of

$$Q_{max} = C_s(V_{DD} - V_{Tn}) \tag{13.10}$$

The hold state is achieved by turning off the access transistor with a word line signal of $WL = 0$. This is shown in Figure 13.24(b).

The dynamic aspect of the cell arises during a data hold time. As discussed in Chapter 9, a MOSFET that is biased into cutoff with $V_G < V_T$ still admits small leakage currents. The DRAM circuit problem is illustrated in Figure 13.25. A logic 1 voltage $V_s = V_{max}$ on the storage capacitors provides the electromotive force for the leakage current I_L flowing away from C_s. This can be described by

$$I_L = -\left(\frac{dQ_s}{dt}\right) \tag{13.11}$$

which shows that the current removes charge from the capacitor. Using equation (13.8) for Q_s gives the capacitor relation

$$I_L = -C_s\left(\frac{dV_s}{dt}\right) \tag{13.12}$$

so that V_s also drops. Assuming an initial voltage of $V_s = V_{max}$ gives the voltage decay illustrated in Figure 13.25. The minimum logic 1 voltage is denoted as V_1 in the drawing. The hold time t_h is defined as the longest period of time that the cell can maintain a voltage large enough to be interpreted as a logic 1; the hold time is also called the **retention time** in the literature. In general, I_L is a function of the voltages and finding $V_s(t)$ requires solving a non-linear equation. However, we may estimate by t_h

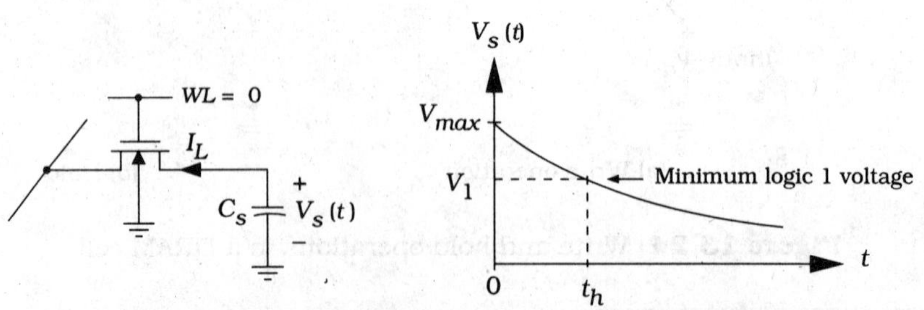

Figure 13.25 Charge leakage in a DRAM cell

assuming that I_L is a constant and writing

$$I_L \approx -C_s\left(\frac{\Delta V_s}{\Delta t}\right) \tag{13.13}$$

where ΔV_s and Δt represent changes in the variables. Rearranging gives the hold time equation

$$t_h = |\Delta t| \approx \left(\frac{C_s}{I_L}\right)(\Delta V_s) \tag{13.14}$$

as a first estimate. This shows that the hold time may be increased by using a large capacitance and minimizing the leakage current. As an example, if $I_L = 1$ nA, $C_s = 50$ fF, and $(\Delta V_s) = 1$ V, the hold time is

$$t_h = \left(\frac{50\times10^{-15}}{1\times10^{-9}}\right)(1) = 0.5\mu s \tag{13.15}$$

This illustrates the short hold time of a DRAM cell, and clearly justifies the use of the adjective "dynamic" for the circuit.

Memory units must be able to hold data so long as the power is applied. To overcome the charge leakage problem, DRAM arrays employ a **refresh operation** where the data is periodically read from every cell, amplified, and then rewritten. The procedure is listed in Figure 13.26. The cycle must be performed on every cell in the array with a minimum refresh frequency of about

$$f_{refresh} \approx \frac{1}{2t_h} \tag{13.16}$$

Refresh circuitry is included in the overhead logic that surrounds the cell array. The refresh cycle is designed to operate in the background and is therefore transparent to the user.

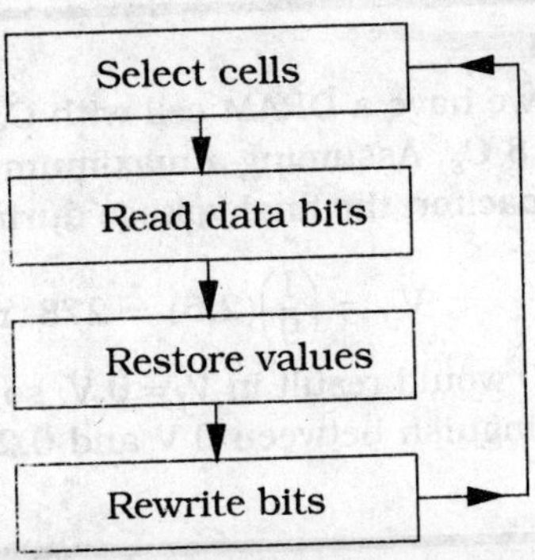

Figure 13.26 Refresh operation summary

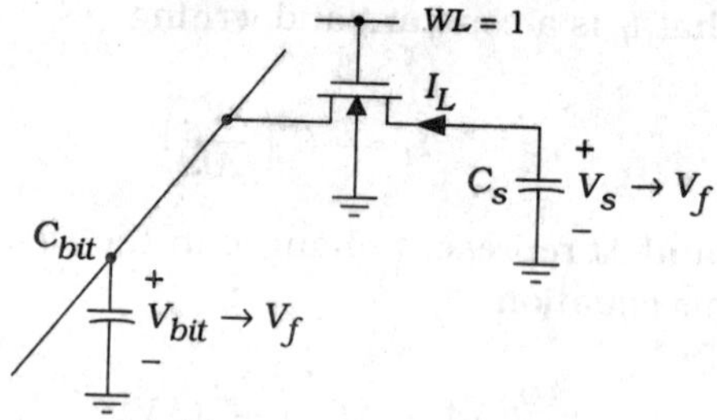

Figure 13.27 Read operation in a DRAM cell

A read operation is shown in Figure 13.27. The voltage V_s on the capacitor at the read time provides the voltage to move charge from C_s to the bit line capacitance C_{bit}, which sets up a charge sharing situation. C_{bit} includes the line capacitance and other parasitic contributions such as the input capacitance of the sense amplifier. The initial charge on the capacitor is

$$Q_s = C_s V_s \tag{13.17}$$

where $V_s = 0$ V for a logic 0, and $V_s > 0$ for a logic 1. Current flow from C_s to C_{bit} continues until the voltages are equal to the final voltage $V_f = V_{bit} = V_s$. The charge is redistributed according to

$$Q_s = C_s V_f + C_{bit} V_f \tag{13.18}$$

The initial and final values of Q_s must be equal by charge conservation, so

$$V_f = \left(\frac{C_s}{C_s + C_{bit}}\right) V_s \tag{13.19}$$

This shows that $V_f < V_s$ for a stored logic 1. In practice, V_f is usually reduced to a few tenths of a volt, so that the design of the sense amplifier becomes a critical factor.

Example 13.3

Suppose that we have a DRAM cell with C_s = 50 fF and a bit line capacitance of C_{bit} = 8 C_s. Assuming a maximum voltage of $V_s = V_{max}$ = 2.5 V on the storage capacitor, the final voltage during a logic 1 read operation is

$$V_f = \left(\frac{1}{9}\right)(2.5) = 278 \text{ mV} \tag{13.20}$$

A stored logic 0 would result in V_f = 0 V, so that the sense amplifier must be able to distinguish between 0 V and 0.28 V to determine the value of the stored bit.

13.3.1 Physical Design of DRAM Cells

Modern DRAM chips have surpassed the 1 Gb density by using novel capacitor structures that are possible with advanced semiconductor processing techniques. The 1T storage cell consists of a single transistor and a storage capacitor. High-density arrays are created by reducing the individual cell area A_{cell} to the smallest size possible. Peripheral circuits for addressing, refresh, and other operations must be added to make the chip functional and can easily consume more than 30% of the total chip area.

In standard MOS processing, the nFET must reside on the silicon wafer; since submicron line widths are standard, the FET area is relatively small. Decreasing the overall cell area usually revolves around the design of the storage capacitor. The value of C_s must be about 40 fF or larger. Using the parallel-plate capacitor formula indicates that we need a plate area A_p of

$$A_p = C_s\left(\frac{t_{ins}}{\varepsilon_{ins}}\right) \tag{13.21}$$

where t_{ins} and ε_{ins} are the insulator thickness and permittivity, respectively. Assuming a silicon dioxide layer that is 50 Å thick implies a plate area of

$$A_p = (40\times10^{-15})\left(\frac{50\times10^{-8}}{3.45\times10^{-13}}\right) = 5.8\times10^{-8}\ \text{cm}^2 \tag{13.22}$$

which is 5.8 μm^2. This is much larger than can be used for large arrays. For example, a 64Mb DRAM usually requires a cell size of about 1.25 μm^2 to meet chip requirements. Much research has been devoted toward building storage capacitors that increase the plate area without increasing the cell surface area A_{cell} (also called the footprint size). There are two main structures in use: trench capacitors and stacked capacitors.

A storage cell that uses a trench capacitor is shown in Figure 13.28. The capacitor is created by using a reactive ion etch (RIE) process to create a deep trench in the silicon. The sides are oxidized to create a glass

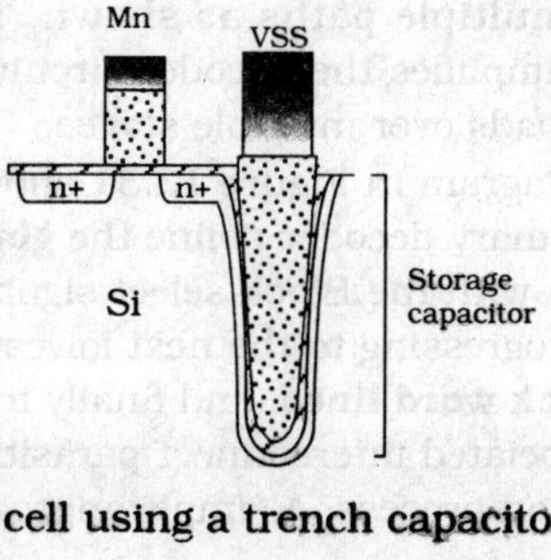

Figure 13.28 A DRAM cell using a trench capacitor

insulator, and then doped polysilicon is used to fill the trench and acts as the upper plate. The lower plate is created by an n+ implant along the entire wall area. The increase in plate area A_p is achieved by using the sidewalls of the trench without increasing the footprint area A_{cell}.

A stacked capacitor design places a polysilicon structure on top of the access transistor as portrayed schematically in Figure 13.29. Advanced 3-dimensional structures can be created to form the upper and lower plates. The plate area A_p depends upon the surface geometry created by the poly plates. Many interesting stacked capacitor designs have been published in the literature. In addition, surface corrugations and "bumps" have been added to further increase the value of C_s. The interested reader is directed to reference [8] for an overview of the subject.

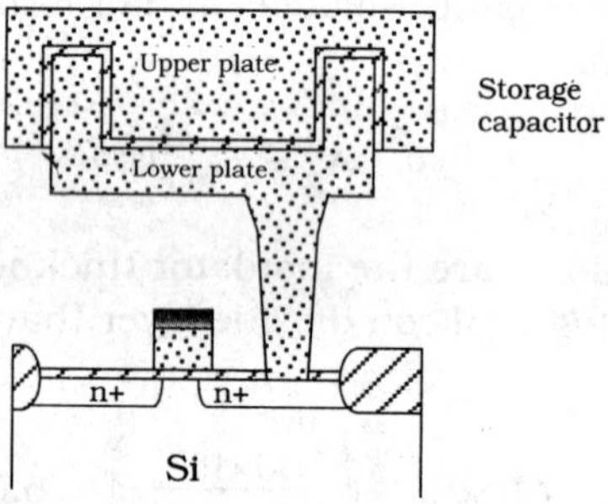

Figure 13.29 Visualization of stacked capacitor structure

13.3.2 Divided-Word Line Architectures

As the capacity of RAMs increases, new layout designs have been developed to reduce the effect of parasitic resistance and capacitance associated with long interconnect lines. Multiple storage cores are used to divide up the total storage area into separate **blocks** of cells; each block defines a specific address range. These can be further divided into **sub-blocks**, **sub-sub-blocks**, and so on, until the size of the cell array can be accessed without excessive delay. Figure 13.30 illustrates the concept. The advantage of dividing the storage array is that the word line routing can be divided into multiple paths as shown. This **divided-word line** (DWL) architecture simplifies the decoder circuits and speeds up access by distributing the loads over multiple stages.

The logic diagram in Figure 13.31 shows the wiring scheme. The outputs of the primary decoder define the **global word lines**. These are used in conjunction with the Block select signals to activate the various **block word lines**. Progressing to the next lowest level in the hierarchy takes us to the **sub-block word lines**, and finally to the **sub-sub-block** signals. The gates and associated interconnect parasitics can be designed to speed up the row selection process. A simple approach is to apply the techniques of

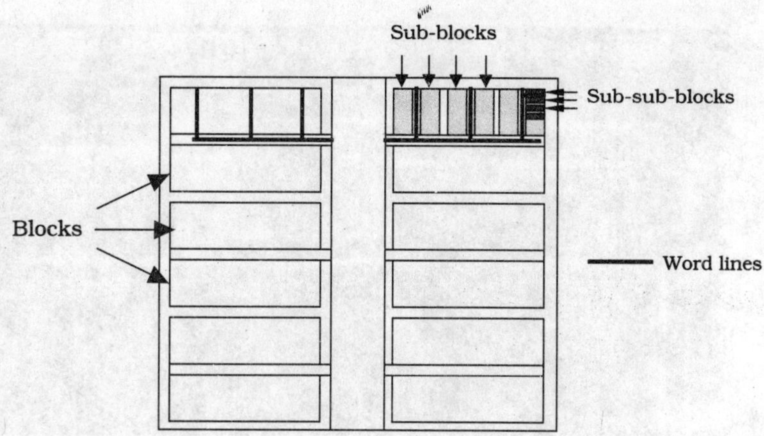

Figure 13.30 Basis for a divided-word line architecture RAM layout

Logical Effort to equalize the delay through each stage (logic gate with output load). This allows one to increase the speed without resorting to excessively large transistors. The technique can be applied to each level in the hierarchy.

Figure 13.32 is a micro-photograph of a portion of an SRAM that uses a multiple-block architecture. Cell block arrays can be identified by their regular layout patterning and are clearly seen in the bottom half. The upper section of the photograph contains decoders, drivers, and auxiliary circuitry including sense amplifiers.

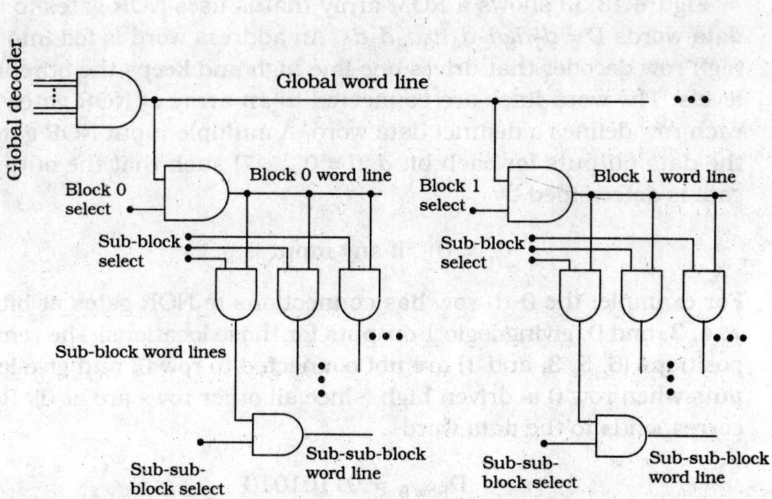

Figure 13.31 Logic for a DWL design

Figure 13.32 Photograph of SRAM blocks and support circuitry

13.4 ROM Arrays

Read-only memories (ROMs) are used for permanent bit storage. The structure of a ROM array is similar to that used for RAMs, but the individual bit cells are much simpler. The data stored in a basic read-only memory is created by the selective placement of FETs. Since this is accomplished in the physical design, the data cannot be altered once the chip is fabricated.

Figure 13.33 shows a ROM array that is uses NOR gates to store 8-bit data words $D = d_7d_6d_5d_4d_3d_2d_1d_0$. An address word is fed into an active-high row decoder that drives one line high and keeps the others at logic 0 levels. The word lines are connected to an array of NOR gates such that each row defines a distinct data word. A multiple-input NOR gate provides the data outputs for each bit d_i ($i = 0,\ldots, 7$) such that the output of each gate is determined by

$$d_i = 0 \quad \text{if any input is a 1} \tag{13.23}$$

For example, the 0-th row has connections to NOR gates at bit positions 7, 4, 2, and 0, giving logic 1 outputs for those locations. The remaining bit positions (6, 5, 3, and 1) are not connected to row 0, and give logic 0 outputs when row 0 is driven high (since all other rows are at 0). Row 0 thus corresponds to the data word

$$D_{\text{Row }0} = 01101010 \tag{13.24}$$

Every row is programmed in the same manner.

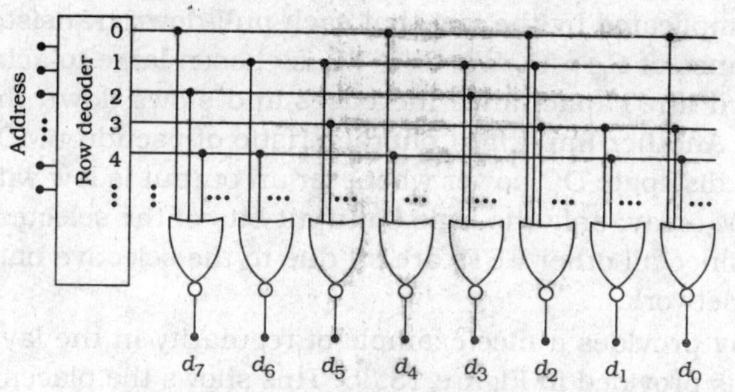

Row	Data
0	01101010
1	10010011
2	01110111
3	11011000
4	00101100

Figure 13.33 Logic diagram for a NOR-based ROM

The pseudo-nMOS implementation of the ROM is shown in Figure 13.34. Since only one pull-up pFET is required for each NOR gate, the task of programming centers on placement of nFETs that act as pull-down devices. A logic 0 output is obtained by providing a FET with its gate connected to the driving word line. This is understood by noting that when a pull-down transistor turns on, it provides a good connection to ground and pulls the output low. Pseudo-nMOS circuits are ratioed, so that the value of the output low voltage V_{OL} is determined by the nFET/pFET ratio $(\beta_n/\beta_p) > 1$ as discussed in Chapter 9. Selecting the nFET

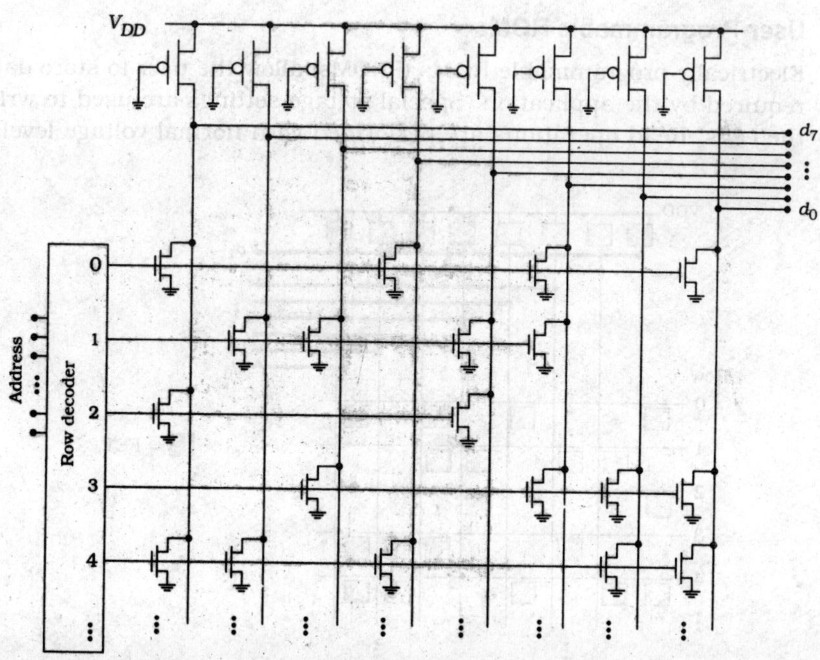

Figure 13.34 ROM array using pseudo-nMOS circuitry

aspect ratio is complicated by the fact that each pull-down transistor has an input capacitance of $C_G = C_{ox}WL$, so if W_n is chosen large to achieve a low V_{OL}, the word line capacitance increases and slows down the row decoder circuits. Another important characteristic of pseudo-nMOS circuits is that they dissipate DC power whenever an output is low with $V_{out} = V_{OL}$. In the ROM array, only the logic 0 output bits of the selected word consume power since all other FETs are off due to the selective nature of the row decoder network.

The ROM array provides a nice example of regularity in the layout. A simple FET map is provided in Figure 13.35. This shows the placement of the FETs relative to the input and output lines. Metal1 is used for the NOR gate connections running vertically (except for the VDD line), and the outputs are taken out on horizontal Metal2 lines. This results in the layout shown in Figure 13.36. Reprogramming is accomplished by adding or removing pull-down transistors.

Various approaches can be used to provide a library-based ROM with this design. One technique is to place a pull-down nFET at every intersection of a word line and a NOR output; a '0' is programmed into that location by connecting the word line to the FET gate with a poly contact. This is an example of a **mask-programmable** ROM where the stored data is defined by the poly-contact mask. Alternately, one can start with a blank nFET array and use a CAD tool to place transistors as needed.

13.4.1 User-Programmable ROMs

Electrically programmable ROMs (PROMs) allow the user to store data as required by the application. Special voltage settings are used to write to the cells. Read operations are performed with normal voltage levels, so

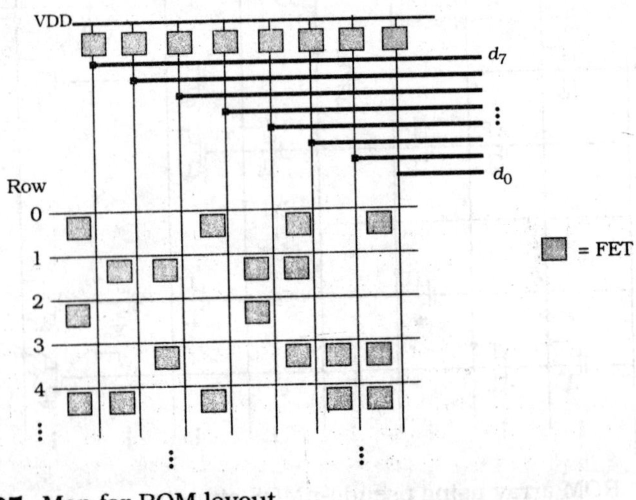

Figure 13.35 Map for ROM layout

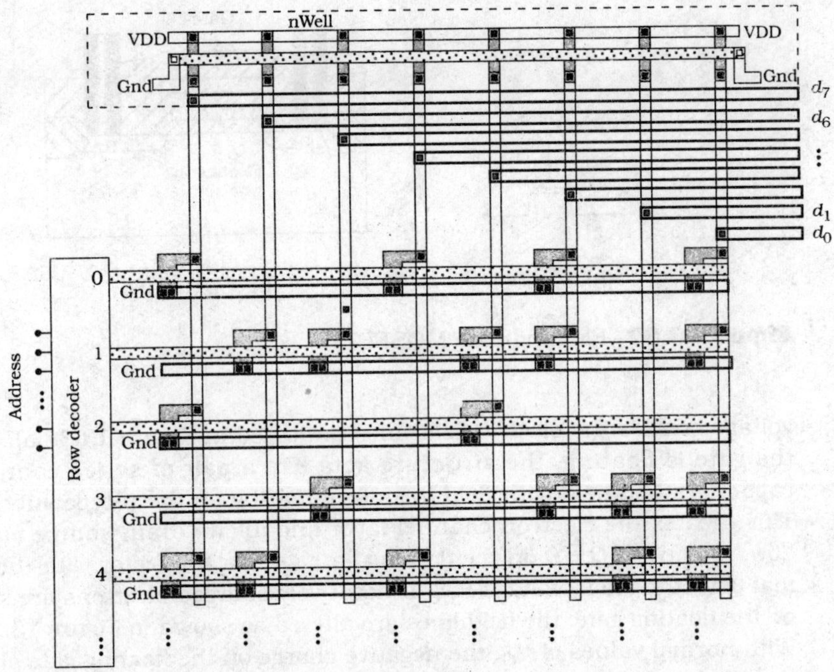

Figure 13.36 ROM layout based on FET map

that the data remains unchanged. Many ROM devices provide for erasure (E) and rewriting the contents of the array. Optical erasure using UV light was used in early EPROM designs, but these have been replaced by electrically alterable devices. Electrically-erasable EPROMs (E^2PROMs) are used to store the BIOS code in personal computers, and allow the user to update the board characteristics for new devices.[3]

A reprogrammable ROM array is built using special FETs that use a pair of stacked poly gates and has the circuit symbol shown in Figure 13.37(a). The topmost gate constitutes the usual gate terminal of the transistor. However, another poly gate layer is sandwiched in between the top poly and the silicon substrate. It is not electrically connected to any part of the transistor or auxiliary circuitry, and is therefore called an electrically **floating** gate. The details are shown in Figure 13.37(b). The floating gate is used to store negative electron charge, which increases the threshold voltage of the transistor above its normal value.

To understand the charge storage mechanism and effects, consider first a transistor that has zero charge on the floating gate. Applying a gate

3 BIOS stands for Basic Input/Output System. The BIOS controls the booting procedure when a PC is powered-up and allows the operating system to be loaded into the system memory.

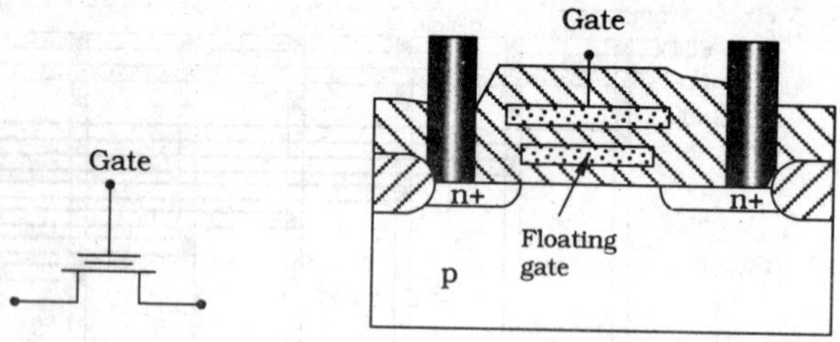

(a) Symbol (b) Structure

Figure 13.37 Floating-gate MOSFET

voltage creates the electric field lines indicated in Figure 13.38(a). Since the gate is floating, the structure acts like a pair of series-connected capacitors and field with field lines terminating on the p-type substrate. This creates the electron channel layer and allows drain-source current flow. A value of $Q = 0$ on the floating gate gives a transistor with the normal (low) threshold voltage V_{Tn}. If negatively charged electrons are stored on the floating gate, the field lines are altered as shown in Figure 13.38(b). With normal values of V_G, the negative charge on the floating gate shields the electric field lines and prevents them from reaching the silicon surface. No channel is formed, and the device remains in cutoff. If we increase the gate voltage to a high value $V_{Tn,H}$, then the FET goes active. However, we can design the transistor so that $V_{Tn,H} > V_{DD}$, which insures that it is always in cutoff when placed in a circuit.

The dual threshold voltage characteristics give rise to the EPROM scheme shown in Figure 13.39 for an 8-bit word; only the read circuitry is shown for simplicity. The nFETs with a low (normal) threshold voltage V_{Tn} are denoted by "L" while high threshold voltage devices with $V_{Tn,H} > V_{DD}$

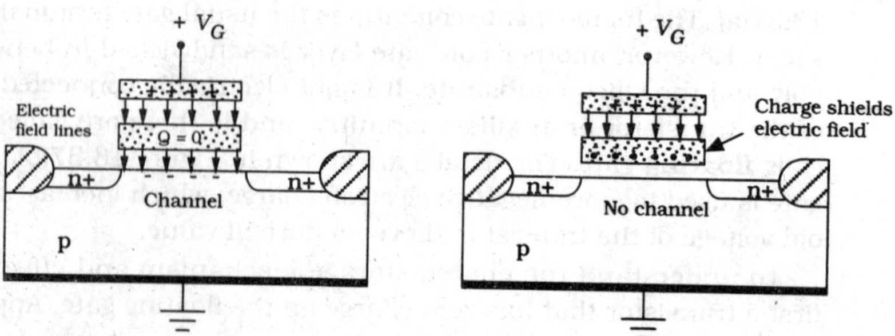

(a) Normal V_{Tn} state (b) Increased $V_{Tn,H}$ state

Figure 13.38 Effect of charge storage on the floating gate

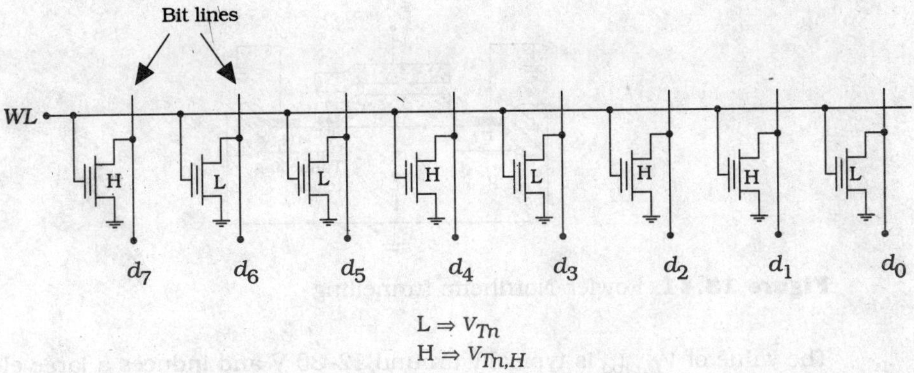

Figure 13.39 A E^2PROM word using floating-gate nFETs

are labeled as "H" in the drawing. The gates of the row are connected to the word line signal *WL* from a row decoder. When the word is accessed with *WL* = 1, a voltage of $V_{WL} = V_{DD}$ is applied to the logic transistors. Low (L) threshold voltage nFETs turn on and pull the output to ground (0 V). High (H) threshold voltage transistors, on the other hand, remain in cutoff and produce logic 1 output voltages using pull-up devices on the bit lines. For the example shown, the output word is

$$d_7d_6d_5d_4d_3d_2d_1d_0 = 10010110 \tag{13.25}$$

To allow the storage of arbitrary data words, a floating-gate FET must be wired as a pull-down device on every NOR gate in the array.

Now that we have seen how the floating-gate FETs are used in circuits, let us examine the programming technique. In the structure we have been studying, electronic charge is transferred to the gate using quantum mechanical tunnelling using **hot electrons**, which are highly energetic channel electrons. The conditions needed to induce the tunnelling are obtained by using a gate voltage $V_{G,prog}$ to create an electron channel, and a programming voltage applied to the drain as shown in Figure 13.40.

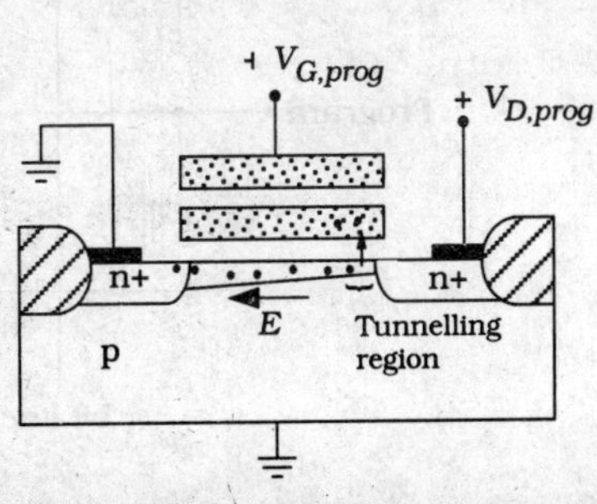

Figure 13.40 Programming a floating-gate FET

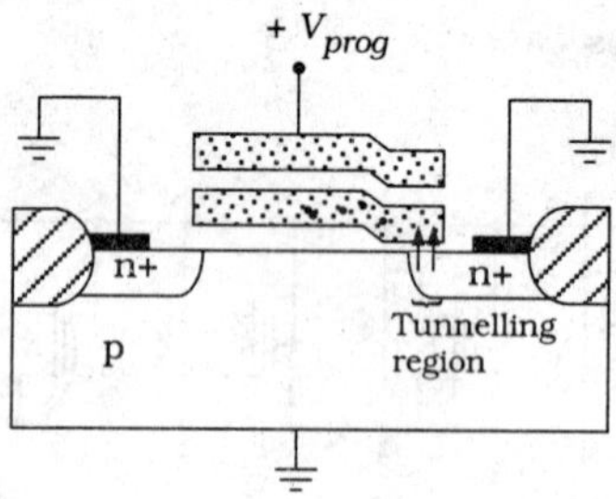

Figure 13.41 Fowler-Nordheim tunnelling

The value of $V_{D,prog}$ is typically around 12–30 V and induces a large electric field that points from the drain to the source. The strong electric field accelerates electrons to the drain side, where they can scatter and tunnel to the floating gate. Retention times are usually estimated to be on the order of 10–20 years in these structures.

Other floating-gate transistor designs use **Fowler-Nordheim emission** where the gate geometry is modified so that a portion of the floating gate extends over the n+ drain as shown in Figure 13.41. A large gate voltage is applied to create a large electric field between the substrate and gate, which enhances the tunnelling through the oxide. Both the drain and source are grounded during the programming operation. The cell circuit in Figure 13.42 accomplishes the write operation by pulsing the Program line while the word line is at $WL = 1$ and the bit and source lines are both grounded.

Erasure is accomplished by reversing the polarity of the applied voltages. General EPROM arrays allow bit erasure, while **flash EPROMs** are wired in a manner that erases large blocks of cells simultaneously. The latter is particularly useful for temporary storage of large data files such as those generated by digital photography.

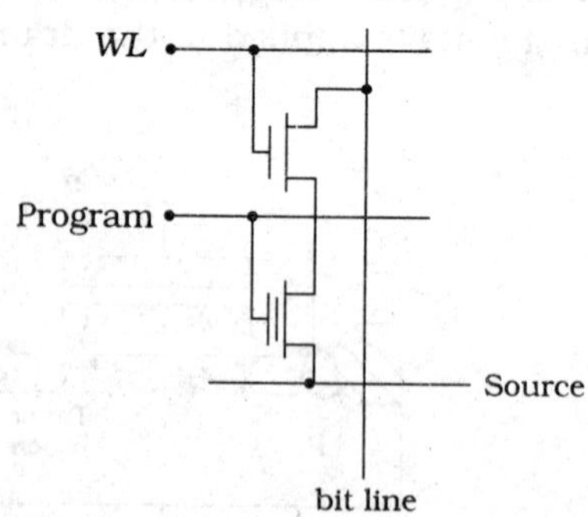

Figure 13.42 EEPROM cell with write line

13.5 Logic Arrays

A **logic array** is usually interpreted as being a structured unit that can be "programmed" to provide various functions and system tasks. They can be similar in structure to ROMs but are usually applied in more general situations. In most cases, the structure of the logic array is invariant. The user programs the array using a defined set of rules.

13.5.1 Programmable Logic Arrays

A useful example is a **programmable logic array** (PLA) for creating SOP (sum-of-product) logic expressions. Consider a group of four input variables a, b, c, d. An SOP function has the form

$$f = \sum_i m_i(a, b, c, d) \tag{13.26}$$

where $m_i(a, b, c, d)$ are minterms, i.e., terms that consist of the input variables or their complements ANDed together. In canonical form, a minterm uses every variable; the numerical value of the subscript i is the equivalent decimal value of the word. For the present case, the lowest order minterms are given by

$$\begin{aligned} m_0 &= \bar{a} \cdot \bar{b} \cdot \bar{c} \cdot \bar{d} \qquad m_1 = \bar{a} \cdot \bar{b} \cdot \bar{c} \cdot d \qquad m_2 = \bar{a} \cdot \bar{b} \cdot c \cdot \bar{d} \\ m_3 &= \bar{a} \cdot \bar{b} \cdot c \cdot d \qquad m_4 = \bar{a} \cdot b \cdot \bar{c} \cdot \bar{d} \qquad m_5 = \bar{a} \cdot b \cdot \bar{c} \cdot d \end{aligned} \tag{13.27}$$

An SOP form is obtained by OR'ing the minterms such as

$$f = \bar{a} \cdot \bar{b} \cdot c \cdot \bar{d} + \bar{a} \cdot b \cdot \bar{c} \cdot d + a \cdot \bar{b} \cdot \bar{c} \cdot \bar{d} + a \cdot \bar{b} \cdot c \cdot d \tag{13.28}$$

This is equivalent to an AND-OR logic sequence, which forms the basis for the VLSI implementation. The general structure of an AND-OR PLA is shown in Figure 13.43. The inputs are fed to the AND-plane that calcu-

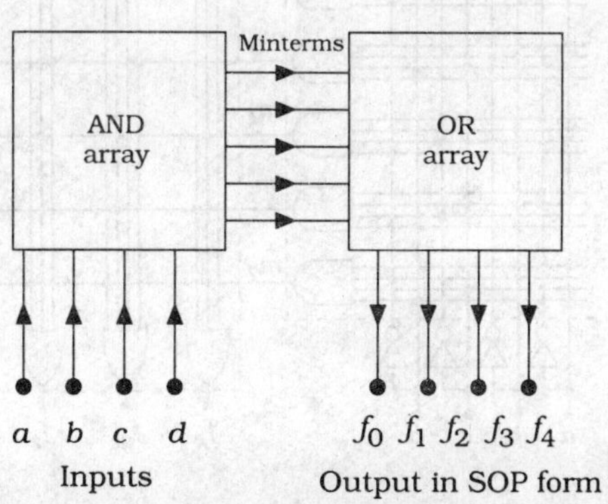

Figure 13.43 Structure of an AND-OR PLA

lates the needed minterms. These are then fed to the OR-plane, where they are OR'ed together. Several SOP functions $f_0, f_1, \ldots,$ can be formed by adding different minterms together. The gate-level diagram for a PLA is shown in Figure 13.44. The AND plane has been chosen to provide five minterms by combining inputs. These feed into the OR plane where they are used to create the output functions shown. Each output can be determined by tracing the inputs. For example

$$f_x = m_0 + m_4 + m_5 \tag{13.29}$$

as can be easily verified by directly reading the connection. Another example is

$$f_y = m_3 + m_4 + m_5 + m_6 \tag{13.30}$$

It is important to remember that each minterm is in canonical form, i.e., every variable appears in either complemented or uncomplemented form. These can be read directly from a function table.

The most usable type of PLA circuit is one that can be easily programmed by placing one FET per connection. Pseudo-nMOS or dynamic logic styles provide the obvious solution, but both have sizing problems with series-connected transistors as needed in NAND gates. One solution is to create the AND-OR array in a NOR-based logic cascade. Figure 13.45(a) shows the desired logic flow. Using complemented inputs to a NOR gate produces the AND function in the schematic of Figure 13.45(b). this can be seen from evaluating

$$\begin{aligned} g &= \overline{(\overline{a} + \overline{b})} \\ &= a \cdot b \end{aligned} \tag{13.31}$$

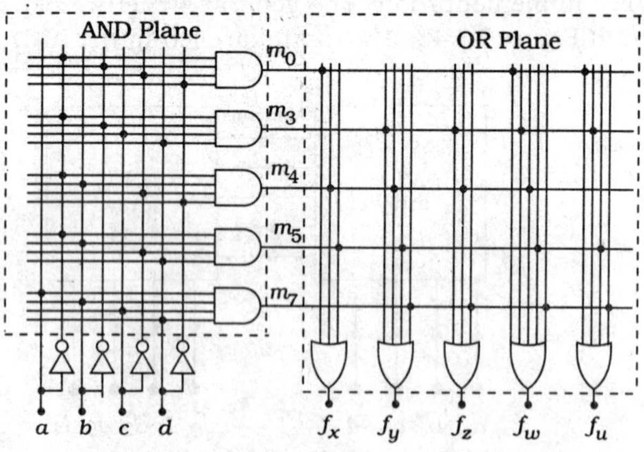

Figure 13.44 Logic gate diagram of the PLA

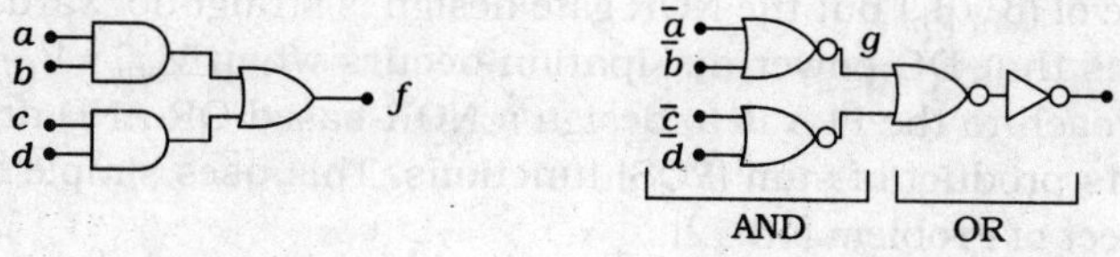

(a) AND-OR logic (b) NOR-based design

Figure 13.45 NOR-gate PLA logic

using a DeMorgan reduction. The OR operation is produced with a NOR-NOT cascade; alternately, the complement $\bar{f}$ may be used as the output. The basic structure for a dynamic cascade based on this equivalence is illustrated by the simple example in Figure 13.46. The AND plane consists of NOR gates, with the outputs fed to the NOR gates. The complemented outputs are

$$\begin{aligned} \bar{f}_1 &= \overline{a \cdot b \cdot c + \bar{a} \cdot \bar{b} \cdot \bar{c}} \\ \bar{f}_2 &= \overline{a \cdot b \cdot c} \end{aligned} \tag{13.32}$$

Other functions can be created by expanding the gates in either (or both) plane(s). A single clock ϕ is used to synchronize the operation of both planes. These can be separated into two different signals where the input plane slightly leads the output plane. In this design, the size of the nFETs determines the high-to-low times of both gates.

A static pseudo-nMOS design is obtained by eliminating the n-channel evaluate FETs, and grounding the gates of the pFETs. Since pseudo-nMOS circuits are ratioed, the output-low voltage V_{OL} depends on the

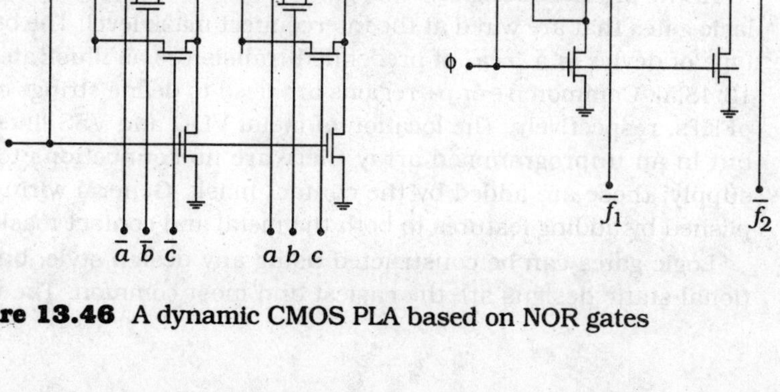

Figure 13.46 A dynamic CMOS PLA based on NOR gates

value of (β_n/β_p) but the NOR gate design is straightforward. Another problem is that DC power dissipation occurs when $V_{out} = V_{OL}$. An alternate approach to the PLA is to design a NOR-based OR-AND array that implements product of sum (POS) functions. This uses simple logic and is the subject of Problem [13.12].

The programmability feature of this approach is seen in Figure 13.47. Either logic plane can be represented as an array of NOR gates. The output for each gate is determined by the nFET logic block. The designer programs the array by placing and wiring FETs into each gate as needed. The physical design is characterized by regular FET patterns placed in between precharge-evaluate transistor pairs.

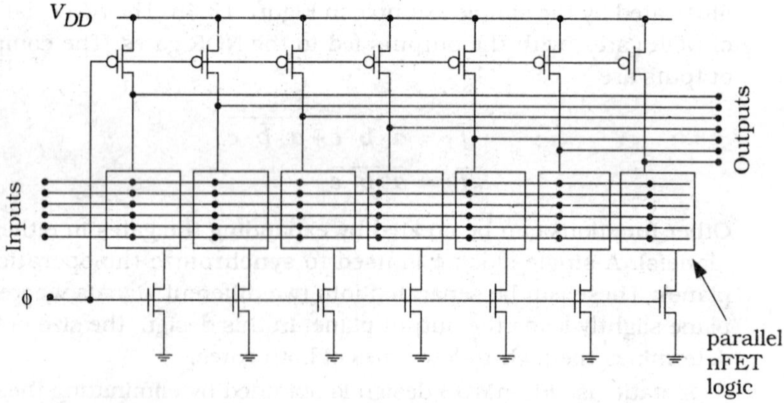

Figure 13.47 Generic NOR-based logic plane

13.5.2 Gate Arrays

The term **gate array** is used to describe an entire class of devices. It usually refers to a user-programmable chip that can be logically configured as needed. Programming techniques vary with the structure of the chip itself.

At the physical design level, a gate array consists literally of an array of logic gates that are wired at the interconnect mask level. The basis for this type of device is a "sea" of predefined transistors as illustrated in Figure 13.48(a). Common n+ or p+ regions are used to define strings of nFET and pFETs, respectively. The location of metal VDD and VSS lines is known, but in an unprogrammed array there are no connections to the power supply; these are added by the contact mask. General wiring is accomplished by adding features to both the metal and contact masks.

Logic gates can be constructed using any design style, but combinational static designs are the easiest and most common. The wiring for a

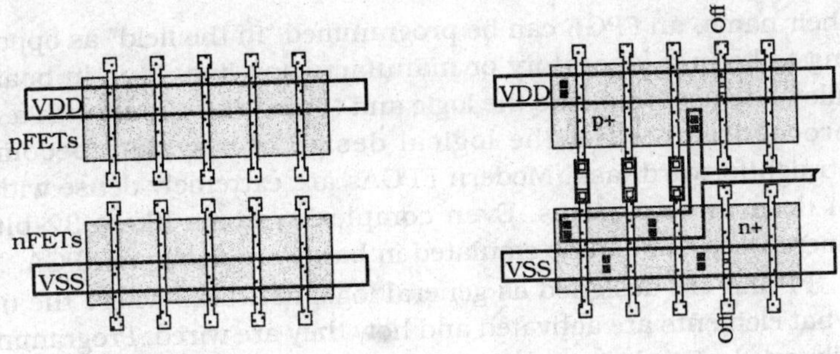

(a) Gate array base (b) NOR3 gate wiring

Figure 13.48 Transistor arrangement in a gate array

NOR3 gate is shown in Figure 13.48(b). Power supply contacts have been added to the VDD and VSS lines. Complementary nFET/pFET pairs are wired using metal and poly contacts. Transistor wiring is attained with metal and active contacts to create three parallel nFETs and three series pFETs. Electrical isolation in this type of array is achieved using cutoff transistors. In the NOR3 gate, the logic pFETs are isolated from the other p-channel transistors by tying a gate to the power supply VDD; an nFET gate connected to VSS provides the same effect on the nFET string. More advanced technologies use oxide isolation to overcome high leakage levels. This leads to more stable electrical characteristics, but makes the wiring more complicated unless common n+/p+ strings are included.

Gate arrays allow rapid prototyping using semicustom logic. Since the designer has access to individual transistors, there are no *a priori* constraints on the circuits or logic. The process can be taken to a high level of automation since the gate map is well defined. One drawback is that the final configuration is from a masking step, so that the wafer must be sent through a portion of the manufacturing line. This adds time to the design cycle. Another problem may arise in the electrical characteristics. In a uniform sea of gates, the aspect ratios are the same for each FET polarity. It is not possible to adjust the sizes, eliminating some circuits that are sensitive to the transistor sizings. Large values of (W/L) are common to allow high-drive circuits to be designed in a simple manner. Since the wiring parasitics, especially the line capacitance, increase with length, switching speed may be a problem.

Gate array designs result in finished die that are larger than could be obtained with custom sizing. This is not usually a concern since the main objective is to provide fast turnaround with semicustom control of the circuitry at the FET level.

Field-programmable gate arrays (FPGAs) have become very popular for testing logic designs and limited production products. As implied by their name, an FPGA can be programmed "in the field" as opposed to having to be in a laboratory or manufacturing line. Plug-in boards for PCs allow the user to define the logic and "burn" the circuit with a fast, simple procedure. Testing the logical design of a system becomes a more straightforward task. Modern FPGAs are extremely dense with hundreds of thousands of gates. Even complex systems like a 32-bit pipelined microprocessor can be emulated in hardware using an FPGA.

FPGAs are designed as general logic networks where the user defines what elements are activated and how they are wired. Programming is usually an electrical process that is achieved using an **antifuse** arrangement that is built into the structure at the physical level. Figure 13.49(a) shows an antifuse. By definition, an antifuse device normally acts like an open circuit between two conducting layers 1 and 2 as shown. The antifuse is "blown" by forcing a high current through it, which melts the antifuse layer and creates a low-resistance contact between the Metal1 and n+. This can be applied between a metal and active n+/p+ region as in Figure 13.49(b), or between a metal and poly layer as in Figure 13.49(c).

Primitive networks in FPGAs vary with the vendor. The most general approach is to provide a logic block that contains elements such as combinational logic, latches and/or flip-flops, multiplexors, and lookup tables. System design is facilitated by a CAD tool set and design notes that cover many standard situations, such as finite state machines or concurrent processors. The details of FPGA design are well beyond the scope of this text, but several excellent books have been written on the subject. Smith's book (Reference [10]) covers many different commercial products, and is almost encyclopedic in nature. Reference [3] is a very readable introduction to the subject. Vendor data from web sites and in hardcopy form is quite abundant, with much tutorial information usually provided to help the potential customer.

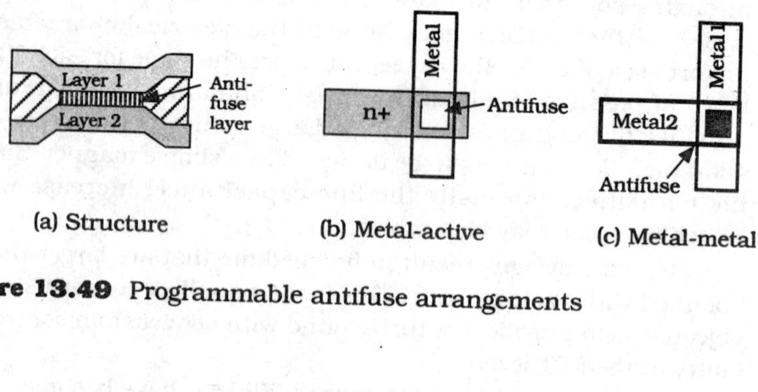

Figure 13.49 Programmable antifuse arrangements

13.6 References for Further Study

[1] R. Jacob Baker, Harry W. Li, and David E. Boyce, **CMOS Circuit Design, Layout and Simulation**, IEEE Press, Piscataway, NJ, 1998.

[2] Abdellatif Bellaouar and Mohamed I. Elmasry, **Low-Power Digital VLSI Design**, Kluwer Academic Publishers, Norwell, MA, 1995.

[3] Stephen D. Brown, Robert J. Francis, Jonathan Rose, and Zvonko G. Vranesic, **Field-Programmable Gate Arrays**, Kluwer Academic Publishers, Norwell, MA, 1992.

[4] Michael D. Ciletti, **Modeling, Synthesis and Rapid Prototyping with the Verilog HDL**, Prentice Hall, Upper Saddle River, NJ, 1999.

[5] Paul R. Gray and Robert G. Meyer, **Analysis and Design of Analog Integrated Circuits**, 3rd ed., John Wiley & Sons, New York, 1993.

[6] Ken Martin, **Digital Integrated Circuit Design**, Oxford University Press, New York, 2000.

[7] Betty Prince, **High Performance Memories**, John Wiley & Sons, Ltd., West Sussex, U.K., 1999.

[8] Betty Prince, **Semiconductor Memories**, 2nd ed., John Wiley & Sons, Ltd., West Sussex, U.K., 1991.

[9] Jan M. Rabaey, **Digital Integrated Circuits**, Prentice-Hall, Upper Saddle River, NY, 1996.

[10] Michael J. S. Smith, **Application-Specific Integrated Circuits**, Addison-Wesley, Reading, MA, 1997.

[11] John P. Uyemura, **A First Course in Digital System Design**, Brooks-Cole, Monterey, CA, 2000.

[12] John P. Uyemura, **CMOS Logic Circuit Design**, Kluwer Academic Publishers, Norwell, MA, 1999.

[13] M. Michael Vai, **VLSI Design**, CRC Press, Boca Raton, FL, 2001.

13.7 Problems

[13.1] Construct switch-level Verilog modules using transistor primitives for the two SRAM cells shown in Figure 13.2.

[13.2] Construct a Verilog descriptions of memory arrays that store

(a) 32-bit words in 2048 locations.

(b) 16K bytes.

(c) 8-bit words in 8K locations.

[13.3] Suppose that a 6T SRAM cell is designed with a β-ratio of 1. Is it still possible to write to the cell even though $R_A = R_n$? Explain your answer in terms of the circuit.

[13.4] Consider the divided-word line architecture illustrated in Figure 13.31. Apply the concepts of Logical Effort to explain why this design can be faster than one that uses a single word line. Present your arguments in a quantitative manner with equations. Make the necessary assumptions about capacitance values, etc., and use a value of $r = 2$ for simplicity in your analysis.

[13.5] Design the logic circuit that will avoid write conflicts in the dual-port SRAM cell in Figure 13.7. Assume that both ports have equal priority, and that emphasis is on avoiding simultaneous write access.

[13.6] The storage capacitor in a DRAM has a value of $C_s = 55$ fF. The circuitry restricts the capacitor voltage to a value of $V_{max} = 3.5$ V. When the access transistor is off, the leakage current off of the cell is estimated to be 75 nA.

(a) How many electrons can be stored on C_s?

(b) How many fundamental charge units q leave the cell in 1 second due to the leakage current?

(c) Calculate the time needed to reduce the number of stored charges to 100.

[13.7] A DRAM cell has a storage capacitance of $C_s = 45$ fF. It is used in a system where $V_{DD} = 3.3$ V and $V_{Tn} = 0.55$ V. The bit line capacitance is $C_{bit} = 250$ fF.

(a) Find the maximum amount of charge that can be stored on C_s.

(b) Suppose that the voltage on the capacitor is charged a level of V_{max}. The word line controlling the access FET is dropped to a value $WL = 0$ at time $t = 0$. The leakage current is estimated to be 50 nA. To detect a logic 1 state, the voltage on the bit line must be at least 1.5 V. Find the hold time.

[13.8] Consider a DRAM cell that has a storage capacitance of $C_s = 55$ fF. The power supply is $V_{DD} = 3.0$ V and the access FET has a threshold voltage of $V_{Tn} = 0.65$ V. The leakage current from the storage capacitor is estimated to be 250 pA, and the bit line capacitance is $C_{bit} = 420$ fF.

The capacitor has a voltage V_{max} across it when the word line is brought low at time $t = 0$. A read operation is initiated at a time $t = 10$ ms by elevating the word line up to a value $WL = 1$. Find the voltage on the bit line.

[13.9] Design an AND-OR PLA that has the following outputs:

$$\begin{aligned} f_1 &= m_0 + m_2 + m_6 \\ f_2 &= m_0 + m_1 + m_4 + m_5 + m_6 \\ f_3 &= m_3 + m_4 + m_7 \end{aligned} \tag{13.33}$$

Construct the logic diagram, and then translate it into a CMOS-compatible circuit.

[13.10] Design an OR-AND PLA that provides the following outputs:

$$
\begin{aligned}
F_1 &= M_2 \cdot M_3 \cdot M_5 \\
F_2 &= M_0 \cdot M_1 \cdot M_4 \\
F_3 &= M_1 \cdot M_2 \cdot M_6 \cdot M_7
\end{aligned}
\tag{13.34}
$$

What common CMOS gates could be used to implement the physical design of the array?

[13.11] Design a dynamic CMOS AND-OR PLA using NOR gates as a basis. Design the circuitry such that the inputs are a, b, c and outputs are

$$
\begin{aligned}
g_1 &= (a \cdot \bar{b} \cdot c) + (\bar{a} \cdot \bar{b} \cdot \bar{c}) \\
g_2 &= (\bar{a} \cdot \bar{b} \cdot c) + (a \cdot b \cdot c) \\
g_3 &= (a \cdot \bar{b} \cdot \bar{c}) + (a \cdot \bar{b} \cdot c) \\
g_4 &= (\bar{a} \cdot \bar{b} \cdot \bar{c}) + (a \cdot \bar{b} \cdot \bar{c})
\end{aligned}
\tag{13.35}
$$

[13.12] Design a dynamic CMOS NOR-NOR PLA that has a, b, c as inputs and outputs the POS functions

$$
\begin{aligned}
f_1 &= (a + \bar{b} + c) \cdot (\bar{a} + b + c) \\
f_2 &= (\bar{a} + b + \bar{c}) \cdot (a + b + c) \\
f_3 &= (a + \bar{b} + \bar{c}) \cdot (\bar{a} + \bar{b} + \bar{c})
\end{aligned}
\tag{13.36}
$$

Construct the logic network first, then design the electronic circuit.

[13.13] Design a FET-programmable ROM that contains the following data.

Address	Data
0	0100
1	1111
2	1010
3	0001
4	1011
5	0111
6	1110
7	1001

(13.37)

[13.14] Consider the general gate array base illustrated in Figure 13.48(a). Design the wiring that would create a complex logic gate for the function

$$
F = \overline{a \cdot b \cdot c + a \cdot e}
\tag{13.38}
$$

with standard series-parallel structuring.

System-Level Physical Design

14

CMOS VLSI design revolves around the physical characteristics of silicon circuitry. A high-level architectural function can be implemented using any of several different circuits, but the blocks must be wired together to complete the design of the chip. In this chapter we will examine aspects of macroscale physical design.

14.1 Large-Scale Physical Design

Up to this point we have concentrated on studying relatively simple logic functions using CMOS technology. Once a cell library is created, it can be used to build a complex VLSI system by instancing logic units into the master design. This bottom-up description represents the transition from low-level primitives to a high-level system. Associated with this change in the design hierarchy are physical considerations that arise when the emphasis evolves from the bit-level micron-sized structures to larger-size units and sections. Characteristics of interconnect lines, signal distribution, and large circuit blocks are only a few of the problems that must be dealt with. Since the performance of the chip is ultimately determined by its components, large-scale aspects of physical design and layout are critically important. Embedding a fast switching network into a slow interconnect mesh neutralizes the speed obtained in the low-level design.

A typical top-down design flow that illustrates some of the important problems is shown in Figure 14.1. After the HDL description of the system has been written and verified, logic synthesis provides a first design. The network is then subjected to simulation with emphasis on insuring that the design behaves as desired. Physical design steps of floorplanning, cell placement, clocking paths, and signal routing are next in the sequence. These are concerned with solving the large-scale problems of chip build-

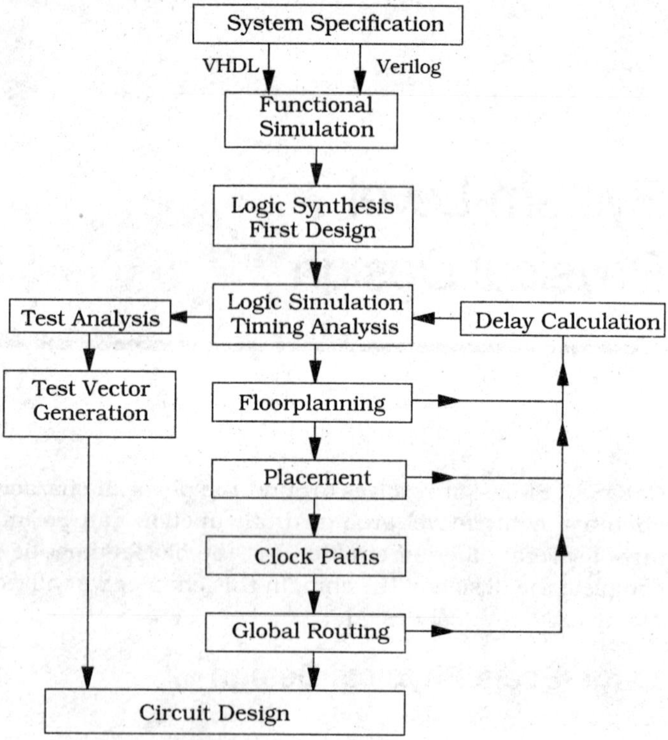

Figure 14.1 Design flow showing chip-level physical design issues

ing, and are the subject of the present chapter. Functionality of a high-speed digital system is closely tied to the overall timing: clock distribution, gate delays, latching, and other considerations. Timing information provides a feedback path that links each successive design level to the next. Testing problems are also examined during the design phase as indicated on the left side of the drawing. The problem of testing VLSI chips is so important to the industry that an entire chapter is dedicated to it later in the book.

The size of a silicon integrated circuit increases with the complexity and transistor count of the network. Overall, our target is to create a chip with the smallest area A_{chip} to increase both the number of die N per wafer and the overall yield Y. For a given process and wafer size, the goal may be to select the smallest standard size die for the design. Regardless of the actual die size, a number of physical design-related problems arise when building the entire chip. The most important issues are introduced in this chapter. Some of the problems, such as interconnect delay and wiring, are considered to be major obstacles for projected designs of the future.

14.2 Interconnect Delay Modeling

One of the most critical problems in high-density VLSI is dealing with interconnect lines. They introduce signal delays that affect system timing, and often lead to extremely complicated layout routing problems. We have seen many examples of the special treatment given to interconnect lines during the fabrication process: silicided poly lines, multiple metal interconnect layers, and the use of copper. In this section, we will build equivalent circuits that provide mathematical models for interconnect delays. Routing techniques are discussed in Section 14.5.

The starting point of our analysis is the simple isolated interconnect line shown in Figure 14.2 that represents an arbitrary material layer on the chip. The dimensions of the line are shown as having a length l, a width w, and a thickness t. The line resistance R_{line} from In to Out is given by the formula

$$R_{line} = R_s\left(\frac{l}{w}\right)\ \Omega \tag{14.1}$$

where R_s is the sheet resistance of the layer and (l/w) is the number of squares with dimensions $(w \times w)$. Defining the resistance per unit length r by

$$r = \frac{R_s}{w}\ \ \Omega/\text{cm} \tag{14.2}$$

the equation becomes

$$R_{line} = rl \tag{14.3}$$

This shows the increase of parasitic resistance with the line length l in a form that is easy to deal with. It is obvious that high-resistivity layers

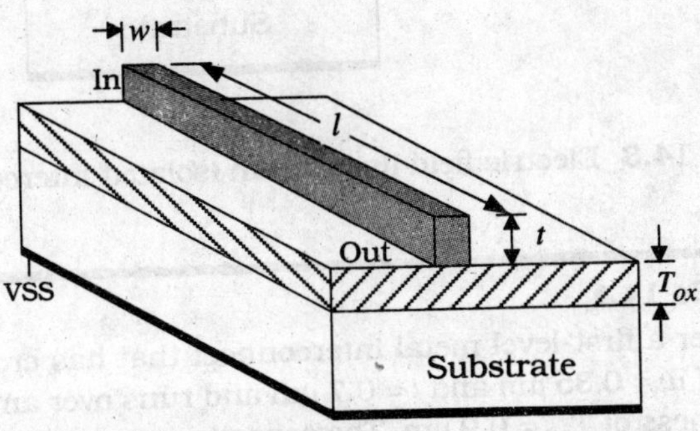

Figure 14.2 Isolated interconnect line

such as polysilicon are more problematic than low-resistivity metals.

The total line capacitance C_{line} can be estimated using the simple parallel-plate formula

$$C_{line} = \frac{\varepsilon_{ox} l w}{T_{ox}} \text{ F} \tag{14.4}$$

where T_{ox} is the thickness of the insulating oxide between the line and the substrate. C_{line} is also called the **self-capacitance** of the line. Although the equation provides a first estimate, it ignores the effects of fringing electric fields from the edges and sides when the line is at a positive voltage; Figure 14.3 illustrates the origin of the fringing field corrections. An empirical equation for the capacitance per unit length c that accounts for these effects is [10]

$$c = \varepsilon_{ox}\left[1.15\left(\frac{w}{T_{ox}}\right) + 2.8\left(\frac{t}{T_{ox}}\right)^{0.222}\right] \text{ F/cm} \tag{14.5}$$

such that

$$C_{line} = cl \tag{14.6}$$

is the total capacitance in farads. Conceptually, the first term accounts for fringing from the bottom side of the line while the second term depends on the thickness t and is due to the sidewall effects. In a CMOS chip, the largest line capacitance values will be on the layers that are the closest to the substrate.

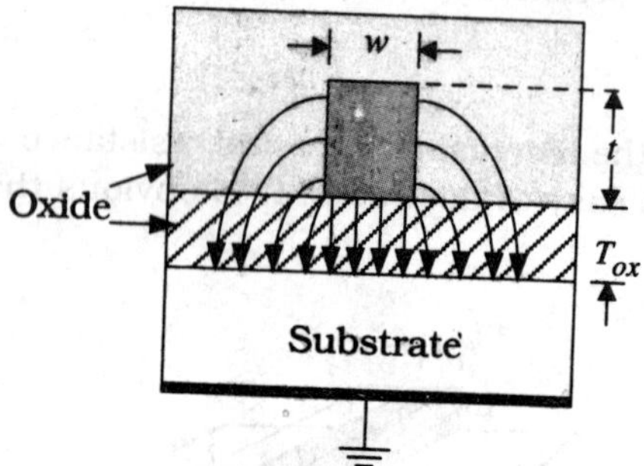

Figure 14.3 Electric field lines for an isolated interconnect

Example 14.1

Consider a first-level metal interconnect that has cross-sectional dimensions of $w = 0.35$ µm and $t = 0.7$ µm and runs over an oxide layer that has a thickness of $T_{ox} = 0.9$ µm. The capacitance per unit length is

$$c = (3.9)(8.854\times10^{-14})\left[1.15\left(\frac{0.35}{0.9}\right) + 2.8\left(\frac{0.7}{0.9}\right)^{0.222}\right] \tag{14.7}$$

$$= 1.07 \text{ pF/cm}$$

If the sheet resistance is $R_s = 0.02\ \Omega$, then the resistance per unit length is

$$r = \frac{0.02}{0.35\times10^{-4}} = 571 \quad \Omega/\text{cm} \tag{14.8}$$

An interconnect with a length of $l = 40\ \mu\text{m}$ is characterized by

$$\begin{aligned} R_{line} &= (571)(40\times10^{-4}) = 2.29 \ \ \Omega \\ C_{line} &= (1.07)(40\times10^{-4}) = 4.28 \text{ fF} \end{aligned} \tag{14.9}$$

Increasing the length to $l = 225\ \mu\text{m}$ gives

$$\begin{aligned} R_{line} &= (571)(225\times10^{-4}) = 12.85 \ \ \Omega \\ C_{line} &= (1.07)(225\times10^{-4}) = 24.1 \text{ fF} \end{aligned} \tag{14.10}$$

While the resistance remains relatively small (because R_s is small), the parasitic line capacitance is on the order of MOSFET values, making it important to the analysis.

Calculating the values of R_{line} and C_{line} allows us to construct circuit models to study the effects of the parasitics. The simplest approach to modeling the line is to construct the simple two-element circuit shown in Figure 14.4 to include the effects of the line from In to Out. This is called a "single-rung ladder circuit" because of the way it is drawn. If the input voltage $v_i(t)$ changes from a 0 to a 1 (or vice versa) in a step-like manner, the change in the output voltage $v_o(t)$ is delayed by the time constant

$$\tau = R_{line}C_{line} \tag{14.11}$$

which constitutes a low-order estimate of how the interconnect line affects the signal transmission. We must delve deeper into the origin of

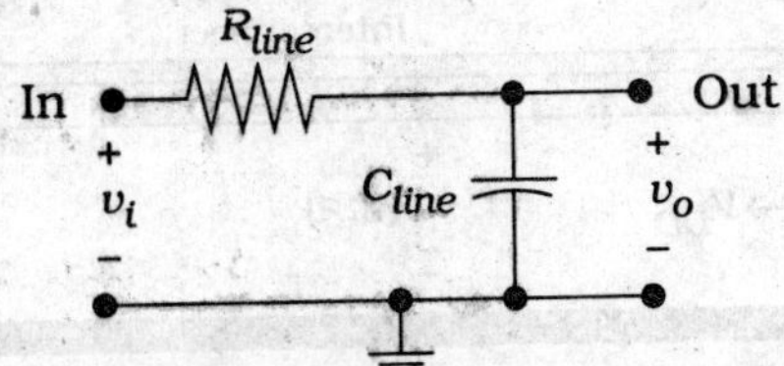

Figure 14.4 Single-rung ladder model

the parasitics in order to understand why this is only a crude approximation. Our study will also allow us to derive a more accurate value for the time constant.

Consider an interconnect line that extends in the z-direction as shown in Figure 14.5. The input voltage $v_i(t)$ is applied at $z = 0$ while the output is at $z = l$. If we pulse the input voltage to $v_i \rightarrow V_{DD}$, then the driving source must move current through the resistive line while simultaneously charging the line capacitance it encounters. Physically, the charging starts at $z = 0$ and progresses to the right with increasing z, so that the line voltage itself is really a function of both position z and time t. We therefore designate it by $v(z, t)$. Both the resistance and capacitance of the line are incremental in nature; approximating their values using the lumped element values R_{line} and C_{line} inherently limits the accuracy of the analysis.

There are two techniques that can be used to model the problem. The first is to divide the line into several RC ladder rungs that approximate the distributed nature of the parasitics. Figure 14.6 illustrates the procedure. The line resistance R_{line} and line capacitance C_{line} are divided into m-segments with values given by

$$R_m = \frac{R_{line}}{m}, \quad C_m = \frac{C_{line}}{m} \tag{14.12}$$

These are then used to construct multirung ladders. The case for $m = 1$ is shown in Figure 14.6(a). This circuit is defined by the reference time constant

$$\begin{aligned}\tau_1 &= R_1 C_1 \\ &= R_{line} C_{line}\end{aligned} \tag{14.13}$$

The time constant τ_2 for the 2-rung ladder in Figure 14.6(b) can be analyzed using the Elmore formula as

$$\tau_2 = C_2(2R_2) + C_2(R_2) = 3R_2C_2 \tag{14.14}$$

Similarly, the 3-rung ladder in Figure 14.6(c) is described by the time constant expression

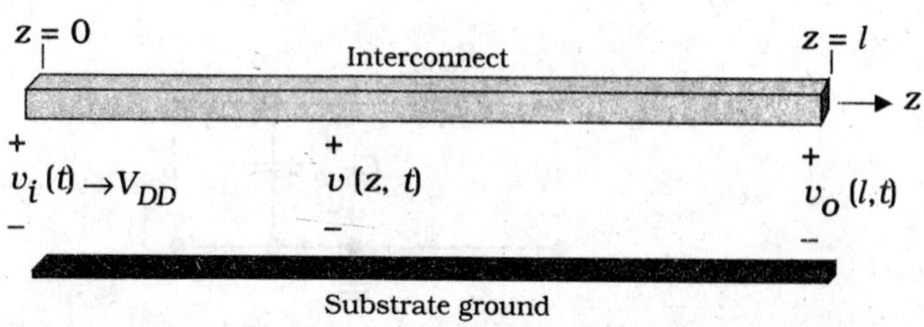

Figure 14.5 Physical model of an interconnect line

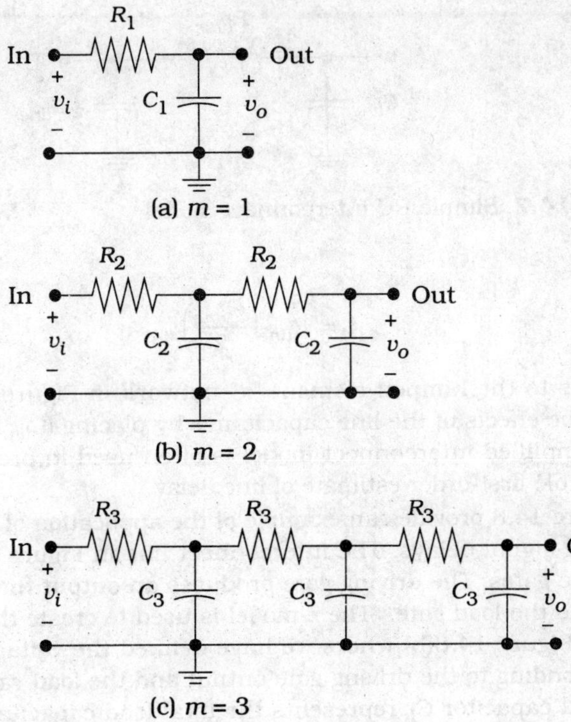

Figure 14.6 Multiple-rung ladder circuits

$$\tau_3 = C_3(3R_3) + C_3(2R_3) + C_3(R_3) = 6R_3C_3 \qquad (14.15)$$

In general, an m-rung ladder has a time constant of

$$\tau_m = \frac{m(m+1)}{2}R_mC_m \qquad (14.16)$$

Substituting the relations in equation (14.12) into this expression gives the time constant in terms of the total line resistance and capacitance in the form

$$\tau_m = \frac{m(m+1)}{2}\left(\frac{R_{line}}{m}\right)\frac{C_{line}}{m} = \frac{m(m+1)}{2m^2}R_{line}C_{line} \qquad (14.17)$$

For large values of m, this has the limiting form

$$\tau = \tau_m \rightarrow \frac{1}{2}R_{line}C_{line} \qquad (14.18)$$

as the total time constant τ for the line. Rewriting this as

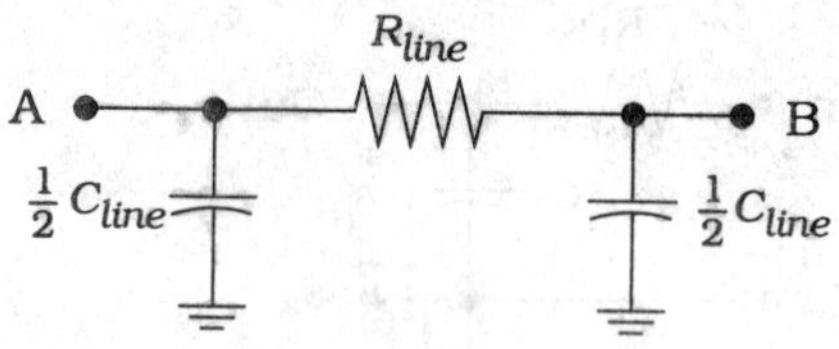

Figure 14.7 Simple RC interconnect model

$$\tau = R_{line}\left(\frac{C_{line}}{2}\right) \tag{14.19}$$

leads us to the lumped-element "π" network in Figure 14.7.[1] This balances the effects of the line capacitance by placing $(C_{line}/2)$ on each end. This simplified interconnect model is often used in practice to obtain a reasonable first-order estimate of line delay.

Figure 14.8 provides an example of the application of the model to calculating signal delays. The interconnect line in Figure 14.8(a) connects two logic gates. The driving gate produces an output that must be transmitted to the load gate. The π-model is used to create the equivalent circuit in Figure 14.8(b) where we have defined the voltages $v(t)$ and $v_L(t)$ corresponding to the driving gate output and the load value, respectively. The load capacitor C_L represents the total load capacitance and is given by

$$C_L = \left(\frac{1}{2}\right)C_{line} + C_{in} \tag{14.20}$$

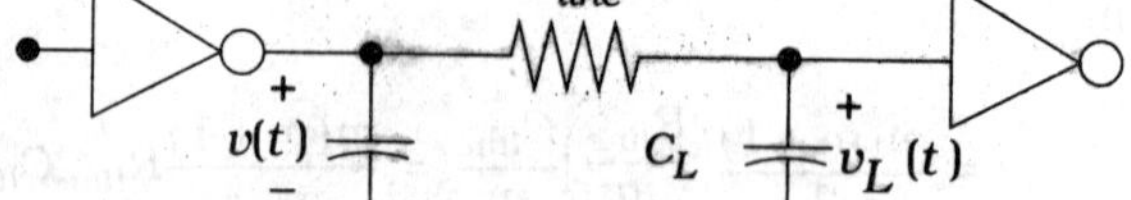

(a) Physical structure

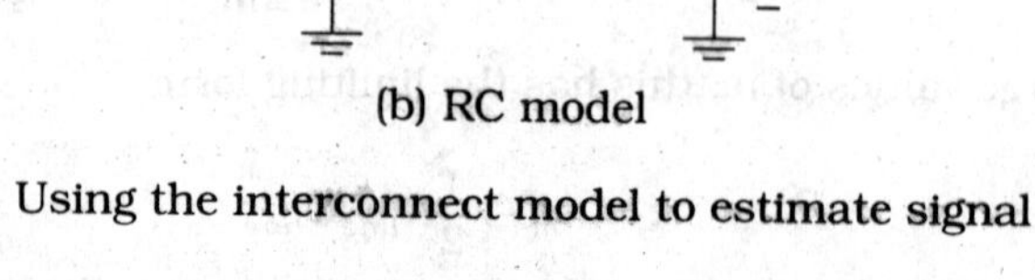

(b) RC model

Figure 14.8 Using the interconnect model to estimate signal delays

[1] This is called a "π" network because the elements are arranged to form the Greek letter.

with C_{in} being the input capacitance of the load gate. Note that $v(t)$ is defined as a driving voltage across the left capacitor. The line-induced signal delay can be calculated by specifying the driving voltage $v(t)$ and then calculating the response $v_L(t)$.

Let us use the pulse input for $v(t)$ shown in Figure 14.9. At time $t = 0$, the voltage changes in a step-like manner from 0 V to V_{DD}. The transition may be written as

$$v(t) = V_{DD}u(t) \tag{14.21}$$

where $u(t)$ is the unit step function defined by

$$u(t) = \begin{cases} 0 & (t < 0) \\ 1 & (t \geq 0) \end{cases} \tag{14.22}$$

The step response for the circuit is

$$v_L(t) = V_{DD}[1 - e^{-(t/\tau)}] \tag{14.23}$$

where the time constant is given by

$$\tau = R_{line}C_L \tag{14.24}$$

The effect of the parasitic is to slow the rise as seen in the $v_L(t)$ plot. The most important point to see is that $v_L(t)$ must reach a reach a logic 1 voltage (shown as "1" in the plot) before the load gate interprets the input as a logic 1 value.

When $v(t)$ makes a transition from V_{DD} to 0 V at time t_1, the reverse situation occurs. Modeling the driving voltage by

$$v(t) = V_{DD}[1 - u(t - t_1)] \tag{14.25}$$

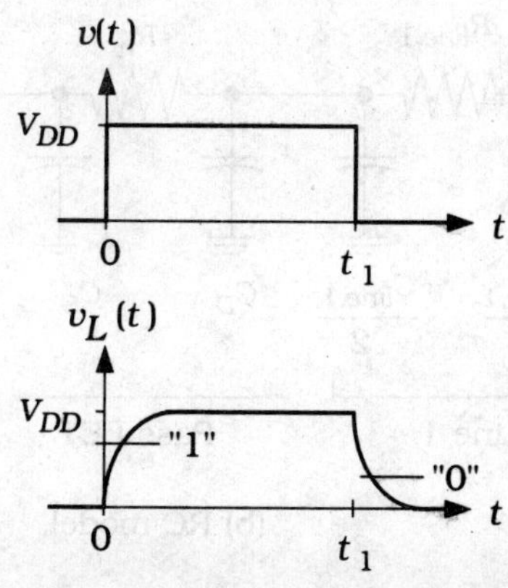

Figure 14.9 Step response for the interconnect circuit

gives a decaying voltage response in the form

$$v_L(t) = V_{DD}e^{-(t-t_1)/\tau} \tag{14.26}$$

which is valid for $t \geq t_1$. The falling value of $v_L(t)$ must reach the "0" voltage shown in the plot before the load gate senses the transition. This simple example illustrates the fact that interconnect parasitics always induce signal delays in VLSI networks.

Another example is shown in Figure 14.10. The original circuit in Figure 14.10(a) consists of two interconnect lines with a pass transistor in between. The circuit equivalent in Figure 14.10(b) is obtained by using the RC π-models for both the wires and transistor. The nFET parasitics R_n, C_D, and C_S are calculated using the equations presented in Chapter 6. Note that parallel capacitors may be combined to give

$$C_1 = C_D + \frac{C_{line,1}}{2}$$
$$C_2 = C_S + \frac{C_{line,2}}{2} \tag{14.27}$$

at the interior node. The simplified model may be analyzed with a circuit simulation program to determine the delay characteristics.

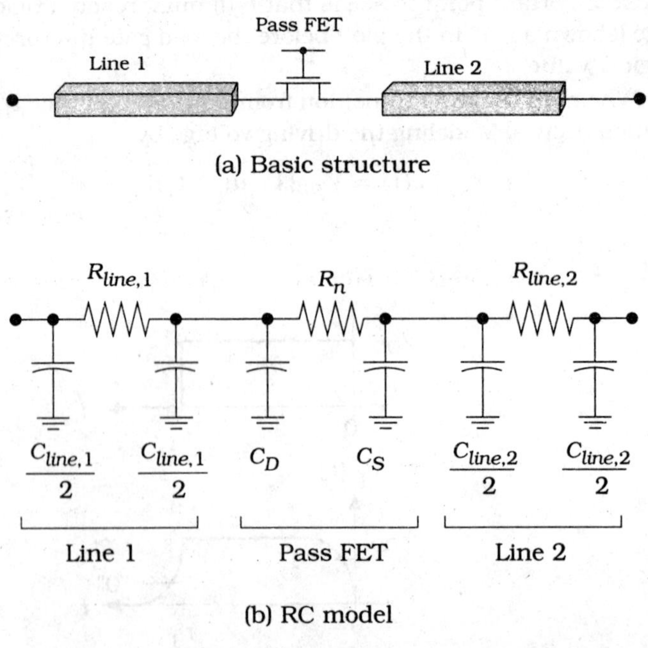

Figure 14.10 Modeling interconnect lines with a series pass FET

A more accurate analysis is obtained by modeling the interconnect using an m-rung ladder network. The drawback of this approach is the increased computational time needed in simulating the circuit. For example, the CPU time required by SPICE to analyze an n-node circuit increases as n^2. Large VLSI circuit sections can take many hours to simulate even if simple interconnect models are used. It is usually more efficient to provide delay models for the individual lines, then use these as macros in the circuit simulation.

While RC ladders are easy to visualize, the problem of modeling the voltage $v(z,t)$ in Figure 14.5 is intrinsically differential in nature. Analyzing the interconnect at this level gives the partial differential equation [14]

$$\frac{\partial^2 v(z,t)}{\partial z^2} = rc\frac{\partial v(z,t)}{\partial t} \tag{14.28}$$

that describes the voltage as a function of position and time. With a step-input voltage of

$$v(z=0,t) = V_{DD}u(t) \tag{14.29}$$

acting as a boundary condition, the voltage on an infinite line is given by

$$v(z,t) = V_{DD}\operatorname{erfc}\left(\sqrt{\frac{rc}{4t}}\,z\right)u(t) \tag{14.30}$$

where erfc(ξ) is the **complementary error function**. In general, erfc(ξ) decreases with increasing argument ξ and is described by the integral representation

$$\operatorname{erfc}(\xi) = \frac{1}{\sqrt{\pi}}\int_{\xi} e^{-\alpha^2}\,d\alpha \tag{14.31}$$

Differentiating gives the slope

$$\frac{d}{d\xi}([\operatorname{erfc}(\xi)]) = -\frac{1}{\sqrt{\pi}}e^{-\xi^2} \tag{14.32}$$

which is of Gaussian form. Note, however, that $\xi = \xi(z,t)$ in this case so that both space and time variations are included in the motion. The differential in equation (14.28) has the same form as the heat diffusion equation of thermodynamics. Because of this, the voltage is viewed as **diffusing** down the line such that the error function argument

$$\xi = \sqrt{\frac{rc}{4t}}\,z \tag{14.33}$$

describes the motion. This can be seen by holding ξ constant: as time t

increases, the position z needed to keep ξ at the same value also increases in a non-linear (squared) manner.

In practice, it is easier to use the numerical values provided in computational programs such as MatLab and MathCad. While the differential equation provides more accurate values for the signal delays, adding realistic constraints such as a finite line with a capacitive node makes the analysis quite complicated. Only a few problems can be solved in closed form, making numerical analysis mandatory. Because of this, VLSI designers tend to prefer the simpler RC models for first estimates in most signal paths.

14.2.1 Signal Delay versus Line Length

One of the most important results of this analysis is the dependence of the delay time constant τ on the length l of the line. Consistent results can be obtained using any of the analysis techniques discussed above.

The simplest estimate for the time constant is from equation (14.11) in the form

$$\tau = R_{line}C_{line} \tag{14.34}$$

Substituting from equations (14.3) and (14.6) gives

$$\tau = Bl^2 \tag{14.35}$$

where $B = rc$ is a constant for the line with units of sec/cm^2. This shows that the signal delay is proportional to the square of the line length. The quadratic dependence is illustrated in Figure 14.11, and has a major effect on the use of long interconnects in VLSI networks. Since unequal line lengths have different signal delays, this requires that interconnect routing be carefully planned, especially in critical datapaths. The system designer must take care to accurately model and design interconnect networks to insure that the system can operate at the desired speed.

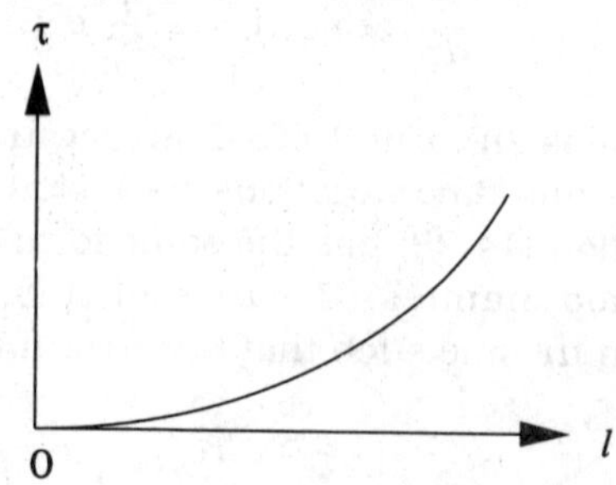

Figure 14.11 Parabolic dependence of the time delay on line length

Example 14.2
Suppose that the signal delay on an interconnect of length 50 μm is known to be 0.13 ps. If the line is increased to 100 μm, the delay rises to a value of

$$\tau = \left(\frac{0.13}{50^2}\right)100^2 = 0.52 \text{ ps} \tag{14.36}$$

where we have used the given data to find *B* in the equation. A line that is 200 μm long has a delay of

$$\tau = \left(\frac{0.13}{50^2}\right)200^2 = 2.08 \text{ ps} \tag{14.37}$$

This shows that the relative lengths of interconnect wires become the important factor.

14.2.2 Dealing with Interconnect Delays

Signal delays along interconnect lines can be limiting factors in high-speed system design. In critical single-bit paths, they must be added to the normal gate delays to obtain an accurate picture of the problem. They become especially important for global distribution of signals such as the clock ϕ in a synchronous system. In word-oriented architectures where every bit of an *n*-bit word must be transmitted from one unit to another, the slowest bit-transmission path determines the data flow speed for the entire word. Careful routing schemes are used in an attempt to equalize the line length for every bit.

Since interconnect delays are intrinsically circuit and layout problems, detailed analyses are usually performed by circuit design groups. They are charged with the task of creating accurate circuit models that can be used in simulation programs without consuming excessive amounts of computer time. Design manuals often provide this type of information in circuit or code form that can be used directly by other designers by inserting numerical values for the parameters. High-level system and logic designers are then able to estimate interconnect delays along all paths for use in architectural verifications.

The importance of interconnect delays cannot be overemphasized. More examples will be presented later in the context of specific problem areas. Of these, the problem of global clock distribution discussed in the next chapter is one of the most critical aspects of high-speed synchronous design.

14.3 Crosstalk

Whenever an interconnect line is placed in close proximity to any other interconnect line, the conductors are coupled by a parasitic capacitance. Pulsing a voltage on one of the lines induces a stray signal on all lines that are coupled to it. This phenomenon is called **crosstalk**. Since a stray signal at the input to a logic gate may cause an incorrect output, dealing with crosstalk problems is a very important aspect of designing high-density VLSI chips.

Consider the layout shown in Figure 14.12 where Line 1 and Line 2 are coupled to each other by parasitic capacitance. Capacitance increases as the distance between two conductors decreases. The strongest coupling thus occurs where the two lines are separated by the minimum distance S. Suppose that Line 1 and Line 2 have voltages of V_1 and V_2, respectively. Let us denote the total coupling capacitance by C_c. The voltage difference $V_{12} = (V_1 - V_2)$ induces a current i_{12} flowing from Line 1 to Line 2 as described by the basic capacitor equation

$$i_{12} = C_c \frac{dV_{12}}{dt} = C_c \frac{d(V_1 - V_2)}{dt} \tag{14.38}$$

This expresses the fundamental basis of capacitive crosstalk. Strong coupling exists if C_c is large, or if the voltage difference $(v_1 - v_2)$ changes very quickly in time. Since high-speed design requires large time derivatives (corresponding to rapidly changing signals), VLSI design usually deals with crosstalk by reducing C_c and then examining the switching limits.

Figure 14.13 provides a cross-sectional view of two adjacent interconnect lines (labeled 1 and 2) that are separated by a spacing S. An empirical equation for the coupling capacitance per unit length is [10]

$$c_c = \varepsilon_{ox}\left[0.03\left(\frac{w}{T_{ox}}\right) + 0.83\left(\frac{t}{T_{ox}}\right) - 0.07\left(\frac{t}{T_{ox}}\right)^{0.222}\right]\left(\frac{S}{T_{ox}}\right)^{-1.34} \tag{14.39}$$

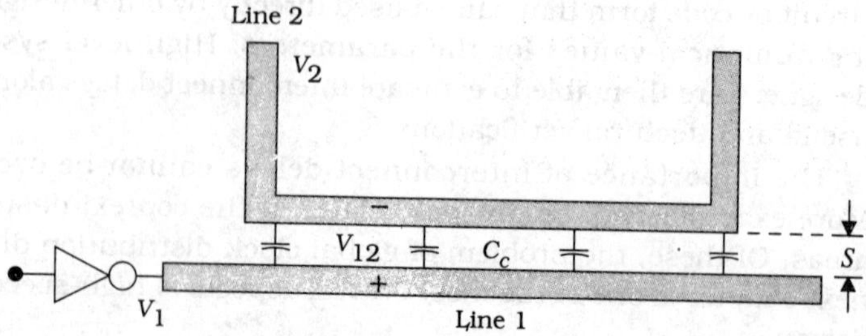

Figure 14.12 Capacitive coupling between two lines

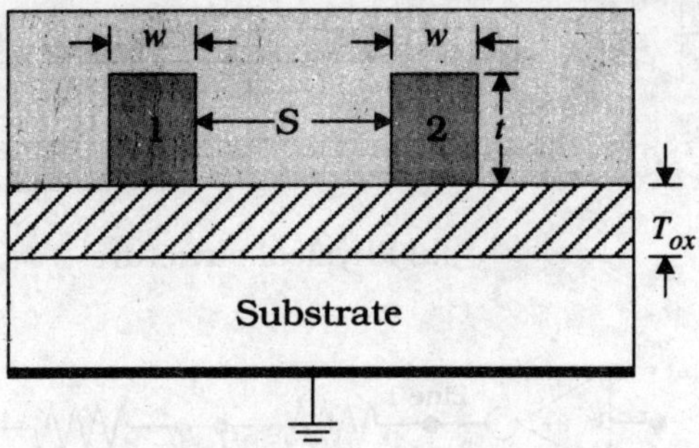

Figure 14.13 Geometry used for coupling capacitance calculation

in units of F/cm. The total coupling capacitance C_c in farads is computed from

$$C_c = c_c l_c \tag{14.40}$$

where l_c is the length of the coupled section with spacing S. This shows that c_c increases as $(1/S)^{4/3}$, so that using a small line spacing increases the coupling capacitance. Layout design rules account for this fact in the numerical values specified for S_{min}. A crosstalk-based design rule overrides the fact that the lithography may be capable of producing finer line spacing.

An example of using the coupling capacitance C_c is shown in Figure 14.14. The original circuit [Figure 14.14(a)] along with the layout dimensions provides the details needed to compute the parasitics including the value of C_c. This can be used in a lumped-element equivalent model such as that shown in Figure 14.14(b). This approach uses a symmetric RC equivalent circuit for each interconnect line and models the coupling using a single capacitor with value C_c in the middle. The alternate model shown in Figure 14.15 divides the coupling into two capacitors ($C_c/2$); other topologies are possible.

The total capacitance of a line consists of the self-capacitance (from the line to ground) and any coupling terms. Let us denote the self-capacitance of Lines 1 and 2 by C_{11} and C_{22}, respectively; these are just the appropriate values of C_{line} for each. The total capacitance seen looking into Line 1 is given by

$$C_1 = C_{11} + C_c \tag{14.41}$$

Similarly, the total capacitance of Line 2 is

$$C_2 = C_{22} + C_c \tag{14.42}$$

These values are important for designing the driving circuits for each line. If an interconnect line is coupled to two adjacent lines, then both lines

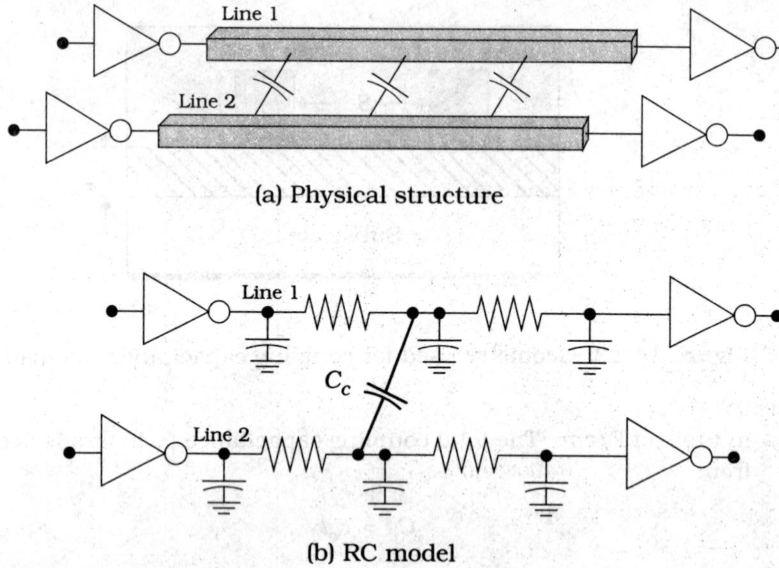

Figure 14.14 Lumped-element coupling circuit model

contribute to the total capacitance. Figure 14.16 illustrates the case where Line 1 interacts with both Line 2 and Line 3. The total capacitance of Line 1 per unit length in the closely spaced sections is

$$c_1 = c + 2c_c \text{ F/cm} \tag{14.43}$$

Multiplying by the length gives the total capacitance. Theoretically, every conductor on the chip interacts with every other conductor. In practice, however, we usually limit ourselves to **nearest-neighbor** coupling; this is justified by the decrease in c_c with increasing S.

Let us examine the physics of the interaction for the 3-line network.

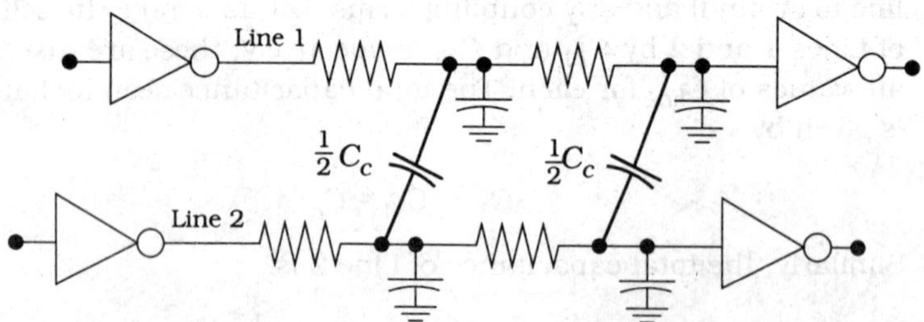

Figure 14.15 Alternate model for coupling circuit

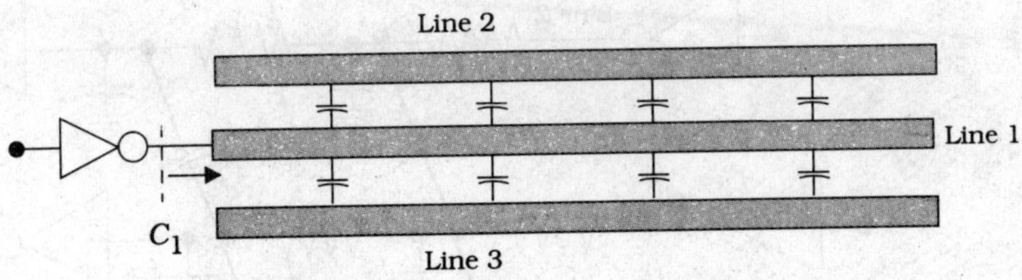

Figure 14.16 Multiple-line coupling

With nearest-neighbor coupling, the charge Q_1 on Line 1 is written as

$$Q_1 = C_{11}V_1 + C_{12}(V_1 - V_2) + C_{13}(V_1 - V_3) \tag{14.44}$$

where C_{12} and C_{13} are the coupling capacitances from lines 1 to 2 and lines 1 to 3, respectively, and V_i is the voltage on the i-th line (i = 1, 2, 3). The charges on Line 2 and Line 3 are

$$\begin{aligned} Q_2 &= C_{21}(V_2 - V_1) + C_{22}V_2 \\ Q_3 &= C_{31}(V_3 - V_1) + C_{33}V_3 \end{aligned} \tag{14.45}$$

since they do not interact with each other. These equations may be combined to give the matrix form

$$\begin{bmatrix} Q_1 \\ Q_2 \\ Q_3 \end{bmatrix} = \begin{bmatrix} (C_{11}+C_{12}+C_{13}) & -C_{12} & -C_{13} \\ -C_{21} & (C_{22}+C_{21}) & 0 \\ -C_{31} & 0 & (C_{33}+C_{31}) \end{bmatrix} \begin{bmatrix} V_1 \\ V_2 \\ V_3 \end{bmatrix} \tag{14.46}$$

We can show that the capacitance matrix is symmetric with $C_{ij} = C_{ji}$. Since current is the time derivative of the charge, we compute

$$\begin{bmatrix} i_1 \\ i_2 \\ i_3 \end{bmatrix} = \frac{d}{dt}\begin{bmatrix} Q_1 \\ Q_2 \\ Q_3 \end{bmatrix} = \begin{bmatrix} (C_{11}+C_{12}+C_{13}) & -C_{12} & -C_{13} \\ -C_{21} & (C_{22}+C_{21}) & 0 \\ -C_{31} & 0 & (C_{33}+C_{31}) \end{bmatrix} \frac{d}{dt}\begin{bmatrix} V_1 \\ V_2 \\ V_3 \end{bmatrix} \tag{14.47}$$

This shows that any change in the line voltage (dV_1/dt) changes both $i_2(t)$ and $i_3(t)$, with the magnitude of the effect dependent on the size of the capacitors and the rate of change of the voltage. Similarly, changing voltages of (dV_2/dt) or (dV_3/dt) causes $i_1(t)$ to flow. A circuit level model for the 3-line network is shown in Figure 14.17. This may be analyzed using a standard circuit simulator.

It is worthwhile at this point to introduce a formula for the capacitance of an isolated plate as shown in Figure 14.18. The total plate capacitance C_p [F] may be estimated using

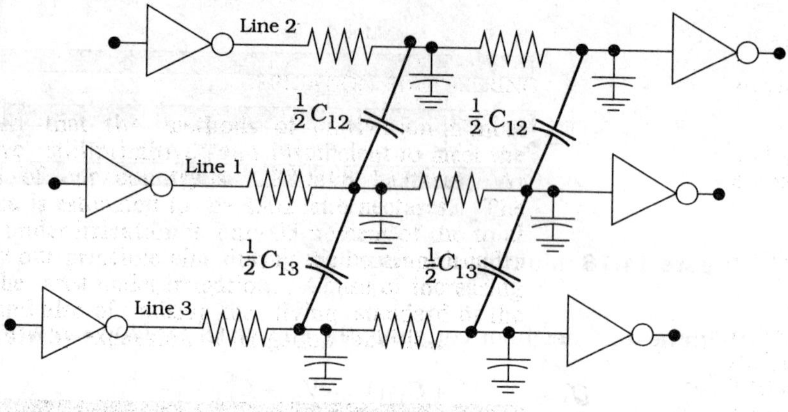

Figure 14.17 Circuit model for 3-line coupling problem

$$C_p = \varepsilon_{ox}\left[1.15\left(\frac{A}{T_{ox}}\right) + 1.40P\left(\frac{t}{T_{ox}}\right)^{0.222} + 4.12T_{ox}\left(\frac{t}{T_{ox}}\right)^{0.728}\right] \quad (14.48)$$

where $A = wl$ is the bottom area of the plate and $P = 2(w+l)$ is the circumference. In this expression, the first term accounts for the bottom and fringing contributions, the second term gives the sidewall additions, and the last term accounts for the corners.

Crosstalk also occurs between overlapping lines on different material layers. Figure 14.19 illustrates the case where Metal2 crosses over a Metal1 line. The critical parameter is the oxide thickness $T_{ox,12}$ between the two layers. The simplest approximation for overlap capacitance $C_{ov,12}$ is the parallel-plate formula

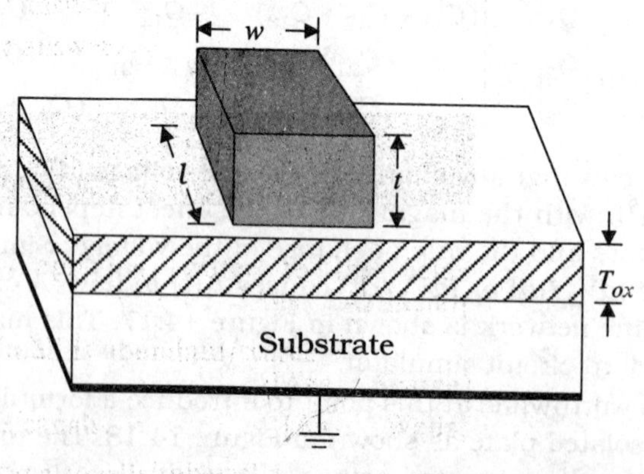

Figure 14.18 Plate geometry

$$C_{ov,12} = \frac{\varepsilon_{ox} A_{ov}}{T_{ox,12}} \tag{14.49}$$

with $A_{ov} = w_1 w_2$ as the overlap area. Although this ignores fringing fields, it is sufficient for small overlap areas. Since we attempt to minimize all capacitances in the circuit, this leads to the layout strategy where we attempt to draw interconnects on a given layer so that they are perpendicular to the lines on the layers directly above and below it. In other words, we try to draw Metal1 lines that are perpendicular to Metal2 lines, Metal2 lines that are perpendicular to Metal3 lines, and so on.

14.3.1 Dealing with Crosstalk

Crosstalk problems can be very involved and often require specialized group studies. While simplified equations are useful for estimating the coupling parameters, computer programs have been developed to calculate 2- and 3-dimensional coupling parameters directly from Maxwell's equations of electromagnetic theory. In addition to detailed information on the field strengths and gradients, these codes provide numerical values for parameters such as c and c_c that can be used directly to calculate the capacitances. The line resistance and capacitances are used to create equivalent circuit models, which are then subjected to simulation studies using programs such as SPICE.

The detailed examination of crosstalk is usually delegated to the domain of circuit designers and electromagnetics specialists. VLSI system designers usually see the results of these studies in various parametric forms such as noise fluctuation levels on the nodes. Other times, the research results in design rule changes at both the device and system level.

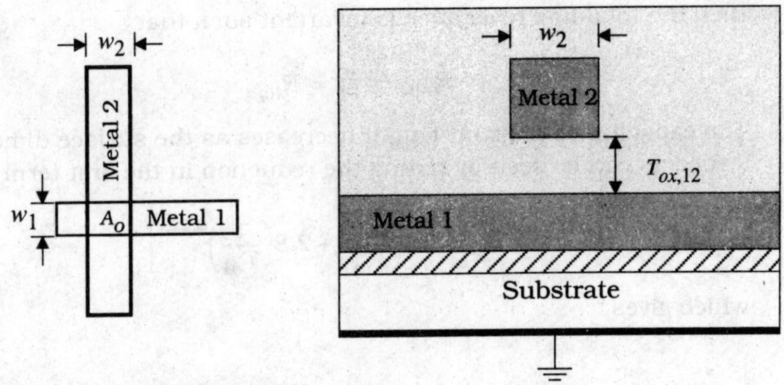

Figure 14.19 Overlap capacitance

14.4 Interconnect Scaling

Although scaling theory[2] was originally introduced to describe FETs, it can be applied to interconnect lines to yield useful results. This is consistent with the view that a shrinking transistor is accompanied by reduced-size interconnects that affect the overall performance of a circuit.

The three geometrical dimensions of an interconnect line that are set in the processing are the width w, the thickness t, and the oxide thickness T_{ox}. Improved lithography allows us to reduce the linewidth to a smaller value

$$\tilde{w} = \frac{w}{s} \tag{14.50}$$

where $s > 1$ is the scaling factor. This is the fundamental effect of scaling the surface geometry of an interconnect line.

To understand how this affects the electrical characteristics, recall that the sheet resistance of a layer is given by

$$R_s = \frac{\rho}{t} \tag{14.51}$$

The resistivity ρ is not changed by shrinking w, so that the line resistance per unit length increases as seen by writing

$$\tilde{r} = \frac{R_s}{\tilde{w}} = sr \tag{14.52}$$

If we assume that the line length l scales according to

$$\tilde{l} = \frac{l}{s} \tag{14.53}$$

then the total line resistance is invariant such that[3]

$$\tilde{R}_{line} = \tilde{r}\tilde{l} = R_{line} \tag{14.54}$$

The capacitance per unit length decreases as the surface dimensions are scaled as can be seen by noting the reduction in the first term of

$$\tilde{c} = \varepsilon_{ox}\left[1.15\left(\frac{\tilde{w}}{T_{ox}}\right) + 2.8\left(\frac{t}{T_{ox}}\right)^{0.222}\right] \tag{14.55}$$

which gives

---

2 Scaling theory was introduced in Section 6.5.1.

3 Note that scaling a layout does not mean that the interconnect lengths scale in the same manner.

$$\tilde{c} = = \varepsilon_{ox}\left[1.15\left(\frac{\tilde{w}}{sT_{ox}}\right) + 2.8\left(\frac{t}{T_{ox}}\right)^{0.222}\right] \quad (14.56)$$

If we can ignore fringing or assume that the first term dominates,

$$\tilde{c} \approx \frac{c}{s} \quad (14.57)$$

Scaling the line length l then approximates the new line capacitance as

$$\tilde{C}_{line} = \tilde{c}\tilde{l} = \frac{C_{line}}{s^2} \quad (14.58)$$

which shows a $1/s^2$ reduction.

A polysilicon line will exhibit the highest sheet resistance in a process even if it is silicided. In this case, it would be important to decrease the line length so as not to increase the value of R_{line}. The time constant for the line scales according to

$$\tilde{\tau} = \tilde{R}_{line}\tilde{C}_{line} = \frac{\tau}{s^2} \quad (14.59)$$

which is due to the reduction in the line capacitance. Note that if l is not scaled, then τ is not affected by the surface scaling. The same comments apply to an arbitrary metal line where the time constant is dominated by the line capacitance.

Let us now examine the situation where the vertical dimensions t and T_{ox} are reduced such that

$$\tilde{t} < t \quad , \quad \tilde{T}_{ox} < T_{ox} \quad (14.60)$$

which are generally valid for any vertical scaling factor $s_v > 1$. Reducing the thickness t has the undesirable effect of increasing the sheet resistance since

$$R_s = \frac{\rho}{t} \quad (14.61)$$

with the resistivity ρ a constant. Similarly, a thinner oxide increases c, so that both R_{line} and C_{line} would increase, leading to longer delays. If we instead **increase** t and T_{ox}, both r and c would be smaller.

As a final case, let us examine how scaling affects the coupling capacitance and, therefore, crosstalk. A brute-force scaling of the surface geometry of neighboring lines would stipulate that

$$\tilde{S} = \frac{S}{s} \quad (14.62)$$

where S is the spacing between the lines as shown previously in Figure

14.13. To see how this affects the coupling, let us examine the basic formula

$$c_c = \varepsilon_{ox}\left[0.03\left(\frac{\tilde{w}}{T_{ox}}\right) + 0.83\left(\frac{t}{T_{ox}}\right) - 0.07\left(\frac{t}{T_{ox}}\right)^{0.222}\right]\left(\frac{S}{T_{ox}}\right)^{-1.34} \tag{14.63}$$

for the coupling capacitance per unit length. The overall multiplying factor

$$c_c \propto \left(\frac{1}{S}\right)^{1.34} \tag{14.64}$$

shows that decreasing S increases the coupling capacitance. While the actual increase may be offset somewhat by scaling other terms such as w and T_{ox}, reducing the crosstalk often dominates all other considerations including real estate consumption. As processes evolve, reduced values of S are possible, but line spacings do not scale as much as FET sizes.

This short discussion of interconnect scaling illustrates how the theory is used to provide ideas for improving performance. By itself, it is a highly idealized approach that cannot be implemented in practice due to processing limitations. However, it does act as a catalyst for future improvements which explains in part why it is still considered worth studying.

14.5 Floorplanning and Routing

Cell-based VLSI design employs predesigned electronic circuit modules that are instanced as needed to create the system. At the chip level, every module is viewed as a block that consumes area and must be wired into the network. This step links the system and subsystem architecture directly to the silicon physical design. At this scale, the physical design problems are very different from those encountered in transistor and gate layout. Long interconnects, complex wiring meshes, and other large-scale factors are critical to the overall performance of the finished design. Many aspects of **design automation** are devoted to these problems.

Floorplanning deals with the placement of the logic blocks into the overall design. This is done very early in the design cycle so that area budgets can be assigned, and the overall size of the chip can be estimated. The initial floorplan can be based on large, complex functional units and how they are wired together by the system architecture. Once an area is allocated, the designs of the subsystems that make up the large units are themselves constrained. Floorplans are drawn before the physical design is even started, so it requires an experienced group of designers to provide guidelines based on previous designs.

When a logic module is placed into the design, it must be wired to other units. While simple point-to-point wires may be easy to add, modern VLSI systems require millions of connections. Interconnect routing schemes have been developed to provide a structured approach to attacking this problem. **Place-and-route** CAD tools are useful for wiring complex systems. The designer specifies the beginning and end points of an interconnect wire, and the tool generates a solution that does not violate any design rules. These codes are based on different types of graph algorithms, and exhibit various degrees of success.

Let us examine the problem of floorplanning first. Any digital system can be decomposed into a set of units that are wired together in a specific manner. A simple example is shown in Figure 14.20(a). The interconnect lines indicate communication between distinct blocks and each carries a different number of bits. If the dimensions of the blocks are scaled according to their actual size in the layout, then we may use the block diagram to create a first-try floorplan as in Figure 14.20(b). Wiring channels are provided in between adjacent blocks to facilitate the wire routing. This is important for minimizing interconnect lengths, and may be mandatory if we are limited to only one or two interconnect layers.

This example can be used to illustrate a **sliceable floorplan**, which is one of the simplest approaches to large-scale layout. A sliceable floorplan is defined as either a single module, or a floorplan that can be partitioned into modules (or module groups) using a vertical or horizontal line that traverses a contiguous group of modules. Let us redraw the floorplan of Figure 14.20(b) into the equivalent representation shown in Figure 14.21(a). A vertical cut line may be used to obtain the first division shown in Figure 14.21(b). The second division into the groups portrayed in Figure 14.21(c) is obtained using two horizontal cut lines. This process may

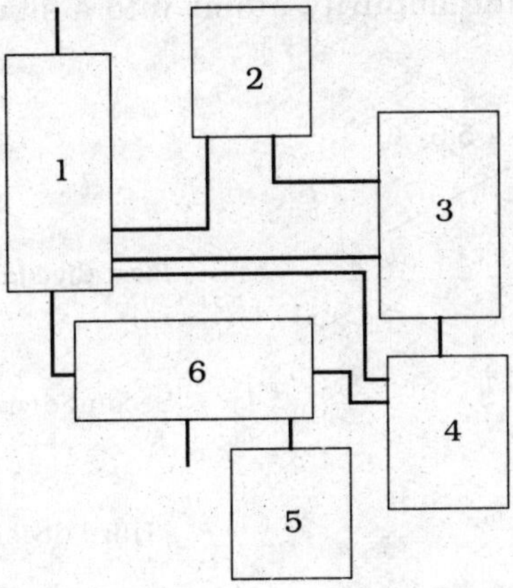

(a) System block diagram

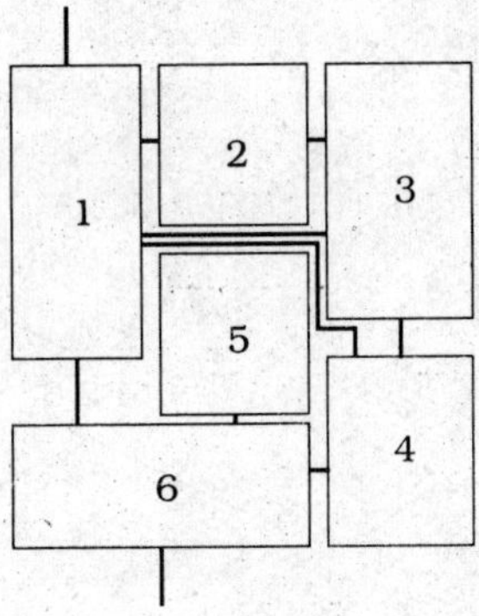

(b) Initial floorplan

Figure 14.20 Using a block diagram for initial floorplanning

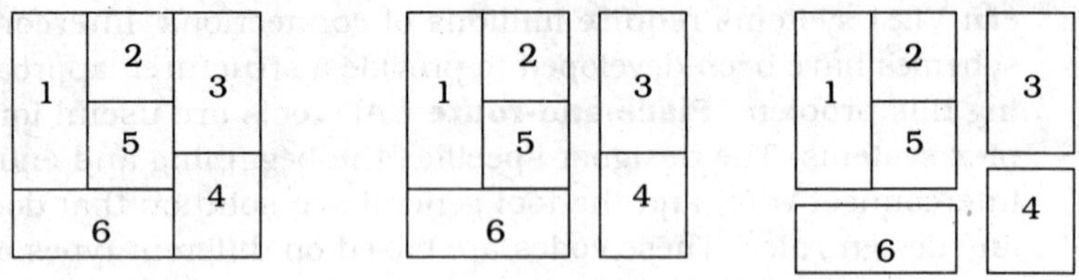

Figure 14.21 Sliceable floorplan example

be continued until only separate modules remain. The divisions can be described using the tree structure drawn in Figure 14.22. The numbers denote the connected module groups, and the slicing process can be seen by the division level indicated at the bottom of each branch. Note that the lowest entries are the basic modules. An example of a non-sliceable floorplan is shown in Figure 14.23(a). This cannot be divided by using either a horizontal or a vertical cut line without dividing up a module. However, it can be used as module in a larger sliceable floorplan, such as illustrated in Figure 14.23(b). The sliceable design can be described by a tree network. If a non-sliceable module is used in a sliceable design, then a tree structure can still be created, but the cut lines must be defined in a more restrictive manner. When a floorplan tree can be constructed, the design is called a **hierarchically defined floorplan**. These are conceptually straightforward to understand and have been used as the basis for deriving layout algorithms that follow the tree structure.

One approach to floorplanning is the concept of **simulated annealing** which is based on the manner in which crystal formation occurs in solid materials. Recall that implanting impurity atoms into a silicon wafer

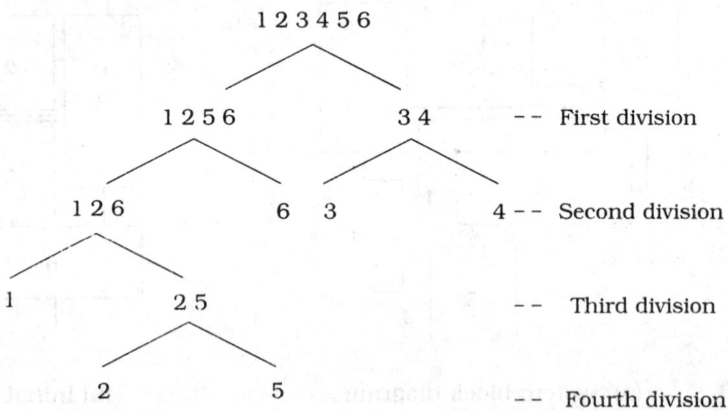

Figure 14.22 Binary division tree

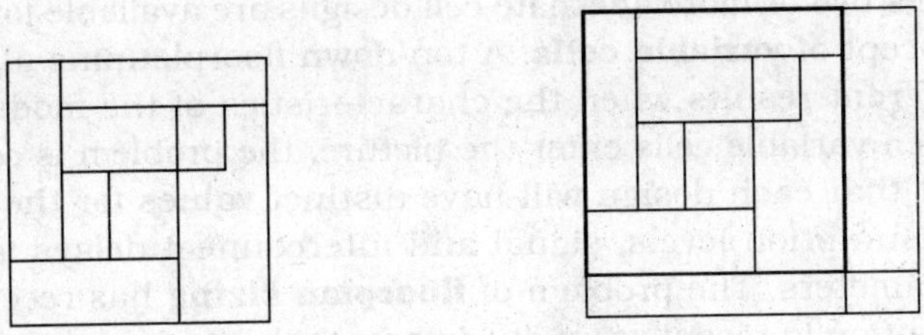

(a) Non-sliceable floorplan (b) Use in a sliceable design

Figure 14.23 Example of a non-sliceable floorplan

causes damage to the crystal structure that is healed using an annealing step. The wafer is heated, imparting thermal energy to the atoms and inducing motion. As the temperature is lowered, the atoms seek minimum energy positions which places them at required points in the crystal lattice. Simulated floorplan annealing is an algorithmic technique that is based on the possible groupings at the vertex of a binary tree; the dependency on the tree structure restricts the approach to sliceable floorplans.

Consider a sliceable floorplan that describes a complex system. It is usually possible to construct more than one binary tree that describes the floorplan. Since every tree is valid, the existence of multiple trees expands the **solution space** of the problem. Simulated annealing takes advantage of multiple solutions by examining different configurations of the problem in a sequential manner. Various solutions are compared at each vertex in the tree, and one is selected as being the best. Moving to the next vertex starts the process over, so that the overall solution has been constructed in a step-by-step manner. Every choice is accompanied by a **cost metric** that is minimized at each vertex. The procedure results in a **global cost function** that can be used to assess the solution.

Other algorithmic approaches to floorplanning have similar characteristics. Each tends to centralize on designing floorplans using a few well-defined constraints and minimization parameters. Iterative routines allow many solutions to be compared. Hierarchical-based approaches have proven useful in many situations, and linear programming packages have also been used for tackling the problem.

An intrinsic problem of floorplanning is defining the modules themselves. The number of modules and the wiring nets establish the structure of the floorplan tree, which in turn affects the design. We know from our studies of the VLSI design hierarchy that cellular-based design yields many different levels of complexity and implementations. Given an arbitrary digital module, it can always be decomposed into smaller units. Or, it can be used as a primitive element in a larger module. If only a single implementation is available for a module, it is called a **fixed cell**. Oftentimes two or more alternate cell designs are available for use, defining the concept of **variable cells**. A top-down floorplanning algorithm produces different results when the characteristics of the modules are changed. When variable cells enter the picture, the problem is complicated by the fact that each design will have distinct values for the chip area, power consumption levels, signal and interconnect delays, and other critical parameters. The problem of **floorplan sizing** has received considerable attention in recent years due in part to the increasing layout complexity of large chips. A common constraint in most approaches is the requirement of rectangular cells since they are the easiest to use in solving the general problems of floorplanning and cell placement in general.

Once the general floorplan has been established, interconnect routing becomes a critical concern. Although routing considerations enter into the choice of floorplan, interconnect-induced delay and timing problems won't be evident until the physical design is well under way. Routing is usually performed in two steps: global and detailed. Global routing deals with finding connection paths at the system level without specifying the actual geometric information. Detailed routing uses the global results and provides the layout specifics such as the layers used and placement of vias. Figure 14.24 provides a conceptual view of the two. Once the detailed routing is completed, the interconnect delays are set and the design must undergo verification.

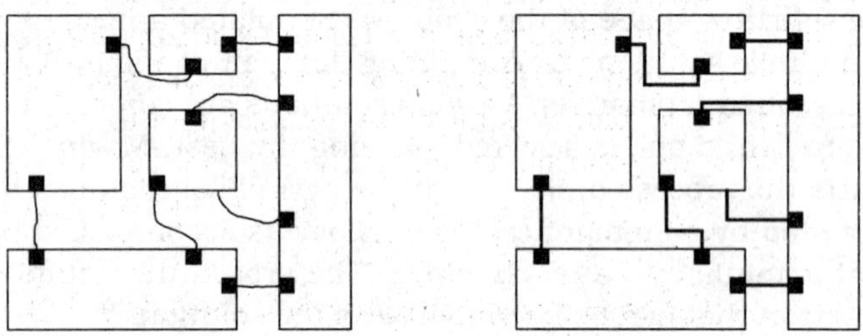

(a) Global (b) Detailed

Figure 14.24 Routing steps

Global routing is usually modeled using graph theory. Many variations have been published in the literature, with some more intuitive than others. The routing model must be closely tied to the processing and the layout design style. Grid models and checkerboard-based algorithms are among the simplest. An example of a grid model is shown in Figure 14.25. Grid points are overlaid on the floorplan; grid points between the modules are used to connect the lines that make up a given net. A **line-probe** algorithm provides a routing path from a source point to a target point by generating lists of lines from both that are not blocked by a module. A

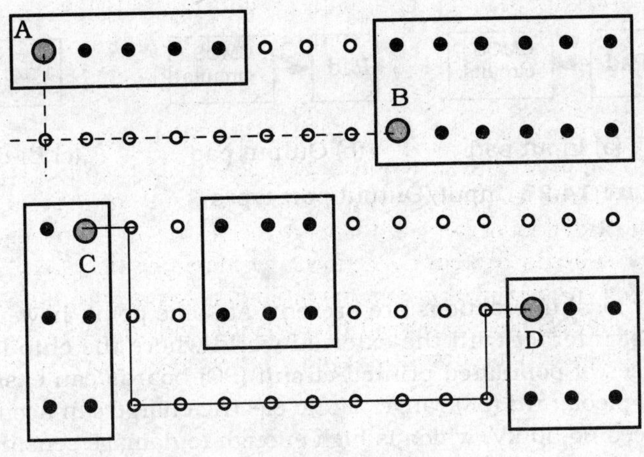

Figure 14.25 Grid model for global routing

solution is found when a line from the source list intersects a line from the target list, as from A to B in the drawing. If a solution is not found, additional lists are created and linked to the source and target data. The **maze-routing** approach is an alternate technique that starts by examining possible paths from the source, and then expanding outward until the target is hit by one. The path drawn from C to D is an example of routing using the maze viewpoint. Many path-finding algorithms have been studied in an effort to find a quick and efficient approach.

Detailed routing is approached by solving one region of the global solution at a time. Routing regions are divided into two types. A **channel** is a rectangular routing area that is bounded on two sides by modules; a **switchbox** region is a rectangular area that is bounded on all four sides. Given a routing region, the solution must lead to a layout that can physically fit into the allocated area. Increasing the number of interconnect metal layers makes the routing easier, but via placement enters the problem.

The field of design automation has evolved to be a very sophisticated discipline within VLSI. New products to deal with complexity and verification issues are announced regularly. As the density and complexity of chips continue to increase, developing better and more efficient CAD tools becomes mandatory for advancement of the discipline.

14.6 Input and Output Circuits

On-chip CMOS circuitry is aided by the fact that it is shielded from the outside world by the packaging. This simplifies the design environment considerably. Capacitance levels are limited to the femtofarad level and

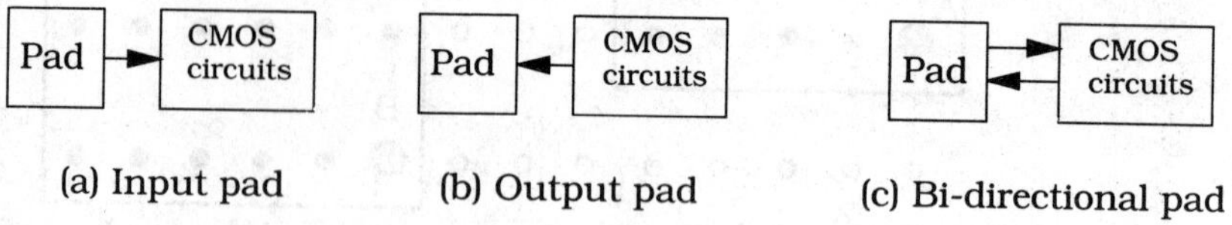

Figure 14.26 Input/Output port types

only local interactions are present. At some point, however, the circuitry must interact with the external world where the chip is used. Capacitances of populated printed-circuit (PC) boards can easily be in the 50-100 picofarard (pF) range. Static electric charge can accumulate to levels exceeding 50 kV, which is high enough to damage sensitive MOS transistors. Input and output circuitry must be designed to shield the internal network from being affected by these types of problems.

The interface ports between VLSI circuits and the outside world can be classified into three types: input, output, or bi-directional. These are shown for pad-type I/O in Figure 14.26. Input port circuits are designed to protect the sensitive CMOS circuits from high voltages, while the design of the output circuits tends to concentrate on driving high-capacitance PC board lines. Bi-directional circuits are combinations of these and allow a port to work as both an input and an output.

14.6.1 Input Circuits

Static electric charges tend to accumulate on dielectric (non-conductive) surfaces and can lead to extremely high voltages. A common cause of static charge buildup is due to frictional interactions.[4] Most people have experienced this in some form or another. Walking across a carpeted room may accumulate enough charge to create a spark when touching a doorknob. Physically, the voltage has grown to a value that is large enough to break down the insulating properties of air.

MOS field-effect transistors are extremely sensitive to electrostatic discharge (ESD) events. Consider the MOS structure shown in Figure 14.27 where the gate oxide thickness t_{ox} is typically less than 100 Å = 10 nm. With a gate voltage of V_G applied, the oxide electric field E_{ox} V/cm can be estimated as

$$E_{ox} \approx \frac{V_G}{t_{ox}} \tag{14.65}$$

With V_G = 3 V and t_{ox} = 0.01 μm, the electric field in the insulating layer has a value of about $E_{ox} \approx 3\times10^6$ V/cm. The maximum electric field that

4 The study of friction is the basis for the field of **tribology**.

can be applied across a silicon dioxide insulator is typically on the order of about $E_{max} \approx 5\text{–}10 \times 10^6$ V/cm, depending upon the processing. If the electric field exceeds this value, breakdown occurs and current flows to the substrate. This destroys the insulating properties of the oxide and, hence, the transistor characteristics. Even a single bad FET renders the entire chip useless so this problem is taken quite seriously.

Static charge sources are encountered throughout the manufacturing process. In addition, the chips must be handled for everyday procedures such as shipping, unpacking, and board insertion. To account for this, modern CMOS chips are designed with circuits to protect the input transistors from excessive charge levels, and the devices are shipped in conductive foam. Even with these safeguards, it is prudent to follow the manufacturers' recommendations when handling chips. These are designed to reduce static charge levels by using grounded work planes and wrist straps that can drain excess charges to ground.

CMOS input protection circuits are designed to provide discharge paths away from the transistors. The network must be relatively transparent to the input signals and operate only when abnormal voltages are applied. The most common designs use resistor-diode networks to provide charge-dumping paths. Figure 14.28 summarizes the relevant characteristics of a pn junction diode structure. The anode and cathode sides of a diode are defined by the p- and n-regions shown in Figure 14.28(a). The voltage V is defined positive with the polarity shown: + on the anode and – on the cathode. Positive current I is defined to flow into the anode and out of the cathode. The circuit symbol for a diode is shown in Figure 14.28(b) with positive voltage and current. The I-V characteristics in Figure 14.28(c) show I as a function of V. A condition of $V > 0$ defines a forward bias where substantial current flows. Reversing the polarity so that the + side is on the cathode and the – side is on the anode defines a state

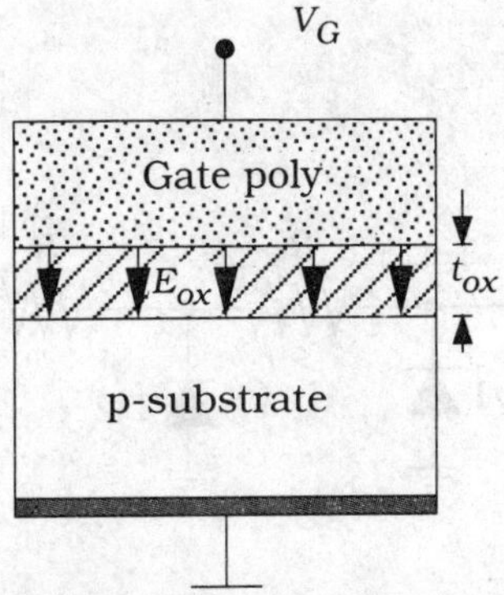

Figure 14.27 Oxide electric field in an MOS structure

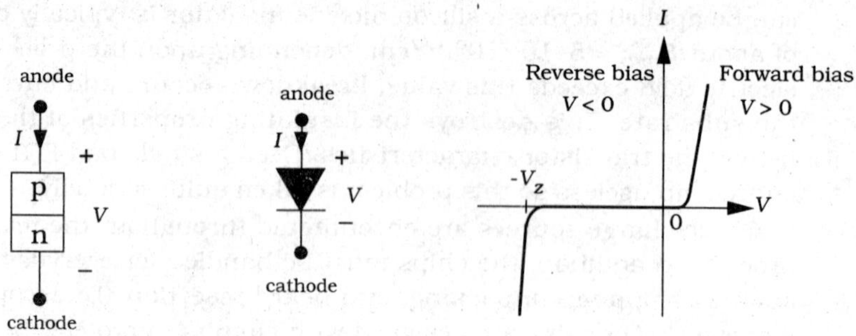

(a) Structure (b) Symbol (c) *I-V* curve

Figure 14.28 Diode characteristics

of reverse bias such that $V < 0$. A reverse-biased diode blocks the current flow for low voltages, but exhibits a breakdown at the **Zener voltage** V_Z when $V = -V_Z$ as shown in the plot. The value of V_Z is set by the doping levels, and is a characteristic of the junction. Junction breakdown is non-destructive: if the voltage is removed and then reapplied, the diode acts in a normal fashion.

Diodes can be used with resistors to form the input protection circuit shown in Figure 14.29. If an excessively large positive voltage is applied to the input pad, the resistors drop the voltage level along the input line. Under these conditions, the diodes D1 and D2 undergo breakdown and steer charge away from the gates of the input stage transistors. The diodes in a CMOS circuit typically have V_Z = 10–12 V or smaller depending upon the junction. At the physical design level, the resistors can be made using n-implanted layers in p-substrate as shown in Figure 14.30(a). The n+ region uses a serpentine pattern to obtain a square layout with a resistance of

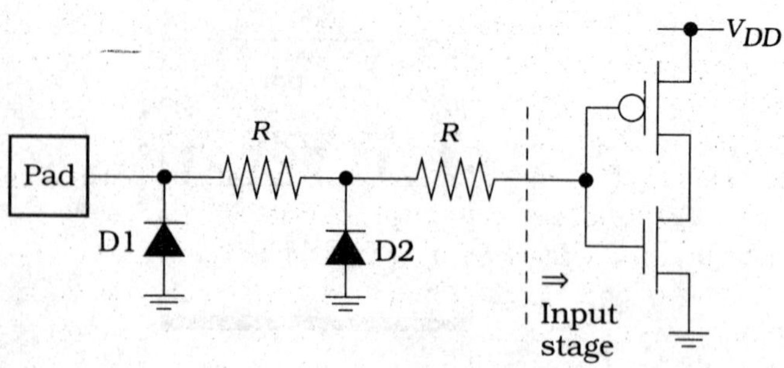

Figure 14.29 Input ESD protection circuit

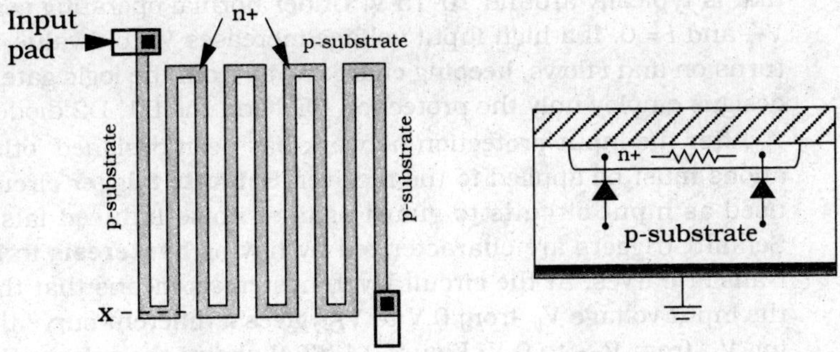

(a) Top view (b) Cross-section at x

Figure 14.30 Input resistor-diode structure

$$R = R_{s,n}n \tag{14.66}$$

where $n = (l/w)$ is the number of squares. In this type of geometry. A 90° corner does not count as full squares; electrostatic analysis shows that $n_{corner} \approx 0.69$ is a reasonable estimate. The reverse-biased pn junction diodes are automatically built into the structure as shown in Figure 14.30(b). Both the resistor and the diode are distributed structures, not discrete devices as in the circuit diagram.

Another input protection scheme is shown in Figure 14.31. This uses diodes D1 and D2, but additional diodes D3 and D4 have been added between the input and the power supply. The circuit keeps the DC voltage reaching the gate in the range $[-V_d, V_{DD} + V_d]$ where $V_d \approx 0.7$ V is the **on-voltage** of the diode, i.e., the value required to induce current flow. A special high-threshold voltage nFET has been included to provide additional charge drainage. This uses the thick isolation field oxide (FOX) as a gate insulator so that the weak field-effect gives a high threshold voltage V_{TF}

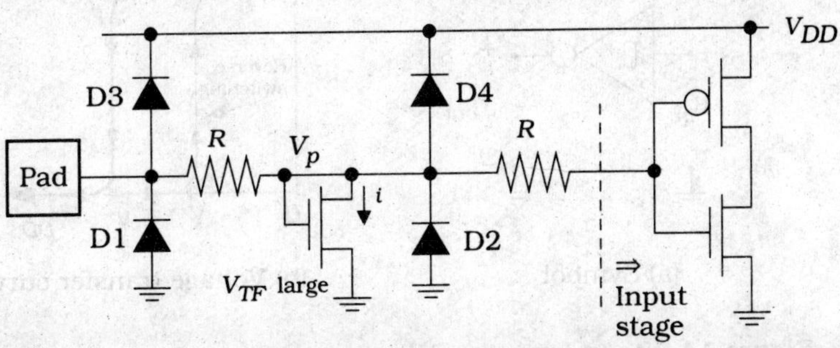

Figure 14.31 Alternate input protection circuits

that is typically around 10–15 V. Under normal operating conditions $V_P < V_{TF}$ and $i = 0$. If a high input voltage increases V_P to a value V_{TF}, the FET turns on and i flows, keeping charge away from the logic gate input. Some designs employ only the protection FET and the D1, D2 diodes.

After the input protection network has been designed, other considerations must be applied to the receiver. **Schmitt trigger** circuits are often used as input circuits to guard against noise-induced false switching. Schmitt triggers are characterized by having **hysteresis** in their voltage transfer curves. At the circuit level, hysteresis means that the increasing the input voltage V_{in} from 0 V to V_{DD} gives a different curve than decreasing V_{in} from V_{DD} to 0 V. Figure 14.32(a) shows the schematic symbol for an inverting Schmitt trigger gate; the icon in the middle of the triangle distinguishes it from a simple inverter. It shows the characteristic shape of the VTC illustrated in Figure 14.32(b). When V_{in} is increased from 0 V, V_{out} stays high at V_{DD} until V_{in} reaches the forward trigger voltage V^+; V_{out} then drops to 0 V. For reverse switching, V_{in} starts at V_{DD} and is decreased giving $V_{out} = 0$ V. The output remains low until V_{in} is decreased to the reverse trigger voltage V^-. For $V_{in} < V^-$, $V_{out} = V_{DD}$. Note that $V^- < V^+$ is required for a functional Schmitt trigger. The hysteresis insures that small fluctuations on the rising or falling edge of the input signal do not induce a false switching event.

A CMOS Schmitt trigger circuit is shown in Figure 14.33. This uses a mirror design where the nFETs determine V^+ and the pFETs determine the value of V^-. Consider the nFET circuits. M1 and M2 are in series and are both driven by the input voltage. When $V_{in} = 0$, $V_{out} = V_{DD}$ and M3 is on. Since the drain of M3 is connected to the power supply, it acts as a feedback path. As V_{in} is increased, it keeps M2 off even after M1 turns on. The analysis shows that the forward trigger voltage is given by

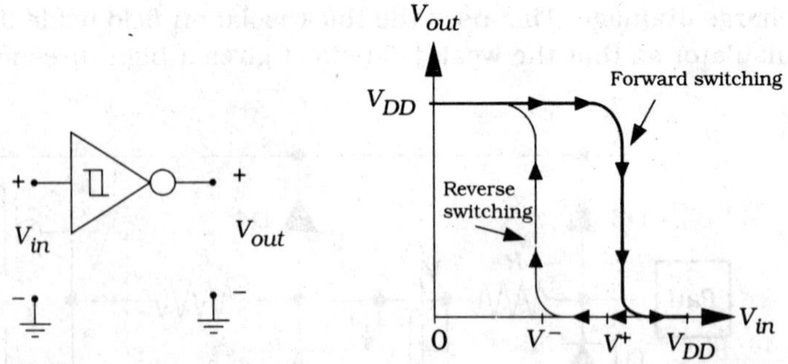

(a) Symbol (b) Voltage transfer curve

Figure 14.32 An inverting Schmitt trigger

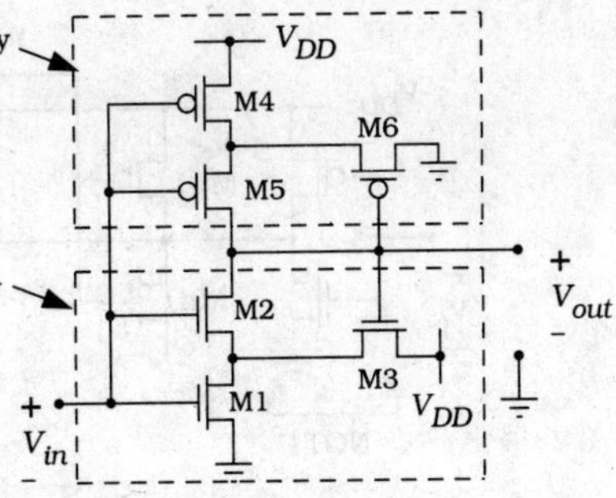

Figure 14.33 A mirror CMOS Schmitt trigger

$$V^{+} = \frac{V_{DD} + \sqrt{\frac{\beta_1}{\beta_3}}\, V_{Tn}}{1 + \sqrt{\frac{\beta_1}{\beta_3}}} \tag{14.67}$$

where the device ratio (β_1/β_3) is the design variable. Since both M1 and M3 are nFETs, this reduces to the ratio of device aspect ratios

$$\frac{\beta_1}{\beta_3} = \frac{(W/L)_1}{(W/L)_3} \tag{14.68}$$

In the same manner, M6 is the feedback transistor for the pFET group. The reverse trigger voltage is found from

$$V^{-} = \frac{\sqrt{\frac{\beta_4}{\beta_6}}\,(V_{DD} - |V_{Tp}|)}{1 + \sqrt{\frac{\beta_4}{\beta_6}}} \tag{14.69}$$

where

$$\frac{\beta_4}{\beta_6} = \frac{(W/L)_4}{(W/L)_6} \tag{14.70}$$

is the pFET ratio. The ratioed characteristic of the circuit can result in a design using relatively large FETs. This is because the series-connected transistors must be made large to compensate for the resistance while the switching voltages are set by the sizes selected for M3 and M6.

A non-inverting Schmitt trigger circuit is shown in Figure 14.34. Transistors Mp1 and Mn1 are used respectively as weak pull-up and pull-down devices that are controlled by the output voltage V_{out} through the feedback connection. Suppose that $V_{in} = 0$ V. The output of the first inverter (NOT1) is high, so that $V_{out} = 0$ V from the second inverter (NOT2).

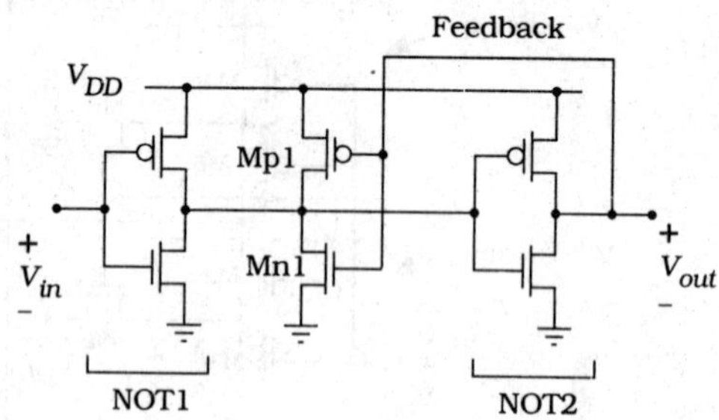

Figure 14.34 A non-inverting Schmitt trigger circuit

This biases Mp1 on and Mn1 off. If we increase V_{in}, the output node of NOT1 is held high by Mp1, which delays the switching. The aspect ratios of Mp1 and Mn1 must be small so as to still allow switching to take place.

14.6.2 Output Drivers

Output circuits must drive the pad capacitance along with the external load connected to the pin. Equation (14.48) may be used to find the capacitance C_{pad} of the pad, but the off-chip load varies with the application. A typical design value is around 80 pF, which is about the loading presented by a test probe. Since this is much larger than the femtofarad levels encountered in normal on-chip design, we must use large output transistors to maintain high speeds.

Example 14.3

Suppose that a 0.5 μm CMOS process uses I/O pads on a Metal3 layer that is characterized by a capacitance of 14 aF/μm^2. The unit aF stands for attofarad such that 1 aF = 10^{-18} F.

If we use pads that have dimensions of 75 μm × 75 μm, the pad capacitance is

$$C_{pad} = (14)(75^2) = 78.75 \quad \text{fF} \tag{14.71}$$

that must be added to the external capacitance contributions.

Off-chip driver design is a critical aspect of CMOS VLSI design. We have seen that on-chip switching times of simple logic gates are in the sub-nanosecond range. Transferring high on-chip data rates to the outside world is complicated by the large capacitance values. Scaled driver chains as discussed in Section 8.3 can be used to deal with the problem. Figure 14.35 shows a 4-stage output circuit that must drive a large picofarad-level load capacitor C_L. Theoretically, the delay minimization analysis specifies the number of stages N in the chain as

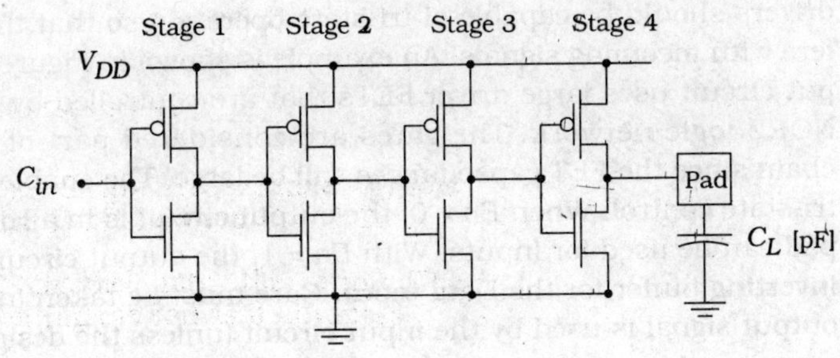

Figure 14.35 Scaled driver chain output circuit

$$N = \frac{\ln\left(\frac{C_L}{C_{in}}\right)}{\ln(S)} \tag{14.72}$$

where S is the scaling factor. However, large output capacitances may lead to large N-values and aspect ratios, so it is sometimes more practical to look at the output switching times requirements. In the example, the output characteristics are determined by Stage 4 since it drives C_L. If we specify values for the rise and fall times t_r and t_f, then the time constant expressions can be used to write

$$\begin{aligned} t_r &= 2.2R_pC_L \\ t_f &= 2.2R_nC_L \end{aligned} \tag{14.73}$$

for the fourth stage. Once the resistances are known, the aspect ratios for can be calculated from

$$\begin{aligned} \left(\frac{W}{L}\right)_{p,4} &= \frac{1}{k'_p(V_{DD}-|V_{Tp}|)R_p} \\ \left(\frac{W}{L}\right)_{n,4} &= \frac{1}{k'_n(V_{DD}-V_{Tn})R_n} \end{aligned} \tag{14.74}$$

The input capacitance into Stage 4 is

$$C_4 = C_{ox}[(WL)_{n4} + (WL)_{p4}] \tag{14.75}$$

which is taken to be the output capacitance seen by Stage 3. Each stage can be designed using the same rise and fall time values, working from the output side toward the interior circuitry. This is repeated until the input capacitance C_{in} is to a "normal" level, which determines the number of stages. Equalizing the stage delays is equivalent to using linear scaling.

A bi-directional pad provides circuitry for both input and output signals. The input circuits are identical to those described above. Output drivers should be capable of tri-state operation so that they do not interfere with incoming signals. An example is shown in Figure 14.36. The output circuit uses large driver FETs that are controlled by the NAND2 and NOR2 logic network. The gates are considered part of a scaled driver chain since the FET capacitances will be large. The enable signal *En* is the tri-state control. When *En* = 0, the output circuit is in a Hi-Z state and the pad can be used for inputs. With *En* = 1, the output circuit acts as a non-inverting buffer for the Data input. Care must be taken to insure that the output signal is used by the input circuit (unless the design requires it).

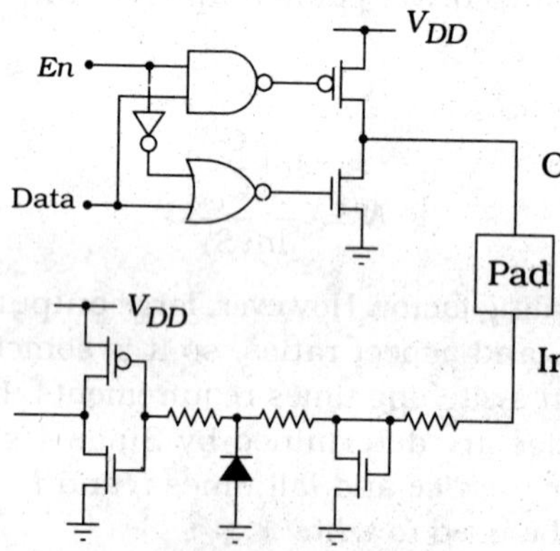

Figure 14.36 A bi-directional I/O circuit

14.7 Power Distribution and Consumption

The power supply values V_{DD} and V_{SS} are externally applied sources that enter the chip environment via two separate pads. The **power supply distribution grid** is a set of metal lines that provide the voltages to every part of the circuit. It must be designed with a geometrical patterning that allows a high packing density while providing the necessary current flow levels.

Two electrical problems tend to dominate the design of power supply grids. The first is electromigration where metal atoms are moved from one end to the other when the current density J [A/cm^2] is large. Since the total current I in amperes is related to the cross-sectional area tw by

$$I = J(tw) \tag{14.76}$$

we can increase the line width w to insure that J remains at acceptable levels. This is usually specified by a design rule table that provides the minimum width w for different ranges of currents. The second problem is the resistance R_{line} of the line. By Ohm's law, the voltage drop across the line is

$$V_{line} = IR_{line} \tag{14.77}$$

so that the voltage that reaches the circuits is altered by this amount. Linewidths, routing, and via placement all contribute to the total resistance between two points. Both problems can be solved by using wide lines, but this brute-force solution consumes excessive area.

Tree-like structures are the most common approach to designing the distribution scheme. The general idea is portrayed in Figure 14.37. The primary V_{DD} line is designed to have a width large enough to carry the total current I for the entire circuit. This is fed into branches, each carrying an average current of I_1 such that

$$I = N_1 I_1 \tag{14.78}$$

where N_1 is the number of secondary lines. Each secondary line feeds into ternary lines that carry a current I_3 and so on, until the individual logic cells are powered. The widths can be calculated once the values of the currents are known. Since digital CMOS circuits have current requirements that vary in time, average values are used to find the widths. Transient characteristics may require widening some lines.

Realistic power grids are designed by routing supply lines and then strapping them together to form a power mesh. Power buses from the pad are usually isolated from those applied to the cells in order to reduce noise problems. The idea can be understood from the drawing in Figure 14.38 where a signal line is placed between two VSS (ground) lines to provide electrical shielding. From the physical viewpoint, isolation is

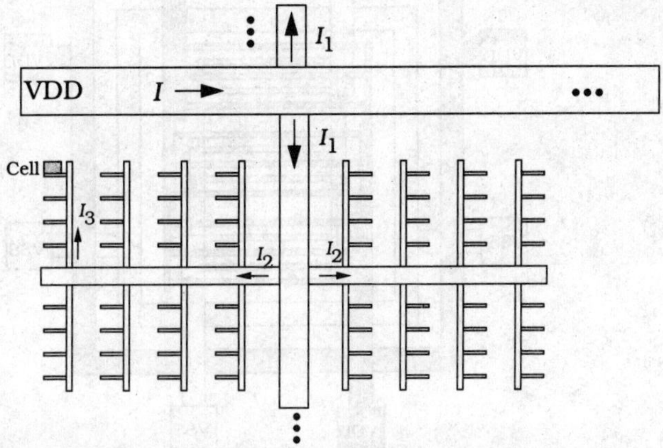

Figure 14.37 Linewidth sizing for power distribution

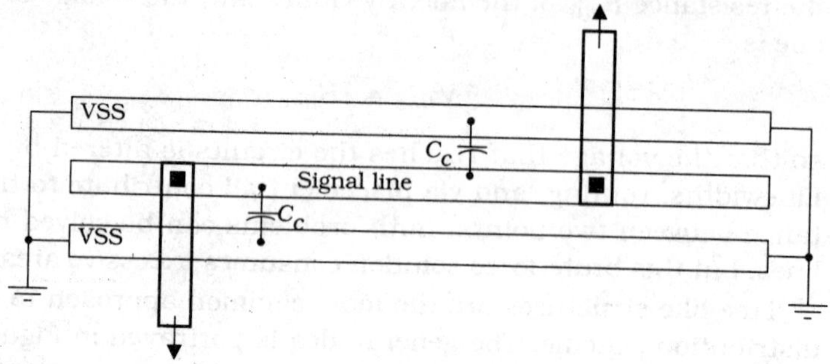

Figure 14.38 Isolation using VSS lines

achieved because coupling capacitances C_c are connected to (relatively) noise-free VSS lines. This can be applied to the 2-metal distribution model presented schematically in Figure 14.39. The power supply voltages VDD and VSS are used to form power rings around the interior circuit regions. The VSS ring acts as an electrical shield for the cells. Metal widths are sized according to the current draw from the logic circuits. Cells are placed between the smallest width VDD and VSS rails.

VLSI design libraries usually have pad frames cells that are used to power the chip, so the design centers on the interior regions. Electrical design rules for widths and loading levels can be applied on a line-by-line basis in critical circuits, but the sheer numbers involved usually mandate an algorithmic approach based on the architecture and circuit design.

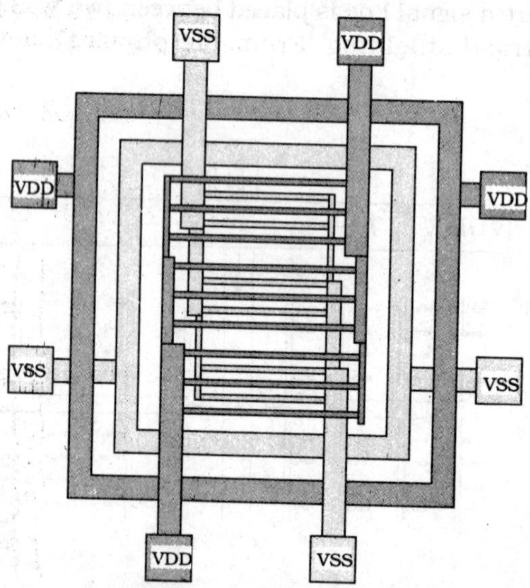

Figure 14.39 Power distribution scheme

14.7.1 Simultaneous Switching Noise

Simultaneous switching noise (SSN) is a fluctuation of the line voltage that is caused by many simultaneous switching events. The effect is also called **delta-I** (ΔI) and **ground-bounce** in the literature. SSN occurs when the current draw on a line changes very rapidly in time. Since high-speed digital networks are designed for fast switching, SSN problems are common in VLSI.

All conductive lines exhibit some **inductance** L that accounts for the magnetic energy storage property of the current flow. Although we usually ignore inductive effects in interconnects that transmit low current signals, they are important to power distribution lines that feed high-speed circuits. The basic problem can be understood from basic electromagnetic theory. An electrical current i produces a magnetic field which gives a magnetic flux Φ such that

$$\Phi = Li \tag{14.79}$$

The inductance L has units of Henrys [H] and has been introduced as a constant of proportionality. Figure 14.40(a) shows the physical model. Faraday's Law of Induction states that a time-varying magnetic flux $\Phi(t)$ induces a voltage $v(t)$ according to

$$v(t) = \frac{d\Phi}{dt} \tag{14.80}$$

Substituting for the flux gives the *I-V* relation

$$v(t) = L\frac{di}{dt} \tag{14.81}$$

for an inductor. The inductor symbol shown in Figure 14.40(b). This equation shows that the induced voltage is proportional to the time rate of change of the current di/dt.

The problems of SSN can be understood by reviewing the switching characteristics of the simple inverter illustrated in Figure 14.41. When the input voltage $V_{in} = 0$, the nFET is off and $i \approx 0$; the zero-current situation also holds for $V_{in} = V_{DD}$ since the pFET is off. Direct current flow from

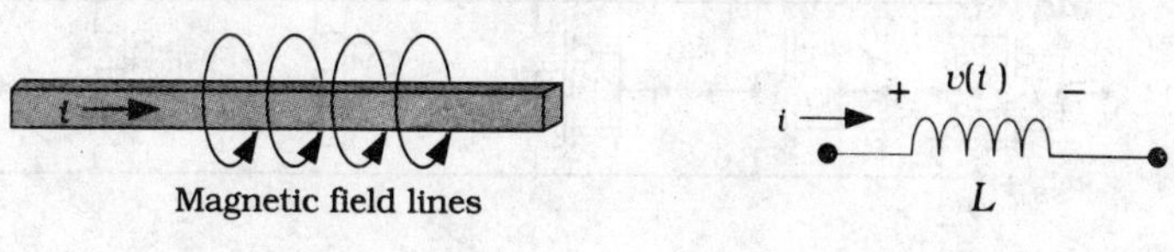

(a) Interconnect line (b) Equivalent element

Figure 14.40 Origin of line inductance

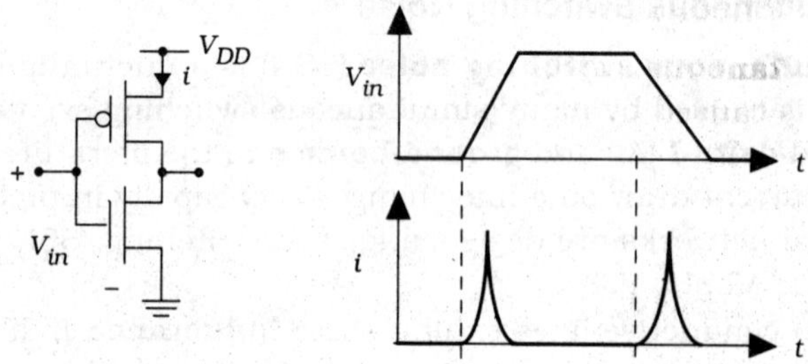

Figure 14.41 Current flow in an inverter circuit

V_{in} to ground occurs only when both transistors are conducting during a change in V_{in}. The plot shows (di/dt) has both positive and negative values. If the input voltage V_{in} can be changed at a faster rate (i.e., a steeper slope), the current derivatives increase.

Consider now the case shown in Figure 14.42 where a common power supply rail is used for a group of circuits. If every gate is switched at the same time, then the current i_1 through the parasitic inductance L_1 is computed from

$$i_1(t) = \sum_j i_j(t) \tag{14.82}$$

where the sum is over the entire collection of gates. The voltage across the inductor is

$$v_1(t) = L_1\left(\frac{di_i}{dt}\right) = L_1\sum_j\left(\frac{di_j}{dt}\right) \tag{14.83}$$

so that the actual value v_{dd} of the voltage applied to the pFETs is

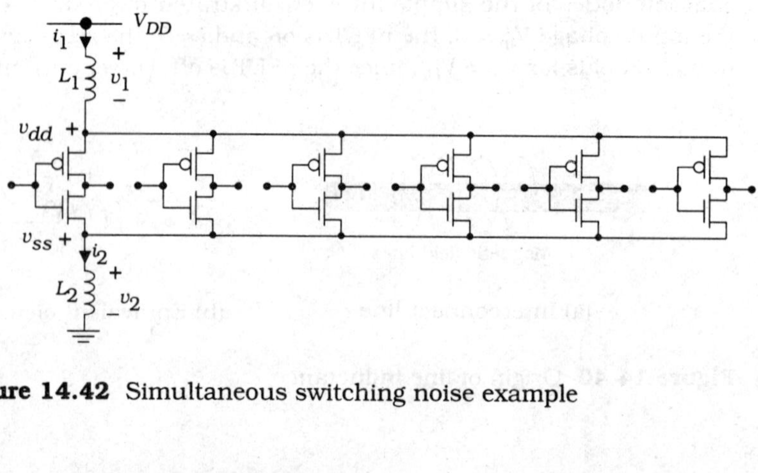

Figure 14.42 Simultaneous switching noise example

$$v_{dd}(t) = V_{DD} - v_1(t) \\ = V_{DD} - L_1 \sum_j \left(\frac{di_j}{dt}\right) \tag{14.84}$$

The voltage v_2 across L_2 is calculated in the same manner so that

$$v_{ss}(t) = v_2(t) = L_2 \sum_k \left(\frac{di_k}{dt}\right) \tag{14.85}$$

is the effective voltage applied to the nFET source terminals. In general, i_1 and i_2 are not equal because of current flow into, or out of, the output nodes of the gates. Note that the derivatives can be positive or negative depending upon the transition. A more complete analysis usually employs the circuit model in Figure 14.43. This adds drain and source resistances R_d and R_s, in addition to line capacitances C_1 and C_2 to obtain higher accuracy in the waveform simulations.

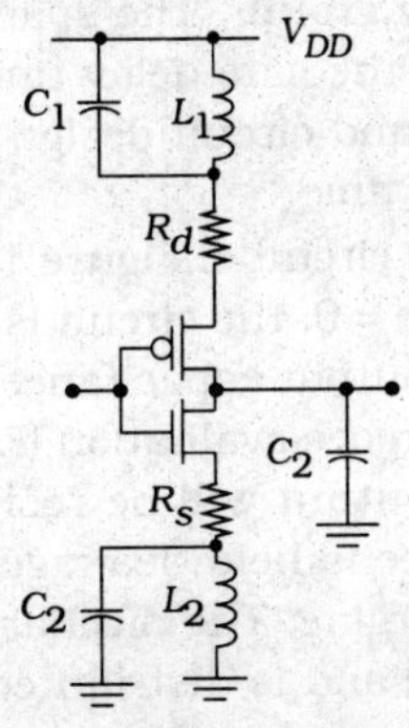

Figure 14.43 Complete circuit model

The main result of this analysis is that the instantaneous values of $v_{dd}(t)$ and $v_{ss}(t)$ establish the output voltages of the logic gates, not the DC values V_{DD} and 0 V. The modified values of the logic voltage levels may cause errors in the logic. Since the total current is a sum of the individual gate contributions, even moderate switching speeds will cause problems if many gates are switched at the same time. A histogram plot such as that shown in Figure 14.44 shows the number of gates that are switching at any given time. The data is usually weighted according to the peak current draw for each gate. Periods of high activity where a large number of gates are changing are of particular concern. The ideal situation is where the number of gates changing at any given time is a constant, since this would allow steady-state voltages to be achieved.

Predicting gate activity is a complicated task due to the complexity of

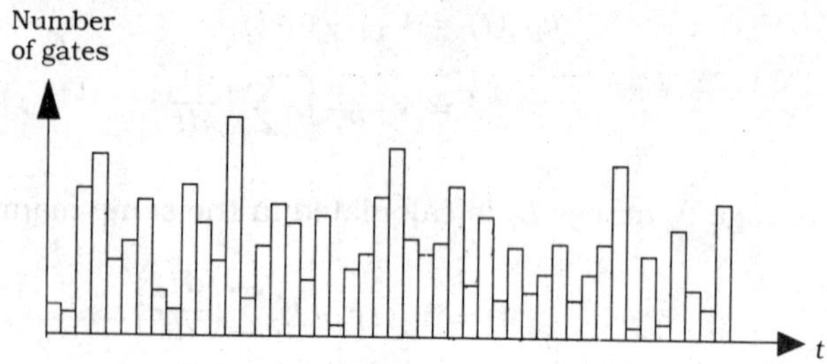

Figure 14.44 Gate switching distribution

modern VLSI systems. Moreover, the circuit design style and clocking have a direct effect on the results. The simplest case of a cascaded chain of random logic gates would be described by the current distribution shown in Figure 14.45. This represents the case where the inputs to the gate 1 logic group are switched at time t_0 and ripple down the chain causing a switch in every group. The spacing in between the current spikes can be estimated as the gate delay times. This type of consideration can be used in the logic and circuit design stage to equalize the current draw level as a function of time.

The dynamic logic circuit of Figure 14.46 is more predictable due to the clock control. When $\phi = 0$, the circuit is in precharge (P) and charging current i_{ch} flows to the output capacitance C_{out}. When the clock changes to $\phi = 1$, the circuit undergoes evaluation (E). If a logic 0 is produced, i_{dis} flows to ground and the output will be recharged during the next precharge interval. If the charge is held, leakage will occur and the stage will still require partial recharging. The discharge current flow takes place during the evaluation phase and is distributed according to the location of each stage in the logic cascade.

SSN levels are also dependent on the packaging and the wiring technique used to connect the die to the pins. In multi-chip modules (MCMs), SSN can cause unwanted interactions among different chips. These and other related problems are the center of many research projects.

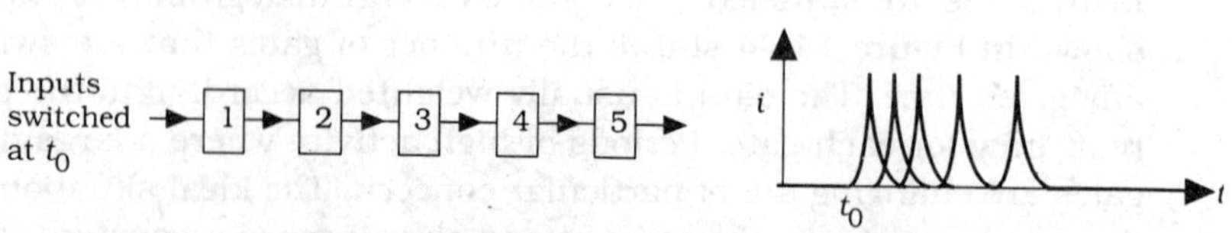

Figure 14.45 Switching current in a random logic chain

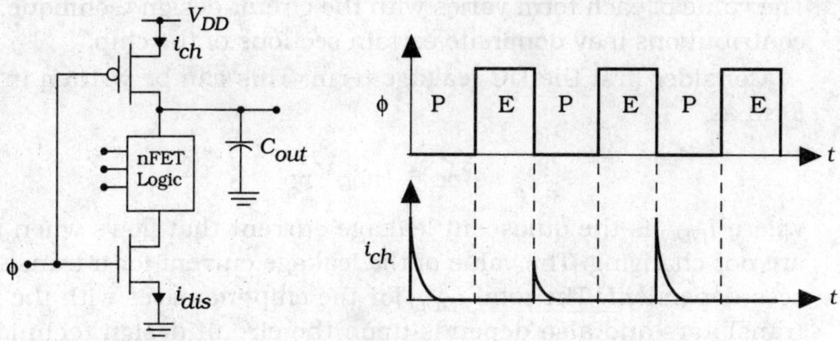

Figure 14.46 Periodic charging current in a dynamic logic gate

14.8 Low-Power Design Considerations

The overall power dissipation P [W] of the chip is critically important in modern VLSI. If a battery power supply is used, then P determines the operating time before a recharge is needed. Even on a desktop system, the power dissipation must be kept low to insure that the silicon doesn't melt (the worst-case situation) and the system cooling scheme is sufficient. Low-power design techniques have been developed at both the circuit and system level. Substantial amounts of research are devoted to studying the problem and solutions.

There are three main sources of power dissipation in a digital CMOS circuit.

- DC power P_{DC} that is due to direct conduction paths from V_{DD} to ground when inputs are stable. Leakage currents are the cause of this component in standard static CMOS logic circuits.
- Switching power P_{sw} that is dissipated when an input change causes the power supply to have a direct current flow path to ground through the transistors. This occurs during the transition portion of a voltage transfer curve (VTC) and was discussed in the previous section as the origin of SSN problems.
- Dynamic switching power P_{dyn} due to charging and discharging capacitive nodes. This is estimated from the general formula

$$P_{dyn} = aCV^2 f \tag{14.86}$$

where C is the capacitance in farads, a is the activity coefficient, V is the voltage swing, and f the signal frequency.

The instantaneous total power dissipation is the sum

$$P = P_{DC} + P_{sw} + P_{dyn} \tag{14.87}$$

The value of each term varies with the circuit design technique, and some contributions may dominate certain sections of the chip.

Consider first the DC leakage term. This can be written in simplified form as

$$P_{DC} = I_{DDQ} V_{DD} \tag{14.88}$$

where I_{DDQ} is the quiescent leakage current that flows when the inputs are not changing. The value of the leakage current for a transistor is process-dependent. The total I_{DDQ} for the chip increases with the number of transistors and also depends upon the circuit design technique. Static CMOS logic gates exhibit the smallest quiescent leakage currents with chip values usually less than 10 μA or so. The resulting power dissipation is on the order of a few tens of microwatts (μW). Although it can be larger in certain designs, it is usually the smallest among the three.

Switching power P_{sw} is a consequence of a gate input signal transition causing a direct current flow path from V_{DD} to ground and is the origin of SSN. It occurs every time the output voltage undergoes a voltage transition, and originates with the circuit design. Static logic gates always dissipate switching power since the conduction path cannot be eliminated. A simple estimate is

$$P_{sw} = \langle I_{sw} \rangle V_{DD} \tag{14.89}$$

where $\langle I_{sw} \rangle$ is the average DC current flow. The contribution from an isolated gate varies with the transistor aspect ratios, since (W/L) determines the current flow level through a FET. The actual magnitude depends upon the shape of the input waveform, making it difficult to calculate using closed-form equations. Circuit simulations are the most accurate.

The dynamic power dissipation is usually considered to be the most difficult to deal with. The general expression

$$P_{dyn} = aCV^2 f \tag{14.90}$$

shows that P_{dyn} increases proportionately with the signal switching frequency f so that it grows with the speed of the circuit. One approach to decreasing the magnitude of this term is to reduce the power supply voltage V_{DD} since it is the maximum (DC) value for V. This also reduces the values of the other contributions. Processor core voltages are currently below 2 V, and the push is on for even lower operating voltages. A reduced power supply voltage is also advantageous in battery-operated units.

Although this may seem like a simple technique, it introduces problems at the circuit level that leads to slower switching speed. This, of course, defeats the purpose of decreasing V_{DD} in the first place. To understand this statement, recall that a non-saturated FET has a current of

$$I_D = \frac{\beta}{2}[2(V_{GS} - V_T)V_{DS} - V_{DS}^2] \quad (14.91)$$

Since the highest voltage in the circuit is V_{DD}, reducing it implies that I_D will also decrease. Reducing I_D implies that it will take longer to charge the output capacitance, increasing both the rise and fall times. This slows down the gate switching speed. To compensate for this effect we can increase the device transconductance term

$$\beta = \mu_n C_{ox}\left(\frac{W}{L}\right) \quad (14.92)$$

Since

$$C_{ox} = \frac{\varepsilon_{ox}}{t_{ox}} \quad (14.93)$$

shrinking the gate oxide thickness t_{ox} increases β so improved processing helps. Otherwise, we must increase the channel width W to maintain the speed.

Many novel and unique approaches for reducing the power dissipation of VLSI chips have been published in the literature. The problem itself is usually tackled at both the circuit design and the architectural level. The interested reader is referred to the literature. Several books on the subject have been listed in the reference section.

14.9 References for Further Study

[1] Abdellatif Bellaouar and Mohamed I. Elmasry, **Low-Power Digital VLSI Design**, Kluwer Academic Publishers, Norwell, MA, 1995.

[2] Anantha P. Chandrakasan and Robert W. Brodersen, **Low Power Digital CMOS Design**, Kluwer Academic Publishers, Norwell, MA, 1995.

[3] Dan Clein, **CMOS IC Layout**, Newnes, Woburn, MA, 2000.

[4] Sabih H. Gerez, **Algorithms for VLSI Design Automation**, John Wiley & Sons, Chichester, England, 1999.

[5] Bryan Preas and Michael Lorenzetti (eds.), **Physical Design Automation of VLSI Systems**, Benjamin-Cummings Publishing Company, Menlo Park, CA, 1988.

[6] Jan M. Rabaey, **Digital Integrated Circuits**, Prentice Hall, Upper Saddle River, NJ, 1996.

[7] Jan M. Rabaey and Massoud Pedram, **Low Power Design Methodologies**, Kluwer Academic Publishers, Norwell, MA, 1996.

[8] Michael Reed and Ron Rohrer, **Applied Introductory Circuit Analysis**, Prentice Hall, Upper Saddle River, NJ, 1999.

[9] Kaushik Roy and Sharat C. Prasad, **Low-Power CMOS VLSI Circuit Design**, John Wiley & Sons, New York, 2000.

[10] T. Sakurai and K. Tamaru, "Simple Formulas for Two- and Three-Dimensional Capacitances," *IEEE Trans. Electron Devices*, vol. ED-30, no. 2, pp. 183-185, Feb. 1983.

[11] M. Sarrafzadeh and C. K. Wong, **An Introduction to VLSI Physical Design**, McGraw-Hill, New York, 1996.

[12] Ramesh Senthinathan and John L. Prince, **Simultaneous Switching Noise of CMOS Devices and Systems**, Kluwer Academic Press, Norwell, MA, 1994.

[13] Naved Sherwani, **Algorithms for VLSI Physical Design Automation**, Kluwer Academic Publishers, Norwell, MA, 1993.

[14] John P. Uyemura, **CMOS Logic Circuit Design**, Kluwer Academic Publishers, Norwell, MA, 1999.

[15] M. Michael Vai, **VLSI Design**, CRC Press, Boca Raton, FL, 2001.

[16] Gary K. Yeep, **Practical Lower Power Digital VLSI Design**, Kluwer Academic Publishers, Norwell, MA, 1998.

14.10 Problems

[14.1] Consider an interconnect with the geometry shown in Figure 14.3 with T_{ox} = 1.10 μm, w = 0.5 μm, and t = 0.90 μm.

(a) Calculate the capacitance per unit length c in [pF/cm] using the simple parallel-plate formula that ignores the fringing capacitance.

(b) Find the value of c predicted by the empirical expression that includes fringing.

(c) Assuming that the result in part (b) is correct, find the percentage error incurred if fringing is neglected.

(d) The interconnect line has a sheet resistance of R_s = 0.08 Ω. Find the values of R_{line} and C_{line} if the line is 100 μm long.

[14.2] An interconnect has the geometry shown in Figure 14.3 with T_{ox} = 0.90 μm, w = 0.35 μm, and t = 1.10 μm.

(a) Find the value of c predicted by the empirical expression that includes fringing.

(b) The interconnect line has a sheet resistance of R_s = 0.04 Ω. Find the values of R_{line} and C_{line} if the line is 48 μm long.

(c) Construct an m = 7 RC ladder equivalent for the line, then use the model to determine the time constant. Compare this with the time constant obtained using the simpler formula $\tau = R_{line} C_{line}$.

[14.3] Consider a CMOS process that has 4 metal layers with thickness shown in Figure P14.1. Find the capacitance per unit length c [pF/cm] (metal-to-substrate capacitance) for each layer. Assume that CMP is performed after every oxide layer is deposited.

[14.4] Calculate the overlap capacitance per μm^2 for the M1-M2 layers shown in Figure P14.1. Repeat for an overlap of M3-M1 patterns.

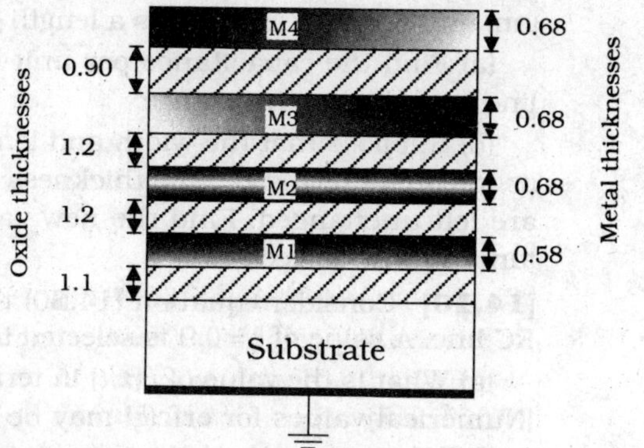

Figure P14.1

[14.5] An interconnect has the geometry shown in Figure 14.3. The values for the important parameters are T_{ox} = 10,000 Å, w = 0.35 μm, t = 0.85 μm, R_s = 0.008 Ω, and l = 122 μm.

(a) Find the values of R_{line} and C_{line}. Always include fringing effects in your calculations.

(b) Create RC ladder equivalents for m = 2 and m = 6 segments. Find the difference in the time constants for the two. Then compare your result for m = 6 with equation (14.19).

[14.6] An interconnect is described by w = 0.35 μm, T_{ox} = 1.20 μm, t = 0.95 μm, and R_s = 0.008 Ω.

(a) Calculate the values of r [Ω/cm] and c [pF/cm].

(b) Suppose that ξ = 0.05 in equation (14.33). Find the equation for the interconnect time delay.

(c) Calculate the delays in ps for line lengths of 100 μm, 200 μm, and 300 μm.

[14.7] Two lines on an interconnect level are separated a spacing of S = 0.50 μm. Each individual line has w = 0.35 μm, T_{ox} = 1.1 μm, and t = 1 μm.

(a) Calculate the coupling capacitance per unit length c_c.

(b) Find the coupling capacitance if the interaction length is 20 μm. Repeat for a 30 μm interaction length.

[14.8] Consider two interconnect lines that are separated a spacing of S = 0.40 μm. Each individual line has w = 0.25 μm, T_{ox} = 1.2 μm, and t = 0.85 μm.

(a) Calculate the self-capacitance per unit length c for a lines.

(b) Calculate the coupling capacitance per unit length c_c between the two lines.

(c) Suppose that the lines are both 18 μm long. Find the total capacitance seen looking into one of the lines.

[14.9] An interconnect is described by $w = 0.4$ μm, $T_{ox} = 1.0$ μm, $t = 0.84$ μm, and $R_s = 0.005$ Ω. It has a length of 50 μm.

(a) Find the capacitance per unit length, the total capacitance of the line, and the line resistance.

(b) Suppose that the width and length are scaled by a factor of $s = 1.5$ as in equation (14.50). The thickness values of the material and the oxide are left unchanged. Find the new values of line capacitance and resistance for the scaled line.

[14.10] Consider equation (14.30) that describes $v(z,t)$ on a distributed RC line. A value of $\xi = 0.9$ is selected in the erfc(ξ) function.

(a) What is the value of $v(z,t)$ in terms of the power supply voltage V_{DD}? [Numerical values for erfc(ξ) may be found in mathematical tables or in most PC computational programs.]

(b) The line has parameters $w = 0.5$ μm, $T_{ox} = 0.9$ μm, $t = 0.90$ μm, and $R_s = 0.04$ Ω. Find the value of B in the delay equation $\tau = Bl^2$ that is implied by this choice for ξ.

[14.11] The Schmitt trigger circuit shown in Figure 14.33 uses a power supply of $V_{DD} = 3.3$ V, and the transistors are defined by the threshold voltage values of $V_{Tn} = 0.7$ V and $|V_{Tp}| = 0.8$V. Calculate V^+ and V^- if the device ratios are $(\beta_1/\beta_3) = 6$ and $(\beta_4/\beta_6) = 4$.

[14.12] Consider the Schmitt trigger circuit shown in Figure 14.33 that is built in a process where $V_{DD} = 5$ V, $V_{Tn} = 0.7$ V, and $|V_{Tp}| = 0.8$V. Design the circuit to give $V^+ = 3.9$ V and $V^- = 1.2$ V.

VLSI Clocking and System Design

15

Synchronous design employs clocking signals to coordinate the movement of data through the system. The overall speed of a VLSI network is often determined by the limitations imposed by the clocking. In this chapter we will examine synchronous logic design in a CMOS environment.

15.1 Clocked Flip-flops

The simplest CMOS clocking scheme is based on a single clock signal $\phi(t)$ that has a period T sec and a frequency $f = (1/T)$ Hz. An ideal clocking waveform is shown in Figure 15.1. The amplitude varies from 0 to 1 with a corresponding voltage range of $[0, V_{DD}]$. In classical digital design, data flow is synchronized by using the clock to control the loading of latches or flip-flops. Figure 15.2 illustrates this for a D-type flip-flop. The data bit D is loaded into the DFF only on a rising clock edge. If the rising edge occurs at a time t_0, then the output has this value after a time delay t_{ff}:

$$Q(t_0 + t_{ff}) = D(t_0) \tag{15.1}$$

The data capture occurs again at the next rising edge (t_1 in the graph) and

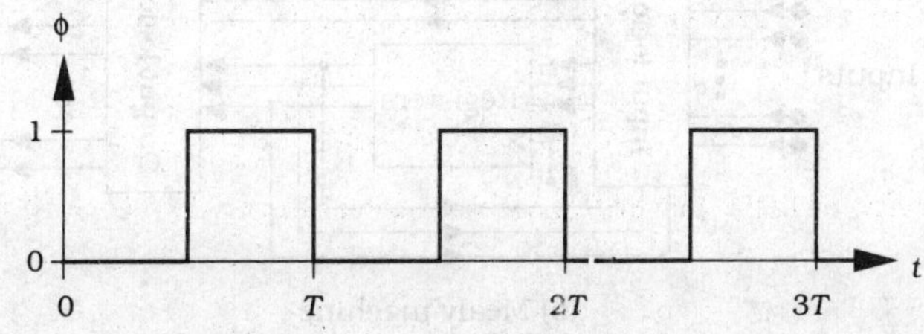

Figure 15.1 Ideal clocking signal

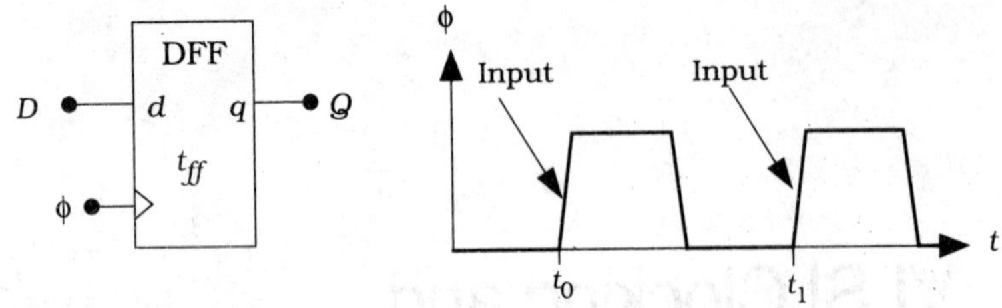

Figure 15.2 Timing in a DFF

on every 0 to 1 clock transition thereafter. The limiting circuit factor is the DFF delay time t_{ff} that is determined by the electronics and the load. Decreasing t_{ff} allows for a higher frequency clock, which increases the data flow rate. The same comment applies to a negative edge-triggered DFF. Other types of flip-flops, such as the JK, can be built in CMOS, but are rarely used in high-density design because they are slower and consume more area.

15.1.1 Classical State Machines

Clocked flip-flops provide the basis for classical sequential logic networks. Two models for state machines that use single-clock timing are shown in Figure 15.3. The Moore machine in Figure 15.3(a) feeds inputs into the Input logic block that consists of combinational logic with outputs that are fed into a register. The register itself is m bits wide and can be built

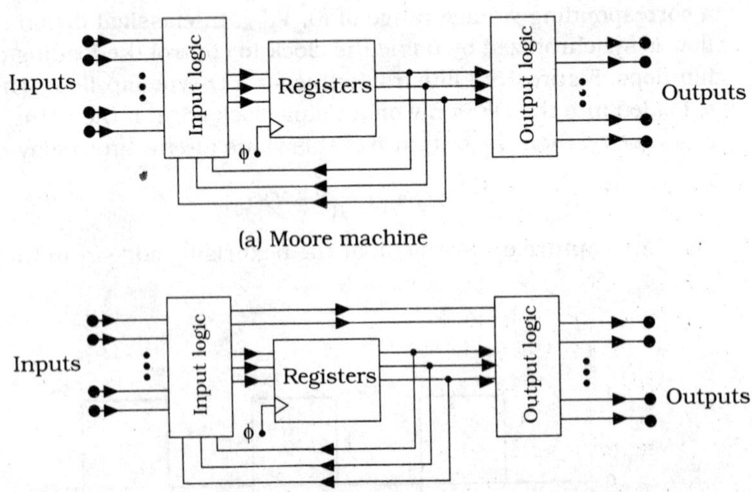

(a) Moore machine

(b) Mealy machine

Figure 15.3 Moore and Mealy state machines

using flip-flops; this results in a machine with 2^m states. The outputs of the state registers are fed to the output logic block to produce the Output data. They are also fed back to the Input logic block. Since the stored data is delayed by one clock cycle, the present state of the machine affects the next state. The Mealy machine in Figure 15.3(b) has the same basic structure, but allows the present inputs to have an immediate effect on the outputs.

The generalized architecture contained in the Huffman model shown in Figure 15.4 contains both the Moore and Mealy models as special cases. The externally applied Primary inputs $x_0, x_1, \ldots, x_n$ are fed into the Combinational logic block along with the Secondary inputs $q_0, \ldots, q_k$ that originate from the Memory unit. The Primary outputs $f_0, f_1, \ldots, f_m$ are taken from the logic block. The Secondary outputs $d_0, d_0, \ldots, d_k$ are used as inputs into the clocked DFFs; these act as Secondary inputs during the next clock cycle.

Sequential logic circuits can be designed and implemented in VLSI using any of these classical models. This is the most common approach for designing cell-based ASICs at the logic level where the engineer works from state diagrams and logic circuits. CAD tools are used to synthesize the circuits, and place-and-route provides floorplanning and interconnect wiring. Details of the silicon are invisible to the designer. Indeed, sophisticated toolsets allow one to design ASICs without requiring any knowledge of physical design. This makes them very useful for quick turnaround prototypes and may be used for volume production if the speeds are acceptable.

FPGAs designs are also heavily based on classical state machine the-

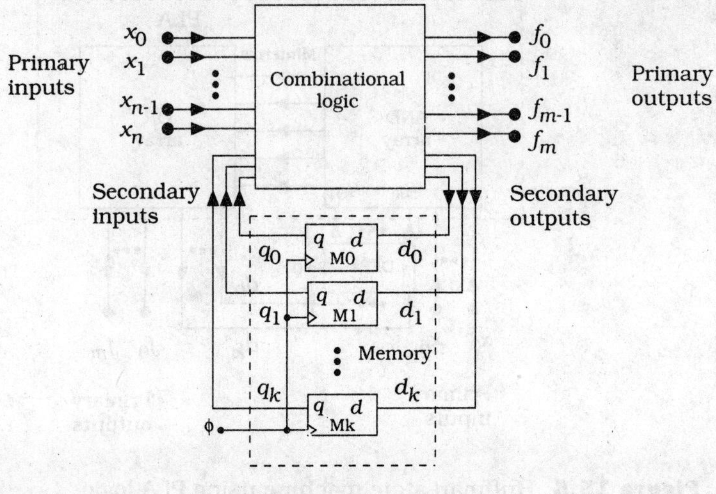

Figure 15.4 Huffman model of a state machine

ory, but usually require a deeper understanding of VLSI concepts. Combinational logic circuits are achieved using circuit styles that vary with the vendor. Individual gates, PLAs, programmable logic devices (PLDs), and groups of multiplexors are common. Programming is achieved with EPROMs, fuses, and SRAM arrays, and some contain lookup tables (LUTs) to aid in the design. While the chip level design is automated, it is possible for the designer to manually perform partitioning, cell placement, and routing to speed up the circuit. Since the specifics of both ASIC and FPGAs design are vendor-dependent, the interested reader is directed to the device-specific literature for the details. We will pursue a more general view of the design problem that illustrates important aspects of the circuit and physical design levels.

Classical state machines are used in large VLSI chips of all types. The design approach itself is based on the standard flow, but is usually more closely linked to the physical design aspects. In cell-based chip-level VLSI, the flip-flops would be instanced from the library as predefined modules; the logic networks can be designed using primitive gates or logic arrays. The critical aspects of the physical design center around gate and interconnect delays, and clocking.

Let us study a structure of a state machine that illustrates the main ideas using the Huffman model as a basis. Figure 15.5 illustrates a block diagram that uses an AND-OR programmable logic array to perform combinational logic. The AND-plane of the PLA can be programmed to produce minterms m_r using the inputs $x_0, \ldots, x_n$ and $q_0, \ldots, q_k$. The OR-plane gives SOP outputs of

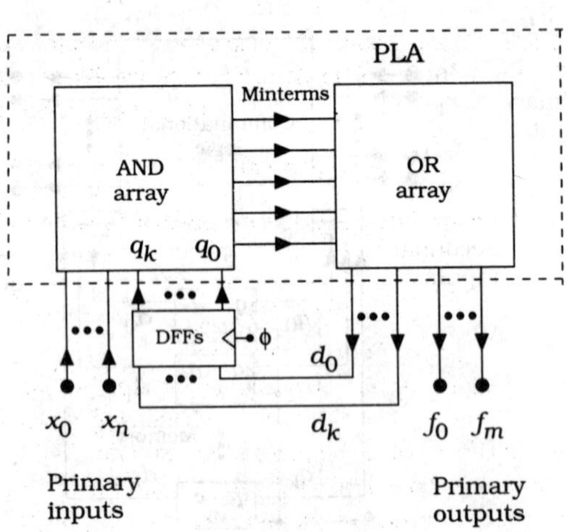

Figure 15.5 Huffman state machine using PLA logic

$$\sum_r m_r \tag{15.2}$$

with some used as primary outputs f_0, ..., f_m and others fed back as secondary outputs d_0, ..., d_k that are stored in the DFFs for use as secondary inputs during the next clock cycle. The details of the logic design are identical to that encountered in standard digital systems; VLSI considerations enter the picture in the PLA. The clock period T must be large enough to allow completion of the logic cycle. This condition can be expressed as

$$T > t_{ff} + t_d + t_{su} \tag{15.3}$$

where

- t_{ff} is the delay time from input to output of the flip-flop,
- t_d is the logic delay time through the PLA,
- t_{su} is the "setup time" of the flip-flop, i.e., the time the input must be stable before being latched into the DFF.

Since the logic array may require relatively long interconnects, the circuit may not be fast enough to meet the desired speed. Random combinational logic will usually produce a faster circuit by reducing t_d, but it takes longer to design. Another approach is to change flip-flops to reduce the values of t_{ff} and t_d.

This simple circuit example illustrates that VLSI can easily be adapted to classical state machine topologies that are based on flip-flop storage. CMOS circuits, however, can provide more complex clocking and timing strategies than are possible in other technologies. The full power of synchronous logic in VLSI becomes apparent when these are studied in more detail.

15.2 CMOS Clocking Styles

CMOS admits to a wide variety of clocking styles. Some are quite general, while others are based on the circuit design techniques. In this section we will examine some of the most common approaches used to synchronize data flow in a VLSI environment.

15.2.1 Clocked Logic Cascades

The simplest is based on using a single-clock ϕ, either by itself (a single-phase design) or in conjunction the complementary signal $\overline{\phi}$ (a dual-phase system) as introduced previously in Chapter 2. Figure 15.6 shows the idealized waveforms for the pair. In terms of voltages, we have

$$V_{\overline{\phi}} = V_{DD} - V_{\phi} \tag{15.4}$$

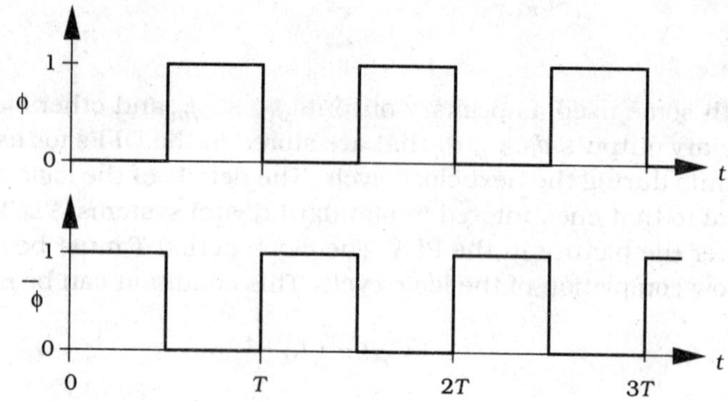

Figure 15.6 Complementary clocks

since both are assumed to be full-rail signals that range from 0 V to V_{DD}. A non-overlapping pair would always satisfy

$$\phi \cdot \bar{\phi} = 0 \tag{15.5}$$

for all times *t*, but this is not entirely true for the complementary signals due to the necessary overlap during the rise/fall times.

Applying the clock signals directly to FETs gives a simple approach for controlling data flow. The three main clocking elements are shown in Figure 15.7. Either of the single-polarity transistor switches in Figure 15.7(a) can be used, but the nFET is generally chosen over the pFET because of its superior conduction characteristics. One problem with using the nFET as a clocking device is that the output is restricted to the voltage range [0, V_{max}] where

$$V_{max} = V_{DD} - V_{Tn} \tag{15.6}$$

due to the threshold loss. In addition to passing a weak logic 1, the tran-

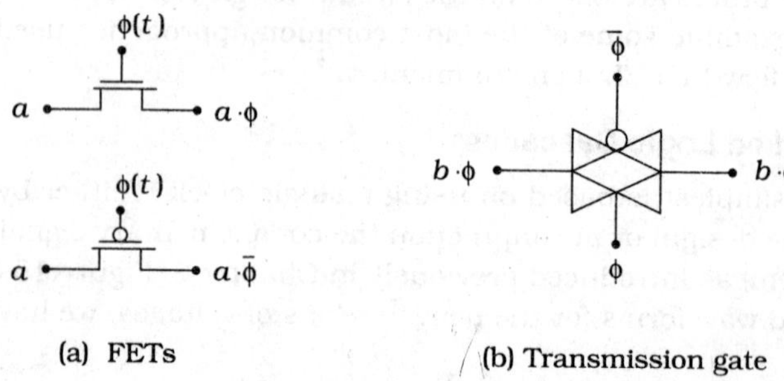

Figure 15.7 Clock-controlled transistors

sition is quite slow. The transmission gate in Figure 15.7(b) passes the full voltage range [0, V_{DD}] with fast switching for both logic 0 and logic 1 inputs. However, a TG requires two transistors (an nFET and a pFET) and the associated parallel wiring increases routing problems. In modern high-speed design, single nFETs tend to be used over TGs due to their simplicity; logic 1 transmission problems are dealt with by careful design of the peripheral circuitry. Since clocked TG circuits have already been studied in Chapter 2, we will concentrate on clocked nFETs here. Every circuit can be redesigned to accommodate TGs if desired.

Consider the logic cascade in Figure 15.8. Data transfer between the combinational logic (C/L) units is through clocked nFETs with ϕ controlling the inputs into Units 1 and 3 while $\overline{\phi}$ controls the bits into Unit 2. Synchronous flow can be seen by starting with a clock of $\phi = 1$ ($\overline{\phi} = 0$). The inputs a_0, ..., a_3 are admitted into Unit 1 and are available at the output of the logic block after the characteristic delay time. When the clock changes to $\phi = 0$ ($\overline{\phi} = 1$), b_0, ..., b_3 is transferred into Unit 2. On the next half-clock cycle ϕ rises to 1 (so that $\overline{\phi} = 0$) and allows c_0, ..., c_3 to enter Unit 3. The output d_0, ..., d_3 of the cascade is available after the logic delay. This clocking style transfers data from one unit to the next unit every half-clock cycle and the data flow is intuitively obvious.

Timing Circles and Clock Skew

Timing circles are simple constructs that can be useful for visualizing data transfer. The timing circle for the clocked cascade is shown in Figure 15.9. Both clock signals ϕ and $\overline{\phi}$ have a period T and are assumed to have a logic 1 value for one-half of the period ($T/2$) and a logic 0 value for one-half the period. This defines what is known as a 50% duty cycle. Since the clocks repeat every period, we construct a timing circle with one complete revolution representing T, and then label each portion of the circle with the clock(s) that is 1 during that time period. The pair (ϕ, $\overline{\phi}$) gives the timing circle shown in the drawing. It is useful to introduce terminology to describe the information contained in the circle. The phrase "during ϕ" will be interpreted to mean the time when $\phi = 1$. Similarly, we will under-

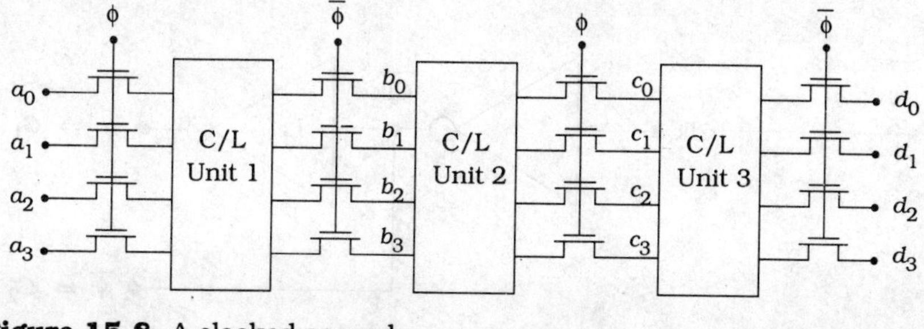

Figure 15.8 A clocked cascade

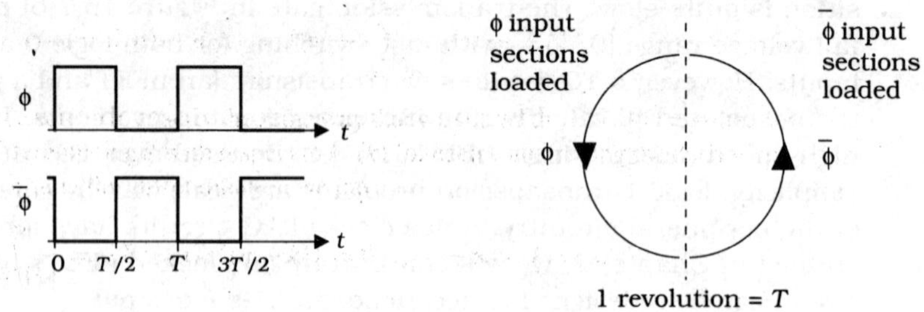

Figure 15.9 Timing circle for a single-clock, dual-phase cascade

stand that "during $\overline{\phi}$" implies the time interval when $\overline{\phi} = 1$. When applied to the clocked cascade in Figure 15.8, the timing circle shows the data flow into each type of unit. During ϕ, inputs are accepted into the odd-numbered units which are called the "ϕ-input sections" in the drawing. Similarly, the even-numbered unit 2 (the "$\overline{\phi}$-section") accepts inputs during $\overline{\phi}$. Alternating ϕ and $\overline{\phi}$ sections in a cascade thus results in a synchronous movement of data through a system.

The simplicity of a clocked cascade makes it attractive as a basic design methodology. When we make the transition to the circuit and physical design levels, however, several complicating factors must be dealt with. A critical problem that affects all clocked systems is that of **clock skew**. Clock skew is where the timing of a clock is out of phase with the system reference. It can originate from different sources, and limits the clock frequency. In a synchronous system, this is equivalent to limiting the data flow rate and the overall speed.

Let us examine the clock skew that originates in the clock generation circuits. The circuit in Figure 15.10 uses inverters to generate ϕ and $\overline{\phi}$ from the clocking signal Clk. If the line capacitances C_1 and C_2 are equal, then ϕ will be slightly delayed from $\overline{\phi}$ by the inverter delay t_d. This defines the skew time $t_s = t_d$ shown in the clocking waveforms of Figure 15.11. Note that the overlap is increased by the clock skew. This causes problems in data synchronization and flow, so much effort is directed toward

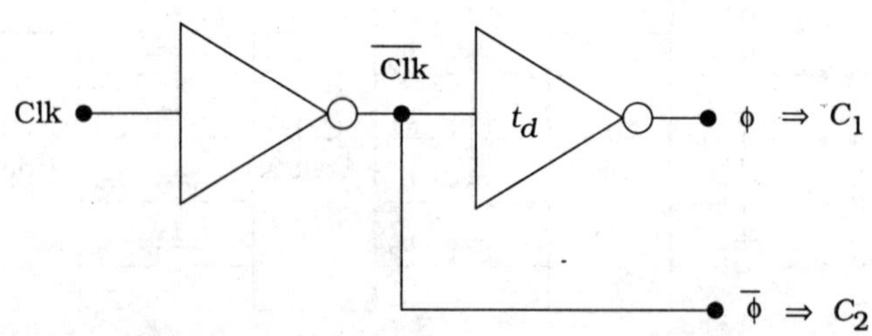

Figure 15.10 Clock generation circuit

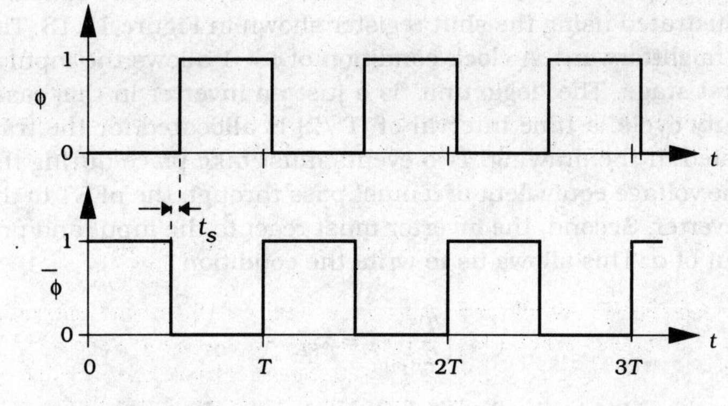

Figure 15.11 Clock skew

minimizing the value of t_s relative to the clock period T. The sensitivity of the circuit to clock skew varies with the design style.

The overall effects of clock skew can be seen by modifying the timing circle to that shown in Figure 15.12. In this approach, the skew has been distributed evenly around the vertical reference axis drawn as a dashed line. The skew time t_s reduces the amount of time for both ϕ and $\overline{\phi}$ data transfer events. This may require that a slower clock frequency be used to allow the logic units to process the data. Skew that originates in the clock generation circuits may be controlled to a limited extent by designing the distribution network. This is equivalent to varying the values of C_1 and C_2 shown in Figure 15.10, and is treated in more detail in the context of the clock distribution problem discussed in Section 15.4.

Circuit Effects and Clock Frequency

The logic-level description of the clocked cascade masks the circuit characteristics that determine the ultimate speed. Since the data transfer rate through the cascade is determined by the clock frequency, it is important

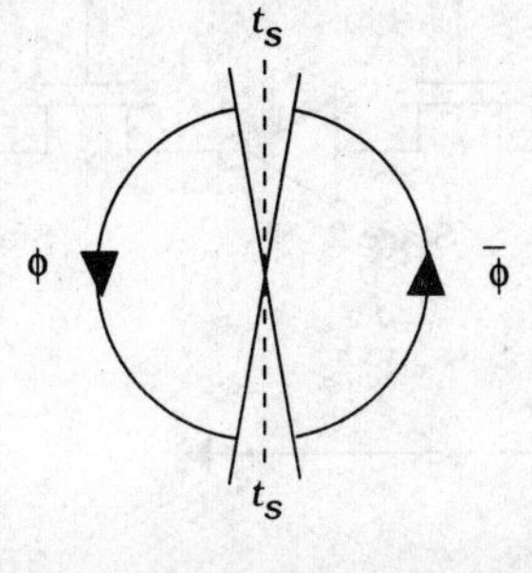

Figure 15.12 Timing circle with clock skew

to understand the electronic factors that limit the speed. These can be illustrated using the shift register shown in Figure 15.13 The operation is straightforward. A clock condition of $\phi = 1$ allows the input a to enter the first stage. The "logic unit" is a just an inverter in this case. With a 50% duty cycle, a time interval of $(T/2)$ is allocated for the transfer as indicated in the drawing. Two events must take place during this time. First, the voltage equivalent of a must pass through the nFET to the input of the inverter. Second, the inverter must react to the input and produce an output of $\overline{a}$. This allows us to write the condition

$$\left(\frac{T}{2}\right)_{min} = t_{FET} + t_{NOT} \tag{15.7}$$

for the minimum allowed half-period. In this equation, t_{FET} is the delay time through the pass transistor while t_{NOT} is the gate delay. Since the worst-case transmission through an nFET is a logic 1 transfer, we have

$$\left(\frac{T}{2}\right)_{min} = t_{r,FET} + t_{HL,NOT} \tag{15.8}$$

where $t_{r,FET} = 18\tau_{FET}$ is the nFET delay and $t_{HL,NOT}$ is the high-to-low time for the NOT gate. The maximum clock frequency for the shift register is thus

$$f_{max} = \frac{1}{T_{min}} = \frac{1}{2(t_{r,FET} + t_{HL,NOT})} \tag{15.9}$$

The shift register result may be extended to an arbitrary logic cascade by writing

$$f_{max} = \frac{1}{T_{min}} = \frac{1}{2(t_{r,FET} + t_{CL})} \tag{15.10}$$

where t_{CL} represents the longest combinational logic delay in the chain.

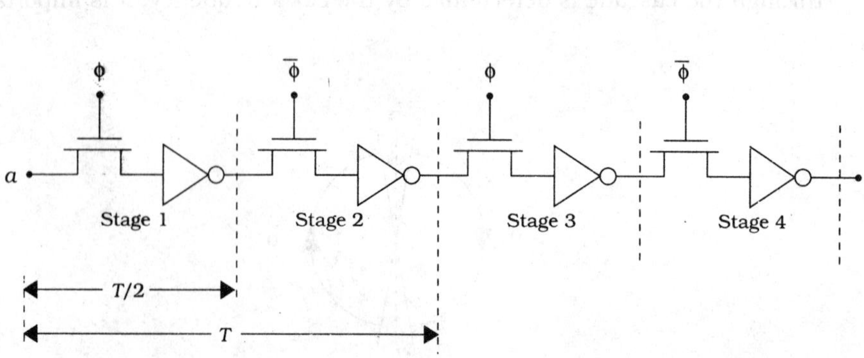

Figure 15.13 Shift register circuit

This clearly illustrates how the system clock speed is determined by the gate delays. Maximum performance requires the worst-case logic path be made as fast as possible. For a set design, this is accomplished at the circuit level by careful selection of the circuit design style or the layout. Alternately, it may be worthwhile to break the logic block into smaller sections and redesign the cascade.

Although it may not be obvious, this CMOS design style is dynamic in nature and exhibits charge leakage problems. Consider the case where a logic 1 voltage is transmitted through the nFET and then the clock goes to a value of $\phi = 0$. This is illustrated in Figure 15.14(a) Although the nFET is in cutoff, a leakage current I_{leak} flows and removes charge from the capacitor C_{in} as described by[1]

$$I_{leak} = -C_{in}\frac{dV_{in}}{dt} \tag{15.11}$$

The leakage current is a function of the voltage V_{in}, making this a non-linear differential equation. Assuming an initial condition of $V_{in}(0) = V_{max}$ gives a voltage decay similar to that shown in Figure 15.14(b). In the plot, V_1 denotes the minimum voltage that is needed for the inverter to recognize the input as a logic 1 value. The hold time t_h is the limiting factor in maintaining this input state. With a 50% duty cycle for the clock, this implies that

$$\left(\frac{T}{2}\right)_{max} = t_h \tag{15.12}$$

since the clock is 0 for half of the clock period. This sets the minimum

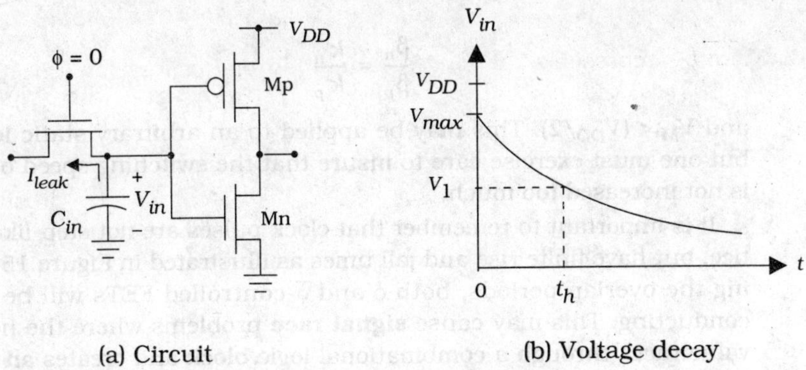

(a) Circuit (b) Voltage decay

Figure 15.14 Charge leakage in the shift register

[1] Charge leakage is treated in Section 9.5 of Chapter 9.

clock frequency f_{min} as

$$f_{min} = \frac{1}{T_{max}} = \frac{1}{2t_h} \tag{15.13}$$

If $f < f_{min}$ then the data will be corrupted. While this consideration does not affect a high-speed network, it does show that there is a minimum clock speed requirement. One consequence of this property is that the clock cannot be stopped for testing as with a static circuit.

The hold time depends on the leakage current, the input capacitance, and the value of V_1. One circuit technique to reduce the charge leakage effect and also increase f_{max} is to design the inverter to have a relatively small value of the midpoint voltage V_M. Since V_M is in between the 0 and 1 voltage ranges, decreasing it also decreases V_1. To design the circuit, recall that

$$V_M = \frac{V_{DD} - |V_{Tp}| + \sqrt{\frac{\beta_n}{\beta_p}} V_{Tn}}{1 + \sqrt{\frac{\beta_n}{\beta_p}}} \tag{15.14}$$

gives V_M for the inverter in terms of the transistor transconductance ratio

$$\frac{\beta_n}{\beta_p} = \frac{k'_n \left(\frac{W}{L}\right)_n}{k'_p \left(\frac{W}{L}\right)_p} \tag{15.15}$$

If we use equal size devices, this reduces to

$$\frac{\beta_n}{\beta_p} = \frac{k'_n}{k'_p} \tag{15.16}$$

and $V_M < (V_{DD}/2)$. This may be applied to an arbitrary static logic gate, but one must exercise care to insure that the switching speed of the gate is not increased too much.

It is important to remember that clock pulses are not step-like in practice, but have finite rise and fall times as illustrated in Figure 15.15. During the overlap periods, both ϕ and $\overline{\phi}$-controlled FETs will be partially conducting. This may cause **signal race** problems where the next input value races through a combinational logic block and creates an incorrect output that is transmitted to the next stage. These situations must be checked in the simulation and verification phase of the design.

The shift register in Figure 15.16 avoids charge leakage by providing a static feedback loop. The drawback of this design is the increased gate count and the routing of the feedback paths. A short inspection of the cir-

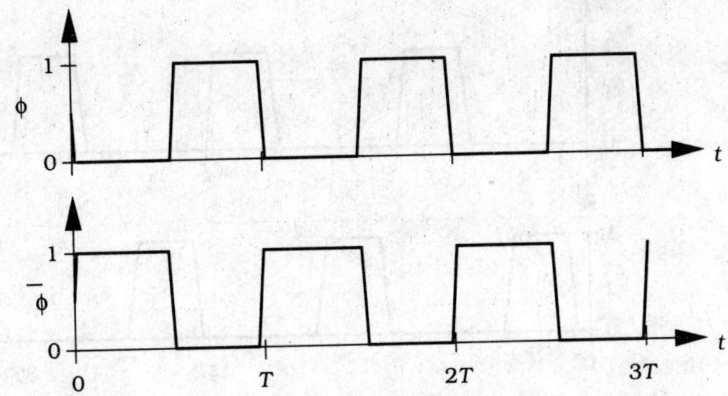

Figure 15.15 Clocking waveforms with finite rise and fall times

cuit will verify that this is just a modified master-slave DFF circuit. Combinational logic blocks can be placed between the stages at the planes defined by the dashed lines to achieve a clocked cascade.

Dual Non-overlapping Clocks

In this technique, two distinct non-overlapping clocks ϕ_1 and ϕ_2 are used such that

$$\phi_1(t) \cdot \phi_2(t) = 0 \tag{15.17}$$

is enforced for all times *t*. This is similar to the single-clock, dual-phase approach except that a duty cycle value of less than 50% is used. Figure 15.17 illustrates a typical set of waveforms. This can be used to control the data flow through a logic cascade in the same manner as with the ϕ, $\overline{\phi}$ pair. The timing circle for a dual-clock system is shown in Figure 15.18; the data transfer time is decreased because of the narrower pulse width. Signal races are eliminated because the non-overlap condition is maintained.

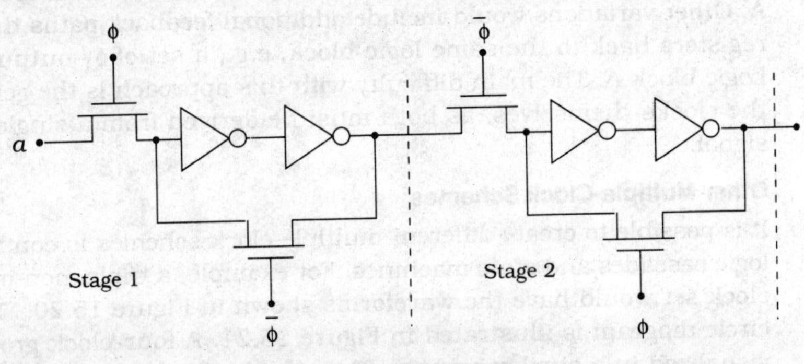

Figure 15.16 Static shift register design

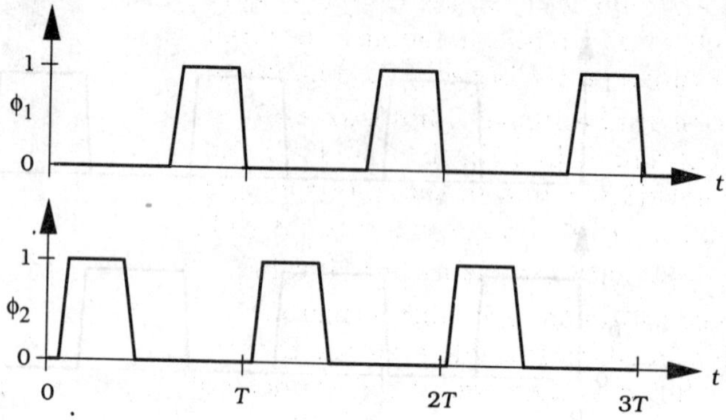

Figure 15.17 Dual non-overlapping clocks

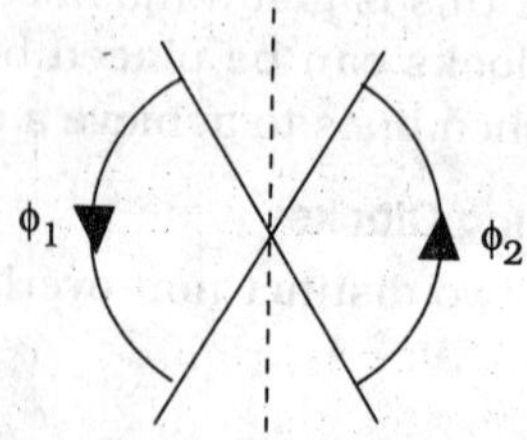

Figure 15.18 Timing circuit for a 2-clock network

Finite-state machines that are based on dual-clock schemes can provide powerful interactive capabilities. One of the simpler topologies is shown in Figure 15.19. This consists of two distinct logic networks Logic block A and Logic block B that are connected via separately controlled feedback registers. In this configuration, the secondary outputs of Logic block A are fed into a ϕ_1-controlled register, which provides secondary inputs into Logic block B. The ϕ_2-controlled register on the left side of the network takes signals from Logic block B and sends them to Logic block A. Other variations would include additional feedback paths that link the registers back to the same logic block, e.g., a set of ϕ_1-outputs back to Logic block A. The main difficulty with this approach is the generation of the clocks themselves, as both must be derived from a single reference signal.

Other Multiple-Clock Schemes

It is possible to create different multiple-clock schemes to control clocked logic cascades and state machines. For example, a triple, non-overlapping clock set would have the waveforms shown in Figure 15.20. The timing circle diagram is illustrated in Figure 15.21. A four-clock group can be visualized in a similar manner. These and other clocking schemes have

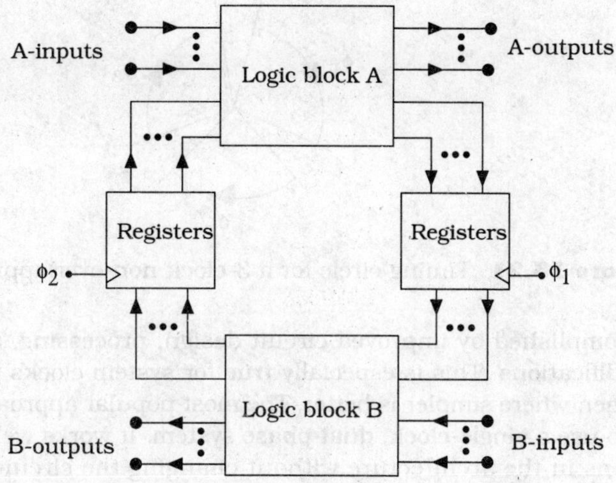

Figure 15.19 A dual-clock finite-state machine design

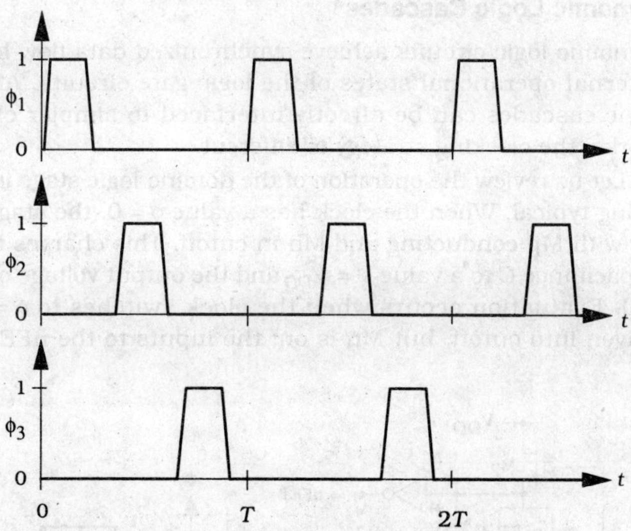

Figure 15.20 Triple, non-overlapping clock signals

been introduced for digital MOS integrated circuit design over the years. Several were successfully used as system control clocks in commercial chips that were based on older digital design styles. Some 3- and 4-phase clocking strategies were used for novel low-frequency pMOS and nMOS-only dynamic logic circuits.

In modern high-speed VLSI, however, complicated clocking schemes introduce too many problems to make them worthwhile. Speed gains are

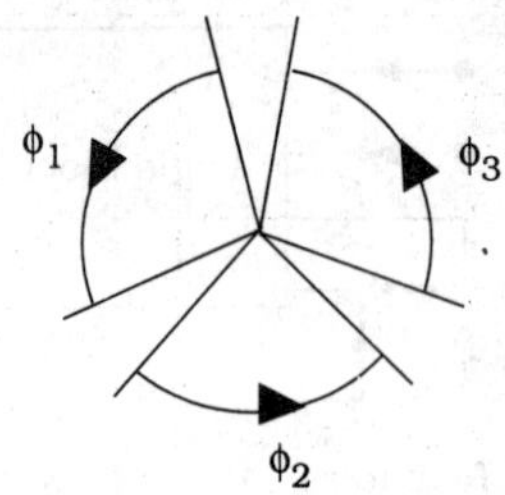

Figure 15.21 Timing circle for a 3-clock non-overlapping network

accomplished by improved circuit design, processing, and architectural modifications. This is especially true for system clocks that are 1 GHz or higher, where simpler is better. The most popular approach in VLSI design is to use a single-clock, dual-phase system. It works well and allows variations in the architecture without changing the circuit design style. We will therefore narrow our study to encompass only simple clocking techniques.

15.2.2 Dynamic Logic Cascades[2]

Dynamic logic circuits achieve synchronized data flow by controlling the internal operational states of the logic gate circuits. Although dynamic logic cascades can be directly interfaced to simpler clocked logic networks, the clocking strategy is different.

Let us review the operation of the domino logic stage in Figure 15.22 as being typical. When the clock has a value $\phi = 0$, the stage is in precharge (P) with Mp conducting and Mn in cutoff. This charges the internal node capacitance C to a value $V = V_{DD}$ and the output voltage of the stage is $V_{out} = 0$. Evaluation occurs when the clock switches to $\phi = 1$. The pFET is driven into cutoff, but Mn is on; the inputs to the nFET logic array are

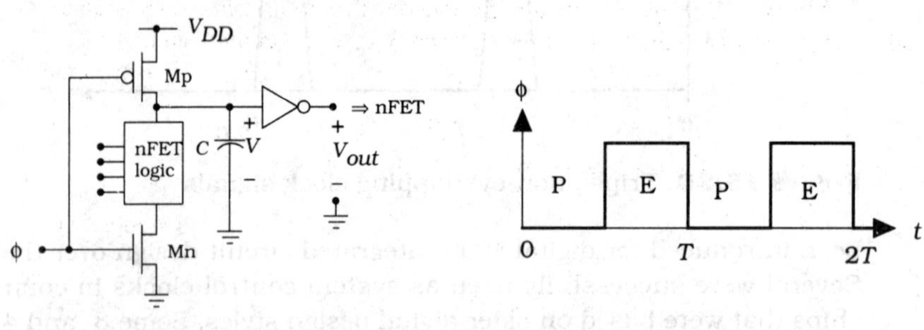

Figure 15.22 Operation of a domino logic stage

[2] This entire section builds on the material presented in Section 9.5 of Chapter 9.

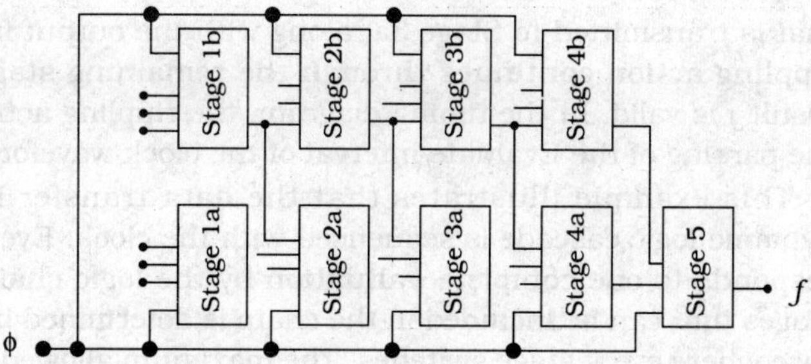

Figure 15.23 A dynamic logic cascade

valid during this time. If the inputs result in an open switch condition for the logic block, V is held high and the output of the gate is 0, both voltage-wise and logically. The output is connected to a logic nFET in the next stage, and $V_{out} = 0$ V will keep it in cutoff. On the other hand, a closed switch condition from the top to the bottom of the logic array allows C to discharge, giving $V = 0$ V. The output then switches to a logic 1 voltage of $V_{out} = V_{DD}$. This drives the logic FET of the next stage into conduction. This serves to illustrate that the clock automatically controls the data flow since inputs and outputs are valid only during the evaluation interval.

Dynamic CMOS system timing can be understood by applying the same analysis to a logic chain. Consider the logic chains illustrated in Figure 15.23 where the stages have the same basic domino structure. Every input to the second through fourth stages is assumed to originate from a domino gate, but they are not shown explicitly. A single clocking waveform ϕ is applied to every stage in the chain, so that the cascade behaves as a single logic group. The waveform in Figure 15.24 shows the behavior of the chain for both the precharge and evaluate intervals. When $\phi = 0$, every stage undergoes precharge at the same time and no data transfer takes place. Evaluation occurs when $\phi = 1$. At this time, the inputs to Stages 1a and 1b are assumed to be valid and result in an output that is fed to Stage 2a along with inputs from other gates. This produces a result

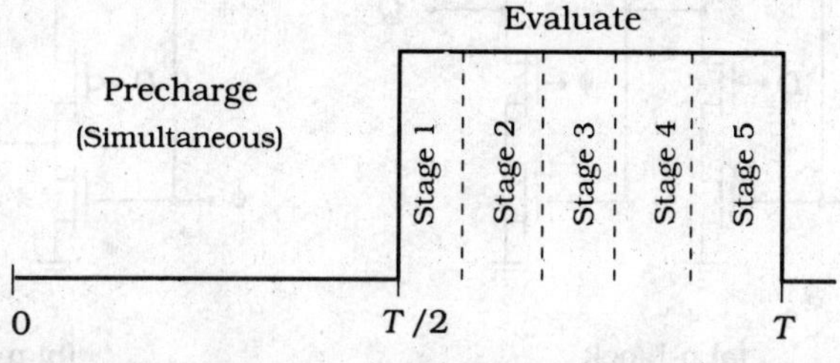

Figure 15.24 Timing sequence in the domino cascade

that is transmitted to Stage 3a, along with the output from Stage 2b. The rippling action continues through the remaining stages until the final result f is valid. In the timing diagram, the rippling action is indicated by the parsing of the Evaluate interval of the clock waveform.

This example illustrates that the data transfer into and out of a dynamic logic cascade is sequenced with the clock. Every clock cycle corresponds to one complete evaluation by the logic chain. The number of stages that can be included in the chain is determined by the delay for the case where every stage switches. The maximum allowed evaluation time is set by the width ($T/2$) of the evaluation pulse. Long logic chains can be accommodated by relatively slow clocks. However, this introduces problems in charge leakage so charge keeper circuits become mandatory.

Although this example has been based on domino circuits, the main results apply to most dynamic CMOS logic families. When dynamic logic cascades are used, they are interfaced to static circuits at both the input and the output sides. Dataflow is thus achieved at the system level.

Many types of dynamic CMOS latches have been published in the literature. While most are single-clock, dual-phase circuits, the TSPC (true single-phase clock) logic design style uses only a single clock ϕ throughout. The single-phase latches can be interfaced with static gate designs for data synchronization. Two TSPC latches are shown in Figure 15.25. The "n-block" circuit in Figure 15.25(a) consists of two stages. The first stage is a simple dynamic inverter, while the second stage provides the latching operation using the middle clock-controlled nFET. Precharge occurs when $\phi = 0$; during this time, the output Q is in a high-impedance state (i.e., an open circuit). When $\phi = 1$, the first stage is in evaluation while the output stage operates as a modified NOT circuit. The data input D is accepted and a buffered value occurs at Q. This value is held by the output capacitance at Q when the clock returns to $\phi = 0$. The p-block latch in Figure 15.25(b) operates in a similar manner. TSPC logic provides

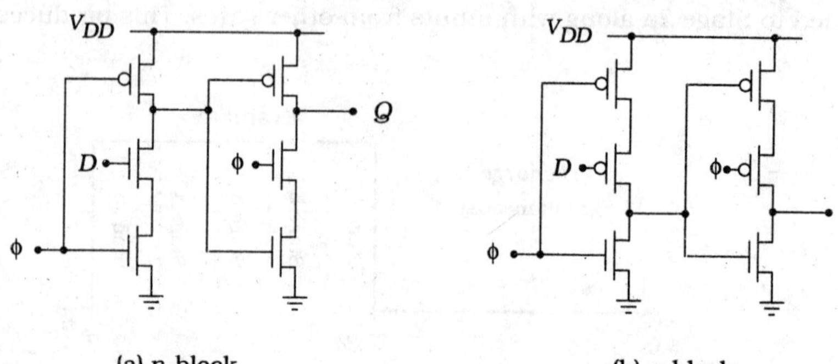

Figure 15.25 True single-phase clock latches

many attractive features for VLSI designs. However, since it is intrinsically dynamic in nature, charge leakage and charge sharing problems will be present.

15.3 Pipelined Systems

Pipelining is a technique that is used to increase the throughput of a sequential set of distinct data inputs through a synchronous logic cascade. Since computer instructions are inherently sequential in nature, pipelining is used in microprocessors to increase the MIPs rating.[3]

Let us analyze the timing requirements for the simple register I/O network shown in Figure 15.26 as a basis for introducing the concepts. The input data bit D_1 is latched into the DFF on a rising clock edge, and is available as Q_1 after the flip-flop delay time t_{ff}. Bit Q_1 enters the combinational logic network (along with other inputs that are not shown) and produces a result D_2 after a delay time t_d. The result D_2 is latched into the output DFF on the next rising clock edge.

This sequence can be used to establish the timing requirements on the clocking waveform. Since data are latched into the FFs on every rising clock edge, we must insure that the clock period T is large enough to allow for normal circuit delays. An example set of waveforms is shown in Figure 15.27. The flip-flop delay t_{ff} and the logic delay time t_d are shown on the Q_1 and D_2 waveforms, respectively. Two FF times are shown on the D_1 plot. The setup time t_{su} is the time prior to the clock edge where the input must stable, while the hold time t_{hold} is the minimum time that the inputs must remain stable after the clock edge to latch the correct value. We have also introduced the possibility of a skew time t_s that separates the input clock ϕ and the output register clock ϕ'. This set of waveforms shows that the clock period T must satisfy

$$T > t_{ff} + t_d + t_{su} + t_s \tag{15.18}$$

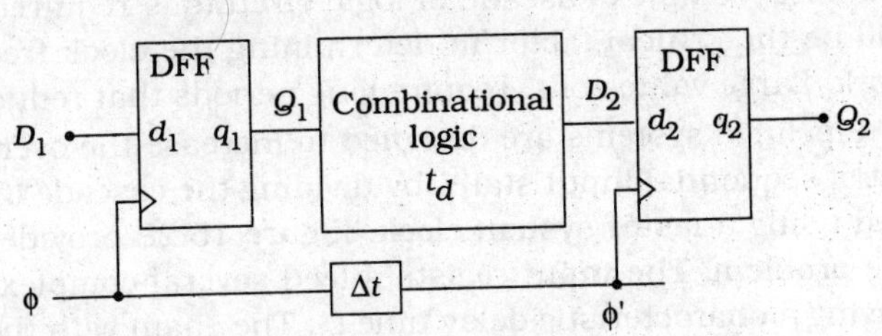

Figure 15.26 Basic pipelined stage for timing analysis

[3] MIPs is an acronym for millions of instructions per second.

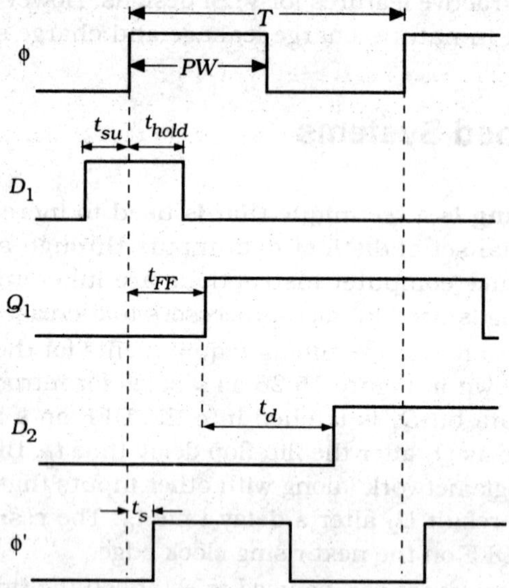

Figure 15.27 Waveform quantities for timing analysis

to allow for all of the circuit delays. The hold time requirement t_{hold} places a restriction on the DFF of

$$t_{hold} < PW \tag{15.19}$$

where *PW* is pulse width of the clock. Equation (15.18) gives the limit on the clock frequency as

$$f < \frac{1}{t_{ff} + t_d + t_{su} + t_s} \tag{15.20}$$

In standard design, a single clocking rate is used for the entire system, so that the system clocking is determined by the slowest subsystem or unit. If a long, complex cascade of logic circuits is required, the delay time t_d will be the critical factor in determining the clock frequency of this network. Large values of t_d require long periods that reduce the value of *f*.

Pipelined systems are designed to increase the overall throughput of a set of sequential input states by dividing the cascade into smaller sections and using a faster system clock. Figure 15.28 provides a visualization of the problem. The input registers feed several complex logic chains, each having a characteristic delay time t_d. The chain with the largest value of t_d determines the clock rate for this unit; the numerical value of the delay time for a logic cascade is established by the circuit design style, the processing parameters, and the physical design.

The idea of pipelining can be understood by noting that logic calcula-

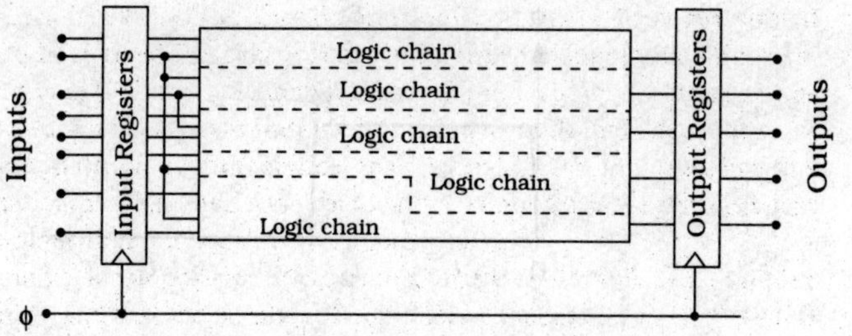

Figure 15.28 Logic chains in a clocked system

tions start at the input sections, and propagate through the chain to the output. Once a circuit completes a calculation and passes the result on to the next stage, it remains idle for the rest of the clock cycle. The progression of circuit usage is shown in Figure 15.29 where we assume that the rising edge of the clock occurs at a time $t = 0$. Successive times are denoted by t_i where $t_{i+1} > t_i$ for $i = 1, 2, 3$. The clocking waveform in Figure 15.30 illustrates the relative time values. Since the delay through a logic

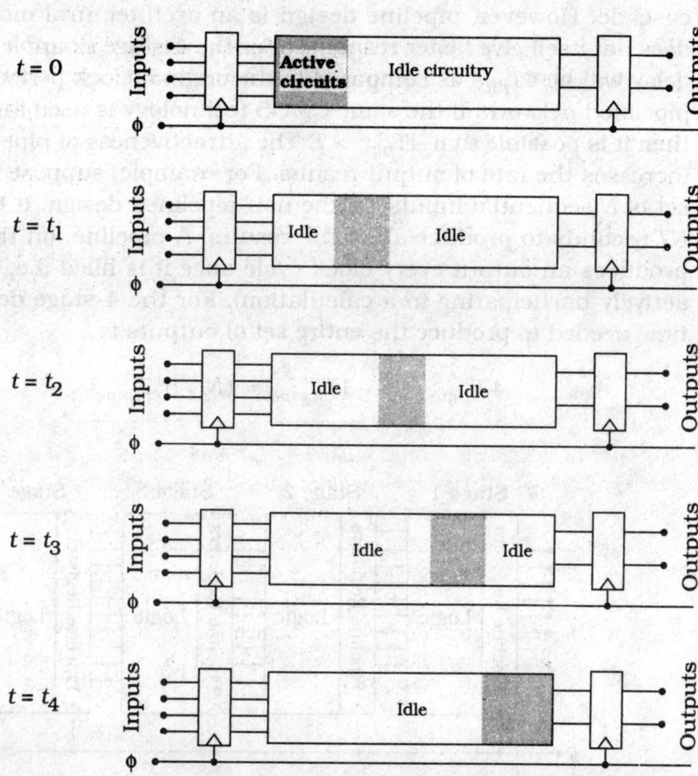

Figure 15.29 Circuit activity in a logic cascade

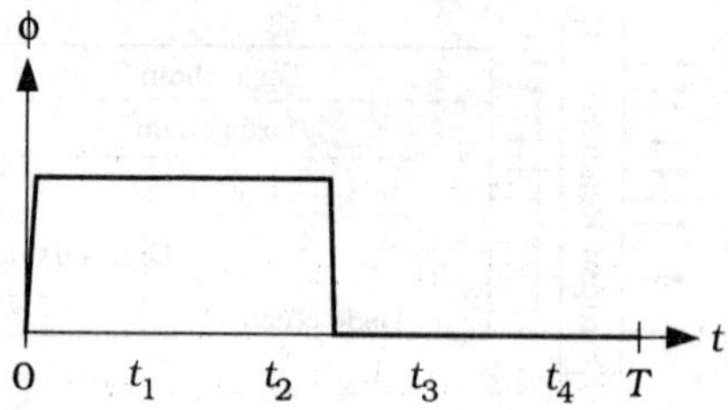

Figure 15.30 Progression times in the logic cascade

gate varies its complexity and parasitics, the logic propagation rate will not be uniform. Some circuits will exhibit longer delays than others.

This example provides the basis for pipelining. If we divide the long logic into smaller groups, add registers between the sections, and use a faster clock, then most of the circuits will be active at any given time. Figure 15.31 shows a 4-stage pipeline. Each stage consists of a set of input registers and a logic network. The clock frequency $f_{pipe} = (1/T_{pipe})$ is set by the slowest stage and is larger than the frequency f used in the original cascade. However, pipeline design is an architectural modification and does not itself give faster response. For the 4-stage example, the total time delay will be $4T_{pipe}$ as compared to the original clock period T in the non-pipelined network. If the same CMOS technology is used for both designs, then it is possible that $4T_{pipe} > T$. The attractiveness of pipelining is that it increases the rate of output results. For example, suppose that we have a set of N sequential inputs. In the non-pipelined design, it takes a total of NT seconds to produce all of the results. A pipeline, on the other hand, produces an output every clock cycle once it is filled (i.e., every stage is actively participating in a calculation). For the 4-stage design, the total time needed to produce the entire set of outputs is

$$4T_{pipe} + (N-1)T_{pipe} = (N+3)T_{pipe} \tag{15.21}$$

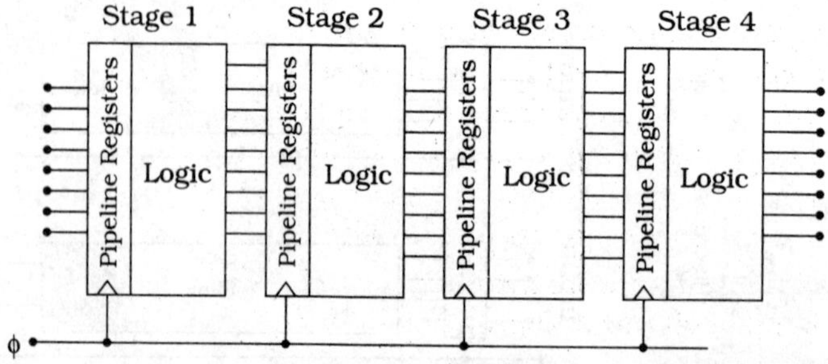

Figure 15.31 A 4-stage pipeline

where the first term on the left side accounts for the time needed to initially fill the pipe. This is an idealized value and assumes that it is possible to maintain a full pipeline. High-level pipelined design of processor data paths is a standard topic in computer architecture, so we will not discuss the details here.[4] Let us instead examine the important aspects of using a pipeline in a VLSI design.

A particularly important problem is selecting the break points in the original system to define smaller pipeline stages. For the i-th stage, the smallest allowable clock period T_i is determined by the condition

$$T_i > t_{ff} + t_{su} + t_{d,i} + t_{s,i+1} \tag{15.22}$$

If we assume that identical pipeline registers are used in every stage, then the terms $(t_{ff} + t_{su})$ are constant; also, the skew time $t_{s,i+1}$ is set by the clock distribution scheme. This leaves the logic delay time $t_{d,i}$ as the critical factor. Once the logical behavior and operations for a pipeline stage have been selected, the logic and circuit design centers on minimizing $t_{d,i}$. For an m-stage pipeline, the period for the pipeline clock is chosen to be

$$T_{pipe} = \max\{T_1, \ldots, T_m\} \tag{15.23}$$

as this allows the slowest unit sufficient time to complete the calculations. Much engineering time and effort is directed toward dealing with the slower units. Various algorithmic solutions are created and implemented using different CMOS circuit design styles. For example, domino logic cascades have been used in critical ALU circuits in chips where most of the circuitry was static in nature.

An architectural approach to increasing the speed is to further subdivide one or more of the units and increase the number of pipeline stages. This decreases the delay time through a unit, and allows a faster clock. However, this **deep pipeline** design requires more clock cycles to complete the entire chain of calculations so the reduction in the clock period may not result in a faster network. Making this design acceptable often requires a substantial increase in engineering effort. Another factor that enters the picture is the desire to associate every pipeline stage with a large-scale behavioral function. For example, breaking up a memory unit into smaller sections will make both the design and verification much more difficult.

Physical design considerations also enter the picture. Real estate costs and power dissipation levels are always important. Every logic/circuit solution has different layout and operational characteristics, and these must be factored into the overall design. Pipeline registers must latch every input bit, and can be quite large in modern systems that use word

4 An excellent discussion of pipelined architectures is contained in Reference [8].

sizes of 32b or more. At the chip level, every register requires a clocking signal and input/output lines, which complicates the interconnect routing problem. Both clock and signal skew problems increase with the layout area. Although pipelining is a rather specific architectural design style, most of the problems mentioned arise in any advanced VLSI design.

A variation of the basic pipeline design is shown in Figure 15.32. Alternating positive edge- and negative edge-triggered input registers are used to latch inputs on every change in the clock. This has the same basic problems as the system that uses only positive-edge triggered flip-flops. Since the physical limit is still the delay through the logic network, one does not gain an increase in speed. It does, however, allow slower clocks to be used which may be easier to generate and distribute.

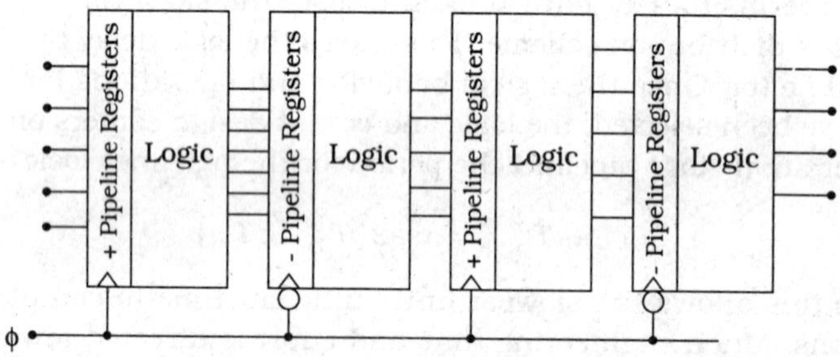

Figure 15.32 Pipeline with positive edge- and negative edge-triggering

15.4 Clock Generation and Distribution

Many large system designs are limited by problems in the clock distribution. This becomes particularly troublesome when clock frequencies f reach the 1 GHz (10^9 Hz) level corresponding to a clock period of

$$T = \frac{1}{f} = 1 \text{ ns} \tag{15.24}$$

as this approaches the switching speed of the gates themselves. Distribution of the clocking signal to various points of the chip is complicated because the intrinsic RC time delay τ increases as the square of the line length l according to

$$\tau = Bl^2 \tag{15.25}$$

with B [sec/cm^2] having a constant value determined by the physical layout and material composition. A simple view of the problem is illustrated by the chip floorplan of Figure 15.33. The input clock signal Clk must be

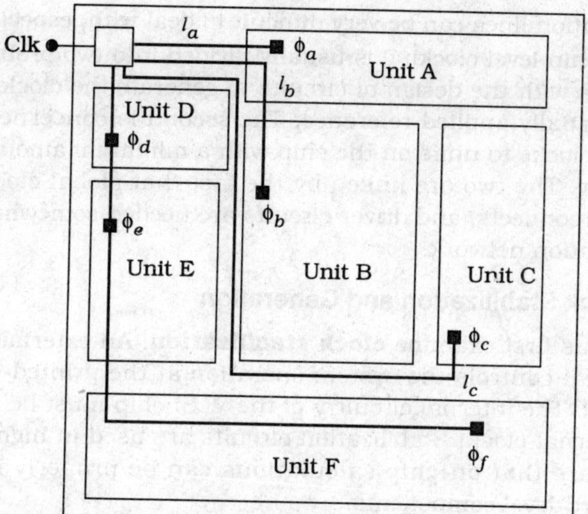

Figure 15.33 Clock distribution to on-chip modules

distributed to every unit. If we employ direct clock distribution lines as shown, the difference in line lengths implies different signal delays. Each unit will operate at the same frequency, but the units would be out of phase with each other. The received clock ϕ_a would be ahead of ϕ_b, which would in turn be ahead of ϕ_c as shown by the waveforms in Figure 15.34. In the present example, ϕ_b is delayed from ϕ_a by a time interval

$$\Delta t_1 = B(l_b^2 - l_a^2) \tag{15.26}$$

while ϕ_c lags behind ϕ_b by

$$\Delta t_2 = B(l_c^2 - l_b^2) \tag{15.27}$$

Similar arguments hold for the single distribution line scheme feeding Units D, E, and F. Clock skew problems that originate from the signal dis-

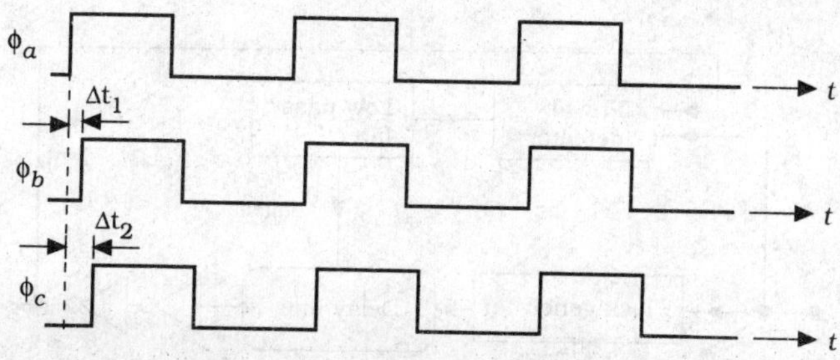

Figure 15.34 Example of clock skew

tribution lines can be very difficult to deal with, especially in large chips.

Chip-level clocking is usually divided into two problem areas. The first deals with the design of circuits to generate the clocking signals from an externally applied reference. The second is concerned with distributing the clocks to units on the chip with a minimum amount of distortion and skew. The two are linked by the fact that global clocking requires long interconnects, and driver circuits are needed somewhere in the clock distribution network.

15.4.1 Clock Stabilization and Generation

Let us first examine **clock stabilization**. An externally applied clocking signal controls the system operation at the printed-circuit board (PCB) level. The internal circuitry of the VLSI chip must be synchronized to the external clock. Stabilization circuits are used in high-speed networks to insure that on-chip calculations can be properly interfaced to other board-level components.

A conceptually simple approach to clock stabilization is shown in Figure 15.35. The external clocking signal is applied to the Clock generator circuit that produces the necessary clocking waveforms. This is fed though a voltage-controlled Delay line unit that can slow down the signal if necessary. A scaled buffer chain provides the drive strength needed to distribute the clock to the chip. The upper part of the network provides frequency stabilization. The output signal is sampled and sent to a phase detector circuit where it is compared with the externally applied clock. The phase detector produces an output that indicates if the output signal is leading or lagging the external clock. This information is used by the Low pass filter circuit to produce a voltage V_{adjust} that controls the RC time constant of the Delay line circuit. The **phase-locked loop** (PLL) scheme in Figure 15.36 operates in a similar manner. A PLL is designed to detect any difference in phase between an input and a reference signal, and produce an output waveform that is properly synchronized to the reference. In this case, the external clock is the input signal while the feedback loop monitors the output and allows for corrections. A PLL is also useful for stabilizing frequency multiplier circuits that produce higher-

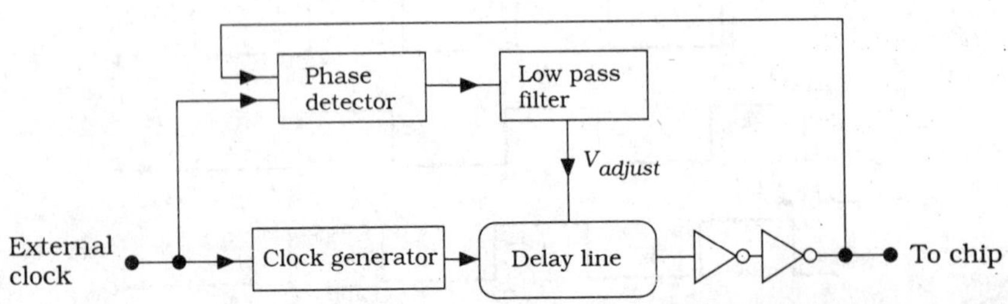

Figure 15.35 A basic clock stabilization network

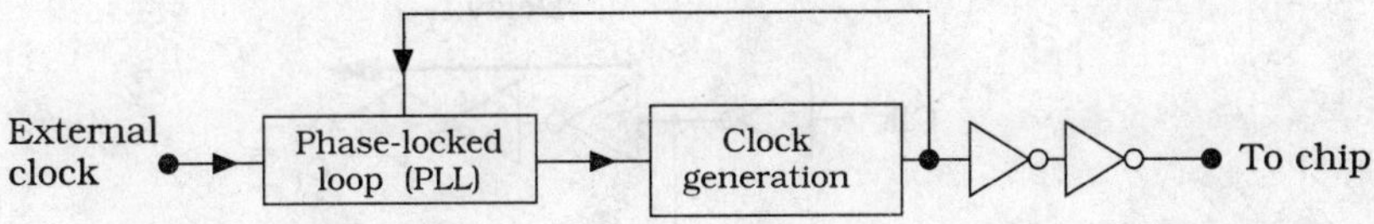

Figure 15.36 Phase-locked loop (PLL) stabilization circuit

order harmonics of the fundamental. Similar circuits are found in **clock-recovery modules** that are used in serial data transmission systems. A clock recovery circuit extracts the bit-time interval and the clocking signal from the data stream. In this type of a system, the data **framing** format provides a short burst of 0's and 1's that provides a synchronization reference.

Clock generator circuits are designed using relatively simple logic configurations. Minimizing the skew from the clocking circuit is accomplished by careful sizing of the transistors. The simple inverter-based cascade in Figure 15.37 produces complementary clocks ϕ and $\overline{\phi}$ from the single input Clk signal. Scaled inverter chains are used to provide the drive currents needed for the capacitances C_1 and C_2 shown in the drawing. Both elements can be written in the generic form

$$C = C_{line} + \sum C_G \tag{15.28}$$

where C_{line} is the line capacitance and the second term sums the gate capacitance for every clocked FET connected to the line.

The skew minimization problem is shown in Figure 15.38. This identifies the two paths for ϕ and $\overline{\phi}$ through the upper and lower chains, respectively. The idea is to design the two circuits such that the delay times t_1 and t_2 are equal. In the technique of Logical Effort, this is called a **fork circuit** due to its topology [13]. Design is based on equalizing the electrical effort of the two chains, noting that C_1 and C_2 are (usually) unequal. This establishes both the number of stages and the relative sizes in each chain.

Another technique for generating two clock phases uses the D-latch circuit in Figure 15.39. Since the core circuitry is just an SR-latch with complemented inputs, the outputs themselves are always complements

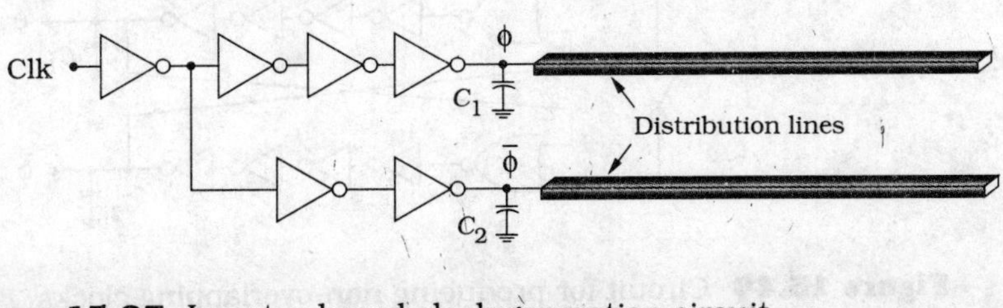

Figure 15.37 Inverter-based clock generation circuit

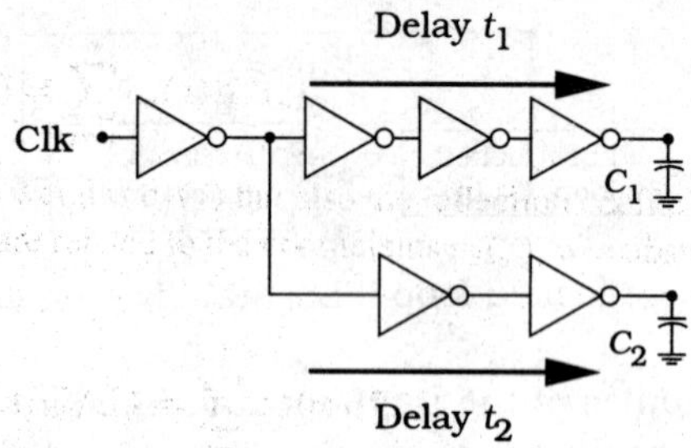

Figure 15.38 Skew minimization circuit

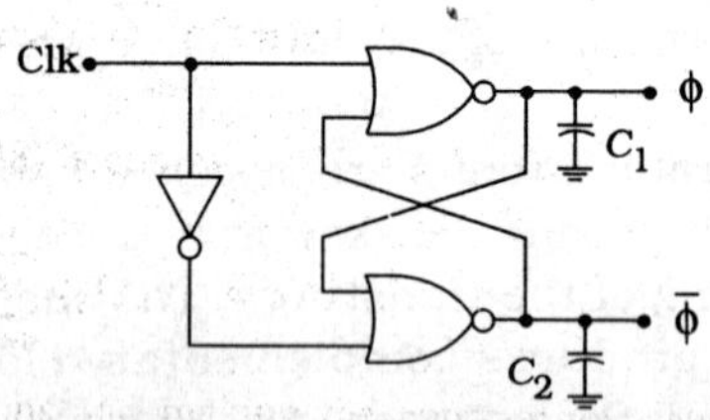

Figure 15.39 Generating complementary clocks using a latch

and can be used as ϕ and $\overline{\phi}$. There are two sources of skew between the output waveforms. A small amount may arise from the inverter, but a difference in output capacitances C_1 and C_2 may cause more problems. The individual NOR gates may be "tweaked" to compensate if necessary. A variation of this circuit is shown in Figure 15.40. Inverter chains are added to delay the feedback signals. This has the effect of producing non-overlapping outputs, with the extent of the separation determined by the chain delay.

15.4.2 Clock Routing and Driver Trees

Once a stable set of clocking signals has been generated, it must be distributed to the subunits on the chip. At the architectural level, clock distribution is usually viewed as a routing problem where the lengths of the line segments are critical. Physical design considerations enter the picture when the large capacitances and layout routes are realized.

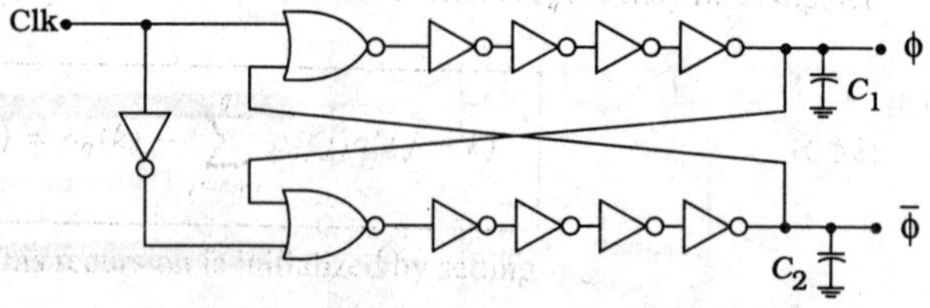

Figure 15.40 Circuit for producing non-overlapping clocks

Let us examine the routing problem first. Suppose that the initial floorplan places clock receivers as shown in Figure 15.41(a). Since these are distributed over the entire chip area, it is useful to group points together; Figure 15.41(b) shows the case where each grouping has been chosen to contain 4 receiver points. If we then wire together the points inside each grouping, then our problem is reduced to driving groups. One approach to the interior routing is shown in Figure 15.41(c). This uses near-vertical lines to connect receiver points; the vertical lines are then connected using wires with a horizontal orientation. Since the placement of the receiver points was random in the example, it is not possible to use strictly vertical or horizontal routings.

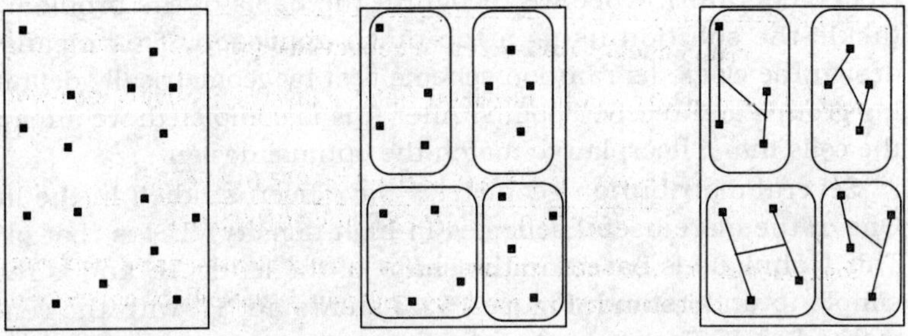

Figure 15.41 Simplified view of the clock routing problem

Other algorithms can be used to define clocking groups. Partitioning is used in Figure 15.42 to obtain a similar result. The first partitioning uses a horizontal line as shown in Figure 15.42(a). The second partitioning step employs a vertical line to divide the chip into quadrants and results in the grouping of Figure 15.42(b). A third step based on horizontal boundaries creates the 2-point groups in Figure 15.42(c); the points

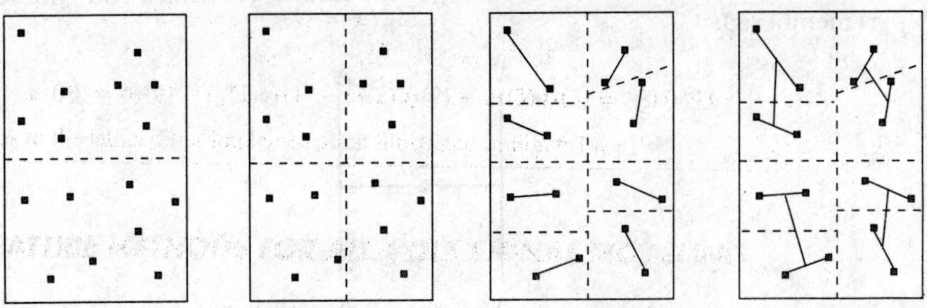

Figure 15.42 Partitioning steps for defining clocking groups

within the group can be wired as shown. Four-point groups are obtained by connected two 2-point groups, and resulting in the initial routing portrayed in Figure 15.42(d). Note that the results of the grouping and partitioning techniques are similar in this example.

The examples above illustrate several important points. Even a moderately complex design will have several thousand clock-driven transistors. To simplify the routing problem, we group closely spaced points together and then concentrate on driving the groups instead of the individual points. It may be necessary to create "super-groups" that consist of many smaller groups; in network theory, this is the same as creating an interconnect wiring tree. The process is simply hierarchical design starting at the bottom (individual receiver points) and working toward the top (the clock generator). Once we recognize the basis of the problem, we may tackle the solution using a top-down approach. This means that we design the clock distribution scheme first by geometrically defining clocking groups and receiver points. After this is completed, we attempt to "fit" the cells into a floorplan to match the optimal design.

Several distribution geometries have been studied in the literature. One of the more useful schemes in high-density VLSI is that of **H-trees**. This technique is based on the shape of the letter "H" and is really quite simple to understand. Figure 15.43 shows an "H" with the center point marked as X. The symmetry of the construction shows that the distance l_{XA} from the center to any tip A is a constant. If we broadcast a signal from X and place receivers at equivalent A-points, then the delay

$$\tau_{XA} = Bl_{XA}^2 \tag{15.29}$$

is the same. In other words, all of the received signals are in phase. An H-tree is built using the H-shape as a primitive on a geometrical grid. A macro-design is shown in Figure 15.44. The master clock is placed at the center point X, and the received signals are taken from the tips of the small H's. These points can be used as sources for local clock distribution that may use smaller H-trees or direct drivers. While H-trees seem to be an obvious solution, there are two important qualifiers that must be remembered:

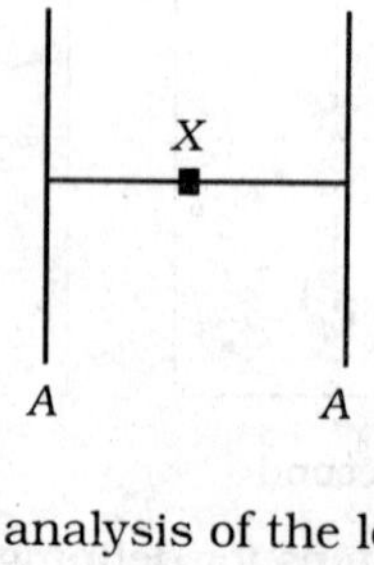

Figure 15.43 Geometrical analysis of the letter "H".

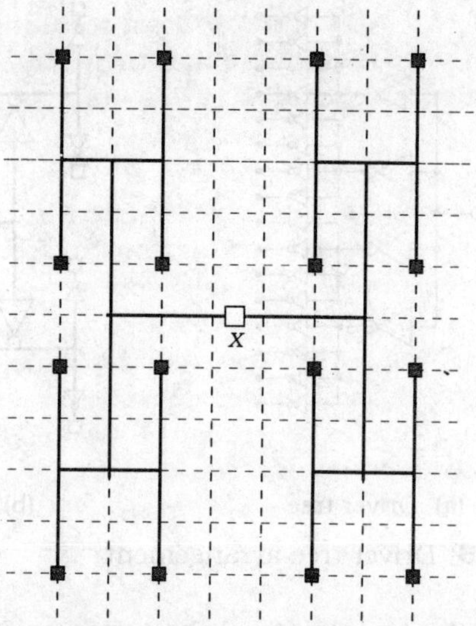

Figure 15.44 Macro-level H-type distribution tree

- The lengths and electrical characteristics must be the same for every clock path to produce the desired effect, and
- The load capacitance at every receiver point must be the same.

Since the symmetry of the tree is critical, it is usually necessary to provide a layer of interconnect metal where the distribution tree has the highest priority. After the tree is constructed, remaining space can be used for general wiring as needed. If this is not possible, then the tree may be designed using different interconnect layers, but the routing paths should be identical. The second item is electronic in nature. It requires us to use electrically identical receiver/driver circuits at every point.

A constraint on the ideal top-down design is that the floorplan and system components must be placed to accommodate the receiver points defined by geometry of the H-tree. This requires an exceptional amount of planning at the circuit and physical design levels, and may not be practical. All is not lost, however, since the main idea is equalizing the signal delay, which is an electrical problem. It is possible to deviate from the exact specification by careful circuit design.

It is generally accepted that the easiest approach to controlling clock skew is to include drivers in the distribution path. A **driver tree** such as that shown in Figure 15.45(a) can be used to break up the interconnect into smaller lengths and provide reasonably sized buffers. Driver placement depends upon the distribution strategy; an H-tree example is shown in Figure 15.45(b). The design of the driver tree circuits is based on the

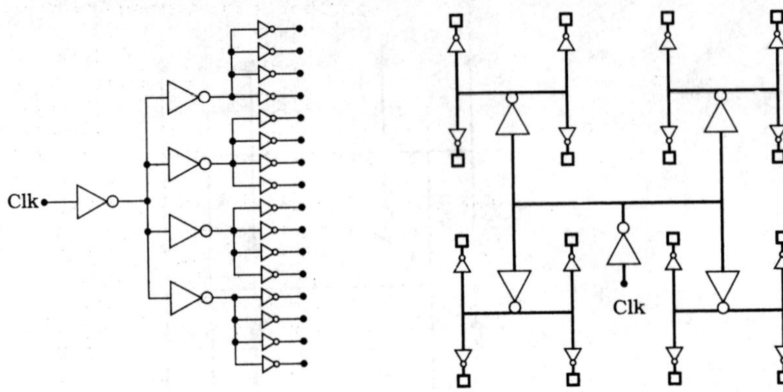

(a) Driver tree (b) Application to H-tree

Figure 15.45 Driver tree arrangement

parasitics of the inverters and the interconnect. In Figure 15.46, π-RC models have been used to include the line resistance and capacitance. Once the values for each segment are known, the drivers can be sized accordingly. This model shows that the actual geometry of the distribution grid is not important since electrical symmetry can be achieved by equalizing the values of the RC components in each set of branches. Many acceptable variations in the H-tree can be created. The key is to maintain equal line lengths in equivalent segments. An example is shown in Figure 15.47 where the routing provides identical distribution paths to every receiver point. Although no large-scale patterning is obvious, the layout

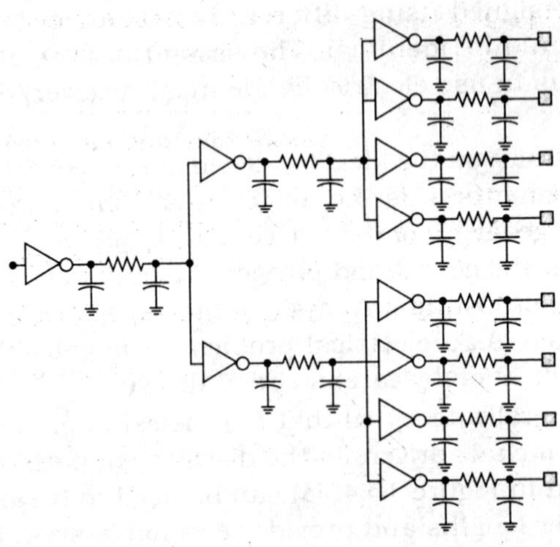

Figure 15.46 Driver tree design with interconnect parasitics

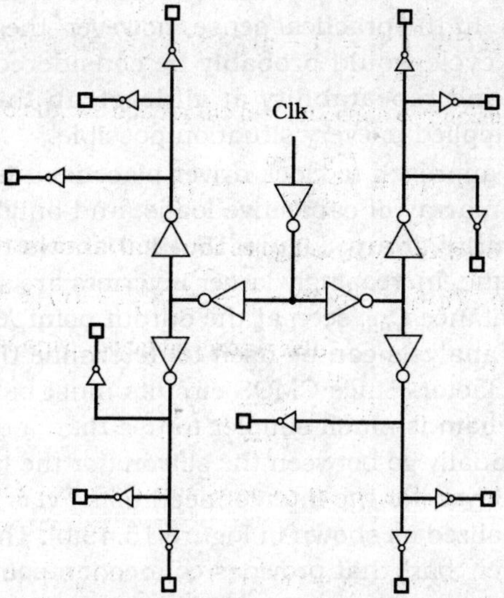

Figure 15.47 Distribution scheme with equivalent driver segments

provides the desired results This flexibility is useful at the physical design level, especially in high-density layouts.

Symmetry in the driver tree simplifies both electrical design and layout. A non-symmetrical scheme with acceptable performance can be created, but is very difficult to design. Figure 15.48 shows a driver network that employs non-symmetrical tree branching. The design problem centers around using the common input Clk to produce aligned (zero-skew) signals at receiver points A, B, and C. Theoretically, this can be accomplished to within a reasonable skew specification. Logical Effort may be

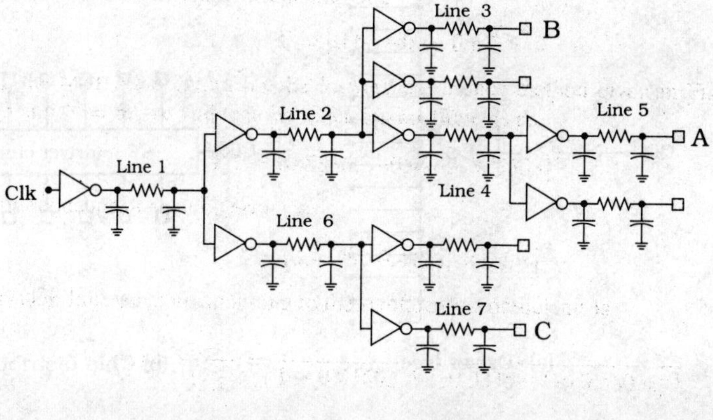

Figure 15.48 Electrical circuit for a non-symmetrical distribution

used for the first sizing, and the resulting design simulated to include the resistances. In the practical sense, however, the additional time added to the design cycle would probably be considered wasted. VLSI relies on symmetry and repeatability at all levels in the design hierarchy, and should be applied to every situation possible.

Another approach to clock driver placement is to treat the distribution points as an array of capacitive loads, and build a single buffer chain to drive the entire group. Figure 15.49(a) shows the equivalent circuit for this technique. Increasingly larger inverters are sized to accommodate the large capacitance C_{out} seen at the output point X of the last stage. Classical scaling analysis can be used to determine the number of stages and the scaling factor. Since CMOS circuits must be built at the silicon level, the driver chain is much simpler to use than a distributed tree where we must continually go between the silicon (for the transistors) and a higher-level metal layer (for the interconnect) using vias and contact cuts. Layout can be visualized as shown in Figure 15.49(b). The driver circuits result in a large driver "bar" that provides connection points; in practice, the "bar" would be the large transistors of the output stage. The lengths of the distribution lines determine the skew relative to the phase at X. Equal-length lines may be used all around, or different lengths may be used to purposely induce time delays on the outer units.

The use of skew-delayed sections brings up an interesting point. All of the clocking schemes assume that we want to synchronize every unit in the system to be in phase with one another. This allows for stable and reliable design, but is not always easy to achieve. Alternate clocking techniques can be used if we are willing to depart from this type of system clocking. One approach is to use **locally synchronous** units where different logic sections operate with a phase that is independent of others. The concept is illustrated in Figure 15.50. The clocks φ, φ', and φ" control the

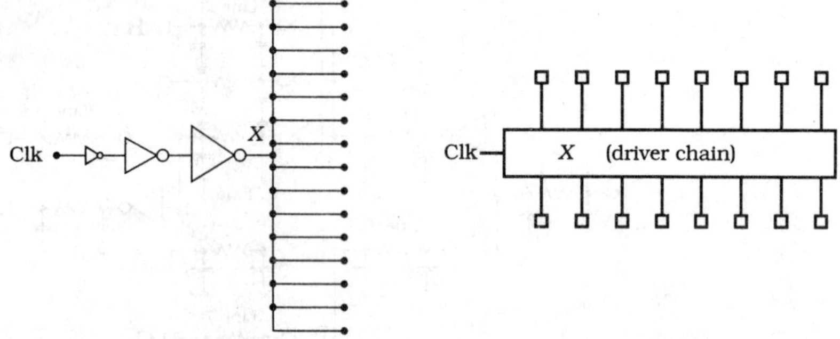

(a) Driver tree (b) Chip distribution

Figure 15.49 Single driver tree with multiple outputs

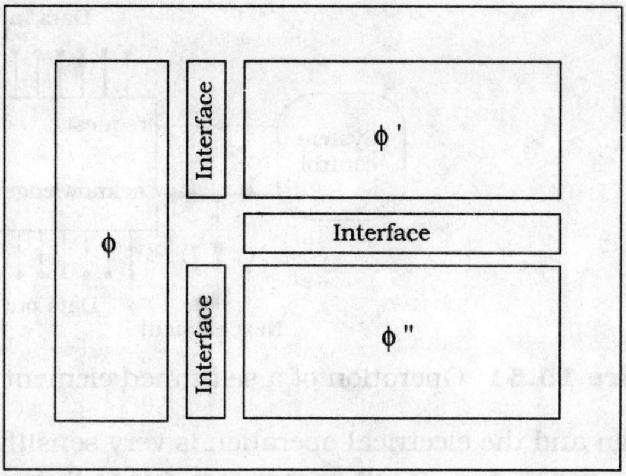

Figure 15.50 Asynchronous system clocking

behavior of the individual units. At the VLSI level, these would have the same frequency but different phases. Communication among the units is achieved using interface logic circuits.

Synchronous system operation can be achieved by using a slower global clock to control the interface circuits. This technique is well known in board-level designs. For example, the internal frequency of a microprocessor chip is much larger than the system clock used on the motherboard of a personal computer. In current designs, the fastest CPU frequencies are above 1 GHz, while board-level operations are around 100 to 200 MHz in the best designs. Clock multiplier circuits are used to take the board frequency up to the processor level. A more efficient approach for single-chip VLSI designs is to use phase-lock techniques to communicate between units. This allows fast system operation so long as the interfacing is relatively infrequent.

An asynchronous technique that has received a lot of attention in the literature is that of **self-timed** systems. A self-timed system is usually totally asynchronous and does not use any externally supplied clocks. Instead, a self-timed element uses internally generated signals to time its operations. A simple view of a self-timed element is shown in Figure 15.51. System interfacing is achieved by handshaking lines that transmit a request signal to the element to perform an operation; the element responds with an acknowledge signal when the operation is completed. The acknowledge signal may be sent back to the control unit, or to the next element to create a self-timed cascade. The acknowledge signal itself is usually an internally generated *analog* delay. Self-timed circuits are used regularly in many applications such as in DRAM array-accessing circuits. They are fast and generally require less area.

Theoretically, an asynchronous machine gives the highest performance, so there is continued interest in using self-timed networks. The drawback to self-timed VLSI systems is that the circuits are difficult to

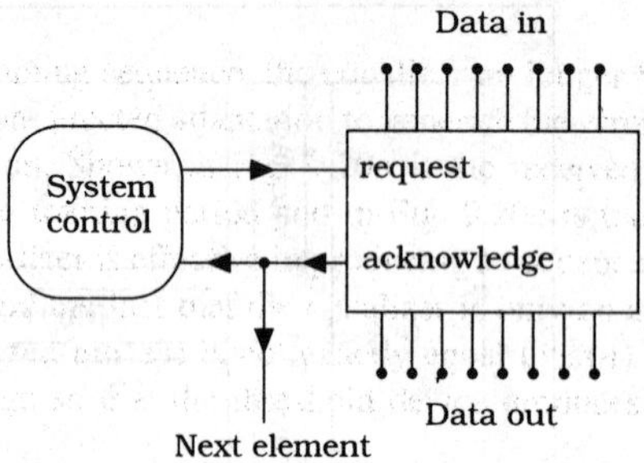

Figure 15.51 Operation of a self-timed element

design and the electrical operation is very sensitive to standard variations of the device and circuit parameters due to fluctuations in the processing. This makes it difficult to achieve a VLSI design that can be mass-produced in a regular manufacturing line. A practical implementation problem with using self-timed circuits in a VLSI design is that the development of adequate CAD toolsets that can handle self-timed circuit-to-system problems has lagged.

15.5 System Design Considerations

A recurring theme in this book is that CMOS VLSI system design should not be viewed as just another digital implementation technology. Although a high-level architectural model is used to initiate the design cycle, the logical algorithms, circuit design styles, and the physical design aspects all contribute to the performance of the final product. This view is important to keep in mind as it provides the key to developing new architectures and design styles for the next generation.

VLSI design choices are often dictated by weighing all known relevant factors and then making selections based on satisfying the most critical specifications. Secondary considerations are examined to insure that they will not have adverse effects. Several design cycles may be needed to work out all of the details to the point where an acceptable design is achieved. In our approach to VLSI we have studied the translation of logic networks from high-level abstract descriptions down to the physical design level, and have discovered that there are a myriad of choices at every level. It is possible, however, to highlight some of the major VLSI-specific design ideas that have evolved over the years. We will be content to restrict our discussion to a few topics. An excellent system-level view of designing an entire microprocessor chip is presented in Reference [12].

15.5.1 Bit Slice Design

Consider an arithmetic and logic unit (ALU) that accepts n-bit words A and B and produces an n-bit result C as in Figure 15.52. At the architec-

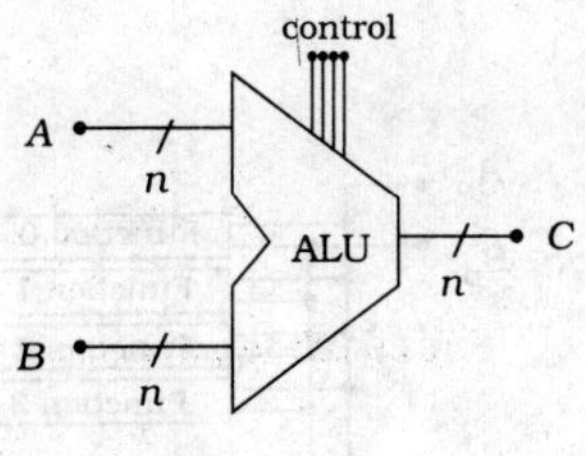

Figure 15.52 An n-bit ALU

tural level, the ALU is described by writing

$$C = C(A, B; \text{control}) \tag{15.30}$$

with the control signals determining the functional dependence of C on the inputs A and B. We may use the block diagram shown in Figure 15.53 to describe the operation of the ALU. In this view, the A and B inputs are fed to different logic sections within the large Logic block. Every section accepts the n-bit words as inputs, but produces a distinct n-bit output f_k where $k = 0, 1, \ldots, (m-1)$ where m is the number of functions. All results are available, but the n-bit ALU output C is determined by the control signals applied to an m:1 MUX.

Bit slice design is based on the fact that logic deals with data at the bit-level. Word-size operations are obtained using individual parallel-wired bit-level circuits that perform the same operation. Figure 15.54 is the bit-level equivalent for the p-th bit of the overall block diagram in Figure 15.53. The inputs are A_p and B_p for p in $[0, n-1]$; the output C_p is the result. With the bit slice philosophy, an n-bit ALU is created by paralleling n identical slices as shown in Figure 15.55. Note that the repetition seen in the block diagram will also appear at the logic, circuit, and silicon levels. The advantage of this approach is that once the slice is designed it can be stored in a library and instanced as needed to create the desired ALU width. This is analogous to creating a group of 1-bit datatypes in Ver-

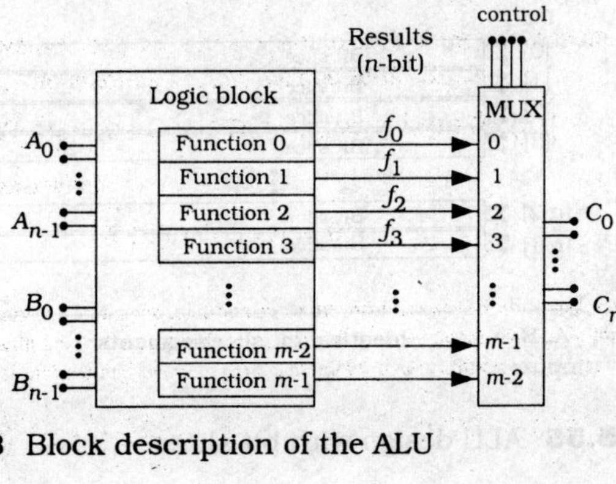

Figure 15.53 Block description of the ALU

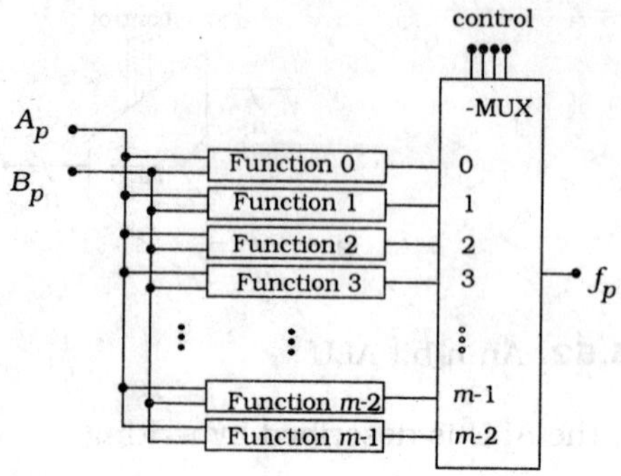

Figure 15.54 An ALU bit slice

ilog of the form

```
reg A, B, C ;
```

and later expanding it to a 32-bit word by changing the code to

```
reg [ 31,0 ] A, B, C ;
```

Bit slice design is very powerful because of this aspect.

At the VLSI level, experience shows that applying the bit slice concept may or may not improve the design. Individual interconnect wiring is much simpler in a single slice layout, and this factor alone makes the technique attractive. However, some circuits, such as CLA networks, work best with 4-bit or larger input groups. This is because the algorithms themselves depend on having simultaneous access to several bit positions in the input words. Dividing up the words into single-bit circuits complicates the wiring among adjacent slices. In these cases, it may be more efficient to use multibit slices in the same manner. An alternate approach is to vary the width of the slice through the datapath.

A problem that enters at the physical design level is the size and shape of the slice when it is built on silicon. The slices should pack together to form a block that fits into the floorplan, but this sometimes leads to indi-

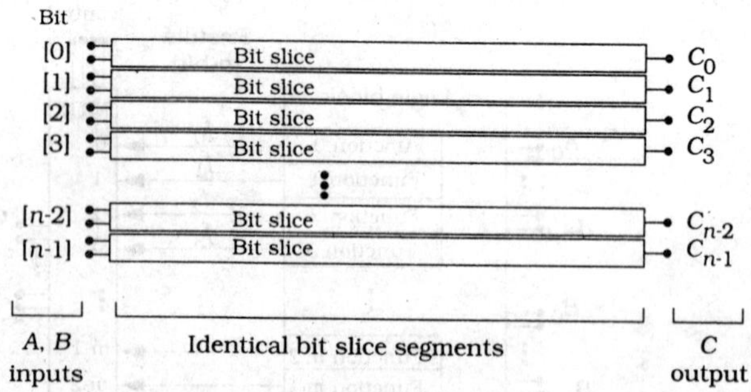

Figure 15.55 ALU design with bit slices

vidual slices that have disproportionate cell aspect ratios *width:height*. If one is much larger than the other, then long interconnects may become a problem. On the other hand, if *width:height* is around 1:1 (indicating a square), then it may be difficult to build up a word-size unit with a convenient shape.

This example illustrates some of the problems that arise when applying a simple architectural idea to a VLSI design. As we move downward in the design hierarchy, the problem remains the same but the viewpoint and consequences change.

15.5.2 Cache Memory

In microcomputer design, large system memories are located on system boards, and are physically and electrically far from the central processor unit (CPU). Moreover, the system board clock is much slower than the internal CPU clock, so that accessing the main memory is usually considered to be one of the slowest operations in the datapath.

Cache memory is designed to be placed in between the CPU and the main memory to speed up the system operation. Cache consists of small segments of fast memory (SRAM) that can be used locally as read/write storage by the processor. It communicates with the system memory and allows blocks of data to be transferred. The concept is illustrated in Figure 15.56. Direct memory access in Figure 15.56(a) can be slow. Adding on-chip cache as shown in Figure 15.56(b) allows the system to maintain chip-level speeds. The closest cache is called L1 (Level 1) cache in the hierarchy; it is usually an on-chip array.[5] The drawing shows two types of local memory that are found in CPU design. I-cache is used to hold instructions that are fetched from the main memory where the program code is stored. Keeping the I-cache as full as possible allows the program to run freely. The data-cache (D-cache) block holds the results of computations; these may be transferred to main memory or held as operands for subsequent instructions.

The I-cache and D-cache units are used in superscalar computer architectures where multiple pipeline datapaths are employed to increase

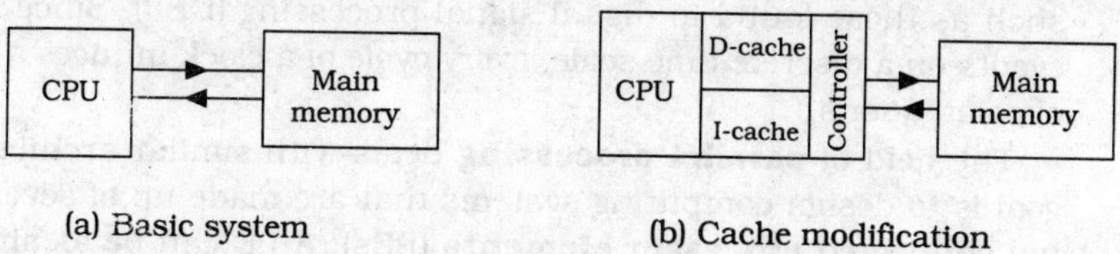

Figure 15.56 Adding cache memory

[5] Level 2 (L2) cache is the next level beyond the L1 cache.

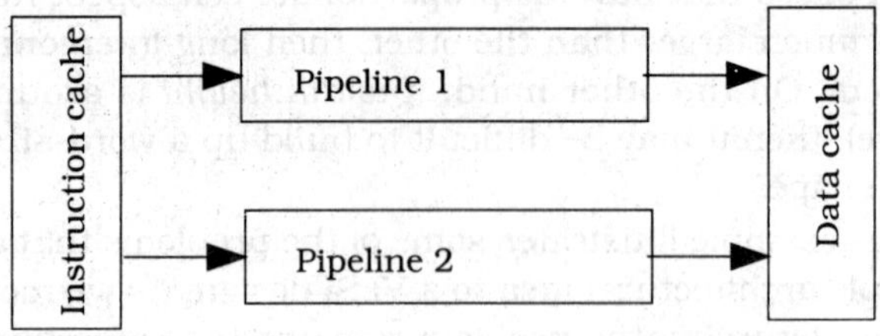

Figure 15.57 Block diagram of a dual-issue superscalar machine

the instruction processing speed. Figure 15.57 shows a simple block diagram for a dual-issue superscalar design. Instructions are stored and sorted in the I-cache, and are directed to either pipe depending on the necessary sequencing. Results are accumulated in the D-cache for use as needed. The cache controller circuits direct the data flow to and from the main memory. It must keep the main memory updated with new results and also keep the instruction cache as full as possible. The system-level view of cache and superscalar design is discussed in many textbooks on computer architecture, and will not be covered here.

In VLSI design, the need for cache is based on physical limitations that affect system performance. It is not possible to provide large amounts of memory on the chip due to the large area needed. Since we expect a slowdown at any input/output port, on-chip memory is highly desirable. The size of a cache unit cannot be too large. However, since it is added to the system to enhance the speed, SRAM circuits are used even though they consume more area than DRAM cells. Library designs are useful since they can be instanced directly onto the chip and are well characterized.

15.5.3 Systolic Systems and Parallel Processing

System-level VLSI allows us to visualize the dataflow through large systems that consist of many components. In a **systolic system**, the movement of the data is controlled by a clock, moving one phase on each cycle. The name is an analogy to the pumping of blood by the heart. Systolic design is ideal for hardwire implementation of various type of algorithms such as those found in digital signal processing (DSP). Since DSP treats events on a discrete time scale, every cycle of a clock induces a new group of calculations.

The field of **parallel processing** deals with similar architectures. Its goal is to design computing systems that are made up of several individual connected **processor elements** (PEs). A PE can be as simple as an AND gate, or as complex as a general-purpose microprocessor. A parallel machine is designed to have many PEs work on different parts of a program in a simultaneous manner. In a VLSI implementation, a PE acts as a unit that can be replicated by instancing. Figure 15.58 provides a visual-

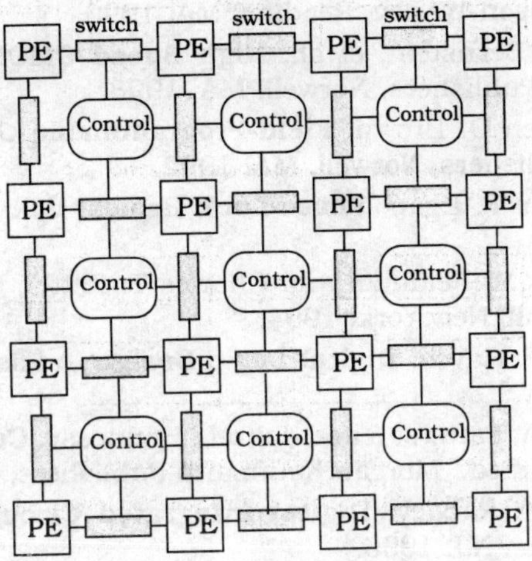

Figure 15.58 Regular patterning in a parallel processing network

ization of the basic features found in a parallel processing network. Individual processor elements communicate via switching arrays, which are controlled by either local or global signals. The communication paths are dictated by the hardwired switches and the control networks. Every base element is repeated throughout the system, making use of one of the strongest aspects of VLSI. The allure of creating a giant processor of this type fostered efforts to achieve wafer scale integration (WSI) designs. This has not yet been achieved because of the difficulty involved in manufacturing a defect-free wafer.

15.5.4 Summary

These short examples were discussed to emphasize the interaction of the different levels within the design hierarchy. Trade-offs must be made to insure that the finished product has the desired characteristics. Altering the design at one level usually induces the need to change the design at most other levels. This is, of course, the viewpoint that has been propagated throughout the book.

VLSI systems design is a challenging field because of the number and types of problems that arise. New products always produce surprises, especially if the chip is based on state-of-the-art considerations.

15.6 References for Advanced Reading

[1] Abdellatif Bellaouar and Mohamed I. Elmasry, **Low-Power Digital VLSI Design**, Kluwer Academic Publishers, Norwell, MA, 1995.

[2] H. B., Bakoglu, **Circuits, Interconnections, and Packaging for VLSI**, Addison-Wesley, Reading, MA, 1990.

[3] Kerry Bernstein, et al., **High Speed CMOS Design Styles**, Kluwer Academic Publishers, Norwell, MA, 1998.

[4] Stephen D. Brown, **Field-Programmable Gate Arrays**, Kluwer Academic Publishers, Norwell, MA, 1992.

[5] William F. Egan, **Phase-Lock Basics**, Wiley-Interscience, New York, 1998.

[6] James M. Feldman and Charles T. Retter, **Computer Architecture**, McGraw-Hill, New York, 1994.

[7] John P. Hayes, **Digital Logic Design**, Addison-Wesley, Reading, MA, 1993.

[8] David A. Patterson and John L. Hennessy, **Computer Organization & Design**, 2nd ed., Morgan Kaufmann Publishers, San Francisco, 1998.

[9] Jan M. Rabaey, **Digital Integrated Circuits**, Prentice-Hall, Upper Saddle River, NJ, 1996.

[10] M. Sarrafzadeh and C. K. Wong, **An Introduction to VLSI Physical Design**, McGraw-Hill, New York, 1996.

[11] Navid Sherwani, **Algorithms for VLSI Physical Design Automation**, Kluwer Academic Press, Norwell, MA, 1993.

[12] Bruce Shriver and Bennett Smith, **The Anatomy of a High-Performance Microprocessor**, IEEE Computer Society Press, Los Alamitos, CA, 1998.

[13] Ivan Sutherland, Bob Sproul, and David Harris, **Logical Effort**, Morgan Kaufmann Publishers, San Francisco, 1999.

[14] John P. Uyemura, **A First Course in Digital Systems Design**, Brooks-Cole Publishers, Pacific Grove, CA, 2000.

[15] John P. Uyemura, **CMOS Logic Circuit Design**, Kluwer Academic Publishers, Norwell, MA, 1999.

Reliability and Testing of VLSI Circuits

16

VLSI testing deals with techniques that are used to determine if a die behaves properly after the fabrication sequence is completed. If a die passes the testing phase, it is packaged and sold. **Reliability** is concerned with projecting the lifetime of a component once it is placed into operation.

16.1 General Concepts

Let us begin our study by examining what happens to a wafer once the fabrication steps have been completed. The end result of weeks of processing is an array of die sites, each of which is potentially a functional circuit that can be packaged and sold. Unfortunately, not every circuit will operate as designed because of problems or random variations that occur during the processing. Every site must be subjected to a round of electrical tests that are chosen to determine whether the circuit is good or bad.

The wafer testing procedure is illustrated schematically in Figure 16.1. A **test probe** head allows electrical contact to the I/O points of a die. Several sets of stimuli are applied to the inputs and the response is taken from the outputs. The system is programmed to accept or reject the die based on this set of measurements. A bad die is marked for future reference. After every die has been tested, the wafer is scored along lines that run in between the individual sites. Applying a little pressure to the wafer induces cracks along the score lines, resulting in individual dice without ruining the circuitry. Good circuits enter the final assembly phase where robotic equipment is used to place the die into a package, connect the die to the package electrodes, and then seal the package.

Once an integrated circuit is used in a board design, the reliability of

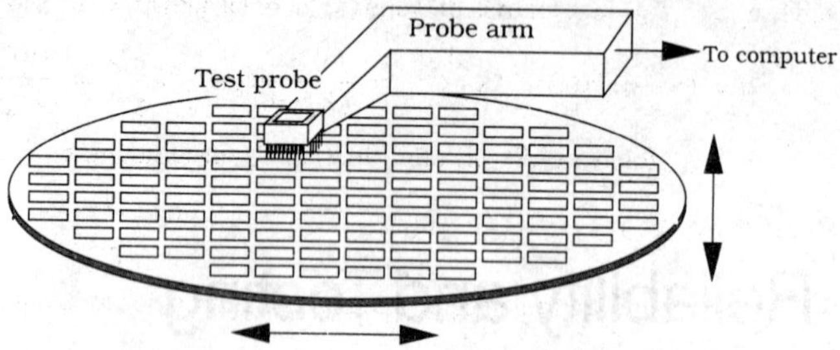

Figure 16.1 Visualization of wafer testing

the chip becomes an important factor. Every electronic circuit eventually "wears out" and VLSI chips are no exception. The projected **lifetime** is the number of hours of operation that can be expected before a failure occurs. This is described graphically using a **bathtub curve**[1] shown in Figure 16.2 which plots the number of failures of a given system as a function of time. Bathtub curves are semilog plots with time t in hours, and points plotted as $\log_{10}(t)$. There are three general regions shown in the curve. **Infant mortalities** are failures that occur after a very short period of time, i.e., early in the life of the system. These tend to arise from manufacturing defects that manifest themselves after a few hours of operation. The central portion of the curve represents **random failures** during normal operation, while **wear-out** describes the end of life.

The number of infant mortalities can be large for any device, especially during the initial months of production when the design and manufacturing are being refined. Once a circuit is mounted on a board, the cost of

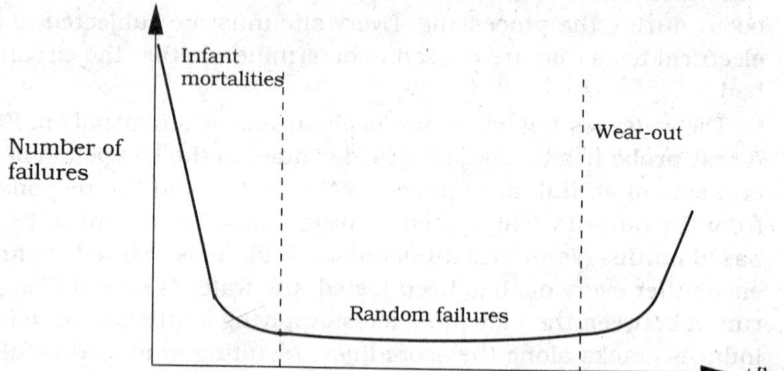

Figure 16.2 Bathtub reliability curve

1 The name "bathtub curve" is obtained from the shape of the reliability plot.

repair exceeds the cost of the IC itself, so it is desirable to weed out as many infant mortalities as possible by performing a **burn-in** operation. During a burn-in, the circuits are operated under stressed conditions with higher-than-normal voltage levels, high temperatures, and high-humidity environments. The idea is to induce potential failures to occur during burn-in and avoid using the devices at the board level. This increases the system reliability considerably. Electronic parts vendors, including VLSI chip manufacturers, usually provide some type of written warranty for their products. It is to both sides' advantage to produce reliable components.

Burn-in systems are also used to obtain reliability data. A large sample of circuits are powered up and placed in oven/humidity chambers. The test group is monitored for degradation and/or failures, with the time noted for each. One year has only 24 × 365 = 8760 hours, so that some test systems are operated continuously for several years! One technique is to measure a critical parameter (or a set of parameters) and track the degradation in time. An idealized example is shown in Figure 16.3 for a situation where the power supply current I of a circuit is found to increase over time. The maximum allowed current consumption is denoted by the Max level. Data taken at regular intervals gives data points that show the high value, the low value, and a mean value for the test group. This can be used to extrapolate the expected time. The simplest straight-line technique approximates the current using the dashed line shown. Extrapolating to the Max level would then give an estimate of the lifetime of the device. One problem with this approach is that a slight change in the slope of the line changes the projected lifetime significantly due to the logarithmic time scale.

16.1.1 Reliability Modeling

Reliability data can be used to construct mathematical models that are useful in predigesting failure rates. Suppose that we test a group of devices starting at time $t = 0$. As time progresses, a device fails at time t_1,

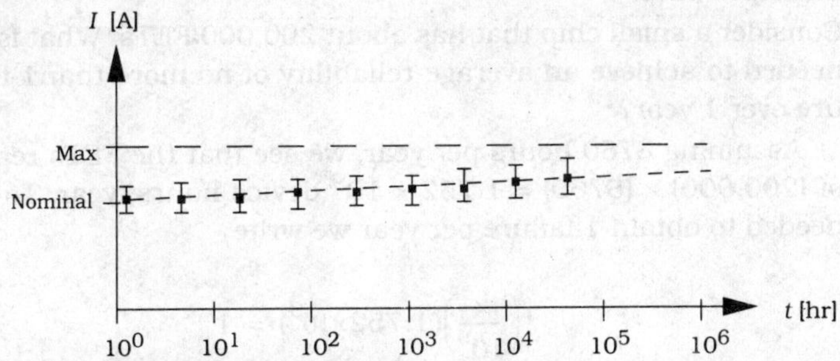

Figure 16.3 Example of test data results

then one at time t_2, and so on. At the end of the test period, we find that N devices fail. The total number of operational hours for the group is the sum

$$T = \sum_{i=1}^{N} t_i \tag{16.1}$$

that we will measure in hours. With this, we can define the **average failure rate** of the test group as the number of failures divided by the total number of operational hours:

$$\lambda_{av} = \frac{N}{T} \tag{16.2}$$

The **mean time to failure** (MTTF) is given by

$$\text{MTTF} = \frac{T}{N} = \frac{1}{\lambda_{av}} \tag{16.3}$$

and represents an average lifetime for the test group.

The units used to describe failure rates acknowledge that the values are very small. One of the more commonly used metrics is the **FIT**, which stands for **failure-in-time**. This is defined by

$$1 \text{ FIT} = 1 \text{ failure in 1 million parts over 1000 hours} \tag{16.4}$$

or

$$1 \text{ FIT} = 1 \text{ ppm/K} \tag{16.5}$$

where **ppm** stands for parts per million. A failure rate of 1 FIT (10^9 operating hours) corresponds to a lifetime of about 125,000 years for one device out of a large sample. **Accelerated-stress life testing** where extreme conditions are used to simulate aging is often employed to determine reasonable FIT values.

Example 16.1

Consider a small chip that has about 200,000 FETs. What is the FIT value needed to achieve an average reliability of no more than1 transistor failure over 1 year?

Assuming 8760 hours per year, we see that the FETs represent a total of (200,000) × (8760) = 1.752×10^9 device hours/year. To find the FITs needed to obtain 1 failure per year we write

$$\left(\frac{x}{10^9}\right)(1.752 \times 10^9) = 1 \tag{16.6}$$

where x is the FIT value. Solving gives $x = 0.67$ FITs as the required rate.

This simple approach to failure analysis provides valuable insight, but is not sufficient to yield accurate estimates. More sophisticated mathematical models provide higher confidence levels, and are used throughout the field. To develop a general approach, suppose that we measure the failures in a test group and find that a function of time t. We introduce a **probability density function** (PDF) $f(t)$ as a function of time such that

$$f(t)\,dt = \text{fraction of failures in the time increment } dt \tag{16.7}$$

A simple model is the exponential function

$$f(t) = \lambda e^{-\lambda t} \tag{16.8}$$

where λ is a constant. The **cumulative distribution function** (CDF) $F(t)$ is related to the PDF by

$$F(t) = \int_0^t f(\eta)\,d\eta \tag{16.9}$$

where η is a dummy integration variable. For the exponential distribution, we may integrate equation (16.8) to obtain

$$F(t) = 1 - e^{-\lambda t} \tag{16.10}$$

Device lifetimes are described by models known as **life distributions**. The CDF is a life distribution that can be interpreted as saying

- $F(t)$ is the probability that a random unit of the test group will fail by t hours.

Alternately,

- $F(t)$ is the fraction of units in the test group that fail by t hours.

Both are useful in practice. The **reliability function** $R(t)$ is defined by

$$R(t) = 1 - F(t) \tag{16.11}$$

and describes the units that have not failed, i.e., the ones that are still functioning after t hours.

These functions allow us to define the **hazard rate** $h(t)$ by

$$h(t) = \frac{1}{R(t)}\left(\frac{dF}{dt}\right) \tag{16.12}$$

This can be interpreted as the probability of a unit failing in a time between t and $(t + dt)$. Evaluating the derivative gives

$$h(t) = \frac{f(t)}{R(t)} \tag{16.13}$$

The hazard rate is also called the **instantaneous failure rate** or just the **failure rate** in the literature.[2] The bathtub curve in Figure 16.2 is an example of a hazard rate plot. Integrating the hazard rate gives the **cumulative hazard function**

$$H(t) = \int_0^t h(\eta)d\eta \tag{16.14}$$

This can be shown to be equal to

$$H(t) = -\ln[R(t)] \tag{16.15}$$

which gives

$$F(t) = 1 - e^{-H(t)} \tag{16.16}$$

as the relation between $F(t)$ and $H(t)$.

The **average failure rate** (AFR) between two times $t_2 > t_1$ is obtained by the ratio of the failure rate to the duration of the time interval. Assuming $t_2 = T$ and $t_1 = 0$ for simplicity gives

$$AFR(T) = \frac{1}{T}\int_0^T h(\eta)d\eta \tag{16.17}$$

This may be expressed in the alternate forms

$$AFR(T) = \frac{H(T)}{T} = \frac{\ln R(T)}{T} \tag{16.18}$$

The AFR is very useful to specify the failure rate of a device that is operated for T hours.

Let us examine the application of these functions with the exponential distribution. Suppose that we perform an accelerated stress life testing experiment on a test group of chips and plot the number of failures to arrive at the histogram shown in Figure 16.4(a). We can model this data with the continuous distribution

$$f(t) = \lambda e^{-\lambda t} \tag{16.19}$$

where the actual value of λ can be estimated from the curve. The curve is shown in Figure 16.4(b). Since

$$F(t) = 1 - e^{-\lambda t} \tag{16.20}$$

the failure rate is

2 We note that some treatments define the failure rate as $f(t)$ without dividing by $R(t)$.

$$h(t) = \frac{1}{R(t)}\left(\frac{dF}{dt}\right) = \lambda \tag{16.21}$$

i.e., the failure rate is a constant in time with a value of λ failures/hour. This is characteristic of the simple modeling provided by the exponential function. The average failure rate is computed as

$$AFR(T) = \frac{\lambda T}{T} = \lambda \tag{16.22}$$

which is also constant. In general, the MTTF is given by moment

$$\text{MTTF} = \int_0^\infty t f(t)\, dt \tag{16.23}$$

For the exponential distribution this becomes

$$\text{MTTF} = \int_0^\infty t\lambda e^{-\lambda t}\, dt = \frac{1}{\lambda} \tag{16.24}$$

using an integration by parts. The exponential distribution gives the same results as our simpler expressions that were written using intuition. The limiting factor of the analysis is the assumption of a constant failure rate.

Different distribution functions $f(t)$ can be introduced to model situations with non-constant failure rates. The **Weibull distribution** employs

$$f(t) = \frac{m}{t}\left(\frac{t}{c}\right)^m e^{-(t/c)^m} \tag{16.25}$$

where $m > 0$, and c (> 0) is called the **characteristic life**. This results in a

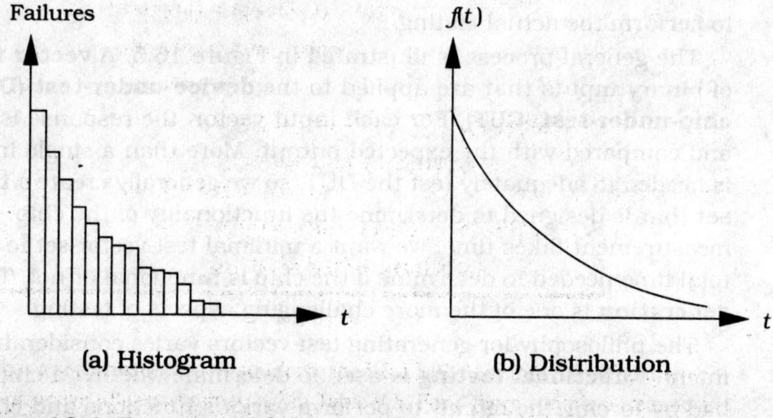

Figure 16.4 Exponential failure model

failure rate of

$$h(t) = \frac{m}{t}\left(\frac{t}{c}\right)^m \tag{16.26}$$

and an average failure rate of

$$AFR(T) = \frac{1}{c}\left(\frac{T}{c}\right)^{m-1} \tag{16.27}$$

The shape of the distribution can be chosen by adjusting m. Odd values of m yield curves that look similar to exponentials, while even values of m produce distributions that are more Gaussian-like, with peaks at certain c-values. For example, $3 \leq m \leq 4$ yields a bell-shaped curve. Other distributions, including normal and log-normal functions, are also used in reliability modeling.

Reliability modeling is a fascinating field of study that employs statistical analysis of data in an effort to determine projected lifetimes and failure rates. A particularly challenging problem is the design of experiments and associated models that provide meaningful data. As the complexity of VLSI processing equipment increases, reliability issues become more critical. Many engineers and scientists with strong backgrounds in physics and statistical modeling pursue careers in this field.

16.2 CMOS Testing

Let us now direct our discussion to the problem of testing digital CMOS circuits. The idea is quite simple, Given a digital integrated circuit, we want to determine whether or not it operates correctly. It is assumed that the logic is correct and that it should be possible to build a working chip. Since the functionality is defined by the design, we can calculate how the network should respond to a set of input stimuli and use this knowledge to perform the actual testing.

The general process is illustrated in Figure 16.5. A **vector** is an array of binary inputs that are applied to the **device-under-test (DUT)** or the **chip-under-test (CUT)**. For each input vector, the response is measured and compared with the expected output. More than a single input vector is needed to adequately test the DUT, so we generally create a **test vector set** that is designed to determine the functionality of the chip. Since each measurement takes time, we want a minimal test vector set to reduce the total time needed to determine if the chip is functional or not. **Test vector generation** is one of the more challenging aspects of testing.

The philosophy for generating test vectors varies considerably with the intent. **Functional testing** is used to determine whether a chip is good or bad by forcing the circuit to perform various functions and checking the response. While this may sound straightforward, brute-force selection of

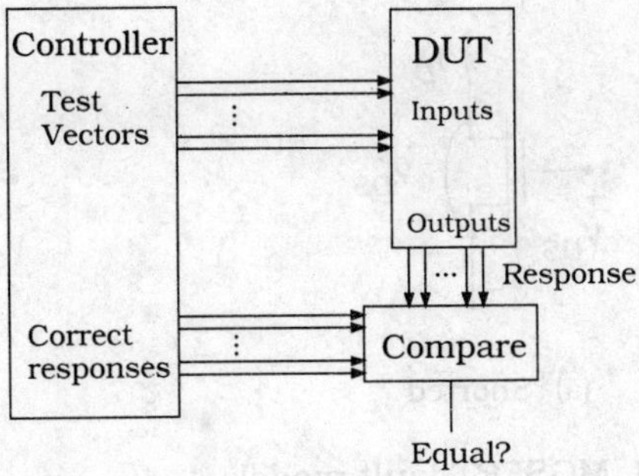

Figure 16.5 Overview of the testing problem

an exhaustive set of test vectors may lead to extremely long test intervals.

Fault modeling is more complicated. Prototype die are used to characterize a fabrication process for problems such as short circuits, open circuits, and bad transistors. Once these faults are identified at the physical level, they are used to generate a test vector set that specifically looks for the problems. Although fault modeling requires a substantial amount of work, it is useful since it searches for known problems and provides feedback to the processing line. Wafer analysis can be performed in a laboratory to verify the cause of the fault, and make corrections in the fabrication step(s) responsible. In the long term, fault modeling helps increase the yield and reliability of the design.

16.2.1 Fault Models in CMOS

Physical-level CMOS faults can usually be modeled with simple equivalent circuits. The models can be used in logic circuits to determine the measurable effects that a particular fault will have on the operation of a gate or logic unit. The basic fault models for FET circuits are relatively simple.

A **short-circuited** FET is one that always conducts drain-source current with an applied drain-source voltage V_{DS}; the gate has no control over the operation. This is also called a **stuck-on** fault. An **open-circuit** or **stuck-off** fault is exactly opposite: current never flows regardless of V_{GS} or V_{DS}. The circuit models for these two faults are somewhat obvious but are shown in Figure 16.6 for completeness. Physically, these problems tend to be due to metallization or etching problems, or mask registration errors.

Two logic-based faults are known as **stuck-at-0 (sa0)** and **stuck-at-1 (sa1)** problems. Stuck-at faults apply to lines that cannot change voltages due to short circuits or other processing-related failures. The simplest visualization is that a node is accidentally connected to the power supply (sa1) or ground (sa0). The effect of a stuck-at fault varies with the loca-

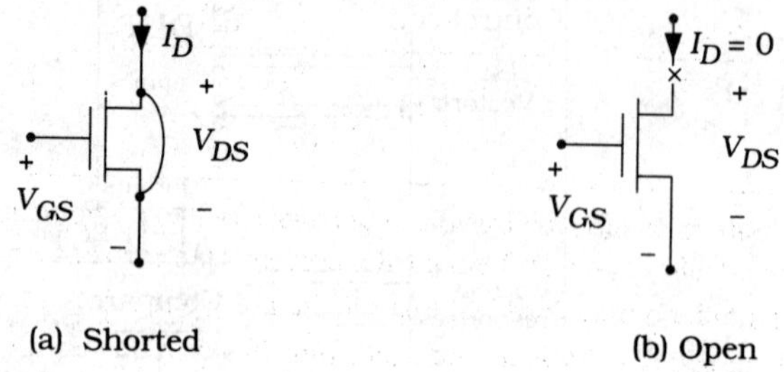

Figure 16.6 MOSFET fault models

tion. For example, an sa0 fault at the gate of an nFET implies that it can never turn on, while an sa0 fault at the gate of a pFET yields a transistor that cannot be shut off. Obviously, these affect the operation of the logic network. A related set of faults are gate-drain and gate-source shorts. Figure 16.7 shows examples of fault models for these defects.

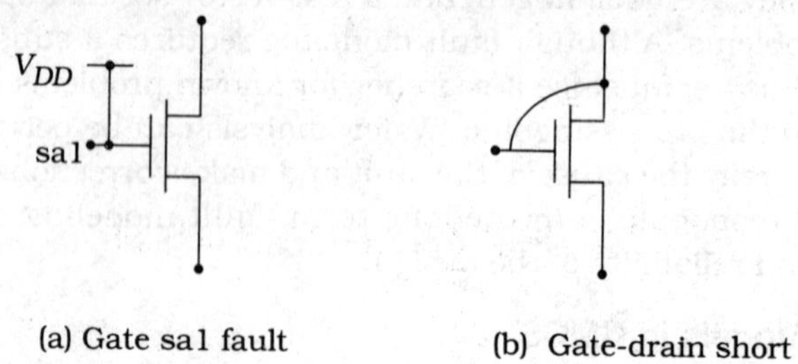

Figure 16.7 Fault model examples

Gate-oxide shorts (GOSs) are specific to MOSFETs. They occur when the insulating gate oxide has a defect and the gate material contacts the substrate as shown in Figure 16.8(a) for an nFET. Assuming an n-doped poly gate, the GOS creates a parasitic pn junction diode between the gate and the p-type substrate. The gate voltage has no control over the drain current which makes the circuit non-functional. Gate-oxide shorts are found in both nFETs and pFETs and tend to originate from nonuniform growth of the gate oxide due to defects on the wafer surface. Many GOS problems are found to occur in local clusters on the wafer. In this case, the affected die(s) tend to have many GOS-defective FETs, which makes them easier to find.

16.2.2 Gate-Level Testing

Fault models are used to characterize failures in logic gates. Creating different fault circuits allows one to find a set of test vectors that can be

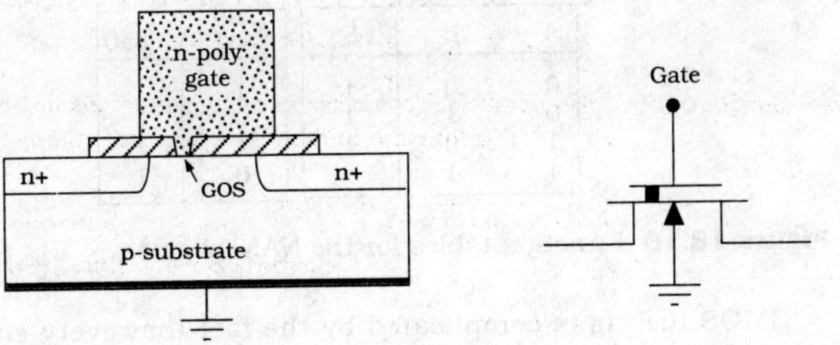

(a) Physical origin

(b) Circuit designation

Figure 16.8 Gate-oxide short in an nFET

applied to the circuit. Let us examine a simple NAND2 gate with inputs *A* and *B* as an example. In Figure 16.9(a), a stuck-at-1 fault at the gate of pFET MpA keeps the transistor in cutoff, while MnA is always conducting. The circuit in Figure 16.9(b) has a stuck-at-0 fault at the *B*-input. This prevents MnB from turning on, and keeps MpB in a conducting state. Both circuits represent distinct cases. The circuits can be used to derive the test vectors needed to find each problem. The function tables in Figure 16.10 provide the necessary information. The normal response of the NAND gate is shown as *F*. The response for the sa1 fault in Figure 16.9(a) is denoted by F_{sa1}. Since MpA never conducts, the output of the gate cannot be pulled to a logic 1 with an input of $(A,B) = (0,1)$. This vector can be used to test for this problem since it should produce a logic 1 output. The sa0 fault in Figure 16.9(b) causes the output to behave as summarized in the F_{sa0} column. In this case, MpB is always on so that the output is stuck at a 1. Using an input vector of $(A,B) = (1,1)$ would find this fault.

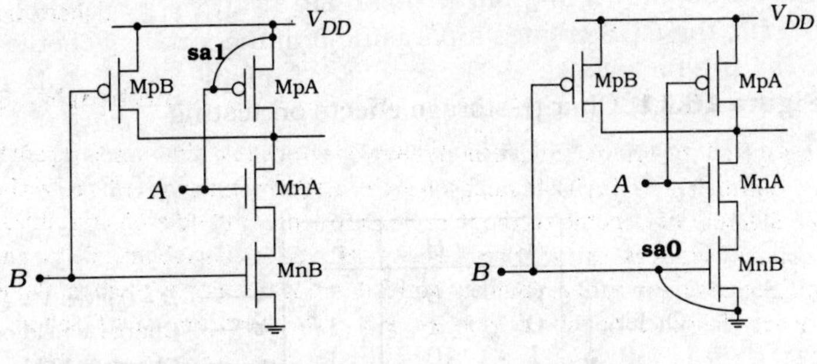

(a) Stuck-at-1 fault

(b) Stuck-at-0 fault

Figure 16.9 NAND2 gate with stuck-at faults

A	B	F	F_{sa1}	F_{sa0}
0	0	1	1	1
0	1	1	0	1
1	0	1	1	1
1	1	0	0	1

Figure 16.10 Function tables for the NAND2 gate

CMOS testing is complicated by the fact that every circuit node is capacitive and has the ability to store charge for a short period of time. If we apply a set of input test vectors to the gate, the response may be affected by this characteristic. Consider the open fault in the NAND2 gate of Figure 16.11. This prevents pFET MpA from conducting and should be detected by the input combination $(A,B) = (0,1)$. Note, however, that the output node has a capacitance C_{out} that cannot be eliminated. If we cycle through the inputs with the sequence $(A,B) = (0,0), (0,1), (1,0), (1,1)$, the stored charge may make it appear that the gate is operating properly. This statement is justified by the function table in Figure 16.12. The first input $(A,B) = (0,0)$ gives a logic 1 output, and the capacitor C_{out} has a voltage $V = V_{DD}$ across it. If we quickly apply the next input $(A,B) = (0,1)$ to insure a short testing cycle, then the output will still look like a logic 1 since C_{out} can hold the charge. Cycling through the remaining inputs gives normal

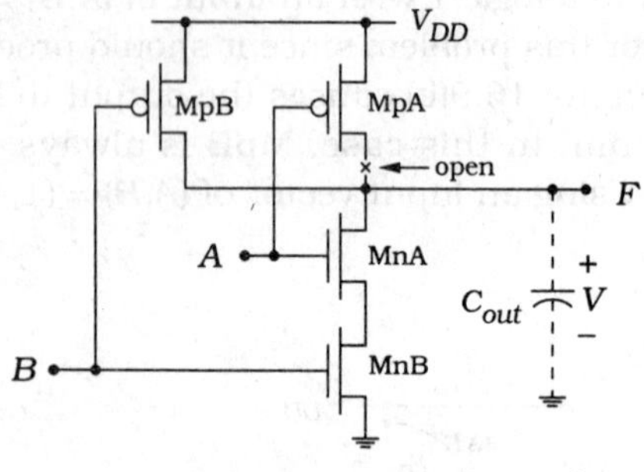

Figure 16.11 Charge-storage effects on testing

Order	A	B	F	
↓	0	0	1	← $B = 0$ gives $V = V_{DD}$
	0	1	1	← Charge is held on C
	1	0	1	← Charge is held on C
	1	1	0	← C discharge to 0 V

Figure 16.12 Function table for charge-storage problem

results, so we have missed the fault entirely.

To compensate for this problem, we use an **initialization vector** that "prepares" the gate for the actual test vector. In the present case, the sequence $(A,B) = (1,1), (0,1)$ would find the fault since the initialization vector $(A,B) = (1,1)$ discharges the output to 0 V, and the fault prevents $(A,B) = (0,1)$ from producing a logic 1 output.

Another type of problem arises with stuck-on and stuck-off faults. Consider the circuit in Figure 16.13(a) where MpA has a stuck-on (shorted) fault. If we apply an input vector of $(A,B) = (1,1)$, then MnA and MnB conduct along with MpA. This leads to the resistor equivalent model shown in Figure 16.13(b). The output voltage is given by the voltage-division rule as

$$V = \left(\frac{R_{nA} + R_{nB}}{R_{nA} + R_{nB} + R_{pA}}\right) V_{DD} \tag{16.28}$$

Since the nFET resistances depend on the aspect ratio while R_{pA} is due to the short, the voltage may give a low value of V which would make it appear that the gate is operating properly. This would be the case if the sum $(R_{nA} + R_{nB})$ is small compared to R_{pA}. If $R_{pA} \approx (R_{nA} + R_{nB})$ then V would be around one-half V_{DD} which may or may not be detected as an incorrect value.

16.2.3 I_{DDQ} Testing

Applying a power supply voltage to a CMOS chip causes a current I_{DD} to flow. When the signal inputs are stable (not switching), the quiescent leakage current I_{DDQ} can be measured. This is illustrated in Figure 16.14. Every chip design is found to have a range of "normal" levels. I_{DDQ} testing is based on the assumption that an abnormal reading of the leakage current indicates a problem on the chip. I_{DDQ} testing is usually performed at

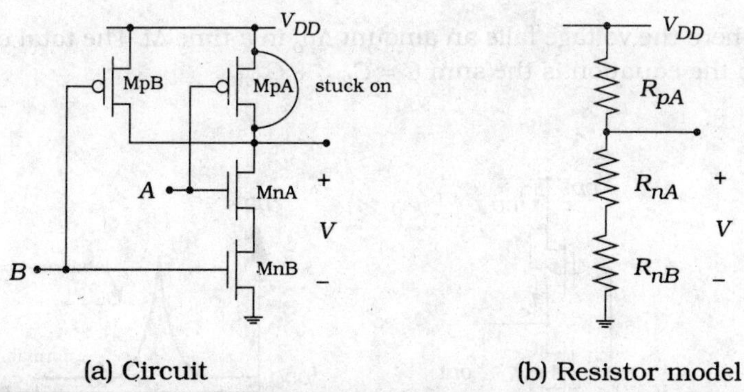

(a) Circuit (b) Resistor model

Figure 16.13 Stuck-on fault in a NAND gate

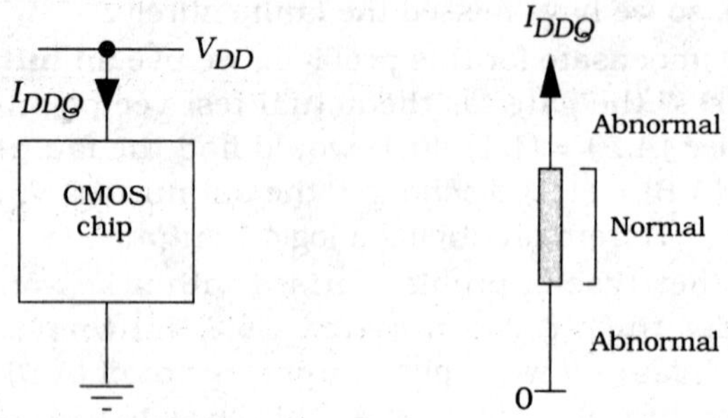

Figure 16.14 Basic I_{DDQ} test

the beginning of the testing cycle. If a die fails, it is rejected and no further tests are performed.

The origin of I_{DDQ} leakage is shown in Figure 16.15. When we sweep the input voltage V_{in} to a NOT gate, the power supply current I_{DD} varies as shown; the peak occurs at the midpoint voltage where $V_{in} = V_{in} = V_{out}$. When the input is in a stable logic 0 or logic 1 voltage range, only the quiescent leakage current I_{DDQ} flows. This consists of reverse-biased pn junction current, subthreshold contributions, and others. If a measurement yields an "abnormal value" of the leakage current, then it is assumed that something is wrong.

The components of a basic measurement system are illustrated in Figure 16.16. The test chip is modeled as being in parallel with the tester capacitance C_{test}. A power supply with a value V_{DD} is connected to the chip by a switch that is momentarily closed at time $t = 0$. The current I_{DD} is monitored by a buffer (a unity-gain amp) and gives the output voltage $v_o(t)$. The value of the current is estimated by

$$I_{DD} \approx C\left(\frac{\Delta v_o}{\Delta t}\right) \tag{16.29}$$

where the voltage falls an amount Δv_o in a time Δt. The total capacitance C in the equation is the sum $C = C_{test} + C_{chip}$.

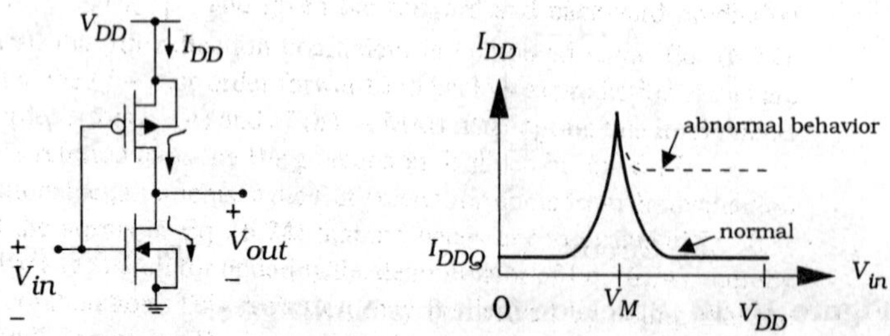

Figure 16.15 Leakage currents in a NOT gate

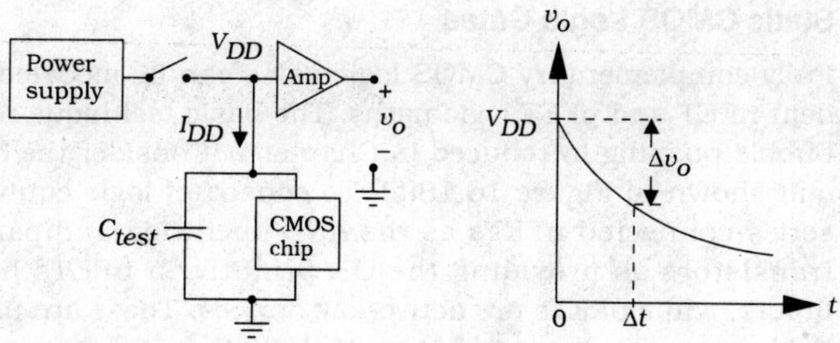

Figure 16.16 Components of an I_{DDQ} measurement system

I_{DDQ} testing can detect clusters of GOSs because they tend to increase the leakage levels. Figure 16.17 shows the situation where an nFET with a GOS is being driven by an inverter circuit. Since the GOS-fault in the nFET is the same as a reverse-biased pn junction, additional leakage currents flow in the circuit with the voltages shown. CMOS chips that are designed with static logic circuits can usually be tested using I_{DDQ} measurements; most ASICs fall under this category. While the technique has been applied to some dynamic circuits, careful attention must be directed toward the formulation of the test vector set and the interpretation of the measurements.

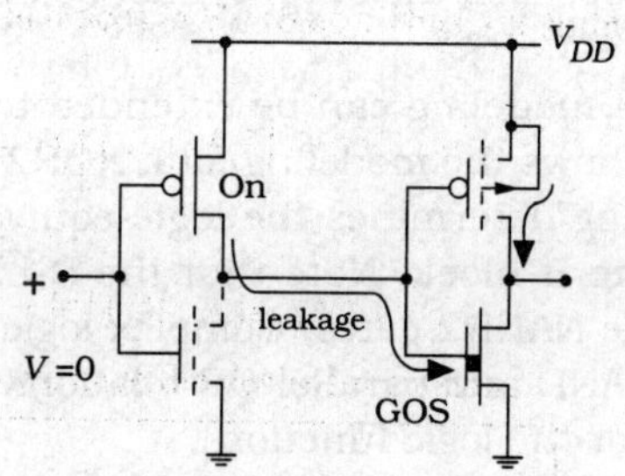

Figure 16.17 I_{DDQ} testing for GOS occurrences

16.3 Test Generation Methods

A physical fault can be transformed into a logical fault model that allows us to develop sets of test vectors. Many techniques have been developed for testing CMOS VLSI chips that use common circuit design styles. The problem can be understood by examining the relationship between the physical circuit and testing by means of examples.

16.3.1 Static CMOS Logic Gates

Fully complementary CMOS logic gates can be modeled using independent nFET and pFET logic paths. The basic technique resembles that of bubble pushing introduced in Chapter 2. Consider the NAND2 logic circuit shown in Figure 16.18(a). To construct logic equivalents, we view series-connected nFETs as the AND operation and parallel-connected transistors as providing the OR primitives; nFETs have active-high inputs, while pFETs are active-low devices. These are used to construct the logic model shown in Figure 16.18(b) that is characterized by separate n- and p-logic paths. The output of the n-path is $S_0 = a \cdot b$ while the p-path produces $S_1 = \overline{a} + \overline{b} = \overline{S_0}$. These are fed in a "B-logic block" that produces the output $f = f(S_0, S_1)$.

The operation of the B-logic block is summarized in the truth table of Figure 16.18(c). For normal NAND operation, $S_0 \neq S_1$ which is equivalent to $S_0 \oplus S_1 = 1$. These are characterized by the second and third lines that result in an output of $f = 1$ or $f = 0$, respectively. If $S_0 = S_1 = 0$, the output is $f = M$ which represents a memory state. In terms of the circuit, this implies that the output is floating due, for example, to FETs in stuck-off states. The last condition $S_0 = S_1 = 1$ is where both the nFETs and pFETs are conducting and the output is being pulled in both directions. The output is designated as "w0" which stands for a weak-0; this assumes that the nFET pull-down strength dominates the pull-up action of the pFETs. If this is not true, then the output must be changed to a weak 1 (w1) or an indeterminate state.

This type of modeling can be extended to arbitrary static logic gates. Figure 16.19 shows the modeling for a NOR2 gate. The series and parallel transistor wiring determines the logic-equivalent gates, and the outputs are fed into the B-block. Note that the B-block logic is taken to be the same as for the NAND2 gates. Complex logic gate models are built applying the series-AND and parallel-OR relations to groups of transistors. The AOI circuitry for the logic function

$$F = \overline{a \cdot b + c \cdot d} \tag{16.30}$$

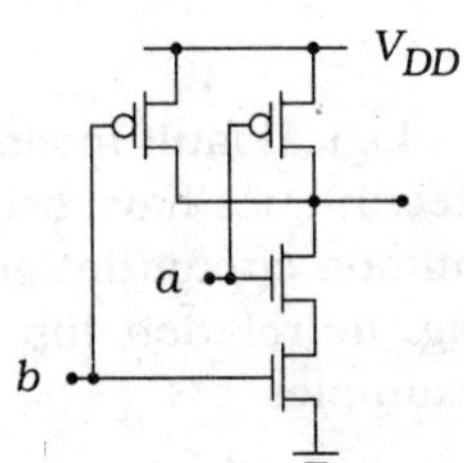

(a) Circuit

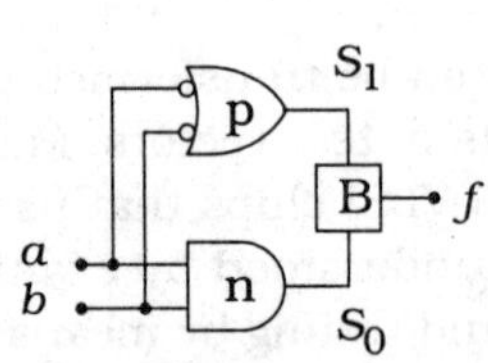

(b) Logic model

S_0	S_1	f
0	0	M
0	1	1
1	0	0
1	1	w0

(c) B-logic

Figure 16.18 NAND2 logic model

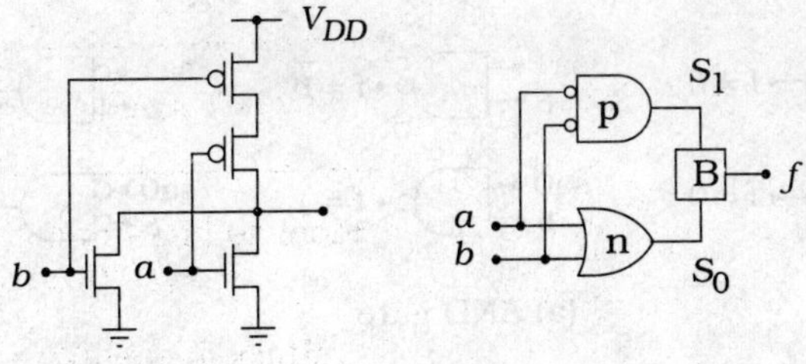

S_0	S_1	f
0	0	M
0	1	1
1	0	0
1	1	w0

(a) Circuit (b) Logic model (c) B-logic

Figure 16.19 Logic modeling of a NOR2 gate

is shown in Figure 16.20(a). The nFET logic-equivalent shows the AO patterning that produces S_0. The p-path uses assert-low inputs into an OA network with an output of $S_1 = \overline{S_0}$. The B-logic block produces an output of f that is equal to F if $S_0 \neq S_1$. A condition of $S_0 = S_1$ results if a fault occurs in one of the logic paths.

16.3.2 Logical Effects of Faults

Once an equivalent logic network has been derived, we can apply fault modeling to various points and analyze the effects at the outputs. The circuit viewpoint has already been discussed in Section 16.2.2 for the case of static CMOS gates. Simple logic-level modeling provides a few more useful pieces of information.

Consider the effect of stuck-at faults sa0 and sa1. Figure 16.21 shows the effects of these faults when they occur at the inputs to basic logic gates. The AND-family response is summarized in Figure 16.21(a). The outputs are calculated using the appropriate logic table, and assume that the input fault shown is the only possible problem. For example, the simple AND operation

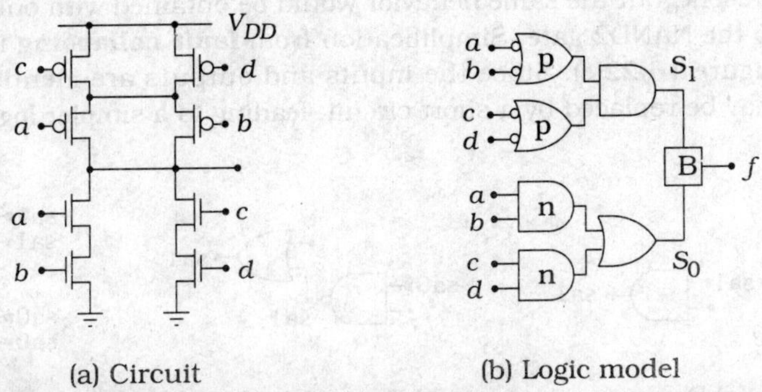

(a) Circuit (b) Logic model

Figure 16.20 Logic modeling for an AOI gate

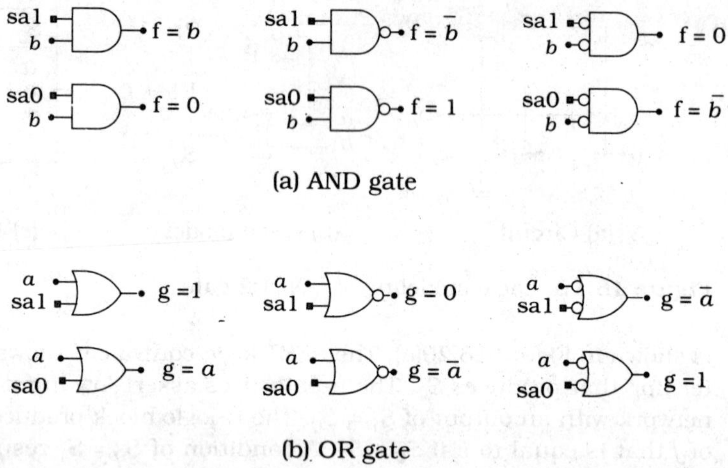

Figure 16.21 Summary of stuck-at fault effects for primitive logic gates

$$f(a, b) = a \cdot b \tag{16.31}$$

shows that a sa0 at either input gives $f = 0$ while a sa1 at a gives $f = \bar{b}$. The stuck-at characteristics of the OR-family of primitive logic gates are summarized in Figure 16.21(b) using the same approach.

Multiple-fault situations can be somewhat tricky to deal with. Figure 16.22 illustrates some ideas in fault simplification. Figure 16.22(a) shows the case where sa1 faults occur at both the input and the output of an AND gate. In this case, the output overrides the input fault, so that anything to the left of the gate may be ignored. This is called **fault dominance** and is useful for simplifying test vector generation. Equivalent faults are shown in Figure 16.22(b). In this case, three distinct faults are present, but the same behavior would be obtained with only the sa0 input to the NAND2 gate. Simplification from **fault collapsing** is illustrated in Figure 16.22(c). Since the inputs and outputs are identical, either gate may be replaced by a short circuit, leading to a simpler logic network.

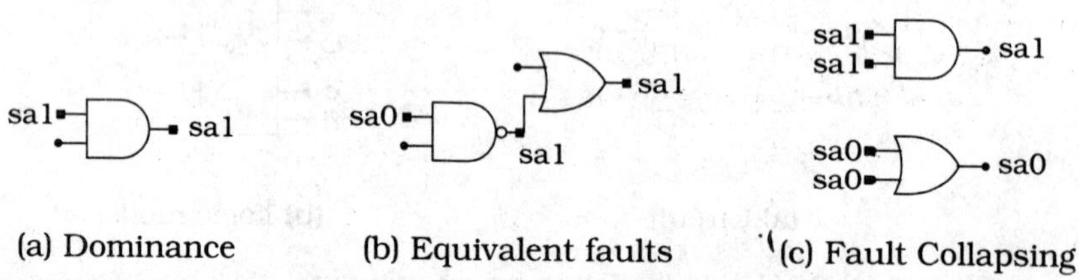

Figure 16.22 Fault simplification

These examples show how logic-level fault models can be used to describe physical circuit flaws. An important technique for generating test vector sets is to place faults at various locations in the network and then calculate the effects. Characterizing the response in this manner allows us to formulate test vectors that target specific faults.

16.3.3 Path Sensitization

When the gate to be tested is embedded within a larger logic network, we can use the existing circuity to create a specific path from the location of the fault to an observable output point. This technique is called **path sensitization**, and the process of creating the path is called **propagation** since the fault is viewed as being propagated through the logic network.

Consider the simple logic circuit in Figure 16.23 that implements the function

$$F = a_1 \cdot a_2 + \bar{a}_2 \cdot a_3 \tag{16.32}$$

We want to determine the inputs to test for a sa0 fault at the input a_3. Path sensitization is performed in two steps. The first is called forward drive. In this step, we want to distinguish the effects of normal operation from the fault. For the case of the sa0 fault, we set $a_3 = 1$ so that it is different from the faulty input. To propagate this value through the AND gate G2 we must have an inverter output of 1. Combined with $a_3 = 1$ then gives the output of G2 as 1. To propagate this value through the OR gate G3, we need the output of G1 to be 0, which completes the forward drive.

The second step is called backwards trace. This uses the results of the forward drive to determine the inputs needed to detect the fault. The first condition is $a_3 = 1$. To insure that the output of the inverter is 1, we must also select $a_2 = 0$. Finally, to insure that the output of G1 is a 0, we need either a_1 or a_2 to be 0. Since we have already selected $a_2 = 0$, a_1 can be a either a 0 or a 1. This gives the test vector for the sa0 fault as

$$(a_1, a_2, a_3) = (d, 0, 1) \tag{16.33}$$

where d is a don't care state. This simple example illustrates the procedure. However, it is not always possible to obtain a realizable test vector for a single path. Multiple-path sensitization is then necessary.

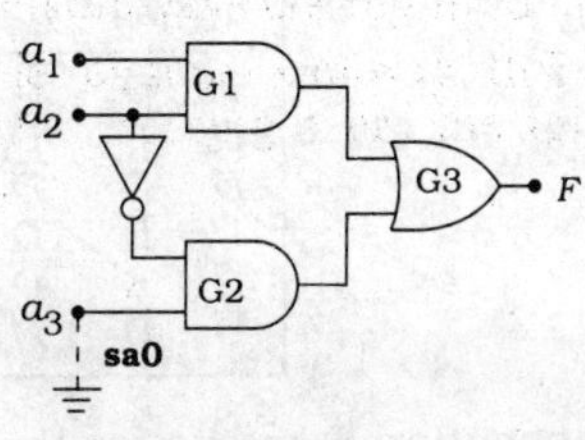

Figure 16.23 Path sensitization example

16.3.4 The D-Algorithm

In this approach, the variable D is introduced to model *discrepancies* between a good circuit and a faulty circuit. By definition, a $D = 1$ indicates a good circuit, while a problem is indicated by $D = 0$. The complement $\overline{D}$ is defined in the opposite manner: $\overline{D} = 0$ is good and $\overline{D} = 1$ is faulty. The D-algorithm provides a technique for deriving test vectors for any observable fault. This power does not come without complexity, and a full treatment is well beyond the scope of this text. It is, however, possible to understand the basic ideas without going into detail.

The first aspect we will examine is that of a **singular cover** for a gate; this is equivalent to a row in the truth table presented in the compact form illustrated by the NAND example in Figure 16.24(b) where d is a don't care condition. In this form, there are three distinct rows that are the cubes of the singular cover. Figure 16.24(c) shows the primitive D-cube of the NAND gate. By definition, the primitive D-cube consists of the input vectors that are required to produce a D or a $\overline{D}$ at the output when there is a fault. In the present case, these correspond to $(a_1, a_2) = (0, d)$ and $(d, 0)$ since these give an output of $a_3 = 1$.

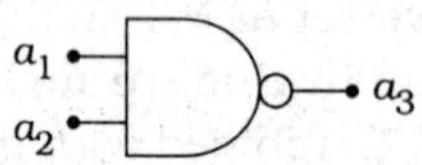

(a) Logic gate

a_1	a_2	a_3
0	d	1
d	0	1
1	1	0

(b) Singular cover

a_1	a_2	a_3
0	d	D
d	0	D

(c) D-cube

Figure 16.24 Singular cover of a NAND2 gate

The propagation D-cubes of a gate are the primitive cubes that are needed to propagate a D or $\overline{D}$ at one or more inputs to the output. Figure 16.25 shows the propagation D-cubes for the NAND2 gate. The D-algorithm then examines intersections of the propagation D-cubes to determine the test vector sets.

The D-algorithm is well known in testing theory. It is a very practical approach because it allows for the sensitization of multiple paths using a

a_1	a_2	a_3
1	D	$\overline{D}$
D	1	$\overline{D}$
D	D	$\overline{D}$
1	$\overline{D}$	D
$\overline{D}$	1	D
$\overline{D}$	$\overline{D}$	D

Figure 16.25 Propagation D-cubes for the NAND2 gate

structured methodology. One drawback of path sensitization is that the process of generating test vectors may become long and involved. The overall time can be reduced by coupling the techniques with **fault simulation**. In this approach, we apply a test vector and then determine what faults can be detected. This is usually less time consuming than solving the inverse problem.

16.3.5 The Boolean Difference

Another approach for test vector generation is **Boolean differences**. Consider the n-input network shown in Figure 16.26. The output is the general function

$$f(a) = f(a_1, a_2, \ldots, a_n) \tag{16.34}$$

Let us select an arbitrary input a_k and define

$$\begin{aligned} f_k &= f(a_1, a_2, \ldots, a_k = 1, \ldots, a_n) \\ f_{\bar{k}} &= f(a_1, a_2, \ldots, a_k = 0, \ldots, a_n) \end{aligned} \tag{16.35}$$

Using Shannon's expansion theorem, we can write the original function as

$$\begin{aligned} f(a) &= a_k \cdot f_k + \bar{a}_k \cdot f_{\bar{k}} \\ &= a_k \cdot f_k \oplus \bar{a}_k \cdot f_{\bar{k}} \end{aligned} \tag{16.36}$$

which provides an expression that deals with the input a_k explicitly.

Suppose that we want to test for a fault at a_α. This gives an output that is described by $f(a)$. Since this is a faulty value, we know that comparing it to the correct output $f(a)$ will give

$$f(a) \neq f_\alpha(a) \tag{16.37}$$

if the inputs a are the same. We may thus write

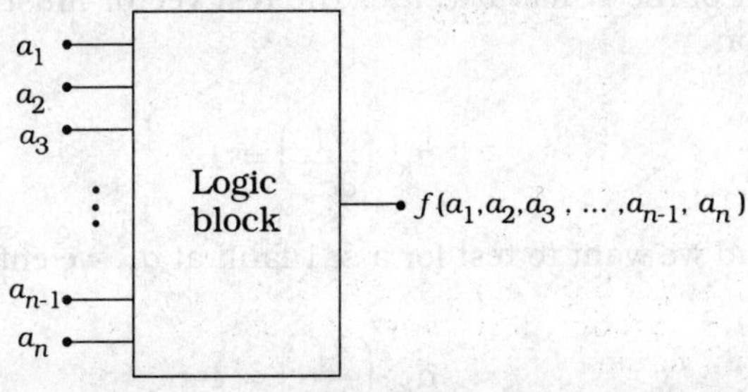

Figure 16.26 Basic network for deriving the Boolean difference

$$f(a) \oplus f_{\alpha}(a) = 1 \tag{16.38}$$

and define a test parameter

$$t_{\alpha} = f(a) \oplus f_{\alpha}(a) \tag{16.39}$$

such that $t_{\alpha} = 1$ indicates a fault.

Now suppose that we have an sa0 fault at a_k. The test parameter for this case is

$$\begin{aligned} t_{\alpha} &= f(a) \oplus f_{\bar{k}}(a) \\ &= [a_k \cdot f_k \oplus \bar{a}_k \cdot f_{\bar{k}}] \oplus f_{\bar{k}} \\ &= a_k \cdot f_k \oplus (\bar{a}_k + 1) f_{\bar{k}} \\ &= a_k \cdot f_k \oplus a_k f_{\bar{k}}(a) \end{aligned} \tag{16.40}$$

or,

$$t_{\alpha} = a_k \cdot (f_k \oplus f_{\bar{k}}) \tag{16.41}$$

The Boolean difference is defined as

$$\frac{\partial f}{\partial a_k} = f_k \oplus f_{\bar{k}} \tag{16.42}$$

which gives

$$t_{\alpha} = a_k \cdot \left(\frac{\partial f}{\partial a_k}\right) \tag{16.43}$$

It is seen that

$$\frac{\partial f}{\partial a_k} = 1 \qquad \text{iff } f_k \neq f_{\bar{k}} \tag{16.44}$$

However, this implies that changing a_k also changes the output, so that the fault is observable. For the sa0 test at a_k we assign the input the complement of the stuck-at value. The test vector must therefore satisfy the condition

$$a_k \cdot \left(\frac{\partial f}{\partial a_k}\right) = 1 \tag{16.45}$$

If instead we want to test for a sa1 fault at a_k, we enforce the condition

$$\bar{a}_k \cdot \left(\frac{\partial f}{\partial a_k}\right) = 1 \tag{16.46}$$

to determine the test vector.

As an example of how the technique works, consider the simple OA network in Figure 16.27. The output function is

$$f(a) = (a_1 + a_2) \cdot (a_3 + a_4) \tag{16.47}$$

Assume that we want to detect faults at a_3. We need

$$\begin{aligned} f_{\bar{3}} &= (a_1 + a_2) \cdot (a_4) \\ f_3 &= (a_1 + a_2) \cdot (1) \end{aligned} \tag{16.48}$$

so that the Boolean difference is

$$\begin{aligned} \frac{\partial f}{\partial a_3} &= f_{\bar{3}} \oplus f_3 \\ &= (a_1 + a_2) \cdot (a_4) \oplus (a_1 + a_2) \\ &= (a_1 + a_2) \cdot \bar{a}_4 \end{aligned} \tag{16.49}$$

For a sa0 fault at a_3 we use the condition

$$a_3 \cdot \left(\frac{\partial f}{\partial a_3}\right) = 1 \tag{16.50}$$

which gives the equation

$$a_3 \cdot (a_1 + a_2) \cdot \bar{a}_4 = 1 \tag{16.51}$$

The test vectors that satisfy this relation are

$$(a_1 a_1 a_3 a_4) = (1d10) \text{ or } (d110) \tag{16.52}$$

where d is a don't care input. Similarly, a sa1 fault at a_3 gives the condition

$$\bar{a}_3 \cdot (a_1 + a_2) \cdot \bar{a}_4 = 1 \tag{16.53}$$

and

$$(a_1 a_1 a_3 a_4) = (1d00) \text{ or } (d100) \tag{16.54}$$

as the test vectors. The Boolean difference technique can also be used to find test vectors for internal nodes of a logic network.

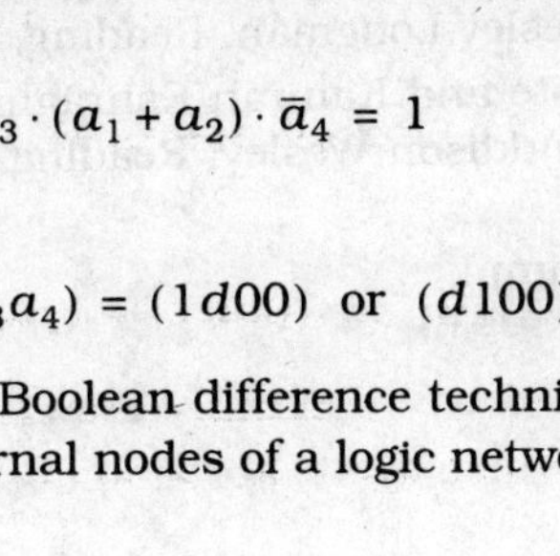

Figure 16.27 Application example of the Boolean difference

16.4 Summary

Reliability and testing are critical aspects of modern VLSI systems. We have only "scratched the surface" in this short treatment but have attempted to illustrate some of the main problems and solutions.

As the complexity of VLSI systems increases, testing becomes more and more difficult. Reliability is a continuing concern as users tend to expect that their systems will last forever. The interested reader who takes time to study further will find that this can be a lifelong career path that gains importance with each successive generation.

16.5 References

[1] Harry Bleeker, Peter van den Eijnden, and Frans de Jong, **Boundary-Scan Test**, Kluwer Academic Publishers, Dordrecht, The Netherlands, 1993.

[2] Niraj K. Jha and Sandip Kunda, **Testing and Reliable Design of CMOS Circuits**, Kluwer Academic Publishers, Norwell, MA, 1990.

[3] Arthur B. Glaser and Gerald E. Subak-Sharpe, **Integrated Circuit Engineering**, Addison-Wesley, Reading, MA, 1977.

[4] Ravi K. Gulati and Charles F. Hawkins (eds), **I_{DDQ} Testing of VLSI Circuits**, Kluwer Academic Publishers, Norwell, MA, 1993.

[5] Kenneth P. Parker, **The Boundary-Scan Handbook**, Kluwer Academic Publishers, Norwell, MA, 1992.

[6] Paul A. Tobias and David Trindade, **Applied Reliability**, Van Nostrand Reinhold, New York, 1986.

[7] Michael John Sebastian Smith, **Application-Specific Integrated Circuits**, Addison-Wesley Longman, Reading, MA, 1997.

[8] Neil H.E. Weste and Kamran Eshraghian, **Principles of CMOS VLSI Design**, 2nd ed., Addison-Wesley, Reading, MA, 1993.

Index

A

B

C

D

E

F

N

O

P

Limited Use License Agreement

Limited Use License Agreement

This is the John Wiley and Sons, Inc. (Wiley) limited use License Agreement, which governs your use of any Wiley proprietary sof ware products (Licensed Program) and User Manual(s) delivered with it.

Your use of the Licensed Program indicates your acceptance of the terms and conditions of this Agreement. If you do not accept c agree with them, you must return the Licensed Program unused within 30 days of receipt or, if purchased, within 30 days, as eviden by a copy of your receipt, in which case, the purchase price will be fully refunded.

License: Wiley hereby grants you, and you accept, a non-exclusive and non-transferable license, to use the Licensed Program and Us Manual(s) on the following terms and conditions only:

a. The Licensed Program and User Manual(s) are for your personal use only.
b. You may use the Licensed Program on a single computer, or on its temporary replacement, or on a subsequent computer only.
c. The Licensed Program may be copied to a single computer hard drive for playing.
d. A backup copy or copies may be made only as provided by the User Manual(s), except as expressly permitted by this Agreen
e. You may not use the Licensed Program on more than one computer system, make or distribute unauthorized copies of the Lic Program or User Manual(s), create by decompilation or otherwise the source code of the Licensed Program or use, copy, mo transfer the Licensed Program, in whole or in part, or User Manual(s), except as expressly permitted by this Agreement. If you transfer possession of any copy or modification of the Licensed Program to any third party, your license is automa terminated. Such termination shall be in addition to and not in lieu of any equitable, civil, or other remedies available to

Term: This License Agreement is effective until terminated. You may terminate it at any time by destroying the Licensed Progr User Manual(s) together with all copies made (with or without authorization).

This Agreement will also terminate upon the conditions discussed elsewhere in this Agreement, or if y u fail to co with any term or condition of this Agreement. Upon such termination, you agree to destroy the Licensed Program, User Manu and any copies made (with or without authorization) of either.

Wiley's Rights: You acknowledge that all rights (including without limitation, copyrights, patents and trade secrets) in the Lice Program (including without limitation, the structure, sequence, organization, flow, logic, source code, object code and all mean forms of operation of the Licensed Program) are the sole and exclusive property of Wiley. By accepting this Agreement, you d become the owner of the Licensed Program, but you do have the right to use it in accordance with the provisions of this Agreen You agree to protect the Licensed Program from unauthorized use, reproduction, or distribution. You further acknowledge that th Licensed Program contains valuable trade secrets and confidential information belonging to Wiley. You may not disclose any com nent of the Licensed Program, whether or not in machine readable form, except as expressly provided in this Agreement.

WARRANTY: TO THE ORIGINAL LICENSEE ONLY, WILEY WARRANTS THAT THE MEDIA ON WHICH THE LICENSEI PROGRAM IS FURNISHED ARE FREE FROM DEFECTS IN THE MATERIAL AND WORKMANSHIP UNDER NORMAL US FOR A PERIOD OF NINETY (90) DAYS FROM THE DATE OF PURCHASE OR RECEIPT AS EVIDENCED BY A COPY OF YOUR RECEIPT. IF DURING THE 90 DAY PERIOD, A DEFECT IN ANY MEDIA OCCURS, YOU MAY RETURN IT. WILI WILL REPLACE THE DEFECTIVE MEDIA WITHOUT CHARGE TO YOU. YOUR SOLE AND EXCLUSIVE REMEDY IN T EVENT OF A DEFECT IS EXPRESSLY LIMITED TO REPLACEMENT OF THE DEFECTIVE MEDIA AT NO ADDITIONA CHARGE. THIS WARRANTY DOES NOT APPLY TO DAMAGE OR DEFECTS DUE TO IMPROPER USE OR NEGLIGEN

THIS LIMITED WARRANTY IS IN LIEU OF ALL OTHER WARRANTIES, EXPRESSED OR IMPLIED, INCLUI ING, WITHOUT LIMITATION, ANY WARRANTIES OF MERCHANTABILITY OR FITNESS FOR A PARTICULAR PURPC

EXCEPT AS SPECIFIED ABOVE, THE LICENSED PROGRAM AND USER MANUAL(S) ARE FURNISHED BY WILEY ON AN "AS IS" BASIS AND WITHOUT WARRANTY AS TO THE PERFORMANCE OR RESULTS YOU MAY OBTA BY USING THE LICENSED PROGRAM AND USER MANUAL(S). THE ENTIRE RISK AS TO THE RESULTS OR PERFOR MANCE, AND THE COST OF ALL NECESSARY SERVICING, REPAIR, OR CORRECTION OF THE LICENSED PROGR M AND USER MANUAL(S) IS ASSUMED BY YOU.

IN NO EVENT WILL WILEY OR THE AUTHOR, BE LIABLE TO YOU FOR ANY DAMAGES, INCLUDING LOS PROFITS, LOST SAVINGS, OR OTHER INCIDENTAL OR CONSEQUENTIAL DAMAGES ARISING OUT OF THE USE OR INABILITY TO USE THE LICENSED PROGRAM OR USER MANUAL(S), EVEN IF WILEY OR AN AUTHORIZED WILEY DEALER HAS BEEN ADVISED OF THE POSSIBILITY OF SUCH DAMAGES.

General: This Limited Warranty gives you specific legal rights. You may have others by operation of law which varies from state to state. If any of the provisions of this Agreement are invalid under any applicable statute or rule of law, they are to that extent deeme omitted.

This Agreement represents the entire agreement between us and supersedes any proposals or prior Agreements, oral or written, and any other communication between us relating to the subject matter of this Agreement.

This Agreement will be governed and construed as if wholly entered into and performed within the State of New York. You acknowledge that you have read this Agreement, and agree to be bound by its terms and conditions.